Lecture Notes in Computer Science 15714

Founding Editors

Gerhard Goos
Juris Hartmanis

AF148241

The series Lecture Notes in Computer Science (LNCS), including its subseries Lecture Notes in Artificial Intelligence (LNAI) and Lecture Notes in Bioinformatics (LNBI), has established itself as a medium for the publication of new developments in computer science and information technology research, teaching, and education.

LNCS enjoys close cooperation with the computer science R & D community, the series counts many renowned academics among its volume editors and paper authors, and collaborates with prestigious societies. Its mission is to serve this international community by providing an invaluable service, mainly focused on the publication of conference and workshop proceedings and postproceedings. LNCS commenced publication in 1973.

Elvio Amparore · Łukasz Mikulski
Editors

Application and Theory of Petri Nets and Concurrency

46th International Conference, PETRI NETS 2025
Paris, France, June 22–27, 2025
Proceedings

 Springer

Editors
Elvio Amparore 🆔
University of Torino
Turin, Italy

Łukasz Mikulski 🆔
Faculty of Mathematics and Computer
Science
Nicolaus Copernicus University
Toruń, Poland

ISSN 0302-9743 ISSN 1611-3349 (electronic)
Lecture Notes in Computer Science
ISBN 978-3-031-94633-2 ISBN 978-3-031-94634-9 (eBook)
https://doi.org/10.1007/978-3-031-94634-9

This Springer imprint is published by the registered company Springer Nature Switzerland AG
The registered company address is: Gewerbestrasse 11, 6330 Cham, Switzerland

If disposing of this product, please recycle the paper.

Preface

This volume contains the proceedings of the 46th International Conference on Application and Theory of PETRI NETS and Concurrency (PETRI NETS 2025). The aim of this series of conferences is to create an annual opportunity to discuss and disseminate the latest results in the field of PETRI NETS and related models of concurrency, including their tools, applications, and theoretical progress. The 46th conference and affiliated events were organized by the SAFER team from Laboratoire d'Informatique de Paris Nord (LIPN), Université Sorbonne Paris Nord, jointly with members of the MeFoSyLoMa group, during June 22–27, 2025.

This year, 48 papers were submitted to PETRI NETS 2025. 41 papers were single-blind reviewed by at least three reviewers. The discussion phase and final selection process by the Program Committee (PC) were supported by the EasyChair conference system. From 38 regular papers and 3 tool papers, the PC selected 22 papers for presentation: 21 regular papers and 1 tool paper. After the conference, some of these authors were invited to submit an extended version of their contribution for consideration in a special issue of a journal.

We thank the PC members and other reviewers for their careful and timely evaluation of the submissions and the fruitful constructive discussions that resulted in the final selection of papers. The Springer LNCS team provided excellent and welcome support in the preparation of this volume.

The keynote presentations were given by

- Dirk Beyer, Ludwig-Maximilians-Universität München (LMU Munich), Germany
- Patricia Bouyer, LMF, ENS Paris-Saclay, France, and
- Marco Montali, Free University of Bozen-Bolzano, Italy.

The conference series is coordinated by a steering committee with the following members: W. van der Aalst (Germany), E. Amparore (Italy), É. André (France), J. Barros (Portugal), I. Brito (Portugal), D. Buchs (Switzerland), G. Ciardo (USA), A. Costa (Portugal), J. Desel (Germany), S. Donatelli (Italy), L. Gomes (Portugal), S. Haddad (France), K. Hiraishi (Japan), J. Kleijn (The Netherlands), F. Kordon (France) (co-chair), M. Koutny (UK), M. Köhler-Bußmeier (Germany), L.M. Kristensen (Norway), C. Lin (China), Ł. Mikulski (Poland), D. Moldt (Germany), C. Olarte (France), W. Penczek (Poland), L. Petrucci (France) (co-chair), L. Pomello (Italy), W. Reisig (Germany), G. Rozenberg (The Netherlands), A. Valmari (Finland), J. van der Werf (The Netherlands), K. Wolf (Germany), and A. Yakovlev (UK).

Alongside PETRI NETS 2025, the following workshops took place:

- Algorithms and Theories for the Analysis of Event Data (ATAED 2025),
- International Workshop on PETRI NETS and Software Engineering (PNSE 2025), and
- Petri Net games, examples, and quizzes for education, contest, and fun (PENGE 2025).

Workshops were co-chaired by Giuliana Franceschinis (Italy) and Karsten Wolf (Germany).

Other collocated events included the Petri Net Course and Tutorials, coordinated by Jörg Desel (Germany) and Lars M. Kristensen (Norway), and the Model Checking Contest, organized by Fabrice Kordon (France), as well as a Tool Exhibition coordinated by Dylan Marinho (France).

We greatly appreciate the efforts of all members of the Local Organizing Committee, chaired by Étienne André and Carlos Olarte, for their time spent in the organization of this event. The local organizers would like to warmly thank Université Sorbonne Paris Nord and LIPN (CNRS UMR 7030) for their support, as well as the whole MeFoSyLoMa group. A special thank you as well to the steering committee co-chairs Fabrice Kordon and Laure Petrucci.

We hope you enjoy reading the contributions in this LNCS volume.

June 2025   Elvio Gilberto Amparore
                                                                    Łukasz Mikulski

Organization

Program Committee

Elvio Amparore (Co-chair)	University of Turin, Italy
Abel Armas Cervantes	University of Melbourne, Australia
João Paulo Barros	Instituto Politécnico de Beja, Portugal
Benoît Delahaye	Université de Nantes, France
João M. Fernandes	University of Minho, Portugal
Giuliana Franceschinis	Università del Piemonte Orientale, Italy
Luis Gomes	Universidade NOVA de Lisboa, Portugal
Xudong He	Florida International University, USA
Loic Helouet	INRIA, France
Wojciech Jamroga	Polish Academy of Sciences, Poland
Gabriel Juhas	Pan-European University, Slovakia
Anna Kalenkova	University of Adelaide, Australia
Michael Köhler-Bußmeier	Hamburg University of Applied Sciences, Germany
Lars Michael Kristensen	Western Norway University of Applied Sciences, Norway
Lisa Luise Mannel	RWTH Aachen University, Germany
Łukasz Mikulski (Co-chair)	Nicolaus Copernicus University in Toruń, Poland
Guillermo Perez	University of Antwerp, Belgium
Marta Pietkiewicz-Koutny	Newcastle University, UK
Artem Polyvyanyy	University of Melbourne, Australia
Lucia Pomello	Università di Milano-Bicocca, Italy
Andrey Rivkin	Technical University of Denmark, Denmark
Natalia Sidorova	Technische Universiteit Eindhoven, The Netherlands
Jeremy Sproston	University of Turin, Italy
Jiri Srba	Aalborg University, Denmark
Nathalie Sznajder	Sorbonne Université, CNRS, LIP6, France
Remigiusz Wisniewski	University of Zielona Góra, Poland
Karsten Wolf	Universität Rostock, Germany

Additional Reviewers

Anastasiadi, Elli
Bernardinello, Luca
Busatto-Gaston, Damien
Chatain, Thomas
Ferigato, Carlo
Guillou, Lucie
Hofman, Piotr
Karatkevich, Andrei
Karunaratne, Anandi
Kim, Yan
Koutny, Maciej
Kurpiewski, Damian

Li, Tian
Lime, Didier
Muniz, Marco
Parrot, Rémi
Petrucci, Laure
Sangnier, Arnaud
Schou, Morten Konggaard
Sidoruk, Teofil
Staquet, Gaëtan
Wojnakowski, Marcin
Zuleger, Florian
Łabiak, Grzegorz

Contents

Automated Reasoning for Data-Aware Petri Nets

Marco Montali[✉]

Free University of Bozen-Bolzano, Bolzano, Italy
marco.montali@unibz.it

Abstract. The focus on work processes in process science is increasingly shifting from a case-centric, pure control-flow perspective, to a data-aware setting where (possibly multiple interrelated) objects and their properties are updated by and influence the process. Data may range from simple attributes (such as strings or numbers) to complex data structures like relational databases. Modelling, analysis, and mining of such data-aware processes call for formalisms and techniques that simultaneously tackle time/dynamics and the interplay with these different forms of data. In this short paper, we focus on data-aware work processes whose underlying control-flow backbone is described as a Petri net. We overview the main modelling requirements and constructs emerging in different proposals, ranging from case- to object-centric processes. We then summarize how artificial intelligence techniques from automated reasoning have been employed and further developed to obtain foundational and practical results in data-aware process analysis and mining.

Keywords: Petri nets · automated reasoning · data-aware processes · object-centric processes

1 Introduction

Within process science, work processes are traditionally modelled under two strong assumptions:

1. *Case-centricity.* The process comes with a clear notion of *case type* (such as order, patient, claim), so that every instance of the case type is in one-to-one correspondence with an instance of the process.
2. *Control-flow focus.* The process describes which activities must/may be executed in a process instance, and in which possible orderings.

Under these assumptions, the most well-established approach for formalising and analysing work processes consists in employing (bounded) Petri nets (or more restricted well-behaved classes such as sound workflow nets) and related state-space exploration and mining techniques [1,25,31,66].

Both assumptions have been recently lifted in a new, process science wave aiming at capturing real-life processes more precisely. Regarding point (2), it has been argued that additional perspectives, beyond the control-flow backbone,

E. Amparore and L. Mikulski (Eds.): PETRI NETS 2025, LNCS 15714, pp. 1–17, 2025.
https://doi.org/10.1007/978-3-031-94634-9_1

need to be considered, in particular that of *data*. Interestingly, this need has emerged not only within process science [57], but also in process mining [51], data management [17], and artificial intelligence [9], spawning a series of key results in formal modelling and analysis [17,18,30,41]. In this setting, *data-awareness* refers to the fact that every case comes with a data payload that influences the process control-flow, and is updated through the execution of activities. Such payload may range from simple case variables storing strings and numbers, to complex structured data such as a full-fledged relational database.

Regarding point (1), many real-life processes cannot be properly described by identifying a single notion of case, as they evolve at once multiple objects, interconnected via complex one-to-many and many-to-many relationships. For example, in an order-to-delivery process each order refers to multiple items, and items are shipped in packages that may combine items from different orders. Straight-jacketing the process representation through a single case notion yields misleading process models [6] and wrong analysis and mining results [2,13]. In this context, data-awareness refers to the need of capturing multiple objects and their evolving relationships. Several process modelling paradigms and corresponding languages emerged to capture these so-called *object-centric processes*, ranging from case-handling [5,46] to artifact-centric [26,63] and object-aware approaches [16,47,61].

From the formal point of view, data-aware processes (both in the case- and object-centric setting) are inherently infinite-state systems, where infinity is not caused by the presence of unboundedly many concurrent threads (like in traditional Petri nets), but is instead due to the presence of data. As already pointed out data, in turn, may range from simple attributes (such as strings or numbers) to complex data structures like relational databases.

This poses three big challenges:

1. How to formally model data-aware processes?
2. How to analyse data-aware processes?
3. How to mine data-aware processes?

In this short paper, we report on how these three questions are being answered in a flourishing stream of research that studies data-aware work processes whose underlying control-flow backbone is described as a Petri net.

As for modelling, we overview the main modelling requirements and constructs emerged in different proposals, ranging from case- to object-centric processes. We do not provide a systematic literature review but only consider some of the most representative data-aware extensions of Petri nets to capture key features.

As for analysis and mining, we summarise how artificial intelligence techniques from automated reasoning have been employed and further developed to obtain foundational and practical results, considering in particular verification of temporal properties and conformance checking.

2   Modelling

In this section, we overview the main data-aware extensions of Petri nets proposed to integrate data and control-flow for work processes. This requires to handle three dimensions:

1. The *control-flow dimension*, dealing with the process dynamics as induced by the classical token game for Petri nets.
2. The *structural dimension of data*, determining which forms of data are supported, and which condition and update languages are provided.
3. The *bidirectional integration* of the control-flow and data dimensions, expressing how data conditions influence the process control-flow, and how the firing of transitions updates the data state; this may also require to enrich tokens with the possibility of referring to data elements or tuples.

We highlight how these three dimensions have been captured in different settings, whose common denominator is the trade-off between expressiveness of representation and the possibility of supporting computation of analysis and mining tasks, which will be the subject of the next sections.

2.1   Case-Centric, Data-Aware Petri Nets

This line comprises processes where it is possible to identify a clear notion of case. Each case is assumed to evolve in isolation, but differently from the classical setting, it is associated to a data payload.

A first wave in this stream deals with case-centric processes operating over an underlying, persistent database. This accounts for the fact that during the execution of a case, some objects and relationships must be persisted beyond the existence of the case itself. For example, in a job hiring process, the applicants must be persisted, together with (the identifier of) the job they applied to. Each case triggers a (bounded) flow of control over activities. Each activity, in turn, is linked to a (parametric) database transaction, which indicates how the underlying database is modified whenever an instance of the activity (with parameters bound to actual object identifiers) is executed. The enablement of activities and the binding of parameters is in turn not only defined by the control-flow of the process, but also conditioned by queries formulated over the database. Even under the assumption of bounded control-flow, in the presence of loops every case can generate unboundedly many data configurations, due to the generation and persistence of unboundedly new identifiers and, possibly, the accumulation of unboundedly many tuples in the database.

A reference approach in this setting is the formalism of *data-centric dynamic systems* (DCDSs [7]), where the process control-flow is not specified using a (bounded) Petri net, but through condition-action rules. This control-flow specification has been related to Petri nets in [8], finally leading to a Petri net version of DCDSs through the formalism of *db-nets* [55]. Essentially, *db-nets* adopt *Petri nets with identifiers* (PNIDs [45,64]) to represent the process control-flow. PNIDs natively support the generation of new identifiers, in a way reminiscent of ν-Petri nets [58], and equip tokens with references to object identifiers and relationships (that is, tuples of object identifiers). Transitions employ the classical inscription approach of colored Petri nets to bind identifiers from consumed tokens and use the binding to produce tokens. Differently from the seminal approach in [45], where patterns in the net together realise database transactions, in db-nets transactions are modelled as external modules a là SQL, and instantiated and invoked when firing a transition with some binding. If the transaction fails, the firing is compensated and the database remains unaltered.

Another reference setting for case-centric, data-centric systems is given by DABs [24]: here, the control flow is represented using block-structured BPMN, and the data dimensions is captured by a read-only and a read-write database. DABs have been instrumental to the introduction of the Petri-net based *catalogue and object-aware nets* (COA nets [37]), further described in Sect. 2.2.

A second line of research considers case-centric processes where the data payload consists of a fixed set of variables. The key ingredient making this setting interesting and challenging is that such variables store values from different data types, thus calling for formalisms dealing with multiple theories at once.

Data Petri nets[1] [49,51] account for case-centric processes with case variables defining a bounded Petri net or workflow net for describing the control-flow of a case, equipped with an external set of variables with primitive datatypes (such as strings, booleans, and numbers). Net transitions are decorated by guards that read and write the variables, thus accounting at once for data conditions on transition enablements, and constrained updates of the variables upon transition firings. Different guard languages for numerical datatypes are considered, ranging from simple variable-to-constant [49] and variable-to-variable conditions [35] to different forms of arithmetic conditions [36].

The formalism of *data-aware processes modulo theories* (DMTs [42]) lifts DPNs to a full "modulo theory" setting. In a DMT, variables range from primitive datatypes to rich, structured datatypes such as lists and uninterpreted functions or, more in general, arbitrary datatypes enjoying some mild, model-theoretic properties.

2.2   Object-Aware Petri Nets

Differently from case-centric processes, so-called *object-centric processes* deal with situations where it is not possible to single out a single notion of case, but rather the process co-evolves multiple objects linked by one-to-many and many-to-many relationships.

This requires to extend the process control-flow with a number of features, such as:

- object creation and removal;
- batch operations, involving (unboundedly) many objects at once;

---

[1] There is a terminological clash with *Petri nets with data* [48], where each token carries a (single) data value from an infinite (ordered or unordered) domain.

- concurrent evolution of distinct objects, tracking the independent progression of objects (such as order payment and loading of its items);
- manipulation of objects and one-to-one and one-to-many relationships;
- object-aware synchronization, imposing that an object of a given type can flow through a transition only if some (*subset synchronization*) or all (*exact synchronization*) related objects of another type simultaneously flow through that transition as well.

An example that uses these constructs is an order-to-delivery process where orders can be created by adding an arbitrary amount of items. Orders can be removed or undergo an approval process, which consists of checking whether *all* contained items are available. Every item is then shipped in one package, while every package can contain items from the same or different orders.

Several proposals have been brought forward in this direction, such as object-centric Petri nets [4], synchronous proclets [32], and Petri nets with identifiers (PNIDs) [43,56,65], possibly equipped with access to an external database – as in COA nets [37]. Such proposals vary in their ability of supporting the features mentioned above, thus providing different levels of precision in capturing object-centric processes.

In particular, object-centric Petri nets [4] contain a dedicate Petri net for every object type involved in the process. Arcs are split into normal and batching - where a batching arc indicates that upon firing, a transition is consumes and/or produces unboundedly many tokens from/to the corresponding place. Finally, some transitions connect different types, indicating a very basic form of synchronization. This approach is further developed in synchronous proclets [32], where transitions involving objects of different types represent forks or joins in the process. Forks establish correlation sets indicating which objects are involving in the firing (e.g., which items are added to an order). Joins use such correlation sets to define a subset synchronization (some items of an order get loaded into a package) or exact synchronization (an order is approved if all its items are available) behaviour.

While in object-centric Petri nets and synchronous proclets the relationships among objects are left implicit, they are explicitly represented and manipulated by the process in approaches based on PNIDs. This makes it possible to capture constructs such as that of reshuffling (where, e.g., items are moved from one order to another), out of reach before [43]. The most recent variant of PNIDs, namely that of *object-centric PNIDs* (OPIDs [43]), infuses PNIDs with batching arcs borrowed from object-centric Petri nets, finally aiming at unifying the two lines.

Finally, notice that although conceptually different, these approaches and those aiming at enrich case-centric processes with an underlying database present many similarities from the formal point of view. In fact, places hosting tokens carrying tuples of identifiers can be seen as relational tables, with the two main differences that they obey to a multiset semantics (hence the same tuple can be present multiple times), and are not explicitly subject to integrity constraints (even though forms of integrity constraints can be implicitly realised depending on which modelling patterns are employed to operate over places).

3 Analysis

Analysis of the data-aware Petri net variants described in Sect. 2 requires to combine queries on the dynamics induced by the process, as well as queries on the involved data. Technically, this translates into queries mixing first-order and temporal quantification.

On the one hand, one may be interested in ascertaining general correctness criteria, such as reachability, safety, and coverability, now expressed over states that combine markings with data configurations. On the other hand, one may want to verify more complex first-order temporal formulae.

Analysing data-aware Petri net variants against these queries poses a foundational and a practical challenge.

From the foundational point of view, even propositional reachability is highly undecidable over severely restricted classes of these models. For example, undecidability of propositional reachability holds for:

- DCDSs equipped with two unary relations and negation within database queries [7];
- DCDSs equipped with one binary relation and positive database queries [53];
- DPNs with two integer case variables and conditions expressing increment, decrement, and test-for-zero [42];
- PNIDs and COA nets with tokens carrying pairs of identifiers [37].

From the practical point of view, there is the need of efficient algorithmic techniques for conducting these reasoning tasks, paired with corresponding effective implementations.

Automated reasoning provides a solid basis to tackle both challenges. We provide next a brief overview of the main results.

3.1 Analysis of State-Bounded Processes Operating over Databases

For processes operating over an underlying relational database, it has been observed that undecidability for DCDSs equipped with two unary relations immediately arises when these relations accumulate unboundedly many identifiers [7]. This leads to a natural strategy to counter this possibility, based on the idea of *state-boundedness* [7,8,12]. In a state-bounded process there is a maximum, pre-defined bound on the maximum number of identifiers that can be accumulated within the database of a single state of the system. Even under state-boundedness, the resulting state-space can still be infinite, due the possibility of encountering, within and across runs, unboundedly many distinct identifiers (which however cannot coexist in the same state).

State-boundedness is a semantic property, decidable to check for a given bound and undecidable to check if the bound is not known [7,27]. Different strategies have been pursued ensure that a process of interest is state-bounded, for different specification languages:

- sufficient, syntactic conditions over the specification of activities, in particular considering how they update the underlying database [7,8];

- situation calculus action theories with fading memory [27];
- DABs operating over a state-bounded read-write database [24];
- resource-constrained ν-Petri nets [54];
- bounded COA nets [37] (where "boundedness" coincides here with the standard notion of boundedness from Petri nets);
- state-bounded db-nets [55] (imposing both state-boundedness of the underlying database and boundedness of the net);
- controlled generation of fresh identifiers and other modeling guidelines [53,62].

For a state-bounded process, fundamental properties such as reachability and safety are decidable. This triggers the question whether verification of first-order temporal formulae is also decidable.

[21] establishes decidability of verification of state-bounded DCDSs against properties expressed in variants of first-order μ-calculus. This is shown constructively, computing a finite state space that faithfully represents the infinite one generated by the process. The abstraction is built by considering the input DCDS, the bound, and the number of variables contained in the formula of interest. Notably, the abstraction can be computed also if the bound is not known: in that case, it is incrementally built, showing that after finitely many steps no new state need to be generated [7].

Among the fragments of first-order μ-calculus originally defined in [7], one of particular interest enforces *persistent quantification*. Intuitively, this enforces that formulae are able to track the identity of objects only if they persist in the databases across consecutive states (whereas objects disappearing from a state are not tracked anymore). For this logic, which can be defined as a syntactic fragment of general first-order μ-calculus, it is shown that the abstraction only depends on the process and the bound, but not on the formula.

The idea of persistent quantification is essential when studying verification against first-order LTL. In fact, linear-time verification turns out to be much more challenging when compared to a branching-time setting: verification of first-order LTL properties is undecidable over state-bounded DCDSs where every state contains at most a single object [21]. This is shown via a reduction from satisfiability of LTL with freeze quantifiers [28]. [22] further explores the reason behind undecidability, singling out that it is due to the ability of the logic to unrestrictedly quantify over objects that may appear in two databases contained in states arbitrarily far away from each other. In fact, decidability of linear-time verification is obtained for the fragment of first-order LTL restricted to persistent quantification, which can be characterised syntactically [22].

Three main questions are left open in this line of research:

1. How to handle multiple datatypes. Positive results beyond databases containing identifiers are only given for ordered domains equipped only with comparison operators [23].
2. How to verify processes with unboundedly growing databases. This is interesting not only for the database setting, but also essential when studying object-centric processes where unboundedly many objects can be generated.

3. How to obtain effective, implementable algorithmic techniques. The construct of the faithful finite-state abstraction is in fact explicit and highly combinatorial, and consequently does not lend itself to be concretely implemented.

These three questions are tackled next.

3.2   Analysis of Processes with Cases Variables

The presence of case variables over multiple datatypes, explored in its full generality in the DMT approach [42], makes the results obtained for state-bounded processes inapplicable. In fact, DMTs easily encode two-counter machines when they are equipped with two numerical variables and conditions expressing increment, decrement, and test-for-zero. Even though processes of this form are state-bounded with a bound of 2, the presence of structure in the underlying datatype makes verification much more difficult than in the case of databases containing identifiers that can only be compared for (in)equality.

Fortunately, dealing with multiple datatypes, which from the logical point of view means dealing with multiple first-order theories, is a well-established problem in the area of automated reasoning known as *satisfiability modulo theories* (SMT [11]). Well-studied SMT theories are the theory of uninterpreted functions, the theory of bitvectors, and the theory of arrays. Widely investigated are also different types of arithmetics for which specific decision procedures are available, like linear arithmetics for integer and rational numbers. Alongside with a rich foundational framework for dealing with such theories, industry-strength SMT solvers exist, such as Yices, CVC5, and MathSAT5. In addition, the SMT-LIB international promotes the adoption of common concrete languages and interfaces for SMT solvers. These aspects are essential towards concrete, feasible implementations of analysis techniques.

While the analysis of case-centric processes operating over numerical variables can take advantage from *quantifier elimination*, as done in [36], dealing with richer datatypes calls for more sophisticated techniques based on *uniform interpolation* [19,20]. This is exploited in [42] to provide a very general framework for linear-time verification of DMTs. The considered logic is a data-aware extension of LTL over finite traces, where state formulae consist of conditions over data variables, referring to their current and possibly next state.[2] For such a logic, the classical automata-theoretic approach to linear-time verification is lifted to the richer case of DMTs, with a fully symbolic approach grounded in SMT: an SMT formula is constructed from the cross-product of the input process and the temporal formula of interest, reducing verification to satisfiability of that formula. Due to undecidability, the corresponding verification procedure is not guaranteed to terminate. In particular, the SMT formula may not be finite.

Interestingly, for the class of so-called *DMTs with finite history*, a semantic notion originally defined for the arithmetic setting in [36], it is shown that the

---

[2] This form of "bounded lookahead" cannot be further relaxed, as comparing data variables in states that are arbitrarily away from each other makes the logic too expressive [22,28].

resulting SMT formula is indeed of a finite size, in turn yielding decidability. While enjoying finite history is a semantic property, undecidable to check, [42] enumerates a series of datatypes with corresponding condition languages that guarantee finite history, reconstructing and subsuming several decidable classes identified in the literature.

Notably, this approach to DMT verification has been fully implemented in the tool lindmt[3], using different SMT solvers as back-end. Feasibility of the implementation in terms of performance is shown in [42].

For a more comprehensive survey of SMT techniques for the analysis of data-aware, case-centric processes, the interested reader can refer to [44].

3.3   Analysis of Unbounded, Object-Centric Processes

The analysis of object-centric processes where only boundedly many objects coexist in every single state of the process can be tackled using the state-bounded setting discussed in Sect. 3.1. However, imposing such a bound may be restrictive in this setting, where inherently unboundedly many objects can be created and co-evolved. In addition, even in a case-centric setting where each case operates over an underlying database, it may be of interest to analyse the process without imposing any restriction on how many tuples can be stored in the database tables.

In spite of being a highly undecidable setting, key results on the safety analysis of these processes have been obtained thanks to a correspondence with array-based systems in SMT [39,40] – which can be seen as a fully symbolic, model-theoretic counterpart of well-structured transition systems. Safety analysis amounts to verifying that the process of interest never reaches a configuration containing undesired patterns of data - something that is intimately connected to a first-order version of coverability for data-aware Petri nets [37]. The core idea in relating array-based systems with object-centric processes (which is essentially the same used to relate array-based systems with case-centric processes operating over a database) is first established in [18]. Essentially, it postulates that a place containing n-tuples of identifiers ranging over possibly different types can be represented using n arrays, where the i-th array stores the elements of the i-th component of the tuples (with matching type). A tuple is retrieved by accessing all such arrays with the same index.

This idea requires to develop genuinely novel model-theoretic results for carrying the standard SMT backward reachability procedure for safety checking of array-based systems to this richer setting [18,41]. Thanks to the major foundational advancements in [18,41], a general sound and complete procedure for backward reachability of processes operating over unbounded relations is obtained, together with the identification of interesting classes for which termination is also guaranteed (in turn witnessing decidability). This has been exploited as the main basis to obtain safety analysis procedures for DABs [24] and COA nets [37] - essentially covering all the features of PNIDs. While this natively supports

---

[3] https://lindmt.unibz.it/.

subset synchronisation (where an object synchronises with one or more related objects), dealing with exact synchronisation (when an object synchronises with all its related objects) requires to introduce universal quantification, a feature not present in standard array-based systems. This aspect has been studied in [38], showing that in this advanced setting, a backward reachability procedure for safety can be defined enjoying completeness and partial soundness. This provides a solid basis to investigate safety analysis of synchronous proclets [32] and OPIDs [43].

Notably, [38] also establishes for the first time that procedures dealing with object identifiers and relationships, as well as those handling numerical data with arithmetics, can be safely combined. This paves the way towards analysis of object-centric processes dealing with objects, relationships, and numerical attributes.

We close this section by pointing out that all the foundational results mentioned above have been correspondingly ported to a working implementation, suitably extending the well-established SMT model checker MCMT for array-based systems[4]. A large experimental evaluation on data-aware processes covering the different features mentioned above shows feasibility of this approach [24,38,41].

4   Conformance Checking

Data-aware process mining, that is, process mining over data-aware processes, is increasingly getting momentum, due to its ability of providing more informative insights on how organisations operate [2,13,52].

We focus here on conformance checking [25]: the task of relating an observed trace with a reference process model, to detect whether the observed trace is a valid behaviour according to the process and, if it not, to what extent it deviates from the model traces (that is, sequences of maximal transition firings supported by the process model).

Among the different conformance checking results, of particular importance are (optimal) *alignments* [15]. An alignment is a sequence of moves relating events in an observed trace and transition firings in a model trace, where every move is of one among three possible types:

- a *log move* if the observed trace progresses of one event, while the model trace does not progress;
- a *model move* if the observed trace does not progress, while the model trace progresses of one firing;
- a *synchronous move* if both the observed trace and the model trace move of one step, agreeing on the involved event/transition.

By defining a cost function that penalises log and model moves, the best alignment is the one that minimises the cost; typically, variants of Levenshtein distance are adopted for defining costs [15]. Given an observed trace and a process

---

[4] https://homes.di.unimi.it/~ghilardi/mcmt/home.html.

model, an optimal alignment consists in retrieving the model trace yielding the alignment with minimal cost overall.

We describe next how alignment-based conformance checking for data-aware processes can be effectively tamed using automated reasoning techniques based on SMT and its optimisation version. In *Optimisation Modulo Theories* (OMT [59, 60]), one is interested in finding, among the different satisfying assignments for the SMT formula of interest, those that are optimal for a given objective function.

4.1   Alignments for Case-Centric Processes with Case Variables

In this setting, observed traces indicate, for every recorded event, the new value of case variables that get updated. At the model level, transition firings come with a binding for written variables. In this light, a synchronous move does not only require that the event of the observed trace matches with the firing transition, but also that they agree on the variable updates [33, 34, 52]. Cost functions can distinguish deviations arising from event-transition mismatches, to deviations where the event and transition agree, but there are mismatches on updated variables (possibly fine-tuning the cost function depending on "how distant" the values are).

The first approach in this direction is [52], where data-aware alignments are defined and computed for DPNs. Better algorithmic bounds and a more general technique that works for multiple datatypes are provided by the conformance-checking approach CoCoMoT [33, 34]. CoCoMot lifts SAT-based approaches for computing pure control-flow alignments [14] to the setting of case variables with multiple datatypes, employing SMT/OMT as the underlying formal and operational framework. This makes the framework directly able to deal with alignments for DMTs that, as pointed out in Sect. 2.1, subsume DPNs.

The main idea behind CoCoMot is the following. When computing alignments with "well-behaved" cost functions, given an observed trace one can derive a maximum bound on the longest model traces to be considered when computing alignments [33, 34]. Intuitively, for a standard edit cost penalising every model and log move by one, this bound corresponds to the sum of the length of the observed trace and the shortest model trace; considering model traces longer than this sum can only degrade the overall alignment cost. Having an upper bound on the length, one can construct an SMT formula symbolically describing all model traces up to that length. Such an SMT formula can then be transformed into one that describes model, log, and synchronous moves with respect to the given observed trace. OMT solving is then employed to find the satisfying assignment of this formula that minimises the objective function computing the alignment cost.

CoCoMot is proven to be correct, and it is accompanied by a working implementation. Even though computing alignments is inherently combinatorial, and dealing with case variables makes it even more difficult, CoCoMot shows good performance thanks to the possibility of discharging the computation to well-established OMT solvers. In addition, the framework is declarative in spirit, and minimal changes to the constructed SMT formula and/or the given objective function make it possible to return conformance checking results that go beyond optimal alignments [34]. The interested reader can refer to [44] for a more detailed, gentle introduction to SMT/OMT-based conformance checking.

4.2 Alignments for Object-Centric Processes

Computing alignments for object-centric processes departs from case-centric approaches, as now executions consist of events that operate over multiple objects at once, and are thus "graph-structured" in nature. A suitable notion of alignment in this richer setting is introduced in [50] for object-centric Petri nets. Object-centric Petri nets inherently under-specify the process, as they are thought as a descriptive, not prescriptive, paradigm [3]. In fact, they strive for simplicity and do not explicitly deal with object identity, relationships, and synchronisation [43]. This may lead to overly optimistic alignments when observed traces contain undesired behaviour related to object reshuffling, such as inadvertently inserting items in wrong packages.

In [43], OPIDs are brought forward as a more precise model for defining prescriptive object-centric processes, merging the ability of object-centric Petri nets to account for batching transitions, with that of PNIDs of tracking in a fine-grained way objects and their possibly evolving relationships. SMT/OMT encodings are provided, lifting the CoCoMot approach to this richer setting. From the foundational point of view, it is shown that also in this object-centric setting, there is an upper bound on the maximal model trace length to consider. At the same time, it is not enough, in the general case, to only consider those objects that explicitly appear in the observed trace. However, another upper bound can be obtained to the maximum number of additional objects to be considered. Thanks to these two upper bounds, an SMT formula can be constructed to symbolically represent the object-centric alignments to be considered for a given observed trace, using OMT solving to return the best one(s). The resulting approach, called oCoCoMot, is proved to be correct. In addition, a working implementation is provided, whose feasibility is also experimentally shown in [43].

5 Conclusions

We have overviewed the main milestones in a decade-long research focussed on enriching Petri net-based models for work processes with different forms of data. We have considered case-centric processes where every process instance comes with a data payload, as well as the more recent "object-centric" wave where a process execution simultaneously operates over multiple interrelated objects. We have considered formal models of data-aware Petri nets in this wide spectrum, and overviewed how techniques stemming from automated reasoning have been adopted and further extended to deal with analysis and process mining tasks.

A number of open problems persist in this fascinating research. We mention three of them: one for modelling, two for analysis and mining. As for modelling, further research is needed to infuse object-centric processes with fine-grained data conditions expressed over data attributes - which is the focus of case-centric processes with case variables. DMTs [41] and OPIDs [43] provide a solid, compatible basis to realise this goal. Analysis and conformance have been so far tackled always assuming an interleaving semantics for concurrency, while the exploration of true concurrent semantics and corresponding algorithmic techniques, widely employed in the pure-control flow setting, is still a green field in the data-aware setting. Finally, object-centric process mining is still in its infancy, with a few algorithmic techniques for discovery [4,29], and seminal works laying formal foundations for this problem [10]. We aim at advancing our understanding of this problem by further exploring the connections between data-aware Petri nets and automated reasoning.

Aknowledgements. This work is partially supported by the NextGenerationEU FAIR PE0000013 project MAIPM (CUP C63C22000770006), and by the PRIN MIUR project PINPOINT Prot. 2020FNEB27.

I am deeply indebted to all the co-authors who accompanied my in this exciting research journey, among which, in particular: Babak Bagheri Hariri, Diego Calvanese, Giuseppe De Giacomo, Alin Deutsch, Dirk Fahland, Paolo Felli, Silvio Ghilardi, Alessandro Gianola, Fabio Patrizi, Andrey Rivkin, Jan Martijn van der Werf, and Sarah Winkler. Additional special thanks to Alessandro Gianola for the valuable inputs to this manuscript.

References

1. van der Aalst, W.: Process Mining - Data Science in Action, 2nd edn. Springer (2016). https://doi.org/10.1007/978-3-662-49851-4
2. van der Aalst, W.M.P.: Object-centric process mining: dealing with divergence and convergence in event data. In: Proceedings of the 17th SEFM (2019). https://doi.org/10.1007/978-3-030-30446-1_1
3. van der Aalst, W.M.P.: Toward more realistic simulation models using object-centric process mining. In: Vicario, E., Bandinelli, R., Fani, V., Mastroianni, M. (eds.) Proceedings of the 37th ECMS International Conference on Modelling and Simulation, pp. 5–13. European Council for Modeling and Simulation (2023). https://doi.org/10.7148/2023-0005
4. van der Aalst, W., Berti, A.: Discovering object-centric Petri nets. Fundam. Inform. **175**(1–4), 1–40 (2020)
5. van der Aalst, W., Weske, M., Grünbauer, D.: Case handling: a new paradigm for business process support. Data Knowl. Eng. **53**(2), 129–162 (2005). https://doi.org/10.1016/J.DATAK.2004.07.003
6. Artale, A., Calvanese, D., Montali, M., van der Aalst, W.M.P.: Enriching data models with behavioral constraints. In: Borgo, S., Ferrario, R., Masolo, C., Vieu, L. (eds.) Ontology Makes Sense - Essays in honor of Nicola Guarino. Frontiers in Artificial Intelligence and Applications, vol. 316, pp. 257–277. IOS Press (2019). https://doi.org/10.3233/978-1-61499-955-3-257

7. Bagheri Hariri, B., Calvanese, D., De Giacomo, G., Deutsch, A., Montali, M.: Verification of relational data-centric dynamic systems with external services, pp. 163–174 (2013)
8. Bagheri Hariri, B., Calvanese, D., Montali, M., Deutsch, A.: State-boundedness in data-aware dynamic systems. In: Baral, C., Giacomo, G.D., Eiter, T. (eds.) Principles of Knowledge Representation and Reasoning: Proceedings of the Fourteenth International Conference, KR 2014, Vienna, Austria, 20–24 July 2014. AAAI Press (2014). http://www.aaai.org/ocs/index.php/KR/KR14/paper/view/8028
9. Baral, C., Giacomo, G.D.: Knowledge representation and reasoning: what's hot. In: Bonet, B., Koenig, S. (eds.) Proceedings of the Twenty-Ninth AAAI Conference on Artificial Intelligence (AAAI 2015), pp. 4316–4317. AAAI Press (2015)
10. Barenholz, D., Montali, M., Polyvyanyy, A., Reijers, H.A., Rivkin, A., van der Werf, J.M.E.M.: There and back again - on the reconstructability and rediscoverability of typed Jackson nets. In: Gomes, L., Lorenz, R. (eds.) PETRI NETS 2023. LNCS, vol. 13929, pp. 37–58. Springer, Cham (2023). https://doi.org/10.1007/978-3-031-33620-1_3
11. Barrett, C.W., Sebastiani, R., Seshia, S.A., Tinelli, C.: Satisfiability modulo theories. In: Biere, A., Heule, M., van Maaren, H., Walsh, T. (eds.) Handbook of Satisfiability. Frontiers in Artificial Intelligence and Applications, 2nd edn., vol. 336, pp. 1267–1329. IOS Press (2021)
12. Belardinelli, F., Lomuscio, A., Patrizi, F.: Verification of agent-based artifact systems. J. Artif. Intell. Res. **51**, 333–376 (2014)
13. Berti, A., Montali, M., van der Aalst, W.M.P.: Advancements and challenges in object-centric process mining: a systematic literature review. CoRR abs/2311.08795 (2023)
14. Boltenhagen, M., Chatain, T., Carmona, J.: Optimized SAT encoding of conformance checking artefacts. Computing **103**(1), 29–50 (2021)
15. Bose, R., van der Aalst, W.: Process diagnostics using trace alignment: opportunities, issues, and challenges. Inf. Syst. **37**(2), 117–141 (2012)
16. Breitmayer, M., Arnold, L., Pejic, M., Reichert, M.: Transforming object-centric process models into BPMN 2.0 models in the philharmonicflows framework. In: Proceedings Modellierung 2024. LNI, vol. P-348, pp. 83–98 (2024). https://doi.org/10.18420/MODELLIERUNG2024_009
17. Calvanese, D., De Giacomo, G., Montali, M.: Foundations of data-aware process analysis: a database theory perspective. In: Proceedings of PODS 2013, pp. 1–12 (2013)
18. Calvanese, D., Ghilardi, S., Gianola, A., Montali, M., Rivkin, A.: SMT-based verification of data-aware processes: a model-theoretic approach. Math. Struct. Comput. Sci. **30**(3), 271–313 (2020). https://doi.org/10.1017/S0960129520000067
19. Calvanese, D., Ghilardi, S., Gianola, A., Montali, M., Rivkin, A.: Model completeness, uniform interpolants and superposition calculus (with applications to verification of data-aware processes). J. Autom. Reason. **65**(7), 941–969 (2021). https://doi.org/10.1007/S10817-021-09596-X
20. Calvanese, D., Ghilardi, S., Gianola, A., Montali, M., Rivkin, A.: Combination of uniform interpolants via Beth definability. J. Autom. Reason. **66**(3) (2022). https://doi.org/10.1007/S10817-022-09627-1
21. Calvanese, D., De Giacomo, G., Montali, M., Patrizi, F.: First-order μ-calculus over generic transition systems and applications to the Situation Calculus. Inform. Comput. **259**(3), 328–347 (2018)

22. Calvanese, D., De Giacomo, G., Montali, M., Patrizi, F.: Verification and monitoring for first-order LTL with persistence-preserving quantification over finite and infinite traces. In: Proceedings of IJCAI. AAAI Press (2022)
23. Calvanese, D., Delzanno, G., Montali, M.: Verification of relational multiagent systems with data types. In: Proceedings of AAAI, pp. 2031–2037. AAAI Press (2015)
24. Calvanese, D., Ghilardi, S., Gianola, A., Montali, M., Rivkin, A.: Formal modeling and SMT-based parameterized verification of data-aware BPMN. In: Hildebrandt, T., van Dongen, B.F., Röglinger, M., Mendling, J. (eds.) BPM 2019. LNCS, vol. 11675, pp. 157–175. Springer, Cham (2019). https://doi.org/10.1007/978-3-030-26619-6_12
25. Carmona, J., van Dongen, B.F., Solti, A., Weidlich, M.: Conformance Checking - Relating Processes and Models. Springer, Heidelberg (2018). https://doi.org/10.1007/978-3-319-99414-7
26. Cohn, D., Hull, R.: Business artifacts: a data-centric approach to modeling business operations and processes. IEEE Data Eng. Bull. **32**(3), 3–9 (2009). http://sites.computer.org/debull/A09sept/david.pdf
27. De Giacomo, G., Lesperance, Y., Patrizi, F.: Bounded Situation Calculus action theories. Artif. Intell. **237**, 172–203 (2016)
28. Demri, S., Lazic, R.: LTL with the freeze quantifier and register automata. ACM TOCL **10**(3) (2009)
29. van Detten, J.N., Schumacher, P., Leemans, S.J.J.: Discovering compact, live and identifier-sound object-centric process models. In: 6th International Conference on Process Mining, ICPM 2024, Kgs. Lyngby, Denmark, 14–18 October 2024. pp. 113–120. IEEE (2024). https://doi.org/10.1109/ICPM63005.2024.10680659
30. Deutsch, A., Hull, R., Li, Y., Vianu, V.: Automatic verification of database-centric systems. ACM SIGLOG News **5**(2), 37–56 (2018). https://doi.org/10.1145/3212019.3212025
31. Dumas, M., Rosa, M.L., Mendling, J., Reijers, H.A.: Fundamentals of Business Process Management, 2nd edn. Springer, Heidelberg (2018). https://doi.org/10.1007/978-3-662-56509-4
32. Fahland, D.: Describing behavior of processes with many-to-many interactions. In: Proceedings of the PETRI NETS (2019)
33. Felli, P., Gianola, A., Montali, M., Rivkin, A., Winkler, S.: CoCoMoT: conformance checking of multi-perspective processes via SMT. In: Polyvyanyy, A., Wynn, M.T., Van Looy, A., Reichert, M. (eds.) BPM 2021. LNCS, vol. 12875, pp. 217–234. Springer, Cham (2021). https://doi.org/10.1007/978-3-030-85469-0_15
34. Felli, P., Gianola, A., Montali, M., Rivkin, A., Winkler, S.: Data-aware conformance checking with SMT. Inf. Syst. **117** (2023). https://doi.org/10.1016/J.IS.2023.102230
35. Felli, P., de Leoni, M., Montali, M.: Soundness verification of data-aware process models with variable-to-variable conditions. Fundam. Inform. **182**(1), 1–29 (2021). https://doi.org/10.3233/FI-2021-2064
36. Felli, P., Montali, M., Winkler, S.: Linear-time verification of data-aware dynamic systems with arithmetic. In: Proceedings of the 36th AAAI Conference on Artificial Intelligence (AAAI 2022), pp. 5642–5650. AAAI Press (2022)
37. Ghilardi, S., Gianola, A., Montali, M., Rivkin, A.: Petri net-based object-centric processes with read-only data. Inf. Syst. **107**, 102011 (2022)

38. Ghilardi, S., Gianola, A., Montali, M., Rivkin, A.: Safety verification and universal invariants for relational action bases. In: Proceedings of the Thirty-Second International Joint Conference on Artificial Intelligence, IJCAI 2023, 19th-25th August 2023, Macao, SAR, China, pp. 3248–3257. ijcai.org (2023). https://doi.org/10.24963/IJCAI.2023/362

39. Ghilardi, S., Nicolini, E., Ranise, S., Zucchelli, D.: Towards SMT model checking of array-based systems. In: Armando, A., Baumgartner, P., Dowek, G. (eds.) IJCAR 2008. LNCS (LNAI), vol. 5195, pp. 67–82. Springer, Heidelberg (2008). https://doi.org/10.1007/978-3-540-71070-7_6

40. Ghilardi, S., Ranise, S.: Backward reachability of array-based systems by SMT solving: termination and invariant synthesis. Log. Methods Comput. Sci. **6**(4) (2010). https://doi.org/10.2168/LMCS-6(4:10)2010

41. Gianola, A.: Verification of Data-Aware Processes via Satisfiability Modulo Theories. LNBIP. Springer (2023). https://doi.org/10.1007/978-3-031-42746-6

42. Gianola, A., Montali, M., Winkler, S.: Linear-time verification of data-aware processes modulo theories via covers and automata. In: Wooldridge, M.J., Dy, J.G., Natarajan, S. (eds.) Thirty-Eighth AAAI Conference on Artificial Intelligence, AAAI 2024, Thirty-Sixth Conference on Innovative Applications of Artificial Intelligence, IAAI 2024, Fourteenth Symposium on Educational Advances in Artificial Intelligence, EAAI 2014, 20–27 February 2024, Vancouver, Canada, pp. 10525–10534. AAAI Press (2024). https://doi.org/10.1609/AAAI.V38I9.28922

43. Gianola, A., Montali, M., Winkler, S.: Object-centric conformance alignments with synchronization. In: Guizzardi, G., Santoro, F., Mouratidis, H., Soffer, P. (eds.) CAiSE 2024. LNCS, vol. 14663, pp. 3–19. Springer, Cham (2024). https://doi.org/10.1007/978-3-031-61057-8_1

44. Gianola, A., Montali, M., Winkler, S.: Smt techniques for data-aware process mining. Künstl Intell. (2025)

45. van Hee, K.M., Sidorova, N., Voorhoeve, M., van der Werf, J.: Generation of database transactions with petri nets. Fundam. Inform. **93**(1–3), 171–184 (2009). https://doi.org/10.3233/FI-2009-0095

46. Hewelt, M., Weske, M.: A hybrid approach for flexible case modeling and execution. In: La Rosa, M., Loos, P., Pastor, O. (eds.) BPM 2016. LNBIP, vol. 260, pp. 38–54. Springer, Cham (2016). https://doi.org/10.1007/978-3-319-45468-9_3

47. Künzle, V., Reichert, M.: PHILharmonicFlows: towards a framework for object-aware process management. J. Softw. Maint. Res. Pract. **23**(4), 205–244 (2011). https://doi.org/10.1002/SMR.524

48. Lasota, S.: Decidability border for petri nets with data: WQO dichotomy conjecture. In: Kordon, F., Moldt, D. (eds.) PETRI NETS 2016. LNCS, vol. 9698, pp. 20–36. Springer, Cham (2016). https://doi.org/10.1007/978-3-319-39086-4_3

49. de Leoni, M., Felli, P., Montali, M.: A holistic approach for soundness verification of decision-aware process models. In: Trujillo, J.C., et al. (eds.) ER 2018. LNCS, vol. 11157, pp. 219–235. Springer, Cham (2018). https://doi.org/10.1007/978-3-030-00847-5_17

50. Liss, L., Adams, J.N., van der Aalst, W.M.P.: Object-centric alignments. In: Proceedings of the ER (2023)

51. Mannhardt, F.: Multi-perspective Process Mining. Ph.D. thesis, Technical University of Eindhoven (2018)

52. Mannhardt, F., de Leoni, M., Reijers, H., van der Aalst, W.: Balanced multi-perspective checking of process conformance. Computing **98**(4), 407–437 (2016)

53. Montali, M., Calvanese, D.: Soundness of data-aware, case-centric processes. Int. J. Softw. Tools Technol. Transf. **18**(5), 535–558 (2016). https://doi.org/10.1007/S10009-016-0417-2

54. Montali, M., Rivkin, A.: Model checking Petri nets with names using data-centric dynamic systems. Formal Aspects Comput. **28**(4), 615–641 (2016). https://doi.org/10.1007/s00165-016-0370-6

55. Montali, M., Rivkin, A.: DB-nets: on the marriage of colored petri nets and relational databases. Trans. Petri Nets Other Model. Concurr. **12**, 91–118 (2017). https://doi.org/10.1007/978-3-662-55862-1_5

56. Polyvyanyy, A., van der Werf, J.M.E.M., Overbeek, S., Brouwers, R.: Information systems modeling: language, verification, and tool support. In: Proceedings of the CAiSE (2019)

57. Reichert, M.: Process and data: two sides of the same coin? In: Meersman, R., et al. (eds.) OTM 2012. LNCS, vol. 7565, pp. 2–19. Springer, Heidelberg (2012). https://doi.org/10.1007/978-3-642-33606-5_2

58. Rosa-Velardo, F., de Frutos-Escrig, D.: Decidability problems in petri nets with names and replication. Fundam. Inform. **105**(3), 291–317 (2010)

59. Sebastiani, R., Tomasi, S.: Optimization modulo theories with linear rational costs. ACM Trans. Comput. Log. **16**(2), 12:1–12:43 (2015). https://doi.org/10.1145/2699915

60. Sebastiani, R., Trentin, P.: OptiMathSAT: a tool for optimization modulo theories. J. Autom. Reason. **64**(3), 423–460 (2020). https://doi.org/10.1007/S10817-018-09508-6

61. Snoeck, M., Verbruggen, C., Smedt, J.D., Weerdt, J.D.: Supporting data-aware processes with MERODE. Softw. Syst. Model. **22**(6), 1779–1802 (2023). https://doi.org/10.1007/S10270-023-01095-4

62. Solomakhin, D., Montali, M., Tessaris, S., De Masellis, R.: Verification of artifact-centric systems: decidability and modeling issues. In: Basu, S., Pautasso, C., Zhang, L., Fu, X. (eds.) ICSOC 2013. LNCS, vol. 8274, pp. 252–266. Springer, Heidelberg (2013). https://doi.org/10.1007/978-3-642-45005-1_18

63. Heath, F.T., et al.: Barcelona: a design and runtime environment for declarative artifact-centric BPM. In: Basu, S., Pautasso, C., Zhang, L., Fu, X. (eds.) ICSOC 2013. LNCS, vol. 8274, pp. 705–709. Springer, Heidelberg (2013). https://doi.org/10.1007/978-3-642-45005-1_65

64. van der Werf, J., Rivkin, A., Montali, M., Polyvyanyy, A.: Correctness notions for petri nets with identifiers. Fundam. Inform. **190**(2–4), 159–207 (2024). https://doi.org/10.3233/FI-242169

65. van der Werf, J.M.E.M., Rivkin, A., Polyvyanyy, A., Montali, M.: Data and process resonance - identifier soundness for models of information systems. In: Proceedings of the PETRI NETS (2022)

66. Weske, M.: Business Process Management - Concepts, Languages, Architectures, 3rd edn. Springer, Heidelberg (2019). https://doi.org/10.1007/978-3-662-59432-2

Petri Nets and Higher-Dimensional Automata

Amazigh Amrane[1] , Hugo Bazille[1][(✉)] , Uli Fahrenberg[1,2] , Loïc Hélouët[4] , and Philipp Schlehuber-Caissier[3]

[1] EPITA Research Lab (LRE), Le Kremlin-Bicêtre, France
hugo1.bazille@epita.fr
[2] IRISA & Inria Rennes, Rennes, France
[3] SAMOVAR, Télécom SudParis, Institut Polytechnique de Paris, Palaiseau, France
[4] IRISA, Rennes, France

Abstract. Petri nets and their variants are often considered through their interleaved semantics, *i.e.* considering executions where, at each step, a single transition fires. This is clearly a miss, as Petri nets are a true concurrency model. This paper revisits the semantics of Petri nets as higher-dimensional automata (HDAs) as introduced by van Glabbeek, which methodically take concurrency into account. We extend the translation to include some common features. We consider nets with inhibitor arcs, under both concurrent semantics used in the literature, and generalized self-modifying nets. Finally, we present a tool that implements our translations.

Keywords: Petri net · Higher-dimensional automaton · Concurrency · Inhibitor arc · Generalized self-modifying net

1 Introduction

We revisit the concurrent semantics of Petri nets as higher-dimensional automata (HDAs). In both Petri nets and HDAs, events may occur simultaneously, and both formalisms make a distinction between parallel composition $a \parallel b$ and choice $a \cdot b + b \cdot a$. However, Petri nets are often considered through their interleaving semantics, annihilating this difference. As an example, Fig. 1 shows two Petri nets and their HDA semantics (see below for definitions): on the left, transitions a and b are mutually exclusive and may be executed in any order but not concurrently; on the right, there is true concurrency between a and b, signified by the filled-in square of the HDA semantics. In interleaving semantics, no distinction is made between the two nets and both give rise to the transition system on the left. To take this into account, *concurrent step semantics* has been introduced [19]. While it captures executions that are impossible when only considering interleavings, it

---

P. Schlehuber-Caissier: Partially funded by the Academic and Research Chair "Architecture des Systèmes Complexes" Dassault Aviation, Naval Group, Dassault Systèmes, KNDS France, Agence de l'Innovation de Défense, Institut Polytechnique de Paris.

E. Amparore and L. Mikulski (Eds.): PETRI NETS 2025, LNCS 15714, pp. 18–40, 2025.
https://doi.org/10.1007/978-3-031-94634-9_2

still misses some possible behaviors, as shown later in the paper. Our goal is thus to introduce a new semantics which encompasses concurrent step semantics and adds the executions that are missed in this one. HDAs are natural candidates for this, and we show that HDA semantics encompasses concurrent step semantics but is more general.

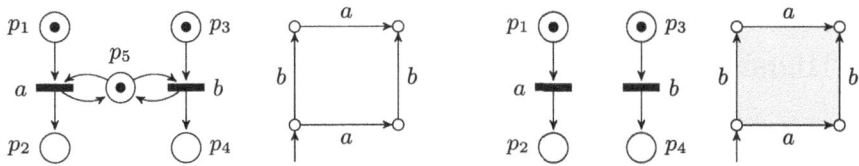

Fig. 1. Petri nets and HDAs for interleaving (left) and true concurrency (right).

The relations between Petri nets and HDAs were first explored by van Glabbeek in [32], where an HDA is defined as a labeled precubical set whose cells are hypercubes of different dimensions. More recently, [14] introduces an event-based setting for HDAs, defining their cells as totally ordered sets of labeled events. This framework has led to a number of new developments in the theory of HDAs [4,6,15,17], so here we set out to update van Glabbeek's translation to this event-based setting.

Petri nets are a powerful model that can represent infinite systems, and yet preserve decidability of reachability [26] and coverability [24]. Despite their expressiveness, Petri nets miss some features that are essential to represent program executions. In [18], the authors introduce inhibitor arcs, which allow preventing a transition t from firing when a place connected to t by an inhibitor arc is not empty. Obviously, this construction allows for the implementation of a zero test, which makes Petri nets with inhibitor arcs Turing powerful.

We investigate the concurrent semantics of Petri nets with inhibitor arcs (PNIs), showing that the a-posteriori semantics of [22] gives again rise to HDAs. For the more liberal a-priori semantics (see again [22]) however, we need to introduce *partial* HDAs in which some cells may be missing, mimicking the fact that some serialisations of concurrent executions are forbidden. We further expand our work to the generalized self-modifying nets of [12], giving their concurrent semantics as *ST-automata* which themselves generalize partial HDAs.

We have developed a prototype tool which implements the translations from Petri nets to HDAs and from PNIs to partial HDAs.[1] Our implementation is able to deal with standard, weighted and inhibitor arcs in a modular fashion.

This article is organised as follows. We begin in Sect. 2 and 3 by recalling HDAs and Petri nets, focusing on their concurrent semantics which allows several transitions to fire concurrently. The following sections present our proper contributions. In Sect. 4, we introduce our translation from Petri nets to HDAs,

---

[1] See https://gitlabev.imtbs-tsp.eu/philipp.schlehuber-caissier/pn2hda.

based on [32]. To overcome the symmetry of the HDAs thus built, Sect. 5 introduces an event order which avoids a factorial blow-up in the construction. We also give several examples to illustrate finer points in HDA semantics.

Then we consider Petri nets with inhibitor arcs in Sect. 6, both under a-posteriori and a-priori semantics, and generalized self-modifying nets in Sect. 7. Section 8 presents our implementation. We refer to the long version [5] for proofs of our results.

2   Higher-Dimensional Automata

Higher-dimensional automata (HDAs) extend finite automata with extra structure which permits to specify independence or concurrency of events. They consist of *cells* which are connected by *face maps*. Each cell has a list of events which are active, and face maps permit to pass from a cell to another in which some events have not yet started or are terminated.

We make this precise. A *conclist* (*concurrency list*) over an alphabet Σ is a tuple $(U, \dashrightarrow, \lambda)$, consisting of a finite set U (of events), a strict total order $\dashrightarrow \ \subseteq U \times U$ (called the *event order*), and a labeling $\lambda : U \to \Sigma$. Conclists represent labeled events running in parallel. If no confusion may arise, we will often refer to conclists by their underlying set only, writing U instead of $(U, \dashrightarrow, \lambda)$, and do the same for other algebraic structures defined throughout. Let $\square = \square(\Sigma)$ denote the set of conclists over Σ.

Conclists $(U_1, \dashrightarrow_1, \lambda_1)$ and $(U_2, \dashrightarrow_2, \lambda_2)$ are *isomorphic* if there is a mapping $\varphi : U_1 \to U_2$ such that $a \dashrightarrow_1 b$ iff $\varphi(a) \dashrightarrow_2 \varphi(b)$ and $\lambda_2 \circ \varphi = \lambda_1$. Isomorphisms between conclists are unique [14], so we may switch freely between conclists and their isomorphism classes in the sequel without mention.

Remark 1. The event order $\dashrightarrow$ is important as a book-keeping device but otherwise carries no computational meaning (see also Sect. 5 below). It plays a key role in distinguishing between events with the same label (autoconcurrency) and is needed for uniqueness of conclist isomorphisms. Conclists without event order are simply multisets, so conclists are multisets totally ordered by the event order, hence lists or words of Σ^*; but we often write them vertically to emphasize that the elements are running in parallel. Event order goes downwards if not indicated.

A *precubical set* $\big(X, \mathrm{ev}, \{\delta_{A,B;U} \mid U \in \square, A, B \subseteq U, A \cap B = \emptyset\}\big)$ consists of a set of *cells* X together with a function $\mathrm{ev} : X \to \square$. For a conclist U we write $X[U] = \{x \in X \mid \mathrm{ev}(x) = U\}$ for the cells of type U. Further, for every $U \in \square$ and $A, B \subseteq U$ with $A \cap B = \emptyset$ there are *face maps* $\delta_{A,B;U} : X[U] \to X[U \setminus (A \cup B)]$ which satisfy

$$\delta_{C,D;U \setminus (A \cup B)} \delta_{A,B;U} = \delta_{A \cup C, B \cup D; U} \tag{1}$$

for every $U \in \square$, $A, B \subseteq U$, and $C, D \subseteq U \setminus (A \cup B)$.

We will omit the extra subscript "U" in the face maps and further write $\delta_A^0 = \delta_{A,\emptyset}$ and $\delta_B^1 = \delta_{\emptyset,B}$. The *upper face maps* δ_B^1 transform a cell x into one in which the events in B have terminated; the *lower* face maps δ_A^0 transform x into a cell where the events in A have not yet started. Every face map $\delta_{A,B}$ can be written as a composition $\delta_{A,B} = \delta_A^0 \delta_B^1 = \delta_B^1 \delta_A^0$, and the *precubical identity* (1) expresses that these transformations commute.

We write $X_n = \{x \in X \mid |\mathrm{ev}(x)| = n\}$ for $n \in \mathbb{N}$ and call elements of X_n *n-cells*. The *dimension* of $x \in X$ is $\dim(x) = |\mathrm{ev}(x)| \in \mathbb{N}$; the dimension of X is $\dim(X) = \sup\{\dim(x) \mid x \in X\} \in \mathbb{N} \cup \{\infty\}$. For $k \in \mathbb{N}$, the *k-truncation* of X is the precubical set $X^{\leq k} = \{x \in X \mid \dim(x) \leq k\}$ with all cells of dimension higher than k removed.

A *higher-dimensional automaton* (*HDA*) $A = (\Sigma, X, \perp)$ consists of a finite alphabet Σ, a precubical set X on Σ, and a subset $\perp \subseteq X$ of initial cells. (We will not need accepting cells in this work.) An HDA may be finite or infinite, or even infinite-dimensional.

Computations of HDAs are *paths*, *i.e.* sequences

$$\alpha = (x_0, \varphi_1, x_1, \ldots, x_{n-1}, \varphi_n, x_n) \tag{2}$$

consisting of cells x_i of X and symbols φ_i which indicate which type of step is taken: for every $i \in \{1, \ldots, n\}$, $(x_{i-1}, \varphi_i, x_i)$ is either

- $(\delta_A^0(x_i), \nearrow^A, x_i)$ for $A \subseteq \mathrm{ev}(x_i)$ (an *upstep*)
- or $(x_{i-1}, \searrow_A, \delta_A^1(x_{i-1}))$ for $A \subseteq \mathrm{ev}(x_{i-1})$ (a *downstep*).

Intuitively, a downstep terminates events in a cell, following an upper face map. This is why downsteps require that $A \subseteq \mathrm{ev}(x_{i-1})$, *i.e.* events that are terminated belong to the cell. Similarly, an upstep starts events by following inverses of lower face maps. The constraints on upsteps require that $A \subseteq \mathrm{ev}(x_i)$, *i.e.* the initiated events belong to the next cell after the step. Both types of steps may be empty.

A cell $x \in X$ is *reachable* if there exists a path α from an initial cell to x, *i.e.* $x_0 \in \perp$ and $x_n = x$ in the notation (2) above. The *essential* part of X is the subset $\mathrm{ess}(X) \subseteq X$ containing only reachable cells. It is not necessarily an HDA, as some faces may be missing.

Example 2. Figure 2 shows a two-dimensional HDA as a combinatorial object (left) and in a geometric realisation (right). It consists of 21 cells: states $X_0 = \{v_1, \ldots, v_8\}$ in which no event is active ($\mathrm{ev}(v_i) = \emptyset$); transitions $X_1 = \{t_1, \ldots, t_{10}\}$ in which one event is active (*e.g.* $\mathrm{ev}(t_3) = \mathrm{ev}(t_4) = c$); and squares $X_2 = \{q_1, q_2, q_3\}$ with $\mathrm{ev}(q_1) = \begin{bmatrix} a \\ c \end{bmatrix}$ and $\mathrm{ev}(q_2) = \mathrm{ev}(q_3) = \begin{bmatrix} a \\ d \end{bmatrix}$.

The arrows between cells in the left representation correspond to the face maps connecting them. For example, the upper face map δ_{ac}^1 maps q_1 to v_4 because the latter is the cell in which the active events a and c of q_1 have been terminated. On the right, face maps are used to glue cells, so that for example $\delta_{ac}^1(q_1)$ is glued to the top right of q_1. In this and other geometric realisations, when we have two concurrent events a and c with $a \dashrightarrow c$, we will draw a horizontally and c vertically.

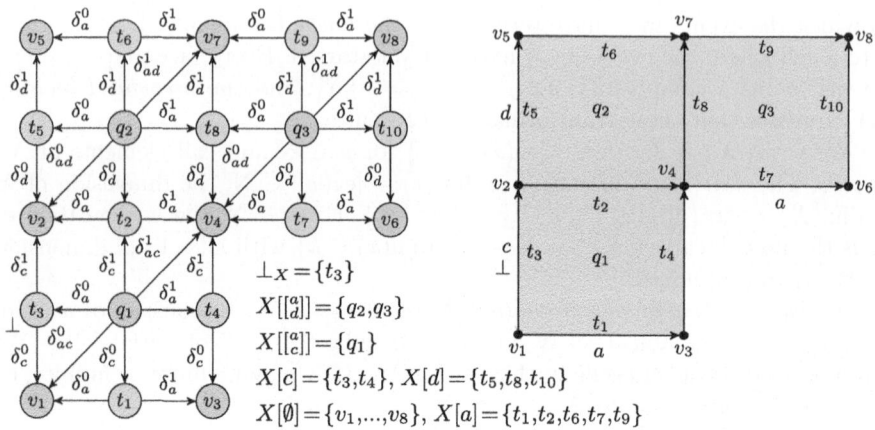

Fig. 2. A two-dimensional HDA X on $\Sigma = \{a, c, d\}$, see Example 2.

The HDA X of Fig. 2 admits several paths, for example $t_3 \nearrow^a q_1 \searrow_c t_2 \nearrow^d q_2 \searrow_a t_8 \nearrow^a q_3 \searrow_{ad} v_8$. Note that $\mathsf{ess}(X) = X \setminus \{v_1, t_1, v_3\}$ is not an HDA, and $X = X^{\leq k}$ for all $k \geq 2$.

Remark 3. We often abuse notation and denote conclists by their labels instead of their events, writing for example $\begin{bmatrix} a \\ c \end{bmatrix}$ for the conclist $(\{e_1 \dashrightarrow e_2\}, \lambda : (e_1 \mapsto a, e_2 \mapsto c))$. (We have already done so in the example above.) As long as there is no autoconcurrency, this abuse of notation is safe and brings no ambiguity; but if we need to assign the same label to several events, we will make events and their labeling explicit in our representation of conclists, writing for instance $\begin{bmatrix} e_1 \mapsto a \\ e_2 \mapsto c \end{bmatrix}$ for the above example.

3   Petri Nets

A *Petri net* $N = (S, T, F)$ consists of a set of places S, a set of transitions T, with $S \cap T = \emptyset$, and a weighted flow relation $F : S \times T \cup T \times S \to \mathbb{N}$. A *marking* of N is a function $m : S \to \mathbb{N}$. Such marking is *k-bounded* if $m(s) \leq k$ for every place s. It is *bounded* if there is a value $k \in \mathbb{N}$ such that m is k-bounded.

Let X be any set. A function $f : X \to \mathbb{N}$ is a *multiset, i.e.* an extension of sets allowing several instances of each element of X. We introduce some notation for these. We write $x \in f$ if $f(x) \geq 1$. Given two multisets f_1, f_2 over X we will write $f_1 \leq f_2$ iff $f_1(x) \leq f_2(x)$ for every element $x \in X$. If $f(x) \in \{0, 1\}$ for all x, then f may be seen as a set, and the notation $x \in f$ agrees with the usual one for sets. The multisets we use will generally be finite in the sense that $\sum_{x \in X} f(x) < \infty$, and in that case we might use additive notation and write $f = \sum_{x \in X} f(x)x$. This notation easily applies to markings of Petri nets, and we will write for instance $m = 2p_1 + p_4$ for a marking such that $m(p_1) = 2, m(p_4) = 1$, and $m(p_i) = 0$ for any other place $p_i \in S \setminus \{p_1, p_4\}$.

For a transition $t \in T$, the *preset* of t is the multiset $^\bullet t : S \to \mathbb{N}$ given by $^\bullet t(s) = F(s,t)$. This preset describes how many tokens are consumed in each place when t fires. The *postset* of t is the multiset $t^\bullet : S \to \mathbb{N}$ such that $t^\bullet(s) = F(t,s)$. It describes how many tokens are produced in each place of the net when firing t.

Petri nets compute by transforming markings. Their standard semantics is an interleaved semantics, where states are markings and a single transition can fire at each step. Let $m : S \to \mathbb{N}$ be a marking and $t \in T$, then t can *fire* in m if $^\bullet t \leq m$. Firing t produces a new marking $m' = m - {}^\bullet t + t^\bullet$.

The *reachability graph* (see for example [9]) of Petri net $N = (S,T,F)$ is the labeled graph $[\![N]\!]_1 = (V,E)$ given by $V = \mathbb{N}^S$ and

$$E = \{(m,t,m') \in V \times T \times V \mid {}^\bullet t \leq m, m' = m - {}^\bullet t + t^\bullet\}.$$

(The reason for the subscript 1 in $[\![N]\!]_1$ will become clear later).

In a reachability graph vertices are markings and edges are labeled by the transition which fires. A computation of a Petri net is a path in its reachability graph. Note that we use *collective token semantics*, *i.e.* tokens in $^\bullet t$ that are consumed by firing t are considered as blind resources. Petri nets also have an *individual token* semantics [20] where transitions distinguish tokens individually by considering their origin. This may be used to model realisation of independent processes; but we will not consider it here.

Let N_1, N_2 be two Petri nets. The reachability graphs $[\![N_1]\!]_1 = (V_1, E_1)$ and $[\![N_2]\!]_1 = (V_2, E_2)$ are *isomorphic*, denoted $[\![N_1]\!]_1 \cong [\![N_2]\!]_1$, if there exist bijections $f : V_1 \to V_2$ and $g : E_1 \to E_2$ such that for all $e_1 = (m_1, t_1, m_1') \in E_1$, $g(e_1) = (m_2, t_2, m_2')$ iff $f(m_1) = m_2$ and $f(m_1') = m_2'$.

Considering Petri nets via their interleaved semantics misses an important point of the model, namely concurrency. Indeed, it does not allow to distinguish between behaviors where a pair of transitions fire in sequence from behaviors where these transitions are independent and can fire concurrently. One way to cope with this issue is to consider executions of Petri nets as *processes* [20], that is, partial orders representing causal dependencies among transitions occurrences. Another possibility is the use of a *concurrent step semantics* [19], where several transitions are allowed to fire concurrently. The concurrent step semantics mimics that of the interleaved semantics, but fires multisets of transitions.

For a multiset $U : T \to \mathbb{N}$ of transitions we write $^\bullet U = \sum_{t \in T} {}^\bullet t\, U(t)$ and $U^\bullet = \sum_{t \in T} t^\bullet\, U(t)$. U is *firable* in marking m if $^\bullet U \leq m$. The *concurrent step reachability graph* [28] of Petri net $N = (S,T,F)$ is the labeled graph $[\![N]\!]_{\mathrm{CS}} = (V,E)$ given by $V = \mathbb{N}^S$ and

$$E = \{(m,U,m') \in V \times \mathbb{N}^T \times V \mid U \neq \emptyset, {}^\bullet U \leq m, m' = m - {}^\bullet U + U^\bullet\}. \quad (3)$$

Figure 3 shows a simple example of a Petri net and its two types of reachability graph. Note that transitions in $[\![N]\!]_{\mathrm{CS}}$ allow multisets of transition rather than only sets, thus several occurrences of a transition may fire in a concurrent step. This feature is called autoconcurrency, and it is well known that allowing autoconcurrency increases the expressive power of Petri nets [31]. Further,

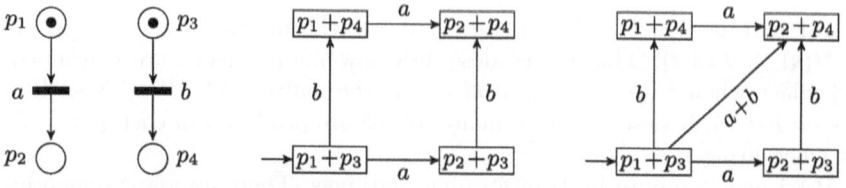

Fig. 3. A Petri net N (left); the reachability graph $[\![N]\!]_1$ (middle); and its concurrent step reachability graph $[\![N]\!]_{\text{CS}}$ (right).

$[\![N]\!]_{\text{CS}}$ is *closed under substeps* in the sense that for all multisets $V \subseteq U$, if $(m, U, m'') \in E$, then we also have $(m, V, m') \in E$ and $(m', U \setminus V, m'') \in E$ for some marking m'.

Notice that our definition of Petri nets allows preset-free transitions t with ${}^\bullet t = \emptyset$. When a transition t has an empty preset, then t is firable from any marking. In an interleaved semantics, allowing preset-free transitions does not change expressive power, so one frequently assumes that ${}^\bullet t \neq \emptyset$ for every $t \in T$. In the setting of a concurrent semantics with autoconcurrency, an arbitrary number of occurrences of each preset-free transitions may fire from any marking, making the transition relation of (3) of infinite degree. We will generally allow preset-free transitions in what follows.

4   From Petri Nets to HDAs

We expand the notion of reachability graph to a higher-dimensional automaton (HDA). The construction is an adaptation of [32, Def. 9] to the event-based setting of HDAs introduced in [14].

Let $N = (S, T, F)$ be a Petri net. Let $\square = \square(T)$ and define $X = \mathbb{N}^S \times \square$ and $\text{ev} : X \to \square$ by $\text{ev}(m, \tau) = \tau$. For $x = (m, \tau) \in X[\tau]$ with $\tau = (t_1, \ldots, t_n)$ non-empty and $i \in \{1, \ldots, n\}$, define

$$\delta_{t_i}^0(x) = (m + {}^\bullet t_i, (t_1, \ldots, t_{i-1}, t_{i+1}, \ldots, t_n)),$$
$$\delta_{t_i}^1(x) = (m + t_i^\bullet, (t_1, \ldots, t_{i-1}, t_{i+1}, \ldots, t_n)).$$

Using (1) to generate the other face maps, this defines a precubical set $[\![N]\!] = X$.

The 0-cells in X are markings of N, and in an n-cell of X, n transitions of N are running concurrently: the events of an n-cell (m, τ) are the elements of the conclist $\tau = (t_1, \ldots, t_n)$ of transitions. If a transition t appears multiple times in this sequence, then it is autoconcurrent. Not all precubical sets are in the image of the translation from Petri nets, see [32, Fig. 11] for an example.

Note that we translate (unlabeled) Petri nets to HDAs with *labeled* events: the labels of the events in $[\![N]\!]$ are the transitions of N. A path in $[\![N]\!]$ is a computation in N in which concurrent transitions can fire concurrently.

$X[\emptyset] = \{p_1+p_3, p_2+p_3, p_1+p_4, p_2+p_4\}$
$X[a] = \{(p_3,a),(p_4,a)\}$
$X[b] = \{(p_1,b),(p_2,b)\}$
$X[[{}^a_b]] = \{(0,[{}^a_b])\}$
$X[[{}^b_a]] = \{(0,[{}^b_a])\}$

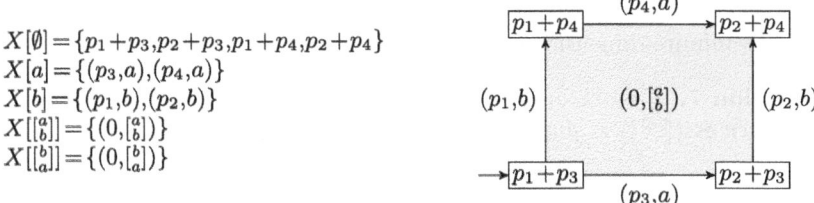

Fig. 4. Higher-dimensional automaton (reachable part only) for the Petri net of Fig. 3. Left: sets of cells; right: geometric realisation (not showing $X[[{}^b_a]]$).

Lemma 4. *The reachability graph of Petri net N is isomorphic to the 1-truncation of $[\![N]\!]$: $[\![N]\!]_1 \cong [\![N]\!]^{\leq 1}$.*

In order for this statement to make sense, we must consider $[\![N]\!]^{\leq 1}$ as a graph: vertices are 0-cells of $[\![N]\!]$, and edges triples of the form $(\delta^0_t(x), x = (m,t), \delta^1_t(x))$.

We can also relate $[\![N]\!]$ to its concurrent step reachability graph, as follows. For a sequence $a = (a_1,\ldots,a_n) \in \Box(\Sigma)$ on some alphabet Σ denote by $\mathrm{pi}(a)$: $\Sigma \to \mathbb{N}$ its *Parikh image*, *i.e.* the multiset given by counting symbols: $\mathrm{pi}(a)(x) = |\{i \mid a_i = x\}|$. For a precubical set X on Σ define a labeled graph $\mathrm{flat}(X) = (V, E)$ (the *flattening* of X) by $V = X_0$ and $E \subseteq V \times \mathbb{N}^\Sigma \times V$ given by

$$E = \{(x, U, z) \mid \exists y \in X : \delta^0_{\mathrm{ev}(y)}(y) = x, \delta^1_{\mathrm{ev}(y)}(y) = z, \mathrm{pi}(\mathrm{ev}(y)) = U\}.$$

That is, edges in $\mathrm{flat}(X)$ are labeled by multisets of events for which there exist corresponding cells in X, identifying all permutations in one edge. We may now reformulate Lemma 4 above to $[\![N]\!]_1 \cong \mathrm{flat}([\![N]\!]^{\leq 1})$, and for $[\![N]\!]_{\mathrm{CS}}$ the following is clear.

Lemma 5. *The concurrent step reachability graph of a Petri net N is isomorphic to the flattening of $[\![N]\!]$: $[\![N]\!]_{\mathrm{CS}} \cong \mathrm{flat}([\![N]\!])$.* □

Note that under this translation, $[\![N]\!]_{\mathrm{CS}}$ being closed under substeps corresponds to the fact that in $[\![N]\!]$, all faces of any cell are also present.

A Petri net $N = (S, T, F)$ together with an initial marking $i : S \to \mathbb{N}$ is called a *marked Petri net*. Now $i \in [\![N]\!][\emptyset]$, so this induces an HDA $[\![N]\!] = (T, X, \bot)$ with $\bot = \{i\}$.

A marked net N is *bounded* if all markings reachable from i in $[\![N]\!]_1$ are k–bounded for some $k \in \mathbb{N}$. Obviously, as firable transitions only depend on the current marking, and as the effect of a firing is deterministic, when a marked net is bounded, the reachable part of $[\![N]\!]_1$, *i.e.* $\mathrm{ess}([\![N]\!]^{\leq 1})$ is finite. However, due to autoconcurrency, this property does not hold for the full $\mathrm{ess}([\![N]\!])$, as show in the following example.

Example 6. Let $N = (\emptyset, \{a\}, F)$ be a Petri net with a single transition a, without places, and with an empty flow relation. With empty initial marking, N is bounded. Now $^\bullet a = \emptyset$, so a is firable in arbitrary autoconcurrency. In $\mathrm{ess}([\![N]\!])$

we get one cell in *every* dimension n: the n-fold autoconcurrency of a. That is, $\text{ess}(\llbracket N \rrbracket)$ is infinite-dimensional (and hence infinite).

Proposition 7. *If marked Petri net N is bounded and has no preset-free transitions, then $\text{ess}(\llbracket N \rrbracket)$ is finite.*

Figure 4 shows the HDA $\text{ess}(\llbracket N \rrbracket)$ for the Petri net N of Fig. 3 with initial marking $i = p_1 + p_3$. In particular, since by construction 0-cells in X are markings of N, $\text{ess}(\llbracket N \rrbracket)$ includes all faces, so is an HDA. (This is a general principle: for any HDA X with $\bot_X \subseteq X_0$, $\text{ess}(X)$ is also an HDA [17].)

Note that the 2-dimensional cell $(0, \left[{}^a_b \right])$ corresponds to the edge of $\llbracket N \rrbracket_{\text{CS}}$ between $p_1 + p_3$ and $p_2 + p_4$ in Fig. 3. Actually, we get *two* 2-dimensional cells, one with event $\left[{}^a_b \right]$ and the other with $\left[{}^b_a \right]$. (We have omitted the second in the geometric realisation.) This is somewhat unfortunate, as they should denote the same concurrent step $\{a, b\}$. We will show in Sect. 5 how to fix this problem.

5   Event Order

The above definition of the HDA $\llbracket N \rrbracket$ is highly symmetric: for a given marking m there is a cell (m, τ) for every sequence $\tau = (t_1, \ldots, t_n)$, even though *in fine* we are only interested in the *multiset* of concurrently active transitions. More precisely, for every permutation $\sigma \in \mathcal{S}_n$ in the nth symmetric group[2] there is a cell $(m, \tau \circ \sigma)$.

That is, $X = \llbracket N \rrbracket$ is a *symmetric* precubical set [21,29]: a precubical set equipped with actions $X_n \times \mathcal{S}_n \to X_n$ of the symmetric groups which are consistent with the face maps, see [21, Sect. 6].

In order to avoid this factorial blow-up, we may fix an arbitrary (non-strict) total order $\preccurlyeq$ on the transitions in T and then instead of $\square(T) \cong T^*$ consider the set

$$T^*_{\preccurlyeq} = \{(t_1, \ldots, t_n) \mid \forall i = 1, \ldots, n - 1 : t_i \preccurlyeq t_{i+1}\}.$$

The definition of the face maps of this reduced $X = \llbracket N \rrbracket$ stays the same, and X is now a (non-symmetric) precubical set with one cell for every marking m and every *multiset* of transitions τ.

The order $\preccurlyeq$ on T may be chosen arbitrarily, and changing it amounts to applying a permutation on X and passing to another, equivalent version of (the symmetrization of) X. Technically speaking, the forgetful functor from symmetric precubical sets to precubical sets is a *geometric morphism* [13,25] in that it has both a left and a right adjoint; this is precisely what is needed to be able to say that the order $\preccurlyeq$ on T is arbitrary and may be chosen and re-chosen at will.

We give some further examples of Petri nets and their HDA semantics, using a lexicographic order on transitions.

---

[2] The group of permutations of $\{1, \ldots, n\}$.

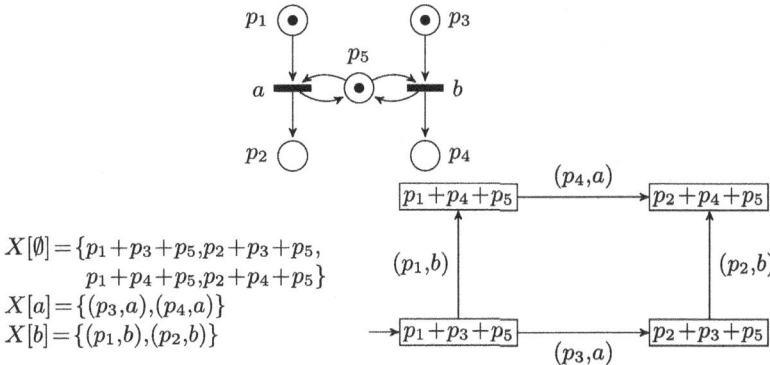

$X[\emptyset] = \{p_1+p_3+p_5, p_2+p_3+p_5,$
$\qquad p_1+p_4+p_5, p_2+p_4+p_5\}$
$X[a] = \{(p_3,a),(p_4,a)\}$
$X[b] = \{(p_1,b),(p_2,b)\}$

Fig. 5. Petri net (top) and HDA semantics (bottom) of Example 8.

Example 8. Figure 5 shows a Petri net N which executes a and b in mutual exclusion. We again show the essential part $X = \mathsf{ess}(\llbracket N \rrbracket)$; the initial cell is $p_1 + p_3 + p_5$. We prove that $X[[\begin{smallmatrix} a \\ b \end{smallmatrix}]] = \emptyset$, so that $\llbracket N \rrbracket$ is in fact isomorphic to the reachability graph $\llbracket N \rrbracket_1$. Assume $x = (m, [\begin{smallmatrix} a \\ b \end{smallmatrix}]) \in X[[\begin{smallmatrix} a \\ b \end{smallmatrix}]]$, then $\delta^0_{ab}(x) = m + p_1 + p_3 + 2p_5$, but there is no reachable marking m' with $m'(p_5) = 2$.

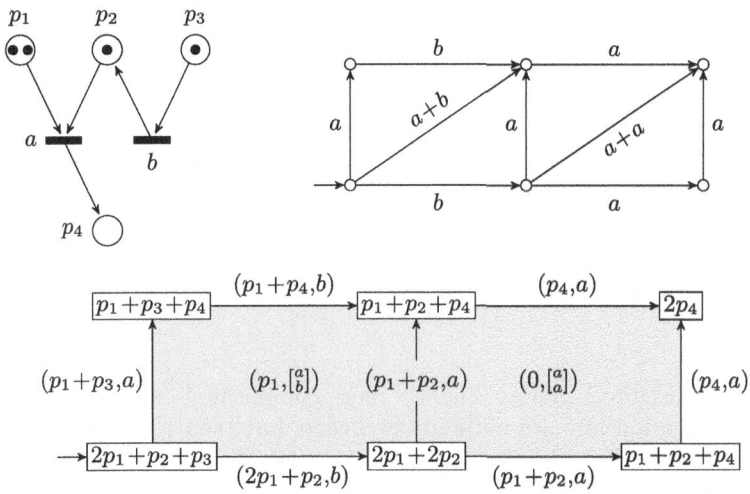

Fig. 6. Petri net (top left), its concurrent step semantics (top right) and HDA semantics (bottom) of Example 9. In the HDA, cells with the same label are identified.

Example 9. Figure 6 shows another Petri net, its concurrent step semantics and its HDA semantics. Note that there is *contact* between the transitions a and b: the pre-place p_2 is modified when firing b. We use this example to pinpoint that some possible behaviors are not captured by the concurrents step semantics. In both semantics, firing b before a enables autoconcurrency of a. However, in concurrent step semantics, as steps are atomic, firing b and a concurrently disables the possibility of autoconcurrency of a, while this is still an accepted behavior in the HDAs semantics. The illustration of the HDA is not quite correct, as the autoconcurrent square $x = (0, \left[\begin{smallmatrix} e_1 \mapsto a \\ e_2 \mapsto a \end{smallmatrix}\right])$ only has *two* different faces:[3] using $\cong$ for conclist isomorphism, we have $\delta^0_{e_1}(x) = (p_1 + p_2, e_2) \cong \delta^0_{e_2}(x) = (p_1 + p_2, e_1)$ and $\delta^1_{e_1}(x) = (p_4, e_2) \cong \delta^1_{e_2}(x) = (p_4, e_1)$.

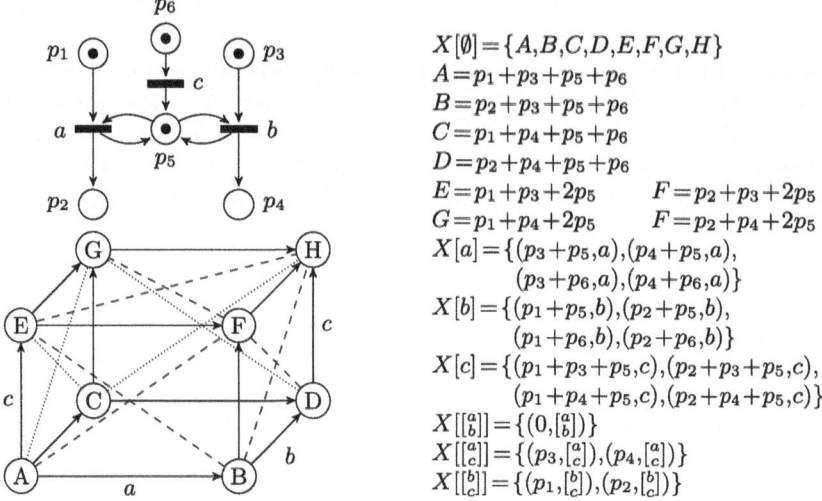

Fig. 7. Petri net and HDA semantics of Example 10. Concurrent squares are indicated using blue crosses instead of filled.

Example 10. Figure 7 shows a slightly more complicated example, where transitions a and b initially are mutually exclusive, but then c introduces independence. Geometrically this is an empty box without bottom face ("Fahrenberg's matchbox" [11]). We have $X\left[\left[\begin{smallmatrix} a \\ b \\ c \end{smallmatrix}\right]\right] = \emptyset$, for if $x = \left(m, \left[\begin{smallmatrix} a \\ b \\ c \end{smallmatrix}\right]\right) \in X\left[\left[\begin{smallmatrix} a \\ b \\ c \end{smallmatrix}\right]\right]$, then $\delta^0_{abc}(x) = m + p_1 + p_3 + 2p_5 + p_6$, which is unreachable.

---

[3] We need to use the extended conclist notation here due to autoconcurrency.

6   Inhibitor Arcs

We now extend our setting to Petri nets with inhibitor arcs. A *Petri net with inhibitor arcs* *(PNI)* $N = (S, T, F, I)$ consists of a Petri net (S, T, F) and a set $I \subseteq S \times T$ of inhibitor arcs. We denote by $°t = \{s \in S \mid (s, t) \in I\}$ the *inhibitor places* of $t \in T$.

The interleaved semantics for PNIs is as follows. Tokens in inhibitor places keep transitions from being firable, so a transition $t \in T$ can fire in marking $m : S \to \mathbb{N}$ if ${}^{\bullet}t \leq m$ and $\forall s \in °t : m(s) = 0$. The reachability graph of a PNI $N = (S, T, F, I)$ is the labeled graph $[\![N]\!]_1 = (V, E)$ given by $V = \mathbb{N}^S$ and $E \subseteq V \times T \times V$ with

$$E = \{(m, t, m') \mid {}^{\bullet}t \leq m, \forall s \in °t : m(s) = 0, m' = m - {}^{\bullet}t + t^{\bullet}\}.$$

PNIs have two different concurrent semantics, one which disables concurrent steps in which transitions may inhibit each other and one which does not. These are called, respectively, *a-posteriori* and *a-priori* semantics in [22], and we treat them both below. We refer to [22, Sect. 2] for an in-depth discussion of these semantics.

6.1   Concurrent a-Posteriori Semantics

In the a-posteriori semantics, a multiset U of transitions is firable in m if

(1) ${}^{\bullet}U \leq m$;
(2) for every $t \in U$ and every place $s \in °t$, $m(s) = 0$;
(3) for every $t_1, t_2 \in U$ such that $t_1 = t_2$ implies $U(t_1) \geq 2$, $t_1^{\bullet} \cap °t_2 = \emptyset$.

The last condition ensures that t_1 cannot produce a token that prevents t_2 from firing, even when t_1 and t_2 are autoconcurrent in U. With this condition, the transitions in U cannot inhibit *each other*. One advantage of a-posteriori semantics is that $[\![N]\!]_{\mathrm{CS}}$ is closed under substeps [8, Prop. 2.8].

Similarly to what we did for Petri nets, we can give an HDA semantics to a PNI $N = (S, T, F, I)$ under a-posteriori concurrent semantics, by restricting the cells to satisfy conditions (2) and (3) above. We let again $\square = \square(T)$ and define

$$X = \{(m, (t_1, \dots, t_n)) \in \mathbb{N}^S \times \square \mid \forall i = 1, \dots, n : \forall s \in °t_i : m(s) = 0,$$
$$\forall i \neq j = 1, \dots, n : t_i^{\bullet} \cap °t_j = \emptyset\}.$$

The rest of the definition of $[\![N]\!]$ now proceeds as before, and [8, Prop. 2.8] ensures that for any $x \in X$ and any $A \subseteq \mathsf{ev}(x)$, we also have $\delta_A^0(x), \delta_A^1(x) \in X$.

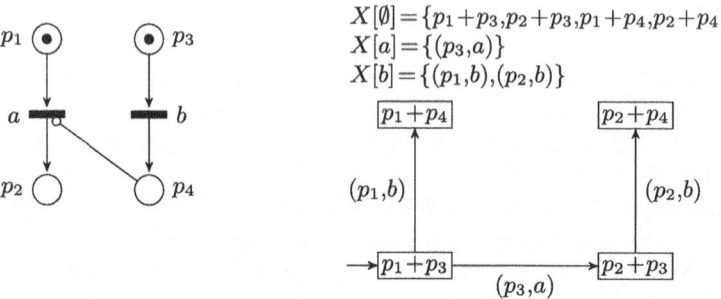

Fig. 8. PNI and HDA semantics of Example 11.

Example 11. Figure 8 shows the Petri net of Fig. 3 with an added inhibitor arc from p_4 to a. That is, transition a is disabled when there is a token in p_4, hence $X[a]$ contains only (p_3, a) but not (p_4, a). Compared to Fig. 4, the 2-cell $(0, \begin{bmatrix} a \\ b \end{bmatrix})$ is also disabled, as $b^\bullet = {}^\circ a = \{p_4\}$: there is no concurrency.

[8] also introduces a subclass of PNIs called *primitive systems* defined as follows. For a marked PNI $N = (S, T, F, I, i)$, let $Inib(N) = \{s \in S \mid \exists(s,t) \in I\}$ be the set of places used to inhibit some transition. Now N is called a *primitive system* if there exists a function $EL : Inib(N) \to \mathbb{N}$ such that for all $s \in Inib(N)$ and all reachable markings m with $m(s) > EL(s)$, if m' is reachable from m, then for all $t \in T$ with $\bullet t \leq m'$, we have $s \notin {}^\circ t$.

Intuitively, in primitive systems, when the bound $EL(s)$ is exceeded in some reachable marking m, then no transition with $s \in {}^\circ t$ is fired in markings that are reachable from m. [8] demonstrates that primitive systems can be simulated by Petri nets (without inhibitor arcs). However, the author shows that while her construction preserves interleaved semantics, it does not preserve concurrent step semantics; it is clear that the same is true for our HDA semantics.

6.2 Concurrent a-Priori Semantics

In the more liberal a-priori concurrent semantics, condition (3) of the multiset firing rules is removed. For intuition, consider again Example 11 and Fig. 8. In a-posteriori semantics, the concurrent step $U = \{a, b\}$ is disabled due to condition (3): b produces a token in inhibitor place p_4 connected to a. This seems rather restrictive: one might argue that *while* the b transition is firing, it has not yet produced a token in p_4, so it should not prevent from starting the firing of a.

On the other hand, if we add the 2-cell $(0, \begin{bmatrix} a \\ b \end{bmatrix})$ to the semantics, we are also forced to add its upper face $\delta_b^1((0, \begin{bmatrix} a \\ b \end{bmatrix})) = (p_4, a)$, given that all faces of cells must be present in HDAs. Now the cell (p_4, a) would have the upper left vertex $p_1 + p_4$ as a lower face, so semantically that means that we can fire a after firing b, which is clearly contrary to what an inhibitor arc should do. To give proper semantics to PNIs we thus must allow HDAs in which some faces are "missing".

These are called *partial HDAs* and have been introduced in [10,16]; we adapt their definition to our event-based setting.

A *partial precubical set* $\left(X, \text{ev}, \{\delta_{A,B;U} \mid U \in \Box, A, B \subseteq U, A \cap B = \emptyset\}\right)$ consists of a set of cells X together with a function $\text{ev} : X \rightarrow \Box$. Further, for every $U \in \Box$ and $A, B \subseteq U$ with $A \cap B = \emptyset$ there are partial face maps $\delta_{A,B;U} : X[U] \rightarrow X[U \setminus (A \cup B)]$ which satisfy

$$\delta_{C,D;U\setminus(A\cup B)}\delta_{A,B;U} \subseteq \delta_{A\cup C, B\cup D;U} \tag{4}$$

for every $U \in \Box$, $A, B \subseteq U$, and $C, D \subseteq U \setminus (A \cup B)$. Except for the face maps being partial, this is the same definition as for HDAs in Sect. 2; we again omit the extra subscripts "U". By the notation $\subseteq$ in (4) we mean that if $\delta_{A,B}$ and $\delta_{C,D}$ are defined, then also $\delta_{A\cup C, B\cup D}$ is defined and equal to the composition $\delta_{C,D}\delta_{A,B}$; but $\delta_{A\cup C, B\cup D}$ may be defined without one or both of the maps on the left-hand side being defined.

A *partial higher-dimensional automaton*, or *pHDA*, $(\Sigma, X, \bot)$ consists of a partial precubical set X on Σ together with a subset $\bot \subseteq X$ of initial cells.

We now give a-priori concurrent semantics to PNIs by translating them to pHDAs. Let $N = (S, T, F, I)$ be a PNI and $X = [\![(S, T, F)]\!]$ its standard HDA semantics, ignoring inhibitor arcs. Define a subset $X' \subseteq X$ by

$$X' = \{(m, (t_1, \ldots, t_n)) \in \mathbb{N}^S \times \Box \mid \forall i = 1, \ldots, n : \forall s \in {}^\circ t_i : m(s) = 0\}$$

and let $[\![N]\!] = X'$.

Hence a cell (m, τ) exists in $[\![N]\!]$ if none of the tokens in m inhibits any transition in τ. Compared to the contact-free semantics, we leave out Busi's last condition (3), thus allowing firing of subsets containing pairs of transitions t_1, t_2 with $t_1^\bullet \cap {}^\circ t_2 \neq \emptyset$.

Lemma 12. $[\![N]\!]$ *is a partial precubical set.*

If we extend the definition of flattening and truncation to partial HDAs and allow the concurrent step semantics to not be closed under substeps, we have the following analogues of Lemmas 4 and 5.

Lemma 13. *The reachability graph of PNI N is isomorphic to the 1-truncation of $[\![N]\!]$: $[\![N]\!]_1 \cong [\![N]\!]^{\leq 1}$.* $\qquad\Box$

Lemma 14. *The concurrent step reachability graph of PNI N is isomorphic to the flattening of $[\![N]\!]$: $[\![N]\!]_{\text{CS}} \cong \text{flat}([\![N]\!])$.* $\qquad\Box$

We also have the following analogue to Prop. 7.

Proposition 15. *If a marked PNI N is bounded and has no preset-free transitions, then $\text{ess}([\![N]\!])$ is finite.* $\qquad\Box$

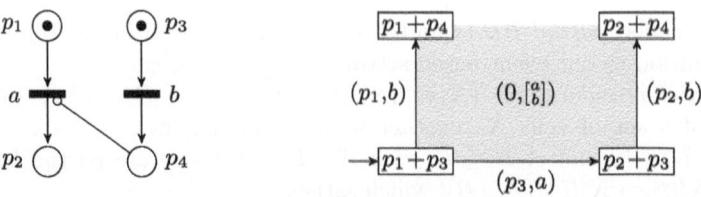

Fig. 9. PNI and partial HDA semantics of Example 16.

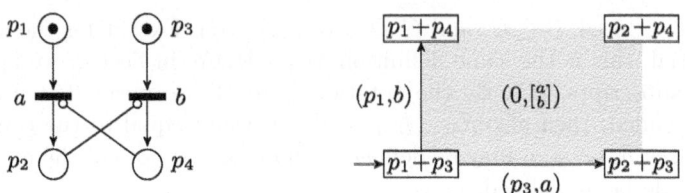

Fig. 10. Second PNI and partial HDA semantics of Example 16.

Example 16. Consider Fig. 9, where Example 11 is interpreted with partial HDA semantics, ignoring condition (3). The 2-cell $(0, \left[\begin{smallmatrix} a \\ b \end{smallmatrix}\right])$ is now present, but its δ_b^1 face is not. There are now two paths from $p_1 + p_3$ to $p_2 + p_4$, passing respectively through $p_2 + p_3$ and $(0, \left[\begin{smallmatrix} a \\ b \end{smallmatrix}\right])$. The latter captures reaching marking $p_2 + p_4$ from $p_1 + p_3$ through the multistep $U = \{a, b\}$ in $[\![N]\!]_{\mathrm{CS}}$.

We may also modify the example by introducing another inhibitor arc from p_2 to b, see Fig. 10. Then transition a (resp. b) is disabled when there is a token in p_4 (resp. p_2). Again, the corresponding partial HDA contains the 2-cell $(0, \left[\begin{smallmatrix} a \\ b \end{smallmatrix}\right])$ even if both its δ_a^1 and δ_b^1 are missing. In the a-posteriori semantics, the marking $p_2 + p_4$ is now unreachable; but in the a-priori semantics, there is a path passing through $(0, \left[\begin{smallmatrix} a \\ b \end{smallmatrix}\right])$ which mimics firing a and b in parallel.

Similarly to [8, Thm. 5.20], it can be proven that the concurrent a-priori semantics of primitive systems may not be simulated by Petri nets without inhibitor arcs.

7 Self-modifying Nets

Instead of considering other extensions one by one, we now pass to generalized self-modifying nets which encompass many other extensions. Recall [12] that a *generalized self-modifying net* (*G-net*) $N = (S, T, F)$ consists of a set of places S, a set of transitions T, with $S \cap \mathrm{N} = S \cap T = \emptyset$, and a weighted flow relation $F : S \times T \cup T \times S \to \mathrm{N}[S]$.

That is, flow arcs are labeled by *polynomials* in place variables; contrary to [12] we do not assume that the labels are sums of monomials. We will propose a concurrent semantics using *ST-graphs*, a generalisation of partial HDAs, see below.

The intuition of the flow polynomials labeling arcs is that when a transition fires, it consumes precisely the number of tokens given by evaluating polynomials of its input arcs, and produces precisely the number of tokens given by evaluating polynomials labeling its output arcs in the current marking. More precisely,

– if $F(s,t) = P$ for an arc (s,t), then firing t consumes $P(m)$ tokens from s, where m is the current marking; so the polynomial P is evaluated by replacing its place variables with the current number of tokens in the respective places;
– if $F(t,s) = P$ for an arc (t,s), then firing t produces $P(m)$ tokens in s, where m is the marking *before* t started firing.

That is, in interleaved semantics the new marking when firing a transition t in marking m is given by $m' = m - {}^{\bullet}t(m) + t^{\bullet}(m)$: ${}^{\bullet}t$ is the function $\varphi : S \to \mathbb{N}[S]$ given by $\varphi(s) = F(s,t)$, so ${}^{\bullet}t(m) : S \to \mathbb{N}$ is given by ${}^{\bullet}t(m)(s) = \varphi(s)(m)$. Now in concurrent semantics, other transitions may fire between starting and terminating t, so we will need to remember the marking before firing t, see below.

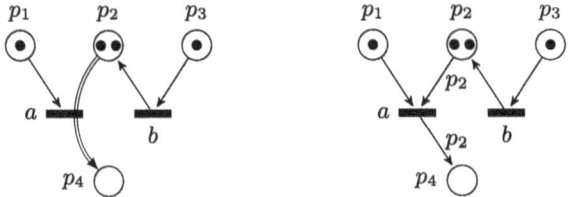

Fig. 11. Petri net with transfer arc (left) and its translation to a G-net (right). Annotations "p_2" at flow arcs indicate that they consume and produce the number of tokens present in p_2. Otherwise, they consume one token as usual.

Example 17. G-nets can model *transfer arcs*, a Petri net extension which, when firing the associated transition, transfers *all* tokens present in the pre-place to the post-place. Figure 11 shows a simple Petri net N with a transfer arc, from p_2 through the a-transition to p_4, and its G-net translation. (In fact, this is a self-modifying net in the sense of [30], a strict subclass of G-nets.) Note that there is contact between the transitions a and b, and the marking $p_2 + 2p_4$ reached by firing first a and then b is not reachable from the marking $p_1 + 3p_3$ obtained after firing b.

The latter is indicated with a "broken" arrow in the intuitive (partial) HDA semantics for N:

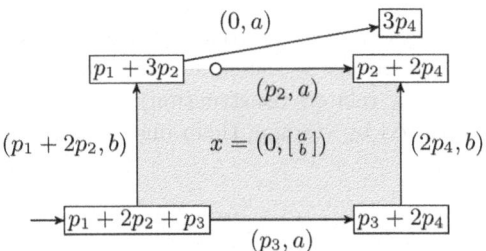

Transitions a and b can fire concurrently, so we have a square $x \in [\![N]\!][\![\begin{smallmatrix} a \\ b \end{smallmatrix}]\!]$, and from here b can terminate, leading to $\delta_b^1(x) = (p_2, a)$. Since $p_2 + 2p_4$ is not reachable from $p_1 + 3p_2$, we must have $\delta_a^0((p_2, a)) \neq p_1 + 3p_2$ if it is defined. In fact, this is unreachable, so it is not defined. We have a partial HDA, with all cells "geometrically present" but one gluing undefined.

In the previous example and Example 21 below, pHDAs allow to capture the concurrent semantics of G-nets, however, we conjecture that they are not sufficient in the general case. Instead, in order to give concurrent semantics to G-nets, we introduce a third kind of automaton, closely related to HDAs and partial HDAs but more general. First, some terminology; again Σ denotes an arbitrary alphabet and $\square = \square(\Sigma)$. A *starter* is a pair (A, U), written ${}_A{\uparrow}U$, consisting of a conclist $U \in \square$ and a subset $A \subseteq U$. A *terminator* is a pair (U, B), written $U{\downarrow}_B$, consisting of a conclist U and a subset $B \subseteq U$. The intuition is that these denote actions of starting resp. terminating subsets of the events in U, passing from $U \setminus A$ to U, resp. from U to $U \setminus B$.

Let $ST = ST(\Sigma)$ denote the (infinite) set of starters and terminators over Σ. An *ST-graph* is a structure (Σ, Q, E, λ) consisting of a set Q of states, a set $E \subseteq Q \times ST \times Q$, and a labeling $\lambda : Q \to \square$ such that for all $(p, x, q) \in E$,

- if $x = {}_A{\uparrow}U$, then $\lambda(p) = U \setminus A$ and $\lambda(q) = U$;
- if $x = U{\downarrow}_B$, then $\lambda(p) = U$ and $\lambda(q) = U \setminus B$.

ST-automata, *i.e.* ST-graphs with initial (and final) states, have been introduced in [3] in order to give operational semantics to HDAs.

Now let $N = (S, T, F)$ be a G-net and define an ST-graph $[\![N]\!]'_{ST} = (T, Q', E', \lambda')$ by $Q' = \mathbb{N}^S \times \square \times (\mathbb{N}^S)^*$, $\lambda'(m, \tau, \mu) = \tau$, and

$$E' = \Big\{ \Big(\big(m + {}^\bullet t_i(\mu_i), (t_1, \ldots, t_{i-1}, t_{i+1}, \ldots, t_n), (\mu_1, \ldots, \mu_{i-1}, \mu_{i+1}, \ldots, \mu_n) \big),$$
$$t_i{\uparrow}(t_1, \ldots, t_n), \big(m, (t_1, \ldots, t_n), \mu \big) \Big) \Big\}$$
$$\cup \Big\{ \Big(\big(m, (t_1, \ldots, t_n), \mu \big), (t_1, \ldots, t_n){\downarrow}_{t_i},$$
$$\big(m + t_i^\bullet(\mu_i), (t_1, \ldots, t_{i-1}, t_{i+1}, \ldots, t_n), (\mu_1, \ldots, \mu_{i-1}, \mu_{i+1}, \ldots, \mu_n) \big) \Big) \Big\}.$$

The intuition is that in a state $x = (m, (t_1, \ldots, t_n), (\mu_1, \ldots, \mu_k)) \in Q'$, $n = k$ if x is reachable, and the marking $\mu_i : S \to \mathbb{N}$ is the memory of how the net was marked before transition t_i started firing.

Remark 18. In the construction of $[\![N]\!]'_{ST}$, one may compose successive starting edges to start multiple transitions at the same time, similarly for terminating edges. (See [3, Sect. 4] for a related construction). Note that concurrency of several transitions is captured by starting them one by one in any order before terminating any of them.

We have primed $[\![N]\!]'_{\mathrm{ST}}$ above, as the memory so-defined remembers too much: in Example 17, transition a only needs to remember the contents of p_2 and b should not require memory at all. We remedy this by introducing a notion of memory equivalence and passing to a quotient. Say that two pairs (t, m), (t, m') are *memory equivalent*, denoted $(t, m) \sim (t, m')$, if ${}^{\bullet}t(m) = {}^{\bullet}t(m')$ and $t^{\bullet}(m) = t^{\bullet}(m')$. Then m and m' have the same *effect* on the net when t is fired.

Now extend $\sim$ to Q' by $\big(m, (t_1, \ldots, t_n), (m_1, \ldots, m_n)\big) \sim \big(m, (t_1, \ldots, t_n),$ $(m'_1, \ldots, m'_n)\big)$ if $(t_i, m_i) \sim (t_i, m'_i)$ for all i. As the memory works by insertion and deletion when starting resp. terminating transitions, $\sim$ is a congruence on the ST-graph $[\![N]\!]'_{\mathrm{ST}}$ and we may form the quotient $[\![N]\!]_{\mathrm{ST}} = [\![N]\!]'_{\mathrm{ST}}/\!\!\sim$.

Similarly to HDAs, we may define the 1-truncation of an ST-graph (Σ, Q, E, λ) as the graph which has as vertices states $q \in Q$ for which $\dim(\lambda(q)) = 0$ and edges (q, a, q') corresponding to states $x \in Q$ with $\dim(\lambda(x)) = 1$. This yields the following analogue of Lemma 4; the relation to the concurrent step reachability graph is left for future work.

Lemma 19. *The reachability graph of G-net N is isomorphic to the 1-truncation of $[\![N]\!]_{\mathrm{ST}}$.* □

We leave open the question of a characterization of G-nets (or subclasses) by means of partial HDAs. They are sufficient in all our examples. On the other hand, note that the ST-automaton semantics we have given conforms with how the reachable part of the semantics is constructed, by starting and terminating events one at a time. (See also Sect. 8).

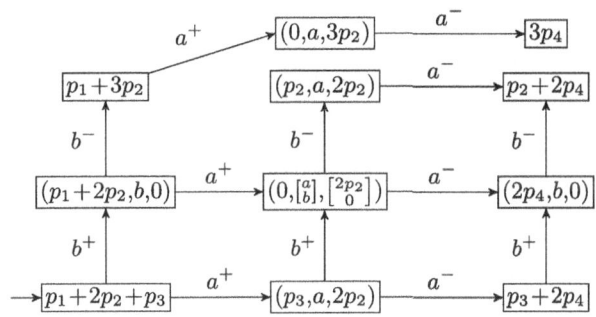

Fig. 12. ST-automaton semantics (reachable part only) of the G-net of Fig. 11, see Example 20. For readability, transitions are labeled with starts (a^{+}) and terminations (a^{-}) of actions rather than starters and terminators.

Example 20. We continue Example 17 by giving $[\![N]\!]_{\mathrm{ST}}$ for the transfer net N in Fig. 12, indicating memory equivalence classes by representatives. We see that, as expected, the sequence $b^{+}b^{-}a^{+}$ leads to a different state than the other permutations.

We show some of the calculations; for readability we make a distinction between a variable p appearing in a polynomial and the current number of tokens in p, denoting the latter by $|p|$. We also write $^\bullet t = \sum_s F(s,t)s$ to denote that t fires by consuming $F(s,t)$ tokens from each place s (and similarly for $t^\bullet$).

We have $^\bullet a = F(p_1,a)p_1 + F(p_2,a)p_2 + F(p_3,a)p_3 = p_1 + |p_2|p_2$ and $a^\bullet = |p_2|p_4$. In addition, the marking before only firing a is $\mu = p_1 + 2p_2 + p_3$. Then $^\bullet a(\mu) = p_1 + \mu(p_2)p_2 = p_1 + 2p_2$. Note also that for $\mu' = 2p_2$, for example, $^\bullet a(\mu) = {}^\bullet a(\mu')$ and $a^\bullet(\mu) = a^\bullet(\mu')$.

More generally, we have $^\bullet a(n_1p_1 + n_2p_2 + n_3p_3) = p_1 + n_2p_2$ for all $n_1, n_2, n_3 \in \mathbb{N}$. Thus $(p_1 + n_2p_2 + p_3, a^+, (p_3, a, p_1 + n_2p_2)) \in E'$. We also have that $(p_3, a, p_1 + n_2p_2) \sim (p_3, a, n_2p_2)$. Indeed, since $F(p_1, a)$ and $F(a, p_1)$ are constants, $^\bullet a(p_1 + n_2p_2) = {}^\bullet a(n_2p_2)$ and $a^\bullet(p_1 + n_2p_2) = a^\bullet(n_2p_2)$.

Thus, for the sequence $a^+b^+b^-$, $((p_3, a, 2p_2), b^+, (0, \left[\begin{smallmatrix}a\\b\end{smallmatrix}\right], \left[\begin{smallmatrix}2p_2\\p_3\end{smallmatrix}\right])) \in E'$ since we had only one token in p_3 before firing b, but $(0, \left[\begin{smallmatrix}a\\b\end{smallmatrix}\right], \left[\begin{smallmatrix}2p_2\\p_3\end{smallmatrix}\right]) \sim (0, \left[\begin{smallmatrix}a\\b\end{smallmatrix}\right], \left[\begin{smallmatrix}2p_2\\0\end{smallmatrix}\right])$. Finally, $((0, \left[\begin{smallmatrix}a\\b\end{smallmatrix}\right], \left[\begin{smallmatrix}2p_2\\0\end{smallmatrix}\right]), b^-, (p_2, a, 2p_2)) \in E'$, leading to a different state than $b^+b^-a^+$.

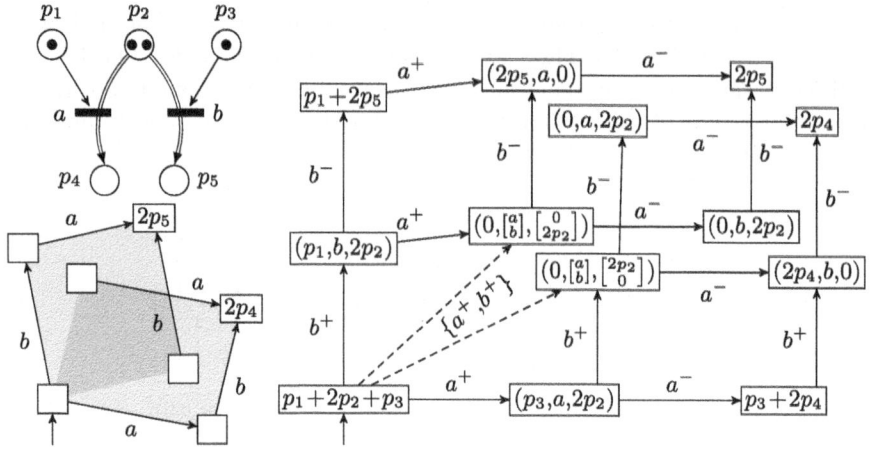

Fig. 13. G-net of Example 21 (top left), ST-automaton semantics (right), and corresponding partial HDA (bottom left).

Example 21. Figure 13 shows a G-net containing two transfer arcs which are in conflict and the corresponding ST-automaton semantics. If we admit that starting a and b at the same time is non-deterministic and may lead to any of the two states in the center (as indicated by the dashed $\{a^+, b^+\}$-labeled transitions in ST-automaton), then this is now a partial HDA: two partial squares labeled $\left[\begin{smallmatrix}a\\b\end{smallmatrix}\right]$ and glued at the initial state.

8   Implementation

We have developed a prototype tool, pn2HDA, written in C++ and implementing our translations from Petri nets to HDAs and from PNIs to partial HDAs. Our implementation is able to deal with standard, weighted and inhibitor arcs in a modular fashion and is available at https://gitlabev.imtbs-tsp.eu/philipp. schlehuber-caissier/pn2hda. Our tool is based on previous work by our student Timothée Fragnaud and on the PNML parser provided by the library Symmetri[4], but any other parser could easily replace this task.

As shown above, Petri nets with or without inhibitor arcs (a-posteriori semantics) can be translated into the same formalism: partial HDAs. The implementation reflects this by having a parametrizable (via templates) representation for Petri nets, which is used in a generic way to build the corresponding pHDA.

For this prototype tool, we have chosen an explicit representation of the pHDA and its face maps as defined at the beginning of Sect. 4. That is each reachable cell $x \in X$ is defined as the tuple (m, τ) corresponding to the conclist and the marking. As event order, the arbitrary total order on transitions introduced in Sect. 5, we have chosen the shortlex order on transition names (*i.e.* names are first sorted by their length, and sequences of identical length are sorted according to the lexicographical order). While this representation is likely not the most efficient, it allows to underpin the correctness of our constructions.

Since we cannot display the constructed pHDAs if their dimension is greater than 3, we have chosen to output them as ST-automata (see Fig. 14). For gathering information about the structure of the automaton like the number of unique conclists or markings we provide a function get_csv_data. In the repository we also provide the PNML files of all Petri nets given as examples in the paper, as well as a selection of models from the Model Checking Contest[5]. This is useful for example for gathering statistics about cells of different dimensions, see Fig. 15.

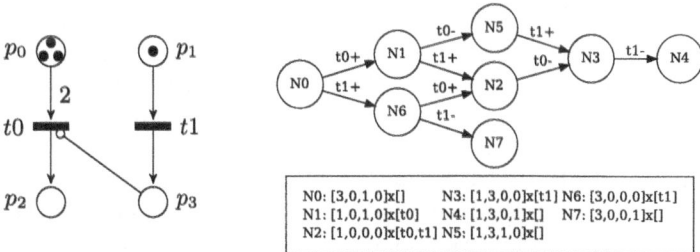

Fig. 14. Petri net (left) and pHDA conversion (right): definition of cells (top) and ST-automaton representation (bottom).

---

[4] See https://github.com/thorstink/Symmetri.
[5] See https://mcc.lip6.fr.

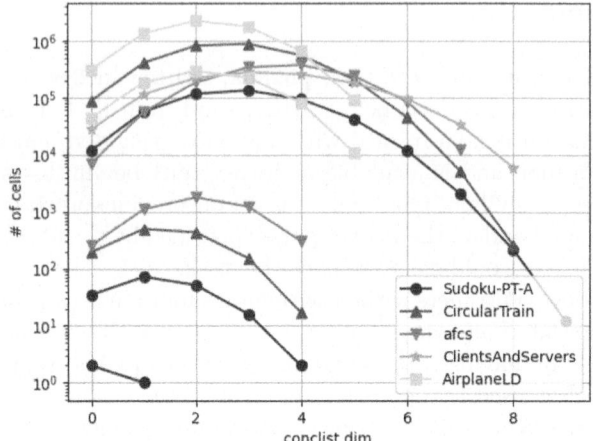

Fig. 15. Statistics on cell distribution for some MCC models. Different models of the same family share a linestyle. We converted the models of a family until either a timeout (1h) or an out-of-memory (32Gb) was reached.

9   Conclusion

We have seen that Petri nets exhibit a natural concurrent semantics as higher-dimensional automata (HDAs) which allows methodical reasoning about the finer points of the semantics of Petri nets and their extensions. The semantics of Petri nets with inhibitors is naturally expressed using partial HDAs in which some faces may be missing.

We have also given concurrent semantics to generalized self-modifying nets (G-nets) which encompass many other extensions. We have given the semantics as ST-automata, a generalization of partial HDAs, using a notion of memory to store the state of the G-net before the starts of transitions. Whether or not the semantics may also be given as partial HDAs, or whether there are interesting subclasses of G-nets which admit partial-HDA semantics, is left open.

Finally, we have presented an implementation of the translations from Petri nets to HDAs and from Petri nets with inhibitors to partial HDAs.

We believe that pHDA and ST-automaton semantics may provide a unifying framework for investigating constructions on Petri nets (such as the removal of inhibitor arcs from primitive systems which we have seen) and their effects on concurrent semantics. This should apply to other well-known simplifications such as removing read arcs; but also for example to unfoldings.

We would also like to apply our setting to other generalizations of Petri nets. Introducing concurrent semantics for affine nets [7], for example, appears difficult; but real-time extensions such as time Petri nets [27] seem natural candidates. Recent work on higher-dimensional timed automata [1,2] will be useful in this context.

Finally, we plan to continue to work on our implementation. We are working on another, implicit, representation of pHDAs which would avoid creating all reachable cells. We would also like to extend our tool to time Petri nets and connect it with Kahl's work on program graphs and homology [23].[6]

Acknowledgements. We are grateful to Eric Lubat for discussions regarding concurrent semantics, to Timothée Fragnaud for the initial implementation of the translation from Petri nets to HDAs, and to Cameron Calk for discovering an error in a earlier version of this work.

References

1. Amrane, A., Bazille, H., Clement, E., Fahrenberg, U.: Languages of higher-dimensional timed automata. In: Kristensen, L.M., van der Werf, J.M. (eds.) PETRI NETS 2024. LNCS, vol. 14628, pp. 197–219. Springer, Cham (2024). https://doi.org/10.1007/978-3-031-61433-0_10
2. Amrane, A., Bazille, H., Clement, E., Fahrenberg, U., Schlehuber-Caissier, P.: Higher-dimensional timed automata for real-time concurrency. CoRR, abs/2401.17444 (2025)
3. Amrane, A., Bazille, H., Clement, E., Fahrenberg, U., Ziemiański, K.: Presenting interval pomsets with interfaces. In: Fahrenberg, U., Fussner, W., Glück, R. (eds.) RAMiCS 2024. LNCS, vol. 14787, pp. 28–45. Springer, Cham (2024). https://doi.org/10.1007/978-3-031-68279-7_3
4. Amrane, A., Bazille, H., Fahrenberg, U., Fortin, M.: Logic and languages of higher-dimensional automata. In: Day, J.D., Manea, F. (eds.) DLT 2024. LNCS, vol. 14791, pp. 51–67. Springer, Cham (2024). https://doi.org/10.1007/978-3-031-66159-4_5
5. Amrane, A., Bazille, H., Fahrenberg, U., Hélouët, L., Schlehuber-Caissier, P.: Petri nets and higher-dimensional automata. CoRR, abs/2502.02354 (2025)
6. Amrane, A., Bazille, H., Fahrenberg, U., Ziemiański, K.: Closure and decision properties for higher-dimensional automata. In: Ábrahám, E., Dubslaff, C., Tarifa, S.L.T. (eds.) ICTAC 2023. LNCS, vol. 14446, pp. 295–312. Springer, Cham (2023). https://doi.org/10.1007/978-3-031-47963-2_18
7. Bonnet, R., Finkel, A., Praveen, M.: Extending the Rackoff technique to affine nets. In: D'Souza, D., Kavitha, T., Radhakrishnan, J. (eds.) FSTTCS. *LIPIcs*, vol. 18, pp. 301–312. Schloss Dagstuhl - Leibniz-Zentrum für Informatik (2012)
8. Busi, N.: Analysis issues in Petri nets with inhibitor arcs. Theoret. Comput. Sci. **275**(1–2), 127–177 (2002)
9. Desel, J., Reisig, W.: Place/transition Petri nets. In: Reisig, W., Rozenberg, G. (eds.) ACPN 1996. LNCS, vol. 1491, pp. 122–173. Springer, Heidelberg (1998). https://doi.org/10.1007/3-540-65306-6_15
10. Dubut, J.: Trees in partial higher dimensional automata. In: Bojańczyk, M., Simpson, A. (eds.) FoSSaCS 2019. LNCS, vol. 11425, pp. 224–241. Springer, Cham (2019). https://doi.org/10.1007/978-3-030-17127-8_13
11. Dubut, J., Goubault, É., Goubault-Larrecq, J.: Natural homology. In: Halldórsson, M.M., Iwama, K., Kobayashi, N., Speckmann, B. (eds.) ICALP 2015. LNCS, vol. 9135, pp. 171–183. Springer, Heidelberg (2015). https://doi.org/10.1007/978-3-662-47666-6_14

---

[6] See https://github.com/twkahl/PG2HDA.

12. Dufourd, C., Finkel, A., Schnoebelen, P.: Reset nets between decidability and undecidability. In: Larsen, K.G., Skyum, S., Winskel, G. (eds.) ICALP 1998. LNCS, vol. 1443, pp. 103–115. Springer, Heidelberg (1998). https://doi.org/10.1007/bfb0055044
13. Fahrenberg, U.: Higher-dimensional automata from a topological viewpoint. Ph.D. thesis, Aalborg University, Denmark (2005)
14. Fahrenberg, U., Johansen, C., Struth, G., Ziemiański, K.: Languages of higher-dimensional automata. Math. Struct. Comput. Sci. **31**(5), 575–613 (2021). https://arxiv.org/abs/2103.07557
15. Fahrenberg, U., Johansen, C., Struth, G., Ziemiański, K.: Kleene theorem for higher-dimensional automata. Log. Methods Comput. Sci. **20**(4) (2024)
16. Fahrenberg, U., Legay, A.: Partial higher-dimensional automata. In: Moss, L.S., Sobocinski, P. (eds.) CALCO. Leibniz International Proceedings in Informatics, vol. 35, pp. 101–115. Schloss Dagstuhl - Leibniz-Zentrum für Informatik (2015)
17. Fahrenberg, U., Ziemiański, K.: Myhill-Nerode theorem for higher-dimensional automata. Fundam. Inform. **192**(3–4), 219–259 (2024)
18. Flynn, M.J., Agerwala, T.: Comments on capabilities, limitations and 'correctness' of Petri nets. Comput. Archit. News **4** (1973)
19. Genrich, H.J., Lautenbach, K., Thiagarajan, P.S.: Elements of general net theory. In: Brauer, W. (ed.) Net Theory and Applications. LNCS, vol. 84, pp. 21–163. Springer, Heidelberg (1980). https://doi.org/10.1007/3-540-10001-6_22
20. Goltz, U., Reisig, W.: The non-sequential behaviour of Petri nets. Inf. Control **57**(2), 125–147 (1983)
21. Grandis, M., Mauri, L.: Cubical sets and their site. Theory Appl. Categories **11**(8), 185–211 (2003)
22. Janicki, R., Koutny, M.: Semantics of inhibitor nets. Inf. Comput. **123**(1), 1–16 (1995)
23. Kahl, T.: On the homology language of HDA models of transition systems. J. Appl. Comput. Topology **8**(4), 859–873 (2024)
24. Karp, R.M., Miller, R.E.: Parallel program schemata. J. Comput. Syst. Sci. **3**(2), 147–195 (1969)
25. Lane, S.M., Moerdijk, I.: Sheaves in Geometry and Logic. Springer (1992)
26. Mayr, E.W.: An algorithm for the general Petri net reachability problem. In: Proceedings of the 13th Annual ACM Symposium on Theory of Computing, 11–13 May 1981, Milwaukee, Wisconsin, USA, pp. 238–246. ACM (1981)
27. Merlin, P.: A study of the recoverability of computer systems. Ph.D. Thesis, Computer Science Department, University of California (1974)
28. Mukund, M.: Petri nets and step transition systems. Int. J. Found. Comput. Sci. **3**(4), 443–478 (1992)
29. Struth, G., Ziemiański, K.: Presheaf automata. CoRR, abs/2409.04612 (2024)
30. Valk, R.: On the computational power of extended petri nets. In: Winkowski, J. (ed.) MFCS 1978. LNCS, vol. 64, pp. 526–535. Springer, Heidelberg (1978). https://doi.org/10.1007/3-540-08921-7_101
31. Glabbeek, R.J.: The individual and collective token interpretations of Petri nets. In: Abadi, M., de Alfaro, L. (eds.) CONCUR 2005. LNCS, vol. 3653, pp. 323–337. Springer, Heidelberg (2005). https://doi.org/10.1007/11539452_26
32. van Glabbeek, R.J.: On the expressiveness of higher dimensional automata. Theor. Comput. Sci. **356**(3), 265–290 (2006). See also [33]
33. van Glabbeek, R.J.: Erratum to "On the expressiveness of higher dimensional automata". Theoret. Comput. Sci. **368**(1–2), 168–194 (2006)

Discovering the Influence of Exogenous Data on Decisions in Processes

Adam Banham[1]([✉])[iD], Yannis Bertrand[2][iD], Robert Andrews[1][iD],
Moe T. Wynn[1][iD], and Sander J. J. Leemans[3,4][iD]

[1] Queensland University of Technology, Brisbane, Australia
adam.banham@hdr.qut.edu.au
[2] KU Leuven, Leuven, Belgium
[3] Fraunhofer FIT, Sankt Augustin, Germany
[4] RWTH Aachen, Aachen, Germany

Abstract. Process mining, when applied to data stored in the information systems of businesses, provides insights into the internal performance of their processes. These insights reveal how the behaviour of processes impacts businesses, and can inform planning for various future scenarios by anticipating how processes will perform in these scenarios. However, these scenarios may be influenced by the context of the process, i.e., external data streams (exogenous data) to the process. In these cases, typical process discovery techniques can produce process models that describe what activities could occur next in a given state, but cannot express the effect of external influences on how likely these are to occur. Our contribution presents an extension of stochastic labelled Petri nets and a discovery technique for our new modelling formalism. The proposed formalism can be used to quantify whether the firing likelihood of a transition is influenced by exogenous data when replaying historical process executions over the net. We compare our approach against existing stochastic techniques over several publicly available event logs. Our results show that our approach can outperform existing data-aware techniques in unstructured processes.

Keywords: Process Mining · Process Enhancement · Stochastic Process Mining · Exogenous Data

1 Introduction

In the field of business process management [8], studying how processes change at runtime [5,10,24,26,29] is important. Studying the dynamics of how processes execute across a collection of varying scenarios enables businesses to plan for future events based on their likely occurrence. Process mining [28] enables businesses to derive the diagnostic information needed to empower such planning by considering historical data about processes. Process discovery techniques [13,14,17] can generate process models that describe when activities occur in a process and how processes are structured [28]. Conformance checking techniques [18,23] can quantify deviations between historical data and a process

© The Author(s), under exclusive license to Springer Nature Switzerland AG 2025
E. Amparore and Ł. Mikulski (Eds.): PETRI NETS 2025, LNCS 15714, pp. 41–62, 2025.
https://doi.org/10.1007/978-3-031-94634-9_3

model to highlight anomalies at runtime [1,6]. However, to understand the likelihood of an activity at runtime, the mining of stochastic processes is required [9].

Stochastic process mining provides a window into understanding process-level decisions by describing which activities could occur next, and how likely these activities are to occur [9]. Various mechanisms for determining next activity likelihood exist [4,5,16,19,20,23,25,27], with recent extensions considering the dependencies within the history of a trace [16] or case/event attributes [23]. Quantifying the likelihood of an activity may require considering many different dependencies, both internal and external, to the process. Addressing this challenge within stochastic process mining [4,9,16,23] is ongoing. Our contribution is a novel approach for determining whether historical process executions were influenced by external data sources, referred to as *exogenous data* [2].

We investigate how stochastic labelled Petri nets [15] (SLPN) could be used to understand if the behaviour of processes has been influenced by exogenous data. SLPNs are themselves extensions of Petri nets, where transitions are extended with a weight function to represent their chance of firing. Various approaches to deriving these weight functions have been used to investigate several aspects of processes [4,16,23], but all have overlooked exogenous factors.

Behaviour in processes may be informed by internal dependencies, or by exogenous influences, i.e. where processes are influenced by their context [29]. Agriculture supply chains can be influenced by climate variability [11,30] and patient care in hospitals frequently involves the continuous observation of the patient's vital signs [2]. Notably, for these cases, existing discovery techniques [4,16,23] cannot capture exogenous influences on the firing likelihoods. In particular, existing techniques overlook temporal factors in exogenous data, where, between discrete actions, new data becomes available, or no data is available.

This paper shares a common aim with recent work [23] focusing on including endogenous data in the firing likelihoods of activities. However, our work focuses on including exogenous data to cater for contextual factors that may influence the behaviour of processes, rather than solely relying on endogenous data. To do so, we consider behaviour in time series data and introduce a temporal component to analysis. Our approach avoids reducing time series data to a static value through the application of a lossy aggregation [2], and considers the condition of having no exogenous data on firing likelihoods. Therefore, this paper investigates:

Problem. *How can exogenous influences on the stochastic behaviour of processes be derived and analysed?*

Our answer to this problem is a stochastic discovery technique which discovers weight functions for transitions in labelled Petri nets. The discovery is presented as an optimisation problem, whereby equalities are derived using diagnostic information, and are solved to form weight functions. The discovered weight functions can be used to investigate exogenous influences on processes.

The paper is organised as follows. Section 2 illustrates existing approaches; Sect. 3 denotes the notation used; Sect. 4 presents our stochastic modelling formalism; Sect. 5 defines our discovery method for such a formalism; Sect. 6 denotes how we quantify our discovered formalism; Sect. 7 evaluates our contributions; and Sect. 8 summarises our work.

Table 1. An extract from the road fines log [7].

trace	event	event attributes or *endogenous data* **process activity**	amount	dismissal	pay amount	timestamp
N74775	1	create fine	35.0	NIL		28.05.2005@00:00
	2	send fine				21.09.2005@00:00
	3	insert fine notification				01.10.2005@00:00
	4	add penalty	71.5			30.11.2005@00:00
	5	insert date appeal to prefecture				30.11.2005@00:00
	6	send appeal to prefecture		#		09.01.2006@00:00
	7	receive result appeal from prefecture				17.02.2006@00:00
	8	notify result appeal to offender				31.03.2006@00:00
A8690	9	create fine	36.0	NIL		08.02.2007@00:00
	10	payment			36.0	21.02.2007@00:00
A21188	11	create fine	22.0	NIL		14.07.2008@00:00
	12	send fine				10.12.2008@00:00
	13	insert fine notification				18.12.2008@00:00
	14	add penalty	44.0			16.02.2009@00:00
	15	send for credit collection				15.10.2010@00:00
⋮	⋮	⋮	⋮	⋮	⋮	⋮

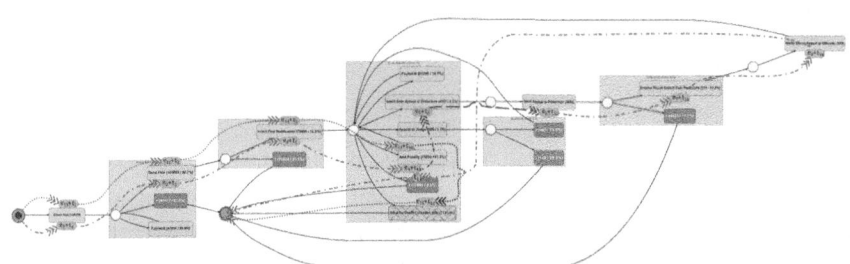

Alignment for trace 'N74775' : $\langle ({e_1 \atop t_1}), ({e_2 \atop t_2}), ({e_3 \atop t_5}), ({e_4 \atop t_{10}}), ({e_5 \atop t_8}), ({e_6 \atop t_{13}}), ({e_7 \atop t_{16}}), ({e_8 \atop t_{18}}), ({\gg \atop t_{11}}) \rangle.$

Alignment for trace 'A21188' : $\langle ({e_{11} \atop t_1}), ({e_{12} \atop t_2}), ({e_{13} \atop t_5}), ({e_{14} \atop t_{10}}), ({e_{15} \atop t_{12}}) \rangle.$

Fig. 1. The normative model capturing the decision points (in green) and credit collection is highlighted. Alignments are captured through dashed directed lines. (Color figure online)

2   Motivating Example

To motivate stochastic process mining and highlight existing techniques [4,16, 23], a running example is introduced. The example uses a snippet of the road fines log [7], which captures the payment and appeal process for fines issued by a police force in Italy (Table 1), and a normative control-flow model [22, Sec. 12.1.3]. Also introduced are optimal alignments [1] for the snippet (Fig. 1). These are important to consider as our approach derives diagnostic information from alignments to inform the firing likelihoods for transitions. This example explores the probability of '*Sent to Credit Collection*' (abbreviated to **credit collection**) occurring. Where the answer for the trace 'A8690' is straightforwardly 0%, as the transition for credit collection was never enabled. However, the question is more challenging for the other traces in Table 1.

In stochastic process mining, transitions have weights and the likelihood of a transition firing is a ratio of its weight over the total weight of all enabled transitions. One approach to assigning weights to transitions is to consider the frequency of the fired transitions [4] and estimate a numerical value for a *base* weight, resulting in Fig. 2a. Where for the trace 'N74775', the discovered net induces a likelihood for credit collection of 31.87% for all three times that it is enabled. The downside of only using base weights is that the likelihood of a transition cannot account for context surrounding the execution.

To contextualise likelihoods for a running execution, base weights can be expanded to *parameterised* weight functions. For example, many parameters on each transition can be introduced allowing for the history of actions in an execution to influence the firing likelihood of the next transition [16] (known as a SLPN-SD). Similarly, parameters can be introduced to account for the value of data attributes as of the last event in an execution [23] (known as a SLDPN). However, these approaches treat executions as discrete changes, and as time passes between actions, the firing likelihoods remain the same.

When we consider trace 'N74775' and the discovered parameterised nets in Fig. 2b and Fig. 2c, then credit collection had firing likelihoods of 20%, 17.3%, and 17.3% or 8.7%, 10.5%, and roughly 0.01% for the respective models. Showing that the dismissal of the fine can be catered for using Fig. 2c. But, if we consider trace 'A21188' where credit collection did occur, where the likelihoods are 8.1% and 9.2% for Fig. 2c, then the batching nature (seen in Fig. 4) of this action is not captured within parameters, as they only capture discrete changes. Moreover, temporal changes in exogenous data are not supported unless we reduce temporal series into static attributes for events, requiring domain knowledge.

This example shows existing techniques [4,16,23] can cater for endogenous factors that may influence the probability of a transition firing. However, without a transition firing, these techniques are unable to adjust the probability of transitions or account for temporal considerations in exogenous data. Thus, we explore how to consider exogenous influences on the firing likelihood of transitions, and the challenges of including exogenous data for stochastic Petri nets.

3   Preliminaries

This section introduces the notation used for event logs, exogenous data, process models and alignments.

Sets. Curly brackets denote sets, e.g. $\Sigma = \{a, b, c\}$. To denote the number of members in a set Σ, $|\Sigma|$ is written. Given some set, say Σ, a multiset $M: \Sigma \mapsto \mathbb{N}_0$ maps the elements of Σ to the natural numbers. Square brackets denote multisets, e.g. $[a^2, b^1]$ s.t. $a, b \in \Sigma$ and all other elements of Σ are mapped to zero. To denote the infinite set of all multisets over a given set Σ, $\mathcal{M}(\Sigma)$ is used. We write $m \in M$ to access all members of a multiset with non-zero frequencies,

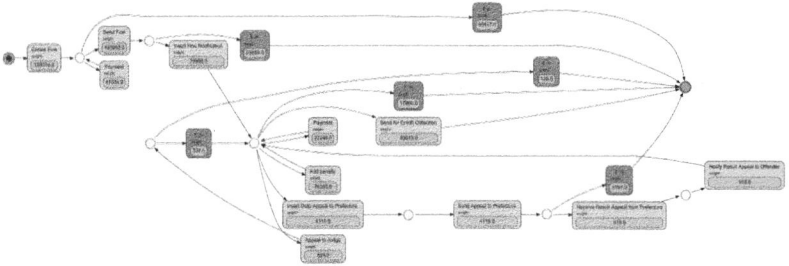

(a) Discovered stochastic labelled Petri net [4, Section 3.6] and only uses base weights.

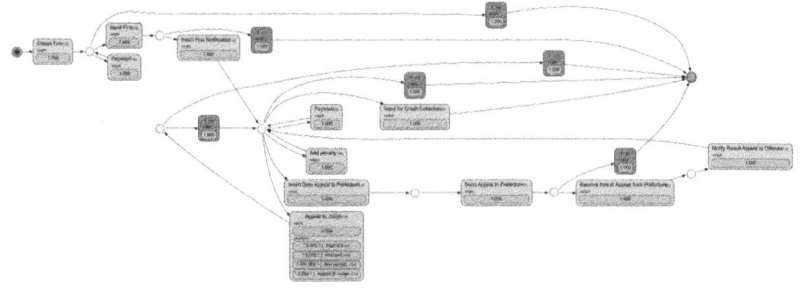

(b) Discovered SLPN-SD [16, Definition 1].

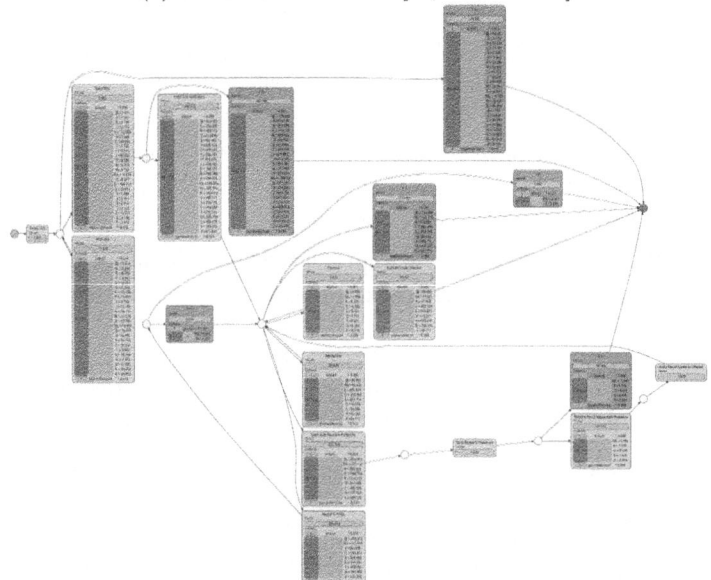

(c) Discovered SLDPN [23] using event attributes in Table 1.

Fig. 2. Existing stochastic techniques applied to the running example. Weights and parameters are denoted below the transition label.

$$\text{uf} = \langle 1^{01-01-2000}, \ldots, 31188^{14-02-2009}, 31222^{15-02-2009}, \ldots, 44884^{18-06-2013} \rangle$$

(a) Number of unpaid fines (unpaid fines).

$$\text{af} = \langle 62^{01-01-2000}, \ldots, 2.007e6^{14-02-2009}, 2.008e6^{15-02-2009}, \ldots, 2.311e6^{18-06-2013} \rangle$$

(b) Total amount of unpaid fines (amount unpaid).

Fig. 3. Exo–series describing inter-case variables for the road fines log.

i.e. $\{m \in \Sigma \mid M(m) > 0\}$. To state relations between a set S and a multiset M, $S \subseteq M$ denotes that S is a subset of M if and only if : $\forall_{s \in S} M(s) > 0$. To denote the cardinality of a multiset M, $||M|| = \sum_{m \in M} M(m)$ is written.

Sequences. Angled brackets denote sequences, e.g. $\langle a, b, c \rangle$. To concatenate two sequences s, s', $s\colon s'$ is written. The length of a sequence s is denoted as $|s|$, and if $|s| = 0$ then s is empty. To access an element of a sequence s, s_i is used to access the i-th element of the sequence s.t. $1 \leq i \leq |s|$. To denote a set of possible sequences over a set Σ, Σ^* is used to denote all possible sequences, while Σ^+ is used to denote all possible non-empty sequences.

Events. An event describes the execution of a process step (activity). Each *event* consists of the activity and a timestamp of the step [28]. The time of event is accessed through the accessor time. A non-empty sequence of events is a *trace* and describes an execution of a process.

Universes. $\mathcal{U}_{ev}, \mathcal{U}_{time}$ denote the sets of identifiers for events, and timestamps. $\mathcal{U}_{att}, \mathcal{U}_{val}$ denote the sets of attributes and all possible values that these can take on. $\mathcal{U}_{ev}, \mathcal{U}_{time}, \mathcal{U}_{att}, \mathcal{U}_{val}$ are pairwise disjoint.

Time Series. Exogenous data will represented be as numerical discrete univariate time series, referred to as an exo–series [2]. Exo–series are sequences of measurements and timestamps, say $x \in (\mathcal{U}_{val} \times \mathcal{U}_{time})^*$, focusing on measuring a single concept. To access the value and time of a measurement, x_i and $\text{time}(x_i)$ is written. $\mathcal{U}_{es} = (\mathcal{U}_{val} \times \mathcal{U}_{time})^*$ denotes the set of all possible exo–series. We call a collection of exo–series measuring the same observation an *exo–panel*.

Example 1. Consider the following exo–series in Fig. 3 for the number of unpaid fines (unpaid fines) and the total amount of unpaid fines (amount unpaid), discussed later in Sect. 7.1. These exo–series will represent exogenous data that is continuously observed to reflect the amount of work currently active in the information system within the running example. The question we investigate within these examples is: "*does the number of unpaid fines influence whether credit collection occurs within executions?*"

Exogenous Annotated Logs. The source of time series data will come from applying the xPM framework [2, Sec. 4], where an event log is annotated with exogenous data through several determinations [2, Sec. 4.1.], creating an *exogenous annotated log* (xlog). This paper solely focuses on the ($\mathcal{L}$) linked exogenous data [2, Sec. 4.1.2].

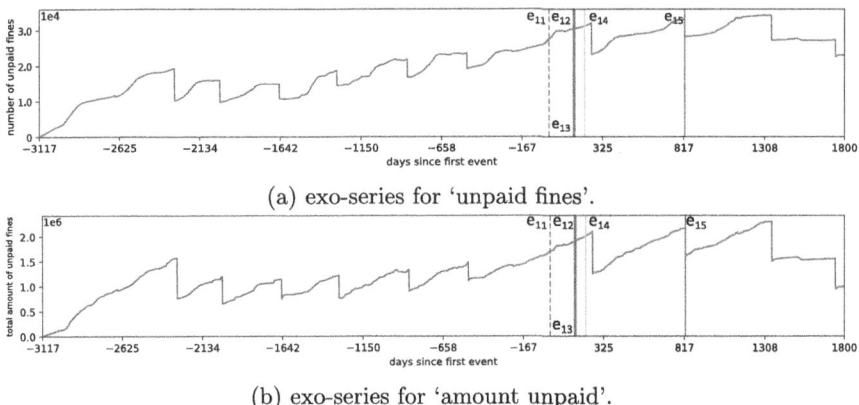

(a) exo-series for 'unpaid fines'.

(b) exo-series for 'amount unpaid'.

Fig. 4. The trace 'A21188' from Table 1 annotated with two exo–series. Verticals denote when events occurred, a dash denotes the first event.

Definition 1 (Exogenous Annotated Logs). *An exogenous-annotated log* $L = (\rho, \Sigma)$ *is a tuple, where ρ is the set of exo-panel names and $\Sigma = \mathcal{M}(\mathcal{U}_{ev}^* \times (\rho \mapsto \mathcal{U}_{es}))$ is a multiset of traces annotated with a mapping of exo-series.*

To consider exo–series x or a mapping of exo–series X, relative to a given timestamp r, referred to as a *truncated* exo-series, the function trun is introduced. These *truncated* exo-series have a relative timestamp denoting the time away from the truncated timestamp r, e.g.. a relative timestamp of a day, $1d$, denotes that a measurement occurred a day before r. Now, trun is formally denoted as:

$$\mathsf{trun}(X, r) = \bigcup_{X(p)=x} \{p \mapsto \mathsf{trun}(x, r)\} \tag{1}$$

$$\mathsf{trun}(\langle (v, t) \rangle : x, r) = \begin{cases} \langle (v, r - \mathsf{time}(t)) \rangle : \mathsf{trun}(x, r) \text{ if } \mathsf{time}(t) \leq r \\ \langle \rangle \text{ otherwise} \end{cases}$$

Example 2. Consider the following visualisation in Fig. 4 of trace 'A21188' from Table 1 annotated with exo–series for 'unpaid fines' (uf) and 'amount unpaid' (uf) from Example 1. To demonstrate trun, consider applying trun in the context of e_{14} (highlighted vertical in Fig. 4) yields:

$$\mathsf{trun}(\{\text{unpaid fines} \mapsto \mathsf{uf}, \text{amount unpaid} \mapsto \mathsf{af}\}, \mathsf{time}(e_{14}))$$
$$= \{\text{unpaid fines} \mapsto \mathsf{trun}(\mathsf{uf}, \mathsf{time}(e_{14})), \text{amount unpaid} \mapsto \mathsf{trun}(\mathsf{af}, \mathsf{time}(e_{14}))\}$$
$$= \{\text{unpaid fines} \mapsto \langle 1^{@4917d}, \dots, 31188^{@2d}, 31222^{@1d}, 31228^{@0d} \rangle,$$
$$\text{amount unpaid} \mapsto \langle 62^{@4917d}, \dots, 2\,007\,000^{@2d}, 2\,008\,000^{@1d}, 2\,009\,000^{@0d} \rangle\}$$

Petri Nets. Petri nets are used as the modelling notation to describe modelled processes. Two such extensions, labelled Petri nets (LPN) and stochastic Petri nets, are considered in this paper.

Definition 2 (Labelled Petri Nets). *A labelled Petri net (LPN) is a tuple $(P, T, F, A, \lambda, M_o)$, where P is a set of places, T is a set of transitions s.t. $P \cap T = \emptyset$, $F \subseteq ((P \times T) \cup (T \times P))$ is a set of directed flows, A is a set of process activities, $\lambda : T \mapsto A \cup \{\tau\}$ is a labelling function, and $M_o \in \mathcal{M}(P)$ is an initial marking.*

The preset of a transition t is denoted as $\bullet t = \{p \mid (p, t) \in F\}$, likewise the postset of a transition is $t^\bullet = \{p \mid (t, p) \in F\}$. A transition is enabled in a marking $M \in \mathcal{M}(P)$ if a marking includes all places with an outgoing arc to the transition, i.e. $\bullet t \subseteq M$. A run of an LPN starts from the initial marking M_o and fires transition until no transitions are enabled. The firing of a transition consumes tokens from the preset and generates fresh tokens in the postset of the transition. The set of enabled transitions from a marking M is denoted as $\mathsf{enbl}(M) = \{t \in T \mid \bullet t \subseteq M\}$.

Alignments [1]. An alignment γ is a sequence of moves, where a move is a combination of an event from the log (or a skip $\gg$) and a transition in the model (or a skip $\gg$). However a move in an alignment cannot consist of two skips $\gg$, i.e. $\left(\begin{smallmatrix}\gg\\\gg\end{smallmatrix}\right)$. A move $\left(\begin{smallmatrix}e\\t\end{smallmatrix}\right)$ is a synchronous move; a move $\left(\begin{smallmatrix}e\\\gg\end{smallmatrix}\right)$ is a log move; and a move $\left(\begin{smallmatrix}\gg\\t\end{smallmatrix}\right)$ is a model move. Also, given a move m, the following functions, $\mathsf{synm}, \mathsf{logm}, \mathsf{modm}$, identify the type of move using the previous cases. Alignments allow for traces to be mapped to a run of an LPN.

Sojourn Times. Lastly, we need to consider when a model move could have occurred. To do so, the sojourn times, the waiting time between two fired transitions, are used to inform when model moves could have occurred. In particular, a histogram of the time between synchronous moves in alignments is computed. For instance, say $\gamma = \langle \left(\begin{smallmatrix}e\\t\end{smallmatrix}\right), \left(\begin{smallmatrix}\gg\\t''\end{smallmatrix}\right), \left(\begin{smallmatrix}e'\\t'\end{smallmatrix}\right) \rangle$ is an alignment, then the sojourn time of $\mathsf{soj}(t')$ is $\mathsf{time}(e') - \mathsf{time}(e)$, and always is towards the transition t'. Formally the function hist computes all sojourn times, defined as:

Definition 3 (Sojourn times). *Let $LPN = (P, T, F, A, \lambda, M_o)$ be a net, let L be a log, let Γ be a multiset of alignments between LPN and L and let $t \in T$ be a transition. Then, hist is a function that takes a transition t to yield a multi-set of sojourn times:*

$$\mathsf{hist}(t) = \bigcup_{\substack{\gamma \in \Gamma, \\ 1 \leq i \leq \Gamma(\gamma)}} \left[\mathsf{time}(e) - \mathsf{time}(e') \mid \forall_{\gamma' : \langle\!\langle\left(\begin{smallmatrix}e'\\t'\end{smallmatrix}\right)\rangle\!\rangle} : \gamma'' : \langle\!\langle\left(\begin{smallmatrix}e\\t\end{smallmatrix}\right)\rangle\!\rangle : \gamma''' = \gamma} \not\exists_{1 \leq j \leq |\gamma''|} \mathsf{synm}(\gamma''_j) \right]$$

4   Stochastic Petri Nets with Exogenous Dependencies

This section introduces our formalism stochastic labelled Petri nets with exogenous dependencies (Exo-SLPN), where weight functions consider exogenous data. Our modelling formalisation for stochastic process mining builds up existing stochastic labelled Petri nets from process mining [3,16,23].

Definition 4 (Stochastic Labelled Petri Nets With Exogenous Dependencies). *An Exo-SLPN is a tuple $(P, T, F, A, \lambda, M_o, \rho, \mathsf{trans}, \omega)$, where $(P, T, F, A, \lambda, M_o)$ is an LPN, ρ is a set of exo-panels, $\mathsf{trans} : (\rho \mapsto \mathcal{U}_{es}) \mapsto (\rho \mapsto (\mathbb{R} \cup \{\bot\}))$ transforms a mapping of exo-panels and exo-series into a mapping of exo-panels and numbers or $\bot$, and $w : (T \times (\rho \mapsto (\mathbb{R} \cup \{\bot\}))) \mapsto \mathbb{R}$ is a weight function.*

The control-flow semantics of an Exo-SLPN is those of the corresponding LPN. Additionally, an Exo-SLPN expresses the likelihood of firing an enabled transition as a ratio of the sum of the weights of all enabled transitions. Formally, given a marking M and a mapping of exo-series E, the probability of firing an enabled transition $t \in \mathsf{enbl}(M)$ is:

$$\mathsf{prob}(M, E, t) = \frac{w(t, \mathsf{trans}(E))}{\sum_{t' \in \mathsf{enbl}(M)} w(t', \mathsf{trans}(E))} \tag{2}$$

While Exo-SLPNs support any representation for a weight function w, we consider forms that use three weights for each transition t. That is, a *base* weight ϕ_t and, for each exo-panel p, two influencing parameters, being an *adjustment* parameter $\varphi_{t,p}$ when exogenous data is available and an *alternate* parameter $\psi_{t,p}$ when exogenous data is unavailable. These parameters quantify how the base weight ϕ_t of a transition t is influenced by exogenous data through $\varphi_{t,p}$ and $\psi_{t,p}$. Given a transition t and a mapping of transforms E, this paper considers several forms for a weight function w:

$$w(t, E) = \phi_t \cdot \prod_{E(p)=e} \begin{cases} (\varphi_{t,p})^e & \text{if } e \neq \perp \\ \psi_{t,p} & \text{otherwise} \end{cases} \qquad \text{with } \phi_t, \varphi_{t,p}, \psi_{t,p} > 0 \tag{3}$$

$$w(t, E) = \phi_t + \sum_{E(p)=e} \begin{cases} e \cdot \varphi_{t,p} & \text{if } e \neq \perp \\ \psi_{t,p} & \text{otherwise} \end{cases} \qquad \text{with } \phi_t, \varphi_{t,p}, \psi_{t,p} > 0 \tag{4}$$

$$w(t, E) = \phi_t + \sum_{E(p)=e} \begin{cases} e \cdot \varphi_{,p} & \text{if } e \neq \perp \\ \psi_{,p} & \text{otherwise} \end{cases} \qquad \text{with } \phi_t, \varphi_{,p}, \psi_{,p} > 0 \tag{5}$$

Equation (3) expresses exogenous influences could be multiplicative with the base weight of a transition, where they can have a dramatic impact on the likelihood of firing. Equation (4) and Eq. (5) express exogenous influences could be additive with the base weight of a transition, where they can only have a positive impact on the likelihood of firing. However, Eq. (5) only introduces one set of parameters per influence (i.e. exo-panel $p : \varphi_{,p}, \psi_{,p}$), rather than individually ($\varphi_{t,p}, \psi_{t,p}$) on transitions. We explore these options to understand which one may have the best generalisation.

Similarly, Exo-SLPNs support any transformation function trans, and the methods in the remainder of this paper do not pose any restrictions on these functions. But for this paper, a default transformation for numerical time series is used. This transformation returns a weighted average of z-scores (Eq. (7)), where measurements further away from the truncation have a smaller influence on the transformed value. Our intuition is that changes that happened furthest away from the decision point will have less impact on the firing likelihood. Z-scores are used to capture the variability and volatility in exogenous data, quantifying exogenous data for firing likelihoods. However, when there is no variance within a panel, the transformation maps to zero to identify that zero change can be observed from this panel. Similarly, when the truncation results in an empty sequence, the transformation maps to $\perp$ to use of the alternate parameter.

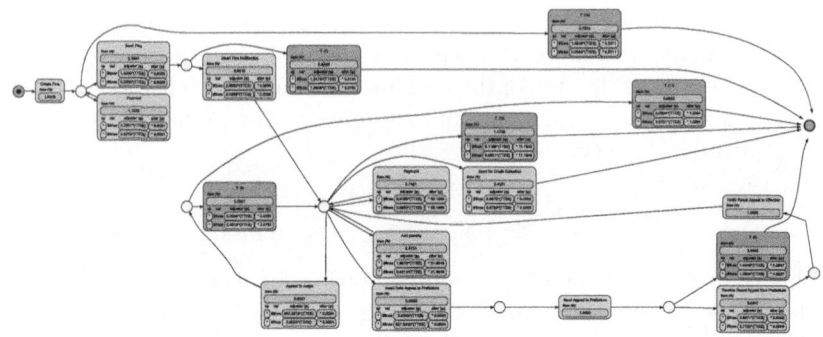

Fig. 5. Exo-SLPN discovered for the road fines log using the normative model.

Definition 5 (Default Numerical Transformation). *Let X be a mapping of panels to truncated exo-series, let $p \in \rho$ be an exo-panel, let $\mu(p)$ be the mean of p, let $\sigma(p)$ be the standard deviation of p, and let $X(p) = x$ be a truncated exo-series for p. Then,* trans$'$ *maps a collection of truncated exo-series into numbers:*

$$\text{trans}'(X)(p) = \begin{cases} \bot & \text{if } X(p) = \langle\rangle \\ 0 & \text{if } \sigma(p) = 0 \wedge X(p) \neq \langle\rangle \\ z & \text{if } \sigma(p) \neq 0 \wedge X(p) = \langle x_1, \ldots, x_n\rangle \end{cases} \tag{6}$$

$$\text{where } z = \frac{\sum\limits_{i=1}^{n} \frac{1}{1+\text{time}(x_i)} \cdot \left|\frac{x_i - \mu(p)}{\sigma(p)}\right|}{\sum\limits_{i=1}^{n} \frac{1}{1+\text{time}(x_i)}} \tag{7}$$

Each transformed-truncated-exo-series is referred to as a TTES.

Example 3. Consider the Exo-SLPN shown in Fig. 5 and trace 'A21188' annotated with exogenous data from Example 2. The goal is to understand the change in probability for credit collection based on exogenous data in the duration between e_{13} and e_{14}. In order to perform transformation, both the mean (mean) and standard deviation (std) of the panels are needed. As such, for unpaid fines these are 20928.975 and 8084.747, for amount unpaid these are 1308182.069 and 479526.257. Therefore, applying trans on the truncated exo-series presented in Example 2, yields a mapping E between panels and TTESes as follows:

$$E = \text{trans}'(\{\text{unpaid fines} \mapsto \langle 1^{@4917d}, \ldots, 31188^{@2d}, 31222^{@1d}, 31228^{@0d}\rangle,$$
$$\text{amount unpaid} \mapsto \langle 62^{@4917d}, \ldots, 2\,007\,000^{@2d}, 2\,008\,000^{@1d}, 2\,009\,000^{@0d}\rangle\})$$
$$= \{\text{unpaid fines} \mapsto 0.864, \text{amount unpaid} \mapsto 0.919\}.$$

Note that a shorthand is used in the following calculation for panels, '#F' for 'unpaid fines' and '$F' for 'amount unpaid'. Then, the probability (using Eq. (3)) of credit collection (**CC**) firing when the trace 'A21188' executed 'add penalty' (**AP**) was:

$$\text{prob}(M, E, \mathbf{CC}) = \frac{w(\mathbf{CC}, E)}{\sum_{t \in \text{enbl}(M)} w(t, E)}$$

$$\approx \frac{3.4321 \cdot \begin{cases} 0.6786^{E(\#F)} & \text{if } E(\#F) \neq \perp \\ 0.0005 & \text{if } E(\#F) = \perp \end{cases} \cdot \begin{cases} 0.8875^{E(\$F)} & \text{if } E(\$F) \neq \perp \\ 0.0005 & \text{if } E(\$F) = \perp \end{cases}}{\cdots + 0.7401 \cdot \begin{cases} 0.0882^{E(\#F)} & \text{if } E(\#F) \neq \perp \\ 59.188 & \text{if } E(\#F) = \perp \end{cases} \cdot \begin{cases} 2.4169^{E(\$F)} & \text{if } E(\$F) \neq \perp \\ 59.188 & \text{if } E(\$F) = \perp \end{cases}} + \cdots$$

$$\approx \frac{3.4321 \cdot 0.6786^{0.864} \cdot 0.8875^{0.919}}{3.4321 \cdot 0.6786^{0.864} \cdot 0.8875^{0.919} + 0.7401 \cdot 0.0882^{0.864} \cdot 2.4169^{0.919} + \cdots}$$

$$\approx 0.4635 \text{ or } 46.35\%.$$

5   Discovery

This section presents how to discover an Exo-SLPN, given an event log and an LPN, by casting the discovery as an optimisation problem. Thereby allowing the identification of exogenous influences on processes from historical data. First, we describe how to reconstruct when choices occurred in the process and extract diagnostic information about these moments, referred to as *choice data*. Then, we present the optimisation problem to discover the weights for Exo-SLPN, whereby values for parameters are obtained, i.e. a base weight ϕ, many adjustment φ parameters and many alternative ψ parameters.

5.1   Constructing Choice Data

After alignments have been obtained for traces in an xlog, the next step is to gather model choices and TTES values for these model choices. That is, for each transition fired in the model from replaying the xlog, the following is recorded: (i) the transitions that were enabled, (ii) the TTES values as they were at the time the transition fired, and (iii) the fired transition. For synchronous moves, TTES can be obtained using the event as they have a timestamp (Eq. (9)). For a log move, no transition is recorded so these moves do not introduce any choice data (Eq. (10)). For model moves, no timestamp is available but could be estimated (discussed in Sect. 5.2) using the function fill (Eq. (11)) to obtain a TTES.

Formally, for a trace, choice data is obtained using Definition 6. Then, choice data for an xlog is then simply the multiset union of its traces.

Definition 6 (Choice data). *Let $(\sigma, \#) \in \Sigma$ be a trace σ annotated with a mapping of exo–series $\#$ from an xlog, let LPN be a net, and let γ be an alignment between σ and LPN. Then, choice takes a marking and an alignment, returning a multiset of choice data:*

$$\text{choice}(M, \langle \rangle) = [\,] \tag{8}$$

$$\text{choice}(M, \langle (\tfrac{e}{t}) \rangle : \gamma') = [\,(\text{enbl}(M), \text{trans}'(\text{trun}(\#, \text{time}(e))), t)\,] \tag{9}$$

$$\cup \, \text{choice}(M \setminus {}^{\bullet}t \cup t^{\bullet}, \gamma')$$

$$\text{choice}(M, \langle (\overset{e}{\gg}) \rangle : \gamma') = \text{choice}(M, \gamma') \tag{10}$$

$$\text{choice}(M, \langle (\overset{\gg}{t}) \rangle : \gamma') = [\,(\text{enbl}(M), \text{fill}(\#, \gamma, |\gamma| - |\gamma'|), t)\,] \tag{11}$$
$$\cup \text{choice}(M \setminus {}^\bullet t \cup t^\bullet, \gamma')$$

Example 4. Consider the snippet of the road fines log (Table 1) and the alignments in Fig. 1, where the abbreviations are used for activities. Then, the recursion of choice is explored for trace 'A21188' to demonstrate how choice data is derived from the diagnostic information stored in the alignment.

$$\text{choice}(M_o, (\overset{e_{11}}{\mathbf{CF}}) : \langle \ldots \rangle) =$$
$$\quad [(\{\mathbf{CF}\}, \{\text{unpaid fines} \mapsto 0.551, \text{amount unpaid} \mapsto 0.529\}, \mathbf{CF})]$$
$$\quad \cup \text{choice}(M', (\overset{e_{12}}{\mathbf{SF}}) : \langle \ldots \rangle)$$
$$= [(\{\mathbf{CF}\}, \{\ldots\}, \mathbf{CF})]$$
$$\quad \cup [(\{\mathbf{SF}, \tau_{16}, \}, \{\text{unpaid fines} \mapsto 0.797, \text{amount unpaid} \mapsto 0.789\}, \mathbf{SF})]$$
$$\ldots$$
$$= [(\{\mathbf{CF}\}, \{\ldots\}, \mathbf{CF}), (\{\mathbf{SF}, \tau_{16}\}, \{\ldots\}, \mathbf{SF}), (\{\mathbf{IFN}, \tau_7\}, \{\ldots\}, \mathbf{IFN}),$$
$$\quad (\{\mathbf{CC}, \mathbf{AP}, \mathbf{PYb}, \mathbf{AtJ}, \tau_{13}\}, \{\ldots\}, \mathbf{AP})]$$
$$\quad \cup [(\{\mathbf{CC}, \mathbf{AP}, \mathbf{PYb}, \mathbf{AtJ}, \tau_{13}\}, \{\text{unpaid fines} \mapsto 1.052,$$
$$\quad \text{amount unpaid} \mapsto 1.174\}, \mathbf{CC})]$$

Thus for trace 'A21188', five triples of choice data are extracted and are combined with choice data from the other traces in the snippet as shown in Table 2.

5.2 Handling Model Moves

To obtain TTES values from model moves, two types of model moves are distinguished: moves on silent transitions, i.e. $\lambda(t) = \tau$, and moves on labelled transitions, i.e. $\lambda(t) \neq \tau$. To define the fill function, first synchronous (which have TTES values) and log moves (which do not have TTES values, as they do not represent transitions) are considered:

$$\text{fill}(\#, \gamma, i) = \begin{cases} \bot & \text{if } i > |\gamma| \tag{12a} \\ \text{trans}'(\text{trun}(\#, \text{time}(e))) & \text{if } \text{synm}(\gamma_i) \, with \, \gamma_i = (\overset{e}{t}) \tag{12b} \\ \text{fill}(\#, \gamma, i+1) & \text{if } \text{logm}(\gamma_i) \tag{12c} \\ \ldots \end{cases}$$

Silent Model Moves. A silent transition models a change in the state of a system without directly visible external consequences. As such, they have neither timestamps nor TTES. Nevertheless, the decision to fire the silent transition and not a potential competing visible transition depends on the TTES value, and as such must be assigned a TTES. As the firing of a silent transition has no corresponding action in the process (i.e. just in the system), this paper posits that the moment they happen is the moment the next labelled transition fires:

$$(12d)$$

$$\text{fill}(\#, \gamma, i) = \begin{cases} \cdots \\ \text{fill}(\#, \gamma, i+1) & \text{if } \mathsf{modm}(\gamma_i) \wedge \lambda(t) = \tau \ \text{with } \gamma_i = (\genfrac{}{}{0pt}{}{\gg}{t}) \\ \cdots \end{cases}$$

Labelled Model Moves. A labelled transition models the execution of a process step; a labelled model move indicates that this step was executed, but not logged, and hence there is no timestamp available to compute the TTES values. Thus, a TTES is obtained through estimation based on other moves with timestamps around the model move.

For the sake of explanation, assume an alignment with two synchronous moves and a model move in between: $\langle \ldots (\genfrac{}{}{0pt}{}{e}{t}), (\genfrac{}{}{0pt}{}{\gg}{t'}), (\genfrac{}{}{0pt}{}{e''}{t''}) \ldots \rangle$. TTES values depend on when the transition t' was fired, and in this case it was between e and e''. Time wise, this means that the time of e plus the sojourn times of t' and t'' is the time of e'' ($\mathsf{time}(e) + \mathsf{soj}(t') + \mathsf{soj}(e'') = \mathsf{time}(e'')$). Neither $\mathsf{soj}(t')$ or $\mathsf{soj}(t'')$ are known, however a distribution of sojourn times for t' and t'' can be observed using the aligned log. Figure 6 illustrates this partially: from the moment that $(\genfrac{}{}{0pt}{}{e}{t})$ happens (vertical line on the left), the distribution of sojourn times of t' (green bars) can be projected. Furthermore, note that $\mathsf{soj}(t')$ cannot have lasted beyond $(\genfrac{}{}{0pt}{}{e''}{t''})$ and that some sojourns times are not appropriate (grayed out bars).

Observe that at each potential timestamp in the green region, the TTES value would be computed using the exo-measures up till the last observed exo-measure. Thus, to estimate the TTES value, the number points in the distribution of $(\genfrac{}{}{0pt}{}{\gg}{t'})$ (the green region) that occurred before to each exo-measure is used to compute a weighted average accordingly:

$$\text{fill}(\#, \gamma, i) = \begin{cases} \cdots \\ \{ p \mapsto \begin{cases} \frac{1}{\|S\|} \sum_{s \in S} E(p)S(s) & \text{if } \|S\| > 0 \\ \bot & \text{otherwise} \end{cases} \mid p \in \# \} \\ \qquad\qquad\qquad \text{if } \mathsf{modm}(\gamma_i) \wedge \lambda(t) \neq \tau \end{cases}$$

$$(12e)$$

$with\,(\genfrac{}{}{0pt}{}{e}{t}), (\genfrac{}{}{0pt}{}{e''}{t''})$ *being the most recent, next sync moves in* γ

and sojourns $S = [s^{\mathsf{hist}(t')(s)} \mid s \in \mathsf{hist}(t') \wedge s \leq \mathsf{time}(e'') - \mathsf{time}(e) \wedge E(p) \neq \bot]$

and transforms of $E = \mathsf{trans}'(\mathsf{trun}(\#, \mathsf{time}(e) + s))$

The edge cases in which no synchronous move precedes or succeeds a model move are handled similarly. Notably two factors are explicitly left out: (i) the distribution over sojourn times of t'' is not leveraged, and (ii) that there may be multiple model moves in between two synchronous moves.

5.3 Solving Equalities

This section outlines how to estimate base weights ϕ, adjustments φ, and alternatives ψ parameters to discover an Exo-SLPN, by casting the problem of estimating these parameters as an linear optimisation problem. To do so, equalities

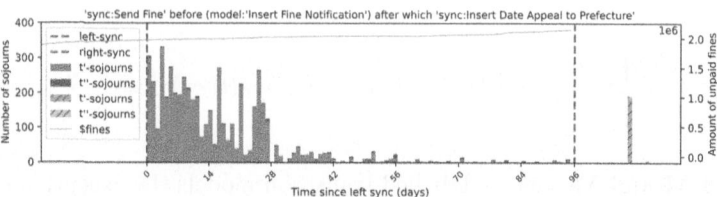

Fig. 6. Illustration of where a model move may have occurred (in green). (Color figure online)

are constructed from the obtained choice data. For each observed choice, the observations in the log are equated with the probability of that choice in the model (expressed in the weight parameters). A standard solver is then applied to minimise the disagreement between the sides of these equations. We considered two solution strategies, one where all the equations are solved in one-shot, and a two-shot approach where base variables are minimised first, then the adjustment and alternative parameters are minimised, both using the following equations.

Recall that the choice data C consists of triples of enabled transitions, the observed TTES values and which transitions fired. For each pair of enabled transitions (T) and TTES values (E) in Eq. (13), an equated fraction is introduced for every observed transition $o \in T$ through Eq. (14). The equated fraction in Eq. (15) states that the relative occurrence of the transition o should match the probability in the model, given the weight parameters denoted in Eq. (16):

$$\forall_{(T,E)\in\{(T,E)|(T,E,t)\in C\}} \tag{13}$$

$$\forall_{o\in\{o|(T,E,o)\in C\}} \tag{14}$$

$$\frac{\|[(T,E,o) \in C]\|}{\|[(T,E,t') \in C]\|} \tag{15}$$

$$= \frac{\phi_o \prod\limits_{E(p)=e} \begin{cases} \varphi_{o,p}^e & \text{if } e \neq \bot \\ \psi_{o,p} & \text{otherwise} \end{cases}}{\sum\limits_{t'\in T} \phi_{t'} \prod\limits_{E(p)=e} \begin{cases} \varphi_{t',p}^e & \text{if } e \neq \bot \\ \psi_{t',p} & \text{otherwise} \end{cases}} \tag{16}$$

If all enabled transitions have been observed at least once when constructing equalities, then one of the equations (an observed transition o from Eq. (14)) for every (T, E) (from Eq. (13)) can be removed without loss of information.

Example 5. For instance, the equalities corresponding to our running example are shown in Table 2. By passing these equalities to a solver, the parameters ϕ, φ, ψ for each transition are assigned a value which minimally reduces the difference between these equalities.

Table 2. The constructed equalities for Example 5. Panel names have been abbreviated. We note the equalities are not necessary for the solver.

enabled (T)	TTESes (E)	fired (t)	equality
{**CF**} (unnecessary)	{...}	[**CF**]	$\frac{1}{1} = \frac{\phi_{\mathbf{CF}} \cdot \varphi_{\mathbf{CF},\#F}^{0.282} \cdot \varphi_{\mathbf{CF},\$F}^{0.357}}{\phi_{\mathbf{CF}} \cdot \varphi_{\mathbf{CF},\#F}^{0.282} \cdot \varphi_{\mathbf{CF},\$F}^{0.357}}$
{**CF**} (unnecessary)	{...}	[**CF**]	$\frac{1}{1} = \frac{\phi_{\mathbf{CF}} \cdot \varphi_{\mathbf{CF},\#F}^{0.530} \cdot \varphi_{\mathbf{CF},\$F}^{0.552}}{\phi_{\mathbf{CF}} \cdot \varphi_{\mathbf{CF},\#F}^{0.530} \cdot \varphi_{\mathbf{CF},\$F}^{0.552}}$
{**CF**} (unnecessary)	{...}	[**CF**]	$\frac{1}{1} = \frac{\phi_{\mathbf{CF}} \cdot \varphi_{\mathbf{CF},\#F}^{0.633} \cdot \varphi_{\mathbf{CF},\$F}^{0.598}}{\phi_{\mathbf{CF}} \cdot \varphi_{\mathbf{CF},\#F}^{0.633} \cdot \varphi_{\mathbf{CF},\$F}^{0.598}}$
...			
{**PYa, SF**, τ_{16}}	{...}	[**PYa**]	$\frac{1}{1} = \frac{\phi_{\mathbf{PYa}} \cdot \varphi_{\mathbf{PYa},\#F}^{0.803} \cdot \varphi_{\mathbf{PYa},\$F}^{0.803}}{\phi_{\mathbf{PYa}} \cdot \varphi_{\mathbf{PYa},\#F}^{0.803} \cdot \varphi_{\mathbf{PYa},\$F}^{0.803} + \cdots}$
{**PYa, SF**, τ_{16}}	{...}	[τ_{16}]	$\frac{1}{1} = \frac{\phi_{\tau_{16}} \cdot \psi_{\tau_{16},\#F} \cdot \psi_{\tau_{16},\$F}}{\phi_{\tau_{16}} \cdot \psi_{\tau_{16},\#F} \cdot \psi_{\tau_{16},\$F} + \cdots}$
{**PYa, SF**, τ_{16}}	{...}	[**SF**]	$\frac{1}{1} = \frac{\phi_{\mathbf{SF}} \cdot \varphi_{\mathbf{SF},\#F}^{0.439} \cdot \varphi_{\mathbf{SF},\$F}^{0.380}}{\phi_{\mathbf{SF}} \cdot \varphi_{\mathbf{SF},\#F}^{0.439} \cdot \varphi_{\mathbf{SF},\$F}^{0.380} + \cdots}$
{**PYa, SF**, τ_{16}}	{...}	[**SF**]	$\frac{1}{1} = \frac{\phi_{\mathbf{SF}} \cdot \varphi_{\mathbf{SF},\#F}^{0.789} \cdot \varphi_{\mathbf{SF},\$F}^{0.798}}{\phi_{\mathbf{SF}} \cdot \varphi_{\mathbf{SF},\#F}^{0.789} \cdot \varphi_{\mathbf{SF},\$F}^{0.798} + \cdots}$
{**CC, PYb, IAP**, τ_{13}, **AP, AtJ**}	{...}	[**AP, IAP**]	$\frac{1}{2} = \frac{\phi_{\mathbf{AP}} \cdot \varphi_{\mathbf{AP},\#F}^{0.310} \cdot \varphi_{\mathbf{AP},\$F}^{0.313}}{\phi_{\mathbf{AP}} \cdot \varphi_{\mathbf{AP},\#F}^{0.310} \cdot \varphi_{\mathbf{AP},\$F}^{0.313} + \cdots}$
			$\frac{1}{2} = \frac{\phi_{\mathbf{IAP}} \cdot \varphi_{\mathbf{IAP},\#F}^{0.310} \cdot \varphi_{\mathbf{IAP},\$F}^{0.313}}{\phi_{\mathbf{IAP}} \cdot \varphi_{\mathbf{IAP},\#F}^{0.310} \cdot \varphi_{\mathbf{IAP},\$F}^{0.313} + \cdots}$
{**CC, PYb, IAP**, τ_{13}, **AP, AtJ**}	{...}	[**AP**]	$\frac{1}{1} = \frac{\phi_{\mathbf{AP}} \cdot \varphi_{\mathbf{AP},\#F}^{0.919} \cdot \varphi_{\mathbf{AP},\$F}^{0.865}}{\phi_{\mathbf{AP}} \cdot \varphi_{\mathbf{AP},\#F}^{0.919} \cdot \varphi_{\mathbf{AP},\$F}^{0.865} + \cdots}$

6 Conformance Checking

This section outlines how we perform conformance checking on an Exo-SLPN, where the data-aware unit Earth mover's distance [23] (duEMSC) is adapted for our purposes. Data-aware unit Earth movers extends the unit Earth-movers [18] by adding the data perspective as an additional dimension when comparing and calculating probability distributions. The measurement of data-aware unit Earth movers is expressed as follows [23]:

$$\text{duEMSC}(L, M) = 1 - \sum_{D \in L_\Delta} \sum_{A \in L_\Sigma} \max(p_L(A \wedge D) - (p_M(A|D) \cdot p_L(D)), 0)$$

where L, M denote an event log and a model, and L_Δ, L_Σ are distributions of data sequences and activity sequences observed in the log. In order to apply duEMSC to a given log L and Exo-SLPN M, we need to define data distribution L_Δ and the joint probabilities p_L, p_M for a given activity and data sequence, but more so in our case a sequence of TTES mappings. The first step is to consider our representation of xlog (Definition 1) as a multiset of pairs for activities and data, stepwise walking traces and generating TTESes for each step.

$$L = (\rho, \Sigma) = \bigcup_{(\sigma, E) \in \Sigma} \bigcup_{k=1}^{\Sigma((\sigma, E))} \left[\left(\sigma, \bigcup_{1 \leq i \leq |\sigma|} \langle \text{trans}'(\text{trun}(E, \text{time}(\sigma_i))) \rangle \right) \right] \quad (17)$$

where Σ is a multiset of traces annotated with a mapping of exo–series E (Definition 1). Using this form the distributions for sequences over activities A and

TTES $\mathcal{E}$ are straightforwardly derived from the multiset representation. Noting that TTES mappings E are continuous, requiring discretisation using a precision reduction factor. Now the probabilities from the log L, given a sequence of activities A and a sequence of TTES mappings $\mathcal{E}$ can be expressed as:

$$p_L(A \wedge \mathcal{E}) = \frac{L((A, \mathcal{E}))}{|L|} \quad p_L(\mathcal{E}) = \frac{|[\mathcal{E}^{L((A,\mathcal{E}))} \mid (A, \mathcal{E}) \in L]|}{|L|} \quad (18)$$

The last step to consider is the probabilities for all paths T through the Exo-SLPN M that induce the activity sequence A under the assumption of the sequence of mapped TTESes $\mathcal{E}$. Note that the sequence $\mathcal{E}$ must be as long as a path $t \in T$, as such padding is applied to ensure that silent transitions use the next following mapping if available, or the last preceding mapping. Now, the model probability for a given activity sequence A under the assumption of $\mathcal{E}$, is the sum probability of all paths T that induce A, expressed as:

$$p_M(A|\mathcal{E}) = \sum_{\langle t_1, \ldots, t_n \rangle \in T} \prod_{i=1}^{n} \frac{w(t_i, \mathcal{E}_i)}{\sum_{t' \in \text{enbl}(M_i)} w(t', \mathcal{E}_i)} \quad (19)$$

Where M_i is the deterministic marking before firing a transition, e.g. $M_1 = M_o$ and $M_o \xrightarrow{t_1} M' \to \ldots \xrightarrow{t_{i-1}} M_i$. Now, duEMSC can be denoted using our definition of an xlog L and Exo-SLPN M, expressed as:

$$\text{duEMSC}(L, M) = 1 - \sum_{\mathcal{E} \in L_\Delta} \sum_{A \in L_\Sigma} \max(p_L(A \wedge \mathcal{E}) - (p_M(A|\mathcal{E}) \cdot p_L(\mathcal{E})), 0) \quad (20)$$

7   Evaluation

This section compares our Exo-SLPN approach against several existing techniques across several publicly available event logs. The proposed technique has been implemented within the ProM framework[1], and evaluation data has been made publicly available to the best of our ability and inline with ethics[2].

7.1   Event Logs and Exogenous Data

This evaluation uses three event logs, from different domains, paired with exogenous data. The first event log comes from the MIMIC III dataset [12] which contains 'MICU' ward admissions focusing on the first 48 h of admission and follows the preparation outlined in [2]. For this log, blood pressure measurements collected from nurse observations are used for exogenous data. These measurements were selected as they may inform what medical interventions should or should not occur with an admission.

---

[1] Found in the ExogenousData Package: github.com/promworkbench/Exogenous Data.

[2] Evaluation data can be found in: github.com/adambanham/exo-slpn-testing.

Table 3. Descriptive qualities of event logs used in our evaluation.

log	#traces	#events	#acts	exo-data	μ	σ	#datapoints
MIMIC [12]	4482	28524	13	blood pressure	8	106	976367
S.Factory [21]	34	418	12	crane jib x	139	128	13038
				crane jib y	43	46	13038
				motorspeed	−21	274	1602128
R. Fines [7]	150370	561470	11	unpaid fines	20928	8084	4829
				amount unpaid	1308182	479526	4898

The second log comes from a smart factory [21], where only the 'WF_101' process is considered and the start events within these executions (as these are associated IoT sensors). This log was paired with IoT sensors used in the factory as exogenous data, where the positioning of a crane and the motor speed of the crane are used. These sensors were selected as their readings may influence when human intervention occurs or trigger further actions.

The third log used is the road fines event log [7]. For this log, inter-case variables from the log were used as exogenous data, i.e. the total number of unpaid fines and the amount of unpaid fines seen in the event log. These inter-case time series were created using data points from all events, and then aggregated into daily observations. These variables are used to understand and answer our running example's question Example 1. All three logs are described in terms of control flow, size and exogenous data in Table 3. Our testing was performed using an AMD Ryzen 9 3900XT with 64GB ram, but we saw excessive amounts of system RAM being required (upwards of 256GB) to compute conformance checking, as such both the MIMIC III and road fines log were sampled[3].

7.2 Log Completeness

To show that our approach can identify the stochastic nature of a process, we considered how the model quality changes as we apply our approach to progressively more complete samples of a given log. The intuition being that with a more complete understanding of the original log, the discovered stochastic nature should better reflect the original or at least not degrade. We created 25 sample logs of the road fines log, where the n-th sample consists of $1000 \cdot n$ traces, and each progressive larger sample contains all samples from previous sample, i.e. given samples s_i, s_{i+1}, s_{i+j} then $s_i \subset s_{i+1}$ and $(s_i \cup s_{i+1}) \subset s_{i+j}$ where $i \geq 1$ and $j > 1$. Note that our approach considers the temporal nature of each trace, so traces with the same control-flow will have a minor temporal difference and these introduce/expand the choice data used for discovery (Sect. 5.1). Furthermore, as the sample gets larger, the number of sojourns considered for model moves (Sect. 5.2) will be increased and will affect the runtime of discovery.

---

[3] See github.com/adambanham/exo-slpn-testing.

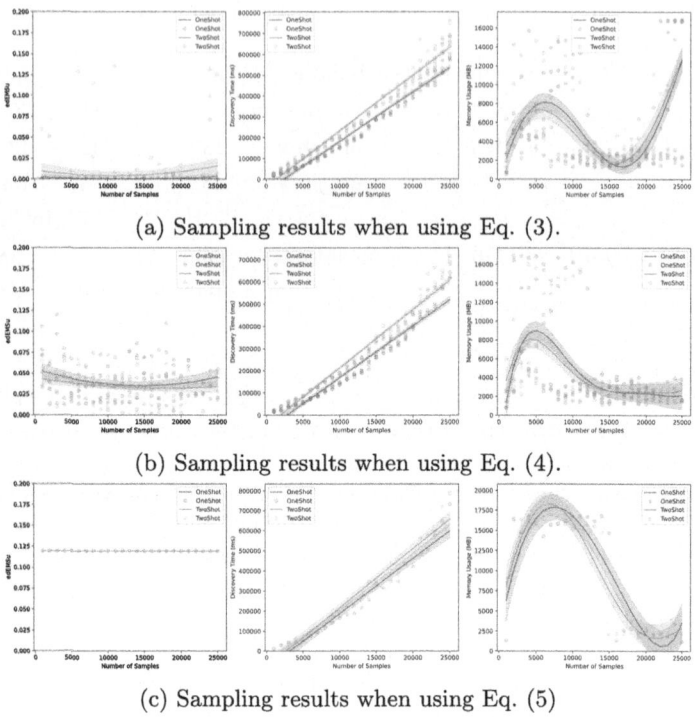

(a) Sampling results when using Eq. (3).

(b) Sampling results when using Eq. (4).

(c) Sampling results when using Eq. (5)

Fig. 7. Results from sampling testing using the road-fines log. The line of best fit and 95% confidence levels are shown to evaluate the trends across runs.

For each sample, we discovered an Exo-SLPN (recording runtime and memory usage) and then using the complete log, we compute duEMSC to quantify the quality of the discovered Exo-SLPN. We sampled the road fines log down to a total of 25,000 traces and used this reduced log as the 'complete log' for testing to avoid excessive system requirements. For a control-flow model the normative model (Fig. 1) was used. Then, our testing was conducted several times for both one-shot and two-shot optimisation (Sect. 5.3). The results of the sampling testing across the alternative weight functions are shown in Fig. 7.

The testing on the multiplicative form (Eq. (3)) in Fig. 7a shows that the equation may overly focus on the exogenous influences and struggle to find the balance between endogenous and exogenous factors. As demonstrated by duEMSC measurements being on average below 0.01 across the sampling with some minor spikes. This struggle may be due to the exogenous influence being modelled as a power of ψ, but interestingly higher quality models were produced by the two-shot approach. Next, the testing for the additive form (Eq. (4)) in Fig. 7b shows that the equation has a higher baseline than Eq. (3) when capturing the stochastic nature, likely due to the exogenous influence being less pronounced. However, Eq. (3) produced a higher quality model than Eq. (4) highest, albeit rather inconsistently, but a large amount of variance can be seen for both equations. The opposite trend occurs for the global additive form (Eq. (5)), which reports no variance at all and a constant reading of 0.119 regardless of sample size. We posit that using Eq. (5) means that exogenous influences are treated as constants and cancel each other out when competition occurs between transitions, which in turn removes all difficulty for optimisation.

Interestingly, the pairwise comparison of two-shot and one-shot optimisation, showed that two-shot has the potential to produce higher quality models, but the one-shot approach was more consistently better in Fig. 7a and Fig. 7b. Unsurprisingly, the discovery time is linear with the size of the log due to the nature of sojourn calculations, and memory usage was similar between all approaches. Furthermore, the additional runtime needed to perform the two-shot optimisation approach is minor in comparison to the possible quality increase, as such we adopted the two-shot optimisation in the following evaluation.

7.3   Model Quality

To compare our approach against existing ones, we discovered a variety of stochastic extensions of Petri nets and considered how close these nets represent the stochastic nature of a log. As existing techniques and the proposed approach require a control-flow model to discover a stochastic Petri net, the following discovery techniques were used: the inductive miner [14] (IM_f), the directly flows miner [17] (DFM), and the POWL miner [13] (IM_{po}) using the default settings. Additionally the normative model [22, Sec 12.1.3] for the road fines log was included, which contains repeated activities which inductive miners cannot find. These techniques were selected as they discover process trees which ensure that the converted Petri net is sound and alignments can be computed. To quantify the quality of a stochastic Petri net, the unit earth movers distance [18] or the data-aware variant [23] were used (see Sect. 6). For these measures, a measurement closer to 1.0 reports that the model perfectly describes the stochastic nature of the log. Note that directly comparing between control flow and data aware stochastic conformance measures is discouraged.

The procedure for discovering and quantifying a model consisted of: (i) discovering a control-flow model with the original log, (ii) sampling the original log with replacement, (iii) discovering stochastic weights using the sampled log and control-flow model, and (iv) measure the discovered stochastic model using the original log. The same sampled log is used across techniques. As noted in Sect. 7.2, the variability of outcomes produced from our approach is high, as such we evaluated them five times and reported on the highest quality outcome.

Results. Table 4 shows the testing results for existing and proposed approaches (far-right columns). Neither control-flow techniques reported 1.0 for any of the used logs, highlighting the need to consider factors outside the control-flow of processes. However, Sign. Deps. [16] was unable to discover a model within 24 h for the mimic log as it was solving over 50,000 parameters. Surprisingly, neither Alig. Est. [4] or Sign. Deps. reported to capture more than 50% of the stochastic nature of the road fines log using the normative model. All techniques favoured

Table 4. Conformance for discovered models with our method highlighted. Highest control-flow scores are bolded. Highest data-aware scores are underlined.

	Disc.	Alig. Est. [4]	Sign. Deps. [16]	Data Deps. [23]	Eq. (3)	Eq. (4)	Eq. (5)
		Unit Earth Movers [18]		Data-Aware Unit Earth Movers [23]			
MIMIC	IM$_f$	**0.0546**	t/o	0.0020	0.0012	0.0001	0.0001
	IM$_{po}$	**0.0031**	t/o	0.0006	0.0187	0.0012	0.0000
	DFM	**0.1713**	t/o	0.0064	0.0147	0.0063	0.0067
S. Factory	IM$_f$	0.0000	0.0000	0.0000	0.0000	0.0000	0.0000
	IM$_{po}$	**0.1205**	0.0000	0.0001	0.1176	0.1149	0.0592
	DFM	**0.1282**	0.1258	0.0006	0.1176	0.1150	0.0295
R. Fines	normative	0.2573	**0.2933**	0.1479	0.0253	0.0540	0.1179
	IM$_f$	0.1003	**0.3865**	0.1654	0.0701	0.0760	0.0413
	IM$_{po}$	0.2201	**0.3955**	0.3326	0.0437	0.0101	0.0100
	DFM	**0.8178**	**0.8178**	0.5100	0.4266	0.3779	0.2148

using a more literal and less structured model, i.e. DFM, for the road fines log. Notably, all measurements for the IM$_f$ model and the S. Factory log reported 0.000 due to the model being unable to replay traces.

Moving to the data-aware setting, across the mimic and factory log, our approach outperforms the existing Data Deps. [23] technique in four out of five cases. Showing that Exo-SLPNs can be competitive with other stochastic nets in unstructured processes. However, Data Deps. outperformed in the more structured case, i.e., the road fines log where there is a clear connection between event attributes and outcomes. Comparing the different weight functions, Eq. (3) produced the highest quality model in seven out of nine cases. These results may indicate that the multiplicative form has a greater capacity to capture the stochastic natures of processes, albeit with greater complexity for optimisation.

8   Conclusion

Understanding how likely are individual actions in processes allows businesses to be flexible and adapt to contextual changes. Thus, techniques providing insights into these likelihoods and how systems have acted in the past are important. As without these insights, ad-hoc changes to make actions more likely may be unsuccessful or harmful to the performance of the process. To this end, we explored how the temporal dimension and contextual factors (i.e. exogenous data) surrounding executions of processes can be included within the analysis of firing likelihoods for actions. Extending the state-of-art techniques and evaluating our approach to understand their capacity to capture firing likelihoods.

The contribution of this paper was a novel means to study whether exogenous data influences the behaviour of processes. A new process modelling formalism Exo-SLPN was introduced to enable the quantification of firing likelihoods based on exogenous data. Then, a discovery technique was developed for Exo-SLPNs, which caters for imperfect alignments between the traces and the control-flow model. Lastly, our evaluation of Exo-SLPNs showed our approach could outperform state-of-art techniques in less structured processes.

In future work, we hope to explore how to visualise complex parameterised stochastic Petri nets within the context of a given event log. Exploring alternatives weight forms (Eq. (3)) and transformations (Definition 5) could advance the expressiveness of Exo-SLPN and allow non-numerical signals to be investigated. Also, considering more scalable and faster computations for conformance checking of data-aware stochastic Petri nets would benefit future research.

Acknowledgments. Adam Banham's work was funded through an Australian Government Research Training Program Scholarship, and a QUT, Centre for Data Science Scholarship. He was also supported by RWTH Aachen, through an Advanced Research Opportunities Program Scholarship. Yannis Bertrand's work was supported by the Flemish Fund for Scientific Research (FWO) with grant number G0B6922N.

References

1. Adriansyah, A.: Aligning observed and modeled behavior. Ph.D. thesis, Mathematics and Computer Science, Eindhoven, The Netherlands (2014)
2. Banham, A., Leemans, S., Wynn, M.T., Andrews, R., Laupland, K.B., Shinners, L.: xPM: enhancing exogenous data visibility. Artif. Intell. Med. (2022)
3. Burke, A.: Process mining with labelled stochastic nets. Ph.D. thesis, Information Systems, Brisbane, Australia (2024)
4. Burke, A., Leemans, S.J.J., Wynn, M.T.: Stochastic process discovery by weight estimation. In: ICPM Workshops. LNBIP (2020)
5. Camargo, M., Dumas, M., González, O.: Automated discovery of business process simulation models from event logs. Decis. Support Syst. (2020)
6. Carmona, J., van Dongen, B.F., Solti, A., Weidlich, M.: Conformance Checking - Relating Processes and Models (2018)
7. de Leoni, M., Mannhardt, F.: Road traffic fine management process (2015)
8. Dumas, M., La Rosa, M., Mendling, J., Reijers, H.A.: Fundamentals of Business Process Management, 2nd edn. (2018)
9. Gal, A.: Everything there is to know about stochastically known logs. In: ICPM (2023)
10. Günther, C.W., Rinderle-Ma, S., Reichert, M., van der Aalst, W.M.P., Recker, J.: Using process mining to learn from process changes in evolutionary systems. Int. J. Bus. Process. Integr. Manag. (1) (2008)
11. Janiesch, C., Koschmider, A., Mecella, M., et al.: The internet of things meets business process management: a manifesto. IEEE Syst. Man Cybern. Mag. (4) (2020)
12. Johnson, A.E., et al.: MIMIC-III, a freely accessible critical care database. Sci. Data (1) (2016)
13. Kourani, H., van Zelst, S.J.: POWL: partially ordered workflow language. In: BPM. LNCS (2023)
14. Leemans, S.: Robust Process Mining with Guarantees - Process Discovery. Conformance Checking and Enhancement. LNBIP (2022)

15. Leemans, S., Maggi, F.M., Montali, M.: Enjoy the silence: analysis of stochastic petri nets with silent transitions. Inf. Syst. **124**, 102383 (2024)
16. Leemans, S., Mannel, L.L., Sidorova, N.: Significant stochastic dependencies in process models. Inf. Syst. (2023)
17. Leemans, S.J.J., Poppe, E., Wynn, M.T.: Directly follows-based process mining: exploration & a case study. In: ICPM (2019)
18. Leemans, S., van der Aalst, W., Brockhoff, T., Polyvyanyy, A.: Stochastic process mining: earth movers' stochastic conformance. Inf. Syst. (2021)
19. Maggi, F.M., Montali, M., Peñaloza, R.: Probabilistic conformance checking based on declarative process models. In: CAiSE Forum. LNBIP. Springer (2020)
20. Maggi, F.M., Montali, M., Peñaloza, R., Alman, A.: Extending temporal business constraints with uncertainty. In: BPM. LNCS. Springer (2020)
21. Malburg, L., Grüger, J., Bergmann, R.: An IoT-enriched event log for process mining in smart factories. Zendo (2023)
22. Mannhardt, F.: Multi-perspective process mining. Ph.D. thesis, Mathematics and Computer Science, Eindhoven, The Netherlands (2018)
23. Mannhardt, F., Leemans, S.J.J., Schwanen, C.T., de Leoni, M.: Modelling data-aware stochastic processes - discovery and conformance checking. In: Petri Nets. LNCS (2023)
24. Rinderle, S., Reichert, M., Dadam, P.: Correctness criteria for dynamic changes in workflow systems - a survey. Data Knowl. Eng. (1) (2004)
25. Rogge-Solti, A., van der Aalst, W.M.P., Weske, M.: Discovering stochastic petri nets with arbitrary delay distributions from event logs. In: Business Process Management Workshops. LNBIP (2013)
26. Scheibel, B., Rinderle-Ma, S.: Decision mining with time series data based on automatic feature generation. In: CAiSE. LNCS (2022)
27. Senderovich, A., Shleyfman, A., Weidlich, M., Gal, A., Mandelbaum, A.: To aggregate or to eliminate? optimal model simplification for improved process performance prediction. Inf. Syst. (2018)
28. van der Aalst, W.M.P.: Process Mining - Data Science in Action, 2nd edn. (2016)
29. vom Brocke, J., Baier, M., Schmiedel, T., Stelzl, K., Röglinger, M., Wehking, C.: Context-aware business process management. Bus. Inf. Syst. Eng. (5) (2021)
30. Yang, J., Ouyang, C., Dik, G., Corry, P., ter Hofstede, A.H.M.: Crop harvest forecast via agronomy-informed process modelling and predictive monitoring. In: CAiSE. LNCS (2022)

Synthesizing Petri Nets from Labelled Petri Nets

Robin Bergenthum[1]([✉])[iD] and Jakub Kovář[2][iD]

[1] Fakultät für Mathematik und Informatik, FernUniversität in Hagen,
Hagen, Germany
robin.bergenthum@fernuni-hagen.de
[2] Lehrgebiet Programmiersysteme, FernUniversität in Hagen, Hagen, Germany
jakub.kovar@fernuni-hagen.de

Abstract. Synthesis automatically generates a process model from a
behavioural specification. If the desired process model is a Petri net,
synthesis is addressed through so-called region theory. Region-based synthesis has been extensively studied for cases where the specification is a
transition system, a step-transition system, a language, or even a partially ordered language. Although the ideas of region-based synthesis are
consistent across different types of specifications, each specification type
has its own definition of regions and uses different representations of the
set of all regions to synthesize a result. Up to this point, state-based and
language-based regions are simply two distinct concepts. In this paper,
we introduce Petri net regions, to synthesize a Petri net from a set of
labelled Petri nets. We synthesize a result that can simulate the specified
behaviour using a set of minimal regions. Thus, we advance region theory
to the next level. Furthermore, we show that both language-based and
state-based regions are Petri net regions as well. There is no longer any
need to distinguish these concepts any more. Using Petri net regions, we
present an implementation of a synthesis algorithm that handles state-
based, language-based, and Petri net-based input.

Keywords: Petri nets · Region theory · Token trails · Labelled Petri
nets · Synthesis · State-based regions · Language-based regions

1 Introduction

Complex systems are often modelled using Petri nets [1,16,34,38]. Petri nets
have formal semantics, an intuitive graphical representation, and the ability to
express concurrency among the actions of a system. However, constructing a
Petri net model for a real-world process is a costly and error-prone task [1,32].
Fortunately, when modelling a system, there are often associated descriptions or
even specifications of the desired process behaviour. These may include log files
of recorded behaviour, example runs, and product specifications describing use
cases. We can model these specifications using a language, a transition system, or
a partial language. If a specification accurately reflects the desired behaviour, we

E. Amparore and Ł. Mikulski (Eds.): PETRI NETS 2025, LNCS 15714, pp. 63–85, 2025.
https://doi.org/10.1007/978-3-031-94634-9_4

can automatically synthesize the most fitting process model. The synthesis problem is to compute a process model such that: (A) the specification is behaviour of the generated model, and (B) the generated model has minimal additional behaviour. Upon reviewing the literature, particularly concerning state-based regions, the synthesis problem is sometimes formulated as determining whether there exists a model with (A) and no additional behaviour. See [2] for a comprehensive introduction into state-based region theory. In this paper, we aim to calculate an upper approximation and consider the check of whether the model perfectly fits the specification as an optional step. In this regard, our formulation is somewhat more liberal and has numerous applications, particularly within the modelling and process mining communities because we always generate a model.

The theory behind Petri net synthesis is known as region theory [18,19]. Region theory has been extensively explored for transition systems [2], languages [15,30], and even partial languages [3,8]. There are many non-trivial theoretical results, concepts, case studies, and tool support from tools such as APT [13], Genet [14], ProM [4,17,41], Viptool [6], and I ♥ Petri nets [11]. Region theory is even extended towards generalised net classes, such as inhibitor nets [29,35], towards more refined firing semantics, like interval semantics [36], and most recently even towards combining both [27]. Nowadays, region theory has two main branches. If the input is state-based, such as a transition system, we apply the theory of state-based regions. In this case, a region is a multi-set of states, and we construct a set of minimal regions to generate a finite set of valid places for the resulting Petri net. If the input is language-based, such as an event log, a language, or a partial language, we apply the theory of language-based regions. Here, a region is a multi-set of tokens produced by prefixes of the language, and we calculate valid places by solving a related integer linear programming (ILP) problem. To generate a finite result, we either calculate a basis of the ILP or use the concept of wrong continuations [7]. Both branches of region theory follow the same general ideas, but use different techniques, definitions, and algorithms.

When working with synthesis in practical examples, we first need to decide on the specific region definition we will use. Only then can we design and formalize our specification. We must restrict ourselves to one of the two techniques. If we work with state-based regions, we need to specify the state space of the desired system, and it is difficult to naturally specify concurrency. When working with language-based regions, we must specify all possible executions of the system, and it becomes challenging to naturally define conflicts. In this community, we know that modelling systems with conflict and concurrency is best done using Petri nets. Therefore, it would be best to develop a region theory for Petri net-based specifications. Let's consider the following example of a simple workflow. After a registration, we must ask for information once, twice, or three times. Simultaneously, we either save and apply, or we stop the process of asking for more information. After all the information is gathered, and after either stop or apply occurred, we check the application. For the sake of this contribution, we present a possible model of this process in Fig. 1. This is possible because we are experienced modellers, well-equipped with the knowledge of how to model using

Petri nets. Nevertheless, it is not a simple model. We use a distributed initial marking, a short-loop and arc weights.

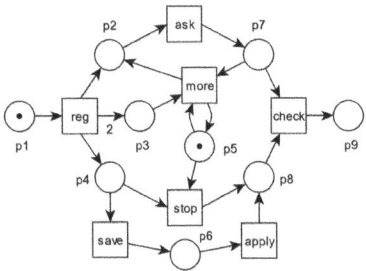

Fig. 1. A Petri net of our example process.

Let's not forget that modelling complex systems is an error-prone task. This is why, in a synthesis-based modelling approach, we do not model from scratch. Instead, we first specify the intended behaviour and then automatically generate the best-fitting model using region theory. We assume it is much easier to model the behaviour, instead of going for the integrated system model. Now, let's assume that we are not able to obtain Fig. 1 directly, but instead rely on synthesis techniques to produce the result.

In state-based region theory, we specify the intended behaviour by modelling the set of reachable states of our system and all transitions between these states. Thus, the correct specification of our workflow is depicted in Fig. 2. Although it is a simple state machine, it is difficult to develop this model of behaviour because, even in this small example, the number of states and transitions is quite large. The reason for this is clear. There is concurrency in our workflow, which results in diamonds in the set of state transitions. Additionally, we have to separate states by counting the number of occurrences of the action *ask*. We claim that we would rather model Fig. 1 from scratch than bother with modelling Fig. 2.

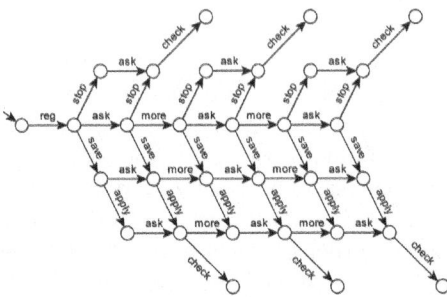

Fig. 2. A state-based specification of our example process.

In language-based region theory, we specify the intended behaviour by modelling the set of runs of our system. A run is a conflict-free set of occurrences of actions, called events, with a later-than relation on these events. A run specifies a single execution of the system but is unable to specify conflict. Thus, a specification of our workflow is depicted in Fig. 3. Although each run itself is easy to understand, it is difficult to keep track of the set of all possible executions and alternatives. Even in this small example, we have to specify six separate runs to capture the behaviour of our workflow system. Adding one more alternative to the start or the end of the workflow would double the number of runs we need to specify. To be fair, in this example, we would prefer Fig. 3 over Fig. 2 in such a workflow-like process example, but it is very easy to see that the size of the transition system, as well as the number of runs, can easily explode if the example increases even slightly in size.

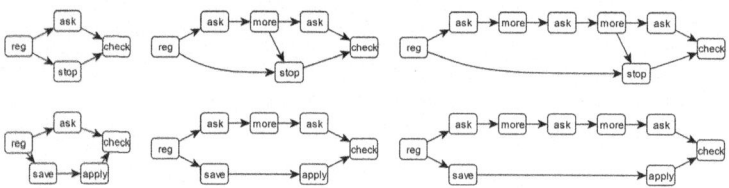

Fig. 3. A language-based specification of our example process.

In this paper, we introduce a new net-based region theory. Using this technique, we can specify the intended behaviour of our system in terms of labelled Petri nets. In our example, we could model the six runs of the system as four labelled nets. Each net specifies exemplary behaviour of our system. One possible specification is presented in Fig. 4. We claim that this is much easier to devise than Figs. 2 or 3, and even easier than the brute-force approach of going to Fig. 1 from scratch. Figure 4 shows four labelled Petri nets. Just like in the transition system in Fig. 2, or the labelled partial orders from Fig. 3, we model the behaviour using labels. Here, different transitions can refer to the same action in the process. Thus, it is not necessary to come up with the integrated model from scratch. The first labelled net of Fig. 4 is a state graph. In fact, it is a one-to-one translation of the diamond on the top left of Fig. 2 into a labelled net. This labelled net has only conflict, there is no concurrency. Note that we could translate the entire Fig. 2 into one big labelled net to get a complete specification of behaviour of or example process, but then again, obviously, this would have the same disadvantages as only using a state-based approach. The second labelled net of Fig. 4 is kind of a marked graph. In fact, it is a one-to-one translation of the fourth partial order of Fig. 3 into a labelled net. This labelled net has only concurrency, there is no conflict. Again, we could translate the entire Fig. 3 into six different conflict-free labelled nets to get a complete specification of behaviour of or example process, but then again, obviously, this would have the same disadvantages as only using a language-based approach.

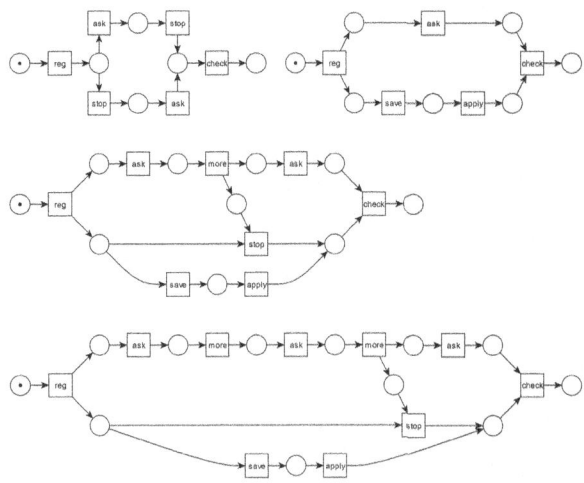

Fig. 4. A Petri net-based specification of our example process.

In our new approach, we are not restricted to state graphs and marked graphs. The third and fourth labelled net of Fig. 4 are general labelled nets with conflict, concurrency, and merging of local states. In this example, we get the third labelled net by kind of folding the partial orders two and five of Fig. 3. We get the fourth labelled net by merging the partial orders three and six of Fig. 3. Remember, we are not restricted to process nets nor branching processes. Both nets, the third and the fourth, model, that we can execute *stop* or *apply* in parallel to *ask* to enable *check*.

Using labelled nets, we can reduce the number of example runs to specify. Adding more alternatives to the start or the end of our example process does not automatically increase the number of nets we need to specify. We can simply add conflict to the labelled nets adding alternative behaviour to the specification. In a net-based specification we can add additional behaviour by either introducing alternative nets or adding conflicts to an already specified labelled net. In that sense, we can break down the entire behaviour into bits, each one an easy-to-handle labelled net. Although, this is a very small example, we claim that it is much easier and more intuitive to model Fig. 4 than to come up with Fig. 2 or 3. We do not claim Fig. 4 is the only right way to specify a workflow. In situations, where state-based modelling fits, we use only state graphs. In examples, where language-based modelling fits, we use only marked graphs. We can mix state-based and language-based models, and we can even model behaviour using general labelled nets. In Sect. 4, we will also see examples, that we can model behaviour using loops and arc weights. Altogether, we claim that a set of labelled nets is flexible and well-suited for modelling a specification of behaviour of a distributed system. Starting from such a specification, this paper addresses the following problem: Let the specification be a set of labelled nets. Synthesize a Petri net that can simulate all nets from the specification.

The new technique is based on the token trail semantics we introduced in [10, 28]. We prove that using this technique, we can define a related theory of regions. We show that this technique is consistent with both state-based and language-based regions. In this sense, it is not only a new region definition for labelled net input but can also serve as meta-theory to unify state-based and language-based concepts. Furthermore, we present an implementation and provide initial experimental results.

2   Preliminaries

Let $\mathbb{N}$ be the non-negative integers. Let f be a function and B be a subset of the domain of f. We write $f|_B$ to denote the restriction of f to B. As usual, we call a function $m : A \to \mathbb{N}$ a multiset and write $m = \sum_{a \in A} m(a) \cdot a$ to denote multiplicities of elements in m. Let $m' : A \to \mathbb{N}$ be another multiset. We write $m \le m'$ iff $\forall a \in A : m(a) \le m'(a)$ holds. We model distributed systems by Petri nets [1, 16, 34, 38].

Definition 1. *A Petri net is a tuple (P, T, W) where P is a finite set of places, T is a finite set of transitions so that $P \cap T = \emptyset$ holds, and $W : (P \times T) \cup (T \times P) \to \mathbb{N}$ is a multiset of arcs. A marking of (P, T, W) is a multiset $m : P \to \mathbb{N}$. Let m_0 be a marking, we call $N = (P, T, W, m_0)$ a marked Petri net and m_0 the initial marking of N.*

Figure 1 depicts a marked Petri net. Transitions are shown as rectangles, places as circles, the multiset of arcs as a set of weighted arcs, and the initial marking as a set of black dots, referred to as tokens. For Petri nets, there is a firing rule. Let t be a transition of a marked Petri net (P, T, W, m_0). We denote $\circ t = \sum_{p \in P} W(p, t) \cdot p$ the weighted pre-set of t. We denote $t \circ = \sum_{p \in P} W(t, p) \cdot p$ the weighted post-set of t. A transition t can fire in marking m iff $m \ge \circ t$ holds. Once transition t fires, the marking of the Petri net changes from m to $m' = m - \circ t + t \circ$. In our example marked Petri net, transition *reg* can fire in the initial marking. If *reg* fires, this removes one token from p_1. Additionally, firing *reg* produces a new token in p_2, two new tokens in p_3, and a new token in p_4. In this new marking transitions *ask*, *stop*, and *save* can fire. *reg* is not enabled any more, because there are no more tokens in p_1. Firing transition *ask* will enable transitions *more*, *stop*, and *save*. Firing transition *stop* will enable transition *ask*. Firing transition *save* will enable transitions *ask* and *apply*. Repeatedly applying the firing rule produces so-called firing sequences. These firing sequences are the most basic behavioural model of Petri nets. For example, the sequence *reg ask stop check* is enabled in the marked Petri net of Fig. 1. The sequence *reg save apply ask check* is another example. Let N be a marked Petri net, the set of all enabled firing sequences of N is the sequential language of N.

Another formalism used to model the behaviour of a Petri net is the reachability graph. A marking is reachable if there is a firing sequence that produces this marking.

Definition 2. *Let $N = (P, T, W, m_0)$ be a marked Petri net. The reachability graph of N is a tuple (R, T, X, m_0), where R is the set of reachable markings of N, T is the set of transitions of N, and X (called transitions as well) is a set of triples in $R \times T \times R$, such that (m, t, m') is in X if and only if t is enabled in (P, T, W, m), and firing t in m leads to the marking m'. We call a tuple (R', T', X', i) a state graph enabled in N if there is an injective function $g : R' \to R$, $g(i) = m_0$, $T' \subseteq T$, $\forall (m, t, m') \in X' : (g(m), t, g(m')) \in X$, and for every $m' \in R'$ there is a directed path from i to m' using the elements of X' as arcs. We call the set of enabled state graphs the state language of N.*

Roughly speaking, every node of a state graph relates to a reachable state, we don't have to include all transitions and states if every node can be reached from the initial node. Thus, a state graph is kind of a prefix of a reachability graph. Figure 2 depicts a state graph modelling the behaviour of the marked Petri net depicted in Fig. 1. The state graph has 31 states and 43 transitions labelled with transitions of the Petri net. The state graph describes the Petri nets behaviour as follows. At first, we must fire transition *reg*. Then, we have some choices. We can execute *stop*, *ask*, or *save*. We have to count occurrences of the action *ask* and after *stop* there is no *more*. Finally, firing *check* always leads to a deadlock. After firing *check*, there are either two, one, or no tokens in p_3, and one or no tokens in p_5. The state language includes firing sequences as the set of all paths through the graph. In this sense, state graphs can merge firing sequences at shared states and may contain loops. However, these graphs are not able to directly express concurrency. Firing *reg* in the initial marking depicted in Fig. 1 leads to the marking $p_2 + 2 \cdot p_3 + p_4 + p_5$. In this marking, transitions *ask* and *stop*, or *ask* and *save* can fire concurrently because they don't share tokens. Neither firing sequences nor state graphs can express this concurrency. Therefore, there are additional semantics for Petri nets in the literature that can explicitly express concurrency. These include step semantics of Petri nets [22], process net semantics of Petri nets [21], token flow semantics of Petri nets [24], and compact token flow semantics of Petri nets [12]. Fortunately, these semantics are equivalent [12,24,25,40] and all define the same partial language. In a partial language, every so-called run is a partially ordered set of events. Obviously, runs can express concurrency and provide an intuitive approach to modelling the behaviour of a distributed system.

Using compact token flow semantics, we can decide if a run is in the partial language of a Petri net in polynomial time [5]. Roughly speaking, a compact token flow is a distribution of tokens on the arcs of a run so that every event receives enough tokens, no event must pass on too many tokens, and all events share tokens from the initial marking.

Definition 3. *Let T be a set of labels. A labelled partial order is a triple $(V, \ll, l)$ where V is a finite set of events, $\ll \subseteq V \times V$ is a transitive and irreflexive relation, and the labelling function $l : V \to T$ assigns a label to every event. A run is a triple $(V, <, l)$, with a relation $< \subseteq V \times V$, iff its irreflexive transitive closure $(V, <^*, l)$ is a labelled partial order. Let $N = (P, T, W, m_0)$ be a marked*

Petri net and $R = (V, <, l)$ *be a run so that* $l(V) \subseteq T$ *holds. Let* $\blacktriangleright, \blacksquare \notin V$ *be two symbols. A compact token flow is a function* $x : ((\{\blacktriangleright\} \times V) \cup < \cup (V \times \{\blacksquare\})) \rightarrow \mathbb{N}$. *Let* $v \in V$ *be an event. We denote* $in_x(v) := x(\blacktriangleright, v) + \sum_{v' < v} x(v', v)$ *the inflow of* v, *and* $out_x(v) := \sum_{v < v'} x(v, v') + x(v, \blacksquare)$ *the outflow of* v. *We define,* x *is valid for* $p \in P$ *iff the following conditions hold:*

(i) $\forall v \in V : in_x(v) \geq W(p, l(v))$,
(ii) $\forall v \in V : out_x(v) = in_x(v) + W(l(v), p) - W(p, l(v))$, *and*
(iii) $\sum_{v \in V} x(\blacktriangleright, v) = m_0(p)$.

   R *is enabled in* N *iff there is a valid compact token flow for every* $p \in P$. *The set of all enabled runs of* N *is the partial language of* N.

Figure 3 depicts six different runs modelling the behaviour of the marked Petri net depicted in Fig. 1. Every run starts with executing transition *reg*. The first run models the concurrent execution of transitions *ask* and *stop* before firing *check*. The second run models the execution of the loop *ask more* before concurrently executing *ask* and *stop*. Transition *stop* can only occur after *more* because of place p_5. The third run models two times the loop. Runs four, five, and six model the execution of the loop of *ask* and *more* in parallel to the execution of the sequence *save apply*. Figure 5 depicts four valid compact token flows for the places p_1, p_2, p_3, and p_5 of Fig. 1 in the last run of Fig. 3. For every token flow, the related place defines the number of tokens each event must receive and the number of tokens an event can pass on to later events. A compact token flow is a distribution of tokens over the arcs of the run, respecting the demand and capability of each event. In addition, a compact token flow distributes the initial marking and collects any superfluous tokens to define the final marking.

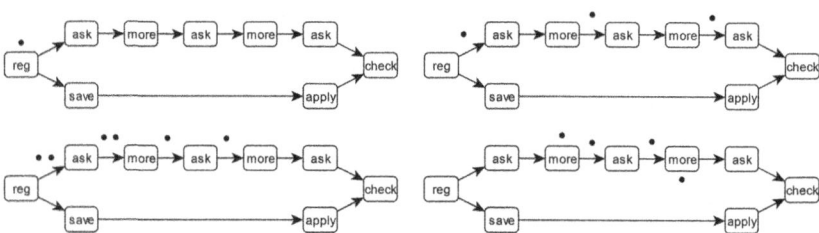

Fig. 5. Four valid compact token flows for places p_1, p_2, p_3, and p_5 in the last run of Fig. 3.

In Fig. 5, in the first token flow, related to place p_1, *reg* consumes from the initial marking. We show this token above the event. In the second token flow, related to place p_2, *reg* produces one token for the first *ask*. Then every occurrence of *more* produces a token for the following *ask*. In the third token flow, related to place p_3, *reg* produces two tokens, one for every occurrence of *more*. Here we clearly see how the token flow is passed to later events. In the fourth token flow, related to place p_5, *more* consumes from the initial marking and produces a token for the second occurrence of *more*. This *more* produces one token for the final marking. We show this token below the event. Figure 3 has six runs, and Fig. 1 has nine places. To show that Fig. 3 is in the partial language of Fig. 1, we need 54 valid compact token flows.

Altogether, Fig. 2 and Fig. 3 model the behaviour of the Petri net in Fig. 1. Figure 2 is unable to express concurrency of transitions. Figure 3 requires six separate runs, because runs cannot contain conflicts. Thus, there is always some trade-off when choosing one semantics over the other.

In the literature, we find other, possibly more advanced techniques for modelling states or executions of Petri nets. Potential candidates include process nets [21], branching processes [21,38], prime event structures [42], oclets [20], and spread nets [37]. These formalisms build upon state graphs and runs but add steps and/or conditions and cuts to enhance the expressiveness of the modelling techniques. However, there is a drawback: as the technique becomes more advanced, modelling behaviour becomes trickier. Specifying all possible steps, combinations of different preconditions, and other factors must be taken into account. Furthermore, by expanding but complicating the semantics, these techniques become just another high-level modelling language to be learned. Although the semantics can often explicitly express conflict and concurrency, they are still not able to intuitively merge local states from different executions. A branching process, for example, is called a branching process because it can only branch, not merge. If we specify behaviour in terms of a branching process, the model will fan out. For this reason, the main application of such techniques is to calculate the complete behaviour of a given model. It is difficult to specify behaviour in terms of these types of structures. To merely illustrate the idea of this problem, Fig. 6 depicts the branching process of Fig. 1. Obviously, we do not want to specify a process in terms of Fig. 6 in order to synthesise Fig. 1.

Figure 6 depicts a branching process. This is a labelled net modelling the behaviour of Fig. 1. Here, there is a one-to-one relation between tokens in Fig. 1 and conditions in Fig. 6. While it is very reasonable to calculate the complete behaviour of Fig. 1 by constructing Fig. 6, it is not sensible to model Fig. 6 in order to generate Fig. 1. The model can express conflict and concurrency, but it must respect the history of every token and can only split, not merge.

To address the problem of intuitively modelling behaviour of a system and to be capable of expressing concurrency, conflict, and merging of local states, we introduced the so-called net language of a Petri net [10,28]. The net language of a Petri net model is the set of all labelled nets such that the model can simulate every labelled net. In our context, simulate means that there is a mapping from every state of the language to every state of the model, respecting the initial marking and preserving the state transition behaviour of enabled multisets of transitions. Roughly speaking, the model behaves just like the specification. Naturally, the net language also includes all enabled firing sequences, state graphs, runs, and branching processes. For these formalisms there is a simple one-to-one translation into a labelled net. It is not surprising that it is easy to prove that

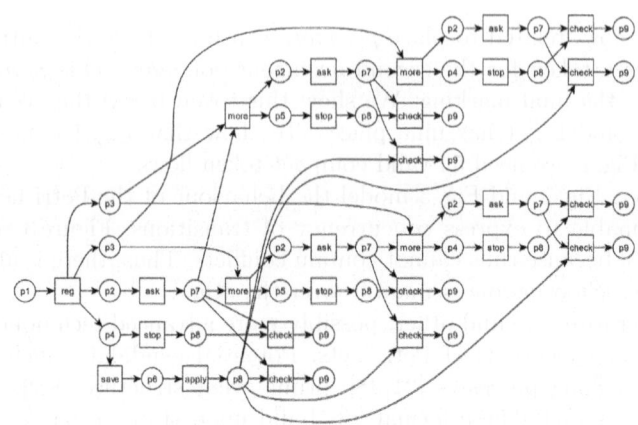

Fig. 6. The branching process of Fig. 1.

the resulting labelled nets are in the net language as well. On the other hand, the net language is not imprecise; the set of all step sequentializations of all the nets in the net language is the step language of the model. In this sense, there is no additional behaviour in the language that is not justified by the model. Furthermore, we already know how to model in terms of Petri nets, so there is no need to learn yet another modelling language. For details, we refer the reader to [10, 28], and simply state the following formal definition of the net language.

Definition 4. *Let T be a set of labels. A labelled net is a tuple (C, E, F, i_0, l) where (C, E, F, i_0) is a marked Petri net and $l : E \to T$ a labelling function assigning a label to every transition of the labelled net.*

Let $N = (P, T, W, m_0)$ be a marked Petri net and $L = (C, E, F, i_0, l)$ be a marked labelled net. A token trail $x : C \to \mathbb{N}$ is a marking of L. Let $e \in E$ be a labelled transition. We denote $in_x(e) := \sum_{(c,e) \in F} F(c, e) \cdot x(c)$ the inflow of tokens to e in x, and the weighted sum $out_x(e) := \sum_{(e,c) \in F} F(e, c) \cdot x(c)$ the outflow of tokens from e in x. We define, x is valid for $p \in P$ iff the following conditions hold:

(I) $\forall e \in E : in_x(e) \geq W(p, l(e))$,
(II) $\forall e \in E : out_x(e) = in_x(e) + W(l(e), p) - W(p, l(e))$, and
(III) $\sum_{c \in C} i_0(c) \cdot x(c) = m_0(p)$.

L is enabled in N iff there is a valid token trail for every $p \in P$. The set of all enabled labelled nets of N is the net language $\mathcal{L}(N)$ of N.

Figure 7 depicts four valid token trails, one in every labelled net of the specification of Fig. 4 for the place p_3 of Fig. 1. A token trail is a marking and, just like a token flow, describes a valid distribution of tokens respecting the demand and capability of each event. A valid token trail shows that, the respective place of the Petri net is able to simulate the labelled net. In this example, regarding place p_3, only transitions labelled *reg* can produce tokens. Thus, in every token trail, the sum of tokens in the post-set of transitions labelled *reg*, minus the sum of tokens in the preset of the same transition, is exactly 2. For every transition labelled *more* there is one less token in the post-set than in the pre-set, because *more* is consuming one token from p_3 in the Petri net model. All the other labels do not interact with p_3, the sum of ingoing tokens is the same as the sum of outgoing tokens.

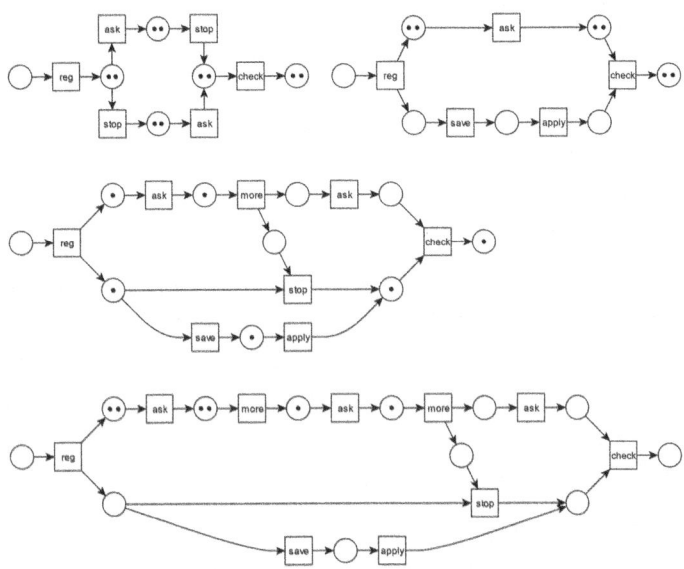

Fig. 7. Four valid token trails for p_3 of Fig. 1 in the four nets of Fig. 4.

If we think of a token trail as a distribution of tokens, in the first net of Fig. 7, we see that if there is a conflict, events kind of share tokens. If there is a merge, events agree on the number of tokens passed. In this example, *reg ask stop* or *reg stop ask* will provide two tokens. In the second net of Fig. 7, we see that if there is a split, events distribute tokens, and if there is a join, events collect tokens. But then again, tokens in a token trail model local states, they don't have to relate to individual tokens of the Petri net. This is why it is much easier to specify behaviour in terms of labelled nets than in terms of process nets or branching processes. Here, we also see that there might be multiple valid token trails for one place because the tokens can move via *ask* or via *save apply*. In nets three and four, there is a mix of splits, joins, conflicts, and merges of tokens. In the third net, one of the tokens is passed to and consumed by *more*. In the fourth net, every *more* receives and consumes one token produced by *reg*.

To prove that the four nets of Fig. 4 are in the net language of Fig. 1, we also have to find valid token trails for the other eight places of Fig. 1. Here, we refer the reader to our web tool at https://www.fernuni-hagen.de/ilovepetrinets/fox/, where we have prepared this example to be dragged to the 🐺 icon. Clicking on places in the Petri net will display the related valid token trails in the labelled nets. For more examples and proofs that token trails indeed cover the state language and the partial language, and that every net of the net language can be simulated by the Petri net, we refer the reader to [10,28]. Note that the net language can also model loops, labelled nets with arc weights, and distributed initial markings.

3 Token Trail Regions

In this section, we address the synthesis problem for a set of labelled nets. The synthesis problem is computing a Petri net from a set of labelled nets such that: (A) every labelled Petri net is in the net language of the synthesised Petri net, and (B) the generated model has minimal additional behaviour. Like mentioned above, to solve this problem. we strictly follow the general ideas of region theory.

Definition 5. *Let $S = \{(C_1, E_1, F_1, i_1, l_1), \ldots, (C_n, E_n, F_n, i_n, l_n)\}$ be a set of marked labelled nets. A token trail region is a marking $r : \bigcup_i C_i \to \mathbb{N}$ iff the following two conditions hold:*

(IV) $\forall \nu, \mu \in \mathbb{N}, e \in E_\nu, e' \in E_\mu, l_\nu(e) = l_\mu(e') : out_r(e) - in_r(e) = out_r(e') - in_r(e')$ and
(V) $\forall \nu, \mu \in \mathbb{N} : \sum_{c \in C_\nu} i_\nu(c) \cdot r(c) = \sum_{c' \in C_\mu} i_\mu(c') \cdot r(c').$

For every labelled transition e, we call $out_r(e) - in_r(e)$ the rise of e. Using this notion, property (IV) ensures: same label, same rise. For every labelled net, we call $\sum_{c \in C_\nu} i_\nu(c) \cdot r(c)$ the initial sum of tokens. Property (V) ensures: all nets have the same initial sum of tokens.

Obviously, the goal of Definition 5 is to define a distribution of tokens on a set of labelled nets such that equally labelled transitions have the same effect on the overall distribution (same label, same rise). If this is the case, for each such distribution, we can directly define a related place in the Petri net to be constructed, such that the arc weights connecting this place to all the transitions directly stem from the effect of every label on the overall distribution. The initial sum of tokens directly defines the initial marking of such a valid place. By construction, the region is a valid token trail for such a place. Thus, adding only places derived from regions to the synthesis result will guarantee (A).

Theorem 1. *Let $S = \{(C_1, E_1, F_1, i_1, l_1), \ldots, (C_n, E_n, F_n, i_n, l_n)\}$ be a set of marked labelled nets, let r be a region in S. We denote $E = \bigcup_i E_i$ and $l = \bigcup_i l_i$. For every $t \in \{l(e)|e \in E\}$ we fix one $e_t \in E$ so that $l(e_t) = t$ and $in_r(e_t) \leq min\{in_r(e)|l(e) = t\}$. The Petri net $N = (P, T, W, m_0)$, where $P = \{p\}$, $T = \{l(e)|e \in E\}$, $W = \sum_{t \in T} in_r(e_t) \cdot (p, t) + \sum_{t \in T} out_r(e_t) \cdot (t, p)$, and $m_0(p) = \sum_{c \in C_1} i_1(c) \cdot r(c)$, is well-defined. For every $(C_\nu, E_\nu, F_\nu, i_\nu, l_\nu) \in S$, $r|_{C_\nu}$ is a valid token trail for p of N. Thus, $S \subseteq \mathcal{L}(N)$ holds.*

Proof. Fix some ν, we prove $r_\nu := r|_{C_\nu}$ is a valid token trail in $(C_\nu, E_\nu, F_\nu, i_\nu, l_\nu)$ for p of N. $\forall e \in E_\nu : W(p, l(e)) = in_r(e_{l(e)}) \leq in_{r_\nu}(e)$ so that (I) holds. $\forall e \in E_\nu : W(l(e), p) - W(p, l(e)) = out_r(e_{l(e)}) - in_r(e_{l(e)}) = rise_r(e_{l(e)}) = rise_{r_\nu}(r)$ so that (II) holds. $m_0(p) = \sum_{c \in C_1} i_1(c) \cdot r(c) = \sum_{c \in C_\nu} i_\nu(c) \cdot r_\nu(c)$ so that (III) holds as well. The region r is a valid token trail in every labelled net of S for p of N. Thus, $S \subseteq \mathcal{L}(N)$ holds.   □

Theorem 1 shows that whenever we have a region, we can construct a related one-place Petri net so that the region is a valid token trail for the place and thus, the net-language includes the specification. Figure 8 depicts markings in our specification so that (IV) and (V) hold. All initially marked places of Fig. 4 have one token each. All transitions labelled *reg* have rise 1. All transitions labelled *stop*, *apply* or *check* have rise −1. All other transitions have rise 0.

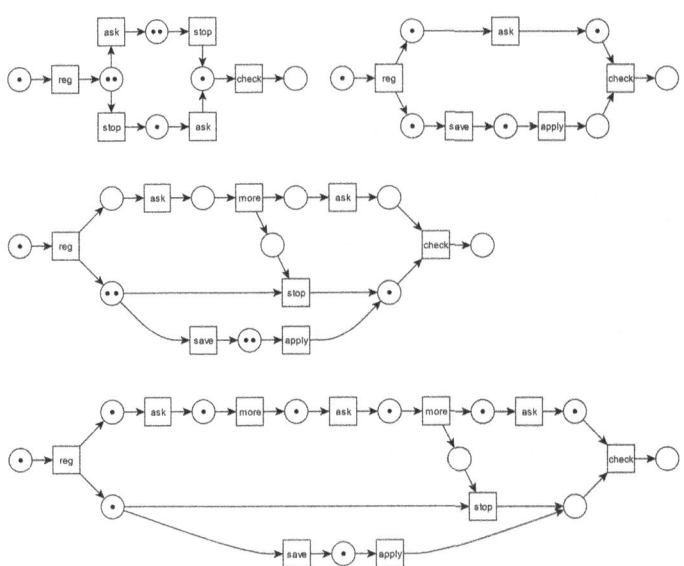

Fig. 8. A token trail region in Fig. 4.

Following Theorem 1, this region is a valid token trail of the place that is initially marked by one token and firing *reg* will consume this token before producing two tokens in this place. Firing *stop*, *apply*, or *check* will consume one token from this place. Transition *save* will have a short-loop to this place, and all other transitions are simply not connected. Adding this place to Fig. 1, the net language of the resulting net will still include Fig. 4.

Corollary 1. *Let S be a set of labelled nets and R be a set of regions in S. As defined in Theorem 1, every region $r \in R$ defines a one-place Petri net N_r. Let N be the union of all Petri nets N_r, $S \subseteq \mathcal{L}(N)$ holds because of Definition 4.*

Corollary 1 provides the first puzzle piece to solve the synthesis problem. We add only places related to regions to our synthesis result to ensure (A). In the next step, we show that we never have to add a place that is not defined by a region. But first, we have to talk about short loops. Figure 9 highlights the idea of the next theorem. In the first row, there is a Petri net and a labelled net of the net language of the Petri net. Thus, there is a valid token trail for p, which is depicted in the second row of the figure. This token trail naturally forms a region defining the Petri net of the second row with matching arc-weights. Every place defined by a region maximizes arc weights but still the token trail for p is also a valid token trail for p'. p' is just more restrictive than p, and there is no need to add places not directly defined by regions.

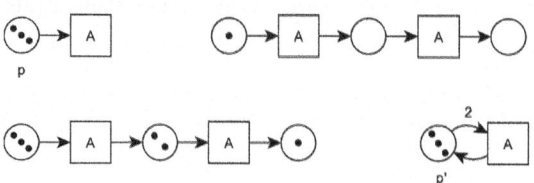

Fig. 9. A Petri net, a marked labelled net, a valid token trail, and a one-place Petri net defined by the token trail region.

Theorem 2. *Let $S = \{(C_1, E_1, F_1, i_1, l_1), \ldots, (C_n, E_n, F_n, i_n, l_n)\}$ be a set of marked labelled nets. Let $N = (\{p\}, T, W, m_0)$ be a one-place Petri net. If $S \subseteq \mathcal{L}(N)$ holds, there is a region r in S defining a one-place net $N' = (\{p'\}, T, W', m_0')$ so that $S \subseteq \mathcal{L}(N') \subseteq \mathcal{L}(N)$ holds.*

Proof. If $S \subseteq \mathcal{L}(N)$ holds, there is a valid token trail for p in every labelled net in S. Together, these markings form a region r in S because: (III) holds for every labelled net so that (V) holds as well. (II) holds for all the labelled transitions so that (IV) holds as well. Because of (I), in every token trail, every labelled transition t has at least $W(l(t), p)$ ingoing tokens, but there might be more. According to Theorem 1, we construct the related one-place net $(\{p'\}, T, W', m_0')$. To define W'. for every t, we take the minimal sum of ingoing tokens of all transitions labelled t. Thus, for every t, $W(p', l(t)) \geq W(p, l(t))$ holds. r is a valid token trail for p' (Theorem 1). Thus, every valid token trail for p' is also a valid token trail for p so that $S \subseteq \mathcal{L}(N') \subseteq \mathcal{L}(N)$ holds. Place p' is just p where short-loops are maximized to get the most restrictive place. □

By combining Theorem 1 and Theorem 2, we conclude that to solve the synthesis problem, we calculate regions and combine the corresponding one-place nets. We never have to add a place not defined by a region.

Corollary 2. *Let S be a set of labelled nets and R be a set of regions in S. Each region r in R defines a one-place Petri net N_r. The possibly infinite union net N of the one-place Petri nets N_r solves the synthesis problem.*

Proof. Let's assume we have the possibly infinite union net N of the one-place Petri nets N_r and a place p, not yet in N. If adding p to N would still ensure (A), there is a valid token trail for p in S. This token trail defines a region and thus, a related place p' is already in N. There is no need to add p because of Theorem 2.   □

In praxis, to solve the synthesis problem, we need to construct a finite Petri net that has the same behaviour as the union Petri net. Upon reviewing the literature related to region theory, there are various approaches to solving this final step. For most state-based regions, when constructing Petri nets without arc weights, the number of regions is simply finite. If we consider state-based regions and Petri nets with arc weights, we typically attempt to solve the synthesis problem under the additional condition that no extra behaviour is introduced. If such a result exists, the state space of the resulting net is the specification, and thus, only minimal regions need to be added. In the case of language-based regions, we aim to achieve the best upper approximation of the specified behaviour. Therefore, even if a region is larger in terms of specified behaviour, it may be smaller regarding unspecified behaviour. Thus, minimal regions might not be sufficient. Here, we calculate the set of wrong continuations, which is the boundary between unspecified and specified behaviour. For partial order-based specifications, this set of wrong continuations is finite, allowing us to determine whether there is an exact solution to the synthesis problem or not. If possible, excluding all wrong continuations while still ensuring (A) solves the synthesis problem.

In this paper, we define regions for a set of labelled nets. Thus, the behaviour is specified in terms of the net language of a Petri net. A net language is always infinite. Adding superfluous additional places, that don't restrict the behaviour or simply copying a net and putting the copy in conflict to the original net, is, and should be, always possible. Thus, because a specification is always finite, and every result will have an infinite net language, technically there is never an exact solution. Altogether, the number of regions is infinite because of arc weights, and the number of wrong continuations is infinite because it's a net language. Thus, still the saturated net having places related to all regions is a solution, but a finite subnet might not be. Although, this sounds like bad news, we are not really interested in the complete, infinite net language of some Petri net. We want to model behaviour in terms of state graphs, partial languages and labelled nets and have a proper synthesis result that can execute the specification but still falls into some reasonable net class. In this paper, we show that a very practical simplification will lead to very good results. As done for state-based regions, we simply restrict the size of a region by some number k. Thus, the number of regions again is finite. We could then start an iterative procedure solving the synthesis problem up to k, checking the result, increasing k, solving again, until we are satisfied. In this procedure every solution will ensure (A), and the size of additional behaviour will decrease.

In the remainder of this section, we show that it is not even necessary to use such a costly iterative approach. We will demonstrate that using state-based or language-based techniques to calculate a selection of token trail regions is sufficient for practical examples. It is easy to translate a state graph into a labelled net. Every state will be a place, every (state) transition will be a labelled transition. The place relating to the initial state will carry one token in the initial marking. By brute force, the reachability graph of the state graph net will be the state graph. The first net of Fig. 4 shows such a labelled net. Its reachability graph is the net itself.

Theorem 3. *Let* $S = (R, T, X, i)$ *be a state graph and* r *be a region in* S. *We construct a labelled net* $N = (C, E, F, m_0, l)$ *by* $C = R$, $E = X$, $F = \sum_{(s,t,s') \in X} (s, (s, t, s')) + ((s, t, s'), s')$, $m_0 = i$, *and* l *maps every* (s, t, s') *to* t. *r is a token trail region in* (C, E, F, m_0, l).

Proof. r is a region in S so that r is a subset of R so that every equally labelled (state) transition either enters, exists, or does not cross r. If a (state) transition (s, t, s') exits r, $s \in r$ and $s' \notin r$ hold. In N, r is a marking, s and s' are places. r marks s with a single token and s' with none. In N, (s, t, s') is a transition connected only to s and s'. The rise of (s, t, s') in r is -1. Using the same arguments, if a (state) transition (s, t, s') enters r, the rise of (s, t, s') in r is 1. If a (state) transition (s, t, s') is non-crossing, the rise of transition (s, t, s') in r is 0. Equally labelled (state) transitions have the same crossing behaviour in S, so that equally labelled transitions have the same rise in N, i.e. (IV) holds. (V) holds, because we only consider one net with an initial sum of tokens 1.   □

Obviously, we can extend Theorem 3 to state-based regions with arc weights. Here, regions are multisets of states, which again perfectly map to markings of a token trail region. The left side of Fig. 10 shows the input to a state-based synthesis algorithm and a region using a multiset of states. Lightly shaded states are in the region once, while darkly shaded states are in the region twice. *reg* enters the region once, *stop* exits the region once, and all other transitions do not cross. On the right, there is the related labelled net with the corresponding token trail region. It is simply the same concept. Assume we have a labelled net, which directly models a state graph, and a token trail region. Using the same relation as defined in Theorem 3, highlighted in Fig. 10, the token trail region in the labelled net corresponds to a region of the state graph. Considering state graphs, every region is a token trail region and vice versa.

In Theorem 4, we obtain the same result for language-based regions. Here, we make use of the one-to-one correspondence between token trails and token flows on partial languages. This result was already introduced in [10]. We repeat the theorem to also demonstrate that both related region definitions are equivalent.

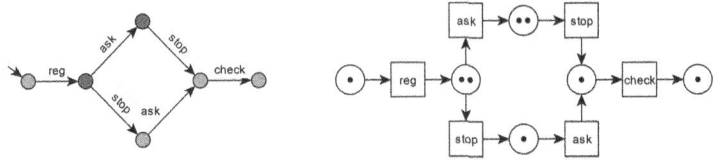

Fig. 10. A state-based region and an identical token trail region.

Theorem 4. *Let $S = \{(V_1, \ll_1, l_1), \ldots, (V_n, \ll_n, l_n)\}$ be a set of labelled partial orders and r be a compact token flow region in S. We construct a set of labelled nets $N = \{(C_1, E_1, F_1, m_1, l_1), \ldots, (C_n, E_n, F_n, m_n, l_n)\}$ by $C_i = (\{\blacktriangleright\} \times V_i) \cup \ll_i \cup (V_i \times \{\blacksquare\})$, $E_i = V_i$, $F_i = \sum_{(v,v') \in \ll_i} (v, (v, v')) + ((v, v'), v') + \sum_{v \in V_i} ((\blacktriangleright, v), v) + (v, (v, \blacksquare))$, and $m_i = \sum_{v \in V_i} (\blacktriangleright, v)$. r is a token trail region in N.*

Proof. We prove the one-to-one relation between token flows and token trails on partial languages. The only difference between (i), (ii), (iii) and (I), (II), (III) is the definition of *in* and *out*. In Definition 4, the sums are weighted by the corresponding arc weights. However, we do not need arc weights to model a labelled partial order as a labelled net. In this case, every weight is 1. In Definition 3 initial and superfluous tokens can be consumed and produced at every event. In this theorem, we add matching places to the pre-set and the post-set of every labelled partial order. Thus, r is a token trail region for S. □

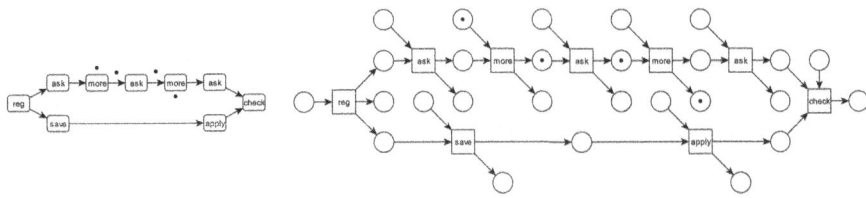

Fig. 11. A language-based region and an identical token trail region.

Figure 11 depicts the relation between a valid compact token flow and a valid token trail on a partially ordered run. It is easy to see, that we could remove a lot of the additional places modelling the initial and final flow. The net is a partial order, so that tokens in the initial marking could spawn in the minimal places and be passed to every event. Similarly, superfluous tokens can be passed to the maximal places. Thus, if there is a token trail on such an extended net, there is also a token trail on the net only modelling the partial order adding minimal and maximal places for the initial and final marking. See the second net of Fig. 8 as an example.

Assume we have a set of labelled nets, which directly model a partial language, and a token trail region. Using the same relation as defined in Theorem 4, highlighted in Fig. 11, the token trail region in the labelled net defines a region of the partial language. Considering partial languages, every region is a token trail region and vice versa.

Concluding this section, as conjectured in [11], token trail regions are a kind of a meta region definition. Instead of state-based regions, we can simply use token trail regions. The same applies to partially ordered regions. Furthermore, we can use minimal regions or wrong continuations to obtain a finite result. However, the goal of this paper is not only to unify both concepts under a meta region definition, but also to introduce a new method for practical applications. By using the new definition, we are not only able to combine state-based and language-based input, but we can also incorporate labelled nets into our specification. These labelled nets can feature distributed initial markings, cycles, and even arc weights. In the following, we present an approach for calculating a set of minimal regions up to a bound of k, in order to address the synthesis problem from a set of labelled nets.

4 Calculating Token Trail Regions

In this section, we introduce a method to calculate all minimal token trail regions up to a given bound k. We present an implementation, a web-tool, and provide examples of the synthesis results to demonstrate that the synthesis is indeed applicable in practical scenarios.

Typically, state-based regions are constructed using a bottom-up approach. Starting from a set of so-called excitation regions, these regions are gradually repaired by incrementally adding more states until they form a set of valid regions. The bottom-up approach generates a set of minimal regions. A similar process could be applied to token trail regions. However, the bottom-up approach can become somewhat unwieldy due to the concurrency in the labelled nets, which introduces many more alternatives when repairing regions.

Most synthesis algorithms based on languages use ILP (Integer Linear Programming) combined with the wrong continuation approach to identify the set of possible regions. For example, see [41] for the most prominent algorithm that applies language-based regions, ILP, and a heuristic based on the directly-follows relation of the input to generate a finite set of regions. The Kokosminer, based on the Master thesis of Karl Heinrichmeyer [23], directly calculates a set of minimal regions with an ILP. In our implementation, we generalize Karl's approach to calculate token trail regions.

An ILP is easy to implement using specialized solvers and is highly flexible. It allows for intuitive implementation of restrictions on the type of regions we aim for, such as one-bound regions, or the type of net class we are targeting, for example, nets with or without arc weights and/or self-loops. For these reasons, we combine the two best proven techniques and introduce a new ILP approach for calculating minimal regions in this paper.

To calculate regions, formally a multiset of places, we implement the conditions of Definition 5 as an ILP. Every place of the specification is a variable of the ILP. For every label of the specification, we fix one transition with the related label. For every other transition carrying the same label, we introduce one equality to the ILP ensuring the rise of both transitions is equal. Let's consider Fig. 4 as an example. The specification has 35 places, thus, the ILP has 35 variables. There are four transitions labelled reg. Let us call them reg^1, reg^2, reg^3, and reg^4. We add three equalities to the ILP. One ensures that the rise of reg^1 is equal to the rise of reg^2. Another, ensuring that the rise of reg^1 is equal to the rise of reg^3. And one more, ensuring the rise of reg^1 is equal to the rise of reg^4. There are eight transitions labelled ask, thus, we add seven more equalities. Altogether we add these equalities for every label. They ensure that every solution of the ILP satisfies (IV).

In the next step, we add equalities to ensure (V). Again, we simply refer to all labelled nets of the specification as net^1, net^2, ..., net^n. We construct one equality such that the sum of the initial tokens of net^1 and net^2 is equal. Another equality ensures that the sum of the initial tokens of net^1 and net^3 is equal, and so on. They ensure that every solution of the ILP satisfies (V).

We add a constraint to each variable, ensuring that it is less than or equal to k, in order to calculate regions only up to k.

Altogether, every non-negative integer solution of this region ILP is a token trail region up to k, and vice versa. This construction is really kind of straight forward, because Definition 5 is an ILP already.

In the next step, we search for a non-empty minimal region. We add the inequality that the sum of all variables is greater than 0, and we also add the same sum, i.e., the sum of all variables, as the objective function to be minimized by the ILP. Now, every optimal solution corresponds to a minimal region, because there is no other region that has fewer or an equal number of tokens in every place, the objective function of such region would be smaller.

With the first minimal region, we build a transition for every label, along with arcs and a place, as defined by Theorem 1. Now, we have to find other minimal regions and related places.

Every region we find is a solution $(s_1, \ldots, s_n)$ assigning values to the variables $p_1, \ldots, p_n$ of the ILP. With every new solution, we extend the ILP. Let's assume we just found a minimal region $r_1 = s_1 \cdot p_1 + \ldots + s_n \cdot p_n$. For every multiplicity $s_i > 0$ of r_1, we add a variable x_{r1pi} to the ILP. The idea is to add inequalities to the ILP so that x_{r1pi} will be 1 if the variable p_i is less than s_i, else 0. Therefore we add the following inequalities:

$$0 \le (p_1 - s_1) + k \cdot x_{r1p1} \le k - 1$$
$$\ldots$$
$$0 \le (p_n - s_n) + k \cdot x_{r1pn} \le k - 1$$

If variable p_i is smaller than the previous solution s_i, the term $(p_i - s_i)$ is less than 0, so that x_{r1pi} must be 1. If p_i is at least s_i, the term $(p_i - s_i)$ is non-negative and x_{r1pi} must be 0. Finally, we add $x_{r1p1} + \ldots + x_{r1pn} > 0$ to our ILP to ensure, that every new solution will be smaller in at least one component.

By iteratively building, solving, and updating the ILP, we can calculate the set of minimal regions up to k for our specification, and subsequently construct the related Petri net.

This synthesis approach is implemented as a new module of the I ♥ Petri nets website. The website is available at www.fernuni-hagen.de/ilovepetrinets/. The 🐎 module, available at www.fernuni-hagen.de/ilovepetrinets/horse, implements the algorithm introduced in this chapter by solving a sequence of ILPs using the GLPK-Solver [31,39]. For this contribution, we prepared four examples that the reader can try to solve using our new region approach.

Example 1, the net based-specification of Fig. 4, synthesises Fig. 1 in 30 seconds. Example 2, the state-based specification of Fig. 2, synthesises Fig. 1, without p_5, instantly. State-based input cannot specify the short-loop. Example 3, four runs of the language-based specification of Fig. 3, synthesises Fig. 1 in 6 minutes. Language-based regions can't handle all six in reasonable time. Example 4 highlights another application of the token trail approach. Obviously, every Petri net is also a labelled net. Thus, the input to the synthesis can be a fully-fledged process model. In Example 4, we input two general labelled nets, one with a distributed initial marking, the other with an arc weight. Here we synthesise Fig. 1 in no-time. This is an exciting feature of token trail regions. We can input sequences, state graphs, partially ordered runs, workflow nets, and complete system models to synthesise a Petri net that simulates the specified behaviour. We encourage the reader to try more examples. Input can be uploaded using the *PNML* standard, or created with an editor available at the I ♥ Petri nets website.

5   Conclusion

This paper introduces a new region definition based on token trails. A token trail region is a label-respecting distribution of local states of a specification, given in terms of a set of labelled nets. We argue that it is very natural to specify behaviour in terms of labelled nets, as they allow for the modelling of conflict and concurrency, specification of different runs of the system separately, and the use of label splitting to unfold complex behaviour, without the need to learn a new modelling technique. Existing literature includes techniques for extending state-based regions by steps [2, 26, 33] or language-based regions by term based representations [9], but these new modelling concepts often feel somewhat forced in practical examples. The concept of token trail regions covers both state-based and language-based regions and can serve as a meta semantic to unify specialised concepts and approaches. We presented an easily accessible web tool, along with initial experiments showing that approximating a synthesis result by a set of minimal regions up to a bound k is a reasonable approach. Despite the ILP being relatively large and synthesis being known as a difficult problem, the JavaScript-based implementation running in the browser can solve the synthesis problem quite efficiently using a specialised solver. Future work will focus on proving more formal guarantees for the generated results. To date, the synthesised result fulfils (A), and no other region is smaller for the specified behaviour. It may be necessary to add non-minimal regions to further exclude non-separable behaviour.

References

1. Van der Aalst, W.M.P., Van Dongen, B.F.: Discovering Petri nets from event logs. In: Transactions on Petri Nets and Other Models of Concurrency VII, pp. 372–422. Springer (2013)
2. Badouel, E., Bernardinello, L., Darondeau, P.: Petri Net Synthesis. Springer, Heidelberg (2015). https://doi.org/10.1007/978-3-662-47967-4
3. Bergenthum, R.: Synthesizing petri nets from hasse diagrams. In: Carmona, J., Engels, G., Kumar, A. (eds.) BPM 2017. LNCS, vol. 10445, pp. 22–39. Springer, Cham (2017). https://doi.org/10.1007/978-3-319-65000-5_2
4. Bergenthum, R.: Prime miner - process discovery using prime event structures. In: Proceedings of ICPM 2019, pp. 41–48 (2019)
5. Bergenthum, R.: Firing partial orders in a petri net. In: Proceedings of PETRI NETS 2021, pp. 399–419. Springer (2021)
6. Bergenthum, R., Desel, J., Lorenz, R., Mauser, S.: Synthesis of petri nets from scenarios with VipTool. In: van Hee, K.M., Valk, R. (eds.) PETRI NETS 2008. LNCS, vol. 5062, pp. 388–398. Springer, Heidelberg (2008). https://doi.org/10.1007/978-3-540-68746-7_25
7. Bergenthum, R., Desel, J., Mauser, S.: Comparison of different algorithms to synthesize a petri net from a partial language. In: Jensen, K., Billington, J., Koutny, M. (eds.) Transactions on Petri Nets and Other Models of Concurrency III. LNCS, vol. 5800, pp. 216–243. Springer, Heidelberg (2009). https://doi.org/10.1007/978-3-642-04856-2_9
8. Bergenthum, R., Desel, J., Lorenz, R., Mauser, S.: Synthesis of petri nets from finite partial languages. Fund. Inform. **88**(4), 437–468 (2008)
9. Bergenthum, R., Desel, J., Mauser, S., Lorenz, R.: Synthesis of petri nets from term based representations of infinite partial languages. Fund. Inform. **95**(1), 187–217 (2009). https://doi.org/10.3233/FI-2009-147
10. Bergenthum, R., Folz-Weinstein, S., Kovář, J.: Token trail semantics - modeling behavior of petri nets with labeled petri nets. In: Proceedings of PETRI NETS 2023, pp. 286–306. Springer (2023)
11. Bergenthum, R., Kovář, J.: A first glimpse at petri net regions. In: Proceedings of ATAED 2022, pp. 60–68. CEUR Workshop Proceedings 3167 (2022)
12. Bergenthum, R., Lorenz, R.: Verification of scenarios in petri nets using compact tokenflows. Fundam. Inform. 117–142 (2015)
13. Best, E., Schlachter, U.: Analysis of petri nets and transition systems. In: Proceedings 8th Interaction and Concurrency Experience (ICE) (2015). https://doi.org/10.4204/EPTCS.189.6. https://github.com/CvO-Theory/apt
14. Carmona, J., Cortadella, J., Kishinevsky, M.: Genet: a tool for the synthesis and mining of petri nets. In: 2009 Ninth International Conference on Application of Concurrency to System Design, pp. 181–185 (2009). https://doi.org/10.1109/ACSD.2009.6
15. Darondeau, P.: Deriving unbounded petri nets from formal languages. In: Sangiorgi, D., de Simone, R. (eds.) CONCUR'98 Concurrency Theory, pp. 533–548. Springer, Heidelberg (1998). https://doi.org/10.1007/BFb0055646

16. Desel, J., Juhás, G.: "What is a petri net?". In: Unifying Petri Nets, pp. 1–25 (2001)
17. van Dongen, B.F., de Medeiros, A., Verbeek, H., Weijters, A., van der Aalst, W.: The ProM framework: a new era in process mining tool support. In: Ciardo, G., Darondeau, P. (eds.) ICATPN 2005. LNCS, vol. 3536, pp. 444–454. Springer, Heidelberg (2005). https://doi.org/10.1007/11494744_25
18. Ehrenfeucht, A., Rozenberg, G.: Partial (set) 2-structures. Part I: basic notions and the representation problem. Acta Informatica 315–342 (1990). https://doi.org/10.1007/BF00264611
19. Ehrenfeucht, A., Rozenberg, G.: Partial (set) 2-structures. Part II: state spaces of concurrent systems. Acta Informatica 343–368 (1990). https://doi.org/10.1007/BF00264612
20. Fahland, D.: Oclets – scenario-based modeling with petri nets. In: Franceschinis, G., Wolf, K. (eds.) PETRI NETS 2009. LNCS, vol. 5606, pp. 223–242. Springer, Heidelberg (2009). https://doi.org/10.1007/978-3-642-02424-5_14
21. Goltz, U., Reisig, W.: The non-sequential behaviour of petri nets. Inf. Control 125–147 (1983)
22. Grabowski, J.: On partial languages. Fundam. Inform. 427–498 (1981)
23. Heinrichmeyer, K.: Process Discovery durch kontinuierliche Kombination minimaler Regionen. Master's thesis, FernUniversität in Hagen (2020)
24. Juhás, G., Lorenz, R., Desel, J.: Can i execute my scenario in your net? In: Proceedings of PETRI NETS, pp. 289–308. Springer (2005)
25. Kiehn, A.: On the interrelation between synchronized and non-synchronized behaviour of petri nets. J. Inf. Process. Cybern. 3–18 (1988)
26. Koutny, M., Pietkiewicz-Koutny, M.: Minimal regions of ENL-transition systems. Fund. Inform. **101**(1–2), 45–58 (2010). https://doi.org/10.3233/FI-2010-274
27. Koutny, M., Pietkiewicz-Koutny, M.: Synthesising ENI-systems with interval order semantics. In: Proceedings of PNSE 2024. CEUR Workshop Proceedings, vol. 3730, pp. 33–52 (2024)
28. Kovář, J., Bergenthum, R.: Token trail semantics II - petri nets and their net language. In: Kristensen, L.M., van der Werf, J.M. (eds.) Application and Theory of Petri Nets and Concurrency, pp. 175–196. Springer (2024)
29. Lorenz, R., Mauser, S., Bergenthum, R.: Theory of regions for the synthesis of inhibitor nets from scenarios. In: Kleijn, J., Yakovlev, A. (eds.) ICATPN 2007. LNCS, vol. 4546, pp. 342–361. Springer, Heidelberg (2007). https://doi.org/10.1007/978-3-540-73094-1_21
30. Lorenz, R., Mauser, S., Juhas, G.: How to synthesize nets from languages - a survey. In: 2007 Winter Simulation Conference, pp. 637–647 (2007). https://doi.org/10.1109/WSC.2007.4419657
31. Makhorin, A.: GLPK (GNU Linear Programming Kit) (2000–2012). https://www.gnu.org/software/glpk/
32. Mayr, H.C., Kop, C., Esberger, D.: Business process modeling and requirements modeling. In: First International Conference on the Digital Society (ICDS 2007), p. 8 (2007). https://doi.org/10.1109/ICDS.2007.9
33. Mukund, M.: Petri nets and step transition systems. Int. J. Found. Comput. Sci. **03**(04), 443–478 (1992). https://doi.org/10.1142/S0129054192000231
34. Peterson, J.L.: Petri Net Theory and the Modeling of Systems. Prentice Hall PTR (1981)
35. Pietkiewicz-Koutny, M.: Synthesising elementary net systems with inhibitor arcs from step transition systems. Fund. Inform. **50**(2), 175–203 (2002)

36. Pietkiewicz-Koutny, M., Koutny, M.: Synthesising elementary net systems with interval order semantics. In: Proceedings of ATAED & PN4TT 2023. CEUR Workshop Proceedings, vol. 3424 (2023)
37. Pinna, G.M., Fabre, E.: Spreading nets: a uniform approach to unfoldings. J. Log. Algebraic Methods Program. 1–33 (2020)
38. Reisig, W.: Understanding Petri Nets: Modeling Techniques, Analysis Methods, Case Studies. Springer (2013)
39. Vaillant, J., Rasulkhani, S., Regue, R., Jones, A.: glpk.js (2014–2023). https://www.npmjs.com/package/glpk.js
40. Vogler, W.: Modular Construction and Partial Order Semantics of Petri Nets. Springer (1992)
41. van der Werf, J., van Dongen, B.F., Hurkens, C., Serebrenik, A.: Process discovery using integer linear programming. In: van Hee, K.M., Valk, R. (eds.) PETRI NETS 2008. LNCS, vol. 5062, pp. 368–387. Springer, Heidelberg (2008). https://doi.org/10.1007/978-3-540-68746-7_24
42. Winskel, G.: Event structures. In: Petri Nets: Applications and Relationships to Other Models of Concurrency, pp. 325–392. Springer, Heidelberg (1986)

Coverability in Well-Formed Free-Choice Nets

Eike Best[1], Raymond Devillers[2], and Petr Jančar[3(✉)]

[1] Department of Computing Science, Carl von Ossietzky Universität Oldenburg, 26111 Oldenburg, Germany
eike.best@informatik.uni-oldenburg.de
[2] Département d'Informatique, Université Libre de Bruxelles, 1050 Brussels, Belgium
raymond.devillers@ulb.be
[3] Department of Computer Science, Faculty of Science, Palacký University, Olomouc, Czechia
petr.jancar@upol.cz

Abstract. The theory of free-choice nets started in the 70's, at the MIT, when F. Commoner and M. Hack discovered that, for this class of Petri nets, it was possible to link their behaviour to the presence of specific substructures. Later their results were extended, and more recently, new light was shed on the theory. The present paper continues in this direction, providing new coverability results, new equivalence properties and a new characterisation of well-formedness.

Keywords: Petri nets · free-choice · well-formedness · coverability · S-component · T-component

1 Introduction

Place/Transition Petri nets [14] (just called "nets" in the following) serve to describe distributed systems in which resources can be absorbed and produced. Formally, a net is a finite, bipartite multigraph in which one type of nodes (called places) models local states while the other type of nodes (called transitions) models local activities. The (multi-) arcs of the graph describe how local activities rely on – and possibly change – local states. Global states may be represented as markings of local states, indicating their multiplicity. The initial state of a system described in this way is called the initial marking, and there is a transition rule governing the reachability of markings from each other by executing transitions.

An initially marked net may, or may not, be bounded or live. Boundedness means that the space of reachable markings is finite. Liveness means that no transition may ever become dead in the sense of no longer being capable of being executed. Viewed as decision problems ("is an initially marked net bounded/live?"), both problems are decidable [9,10,12] but exceedingly complex, liveness (Ackermann complexity [4,11]) even more so than boundedness (exponential space complexity [13]).

E. Amparore and Ł. Mikulski (Eds.): PETRI NETS 2025, LNCS 15714, pp. 86–108, 2025.
https://doi.org/10.1007/978-3-031-94634-9_5

The graph-theoretical structure of such nets allows convenient and useful ways of defining subclasses. The class of nets this paper focusses on is called "free-choice nets" [5,8]. Its defining properties are that no proper arc weights are allowed (thus, the graphs are plain graphs, rather than proper multigraphs), and that there is a restriction concerning local activities in case they compete for resources (it is never the case that one of them can proceed while a competitor cannot). For free-choice nets, the liveness and boundedness problems are easier to solve (liveness is co-NP-complete and boundedness is polynomial in the presence of liveness [6]).

Let a net be called well-formed if it has a live and bounded initial marking. It has been known for a long time (more than 50 years) that well-formed free-choice nets have a very interesting and satisfactory structural theory. We are interested in two results that are collectively known as coverability results [8]. It turns out that in a well-formed free-choice net, all places are covered by special structures called S-components and all transitions are covered by other structures called T-components.

In a formal sense, S-components and T-components are duals of each other. They correspond to each other by exchanging the roles of places and transitions. However, because of the asymmetric nature of the transition rule, there is no straightforward way of deriving the proofs of these coverability results from each other. In particular, to the authors' knowledge, all proofs of S-component coverability known in the literature rely on an auxiliary result characterising the liveness of free-choice nets attributed to F. Commoner [3], while no similar result is known (nor needed in the literature) for T-component coverability.

The starting point of the present paper was to search for more symmetric proofs of these properties. One could wonder about the true interest of re-proving properties with well-known and correct proofs. We feel that this may be justified if this leads to new and interesting results. This will be the case here. We shall in particular carefully delineate, and generalise, the part of the proofs which are perfectly symmetric, in the form of two new coverability results (Theorems 3 and 4). Then, two equivalence results will gather the parts which are not completely symmetric (Theorems 5 and 6), without needing the use of Commoner's characterisation of liveness. The combinations of these four results give us new proofs of the old coverability results. As a bonus, this will also lead us to new characterisations of well-formedness for free-choice nets (Theorems 7 and 8).

The paper's structure is as follows. Sections 2 and 3 set the scene by defining some basic notions about nets and free-choice nets and by stating some of their properties. Section 4 then presents our new coverability results and improved proofs. The new characterisation of well-formedness for free-choice nets can be found in Sect. 5. The last section as usual discusses and concludes the paper.

2   Petri Nets and Their Behaviour

By $\mathbb{N}$ we denote the set $\{0, 1, 2, \dots\}$ of nonnegative integers.

Definition 1. PLACE/TRANSITION PETRI NET, PRESET, POSTSET
*A Place/Transition Petri net, or just a net for short, is a triple $N = (S, T, F)$
where S and T are finite disjoint sets of places (or "local states", "Stellen" in
German from where the letter S originates) and transitions, respectively, and
$F \subseteq (S \times T) \cup (T \times S)$ is a flow relation. For $x \in S \cup T$ and $X \subseteq S \cup T$ we put
$^\bullet x = \{y \mid (y, x) \in F\}$ (the preset of x), $^\bullet X = \bigcup_{x \in X} {}^\bullet x$, and $x^\bullet = \{y \mid (x, y) \in F\}$
(the postset of x), $X^\bullet = \bigcup_{x \in X} x^\bullet$.*   ■ 1

A place will be considered as representing a kind of resource; many interchangeable copies of the resource may coexist in the system, represented by
black tokens[1]; an arc $(s, t) \in F$ represents the fact that transition t needs a
resource from s to be executable (and absorbs it during an execution of t), and
an arc (t, s) represents the fact that an execution of t produces a resource of
kind s. Even though many interchangeable copies of a resource may be present
on some place, a transition may take at most one of them and produce at most
one of them. That is, we are here limiting ourselves to so-called *plain* (or sometimes also named *ordinary*) nets where arcs in F have at most weight 1 and can
therefore be described by a relation, as in Definition 1.

Definition 2. MARKING, ENABLING, FIRING, INFINITE EXECUTION, REACHA-
BILITY SET
*A marking M of N is a finite multiset of places, hence $M \in \mathbb{N}^S$; by $\mathbf{0}$ we denote
the empty multiset. A marked net (N, M_0) is an (unmarked) net N provided with
an initial marking M_0. As usual, (sub)sets may be considered as special cases of
multisets, through their characteristic functions; hence we shall allow notations
like $X \subseteq M$ (or $X \leq M$) and $M + X$ when $X \subseteq S$. For $R \subseteq S$, $M|_R \in \mathbb{N}^R$
denotes the restriction of M to R, and we put $M(R) = \sum_{s \in R} M(s)$ (the total
number of resources in R). A transition $t \in T$ is enabled at M, which is written
$M > t$, if $^\bullet t \subseteq M$. We have $M > tM'$ (M' can be reached from M by exe-
cuting, or firing, t) if $^\bullet t \subseteq M$ and $M' = M - {}^\bullet t + t^\bullet$. For $\sigma \in T^*$ we define
$M > \sigma M'$ inductively: $M > \varepsilon M$ (where ε is the empty word), and if $M > tM'$
and $M' > \sigma M''$ then $M > t\sigma M''$. $M > \sigma M'$ means that M' can be reached
from M by executing σ and we shall say that σ is enabled at M, leading to an
execution $M > \sigma$, if $M > \sigma M'$ for some M'. We shall also consider infinite
executions $M > \tau$, with $\tau \in T^\infty$, when $M > \sigma$ for every finite prefix σ of τ. The
reachability set of M is $[M\rangle = \{M' \mid M > \sigma M'$ for some $\sigma \in T^*\}$.*   ■ 2

Note that firing is *monotonic*: if $M > \sigma M'$, then $M + \overline{M} > \sigma M' + \overline{M}$ for any
$\overline{M} \in \mathbb{N}^S$.

Definition 3. BOUNDEDNESS
*Given a net $N = (S, T, F)$, a place $s \in S$ is bounded at a marking M if there is
a number $b \in \mathbb{N}$ such that $M'(s) \leq b$ for all $M' \in [M\rangle$; it is safe if $M'(s) \leq 1$
for all $M' \in [M\rangle$.*

---

[1] "Black" here means that no additional information is carried by the tokens to drive
   the control flow; the exact colour used in figures is thus irrelevant.

A marking M is bounded if all places are bounded at M (or, equivalently, if the set $[M\rangle$ is finite). In order to highlight the underlying net N, we can also say that (N, M) is bounded, provided M is bounded in N. Similarly, it is safe if each place is safe at M. ■ 3

It may be observed that safeness is a strong form of boundedness, and indeed it was used as such in the first version of the theory [8].

Definition 4. DEAD, LIVE, WELL-FORMED
Given a net $N = (S, T, F)$, a transition t is dead at a marking M if $\forall M' \in [M\rangle$: $^\bullet t \not\subseteq M'$ (t is not enabled at any marking reachable from M). A transition t is live at M if it can never become dead from M: $\forall M' \in [M\rangle\, \exists M'' \in [M'\rangle : M'' > t$.

A marking M is live if all transitions are live at M. In order to highlight the underlying net N, we can also say that (N, M) is live, provided M is live in N.

A net $N = (S, T, F)$ is called well-formed if it has a marking M that is both live and bounded[2]; it is strongly well-formed[3] if it has a marking M that is both live and safe. ■ 4

Note that (strong) well-formedness refers to a net without a marking, while all other notions introduced in Definition 4 are relative to some markings.

Since a net $N = (S, T, F)$ is, in fact, a bipartite directed graph, $S \cup T$ being the set of nodes and F being the set of arcs, all the usual graph-theoretical concepts are applicable.

Definition 5. (SIMPLE) PATH, CYCLE, DISTANCE, (STRONGLY) CONNECTED, ISOLATED
By paths and cycles we mean directed paths and directed cycles. They are simple if nodes are not repeated. The distance of a node y from a node x is the length of a shortest path from x to y (if such a path exists, otherwise it is ∞).

We recall (from graph theory) that by a connected component of a graph we mean any maximal subgraph that is connected when the arc directions are ignored. A strongly connected component is any maximal subgraph that is strongly connected (hence all its nodes lie on a directed cycle; some nodes might have more occurrences in this cycle, i.e., it is not necessarily a simple cycle). An isolated node, which for $N = (S, T, F)$ is a place or a transition x such that $^\bullet x \cup x^\bullet = \emptyset$, is a strongly connected component. ■ 5

Definition 6. SUBNET
By a subnet of $N = (S, T, F)$ we mean any net $N' = (S', T', F|_{S' \cup T'})$ where $S' \subseteq S$ and $T' \subseteq T$. Hence a subnet arises by possibly removing some places and/or transitions and (only) their adjacent arcs. Such a subnet N' is said generated from its underlying net N by S' and T', or by S' alone if $T' = {}^\bullet S' \cap S'^\bullet$, or by T' alone if $S' = {}^\bullet T' \cap T'^\bullet$. ■ 6

---

[2] There may be many such markings; it is not necessary to choose and exhibit one of them; only its existence is required.

[3] Initially, in [8], only this form was considered and simply called well-formed.

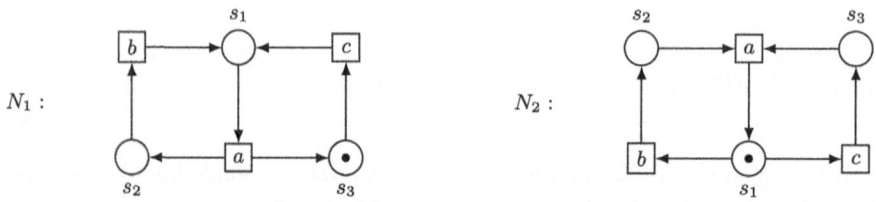

Fig. 1. Two strongly connected, marked, nets N_1 and N_2. As usual, the initial markings are represented by *tokens* inscribed in places.

Example 1. SOME NETS AND SOME SUBNETS (FIG. 1)
Figure 1 illustrates some of the basic notions defined so far. The initial marking of N_1 is $M_0^1 = \{s_3\}$, and the initial marking of N_2 is $M_0^2 = \{s_1\}$. M_0^1 is live, but not bounded. At M_0^1, the infinite sequence $c\,(abc)^\infty$ can be fired (and many others, too), showing that, in fact, all places are unbounded at M_0^1. M_0^2 is bounded, but not live. In N_2, transition a is dead at M_0^2, but neither b nor c are dead at M_0^2; b or c can be fired at M_0^2, and after either firing, all transitions are dead. The subnet generated by $\{s_1, s_2\}$ is the simple cycle $s_1\,a\,s_2\,b\,s_1$ (consisting of four nodes) in N_1 and the cycle $s_1\,b\,s_2\,a\,s_1$ in N_2. ■ 1

Example 2. NON-WELL-FORMEDNESS OF THE NETS IN FIG. 1
If we disregard the two initial markings in Fig. 1, then none of the two resulting unmarked nets is well-formed, since not only the initial markings depicted in the figure, but any arbitrary initial markings, fail to be both live and bounded. ■ 2

The following observation is immediate.

Proposition 1.
A net N is (strongly) well-formed iff so are each of its connected components.
■ 1

The next proposition is not so immediate, but it is well-known; we give no new proof.

Proposition 2. WELL-FORMEDNESS AND STRONG CONNECTEDNESS ([5], THEOREM 2.25)
Every connected component of a (strongly) well-formed net is strongly connected.
■ 2

As a consequence of Propositions 1 and 2, any well-formed net N consists of zero or more pairwise disjoint well-formed strongly connected subnets. Some of them might be isolated places or isolated transitions.

3   Free-Choice Nets, Clusters, and Allocations

The free-choice property of nets captures the property that of two conflicting transitions, either both are enabled or none is enabled at any marking.

Definition 7. FREE-CHOICE AND CHOICE-FREE
A net $N = (S, T, F)$ is a free-choice net if for all $t, t' \in T$, $^\bullet t \cap {}^\bullet t' \neq \emptyset$ entails $^\bullet t = {}^\bullet t'$.
It is restricted free-choice if whenever $(s, t) \in F$, either $|s^\bullet| \leq 1$ or $|{}^\bullet t| \leq 1$.
It is choice-free if for all $t, t' \in T$, $^\bullet t \cap {}^\bullet t' \neq \emptyset$ entails $t = t'$.   ■ 7

Historically, the theory [8] was developed for restricted free-choice nets (then simply called free-choice), then the theory was extended [5] to what we here simply call free-choice, and was at some point called "extended" free-choice [14]. Choice-freeness is still a stronger version of (restricted) free-choiceness, since no two transitions can then be in conflict.

Proposition 3. SUBNETS OF FREE-CHOICE NETS ARE FREE-CHOICE
Any subnet of a free-choice net is a free-choice net. The same is true for the subclasses of restricted free-choice nets and choice-free nets.

Proof: Erasing a transition in a free-choice net does not alter the sets $^\bullet t$ and $^\bullet t'$ used in the definition, thus keeping the free-choice property invariant. Erasing a place might decrease the preset of some transition t, but (by the free-choice property) it similarly decreases the presets of all transitions which share some input place with t, thus, again, not destroying the free-choice-property. The reasoning is similar (sometimes simpler) for restricted free-choice nets or choice-free nets.   ■ 3

This means in particular that any connected component of a free-choice net is itself free-choice. Since we are here essentially interested in properties which are valid for a free-choice net if and only if they are valid for each connected component, in the following, without loss of generality, we shall only consider connected nets.

Some place subsets are of special interest in the theory of free-choice nets.

Definition 8. SIPHONS ANS TRAPS
A subset $D \subseteq S$ of places in a net is called a siphon if $^\bullet D \subseteq D^\bullet$, and a subset $Q \subseteq S$ is called a trap if $Q^\bullet \subseteq {}^\bullet Q$. A siphon is called proper if it is nonempty, and minimal if it is proper and has no other proper siphon in it.   ■ 8

The following characterisation of a live marking in a restricted free-choice net has been introduced in [3], and generalised to free-choice nets in [5].

Theorem 1. COMMONER'S LIVENESS THEOREM
A free-choice net N without isolated places is live under some initial marking M_0 iff every minimal siphon D contains a trap $Q \subseteq D$ with $M_0(Q) > 0$.

The set of nodes $S \cup T$ of a net can be partitioned as follows:

Definition 9. CLUSTER, SIBLING, RIVAL
Let $N = (S, T, F)$ be a net. The clusters of N are the equivalence classes of the relation[4]

$$((F \cap (S \times T)) \cup (F \cap (S \times T))^{-1})^*$$

---

[4] If R is a relation on a set D, we denote by R^{-1} the inverse of R, i.e., $(x, y) \in R^{-1}$ iff $(y, x) \in R$ for any $x, y \in D$.

which is the symmetric, reflexive and transitive closure of $F \cap (S \times T)$. *For a cluster* C, *the set of places (transitions) in* C *is denoted by* S_C *(respectively, T_C), so* $C = S_C \cup T_C$. *A cluster* C *is called nonempty if* $T_C \neq \emptyset$, *i.e.,* C *contains at least one transition. The set of nonempty clusters of* N *is denoted by* $\mathbb{C}$. *The places of a cluster are called* siblings *and the transitions of a cluster are called* rivals.   ■ 9

For free-choice nets, clusters C have a particularly simple form: there is always an F-arc from any place in S_C to any transition in T_C. Note that, if a place has no output, it is a cluster by itself, and the same is true for a transition without input. Note however that the emptiness of a cluster is only based on its set of transitions; this is due to the fact that we will essentially be interested in the enabling notion, which is itself asymmetric (see for instance Lemma 2 below).

In the context of (free-choice) nets, various symmetries may be considered, such as reversing the arcs or exchanging the roles of places and transitions [5,8]. In this paper, we will concentrate on *reverse-duality*, defined as follows.

Definition 10. REVERSE-DUAL NETS
The reverse-dual $rd(N)$ *of a net* $N = (S, T, F)$ *is obtained by exchanging the roles of places and transitions, and by reversing the arcs between them. Formally, $rd(N) = (T, S, F^{-1})$.*   ■ 10

The next properties are easy to verify.

Lemma 1. SOME PROPERTIES OF REVERSE-DUALITY
If N *is a net,*

1. $rd(rd(N)) = N$ *(involution);*
2. *if* N *is (strongly) connected, so is* $rd(N)$ *(preservation of connectivity);*
3. *if* N *is (restricted) free-choice so is* $rd(N)$ *(preservation of free-choiceness);*
4. *if* C *is a cluster of* N, $rd(C)$ *is a cluster of* $rd(N)$, $rd(S_C) = T_{rd(C)}$ *and $rd(T_C) = S_{rd(C)}$ (preservation of clusters).*   ■ 1

Example 3. SOME CLUSTERS IN FIG. 1
N_1 has three clusters $\{s_1, a\}$, $\{s_2, b\}$, and $\{s_3, c\}$. N_2 has two clusters, $\{s_1, b, c\}$ (with b, c being rivals) and $\{s_2, s_3, a\}$ (with s_2, s_3 being siblings). They are shown in green in Fig. 2.   ■ 3

It may be the case that a cluster contains no places (if there are transitions without pre-places), or that it contains no transitions (if there are places without post-transitions). But if a net is strongly connected and contains at least a place and a transition, then a cluster always contains at least one place and at least one transition.

Clusters are particularly interesting in free-choice nets because their transitions enjoy an all-or-nothing property.

Lemma 2. ALL-OR-NOTHING PROPERTY OF CLUSTERS
Let (S, T, F) *be a free-choice net, let* C *be a nonempty cluster of* N, *and let* M *be a marking of* N. *Then* $(\exists t \in T_C : M \xrightarrow{t})$ *if, and only if,* $(\forall t \in T_C : M \xrightarrow{t})$.

Proof: This follows from the property $^\bullet t = {}^\bullet t'$ for any two transition $t, t' \in T_C$. Note however that this is not true for empty clusters, since $(\exists t \in \emptyset$ is always false while $\forall t \in \emptyset$ is always true. ■ 2

Hence if (some transition of) a cluster is enabled by a marking M, an arbitrary transition of the cluster can be chosen to be fired from M. Taking advantage of this property, we shall be interested in executions of free-choice nets that are confined to using only a single transition in each cluster. This is formalised, and elaborated on, in the remainder of this section.

Definition 11. ALLOCATION, α-ALLOCATED
Let $N = (S, T, F)$ be a free-choice net and let $\mathbb{C}$ be the set of its nonempty clusters. A (partial) allocation is a nonempty partial function $\alpha \colon \mathbb{C} \to T$ which selects transitions from clusters, i.e., $\alpha(C) \in T_C$ for all C in the domain of α, and α is total if it is a total function. A cluster is said α-allocated, if it belongs to the domain of α, and a transition t is called α-allocated if there is a cluster C in the domain of α such that $t = \alpha(C)$. ■ 11

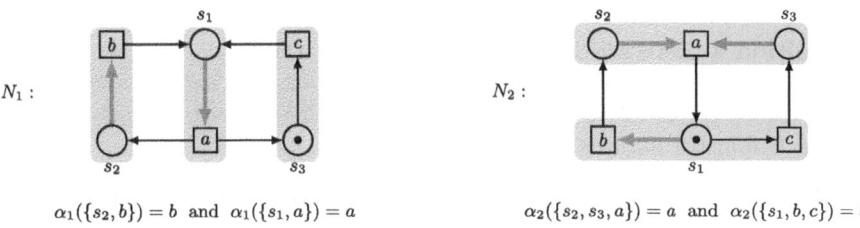

$\alpha_1(\{s_2, b\}) = b$ and $\alpha_1(\{s_1, a\}) = a$ $\alpha_2(\{s_2, s_3, a\}) = a$ and $\alpha_2(\{s_1, b, c\}) = b$

Fig. 2. A partial allocation α_1 (in N_1) and a total allocation α_2 (in N_2), both indicated by bold arcs. The clusters of N_1 and N_2 are shown in green. (Color figure online)

Example 4. ALLOCATIONS IN THE RUNNING EXAMPLE (FIGS. 1 AND 2)
Figure 2 depicts two allocations α_1 in N_1 and α_2 in N_2; α_1 is not total while α_2 is total. In N_1, transition b is chosen in cluster $\{s_2, b\}$; transition a is chosen in cluster $\{s_1, a\}$; and no transition is chosen in cluster $\{s_3, c\}$. In N_2, transition a is chosen in cluster $\{s_2, s_3, a\}$ and transition b is chosen in cluster $\{s_1, b, c\}$. ■ 4

Note that allocations define choice-free subnets of the original free-choice net, compatible with the latter in the sense that all the places belonging to an allocated cluste have the same (unique) successor (and the places of a non-allocated cluster have no successor). For a given allocation α, we may define a useful ordering relation on the transitions of N', as well as a derived equivalence relation.

Definition 12. $\prec$, $\preceq$, AND $\equiv$
Let α be an allocation of a free-choice net $N = (S, T, F)$. Let $x, y \in T$ be two α-allocated transitions. Then $x \prec_\alpha y$ if $x^\bullet \cap {}^\bullet y \neq \emptyset$ (in words: x

immediately precedes y *if there is a place* s *which leads from* x *to* y: $x\,F\,s\,F\,y$*);* $\preceq_\alpha$ *is the reflexive and transitive closure of* $\twoheadleftarrow_\alpha$*;* $x \equiv_\alpha y$ *if* $x \preceq_\alpha y$ *and* $y \preceq_\alpha x$*; and* $[x]_{\equiv_\alpha}$ *is the equivalence class of* x *with respect to* $\equiv_\alpha$*.*   ∎ 12

Example 5. SOME ORDERING RELATIONS REFERRING TO FIG. 2
In N_1, , $a \twoheadleftarrow_{\alpha_1} b$ and $b \twoheadleftarrow_{\alpha_1} a$, as well as $x \preceq_{\alpha_1} y$ for all $x, y \in \{a, b\}$. Also, $[a]_{\equiv_{\alpha_1}} = \{a, b\}$.
    Similarly, in N_2, $a \twoheadleftarrow_{\alpha_2} b$ and $b \twoheadleftarrow_{\alpha_2} a$; $x \preceq_{\alpha_2} y$ for all $x, y \in \{a, b\}$; and $[a]_{\equiv_{\alpha_2}} = \{a, b\}$.   ∎ 5

We now identify a class of allocations associated with transitions, and some of their properties.

Definition 13. DIRECTED ALLOCATION
Let $N = (S, T, F)$ *be a free-choice net, let* α *be an allocation of* N *and let* t *be an* α*-allocated transition. Then* α *is called directed towards* t *if it is total and if* $x \preceq_\alpha t$ *for every* α*-allocated* $x \in T$ *(i.e., there is a directed path from* x *to* t *only visiting allocated transitions).*   ∎ 13

Such allocations always exist, in the following sense.

Proposition 4. EXISTENCE OF DIRECTED ALLOCATIONS
Let $N = (S, T, F)$ *be a strongly connected free-choice net, with at least a place and a transition, and let* $t \in T$ *be a transition of* N*. There exists a total allocation* α *which is directed towards* t*.*

Proof: With each $x \in T$ we associate the distance $d_x(t) \in \mathbb{N}$ of x to t in (the graph of) N. Let α be such that for each nonempty cluster C of N it allocates one $x \in T_C$ with the least $d_x(t)$ (which is finite due to strong connectedness). We have $x \preceq_\alpha t$ for all α-allocated x, as can be shown by induction on $d_x(t)$.   ∎ 4

Next we show that if N has live (and bounded) markings, and if an allocation is fixed, then the existence of special infinite executions which are consistent with this allocation are guaranteed. This proposition is a key ingredient of both main proofs that follow.

Proposition 5. EXISTENCE OF INFINITE EXECUTIONS
Let $N = (S, T, F)$ *be a free-choice net.*

(i) *Let* M_0 *be a live marking of* N *and let* α *be a nonempty (possibly partial) allocation of* N*. There is an infinite firing sequence* $M_0 \xrightarrow{\tau}$ *which (1) agrees with* α *(i.e., for each cluster* C *in the domain of* α*, no transition except possibly* $\alpha(C)$ *is fired), and (2) contains at least one* α*-allocated transition infinitely often.*

(ii) *Let* M_0 *be a live and bounded marking of* N*, let* $t \in T$ *be non-isolated, and let* α *be a total allocation of* N *which is directed towards* t *in its (strongly, from Proposition 2) connected component. There is an infinite firing sequence* $M_0 \xrightarrow{\tau}$ *which (1) agrees with* α *and (2) satisfies* $T_\infty^\tau = [t]_{\equiv_\alpha}$*, where* T_∞^τ *is the set of transitions occurring infinitely often in* τ*.*

Proof:

(i) Let $M_0 \xrightarrow{\tau_1} M_1'$ be a shortest sequence which is firable from M_0 such that M_1' enables a transition contained in a cluster allocated by α. (τ_1 may be empty, but not infinite since M_0 is live and α is nonempty.) This implies that M_1' enables a transition of some allocated cluster C. By the free-choice property, $t_1 = \alpha(C)$ is also enabled by M_1'. Let $M_0 \xrightarrow{\tau_1 t_1} M_1$ and continue from M_1 (which is, of course, also live). This proves Part (1) of Claim (i). Since T is finite, the infinite sequence $\tau_1 t_1 \tau_2 t_2 \ldots$ so constructed contains at least one allocated transition infinitely often, proving Part (2) of Claim (i).

(ii) Using Part (i) of the proposition, we can fix an infinite execution

$$\textsc{Exec} \;=\; M_0 \xrightarrow{t_1 t_2 t_3 \ldots}$$

which agrees with α and contains at least one α-allocated transition, say u, infinitely often. This already proves Part (1) of Claim (ii); we now show that Part (2) is true for $\textsc{Exec}$.

First, we observe that if $x \in T_\infty^{\textsc{Exec}}$ and $x \, F \, s \, F \, y$ for an α-allocated transition y, then also $y \in T_\infty^{\textsc{Exec}}$: otherwise, using the fact that no other y' in the same cluster as y occurs in $\textsc{Exec}$, the number of tokens on s is infinitely often increased, but only finitely many times decreased, in $\textsc{Exec}$, contradicting the boundedness of (N, M_0). Since $T_\infty^{\textsc{Exec}} \neq \emptyset$ and $u \preceq_\alpha t$ by the directedness of α towards t, we have $t \in T_\infty^{\textsc{Exec}}$, and $[t]_{\equiv_\alpha} \subseteq T_\infty^{\textsc{Exec}}$.

Let T_α be the set of α-allocated transitions and $\overline{T} = T_\alpha \backslash [t]_{\equiv_\alpha}$. Then $([t]_{\equiv_\alpha})^\bullet \cap {}^\bullet \overline{T} = \emptyset$, since α is directed towards t; hence the sequence σ arising by projecting $t_1 t_2 t_3 \cdots$ from $\textsc{Exec}$ to $\overline{T}$ is also performable from M_0 (since transitions from $[t]_{\equiv_\alpha}$ do not add tokens to the clusters of $\overline{T}$). But σ cannot be infinite (we would get an infinite execution $\textsc{Exec}'$ of N from M_0 with $t \notin T_\infty^{\textsc{Exec}'}$), which implies that $T_\infty^{\textsc{Exec}} \subseteq [t]_{\equiv_\alpha}$.

Thus $T_\infty^{\textsc{Exec}} = [t]_{\equiv_\alpha}$, proving Part (2) of Claim (ii).   ∎ 5

Example 6. APPLICATION OF THE PROOF OF PROPOSITION 5(i) TO N_1 IN FIG. 2

The shortest firable sequence enabling an allocated transition is c. So, we have $\tau_1 = c$. After firing c, the sequence $(ab)^\infty$ can be executed; i.e., we have $\tau_j = \varepsilon$ for every $j > 1$. Proposition 5(i) cannot be applied to N_2, since the initial marking is not live.   ∎ 6

For Proposition 5(ii), we need a well-formed free-choice net with an initial marking that is live and bounded. None of the nets considered so far satisfy this requirement. Hence we now consider such a net in Fig. 3.

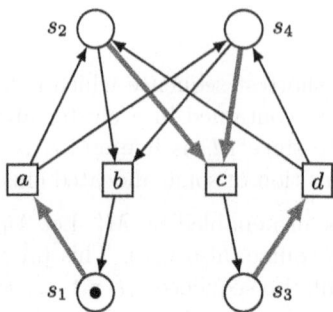

Fig. 3. A well-formed free-choice net with a live and bounded initial marking $M_0 = \{s_1\}$ and a total allocation α shown in bold: from the cluster $\{s_1, a\}$, a is chosen; from $\{s_2, s_4, b, c\}$, c is chosen, and from $\{s_3, d\}$, d is chosen. α is directed towards c and towards d, but not towards a.

Example 7. ILLUSTRATION OF THE PROOF OF PROPOSITION 5(ii) ON FIG. 3

In Fig. 3, there are many infinite firing sequences, for instance $M_0 \xrightarrow{ababa...}$, but there is only one infinite firing sequence agreeing with the allocation α shown in bold in the figure:

$$\mathsf{Exec} \;=\; M_0 \xrightarrow{acdcdc...}$$

Allocation α is directed towards c and towards d, but not towards a. We have $[a]_{\equiv_\alpha} = \{a\}$ and $[c]_{\equiv_\alpha} = [d]_{\equiv_\alpha} = \{c, d\}$. So, $\overline{T} = T_\alpha \setminus [c]_{\equiv_\alpha} = T_\alpha \setminus [d]_{\equiv_\alpha} = \{a\}$, and the projection of Exec onto $\overline{T}$ is the finite, one-letter sequence a. It can also be seen that indeed, $T_\infty^{\mathsf{Exec}} = \{c, d\}$, as claimed in Part (2) of Proposition 5(ii).

∎ 7

4   Coverability Results

The main aim of the present paper is to determine when the places and transitions of a free-choice net are covered by special substructures. A first pair of such results was presented in [8], for strongly well-formed restricted free-choice nets; interestingly, while the two statements were exactly symmetric (with respect to reverse-duality), the proofs have a strong symmetric flavour, but not completely; in particular one of them makes use of the liveness criterion (Theorem 1) but not the other one. The property was generalised to well-formed free-choice nets in [5], but curiously the two proofs were completely different there. The situation was partly recovered in [1], still with the use of the liveness criterion. We shall now pursue in this direction, improving and generalising some statements and proofs.

First, the central concepts of S-components and T-components are defined[5], explained and illustrated.

Definition 14. S-COMPONENT
Let $N = (S, T, F)$ be a net and let $X = (S_X, T_X, F|_{S_X \cup T_X})$ be a subnet of N. Then X is an S-component of N iff

A) *X is strongly connected;*
B) *$|^\bullet t \cap S_X| = 1 = |t^\bullet \cap S_X|$ for all $t \in T_X$;*
C) *$^\bullet(S_X) = T_X = (S_X)^\bullet$.*   ■ 14

Definition 15. T-COMPONENT *Let $N = (S, T, F)$ be a net and let $Y = (S_Y, T_Y, F|_{S_Y \cup T_Y})$ be a subnet of N. Then Y is a T-component of N iff*

A) *Y is strongly connected;*
B) *$|s^\bullet \cap T_Y| = 1 = |^\bullet s \cap T_Y|$ for all $s \in S_Y$;*
C) *$^\bullet(T_Y) = S_Y = (T_Y)^\bullet$.*   ■ 15

If such components are present in a net N, they entail useful behavioural properties of the reachable markings of N when it is provided with an initial marking. For instance, the number of tokens on an S-component always stays constant throughout any firings of its transitions. As a consequence, all the places of an S-component are structurally bounded (i.e., bounded whatever the initial marking), and a net covered by S-components is structurally bounded. On the other hand, if all transitions of a T-component (and no others) are executed exactly once, then the marking is reproduced.

Example 8. S- AND T-COMPONENTS IN A WELL-FORMED FREE-CHOICE NET
As shown in Fig. 4, the well-formed free-choice net in Fig. 3 has two S-components. One of them is spanned by the set of places $\{s_1, s_2, s_3\}$ (drawn in green). The other one is spanned by $\{s_1, s_3, s_4\}$ (drawn in red). Both have exactly one token initially (in the grey marking), and this property remains true for all markings of the corresponding reachability graph.

Likewise, the net has two T-components, spanned, respectively, by the set of transitions $\{a, b\}$ (drawn in blue) and $\{c, d\}$ (drawn in orange). Both T-components can be executed fully, the first by ab from the initial marking (reproducing the marking $\{s_1\}$), and the second, after executing a first, by cd (which reproduces the marking $\{s_2, s_4\}$).   ■ 8

---

[5] In the following, pre- and post-sets are taken in the original net N.

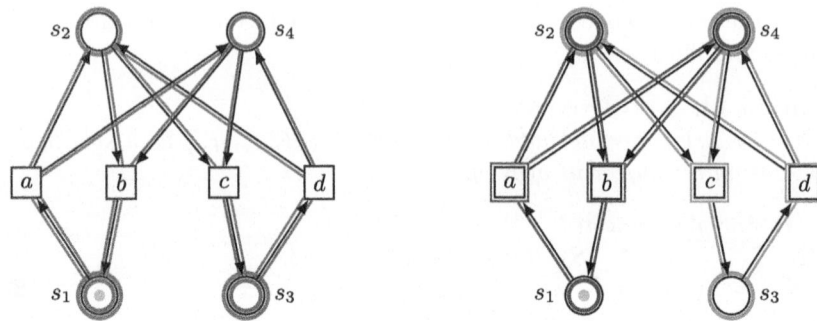

Fig. 4. A well-formed free-choice net with a live and bounded marking $\{s_1\}$ (shown in light grey). It is covered by two S-components (left-hand side, green and red) and by two T-components (right-hand side, blue and orange).

We shall also make use of a slightly relaxed version of S- and T-components. Since we shall only keep about half of the constraints defining them, we shall call them *semi-S-components* and *semi-T-components*.

Definition 16. SEMI-S-COMPONENT
If the following three conditions are satisfied for a subnet $X = (S_X, T_X, F|_{S_X \cup T_X})$ of a net N, then X will be called a semi-S-component of N.

a) *X is strongly connected;*
b) *$|{}^\bullet t \cap S_X| = 1$ for all $t \in T_X$ (and $|t^\bullet \cap S_X| \geq 1$ due to strong connectedness);*
c) *$T_X = {}^\bullet S_X$ (and $T_X \subseteq S_X{}^\bullet$ due to b), so that $T_X = {}^\bullet S_X \subseteq S_X{}^\bullet$).*   ■ 16

Note that, from this definition, the total number of tokens in a semi-S-component may not be decreased by a transition t, since $|t^\bullet \cap S_X| \geq |{}^\bullet t \cap S_X|$, unless t is chosen in $S_X{}^\bullet \setminus {}^\bullet S_X$.

If, in addition to a), b) and c), the following properties are satisfied,

bb) $|t^\bullet \cap S_X| = 1$ for all $t \in T_X$;
cc) $T_X = S_X{}^\bullet$,

then X is an S-component of N (since bb) and cc) are the missing parts of Definition 14).

In fact, semi-S-components are not a really new notion: they coincide with minimal siphons for free-choice nets:

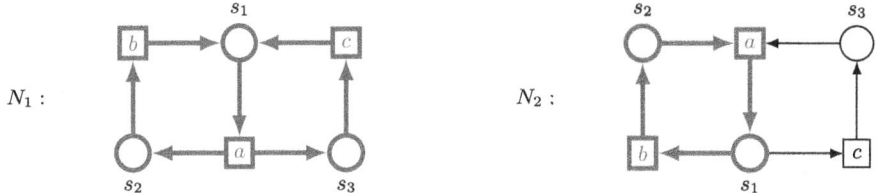

Fig. 5. Semi-S-components of N_1 and of N_2 shown in bold green. (Color figure online)

Theorem 2. CHARACTERISATION OF MINIMAL SIPHONS (THEOREM 4.30 OF [5])
A nonempty set of places R of a free-choice net N is a minimal siphon iff

- *every cluster C of N contains at most one place of R, and*
- *the subnet generated by $R \cup {}^{\bullet}R$ is strongly connected.*

Comparing this with Definition 16, we get the following

Corollary 1. MINIMAL SIPHONS COINCIDE WITH SEMI-S-COMPONENTS
A nonempty set of places R in a free-choice net $N = (S, T, F)$ is a minimal siphon iff $(R, {}^{\bullet}R, F|_{R \cup {}^{\bullet}R})$ is a semi-S-component.   ■ 1

Definition 17. SEMI-T-COMPONENT
If the following three conditions are satisfied for a subnet $Y = (S_Y, T_Y, F|_{S_Y \cup T_Y})$ of a net N, then Y will be called a semi-T-component *of N.*

a) *Y is strongly connected;*
b) *$|s^{\bullet} \cap T_Y| = 1$ for all $s \in S_Y$ (and $|{}^{\bullet}s \cap T_Y| \geq 1$ due to strong connectedness);*
c) *$S_Y = T_Y{}^{\bullet}$ in N (and $S_Y \subseteq {}^{\bullet}T_Y$ due to b), so that $S_Y = T_Y{}^{\bullet} \subseteq {}^{\bullet}T_Y$).*   ■ 17

If, in addition to a), b) and c), the following properties are satisfied,

bb) *$|{}^{\bullet}s \cap T_Y| = 1$ for all $s \in S_Y$;*
cc) *$S_Y = {}^{\bullet}T_Y$,*

then Y is a T-component of N.

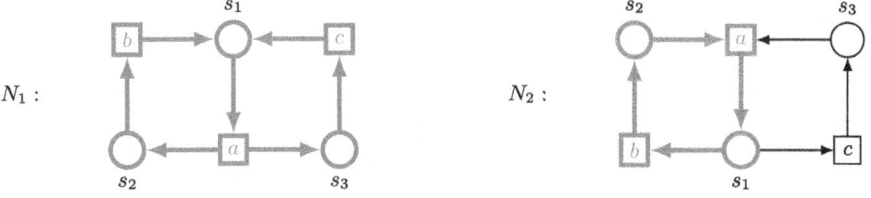

Fig. 6. Semi-T-components of N_1 and of N_2 shown in bold orange. (Color figure online)

Example 9. THE DIFFERENCE BETWEEN SEMI-T-COMPONENTS AND T-COMPONENTS

Figure 6 shows two semi-T-components which are not also T-components for our first running example. Let Y be the semi-T-component shown in N_1. We have $|{}^\bullet s_1 \cap T_Y| = 2 > 1$, hence Part bb) of the definition of a T-component (which requires $|{}^\bullet s \cap T_Y| = 1$ for all $s \in S_Y$) is not satisfied. Let Y be the semi-T-component shown in N_2. We have $s_3 \in {}^\bullet T_Y \setminus S_Y$, hence Part cc) of the definition of a T-component (which requires ${}^\bullet T_Y \subseteq S_Y$) is not satisfied.   ∎ 9

From these definitions we get immediately:

Lemma 3. SYMMETRIC COMPONENTS
N' is a (semi-)S-component of a net N iff $rd(N')$ is a (semi-)T-component of $rd(N)$ (exchange of (semi-)S- and (semi-)T-components).   ∎ 3

Our first pair of coverability results, whose statements *and* proofs are completely symmetric is thus:

Theorem 3. SEMI-S-COMPONENT COVERING
For any strongly connected free-choice net N, each place of N belongs to a semi-S-component of N.

Proof: Let $N = (S, T, F)$ be a strongly connected free-choice net, and $s_0 \in S$. We build progressively a semi-S-component of N that contains s_0; it is given by the final value of the "program variable" $X = (S_X, T_X)$ of the following algorithm (when adding the arcs $F|_{S_X \cup T_X}$):

$X := (\{s_0\}, \emptyset)$;
while ${}^\bullet S_X \cap (T \setminus T_X) \neq \emptyset$ **do**
extend X with places and transitions on a shortest path from S_X to ${}^\bullet S_X \cap (T \setminus T_X)$ (such a path exists since N is strongly connected).

This construction is illustrated in Fig. 7. We claim that the algorithm keeps the following invariant: the subnet $(S_X, T_X, F|_{S_X \cup T_X})$ of N always satisfies the conditions a) and b) of semi-S-components, and the condition c'): $T_X \subseteq {}^\bullet S_X$; this entails that at the end the condition c) $(T_X = {}^\bullet S_X)$ is satisfied.

To verify the claim, we observe that for the initial value $X = (\{s_0\}, \emptyset)$ we clearly have a),b),c'), and then we consider the extension step: it clearly keeps the conditions a) and c'), so it remains to verify that it also keeps the condition b). Let the chosen shortest path be $s_1 F t_1 F s_2 \cdots F t_k$ where $s_1 \in S_X$ and $t_k \in {}^\bullet S_X \cap (T \setminus T_X)$. For each $j, 1 < j \leq k$, the place s_j is in a "fresh" cluster, i.e., in N we have $C(s_j) \cap (S_X \cup \{s_2, s_3, \ldots, s_{j-1}\}) = \emptyset$; indeed: if $s \in S_X \cap C(s_j)$, then $sF t_j F s_{j+1} \cdots F t_k$ is a shorter path that we could choose, and if $s_i \in C(s_j)$ for $2 \leq i < j$, then $s_1 F t_1 F \cdots s_i F t_j F s_{j+1} \cdots F t_k$ is also a shorter path that we could choose. Hence the condition b) is also kept.   ∎ 3

Theorem 4. SEMI-T-COMPONENT COVERING
For any strongly connected free-choice net N, each transition of N belongs to a semi-T-component of N.

Proof: This immediately results from Lemmata 1 and 3, and Theorem 3 applied to $rd(N)$, hence from a symmetry argument applied to the previous result. ∎ 4

We shall now examine the impact of strenghtening the preconditions from strong connectedness to well-formedness (recall Proposition 2), without needing to use the liveness criterion in Theorem 1.

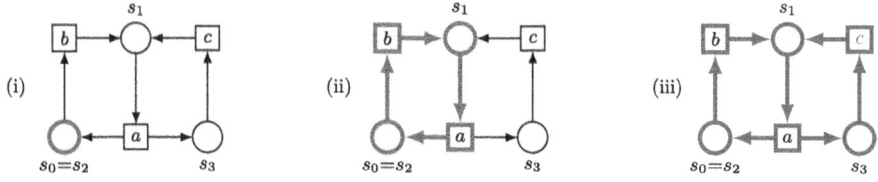

Fig. 7. A construction on the strongly connected net N_1.

Theorem 5. SEMI-S-COMPONENTS ARE S-COMPONENTS IN WELL-FORMED FREE-CHOICE NETS

In any well-formed free-choice net N, every semi-S-component is an S-component.

Proof: Let $N = (S, T, F)$ be a well-formed free-choice net. Propositions 1 and 2 allow us to assume, without loss of generality, that N is strongly connected. Let M_0 be a live and bounded marking of N and let $X = (S_X, T_X, F|_{S_X \cup T_X})$ be a semi-S-component in N, thus satisfying Conditions a), b) and c). The proof is done in two steps, (i) and (ii). In Step (i), we prove that Condition bb) also holds for X, i.e., transitions such as a in N_1 (Fig. 5, left-hand side) cannot occur. In Step (ii), we prove that Condition cc) also holds for X, i.e., transitions such as c in N_2 (Fig. 5, right-hand side) cannot occur.

(i) To show bb), assume that there are transitions $u \in T_X$ for which $|u^\bullet \cap S_X| \geq 2$; we fix such a u and derive a contradiction.

Let $\overline{T}$ be the set of transitions from non-X-clusters, i.e. from the clusters C for which $C \cap T_X = \emptyset$, and let $T' = \overline{T} \cup T_X$. Note that T' contains a transition from every cluster.

First, we claim that

$$\forall t \in T': \; |{}^\bullet t \cap S_X| \leq |t^\bullet \cap S_X| \tag{1}$$

(i.e., intuitively speaking, no transition in T' decreases the number of tokens on S_X). This is because if $t \in \overline{T}$, then $|{}^\bullet t \cap S_X| = 0$ by definition. And if $t \in T_X$, then Condition b) of a semi-S-component yields $|{}^\bullet t \cap S_X| = 1$, hence also $|t^\bullet \cap S_X| \geq 1$ due to the strong connectedness of X. In both cases, the inequality in (1) is valid.

Secondly, we also claim that

$$u \in T' \; \wedge \; |{}^\bullet u \cap S_X| < |u^\bullet \cap S_X| \tag{2}$$

(i.e., intuitively speaking, u properly increases the number of tokens on S_X). We have $u \in T'$ by $u \in T_X$ and $T_X \subseteq T'$, and the inequality in (2) comes from (1) and the definition of u.

Now consider the subnet $N' = (S, T', F|_{S \cup T'})$. It is just so defined that transitions $x \notin T_X$ satisfying $|{}^\bullet x \cap S_X| > |x^\bullet \cap S_X|$ (which reduce the number of tokens on S_X) are *not* contained in it; such transitions x always have a rival in T_X which *is* contained in N' (which follows from the strong connectedness of X, since x has an input place in S_X). Proposition 3 tells us that N' is again free-choice, and it is strongly connected by the strong connectedness of both N and X. Thus, u is a successor of each $t \in T'$ in the subnet N', hence also in N, and it is itself in T'. By Proposition 4, there is a total allocation α of N' which is directed towards u; since N' contains a transition from every cluster of N, α is also a total allocation of N directed towards u. Crucially, transitions x satisfying $|{}^\bullet x \cap S_X| > |x^\bullet \cap S_X|$ are not in T' and hence *not* α-allocated.

By Proposition 5(i), and because M_0 is live, there is an infinite execution M_0 which agrees with α and in which u occurs infinitely often (since α is directed towards u). During this execution, the sum of tokens on S_X necessarily grows above any bound, contradicting the boundedness of M_0.

This contradiction shows that a transition such as u cannot exist and that, instead, any semi-S-component of a well-formed free-choice net satisfies bb), i.e., $\forall t \in T_X: |t^\bullet \cap S_X| = 1$.

(ii) To show cc), assume that there are transitions $v \in (S_X{}^\bullet) \setminus T_X$; we fix such a v and derive a contradiction.

We thus have $|{}^\bullet v \cap S_X| > |v^\bullet \cap S_X| = 0$, where $|{}^\bullet v \cap S_X| > 0$ follows from $v \in S_X{}^\bullet$ and $|v^\bullet \cap S_X| = 0$ follows from $v \notin T_X$ and from $T_X = {}^\bullet S_X$ (Part c) of the definition of a semi-S-component).

Moreover, for each $t \in T$, $|{}^\bullet t \cap S_X| \geq |t^\bullet \cap S_X|$. This follows because if $t \in T_X$, then $|{}^\bullet t \cap S_X| = 1$ by b) and $|t^\bullet \cap S_X| = 1$ by Step (i). And if $t \notin T_X$, then $|t^\bullet \cap S_X| = 0$ as a consequence of Part c) of the definition of a semi-S-component.

Intuitively speaking, the above means that v properly decreases the number of tokens on S_X while no other transition increases this number. Now consider the partial allocation which just allocates v and no other transition. By Proposition 5(i), there is an infinite execution in which v occurs infinitely often. But such an execution decreases the number of tokens on S_X below any bound—a contradiction. This contradiction shows that a transition such as v cannot exist and that, instead, any semi-S-component of a well-formed free-choice net also satisfies cc), ending the proof.

Note that, if N is only composed of an isolated node, the proof is valid, but in this case there is a simpler argument: if N is an isolated place, N is both a semi-S-component and an S-component; if N is an isolated transition, there is no place to cover.

Corollary 2. S-COMPONENT COVERINGS EXIST FOR WELL-FORMED FREE-CHOICE NETS
In a well-formed free-choice net, every place belongs to an S-component.

Proof: Using Theorems 3 and 5.   ■ 2

Corollary 3. WHEN A SEMI-S-COMPONENT IS NOT AN S-COMPONENT
If a free-choice net exhibits a semi-S-component which is not an S-component, then this net is not well-formed.   ■ 3

In order to prove similar results about (semi-)T-components, we could try to apply Theorem 5 to the reduce-dual of the original net and use the next result:

Lemma 4. PRESERVATION OF WELL-FORMEDNESS
A free-choice net N is well-formed iff $rd(N)$ is well-formed (preservation of well-formedness).

Proof:
While the other properties of reverse-duality we have seen are purely structural and easy to check, the present one is behavioural and not trivial; its proof (when applied to each connected component) may be found in [5] (Theorem 6.21) and in [2] (Theorem 6.21 as well)   ■ 4

However, this is not possible since the proof of Lemma 4 uses both results on the coverability of well-formed free-choice-nets by S- and T-components, of which the next result is an important ingredient. Hence, we shall develop a distinct proof in this case.

Theorem 6. SEMI-T-COMPONENTS ARE T-COMPONENTS IN WELL-FORMED FREE-CHOICE NETS
In any well-formed free-choice net N, every semi-T-component is a T-component.

Proof:
Let $N = (S, T, F)$ be a (connected) well-formed free-choice net. Let M_0 be a live and bounded marking of N and let $Y = (S_Y, T_Y, F|_{S_Y \cup T_Y})$ be a semi-T-component in N, satisfying Conditions a), b) and c). We will show that Y also satisfies the conditions bb) and cc) of T-components.

For the sake of contradiction, we assume that bb) or cc) (or both) are violated, and we fix some $s_b \in S_Y$ such that $|{}^\bullet s_b \cap T_Y| \geq 2$ or some $s_c \in {}^\bullet T_Y \smallsetminus S_Y$; this also entails that $T_Y \neq \emptyset$.

Let us first assume that cc) is violated, i.e., there is some $s_c \in {}^\bullet T_Y \smallsetminus S_Y$, and some[6] $t_0 \in s_c{}^\bullet \cap T_Y$. Let α be an allocation directed towards t_0 which is total in Y. Such an allocation exists by Proposition 4 since by hypothesis Y is strongly connected. Since by hypothesis $|s^\bullet \cap T_Y| = 1$ for each $s \in S_Y$, this allocation is unique and goes to any transition in T_Y. Then, it is easy to extend α to a total allocation directed towards t_0 in N; first, if C is a cluster of N intersecting

---
[6] If we refer to net N_2 in Fig. 6, $s_c = s_3$ and $t_0 = a$.

a cluster C' of Y, we choose of course $\alpha(C) = \alpha(C')$; then, still following the construction in the proof of Proposition 4, we construct α for the remaining non-allocated clusters. By construction, since $S_Y = T_Y{}^\bullet$, there is no way to get out of Y while following α, so that $[t_0]_{\equiv_\alpha} = T_Y$. By Proposition 5(ii), there is an infinite execution EXEC of N' from M_0, for which $T_\infty^{\text{EXEC}} = [t_0]_{\equiv_\alpha}$, so that s_c would be infinitely often decreased while the transitions in ${}^\bullet s_c$ may only fire finitely often to increase the marking of s_c. This excludes the existence of the above s_c; hence cc) holds, and we now concentrate on s_b (with $|{}^\bullet s_b \cap T_Y| \geq 2$).[7]

Since Y is strongly connected, in Y we may fix a simple cycle containing s_b; let $S' \subseteq S_Y$ be the set of places in this cycle. Like for the previous point, there is a total allocation α towards each transition in Y and an infinite execution EXEC compatible with α such that $T_Y = T_\infty^{\text{EXEC}}$. Since $|s^\bullet| = 1$ for each $s \in S_Y$, each transition $t \in T_Y$ satisfies $|{}^\bullet t \cap S'| = 1 = |t^\bullet \cap S'|$ (inside the cycle) or $|{}^\bullet t \cap S'| = 0 \leq |t^\bullet \cap S'|$ (outside the cycle); one of them must satisfy $|{}^\bullet t \cap S'| < |t^\bullet \cap S'|$, since $|{}^\bullet s_b \cap T_Y| \geq 2$. Hence the sum of tokens on S' grows above any bound in EXEC—a contradiction with the boundedness of (N, M_0). Hence the existence of s_b is excluded as well, which means that bb) also holds (and thus Y is a T-component, as we aimed to show).   ∎ 6

Corollary 4. T-COMPONENT COVERINGS EXIST FOR WELL-FORMED FREE-CHOICE NETS
In a well-formed free-choice net, every transition belongs to a T-component.

Proof: Using Theorems 4 and 6.   ∎ 4

Corollary 5. WHEN A SEMI-T-COMPONENT IS NOT A T-COMPONENT
If a free-choice net exhibits a semi-T-component which is not a T-component, then it is not well-formed.   ∎ 5

5   Characterisation of Well-Formedness

Unfortunately, the coverability properties by S- and T-components are necessary but not sufficient to characterise the well-formedness of a strongly connected free-choice net. For instance, the strongly connected free-choice net shown in Fig. 8 is covered both by S-components (generated by $\{s_1, s_3, s_4, s_7\}$ and $\{s_2, s_5, s_6, s_7\}$) and by T-components (generated by $\{t_1, t_2, t_4, t_6\}$ and $\{t_1, t_3, t_5, t_7\}$). Yet it is not well-formed, due to Corollary 3, since there is a semi-S-component, shown in bold, with two transitions violating Condition cc) (transitions t_3 and t_4) and a transition violating Condition bb) (transition t_1). For any initial marking of this net, transitions t_3 and t_4 can be fired sufficiently often to create a non-live follower marking (a deadlock, actually). The proof of Theorem 5 fails at the end of Part (i) where Proposition 5(i) is invoked, because of the missing liveness assumption.

---

[7] E.g., $s_b = s_1$ in N_1 in Fig. 6.

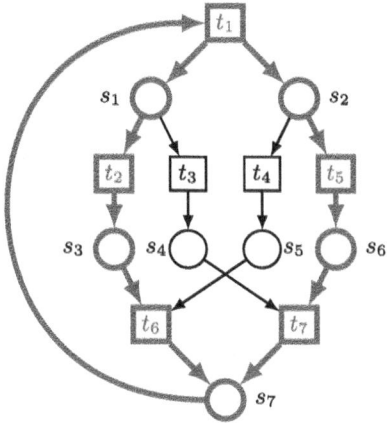

Fig. 8. A non-well-formed free-choice net. A semi-S-component which is not an S-component is shown in red. (Color figure online)

In order to induce well-formedness from these coverability properties, it is necessary to add other constraints, like the one[8] linking the rank of the incidence matrix defined by the arc set to the number of clusters (see Theorem 6.11 in [2]). By contrast, Lemma 3 leads to a (new) full characterisation of well-formedness for free-choice nets.

Theorem 7. WELL-FORMEDNESS CHARACTERISATION
A free-choice net N is well-formed iff each connected component is strongly connected and each semi-S-component is an S-component.

Proof:
$\Rightarrow$
Results from Proposition 1, Proposition 2 and Theorem 5.
$\Leftarrow$
Proposition 1 allows to concentrate on each connected component. Lemma 3 then implies that each place s belongs to a semi-S-component, hence to an S-component (let us call it S_s) by hypothesis. Since the number of tokens in an S-component is constant, for each marking M reachable from an initial marking M_0, we have $M(s) \leq M_0(S_s)$, hence the boundedness (for any initial marking). Since each semi-S-component is an S-component, from Corollary 1 each minimal siphon X is an S-component hence a trap (since $^\bullet S_X = S_X^\bullet$), so that if we start from an initial marking M_0 marking each place ($M_0(s) > 0$ for each $s \in S$), by Theorem 1 M_0 is live and the net is well-formed.   ■ 7

The symmetric property, based on (semi-)T-components, may be easily obtained from the mentioned properties of reverse-dual nets.

Theorem 8. SYMMETRIC WELL-FORMEDNESS CHARACTERISATION
A free-choice net N is well-formed iff each connected component is strongly connected and each semi-T-component is a T-component.

---

[8] Other ones may be found for instance in [5,7].

Proof:
This is a direct consequence of the properties of Lemmata 1, 3, 4, and of Theorem 7 (applied to $rd(N)$).   ■ 8

6 Concluding Remarks

The paper [1] and the book [2] describe two matching algorithms for computing (semi-) S-components covering a given place and (semi-) T-components covering a given transition of a free-choice net. While the justification of the latter is both short and transparent, the justification of the former is not, since it depends circumstantially on auxiliary results such as Commoner's theorem and is thus much longer.

In the present paper, this imbalance has been removed, and we now have two rather concise coverability proofs (Corollaries 2 and 4) which are also almost symmetric with respect to each other. The only source of non-symmetry can be found in the behavioural parts of Theorems 5 and 6. The arguments used there are necessarily different because of the asymmetric role played by places and transitions with respect to transition firings.

It has nevertheless been possible to distil a common argument which has been isolated in the form of Proposition 5, as well as in the pair of Theorems 3 and 4 obtained from the weakening of S-components and T-components into their "semi-"versions.

While the results of Sect. 4 are inspired from [1], the presentation, and especially the proofs, have been considerably reworked; in particular, it is no longer necessary to rely on the Commoner's liveness criterion. The characterisations of well-formedness described in Sect. 5 are novel to the best of our knowledge. In this spirit, in future works, we intend to search for efficient algorithms to build the set of semi-S-components of a net. This could be interesting if there are not too many of them; unfortunately, there are cases where the number of minimal siphons (i.e. of semi-S-components for free-choice nets) is exponential in the size of the net (see for instance the reverse dual of Fig. 9.4 in [2]). Another possibility would then be to search for efficient algorithms to find a semi-S-component which is not an S-component in a strongly connected free-choice net. There could then be a possibility to compete with the known polynomial algorithms checking the well-formedness of a free-choice net (like the one described in Corollary 6.16 of [5]), or with the techniques based on comparability graphs and perfect graphs to decompose a net into State Machines (like [18]).

We also plan to explore whether the concise and symmetric approach described in this paper can be extended to other known (and unknown) results within the rich theory of free-choice nets, and to other classes of nets (weighted marked graphs [15], choice-free nets [16] and equal conflict nets [17] being on top of the list).

Acknowledgements. The authors appreciated the constructive remarks and encouragements of the anonymous referees.

References

1. Best, E., Devillers, R.: Coverability in well-formed free-choice Petri nets. In: Kiefer, S., Křetínský, J., Kučera, A. (eds.) Taming the Infinities of Concurrency: Essays Dedicated to Javier Esparza on the Occasion of His 60th Birthday, pp. 122–132. Springer, Cham (2024). https://doi.org/10.1007/978-3-031-56222-8_6. ISBN 978-3-031-56222-8
2. Best, E., Devillers, R.: Petri Net Primer - A Compendium on the Core Model, Analysis, and Synthesis. Springer, Cham (2024). https://doi.org/10.1007/978-3-031-48278-6. ISBN 978-3-031-48277-9
3. Commoner, F.G.: Deadlocks in Petri Nets. Technical Report CA-7206-2311. Applied Data Research, Wakefield (1972)
4. Czerwinski, W., Orlikowski, L.: Reachability in vector addition systems is Ackermann-complete. In: 62nd IEEE Annual Symposium on Foundations of Computer Science, FOCS 2021, Denver, CO, USA, 7–10 February 2022, pp. 1229–1240. IEEE (2021). https://doi.org/10.1109/FOCS52979.2021.00120
5. Desel, J., Esparza, J.: Free Choice Petri Nets, vol. 40. Cambridge Tracts in Theoretical Computer Science. Cambridge University Press (1995)
6. Esparza, J.: Decidability and complexity of Petri net problems-an introduction. In: Advanced Course on Petri Nets, pp. 374–428. Springer, Cham (1996)
7. Esparza, J., Silva, M.: Circuits, handles, bridges and nets. In: Rozenberg, G. (ed.) ICATPN 1989. LNCS, vol. 483, pp. 210–242. Springer, Heidelberg (1991). https://doi.org/10.1007/3-540-53863-1_27
8. Hack, M.H.Th.: Analysis of production schemata by Petri nets. Technical report, Massachussetts Institute of Technology, MAC TR-94, 1974 (based on his MSc thesis, 1972)
9. Karp, R.M., Miller, R.E.: Parallel program schemata. J. Comput. Syst. Sci. 3(2), 147–195 (1969). https://doi.org/10.1016/S0022-0000(69)80011-5
10. Rao Kosaraju, S.: Decidability of reachability in vector addition systems (preliminary version). In: Lewis, H.R., Simons, B.B., Burkhard, W.A., Landweber, L.H. (eds.) Proceedings of the 14th Annual ACM Symposium on Theory of Computing, San Francisco, California, USA, 5–7 May 1982, pp. 267-281. ACM (1982). https://doi.org/10.1145/800070.802201
11. Leroux, J.: The reachability problem for Petri nets is not primitive recursive. In: 62nd IEEE Annual Symposium on Foundations of Computer Science, FOCS 2021, Denver, CO, USA, 7–10 February 2022, pp. 1241–1252 (2021). https://doi.org/10.1109/FOCS52979.2021.00121
12. Mayr, E.W.: An algorithm for the general Petri net reachability problem. SIAM J. Comput. 13(3), 441–460 (1984). https://doi.org/10.1137/0213029
13. Rackoff, C.: The covering and boundedness problems for vector addition systems. Theor. Comput. Sci. 6, 223–231 (1978). https://doi.org/10.1016/0304-3975(78)90036-1
14. Reisig, W.: Petri Nets: An Introduction, vol. 4. EATCS Monographs on Theoretical Computer Science. Springer, Cham (1985). https://doi.org/10.1007/978-3-642-69968-9. ISBN 3-540-13723-8
15. Teruel, E., Chrzastowski-Wachtel, P., Colom, J.M., Suárez, M.S.: On weighted T-systems. In: Application and Theory of Petri Nets 1992, 13th International Conference, Sheffield, UK, 22–26 June 1992, Proceedings, pp. 348–367 (1992). https://doi.org/10.1007/3-540-55676-1_20

16. Teruel, E., Colom, J.M., Suárez, M.S.: Choice-free Petri nets: a model for deterministic concurrent systems with bulk services and arrivals. IEEE Trans. Syst. Man Cybern. Part A **27**(1), 73–83 (1997). https://doi.org/10.1109/3468.553226
17. Teruel, E., Suárez, M.S.: Structure theory of equal conflict systems. Theor. Comput. Sci. **153**(1&2), 271–300 (1996). https://doi.org/10.1016/0304-3975(95)00124-7
18. Wisniewski, R., Karatkevich, A., Adamski, M., Kur, D.: Application of comparability graphs in decomposition of Petri nets. In: 7th International Conference on Human System Interactions, HSI 2014, Costa da Caparica, Portugal, 16–18 June 2014, pp. 216–220 (2014). https://doi.org/10.1109/HSI.2014.6860478

High-Level Message Sequence Charts: Satisfiability and Realizability Revisited

Benedikt Bollig[1]([✉])[iD], Marie Fortin[2][iD], and Paul Gastin[1][iD]

[1] Université Paris-Saclay, CNRS, ENS Paris-Saclay, LMF, Gif-sur-Yvette, France
{bollig,gastin}@lmf.cnrs.fr
[2] Université Paris Cité, CNRS, IRIF, Paris, France
marie.fortin@irif.fr

Abstract. Message sequence charts (MSCs) visually represent interactions in distributed systems that communicate through FIFO channels. High-level MSCs (HMSCs) extend MSCs with choice, concatenation, and iteration, allowing for the specification of complex behaviors. This paper revisits two classical problems for HMSCs: satisfiability and realizability. Satisfiability (also known as reachability or nonemptiness) asks whether there exists a path in the HMSC that gives rise to a valid behavior. Realizability concerns translating HMSCs into communicating finite-state machines to ensure correct system implementations.

While most positive results assume bounded channels, we introduce a class of HMSCs that allows for unbounded channelswhile maintaining effective implementations. On the other hand, we show that the corresponding satisfiability problem is still undecidable.

Keywords: High-level message sequence charts · Communicating finite-state machines · Satisfiability · Realizability · Synthesis

1 Introduction

Message Sequence Charts (MSCs) provide a visual formalism for representing interactions between processes in concurrent or distributed systems that communicate through FIFO channels. An MSC illustrates a potential execution by recording send, receive, and local events along the relative timelines of each process, with each pair of matching send and receive events connected by an arrow. This intuitive representation makes MSCs appealing for specifying the behavior of such systems.

Specification languages that extend MSCs, such as high-level message sequence charts (HMSCs), also known as message sequence graphs (MSGs), have been extensively studied. HMSCs enrich MSCs by introducing constructs like choice, concatenation, and iteration, similar to automata or regular expressions. This extension allows for specifying a set of behaviors, whether desired (positive specification) or prohibited (negative specification), while maintaining a global view of the system. Despite the high-level, declarative view provided

E. Amparore and L. Mikulski (Eds.): PETRI NETS 2025, LNCS 15714, pp. 109–129, 2025.
https://doi.org/10.1007/978-3-031-94634-9_6

by HMSCs, actual system implementations operate at a lower, more granular level. Automata models, mainly communicating finite-state machines (CFMs), are used to abstract these implementations. In these models, transitions are no longer labeled by MSCs but by atomic actions such as send and receive events. The implementation must ensure that these atomic actions align correctly to meet the global specifications.

A significant amount of research has studied the relationships between different specification languages and their translations into CFMs, e.g., [1,2,5–8,10,12,14–16]. While translating specifications directly into CFMs is referred to as the *synthesis problem*, we will refer to the existence of an implementation as *realizability*. Realizability/synthesis are particularly valuable because they ensure implementations that are correct by design. Early results focused on HMSCs as the specification language and communicating finite-state machines as the implementation model, under the constraint that the automata could not carry additional message contents beyond those specified in the HMSC [1,2,15].

When CFMs are allowed to include additional message contents, they become more expressive, leading to results analogous to Büchi and Kleene theorems.[1] These results establish an expressive equivalence between HMSCs, monadic second-order logic, and CFMs, assuming bounded channels. There is, however, an important distinction in the type of channel bounds considered: universally bounded channels require that no execution exceeds a specified bound [12,14]. In contrast, existentially bounded channels allow for executions with an equivalent bounded one, meaning they can be reordered to fit within the bound [8]. Note that these results have interesting connections to the theory of star-connected rational expressions over Mazurkiewicz traces and their implementations via Zielonka automata in shared-memory systems [6,8,14]. Moreover, they have been used as a framework to establish corresponding results in the realm of multiparty session types [13,17]. While research on HMSCs has primarily addressed MSC languages with bounded channels, it was shown in [5] that any first-order definable property, giving access to the "happen-before" relation and without assuming any channel capacity, can be implemented as a CFM.

In this paper, we identify a class of HMSCs that define languages allowing for unbounded channels, extending beyond existentially bounded channels while guaranteeing effective translation into a CFM implementation. Similarly to globally cooperative HMSCs [8,10], this class is characterized by a restriction on the iteration to sets of connected MSCs, inspired by Mazurkiewicz trace theory. Intuitively, this restriction facilitates process communication and prevents the global HMSC specification from enforcing patterns, like an equal number of messages, between groups of processes that do not communicate sufficiently. However, our approach introduces a novel scope for iteration, relying on a graph-based view of MSC connectedness (nodes are events of the MSC and edges are induced by process successor and matched send-receive pairs) rather than a

---

[1] Another variant of the realizability problem permits implementations to include additional communication, allowing for send and receive events not specified initially [7].

traditional communication graph (nodes are processes and edges are induced by possibly unmatched send-receive events). To establish our main result, we translate HMSCs into existential monadic second-order logic and leverage [5, Theorem 4].

It is important to note that, like in [8] over existentially bounded channels, the implementations derived from our approach are inherently non-deterministic. Deterministic machines can only be obtained under universally bounded channels [9,12]. Moreover, for deadlock-free implementations, additional constraints are required (e.g., [3,10]).

Besides the realizability problem, i.e., the problem of translating HMSCs into CFMs, we also address another fundamental problem: *satisfiability*. Satisfiability consists of asking whether a given HMSC defines *some* behavior at all. It is akin to reachability or nonemptiness problems (e.g., in Petri nets or automata). However, it turns out that these problems are undecidable for unbounded channels already under strong restrictions on HMSCs. This also underpins the importance of positive findings concerning the realizability problem.

Outline. Section 2 introduces the fundamental concepts, including (high-level) message sequence charts (HMSCs), communicating finite-state machines, and (existential) monadic second-order logic. Section 3 states our main results for both realizability and satisfiability and highlights how they compare with previous work. The detailed proofs are presented in subsequent sections. The positive realizability result is obtained in several steps. In Sect. 4, we provide closure properties of MSC languages definable in existential monadic second-order logic. In Sect. 5, these closure properties are exploited for an inductive translation of HMSCs into CFMs. The undecidability proofs concerning satisfiability are presented in Sect. 6. Finally, Sect. 7 discusses open problems and potential directions for future research.

2   Preliminary Definitions

We fix a nonempty finite set of *processes* $\mathscr{P}$. We assume P2P communication through *channels* from $Ch = \{(p,q) \in \mathscr{P} \times \mathscr{P} \mid p \neq q\}$. The set of *actions* (or *action types*) of process $p \in \mathscr{P}$ is $Act_p = Send_p \cup Rec_p$ where $Send_p = \{p!q \mid (p,q) \in Ch\}$ is the set of send actions and $Rec_p = \{p?q \mid (q,p) \in Ch\}$ is the set of receive actions. Here, $p!q$ ($p?q$, respectively) denotes that process p sends a message to (receives a message from, respectively) process q. We let $Send = \bigcup_{p \in \mathscr{P}} Send_p$ and $Rec = \bigcup_{p \in \mathscr{P}} Rec_p$. Finally, $Act = Send \cup Rec$ is the set of all actions.

2.1   Message Sequence Charts

The atomic building blocks of HMSCs are MSCs. An MSC denotes a fragment of a communication scenario. As a fragment, it can be *compositional* [11] (also referred to as partial) in the sense that a send or receive event is not necessarily matched by a corresponding communication event. However, such an event may later be matched when combined with other compositional MSCs within an HMSC. The following definition is illustrated by Example 1 below.

Definition 1 (compositional MSC). *Let Msg be a finite set of* messages. *A compositional MSC (cMSC) over $\mathscr{P}$ and Msg is a tuple $M = (E, \leq, \vartriangleleft, \lambda, \mu)$. Here, E is the nonempty, finite or countably infinite, set of events, which is equipped with a partial order $\leq \subseteq E \times E$. Furthermore, $\lambda \colon E \to Act$ is a labeling function, and $\mu \colon E \to Msg$ is a partial function.*

For $p \in \mathscr{P}$, $E_p = \{e \in E \mid \lambda(e) \in Act_p\}$ is the set of events executed by process p. We denote the restriction of $\leq$ to E_p by $\leq_p \subseteq E_p \times E_p$ and require that $(E_p, \leq_p)$ be a total order that is finite or isomorphic to $(\mathbb{N}, \leq)$. By $\to_p$, we denote the direct-successor relation of $\leq_p$, and we let $\to = \bigcup_{p \in \mathscr{P}} \to_p$. The relation $\vartriangleleft \subseteq \leq$ matches send and receive events according to a first-in-first-out (FIFO) policy: (i) for all $(e, f) \in \vartriangleleft$, there is $(p, q) \in Ch$ such that $\lambda(e) = p!q$ and $\lambda(f) = q?p$; (ii) for all $(e, f), (e', f') \in \vartriangleleft$ and $(p, q) \in Ch$ such that $e, e' \in E_p$ and $f, f' \in E_q$, we have $e \leq_p e'$ iff $f \leq_q f'$. The partial order $\leq$ has to be equal to the reflexive transitive closure of $\to \cup \vartriangleleft$.

Note that there may be unmatched send and receive events, which we gather in the set $\mathsf{unm}(M) = \{e \in E \mid$ there is no $f \in E$ such that $e \vartriangleleft f$ or $f \vartriangleleft e\}$. Messages from Msg are used to label unmatched send and receive events[2], which will allow us to selectively match send and receive events when concatenating several cMSCs. We require that $\mathsf{dom}(\mu) = \mathsf{unm}(M)$.

Finally, we require that, for all $(e, f) \in \vartriangleleft$ and $g \in \mathsf{unm}(M)$, $\lambda(g) = \lambda(e)$ implies $e < g$, and $\lambda(g) = \lambda(f)$ implies $g < f$ (which avoids unmatched events that cannot be matched later on). □

We call a cMSC $M = (E, \leq, \vartriangleleft, \lambda, \mu)$ an MSC if $\mathsf{unm}(M) = \emptyset$. We may then omit μ and just write $(E, \leq, \vartriangleleft, \lambda)$. We call M *finite* if E is finite. By cMSC, fcMSC, MSC and fMSC, we denote the sets of compositional MSCs, finite compositional MSCs, MSCs, and finite MSCs, respectively.

Example 1. Figure 1 depicts six finite cMSCs over $\mathscr{P} = \{p, q\}$ and $Msg = \{a\}$. cMSC M_1 has the three events e, e', and f. We have, e.g., $E_p = \{e, e'\}$, $e \leq_p e'$, $e \to e'$, $e \vartriangleleft f$, $\lambda(e) = \lambda(e') = p!q$, and $\mu(e') = a$. Moreover, $\mathsf{unm}(M_1) = \{e'\}$.

Remark 1. All results presented in this paper hold when extending the definitions in two respects: First, we can add finitely many *attributes*, where an attribute may provide additional information about an event. Second, we can include another type of events, *internal* events, which are exclusive to a process and neither emit nor receive messages (such as "enters critical region").

Concatenation. We define concatenation of cMSCs following [8] but taking into account messages. For cMSCs $M_1 = (E^1, \leq^1, \vartriangleleft^1, \lambda^1, \mu^1) \in$ fcMSC and

---

[2] The use of messages from *Msg* to identify suitable communication events is an extension with respect to previous work. It increases the expressive power of HMSCs but does not introduce considerable additional technical complexity.

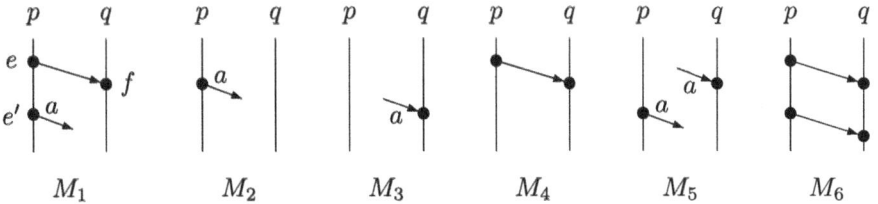

Fig. 1. Example cMSCs.

$M_2 = (E^2, \leq^2, \lhd^2, \lambda^2, \mu^2) \in$ cMSC, the *concatenation* $M_1 \circ M_2 \subseteq$ cMSC is defined as the *set* of cMSCs $M = (E, \leq, \lhd, \lambda)$ such that the following hold:

- $E = E^1 \uplus E^2$,
- $\leq_p = \leq_p^1 \cup \leq_p^2 \cup (E_p^1 \times E_p^2)$ for all $p \in \mathscr{P}$,
- $(\lhd^1 \cup \lhd^2) \subseteq \lhd$ and for all $(e, f) \in \lhd \setminus (\lhd^1 \cup \lhd^2)$, we have $e \in E^1$, $f \in E^2$, and $\mu^1(e) = \mu^2(f)$,
- for all $e \in E^i$ ($i \in \{1, 2\}$), we have $\lambda(e) = \lambda^i(e)$ and, if $e \in$ unm(M), then $\mu(e) = \mu^i(e)$.

In other words, the concatenation of two cMSCs is their "vertical stacking", where one is written below the other while possibly matching send events from the first with receive events from the second.

Example 2. Consider once again Fig. 1. The concatenation $M_2 \circ M_3$ equals $\{M_4, M_5\}$. In M_4, a message edge has been added from the send to the receive event (note that synchronization message a disappears), while in M_5, the edge has not been added. On the other hand, $M_3 \circ M_2$ is the singleton set $\{M_5\}$ (a send event can only be matched with a receive event from a *subsequent* cMSC). The cMSC M_6 is contained in the concatenation $M_1 \circ M_3$. In fact, M_6 is the only cMSC in $M_1 \circ M_3$: leaving the open receive event in M_3 unmatched would result in a structure where this event can never be matched via a concatenation, and which is, therefore, excluded according to Definition 1 (cf. the very last condition).

We extend concatenation to sets $L_1 \subseteq$ fcMSC and $L_2 \subseteq$ cMSC via $L_1 \circ L_2 = \bigcup_{M_1 \in L_1, M_2 \in L_2} M_1 \circ M_2$. Abusing notation, we abbreviate $\{M\} \circ L$ and $L \circ \{M\}$ by $M \circ L$ and $L \circ M$, respectively. Note that concatenation is associative.

Infinite Product. We extend concatenation to infinite sequences. For all $i \geq 1$, let $M_i = (E^i, \leq^i, \lhd^i, \lambda^i, \mu^i) \in$ fcMSC be *finite* cMSCs. The *product* $\prod_{i \geq 1} M_i \subseteq$ cMSC is defined as the *set* of cMSCs $M = (E, \leq, \lhd, \lambda, \mu)$ such that the following hold:

- $E = \biguplus_{i \geq 1} E^i$,
- $\leq_p = \left(\bigcup_{i \geq 1} \leq_p^i\right) \cup \left(\bigcup_{1 \leq i < j} (E_p^i \times E_p^j)\right)$ for all $p \in \mathscr{P}$,
- $\left(\bigcup_{i \geq 1} \lhd^i\right) \subseteq \lhd$ and for all $(e, f) \in \lhd \setminus \left(\bigcup_{i \geq 1} \lhd^i\right)$, we have $e \in E^i$, $f \in E^j$ for some $1 \leq i < j$, and $\mu^i(e) = \mu^j(f)$,

– for all $e \in E^i$ $(i \geq 1)$, we have $\lambda(e) = \lambda^i(e)$ and, if $e \in \mathrm{unm}(M)$, then $\mu(e) = \mu^i(e)$.

The infinite product is extended to languages $(L_i)_{i \geq 1} \subseteq \mathrm{fcMSC}$ as expected. The infinite product is also associative.

2.2   High-Level Message Sequence Charts

To obtain a convenient specification language, cMSCs are combined towards HMSCs using constructs such as choice, concatenation, and iteration. This gives rise to the notion of (compositional) high-level message sequence charts [11].

Definition 2. *A high-level message sequence chart (HMSC) over $\mathscr{P}$ is a tuple $\mathcal{H} = (S, \iota, Msg, R, F, F_\omega)$ where S is a finite set of states, ι is the initial state, Msg is a finite set of messages, $R \subseteq S \times \mathrm{fcMSC} \times S$ is the finite transition relation (recall that fcMSC is the set of finite cMSC over $\mathscr{P}$ and Msg), and $F, F_\omega \subseteq S$ are sets of accepting states (one for finite paths, and one for infinite paths).*

A finite path in $\mathcal{H}$ is a sequence $\rho = s_0 M_1 s_1 \ldots M_n s_n \in S(\mathrm{fcMSC}\, S)^*$ with $n \geq 1$, such that, for all $i \in \{1, \ldots, n\}$, we have $(s_{i-1}, M_i, s_i) \in R$. We say that ρ is a path from s_0 to s_n. Infinite paths are defined similarly. A path is accepting if it starts in the initial state and, (i) if it is finite and ends in an accepting state from F, (ii) if it is infinite and visits an accepting state from F_ω infinitely often. The concatenation of all cMSCs in a path ρ is denoted $cmscs(\rho)$. Recall that $cmscs(\rho)$ is a *set* of cMSCs. The language of $\mathcal{H}$ is the set $L(\mathcal{H}) = \bigcup \{cmscs(\rho) \mid \rho \text{ is an accepting path of } \mathcal{H}\} \cap \mathrm{MSC}$. Note that $L(\mathcal{H})$ contains only MSCs, rather than all cMSCs.

Example 3. Figure 2 illustrates an HMSC $\mathcal{H}$ over $\mathscr{P} = \{p, q\}$. In particular, we have $Msg = \{a, b\}$ and $F = \{5\}$. The figure also depicts an MSC $M \in L(\mathcal{H})$.

It should be noted that, compared to previous definitions, adding messages to HMSCs in terms of Msg increases their expressive power when not restricting to channel-bounded languages. Messages in HMSCs are akin to messages in communicating finite-state machines, as defined below. In fact, with the use of messages, every communicating finite-state machine can be translated into an equivalent HMSC (though we do not delve into the details in this paper), while this is in general not possible without messages [4, Proposition 5.5.1]. Moreover, it is easy to conceive an HMSC that generates a language that is non-regular in a certain sense. Consider a single-state HMSC with $\mathscr{P} = \{p_1, p_2, p_3, p_4\}$, looping over an MSC containing two complete message edges: one from process p_1 to process p_2, and another from process p_3 to process p_4. The resulting language consists of MSCs in which the number of messages from p_1 to p_2 matches the number of messages from p_3 to p_4. This language is not realizable by communicating finite-state machines[3]. Since there is no communication between the

---

[3] In fact, it is not even definable in monadic second-order logic.

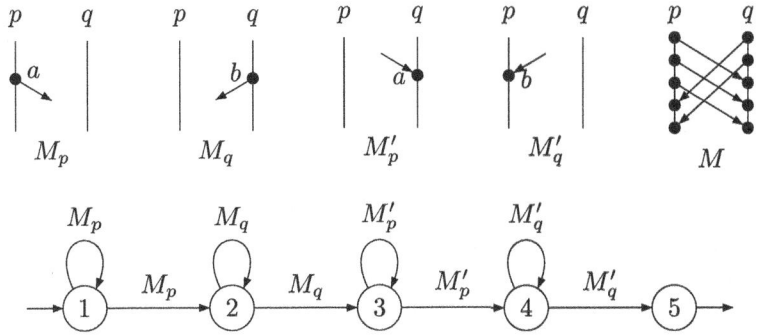

Fig. 2. Example HMSC that is loop-connected but not safe

groups of processes $\{p_1, p_2\}$ and $\{p_3, p_4\}$, any asynchronous implementation of this language would require a counting mechanism to compare the unbounded number of messages.

To define "feasible" classes of HMSCs, we restrict loops to connected cMSCs. We call a cMSC $M = (E, \leq, \lhd, \lambda, \mu)$ *connected* if the undirected graph $(E, \leq \cup \leq^{-1})$ is connected.

Definition 3. *An HMSC is called* loop-connected *if, for all $s \in S$ and all finite paths ρ from s to s, all cMSCs in $cmscs(\rho)$ are connected.*

Example 4. Every cMSC in Fig. 1 is connected, except M_5. The HMSC depicted in Fig. 2 is loop-connected: All loops generate cMSCs where all events are on p or all events are on q. Any such cMSC is connected.

2.3 Communicating Finite-State Machines

Communicating finite-state machines (CFMs) are the operational counterpart to HMSCs and represent a low-level model of distributed systems. While HMSCs offer the possibility to connect specific send and receive events, synchronization in CFMs is solely accomplished via FIFO channels.

We give here an informal account of CFMs, as they are not needed for our technical developments. A formal definition can be found in [5]. Recall that we fixed the set $\mathscr{P}$ of processes. In a CFM, every process $p \in \mathscr{P}$ is represented by a finite-state machine $\mathcal{A}_p$, which allows it to execute actions from Act_p and to send and receive messages from a set Msg. Messages in transit from process p to process q are stored in the (unbounded) FIFO channel (p, q). However, like for HMSCs, they do not occur in (complete) MSCs. In addition, we have simple global acceptance conditions, which take into account finite and/or infinite behaviors. The language of a CFM $\mathcal{A}$ is denoted by $L(\mathcal{A})$.

Example 5. Figure 3 shows a CFM $\mathcal{A}$ over $\mathscr{P} = \{p, q\}$ that is equivalent to the HMSC $\mathcal{H}$ from Fig. 2, i.e., $L(\mathcal{A}) = L(\mathcal{H})$.

Fig. 3. An example CFM

2.4 (Existential) Monadic Second-Order Logic

It has been shown in [5] that CFMs are expressively equivalent to existential monadic second-order logic (EMSO). EMSO formulas over given sets $\mathscr{P}$ of processes and *Msg* of messages are interpreted over cMSCs over $\mathscr{P}$ and *Msg*. They are of the form $\varphi = \exists X_1 \ldots \exists X_k.\psi$. Here, the X_i are existentially quantified second-order variables, interpreted as sets of events of a given cMSC. Moreover, ψ is a first-order formula that (i) can make use of boolean connectives (disjunction, conjunction, and negation), (ii) has access to the process order, direct process-successor relation, and message relation in terms of formulas $x \leq_p y$, $x \rightarrow y$, and $x \lhd y$ (x, y being first-order variables interpreted as events), (iii) can determine the type or message of an event in terms of formulas $p!q(x)$, $p?q(x)$, and $a(x)$ with $a \in Msg$ (which is satisfied if the event representing x is unmatched and carries message a), (iv) uses first-order quantification $\exists x$ and $\forall x$, and (v) can relate first-order and second-order variables using atomic formulas of the form $x \in X_i$.

If φ is a sentence, i.e., no variable is free in φ, we let $L(\varphi)$ denote the set of MSCs $M \in \mathbb{MSC}$ that satisfy φ (written $M \models \varphi$).

Example 6. Below is an example EMSO sentence φ over $\mathscr{P} = \{p, q\}$ (and no messages) such that $L(\varphi) = L(\mathcal{H}) = L(\mathcal{A})$, where $\mathcal{H}$ is the HMSC from Fig. 2 and $\mathcal{A}$ is the CFM from Fig. 3:

$$\varphi = \left(\begin{array}{l} \exists x, y.\big(p?q(x) \land \max(x) \land q?p(y) \land \max(y)\big) \\ \land \,\neg\exists x, y.\big(p?q(x) \land p!q(y) \land x \leq_p y\big) \\ \land \,\neg\exists x, y.\big(q?p(x) \land q!p(y) \land x \leq_q y\big) \end{array} \right)$$

Here, the abbreviation $\max(z) = \neg\exists z'.z \rightarrow z'$ states that event z is maximal on its process. Thus, φ says that a given MSC

- is finite,
- has at least one event of the form $p?q$ (thus, at least one of the form $q!p$),
- has at least one event of the form $q?p$ (thus, at least one of the form $p!q$),
- on both processes, all send events are scheduled before all receive events.

Theorem 1 ([5]). *For every EMSO sentence φ, one can effectively construct a CFM $\mathcal{A}$ such that $L(\varphi) = L(\mathcal{A})$.*

3   Main Results

We consider HMSCs as a specification language and CFMs as a model of implementation. Accordingly, we address two types of questions: *satisfiability* of an HMSC (does a given HMSC produce any behavior?) and *realizability* of an HMSC language as a CFM (sometimes also referred to as *implementability*). While satisfiability is formulated as a decision problem, realizability aims to identify classes of HMSCs that *guarantee* the existence of a corresponding CFM. In this section, we summarize our results for both problems. The proofs are then spread over subsequent sections.

3.1   Realizability of HMSCs

Realizability results identify classes of HMSCs that allow for an effective translation into equivalent CFMs. In [8], it was shown that all *globally cooperative* HMSCs are realizable in that sense. While globally cooperative HMSCs specify existentially bounded behaviors (where each MSC has some linearized execution that does not exceed a given channel capacity [8]), our main result addresses realizability for the class of loop-connected HMSCs, which allows for unbounded behaviors (like the language generated by the HMSC from Fig. 2).

Theorem 2. *Given a loop-connected HMSC* $\mathcal{H}$*, we can effectively construct a CFM* $\mathcal{A}$ *such that* $L(\mathcal{H}) = L(\mathcal{A})$*.*

The proof of Theorem 2 relies on Theorem 1: we only need to prove that any loop-connected HSMC can be translated into an equivalent EMSO formula. We first show in Sect. 4 that EMSO-definable languages of cMSCs are closed under union and concatenation, and in the case where all cMSCs in the language are connected, under iteration. We then use these closure properties together with standard state-elimination techniques in Sect. 5 to construct, from a loop-connected HMSC $\mathcal{H}$, an EMSO formula φ such that $L(\varphi) = L(\mathcal{H})$.

The Case of Bounded Channels. We will now compare Theorem 2 to [8], which shows an analogous result for *globally cooperative* HMSCs over finite cMSCs. Globally cooperative HMSCs are based on the notion of *weakly connected* cMSCs.[4]

A cMSC $M = (E, \leq, \lhd, \lambda, \mu)$ is *weakly connected* if it has a connected (undirected) communication graph. Here, the communication graph of M has $\{p \in \mathcal{P} \mid E_p \neq \emptyset\}$ as set of nodes and it has an (undirected) edge between p and q if, for some $e, f \in E$, we have $\lambda(e) = p!q$ and $\lambda(f) = q?p$ or $\lambda(e) = q!p$ and $\lambda(f) = p?q$. Note that *connected* implies *weakly connected*, but not the other way around. For example, the cMSC M_5 in Fig. 1 is weakly connected, but not connected.

Analogously to the definition of loop-connected, an HMSC is called *weakly loop-connected* if, for all $s \in S$ and all finite paths ρ from s to s, all cMSCs in $cmscs(\rho)$ are weakly connected.

---

[4] The property of being weakly loop-connected is referred to as *loop-connected* in [8].

Let $\mathcal{H} = (S, \iota, Msg, R, F, F_\omega)$ be an HMSC such that $F_\omega = \emptyset$. We call $\mathcal{H}$ *safe* if, for all accepting paths ρ of $\mathcal{H}$, $cmscs(\rho)$ contains an MSC [8]. We call $\mathcal{H}$ *globally cooperative* if it is weakly loop-connected and safe.[5]

Example 7. The HMSC depicted in Fig. 2 is not safe and, therefore, not globally cooperative. In particular, there is no uniform bound on the channel capacity that would allow all behaviors to be scheduled in a way such that no channel exceeds the bound. That is, there is no equivalent HMSC that is globally cooperative.

Theorem 3 (Genest, Kuske, and Muscholl [8]). *Every globally cooperative HMSC $\mathcal{H}$ (with $F_\omega = \emptyset$) can be effectively translated into a CFM $\mathcal{A}$ such that $L(\mathcal{H}) = L(\mathcal{A})$.*

We leave it as an open problem whether, for every globally cooperative HMSC, there is an equivalent loop-connected HMSC.

3.2   Satisfiability of HMSCs

Deciding whether the language of a given HMSC is nonempty turns out to be undecidable, even under the structural restriction of loop-connected HMSCs.

Theorem 4. *The following decision problem is undecidable: Given a loop-connected HMSC $\mathcal{H}$ (with at least two processes), do we have $L(\mathcal{H}) \neq \emptyset$?*

We give two proofs of this result: the first one, in Sect. 6.1, goes through a reduction from Post correspondence problem, and shows that satisfiability is undecidable even for *flat* HMSCs, but uses three processes. The second proof, in Sect. 6.2 is a reduction to the halting problem, and proves that satisfiability is undecidable even with *two* processes – but it involves a non-flat HSMC.

We also show in Sect. 6.3 that restricting to HMSCs where Msg is a singleton set results in an undecidable satisfiability problem (though here we do not assume that the HMSC is loop-connected). Note that this corresponds to the standard definition of HMSCs (cf., for example, [8]).

Theorem 5. *The following decision problem is undecidable: Given an HMSC $\mathcal{H}$ such that Msg is a singleton set, do we have $L(\mathcal{H}) \neq \emptyset$?*

4   Some Closure Properties of EMSO Languages

In this section, we fix a finite set $Msg = \{m_1, \ldots, m_\ell\}$ of messages and we let $n = |\mathscr{P}|$ be the number of processes. We consider (languages of) cMSCs over $\mathscr{P}$ and Msg.

Lemma 1. *EMSO languages of cMSCs are closed under union and concatenation.*

---

[5] The definition can be extended to include infinite cMSCs, but it is more technical.

Proof. Let L_1, L_2 be languages of cMSCs defined by EMSO sentences Φ_1, Φ_2. Without loss of generality, we may assume that Φ_1, Φ_2 use the same *set* variables: $\Phi_1 = \exists X_1 \ldots \exists X_k.\varphi_1$ and $\Phi_2 = \exists X_1 \ldots \exists X_k.\varphi_2$ where φ_1, φ_2 are first-order formulas.

Clearly, the language $L_1 \cup L_2$ is defined by the formula $\exists X_1 \ldots \exists X_k.(\varphi_1 \vee \varphi_2)$.

We show that the concatenation $L_1 \circ L_2$ is defined by a sentence Ψ of the form

$$\Psi = \exists W. \exists Y_{m_1} \ldots \exists Y_{m_\ell}. \exists X_1 \ldots \exists X_k. \psi.$$

The intuition is that variable W identifies the prefix of the composed cMSC and variables $Y_{m_1}, \ldots, Y_{m_\ell}$ are used to guess which messages are "new" in the composition, and what the original message labels were. The first-order part ψ of φ is a conjunction $\psi_1 \wedge \psi_2 \wedge \psi_3 \wedge \psi_4$ checking several conditions. The first requirement in ψ makes sure that W identifies a *finite* (nonempty) prefix of the cMSC:

$$\psi_1 = \exists x_1 \ldots \exists x_n \forall y. \, y \in W \iff \bigvee_{1 \leq i \leq n} y \leq x_i.$$

The second condition says that (only) newly matched send events in W and receive events not in W can be in some Y_m, and the message must be unique.

$$\psi_2 = \forall x \forall y. \, x \lhd y \implies \left(x, y \in W \vee x, y \notin W \vee \bigvee_{m \in Msg} x, y \in Y_m \right)$$

$$\wedge \bigwedge_{m \in Msg} \forall x. \, x \in Y_m \implies \bigwedge_{m' \neq m} x \notin Y_{m'} \wedge \exists y.$$

$$(x \lhd y \wedge x \in W \wedge y \notin W) \vee (y \lhd x \wedge y \in W \wedge x \notin W).$$

The third condition says that the prefix of the cMSC identified by W satisfies Φ_1. To do so, we define a relativisation ξ^W of a first-order formula ξ to elements in W. Message edges with endpoints in Y_m should also be ignored, and the label m added. This is defined inductively as follows:

$$(\exists y. \xi)^W = \exists y. \, y \in W \wedge \xi^W$$
$$m(y)^W = m(y) \vee y \in Y_m \qquad (m \in Msg)$$

The other cases are straightforward:

$$(y \leq z)^W = y \leq z \qquad\qquad\qquad (y \lhd z)^W = y \lhd z$$
$$p(y)^W = p(y) \quad (p \in \mathscr{P}) \qquad\qquad (y \in X)^W = y \in X$$
$$(\xi_1 \vee \xi_2)^W = \xi_1^W \vee \xi_2^W \qquad\qquad\qquad (\neg \xi)^W = \neg \xi^W.$$

With this relativisation, the third formula is $\psi_3 = \varphi_1^W$. We define similarly the relativisation $\xi^{\neg W}$ of a first-order formula ξ to the suffix identified by the complement of W. The last condition is $\psi_4 = \varphi_2^{\neg W}$. $\qquad\square$

Let L be a language of finite cMSCs. We say that L is *connected* if all cMSCs in L are connected.

Lemma 2. *Let L be a language of* finite *cMSCs which is definable in EMSO and* connected. *Then, L^+ and L^ω are definable in EMSO.*

Proof. Assume that L is defined by the sentence $\Phi = \exists X_1 \ldots \exists X_k.\, \varphi$ where φ is a first-order formula. We show that $L^+ \cup L^\omega$ is defined by a sentence of the form

$$\Psi = \exists W.\, \exists Y_{m_1} \ldots \exists Y_{m_\ell}.\, \exists X_1 \ldots \exists X_k.\, \psi.$$

As in the proof of Lemma 1, variables $Y_{m_1}, \ldots, Y_{m_\ell}$ are used to guess which messages are "new" in the composition, and what the original message labels were. The main difference is with variable W since, in an iteration, we may have an arbitrary number of factors and not only a prefix and a suffix. The value of variable W alternates on each process to identify factors of the composition.

More precisely, consider a cMSC $M \in M_1 \circ M_2 \circ \cdots$ with $M_i \in L$ for all $i \geq 1$. The interpretation of the set variables will be as follows:

1. variables $X_1, \ldots, X_k$ are interpreted as the union of the interpretations witnessing the fact that $M_i \models \Phi$ for all i;
2. variable Y_m is interpreted as the set of send or receive events labeled with message m in one of the M_i, and matched in M;
3. variable W is interpreted so that for all processes $p \in \mathscr{P}$, and all $i \geq 1$, either all p-events in M_i are in W or no p-events in M_i are in W. Moreover, if M_i and M_j ($i < j$) are consecutive factors with some p-events (M_k has no p-events for $i < k < j$), then either all p-events in M_i are in W and no p-events in M_j are in W, or the other way around.

The first-order part ψ of Ψ is a conjunction $\psi_1 \wedge \psi_2 \wedge \psi_2' \wedge \psi_3 \wedge \psi_4$ checking several conditions. In the following, we write $x \equiv_W y$ for $(x \in W \iff y \in W)$, and similarly $x \equiv_Y y$ for the formula $\bigwedge_{m \in Msg}(x \in Y_m \iff y \in Y_m)$.

The first requirement in ψ is a simple coherence condition on $Y_{m_1}, \ldots, Y_{m_\ell}$. Only matched send or receive events can be in some Y_m, and the message has to be unique.

$$\psi_1 = \bigwedge_{m \in Msg} \forall x.\, x \in Y_m \implies \exists y.\, (x \lhd y \vee y \lhd x) \wedge y \in Y_m \wedge \bigwedge_{m' \neq m} x \notin Y_{m'}.$$

Notice that under the constraint ψ_1, a matched pair $e \lhd f$ of send/receive events in M are either both in the same Y_m (the matching between them is added in the composition) or neither are in any Y_m (they come from the same component M_i and the matching was already there).

To define the other conjuncts in ψ, we need to introduce a few more notations. The main ingredient is a formula $x \sim y$ meant to be interpreted as x and y being part of the same factor in L. This is where we use the fact that all cMSCs in L are *connected*. Two events come from the same factor if ($\sim_{\mathsf{proc}}$) they are on the same process and in the same W-block, or ($\sim_{\mathsf{msg}}$) they are related by a message which is not added in the composition (which can be checked using the Y_m's), or they are in the transitive closure of these two basic relations. We write $\leq_{\mathsf{proc}} = \bigcup_{p \in \mathscr{P}} \leq_p$ for the partial order restricted to events on the same process.

$$x \sim_{\mathsf{proc}} y = (x \leq_{\mathsf{proc}} y \vee y \leq_{\mathsf{proc}} x) \wedge \forall z.$$

$$(x \leq_{\mathsf{proc}} z \leq_{\mathsf{proc}} y \vee y \leq_{\mathsf{proc}} z \leq_{\mathsf{proc}} x) \implies z \equiv_W x$$

$$x \sim_{\mathsf{msg}} y = (x \lhd y \vee y \lhd x) \wedge \bigwedge_{m \in Msg} \neg(x \in Y_m \vee y \in Y_m)$$

$$x \sim y = \exists x_1, \ldots, x_{2n}. \, x = x_1 \wedge y = x_{2n} \wedge \bigwedge_{1 \leq i < 2n} x_i \sim_{\mathsf{proc}} x_{i+1} \vee x_i \sim_{\mathsf{msg}} x_{i+1}.$$

It is easy to see that $\sim_{\mathsf{proc}}$ is an *equivalence* relation. The relation $\sim$ is clearly *reflexive* and *symmetric*, but without further constraints, it need not be *transitive* (notice that the path chosen between x and y is of length $2n$ where $n = |\mathscr{P}|$ is the number of processes). To ensure transitivity, we request that, if two similar events $x \sim y$ are on the same process, then all events between x and y are in the same W-block, i.e., $x \sim_{\mathsf{proc}} y$ (components are continuous on any given process).

$$\psi_2 = \forall x \forall y. \, (x \sim y \wedge x \leq_{\mathsf{proc}} y) \implies (\forall z. x \leq_{\mathsf{proc}} z \leq_{\mathsf{proc}} y \implies z \equiv_W x).$$

Under the constraint ψ_2, we can check that $\sim$ is transitive, hence it is an equivalence relation. Indeed, assume that $x \sim y \sim z$ and let $x = x_1, \ldots, x_{2n} = y, \ldots, x_{4n-1} = z$ be a witnessing path, i.e., $x_i \sim_{\mathsf{proc}} x_{i+1} \vee x_i \sim_{\mathsf{msg}} x_{i+1}$ for all $1 \leq i < 4n - 1$. Using the reflexivity and transitivity of $\sim_{\mathsf{proc}}$, we can get a (possibly shorter) path alternating between $\sim_{\mathsf{proc}}$ and $\sim_{\mathsf{msg}}$:

$$x = y_1 \sim_{\mathsf{proc}} y_2 \sim_{\mathsf{msg}} y_3 \sim_{\mathsf{proc}} y_4 \cdots \sim_{\mathsf{msg}} y_{2n'-1} \sim_{\mathsf{proc}} y_{2n'} = z.$$

If $n' > n$ then among $y_2, y_4, \ldots, y_{2n+2}$ we find two events y_{2i}, y_{2j} with $i < j$ on the same process. We have $y_{2i} \sim y_{2j-1}$ and y_{2i}, y_{2j-1} on the same process. By ψ_2 we get $y_{2i} \sim_{\mathsf{proc}} y_{2j-1}$. Hence, we may shorten the path by removing $y_{2i}, \ldots, y_{2j-1}$. Repeating this shortening if needed, we end up with a path with $n' \leq n$. In case $n' < n$, since $\sim_{\mathsf{proc}}$ is reflexive, we may extend the path by stuttering the last event $z = y_{2n'} = y_{2n'+1} = \cdots = y_{2n}$. Therefore, $x \sim z$.

We also require that the Y_m variables precisely correspond to messages between different $\sim$-classes:

$$\psi_2' = \forall x \forall y. x \lhd y \implies \left(x \not\sim y \iff \bigvee_{m \in Msg} x \in Y_m \wedge y \in Y_m \right).$$

Now that the equivalence relation $\sim$ allows us to identify events in the same factor, we can state the next constraint: the process order and the message relation should induce a partial order on the set of all components. This is because if $M \in M_1 \circ M_2 \circ \cdots$, process edges or message edges in M which are not already in one of the components must go from some M_i to some M_j with $i < j$. We let

$$[x] \rightsquigarrow [y] = \neg(x \sim y) \wedge \exists x', y'. x \sim x' \wedge y \sim y' \wedge (x' \lhd y' \vee x' <_{\mathsf{proc}} y'),$$

which leads us to the next conjunct in ψ: the relation $\rightsquigarrow$ is acyclic. Any $\rightsquigarrow$-cycle going through the same process more than twice can be shortened, so acyclicity can be expressed as:

$$\psi_3 = \bigwedge_{1 \le r \le 2n} \neg \exists x_0, \dots, x_r.\, x_0 = x_r \wedge \bigwedge_{0 \le i < r} [x_i] \rightsquigarrow [x_{i+1}].$$

Finally, we need to check that every sub-cMSC induced by $\sim$ is indeed in L. To do so, we define a relativisation $\xi^{\sim x}$ of ξ to elements within the $\sim$-equivalence class of some event x. Message edges with endpoints in Y_m should also be ignored, and the label m added. This is defined inductively as follows:

$$(\exists y.\, \xi)^{\sim x} = \exists y.\, y \sim x \wedge \xi^{\sim x}$$
$$m(y)^{\sim x} = m(y) \vee y \in Y_m \qquad (m \in Msg)$$

The other cases are straightforward:

$$(y \le z)^{\sim x} = y \le z \qquad\qquad (y \lhd z)^{\sim x} = y \lhd z$$
$$p(y)^{\sim x} = p(y) \qquad (p \in \mathscr{P}) \qquad\qquad (y \in X)^{\sim x} = y \in X$$
$$(\xi_1 \vee \xi_2)^{\sim x} = \xi_1^{\sim x} \vee \xi_2^{\sim x} \qquad\qquad (\neg \xi)^{\sim x} = \neg \xi^{\sim x}.$$

The last constraint is that every sub-cMSC induced by $\sim$, with the $X_1, \dots, X_n$-labeling inherited from the full cMSC, satisfies φ (and therefore is in L):

$$\psi_4 = \forall x.\, \varphi^{\sim x}.$$

We turn to the proof of correctness, i.e., $L(\Psi) = L^+ \cup L^\omega$.

First, let $M \in M_1 \circ M_2 \circ \cdots$ with $M_i \in L$ for all $i \ge 1$. We show that the cMSC M satisfies ψ when choosing an interpretation of the second-order variables as explained at the beginning of the proof. With the interpretation chosen for Y_m's, it is clear that ψ_1 is satisfied. The crucial part is to verify that, with this interpretation, the relation $x \sim y$ indeed coincide with "x, y are in the same factor" of the composition. First, since MSCs in L are *connected*, it is easy to see that if x, y are in the same (connected) factor M_i then there is a path from x to y of length less than $2n$ using the relation $\sim_{\mathsf{proc}} \cup \sim_{\mathsf{msg}}$. Hence $x \sim y$. Conversely, assume that $x \sim_{\mathsf{proc}} y$. From the property of W, we deduce that x, y must be in the same factor. Now, assume that $x \sim_{\mathsf{msg}} y$. From the property of the Y_m's, we know that the message edge between x and y is not added by the concatenation. Hence, x, y are in the same factor. Therefore, $x \sim y$ implies that x, y are in the same factor. From the property of W, it is now easy to check that ψ_2 is satisfied. The interpretation chosen for Y_m's implies that ψ_2' is also satisfied. By definition of the concatenation, if $x \lhd y \vee x <_{\mathsf{proc}} y$ where x is in M_i and y is in M_j with $i \ne j$ then $i < j$. We deduce that $\rightsquigarrow$ is acyclic and ψ_3 is satisfied. Finally, given the interpretation of the X_i's and Y_m's, the formula ψ_4 is also satisfied. This proves that $L^+ \cup L^\omega \subseteq L(\Psi)$.

Conversely, assume that $M = (E, \le, \lhd, \lambda, \mu)$ satisfies Ψ. Consider an interpretation of the set variables W, Y_m's and X_i's such that ψ is satisfied. By ψ_2,

$\sim$ is an equivalence relation on the set of events. Hence, we find sub-cMSCs $M_1, M_2, \ldots$ of M induced by the equivalence relation $\sim$ and the Y_m's. More precisely, if e is an event in M, then we let $M^e = ([e], \leq, \lhd, \lambda, \mu')$ be the cMSC with $[e] = \{f \mid f \sim e\}$ as set of events, $\leq, \lhd, \lambda$ restricted to $[e]$, and $f \in \mathrm{dom}(\mu')$ if either $f \in [e] \cap \mathrm{dom}(\mu)$ (in which case $\mu'(f) = \mu(f)$) or if $f \in [e] \cap Y_m$ for some $m \in Msg$ (in which case $\mu'(f) = m$). By ψ_1 these cases are exclusive, hence μ' is well-defined. Notice that by ψ_2', if $f \lhd g$ with $f, g \in [e]$ then $f, g \notin \mathrm{dom}(\mu')$. By ψ_3 these sub-cMSCs are partially ordered by $\leadsto$. Hence, without loss of generality we may assume that if $[M_i] \leadsto [M_j]$ then $i < j$. By ψ_4, we have $M_i \in L$ for all $i \geq 1$. One can then verify that $M \in M_1 \circ M_2 \circ \cdots$.

Finally, let finite $= \exists x_1 \cdots \exists x_n \forall y. \bigvee_{i=1}^{n} y \leq x_i$ be a sentence defining finite nonempty cMSCs. We see that L^+ and L^ω are respectively defined by $\Psi \wedge$ finite and $\Psi \wedge \neg$finite.   □

5   Realizability of HMSCs

We now use the closure properties established in the previous section to prove Theorem 2. This is done, essentially, by applying a standard automata-to-regular-expressions translation to HMSCs, and observing that the loop-connected assumption guarantees that iteration is only applied to connected languages.

To that end, we introduce *generalized HMSCs*, which are defined as HMSCs except that transitions are labeled by *languages* of finite cMSCs. More precisely, a generalized HMSC over $\mathscr{P}$ is a tuple $\mathcal{H} = (S, \iota, Msg, R, F, F_\omega)$ where S, ι, Msg, F and F_ω are as before, but with a transition function $R : S \times S \to 2^{\mathrm{fcMSC}}$.

A finite path in $\mathcal{H}$ is a sequence of states $\rho = s_0 s_1 \ldots s_n \in S^+$ with $n \geq 1$. The set $cmscs(\rho)$ is defined as the union of all concatenations $M_1 \circ \cdots \circ M_n$ with $M_i \in R(s_{i-1}, s_i)$ for all $1 \leq i \leq n$. Infinite paths and associated sets $cmscs(\rho)$ are defined similarly. We also write $cmscs_{\mathcal{H}}(s, s') = \bigcup \{cmscs(\rho) \mid \rho$ is a finite path of $\mathcal{H}$ from s to $s'\}$. The language $L(\mathcal{H})$ of a generalized MSC is then defined, as before, as $L(\mathcal{H}) = \bigcup \{cmscs(\rho) \mid \rho$ is an accepting path of $\mathcal{H}\} \cap$ MSC.

A generalized HMSC is called *loop-connected* if for all $s \in S$, all cMSCs in $cmscs_{\mathcal{H}}(s, s)$ are connected. It is called EMSO-definable if for all states s and s', there is an EMSO formula $\varphi_{s,s'}$ such that $R(s, s') = L(\varphi_{s,s'})$.

Note that any (loop-connected) HMSC can be seen as a generalized (loop-connected) HMSC where for all pairs of states (s, s'), $R(s, s')$ is a finite language.

We can now describe a state elimination procedure for (generalized) HMSCs, analogous to state elimination for standard automata. Given a generalized HMSC $\mathcal{H} = (S, \iota, Msg, R, F, F_\omega)$ and a state $s \in S \setminus (F \cup F_\omega \cup \{\iota\})$, we define $\mathcal{H}_s = (S \setminus \{s\}, \iota, Msg, R_s, F, F_\omega)$ as

$$R_s(s_1, s_2) = R(s_1, s_2) \cup \big(R(s_1, s) \circ R(s, s_2)\big) \cup \big(R(s_1, s) \circ R(s, s)^+ \circ R(s, s_2)\big).$$

Lemma 3. *We have $L(\mathcal{H}) = L(\mathcal{H}_s)$. In addition, if $\mathcal{H}$ is loop-connected and EMSO definable, then $\mathcal{H}_s$ is also loop-connected and EMSO definable.*

Proof. It is easy to verify that for all states $s_1, s_2 \in S \setminus \{s\}$, $cmscs_{\mathcal{H}}(s_1, s_2) = cmscs_{\mathcal{H}_s}(s_1, s_2)$. This implies $L(\mathcal{H}) = L(\mathcal{H}_s)$, as well as the fact that if $\mathcal{H}$ is loop-connected, then so is $\mathcal{H}_s$.

Together with Lemmas 1 and 2, this also implies that if $\mathcal{H}$ is both loop-connected and EMSO-definable, then $\mathcal{H}_s$ is as well.   □

To prove Theorem 2, we only need to show that every loop-connected HMSC $\mathcal{H}$ can be translated into an EMSO formula φ with the same language. Without loss of generality, we may assume that $\mathcal{H} = (S, \iota, Msg, R, F, F_\omega)$ is such that $F \cup F_\omega$ is a singleton (as the language of $\mathcal{H}$ is the finite union of all the languages obtained by keeping only one accepting state), and that ι has no incoming transition.

Assume that $F \cup F_\omega = \{s\}$. We successively eliminate all states in $S \setminus \{\iota, s\}$. By Lemma 3, every intermediate generalized HMSC thus obtained is loop-connected and EMSO-definable, and such that $R(s', \iota) = \emptyset$ for all states s', as ι has no incoming transition in $\mathcal{H}$. Let $\mathcal{H}' = (\{\iota, s\}, \iota, Msg, R', F, F_\omega)$ be the HSMC obtained at the end. We then have

$$L(\mathcal{H}') = R'(\iota, s) \cup \left(R'(\iota, s) \circ R'(s, s)^+ \right) \qquad \text{if } s \in F$$
$$L(\mathcal{H}') = R'(\iota, s) \circ R'(s, s)^\omega \qquad \text{if } s \in F_\omega .$$

Since we have EMSO formulas for $R'(\iota, s)$ and $R'(s, s)$, and $R'(s, s)$ is connected, we can construct an EMSO formula φ such that

$$L(\varphi) = L(\mathcal{H}') = L(\mathcal{H}).$$

By Theorem 1, the formula φ can in turn be translated into an equivalent CFM.

6   Satisfiability of HMSCs

In this section, we present the proofs related to satisfiability results. Specifically, for the satisfiability problem of loop-connected HMSCs, we provide two proofs of undecidability. In the first one, we reduce the Post correspondence problem to the satisfiability problem of a loop-connected *flat* HMSC over three processes. By *flat* we mean that it contains only self-loops: if there is a path from s to s' and a path from s' to s then $s = s'$. In the second proof, we reduce the halting problem of a Turing machine on the empty word to the satisfiability problem of a loop-connected HMSC over two processes. Finally, we also reduce the nonemptiness problem of a two-counter machines to the satisfiability problem of HMSCs over four processes and a singleton message set.

6.1   Undecidability for Loop-Connected Flat HMSCs

We start with the reduction from the PCP problem. Consider two finite alphabets A and B and two morphisms $f, g: A^* \rightarrow B^*$. The PCP problem asks

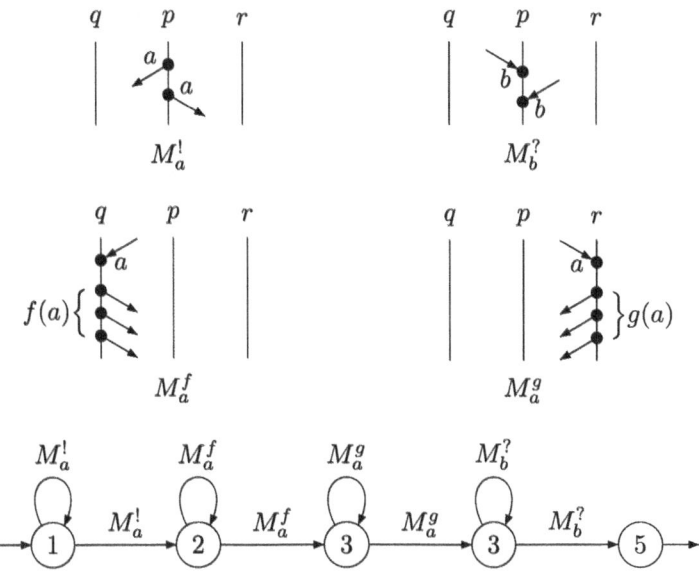

Fig. 4. Connected and flat HMSC used for the reduction from the PCP problem.

whether there is a nonempty word $u \in A^+$ such that $f(u) = g(u)$. Without loss of generality, we assume that $A \cap B = \emptyset$ and $f(A), g(A) \subseteq B^+$.

There are three processes $\mathscr{P} = \{p, q, r\}$ and we will only send messages between p and q and between p and r (no messages are exchanged between q and r). The set of messages is $Msg = A \cup B$. We first define the cMSCs used to label the transitions of the HMSC, see Fig. 4. For each letter $a \in A$, we define the cMSC $M_a^!$ which consists only of two events $e < f$ on process p which are unmatched sends labeled $\lambda(e) = p!q$, $\lambda(f) = p!r$ and $\mu(e) = \mu(f) = a$. Assume that $f(a) = b_1 b_2 \cdots b_k$ with $b_i \in B$ for $1 \le i \le k$. We also define the cMSC M_a^f which consists of $k + 1$ events $e < f_1 < \cdots < f_k$ on process q which are labeled $\lambda(e) = q?p$, $\mu(e) = a$ and $\lambda(f_i) = q!p$, $\mu(f_i) = b_i$ for $1 \le i \le k$. We define similarly the cMSC M_a^g which consists of $1 + |g(a)|$ events on process r, first receiving the message a sent by p and then sending to p the sequence $g(a)$. Finally, for each letter $b \in B$ we have a cMSC $M_b^?$ with two unmatched receives on process p with message b. All these cMSMs are *connected* since in each of them only one process has events.

Now, we define a *flat* and connected HMSC $\mathcal{H}$ corresponding to the following rational expression (see Fig. 4).

$$\left(\sum_{a \in A} M_a^! \right)^+ \cdot \left(\sum_{a \in A} M_a^f \right)^+ \cdot \left(\sum_{a \in A} M_a^g \right)^+ \cdot \left(\sum_{b \in B} M_b^? \right)^+.$$

We show that the PCP problem has a solution if and only if the language of $\mathcal{H}$ is nonempty.

Let $u = a_1 a_2 \cdots a_n \in A^+$ with $a_i \in A$ for $1 \le i \le n$ be a solution of the PCP problem, i.e., $f(u) = g(u) = b_1 b_2 \cdots b_m$ with $b_i \in B$ for $1 \le i \le m$. Consider the accepting path ρ of $\mathcal{H}$ reading the sequence

$$M_{a_1}^! M_{a_2}^! \cdots M_{a_n}^! M_{a_1}^f M_{a_2}^f \cdots M_{a_n}^f M_{a_1}^g M_{a_2}^g \cdots M_{a_n}^g M_{b_1}^? M_{b_2}^? \cdots M_{b_m}^? .$$

It is easy to see that $cmscs(\rho) \cap \mathbb{MSC}$ is nonempty and contains a unique MSC. Hence the flat and connected HMSC $\mathcal{H}$ has a nonemtpy language.

Conversely, assume that $L(\mathcal{H})$ is nonempty. Consider an accepting path ρ of $\mathcal{H}$ such that $cmscs(\rho) \cap \mathbb{MSC} \ne \emptyset$. Fix some $M \in cmscs(\rho) \cap \mathbb{MSC}$. Let $\sigma = \sigma_1 \sigma_2 \sigma_3 \sigma_4$ be the sequence of MSCs read by ρ with $\sigma_1 \in \left(\sum_{a \in A} M_a^! \right)^+$, $\sigma_2 \in \left(\sum_{a \in A} M_a^f \right)^+$, $\sigma_3 \in \left(\sum_{a \in A} M_a^g \right)^+$, and $\sigma_4 \in \left(\sum_{b \in B} M_b^? \right)^+$. Let $u = a_1 a_2 \cdots a_n \in A^+$ be such that $\sigma_1 = M_{a_1}^! M_{a_2}^! \cdots M_{a_n}^!$ and let $v = b_1 b_2 \cdots b_m \in B^+$ be such that $\sigma_4 = M_{b_1}^? M_{b_2}^? \cdots M_{b_m}^?$. Since all messages sent from p to q must be matched in M, we get $\sigma_2 = M_{a_1}^f M_{a_2}^f \cdots M_{a_n}^f$, and since all messages sent back from q to p must be matched in M, we get $v = f(u)$. Similarly, $\sigma_3 = M_{a_1}^g M_{a_2}^g \cdots M_{a_n}^g$ and $v = g(u)$. Therefore, u is a solution of the PCP problem.

6.2   Undecidability for Loop-Connected HMSCs over Two Processes

Next, we move to the reduction from the halting problem of a Turing machine $\mathcal{M}$ on the empty word to the nonemptiness problem of a loop-connected HMSC $\mathcal{H}$ over two processes p and q. Intuitively, process p first guesses an initial configuration $\triangleright s_0 \flat^N$ of $\mathcal{M}$ over the empty word. Here $\triangleright$ is the left endmarker, s_0 is the initial state of $\mathcal{M}$ and $\flat$ is the blank tape symbol. When gessing this initial configuration, process p uses sufficiently many (N) blank symbols for the space needed by the computation until the halting configuration with state s_h is reached. Process p sends its configuration C to process q, which sends back to p the successor C' of C in the computation of $\mathcal{M}$. This is repeated until the halting state is seen by process p, in which case the HMSC $\mathcal{H}$ accepts. An MSC simulating a run of a Turing machine is depicted in Fig. 5.

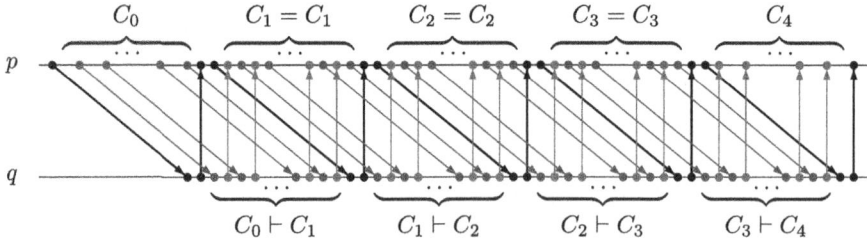

Fig. 5. Simulation of a run of a Turing machine with an MSC.

Formally, let S be the set of states of the Turing machine $\mathcal{M}$ and Γ be the set of tape symbols. We have $s_0, s_h \in S$ and $\triangleright, \flat \in \Gamma$. The transition function of $\mathcal{M}$ is given by a subset $\Delta \subseteq (\Gamma \times S \times \Gamma) \times (S \cup \Gamma)^3$. For instance, $\delta = (\alpha s \beta, s' \alpha \gamma) \in \Delta$ means that in state s when reading β the Turing machine $\mathcal{M}$ goes to state s', replaces β with γ and moves left (a left move is not possible if $\alpha = \triangleright$ is the left endmarker).

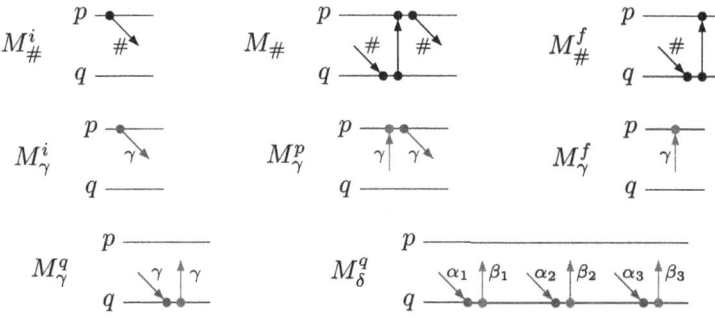

Fig. 6. Basic cMSCs for the simulation of a Turing machine. Here, $\# \notin S \cup \Gamma$ is a new (marker) symbol, $\gamma \in S \cup \Gamma$ and $\delta = (\alpha_1 \alpha_2 \alpha_3, \beta_1 \beta_2 \beta_3) \in \Delta$.

The cMSCs used by the HMSC $\mathcal{H}$ are depicted in Fig. 6. Notice that these cMSCs are all *connected*. We describe the HMSC $\mathcal{H}$ with the following rational expression:

$$\mathsf{Init} = M_\#^i\, M_\triangleright^i\, M_{s_0}^i \left(M_\flat^i\right)^+ \quad \mathsf{Succ} = M_\# \cdot \left(\sum_{\gamma \in \Gamma} M_\gamma^q\right)^* \cdot \left(\sum_{\delta \in \Delta} M_\delta^q\right) \cdot \left(\sum_{\gamma \in \Gamma} M_\gamma^q\right)^*$$

$$\mathsf{Copy} = \left(\sum_{\gamma \in S \cup \Gamma} M_\gamma^p\right)^+ \quad \mathsf{Halt} = \left(\sum_{\gamma \in \Gamma} M_\gamma^f\right)^+ \cdot M_{s_h}^f \cdot \left(\sum_{\gamma \in \Gamma} M_\gamma^f\right)^* \cdot M_\#^f$$

$$\mathcal{H} = \mathsf{Init} \cdot \left(\mathsf{Succ} \cdot \mathsf{Copy}\right)^* \cdot \mathsf{Succ} \cdot \mathsf{Halt}$$

Notice that the HMSC $\mathcal{H}$ is loop-connected. In particular, the loop corresponding to $\left(\mathsf{Succ} \cdot \mathsf{Copy}\right)^*$ is connected since it starts with $M_\#$ which connects the two processes. The MSCs accepted by $\mathcal{H}$ are of the form of Fig. 5 and correspond to halting computations of the Turing machine $\mathcal{M}$ on the empty word, i.e., starting from an initial configuration in $\triangleright s_0 \flat^+$. Therefore, the computation of $\mathcal{M}$ on the empty word halts if and only if $L(\mathcal{H}) \neq \emptyset$. This completes the reduction and the undecidability proof for two processes.

6.3 Undecidability for HMSCs With a Singleton Message Alphabet

We reduce nonemptiness (with counter values 0) in two-counter machines to the corresponding satisfiability problem for HMSCs over $\mathscr{P} = \{p_1, p_1', p_2, p_2'\}$ with the singleton message alphabet $Msg = \{a\}$. As we deal with a singleton set, we henceforth omit any mention of the message. Note that our reduction does not allow us to restrict to loop-connected HMSCs.

Just like an HMSC, a two-counter machine $\mathcal{M}$ has a finite state space with a distinguished initial state and a set of final states (corresponding to F in HMSCs). Moreover, it has two counters, c_1 and c_2, whose values range over $\mathbb{N}$. In a transition of $\mathcal{M}$, any counter can be incremented by one, be decremented by one (provided its current value is positive), or be tested for zero (the transition can only be taken if the current counter value is 0). The nonemptiness problem asks whether we can reach a final states when both counters have value 0.

We construct an HMSC $\mathcal{H}$ such that a final state is reachable in $\mathcal{M}$ iff the language of $\mathcal{H}$ is nonempty. The idea is that processes p_1 and p_1' together simulate counter c_1, and p_2 and p_2' simulate counter c_2. The state-transition structure of $\mathcal{H}$ is exactly the same as that of $\mathcal{M}$. However, instead of incrementing c_1, the HMSC will write a message into the channel (p_1, p_1'), using the cMSC with a single event labeled $p_1!p_1'$. Accordingly, decrementing c_1 corresponds to appending the cMSC whose only event is of type $p_1'?p_1$, thus removing a message from the channel. We proceed analogously for counter c_2. In doing so, the HMSC faithfully simulates the two-counter machine: at any time in an execution, the number of messages in channel (p_1, p_1') corresponds to the value of c_1 in $\mathcal{M}$, and analogously for channel (p_2, p_2') and counter c_2.

It remains to simulate zero tests. A zero test for c_1 is simply replaced, in $\mathcal{H}$, by an MSC containing a single *complete* message from p_1 to p_1'. Such a transition contributes to a run of $\mathcal{H}$ iff, previously, the number of $p_1!p_1'$-events matches the number of $p_1'?p_1$-events, i.e., iff in the simulated run of the two-counter machine, the value of c_1 is 0. The zero test for c_2 is simulated accordingly.

7   Conclusion and Future Work

While we showed that satisfiability problems are undecidable even for restricted versions of HMSCs, we solved the synthesis/realizability problem for the class of loop-connected HMSCs, which allows for unbounded-channel behavior. This realizability result is orthogonal to the known result for globally cooperative HMSCs, which define channel-bounded MSC languages. An interesting open problem is whether, over finite MSCs, loop-connected HMSCs are strictly more expressive than globally cooperative HMSCs:

Problem 1. Can every globally cooperative HMSC $\mathcal{H}$ be translated into a loop-connected HMSC $\mathcal{H}'$ such that $L(\mathcal{H}) = L(\mathcal{H}')$?

Apart from that problem, we leave the following questions for future work:

- Does undecidability of satisfiability hold for loop-connected and flat HMSCs with *two* processes?
- Is there an interesting subclass of (loop-connected or not) HMSCs beyond existentially bounded MSCs with decidable satisfiability/model checking problems?
- What are the CFMs that correspond to loop-connected HMSCs?

References

1. Alur, R., Etessami, K., Yannakakis, M.: Inference of message sequence charts. IEEE Trans. Softw. Eng. **29**(7), 623–633 (2003). https://doi.org/10.1109/TSE. 2003.1214326
2. Alur, R., Etessami, K., Yannakakis, M.: Realizability and verification of MSC graphs. Theor. Comput. Sci. **331**(1), 97–114 (2005). https://doi.org/10.1016/j.tcs. 2004.09.034
3. Baudru, N., Morin, R.: Synthesis of safe message-passing systems. In: Arvind, V., Prasad, S. (eds.) FSTTCS 2007. LNCS, vol. 4855, pp. 277–289. Springer, Heidelberg (2007). https://doi.org/10.1007/978-3-540-77050-3_23
4. Bollig, B.: Automata and logics for message sequence charts. Ph.D. thesis, RWTH Aachen University (2005). https://d-nb.info/975254936
5. Bollig, B., Fortin, M., Gastin, P.: Communicating finite-state machines, first-order logic, and star-free propositional dynamic logic. J. Comput. Syst. Sci. **115**, 22–53 (2021). https://doi.org/10.1016/j.jcss.2020.06.006
6. Diekert, V., Rozenberg, G. (eds.): The Book of Traces. World Scientific (1995). https://doi.org/10.1142/2563
7. Genest, B.: On implementation of global concurrent systems with local asynchronous controllers. In: Abadi, M., de Alfaro, L. (eds.) CONCUR 2005. LNCS, vol. 3653, pp. 443–457. Springer, Heidelberg (2005). https://doi.org/10.1007/ 11539452_34
8. Genest, B., Kuske, D., Muscholl, A.: A Kleene theorem and model checking algorithms for existentially bounded communicating automata. Inf. Comput. **204**(6), 920–956 (2006). https://doi.org/10.1016/j.ic.2006.01.005
9. Genest, B., Kuske, D., Muscholl, A.: On communicating automata with bounded channels. Fundam. Informaticae **80**(1-3), 147–167 (2007). http://content.iospress. com/articles/fundamenta-informaticae/fi80-1-3-09
10. Genest, B., Muscholl, A., Seidl, H., Zeitoun, M.: Infinite-state high-level MSCs: model-checking and realizability. J. Comput. Syst. Sci. **72**(4), 617–647 (2006). https://doi.org/10.1016/j.jcss.2005.09.007
11. Gunter, E.L., Muscholl, A., Peled, D.: Compositional message sequence charts. Int. J. Softw. Tools Technol. Transfer **5**(1), 78–89 (2002). https://doi.org/10.1007/ s10009-002-0085-2
12. Henriksen, J.G., Mukund, M., NarayanKumar, K., Sohoni, M.A., Thiagarajan, P.S.: A theory of regular MSC languages. Inf. Comput. **202**(1), 1–38. https://doi. org/10.1016/j.ic.2004.08.004
13. Honda, K., Yoshida, N., Carbone, M.: Multiparty asynchronous session types. J. ACM **63**(1), 9:1–9:67 (2016). https://doi.org/10.1145/2827695
14. Kuske, D.: Regular sets of infinite message sequence charts. Inf. Comput. **187**(1), 80–109 (2003). https://doi.org/10.1016/S0890-5401(03)00123-8
15. Lohrey, M.: Realizability of high-level message sequence charts: closing the gaps. Theor. Comput. Sci. **309**(1–3), 529–554 (2003). https://doi.org/10.1016/j.tcs.2003. 08.002
16. Morin, R.: Recognizable sets of message sequence charts. In: Alt, H., Ferreira, A. (eds.) STACS 2002. LNCS, vol. 2285, pp. 523–534. Springer, Heidelberg (2002). https://doi.org/10.1007/3-540-45841-7_43
17. Stutz, F., Zufferey, D.: Comparing channel restrictions of communicating state machines, high-level message sequence charts, and multiparty session types. In: GandALF 2022. EPTCS, vol. 370, pp. 194–212 (2022). https://doi.org/10.4204/ EPTCS.370.13

Distributed Reference Net Simulation Based on Event Streaming

Laif-Oke Clasen$^{(\boxtimes)}$, Can Nayci$^{(\boxtimes)}$, and Daniel Moldt$^{(\boxtimes)}$

Faculty of Mathematics, Informatics and Natural Sciences, Department of
Informatics, University of Hamburg, Hamburg, Germany
laif-oke.clasen@uni-hamburg.de, can.nayci@studium.uni-hamburg.de,
Moldt@Informatik.Uni-Hamburg.DE
http://www.paose.de

Abstract. Developing complex systems relies heavily on using models
to analyze components and validate system behavior. Simulation and verification are essential for advancing those systems. Petri nets, a widely
used modeling technique, excel in representing such systems. However,
they face scalability and performance limitations when single computing
platforms confine the simulations.

This paper proposes a novel approach for the distributed simulation
of Reference Nets, overcoming the constraints of single computing platforms. The proposed solution employs distributed synchronization via
synchronous channels and integrates event streaming within computing
clusters. A prototyping methodology grounded in constructivist principles validates the approach.

The primary contribution is a distributed simulator for Reference Nets
as an extension of RENEW. The simulator leverages KAFKAREGISTRY to
facilitate event-driven, loosely coupled communication. This architecture
enhances scalability and robustness, tackling computational challenges in
modern distributed systems.

The proposed solution overcomes bottlenecks of single-platform simulations. It provides a scalable simulator for Reference Nets in distributed
environments, enabling efficient and high-performance system simulation.

Keywords: Petri Nets · Reference Nets · Distributed Systems ·
Event-Streaming · Apache Kafka

1 Introduction

Petri nets are foundational tools in the modeling and analysis of complex systems. Their unique simplicity and expressive power combination have made them
indispensable in fields such as organizational process modeling [30]. Furthermore, their application extends to modern challenges, including the design of

---

Supported by participants of our teaching project classes and many student theses.

E. Amparore and Ł. Mikulski (Eds.): PETRI NETS 2025, LNCS 15714, pp. 130–154, 2025.
https://doi.org/10.1007/978-3-031-94634-9_7

distributed and intelligent systems [21]. Despite their versatility, Petri nets' scalability and computational efficiency remain critical research areas, particularly in the context of distributed environments.

The simulation of Petri nets is fundamental to modeling and analyzing complex systems. However, the computational demands often surpass the capabilities of single platforms, leading to performance bottlenecks. Distributed simulation addresses these challenges by balancing network communication and local computation. Selecting effective communication strategies and technologies is crucial for ensuring scalability and performance.

This research focuses on distributed computing and simulation of Petri nets. It provides scalable solutions for high-performance simulations, enabling efficient simulation of large-scale distributed systems and advancing methodology and implementation.

The central research question is: *How can the concept of distributed and event-driven synchronization through synchronous channels enhance the scalability and robustness of Reference Net simulations in distributed systems?*

To address this research question, a formalism that enables distributed and event-driven synchronization within Petri nets operating in distributed environments is required. The Reference Net formalism, developed by Kummer [36], is the foundation and extended by a distributed and event-driven synchronization via synchronous channels. This formalism, integral to RENEW [37,39], combines the principles of nets-within-nets [59,60], synchronous channels [14], reference semantics, and Java as an inscription language (Sect. 2.4).

This paper's objective is to design and implement a distributed simulator based on this formalism and the extension of the distributed and event-driven synchronization option. The simulator must enable efficient communication between distributed Reference Nets while ensuring robust and scalable execution in a distributed environment.

The hypothesis underpinning this work is that distributed and event-driven synchronization via synchronous channels significantly improves Reference Net simulations' performance by addressing key scalability and robustness limitations. This paper investigates this hypothesis by designing and implementing a distributed simulation framework.

Our approach employs a prototyping methodology [8,45,63] to validate the feasibility and effectiveness of the proposed concepts. This methodology adheres to constructivist principles by leveraging existing research and tools while addressing critical gaps in the current literature. The approach integrates state-of-the-art technologies to implement the proposed concept. This approach enables the practical realization of scalable and robust simulations for Reference Nets in distributed systems.

The paper will be structured as follows, introducing the foundational concepts of the Reference Net formalism, distributed simulation, and the technology stack used in Sect. 2. The objectives of this research are outlined in Sect. 3, followed by the central part. Section 4 describes the distributed system, which is the basis for the distributed simulation environment. The development of

KAFKAREGISTRY and its prototyping phases are presented in Sect. 5. The integration of KAFKAREGISTRY in the RENEWDISTRIBUTE plugin is described in Sect. 6. Our research and produced artifacts will then be evaluated in Sect. 7, related work will be shown in Sect. 8, and the conclusion will be drawn in Sect. 9.

2   Foundations

In this section, we look at the basics of the contribution. We address the topics of event streaming (Sect. 2.1), remote method invocation (Sect. 2.2), synchronous channels (Sect. 2.3), Reference nets (Sect. 2.4) and RENEW (Sect. 2.5).

2.1   Event Streaming

Event streaming denotes the continuous processing of data represented as discrete, immutable events, each with a timestamp and sequence number. These can be stored and reused, enabling efficient analysis and processing.

Apache Kafka is an open-source, distributed event-streaming platform designed for scalability and high performance [23,33]. It is widely used in distributed systems for real-time data processing and transmission.

Kafka's core abstractions are events (records) and topics. Each event includes a key (unique identifier), a value (payload), and metadata such as timestamps [25, p. 23]. Events are assigned to exactly one topic, which serves as a logical channel for related data.

Kafka's architecture comprises brokers, producers, and consumers. Brokers store and distribute events; producers generate and publish events to topics; consumers retrieve and process events. Configurations determine whether all or selected consumers receive specific events, offering flexibility for various processing needs.

A Kafka cluster consists of multiple brokers to ensure scalability and high availability [25, p. 17]. If one broker fails, others take over seamlessly. Partitioning spreads the load across brokers. Figure 1 illustrates this architecture [24, p. 424].

Partitioning improves scalability but reduces availability unless records are replicated. To ensure fault tolerance, each partition must be replicated across at least three nodes, mitigating risks like split-brain and single points of failure. Split-brain occurs when redundant nodes lose synchronization, leading to inconsistent states and potential data loss.

Kafka offers persistence, high-throughput, real-time processing, and support for diverse architectures and languages [25, p. 6f]. Decoupling producers and consumers facilitates loosely coupled system architectures, making it a scalable and robust solution for modern distributed systems when combined with a high-availability setup.

2.2   Remote Method Invocation (RMI)

The concept of Remote Method Invocation (RMI) or Remote Procedure Calls (RPC) describes the invocation of methods or procedures via a network. A con-

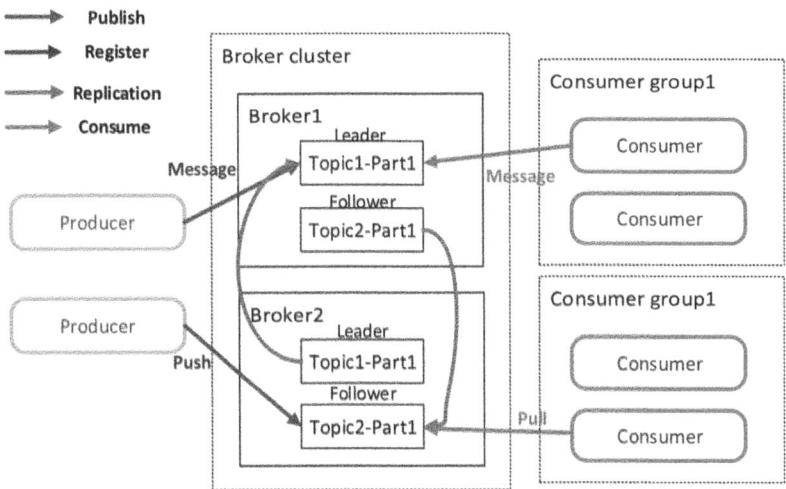

Fig. 1. Infrastructure of a Kafka Cluster (based on Fu et al. [24, p. 424])

crete implementation of this concept is Java's RMI library by allowing method calls on remote objects [64].

The architecture of Java RMI requires a centralized Java RMI Registry operating within its own Java Virtual Machine (JVM). This registry acts as a coordination hub, managing the registration and discovery of distributed services. While effective, the reliance on a centralized registry introduces potential scalability and robustness limitations, particularly in highly dynamic or large-scale environments.

Java RMI employs stubs and skeletons to facilitate remote method calls [2]: Stubs act as proxies for remote objects, enabling local method invocations that are transparently forwarded to the remote object. Skeletons, on the other hand, serve as intermediaries on the server-side, deserializing incoming requests and forwarding them to the real object.

2.3 Synchronous Channels

Synchronous channels in Petri nets enable direct communication between different nets through rendezvous synchronization. This concept was introduced to colored Petri nets by Christensen [14, p. 165]. A synchronous channel is defined by its type, an identifier, and parameters, which collectively determine its function and scope of interaction.

Unlike communication through shared places, synchronous channels establish interactions via interconnected transitions, referred to as uplinks and downlinks [35,36]. Downlinks are active transitions that initiate communication by calling uplinks, which are passive transitions that respond to these calls. This structure facilitates bidirectional communication, allowing parameters to be exchanged between transitions.

An example of this mechanism is depicted in Fig. 2. The uplink named "synchronize" is defined as `:synchronize(a)`, while a corresponding downlink capable of invoking it is specified as `this:synchronize(x)`. The downlink inscription requires a reference to the net where the uplink is defined, and the keyword `this` serves as a self-reference within the same net. This approach provides a robust framework for clear, modular, and scalable communication between Petri nets.

2.4   Reference Nets

Reference Nets, introduced by Olaf Kummer, extend Colored Petri Nets by incorporating advanced modeling concepts such as nets-within-nets [59, 60], synchronous channels [14] (Sect. 2.3), reference semantics, and the use of Java as an inscription language (Sect. 2.4). [36] This integration builds upon the foundational work of Valk on nets-within-nets and Christensen's exploration of synchronous communication mechanisms. Reference Nets enable hierarchical modeling and dynamic token structures while supporting complex system behavior through direct Java integration.

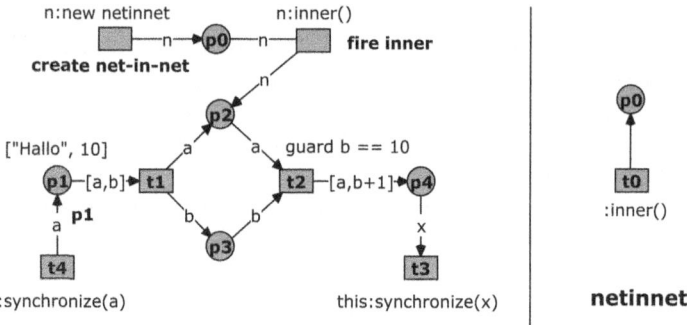

Fig. 2. Two example nets with *Guard*, synchronous channel and nets-within-nets

Figure 2 provides an illustrative example, highlighting their capacity to model intricate system interactions effectively. In the example, the net on the left contains a net-within-net structure, where the inner net displayed on the right is created and fired by a downlink. This hierarchical composition allows for modular and scalable modeling, enabling the encapsulation of complex behavior within individual nets. References from uplink to downlink are established here through the syntax `n:inner()` and `inner()`.

2.5   RENEW

RENEW[1] [37] is an advanced Petri net tool developed by the Algorithms, Randomization, and Theory (ART) research group (formerly Theoretical Foundations of Computer Science, TGI) at the University of Hamburg. Initially

---

[1] **Reference Net Workshop**, available at http://renew.de,.

created for Reference Nets [36], it now supports various formalisms such as Agent Interaction Protocol Diagrams (AIPs) [10–12], Business Process Modelling Notation (BPMN) [29,41], and (distributed) P/T nets with synchronous channels [5,16,58,61,62], making it a versatile platform for modeling and simulation.

RENEW is implemented in Java, leveraging its robustness and platform independence. Its architecture is designed as a plugin system, as described by Duvigneau [20]. Recent updates have enhanced its modularity and maintainability by adopting the Java Platform Module System (JPMS) [17,39].

This modular design enables seamless integration of new plugins. Developers can easily extend RENEW with additional features, including novel net formalisms, advanced simulation algorithms, and enhanced modeling tools. These capabilities make RENEW a flexible and extensible platform, supporting iterative development, experimentation, and adaptation to diverse application domains and evolving research requirements. Its alignment with modern software engineering practices ensures scalability and adaptability for future innovations.

The DISTRIBUTE plugin, developed by Michael Simon [56,57], introduced distributed and method-based synchronization for the Reference Net formalism. It includes a distributed firing algorithm and employs Java's Remote Method Invocation (RMI) for communication between different RENEW's.

The DISTRIBUTE plugin facilitates the distribution of so-called net handles across a network. These handles enable remote access to uplinks on other computers. A downlink referencing a remote uplink is expressed using the syntax handle!link(1,2). In this syntax, the colon, typically used in synchronous channels, is replaced by an exclamation mark. This notation facilitates the straightforward implementation of distributed synchronization.

Figure 3 shows an example of this. As shown in this example, the left net registers itself globally with the *register*-transition and shares its handle with the registry of RMI. When firing the *synchronize*-transition, the right net retrieves the handle from the registry and synchronizes with the left net. In this example, also a String-value is passed to the uplink, which is bound to str="foo".

3   Objectives

This study aims to enhance the scalability and robustness of distributed simulations based on Reference Nets. To this end, a communication medium that inherently supports both properties is required.

Robustness necessitates loosely coupled communication partners to prevent fault propagation. Messages must be processable by multiple consumers and reprocessable by individual ones to increase fault tolerance.

The medium must support concurrent producers and consumers and allow horizontal scaling. Distributed synchronous channels involve multiple message exchanges per firing. Therefore, systems with many such channels require high message throughput.

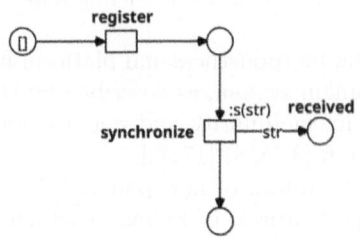 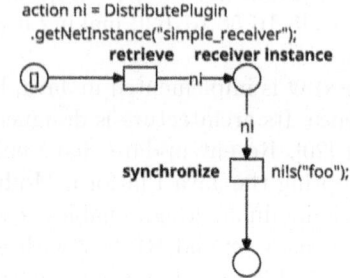

Fig. 3. Two DISTRIBUTE example nets [56].

RMI (Sect. 2.2) lacks scalability due to tightly coupled point-to-point communication. It does not natively support multi-producer or multi-consumer models, limiting robustness, and introduces communication overhead through object-oriented abstractions, reducing throughput.

Message queues are typically centralized and non-partitionable, limiting parallelism and scalability. Robustness is constrained by single-consumption semantics and limited multi-consumer support (e.g., round-robin). Centralization also restricts throughput.

Event streaming (Sect. 2.1) enables parallelism through multi-producer/-consumer support and topic partitioning, improving scalability. Loose coupling and event replay enhance robustness. Kafka, the industry standard, is optimized for processing millions of events per second.

A distributed and event-driven synchronization mechanism based on synchronous channels achieves these properties. The following objectives are defined:

1. Design and implement distributed and event-driven synchronization through synchronous channels for a Reference Net simulator, enabling efficient communication between different simulators.
2. Leveraging a scalable, robust, and event-driven communication medium to support high-performance operations.
3. Conduct distributed simulations within a cluster environment.

These objectives address the limitations of single-platform simulations by introducing mechanisms that ensure robustness and scalability. The proposed solution establishes a reliable foundation for efficient and scalable distributed simulations of Reference Nets.

4   Distributed System

A distributed system addressing the stated objectives (Sect. 3) requires three fundamental components: a distributed environment, an event-based communication medium, and a mechanism for distributed, event-based synchronization of synchronous channels for Reference Nets. Figure 4 depicts the conceptual architecture. This architecture supports multiple simulators and communication media operating concurrently within the distributed environment. Moreover, the system ensures seamless synchronization of synchronous channels through event-based techniques tailored for Reference Nets.

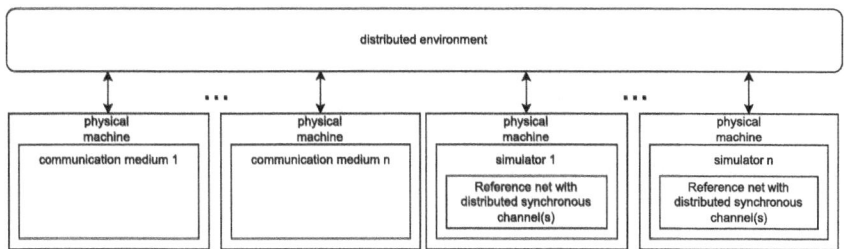

Fig. 4. System for distributed Reference net simulations

Modern container orchestration technologies enable the creation and management of distributed environments. We chose Kubernetes for its robust ecosystem, scalability, and efficient handling of containerized applications. It ensures dynamic deployment and scaling of system components, promoting robustness and flexibility. Kubernetes' inherent capabilities make it suitable for managing complex, distributed systems.

The components of Kafka, such as the broker, will be containerized to enable platform-independency, so they can be easily deployed in a Kubernetes Cluster [6], which is our way of application in the context of our work. Kubernetes [34], the open-source container-orchestration system, is especially suited for the deployment of Kafka Brokers since it offers automatic checks for the health of the broker, may directly restart it and is able to automatically scale the amount of brokers up [6].

The communication medium must meet stringent requirements for robustness, scalability, and efficiency while adhering to an event-driven architecture. Apache Kafka (Sect. 2.1) was selected for this purpose due to its fault-tolerant design and high throughput. Kafka's capabilities as an event streaming platform enable efficient, loosely coupled communication between distributed components. Its architecture aligns with the demands of distributed, event-based systems, ensuring robustness and scalability.

The choice of simulator depends on the specific formalism. For Reference Nets, RENEW (Sect. 2.5) was selected, as it provides exclusive support for this formalism. RENEW's modular plugin architecture facilitates extensions and integrations, making it well-suited for distributed environments. Its design ensures compatibility with the proposed system's objectives.

The existing DISTRIBUTE plugin in RENEW provides a distributed, method-based synchronization of synchronous channels. However, its reliance on Java

RMI (Sect. 2.2) imposes limitations on robustness and scalability. The synchronization algorithm was re-implemented to overcome these constraints using an event-based approach. The newly developed KAFKAREGISTRY plugin replaces Java RMI, leveraging Kafka's event-based architecture to enable distributed synchronization. This enhancement eliminates the bottlenecks associated with RMI, significantly improving scalability and robustness.

The DISTRIBUTE plugin is modified to integrate KAFKAREGISTRY, facilitating seamless event-based communication. Additionally, the existing KAFKA-CLIENT plugin in RENEW, developed by Clasen et al. [16], was utilized to interface with any Kafka broker.

Combining Kubernetes, Apache Kafka, and RENEW creates a robust and scalable distributed system for Reference Nets. The system addresses critical limitations by replacing Java RMI with event-based synchronization via KAFKAREGISTRY, enabling high-performance simulations in distributed environments.

5   KAFKAREGISTRY

This section presents the first prototype of the contribution: KAFKAREGISTRY, which enables RMI functionalities in an event-driven manner. It covers the requirements (Sect. 5.1), specification (Sect. 5.2), design (Sect. 5.3), implementation (Sect. 5.4), and evaluation (Sect. 5.5) of the plugin.

5.1   Requirements

A scalable and robust communication infrastructure is critical to achieving the objectives outlined in Sect. 3. As analyzed in Sect. 4, Kafka meets these requirements by enabling efficient event-driven communication within distributed systems. The KAFKAREGISTRY facilitates seamless interaction with a Kafka broker, ensuring Kafka records' reliable transmission and reception.

The KAFKAREGISTRY implements event-driven Remote Method Invocation (RMI) functionalities, enabling distributed synchronization through Kafka's high performance event-streaming capabilities. This design aligns with the system's architecture, leveraging Kafka's strengths to ensure scalability and robustness.

To maximize performance, the KAFKAREGISTRY retains Kafka's advantages, including fault tolerance, loose coupling, and minimal processing overhead. It avoids introducing bottlenecks or inefficiencies that could undermine Kafka's inherent benefits, ensuring efficient communication across distributed environments.

The KAFKAREGISTRY must seamlessly integrate with RENEW's plugin architecture. This integration enables the KAFKAREGISTRY to function as a core component of the DISTRIBUTE plugin, ensuring compatibility with the broader system. Embedding the KAFKAREGISTRY within RENEW's modular architecture enhances both extensibility and maintainability, adhering to RENEW's established design principles.

The KAFKAREGISTRY establishes a robust foundation for distributed, event-driven synchronization in Reference Net simulations. Its integration with RENEW and alignment with Kafka's strengths provide a scalable, efficient, and reliable solution for modern distributed systems.

5.2   Specification

For an event-based implementation with Kafka, the KAFKAREGISTRY must be able to communicate with Kafka Brokers. Additionally, the Remote Method Invocation (RMI) process encompasses three core functionalities that need to be realized event-driven by the KAFKAREGISTRY. First is the object export, in which the object owner serializes and transmits the object with a global identifier to participants who request it.

Second is the method call, where the recipient sends a serialized method call request, including the global identifier and parameters, to the object owner. The owner executes the method, updates the object state, and returns the result. Finally, the method return, where the object owner encodes the return value and sends it back to the recipient, referencing the global identifier.

The design leverages Kafka's partitioned publish/subscribe model to decouple object state synchronization from method execution logic. By isolating ownership management within topic routing, the architecture achieves loose coupling while preserving consistency through Kafka's ordered event delivery. Scalability emerges from the inherent parallelism of topic partitions, enabling concurrent processing of method invocations across distributed simulator nodes.

In addition, the RENEW simulator will integrate the KAFKAREGISTRY. For this, it is necessary to implement KAFKAREGISTRY as a RENEW plugin.

5.3   Design

The new plugin KAFKAREGISTRY realizes the integration into RENEW. For this purpose, we implement the plugin using two Java modules according to [15, 22], utilizing both an interface and an implementation module. In addition, we use common types across plugin boundaries and extract them into independent ontology modules. Figure 5 shows the plugin's architecture and integration to other plugins within RENEW.

The KAFKAREGISTRY leverages the KAFKACLIENT plugin to enable seamless interaction with the broker. Clasen et al. [16] developed the plugin to facilitate efficient message handling, allowing the production of Kafka records and automatic consumption via parallel threads. This approach decouples production and consumption, ensuring robust event processing. The plugin establishes the connection to the Kafka broker, which each RENEW simulator must predefine during configuration.

We introduce a unique user identifier for each participant to distinguish object owners clearly and enable the transmission of method return values to the method call initiator. Unlike traditional RMI, this approach extends functionality in accessing shared objects. It eliminates the reliance on method proxies, such as RMI stubs, by introducing a novel mechanism where the object accessing a remotely shared object is a complete copy. This copy retains the fields present at the time of export.

Additionally, inspecting remote objects at the receiving end no longer requires implementing getter methods or transmitting remote method calls over the network. This design significantly reduces network overhead and simplifies the interaction with shared objects.

The event design relies on defining distinct events and topics for each RMI functionality. A dedicated event specifically addresses the object export functionality, paired with a unique topic referred to as a class topic. Constructed by combining a prefix with the fully qualified class name of any `Referenceable` object, this topic ensures clarity and organization for network data exchange.

The event key guarantees uniqueness and traceability across the distributed system, which serves as the global identifier of the `Referenceable` object. Within the event value lies the serialized representation of the object, facilitating its efficient transfer and seamless reconstruction in the simulation environment. By adopting this design, a structured and scalable mechanism for exporting objects in distributed Reference Net simulations is achieved.

A structured topic and event schema is the foundation for method call functionality within the RMI paradigm. Each topic integrates a prefix, the class name, and the object's global identifier, ensuring precise and unique addressing. By encapsulating the user identifier, the key provides a mechanism to identify the source of an invocation.

Essential details for method invocation are encapsulated within the event value, a composite data structure. This structure comprises the method signature, name, parameter types, and serialized arguments. The system uses an extended delimiter to serialize and concatenate these components to avoid parsing conflicts or collisions. A long delimiter guarantees unambiguous separation of fields, even for complex payloads. This approach facilitates robust, scalable, and loosely coupled method invocation within the KAFKAREGISTRY.

The event design takes advantage of the method return mechanism offered by the RMI functionality, utilizing the same topic for both method returns and invocation. The event key is the user identifier, and the event's value is the method return value as a serialized object. By integrating this design, return data is seamlessly embedded into the event-driven communication framework, promoting consistency and efficiency in distributed synchronization. This approach facilitates robust, scalable, and loosely coupled method return within the KAFKAREGISTRY.

5.4   Implementation

The KAFKAREGISTRY is implemented in Java [43], leveraging the features of Java 17 to ensure compatibility with modern language enhancements and performance optimizations. The team manages the build process using Gradle [28] 8.4, facilitating dependency resolution, modularization, and reproducible builds.

KAFKAREGISTRY was developed as a dedicated plugin for RENEW, leveraging its modular architecture. This integration ensures seamless compatibility while preserving the extensibility of the RENEW tool.

Integration within RENEW enables seamless reuse of existing plugins, including KAFKACLIENT [5,16,58]. This plugin facilitates interaction with Kafka Brokers by providing essential functionalities for producing and consuming Kafka Records. By leveraging Kafka's publish-subscribe model, KAFKACLIENT enhances system robustness and scalability, ensuring reliable event propagation across computing nodes.

Java requires serialization and deserialization to transmit objects over a network. The implementation employs Java's native serialization mechanism to encode and decode objects, method invocations, and return values into byte streams for transmission via Kafka. Consequently, all inherent limitations and inefficiencies of Java's native serialization persist.

To ensure the reliability of the plugin, unit and integration tests were conducted using JUnit 5 [31] and Mockito [38]. These tests focused on thread safety, serialization accuracy, and Kafka interactions.

We managed version control for the KAFKAREGISTRY integration using Git and GitLab [26]. The repository maintained a structured branching model to ensure code integrity and traceability. Feature branches facilitated the development of isolated enhancements while merging requests enforced systematic peer reviews. Versioned artifacts enabled reproducibility and rollback capabilities, ensuring stability.

Configuration parameters for the plugin, including broker addresses and topic naming conventions, were managed through RENEW's property file system. This setup allowed for modifications without necessitating recompilation.

5.5   Evaluation

The design of KAFKAREGISTRY prioritizes the fundamental advantages of Kafka, including fault tolerance, loose coupling, and minimal processing overhead. These properties are ensured by abstaining from stateful broker management and avoiding additional computational overhead from auxiliary features. As a result, the system maintains Kafka's efficiency while enabling seamless integration into distributed Reference Net simulations.

Integrating with RENEW ensures seamless interoperability with the DISTRIBUTE and KAFKACLIENT plugins. This integration facilitates direct communication with Kafka brokers while leveraging the DISTRIBUTE algorithm for distributed synchronization.

Additionally, the implementation of KAFKAREGISTRY preserves an event-based variant of the three fundamental Remote Method Invocation (RMI) functionalities: object export, method call, and return. This realization makes it possible to integrate the event-driven RMI functionalities into the algorithm for the distributed synchronization of Reference nets in the DISTRIBUTE plugin.

The prototype successfully meets the defined requirements. The KAFKAREGISTRY plugin provides event-driven RMI functionalities. Whereas the DIS-

TRIBUTE plugin implements the core algorithm for distributed synchronization of Reference Nets. The system leverages event-driven communication by combining the plugins KAFKAREGISTRY and DISTRIBUTE to enable distributed and event-driven synchronization through synchronous channels for Reference Nets.

This approach ensures robustness and improves scalability through event streaming. Integrating both plugins is crucial to achieving a high-performance distributed simulation for Reference Nets.

6   Integration of KAFKAREGISTRY and DISTRIBUTE

In this section, we address the second prototype of the contribution. This prototype involves integrating the KAFKAREGISTRY with the DISTRIBUTE. The DISTRIBUTE provides an algorithm for the distributed synchronization of distributed synchronous channels for reference networks. However, the algorithm is based on Java RMI functionalities, which will be replaced by event-based RMI functionalities during the integration. To this end, this section addresses the requirements (Sect. 6.1), specifications (Sect. 6.2), design (Sect. 6.3), implementation (Sect. 6.4), and evaluation (Sect. 6.5) of this integration.

6.1   Requirements

Integrating KAFKAREGISTRY and DISTRIBUTE requires replacing method-based communication with an event-driven paradigm, representing the prototype's core requirement. DISTRIBUTE realizes this transformation by employing the RMI functionalities of KAFKAREGISTRY to achieve distributed synchronization of synchronous channels in Reference Nets.

Additionally, the technical realization of the firing mechanism in the DISTRIBUTE is optimized to minimize execution time and reduce network traffic. Optimizations focus on streamlining token processing and minimizing synchronization overhead.

The integration of KAFKAREGISTRY with DISTRIBUTE must preserve the syntax of Reference nets to maintain backward compatibility. Any modifications must align with the syntactic and behavioral consistency defined in [57], preventing disruptions or the need for adaptation.

6.2   Specification

To enable event-driven communication in the DISTRIBUTE plugin, its Java RMI dependencies must be replaced with KAFKAREGISTRY functionalities. This transition necessitates the complete removal of RMI-based interactions. Adopting this event-driven approach enhances scalability and robustness, mitigating the constraints imposed by method-based communication.

Additionally, we must restructure the network access mechanism for transmitted objects. Traditionally, Remote Method Invocation (RMI) uses a dedicated registry for the remote objects, introducing an intermediary access layer. We must remove this layer to improve efficiency and reduce complexity.

The KAFKAREGISTRY plugin will manage all remote object references in its own registry, aligning with standard RMI behavior. This shift replaces method-based remote communication with an event-driven model. Additionally, build configurations need to be adapted to eliminate any residual components.

The algorithmic design of the firing mechanism must eliminate redundant remote queries for various identifier types. These identifiers remain constant and are shared when exporting remote objects. Efficient identifier management strengthens scalability by preventing unnecessary synchronization delays.

The syntax must remain unchanged to ensure backward compatibility. The event-driven variant adopts the same syntax as the Java RMI-based implementation. This consistency preserves existing specifications and prevents compatibility issues.

6.3   Design

Replacing Java RMI with KAFKAREGISTRY necessitates eliminating Java dependencies and removing the centralized Java RMI registry. This transformation requires a fundamental shift in the system architecture from a centralized coordination model to a decentralized, event-driven paradigm. Based on this integration and the implementation of the KAFKAREGISTRY plugin, we create the section of the module architecture in Fig. 5 for the KAFKAREGISTRY plugin. Instead of maintaining a static registry, shared objects propagate dynamically through Kafka topics.

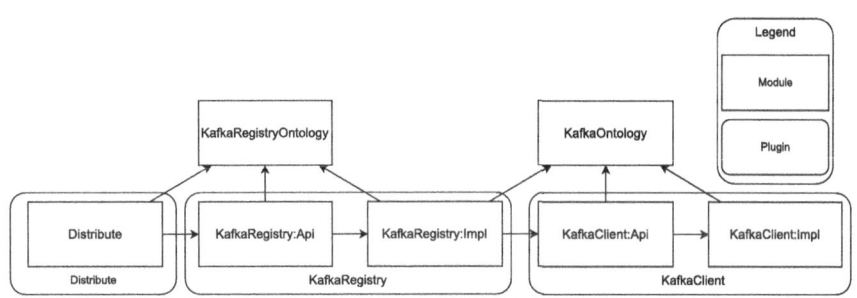

Fig. 5. Module Architecture of the KAFKAREGISTRY Plugin

Object ownership dictates method invocation semantics. Every exported object receives a globally unique identifier, enabling precise routing of requests to the simulator instance responsible for that object. This design ensures that synchronization remains consistent while eliminating the bottlenecks of centralized coordination. By leveraging event-driven communication, the system achieves higher scalability and robustness, addressing the limitations of conventional method-based approaches.

Minimizing network overhead requires embedding critical metadata, such as hash codes and simulation-specific identifiers, directly within objects at the point of initial distribution. This design ensures that remote instances can execute fundamental operations—including equivalence verification and hash map lookups with locally cached data. By eliminating unnecessary remote queries, the system significantly reduces synchronization latency.

Integrating KAFKAREGISTRY and DISTRIBUTE ensure backward compatibility by preserving the established syntax of Reference Nets. The implementation of distributed synchronous channels relies on the DISTRIBUTE plugin, defining the syntax *net-name!channel-name(parameters)* for downlinks and *:channel-name(parameters)* for uplinks. This approach maintains the structural integrity of existing models while enabling seamless migration to distributed execution. By aligning with established syntactic conventions, the solution prevents disruptions in model behavior and preserves compatibility with prior implementations.

6.4   Implementation

Integrating the KAFKAREGISTRY and the DISTRIBUTE plugin replaced RMI dependencies with Apache Kafka-based communication. The `Referenceable` interface and `RemoteReference` class from KAFKAREGISTRY supplanted RMI's `Remote` and stubs. KAFKAREGISTRY handled the caching of shared remote objects, functioning as a registry, eliminating RMI's centralized registry approach. RMI libraries were removed from the codebase to transition to event-driven communication. Firing mechanisms were optimized by locally caching remote identifiers (e.g., hash codes) after initial object export.

Also, the firing mechanism was optimized by caching remote hash codes and simulation identifiers locally after the initial object export. This optimization was done by saving hashcodes and ephemeral inner-simulation identifiers as fields in the object, allowing direct access without remote method calls.

6.5   Evaluation

The evaluation focused on validating the integration of KAFKAREGISTRY and Distribute within the distributed simulation environment. Implementing integration tests ensured the correctness of remote procedure calls, the preservation of net formalisms, and the accuracy of the firing algorithm. Each test examined whether the synchronization mechanisms operated as intended under varying computational loads.

A Dockerfile facilitated the construction of the container image and its deployment to a private registry. The multistage build process optimized the container for execution in a Kubernetes cluster, minimizing dependencies and improving resource efficiency. A Kafka broker was provisioned within this deployment to enable event-driven communication between distributed components.

The experimental setup verified that synchronous channels effectively coordinated transitions across distributed nodes. The event streaming mechanism, implemented via KAFKAREGISTRY, ensured consistent state propagation while maintaining net invariants. The integration of DISTRIBUTE enabled the seamless execution of Reference Net simulations across computational boundaries. The results confirmed that the proposed architecture supports scalable and robust simulation, mitigating the limitations of single-platform execution.

In the following Fig. 6, the virtualized graphical interfaces can be seen side by side, which reside on different nodes in the cluster:

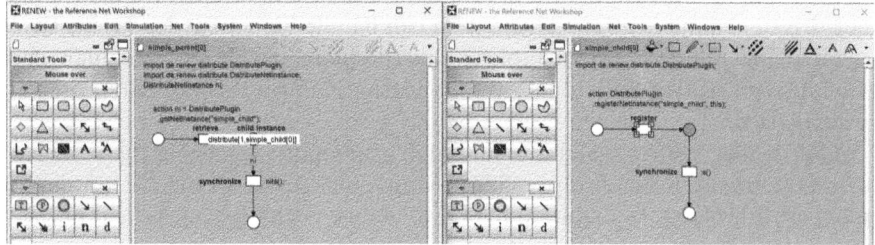

Fig. 6. Deployment in our Kubernetes cluster with sample nets from Simon [56].

As the figure visualizes, sharing the net handle through the KAFKAREGISTRY plugin and the Kafka Broker works as intended. The sharing verifies that the update of the DISTRIBUTE plugin successfully passes the evaluation.

7   Discussion

We can fulfill all three objectives by evaluating the integration of KAFKAREG-ISTRY and DISTRIBUTE (Sect. 6.5) in our Kubernetes cluster. With the example, a distributed and event-driven synchronization could be performed through synchronous channels for reference networks in a distributed environment. We also used Kafka as a scalable, robust, and event-driven communication medium.

The prototypes presented in this study confirm the research question by demonstrating that event-driven synchronization enhances the scalability and robustness of distributed Reference Net simulations. Previously, a centralized Java RMI registry imposed significant communication medium limitations, restricting scalability and fault tolerance. The proposed solution overcomes these constraints by enabling distributed and event-driven synchronization through synchronous channels. This architectural shift ensures efficient communication, reduces bottlenecks and improves resilience in distributed environments. By decoupling system components and leveraging event streaming, the new approach supports high-performance simulations while maintaining the model's structural integrity.

The proposed distributed system supports the integration of any number of simulators. This flexibility enables scalable simulations, provided the subnets' interfaces are defined using distributed synchronous channels. These channels act as connectors between subnets, ensuring proper communication and coordination.

Defining interfaces through synchronous channels supports collaborative modeling. During the modeling phase, teams need only to predefine the interfaces. This approach enables different teams to model their respective subnets independently while ensuring seamless integration during simulation.

A key strength of the proposed solution lies in its use of Kafka as the communication medium. Kafka provides a robust, scalable, and efficient mechanism for event-driven communication. The system can handle high event loads effectively by employing multiple Kafka brokers. Furthermore, Kafka ensures high data availability, tolerating individual communication failures without compromising the simulation.

While Kafka ensures scalability and high availability, achieving these benefits may require many Kafka brokers. This requirement is due to the partitioning of topics across brokers. To maintain high availability, each topic must be replicated at least three times. Partitioning further increases the replication requirements, which can result in a significant number of Kafka brokers being needed.

Distributed synchronization through synchronous channels requires network communication. This synchronization introduces latency compared to local synchronization within a single simulator. Consequently, distributed synchronous channels are inherently slower expressive than their local counterparts. The number of distributed synchronous channels used in a Reference Net should be minimized to mitigate this impact.

The distribution of subnets across simulators must be defined before the simulation begins. This static assignment can lead to uneven workload distribution, where some simulators experience minimal load while others are overloaded. Load balancing of simulations could address this by executing a large number of net instances on each machine. Nevertheless, careful planning is required to optimize the distribution of subnets.

Due to the complexity of general synchronous channels of Reference Nets the expressivity of the channels used for remote synchronization is restricted. However, modeling complex systems follow the idea of loose coupling. Therefore, the channels are usually small and of low complexity already due to appropriate modeling.

The current implementation of distributed synchronous channels assumes a homogeneous simulator environment. This environment is necessary because objects exchanged through these channels require consistent class definitions across simulators. No mechanism exists to dynamically synchronize class definitions between simulators, representing a limitation and an area for future research. A further limitation of the event-based synchronous channels is that only serializable objects can be used for these channels.

The simulators used in the proposed approach are not resilient. If a simulator fails, the simulation cannot recover and must be restarted. Achieving resilience requires storing partial states of simulators. This storing would enable failed simulators to resume from their last saved state. The development of such mechanisms is an avenue for future research.

The proposed distributed simulation for Reference Nets offers significant advantages in scalability, robustness, and collaborative modeling. However, it also introduces challenges such as performance trade-offs, increased infrastructure requirements, and limitations in resilience and heterogeneity. Addressing these challenges in future research will further enhance the applicability and efficiency of distributed Reference Net simulations.

8 Related Work

Several studies explore distributed Petri net simulation but lack practical simulators for any custom net file. Instead, these studies often rely on hardcoded nets to evaluate performance. Examples include Moutinho et al. [42], Gjesdal [27], Bell et al. [7], Buetler et al. [9], and Djemame [19]. While valuable for comparative studies, these approaches are not directly applicable to flexible and scalable Petri nets simulation.

Other approaches introduce custom simulators but depend on global synchronizers. These methods often involve hosting all globally shared places or acting as a global scheduler for transition firing. Examples include Chiola et al. [13], Apaydin-Özkan et al. [4], and Apaydin [3]. Such methods impose centralization, which fundamentally differs from the decentralized solution presented in this paper.

Some simulators operate on local, per-thread simulation sharing to achieve parallelism but lack network-level distribution. These simulators share the same process memory, allowing complete access to net data. Examples include the simulators by Pommereau et al. [46] and Gjesdal [27]. Pre-calculated and pre-split nets, as seen in Knoke et al. [32], also fall into this category. These methods differ from our approach, which enables concurrency across distributed nodes rather than threads.

In Colom [18], distributed P/T nets are simulated. Apart from the different net classes, one of the main differences between our approaches is that Colom [18] does not use places. Instead, a so-called enabling function is used, which is formed from the preconditions and the marking of the net.

Also, in Schmidt [55], distributed simulation of Petri net based Agents is presented with the use of ActiveMQ [1]. This is a different approach than ours since ActiveMQ is a message-queue system built on the Java Message Service (JMS) API [44], which is a different technology than Kafka and more similar to RMI. The fundamental difference is here the message type, where Kafka is an event broker while ActiveMQ is a message broker.

The work by Clasen et al. [16] provides detailed insights into distributed Petri net simulations. Their approach employs event-based communication to synchronize distributed synchronous channels. Specifically, Kafka is used as an event broker, demonstrating its effectiveness in high-throughput, loosely coupled systems.

However, their focus belongs to Place/Transition (P/T) nets, not Reference Nets, and their synchronization method is centralized. They utilize a synchronization service based on a unification strategy to coordinate synchronous channels. This synchronization differs from the distributed algorithm proposed by Simon [56,57], which is implemented in this paper through event-driven and distributed synchronization of synchronous channels.

Simon et al. [56,57] introduce distributed algorithms for Reference Net simulation. While their work shares similarities in synchronization algorithms, the communication strategies and execution environments differ. Simon et al. rely on remote procedure calls (RPCs) for synchronization. While effective, RPCs limit scalability and fault tolerance due to their synchronous, tightly coupled nature.

In contrast, this paper employs event-driven communication, decoupling components, and enhancing scalability and robustness. Additionally, Simon et al. do not utilize containerized orchestration. Our Kubernetes-based deployment enables dynamic scaling, fault tolerance, and optimized resource allocation for distributed computation.

Research by Moldt et al. [40] and Röwekamp et al. [48–54] focuses on distributed Reference Net simulations with platform management as the primary concern. These studies incorporate MULAN agent concepts [47] and use Spring Boot for basic experiments. While these efforts provide foundational insights, they fail to exploit distributed architectures' advantages fully. Consequently, their applicability to complex real-world scenarios is limited.

Recent research addresses these gaps by leveraging Kubernetes orchestration for Reference Nets. This approach improves scalability and robustness, meeting the demands of distributed simulation environments.

The reviewed works highlight the evolution of distributed Petri net simulations. While prior research provides essential foundations, significant gaps remain, particularly in the scalability and robustness of Reference Nets. This paper advances the field by addressing these challenges through event-driven synchronization and distributed architectures. The proposed approach offers a scalable and robust solution that overcomes the limitations of centralized methods, enabling efficient and reliable distributed simulations.

9   Conclusion

9.1   Summary

After introducing the problem in Sect. 1, we explained the foundations to understand event streaming, RMI, synchronous channels, Reference nets, and the tool RENEW in Sect. 2. Then, the objectives in Sect. 3 were defined, which were to design and implement a distributed and event-driven synchronization through synchronous channels for Reference nets, a scalable, robust, and event-driven communication medium, and a distributed simulation within a cluster environment.

Subsequently, Sect. 4 describes the intended distributed system. This system has many communication media and simulators in a distributed environment. Specifically, the system used Kafka as the communication medium, RENEW as the simulator, and Kubernetes as the distributed environment.

This contribution's first prototype in Sect. 5 deals with the KAFKAREGISTRY plugin. This plugin recreates the method-based Java RMI functionalities in an event-driven way. This realization makes it possible to reuse the algorithm of the DISTRIBUTE plugin, which uses the Java RMI functionalities.

However, the KAFKAREGISTRY and the DISTRIBUTE plugin first had to be integrated with the second prototype in Sect. 6. In particular, this replaced the centralized Java RMI registry with a decentralized event-driven variant. We conducted a distributed and event-driven simulation for evaluation purposes within our Kubernetes cluster. Overall the outdated RMI is now replaced by the more modern and better Kafka implementation offering way more modeling and execution options.

Finally, we held the discussion in Sect. 7 on the objectives, research question, and the advantages, disadvantages, and limitations of the concept presented here, and we introduced the related work in Sect. 8.

9.2   Future Work

The KAFKAREGISTRY plugin could support asynchronously executed methods, since currently the method call is blocking until the return value is received. This was omitted here to keep the complexity low, but it could be a valuable addition in the future.

Part of further research will also be to use the synchronization presented here for heterogeneous simulators. For this, we must create an exchange mechanism of shared knowledge so that the simulators can learn new types at runtime.

Furthermore, certain factors have so far limited the resilience of the simulators. Further research will focus on removing these limitations.

In addition, it is currently necessary to define the distribution of the subnetworks to simulators in advance. This predefinition means that the system does not optimally utilize the available resources, as the distribution can be poor, causing some simulators to carry a low load while others carry a high load. Part of further research is to make this distribution resource-driven at runtime.

References

1. Apache ActiveMQ: Apache ActiveMQ Documentation (2025). https://activemq. apache.org/components/classic/documentation/. Accessed 15 Jan 2025
2. Alt, M., Gorlatch, S.: Adapting Java RMI for grid computing. Futur. Gener. Comput. Syst. **21**(5), 699–707 (2005)
3. Apaydin-Özkan, H., Mahulea, C., Júlvez, J., Silva, M.: A control method for distributed continuous mono-T-semiflow Petri nets. Int. J. Control **87**(2), 223–234 (2014)
4. Apaydin-Özkan, H., Júlvez, J., Mahulea, C., Silva, M.: A control method for timed distributed continuous Petri nets. In: Proceedings of the 2010 American Control Conference (ACC), pp. 2593–2600 (2010). https://doi.org/10.1109/ACC.2010. 5530546

5. Bartelt, S.: Prototypische Umsetzung eines Simulators als Komponente von verteil-ter Simulation von P/T Netzen in Renew. Bachelor thesis, University of Hamburg, Department of Informatics, Vogt-Kölln Str. 30, D-22527 Hamburg (2024)
6. Baur, A.: Packaging of kubernetes applications. In: Proceedings of the 2020 OMI Seminars (PROMIS 2020), vol. 1, p. 1. Universität Ulm (2021)
7. Bell, A., Haverkort, B.R.: Sequential and distributed model checking of Petri nets. Int. J. Softw. Tools Technol. Transfer **7**, 43–60 (2005)
8. Budde, R., Kautz, K., Kuhlenkamp, K., Züllighoven, H.: What is prototyping? Inf. Technol. People **6**(2/3), 89–95 (1990)
9. Bütler, B., Esser, R., Mattmann, R.: A distributed simulator for high order Petri nets. In: Rozenberg, G. (ed.) ICATPN 1989. LNCS, vol. 483, pp. 47–63. Springer, Heidelberg (1991). https://doi.org/10.1007/3-540-53863-1_20
10. Cabac, L.: Modeling Agent Interaction Protocols with AUML Diagrams and Petri Nets. Diploma thesis, University of Hamburg, Department of Computer Science, Vogt-Kölln Str. 30, D-22527 Hamburg (2003)
11. Cabac, L.: Modeling Petri Net-Based Multi-Agent Applications. Dissertation, University of Hamburg, Department of Informatics, Vogt-Kölln Str. 30, D-22527 Hamburg (2010). https://ediss.sub.uni-hamburg.de/handle/ediss/3691
12. Cabac, L., Moldt, D.: Formal semantics for AUML agent interaction protocol diagrams. In: Odell, J., Giorgini, P., Müller, J.P. (eds.) AOSE 2004. LNCS, vol. 3382, pp. 47–61. Springer, Heidelberg (2005). https://doi.org/10.1007/978-3-540-30578-1_4
13. Chiola, G., Ferscha, A.: Distributed simulation of Petri nets. IEEE Parallel Distrib. Technol. **1**(3), 33–50 (1993)
14. Christensen, S., Hansen, N.D.: Coloured Petri nets extended with channels for synchronous communication. Technical report DAIMI PB-390, Aarhus University (1992)
15. Clasen, L.: Untersuchung von Softwarearchitektur und Projektorganisation unter Berücksichtigung der Wechselwirkungen im Kontext von Lehrprojekten am Beispiel verteilter Entwicklung der Open-Source Software Renew. Master thesis, University of Hamburg, Department of Informatics, Vogt-Kölln Str. 30, D-22527 Hamburg (2023)
16. Clasen, L., Bartelt, S., Stahl, Y., Moldt, D.: Distributed P/T net simulation prototypes based on event streaming. In: Köhler-Bußmeier, M., Moldt, D., Rölke, H. (eds.) Proceedings of the International Workshop on Petri Nets and Software Engineering 2024 co-located with the 45th International Conference on Application and Theory of Petri Nets and Concurrency (PETRI NETS 2024), 24–25 June 2024, Geneva, Switzerland. CEUR Workshop Proceedings, vol. 3730, pp. 192–216. CEUR-WS.org (2024). https://ceur-ws.org/Vol-3730
17. Clasen, L., Moldt, D., Hansson, M., Willrodt, S., Voß, L.: Enhancement of renew to version 4.0 using JPMS. In: Köhler-Bußmeier, M., Moldt, D., Rölke, H. (eds.) Proceedings of the International Workshop on Petri Nets and Software Engineering 2022 co-located with the 43rd International Conference on Application and Theory of Petri Nets and Concurrency (PETRI NETS 2022), Bergen, Norway, June 20th, 2022. CEUR Workshop Proceedings, vol. 3170, pp. 165–176. CEUR-WS.org (2022). https://ceur-ws.org/Vol-3170
18. Colom, J.M.: Harnessing structure theory of Petri nets in discrete event system simulation. In: Kristensen, L.M., van der Werf, J.M.E.M. (eds.) Application and Theory of Petri Nets and Concurrency - 45th International Conference, Petri NETS

2024, Geneva, Switzerland, 26–28 June 2024, Proceedings. Lecture Notes in Computer Science, vol. 14628, pp. 3–23. Springer (2024). https://doi.org/10.1007/978-3-031-61433-0_1

19. Djemame, K.: Distributed simulation of high-level algebraic Petri nets. University of Glasgow (United Kingdom) (1999)
20. Duvigneau, M.: Konzeptionelle Modellierung von Plugin-Systemen mit Petrinetzen, Agent Technology – Theory and Applications, vol. 4. Logos Verlag, Berlin (2010). http://www.logos-verlag.de/cgi-bin/engbuchmid?isbn=2561&lng=eng&id=
21. Ebert, S., Mey, J., Schöne, R., Götz, S., Aßmann, U.: Distributed Petri nets for model-driven verifiable robotic applications in ROS. In: Innovations in Systems and Software Engineering, pp. 1–27 (2024)
22. Ehlers, K.: Untersuchung des JPMS für Java Programmsysteme zur Dekomposition mittels Modulschnittstellen für einzelne Plugins der Open-Source Software Renew. Bachelor thesis, University of Hamburg, Department of Informatics, Vogt-Kölln Str. 30, D-22527 Hamburg (2023)
23. Apache Software Foundation: Apache Kafka Documentation (2025). https://kafka.apache.org/documentation/. Accessed 21 Jan 2025
24. Fu, G., Zhang, Y., Yu, G.: A fair comparison of message queuing systems. IEEE Access 9, 421–432 (2020)
25. Garg, N.: Apache Kafka. Packt Publishing Birmingham, UK (2013)
26. GitLab: GitLab Documentation (2025). https://docs.gitlab.com/. Accessed 15 Jan 2025
27. Gjesdal, S.L.: A Modular Integrated Development Environment for Coloured Petri Net Models. Master's thesis, The University of Bergen (2023)
28. Gradle: Gradle Documentation (2025). https://docs.gradle.org/. Accessed 15 Jan 2025
29. Haustermann, M.: BPMN-Modelle für petrinetzbasierte agentenorientierte Softwaresysteme auf Basis von Mulan/Capa. Master thesis, University of Hamburg, Department of Informatics, Vogt-Kölln Str. 30, D-22527 Hamburg (2014)
30. van Hee, K.M., Sidorova, N., van der Werf, J.M.: Business process modeling using petri nets. In: Transactions on Petri Nets and Other Models of Concurrency VII, pp. 116–161. Springer (2013)
31. JUnit: JUnit Documentation (2025). https://junit.org/junit5/docs/current/user-guide/. Accessed 15 Jan 2025
32. Knoke, M., Kühling, F., Zimmermann, A., Hommel, G.: Performance of a distributed simulation of timed colored Petri nets with fine-grained partitioning. In: Proceedings of the 2005 Design, Analysis, and Simulation of Distributed Systems Symposium (DASD 2005), San Diego, USA, pp. 63–71 (2005)
33. Kreps, J., Narkhede, N., Rao, J., et al.: Kafka: a distributed messaging system for log processing. In: NetDB 2011: 6th Workshop on Networking meets Databases, Athens, Greece, vol. 11, pp. 1–7 (2011)
34. Kubernetes, A.: Kubernetes Documentation (2025). https://kubernetes.io/docs/home/. Accessed 15 Jan 2025
35. Kummer, O.: Introduction to Petri nets and reference nets. Sozionik Aktuell 1, 1–9 (2001). ISSN 1617-2477
36. Kummer, O.: Referenznetze. Logos Verlag, Berlin (2002). http://www.logos-verlag.de/cgi-bin/engbuchmid?isbn=0035&lng=eng&id=
37. Kummer, O., Wienberg, F., Duvigneau, M., Cabac, L., Haustermann, M., Mosteller, D.: Renew – the Reference Net Workshop (2023). http://www.renew.de/, release 4.1

38. Mockito: Mockito Documentation (2025). https://javadoc.io/doc/org.mockito/ mockito-core/latest/org/mockito/Mockito.html. Accessed 15 Jan 2025
39. Moldt, D., et al.: RENEW: modularized architecture and new features. In: Gomes, L., Lorenz, R. (eds.) Application and Theory of Petri Nets and Concurrency - 44th International Conference, PETRI NETS 2023, Lisbon, Portugal, 25–30 June 2023, Proceedings. Lecture Notes in Computer Science, vol. 13929, pp. 217–228. Springer, Cham (2023). https://doi.org/10.1007/978-3-031-33620-1_12
40. Moldt, D., Röwekamp, J.H., Simon, M.: A simple prototype of distributed execution of reference nets based on virtual machines. In: Bergenthum, R., Kindler, E. (eds.) Algorithms and Tools for Petri Nets Proceedings of the Workshop AWPN 2017, Kgs. Lyngby, Denmark, 19–20 October 2017, pp. 51–57. DTU Compute Technical Report 2017-06 (2017)
41. Mosteller, D., Haustermann, M., Moldt, D.: Prototypical graphical simulation feedback in reference net-based domain-specific languages within a meta-modeling environment. In: Bergenthum, R., Kindler, E. (eds.) Algorithms and Tools for Petri Nets Proceedings of the Workshop AWPN 2017, Kgs. Lyngby, Denmark, 19–20 October 2017, pp. 58–63. DTU Compute Technical Report 2017-06 (2017)
42. Moutinho, F., Gomes, L.: Asynchronous-channels within Petri net-based GALS distributed embedded systems modeling. IEEE Trans. Industr. Inf. **10**(4), 2024–2033 (2014)
43. Oracle: Java Documentation (2025). https://docs.oracle.com/en/java/. Accessed 15 Jan 2025
44. Oracle: Java Messaging Service Documentation (2025). https://docs.oracle.com/ javaee/6/tutorial/doc/bncdr.html. Accessed 15 Jan 2025
45. Pomberger, G., Pree, W., Stritzinger, A.: Methoden und Werkzeuge für das Prototyping und IHRE Integration. Inform. Forsch. Entwickl. **7**(2), 49–61 (1992)
46. Pommereau, F., de La Houssaye, J.: Faster simulation of (Coloured) Petri nets using parallel computing. In: Application and Theory of Petri Nets and Concurrency: 38th International Conference, Petri Nets 2017, Zaragoza, Spain, 25–30 June 2017, Proceedings 38, pp. 37–56. Springer (2017)
47. Rölke, H.: Modellierung von Agenten und Multiagentensystemen – Grundlagen und Anwendungen, Agent Technology – Theory and Applications, vol. 2. Logos Verlag, Berlin (2004). http://logos-verlag.de/cgi-bin/engbuchmid?isbn=0768&lng=eng& id=
48. Röwekamp, J.H.: Investigating the java spring framework to simulate reference nets with RENEW. In: Lorenz, R., Metzger, J. (eds.) Algorithms and Tools for Petri Nets, Proceedings of the 21th Workshop AWPN 2018, Augsburg, Germany, pp. 41–46. No. 2018-02 in Reports/Technische Berichte der Fakultät für Angewandte Informatik der Universität Augsburg (2018). https://opus.bibliothek.uni-augsburg.de/opus4/41861
49. Röwekamp, J.H.: Skalierung von nebenläufigen und verteilten Simulationssystemen für interagierende Agenten. Ph.D. thesis, University of Hamburg, Department of Informatics, Vogt-Kölln Str. 30, D-22527 Hamburg (2023). https://ediss.sub.uni-hamburg.de/handle/ediss/10040
50. Röwekamp, J.H., Buchholz, M., Moldt, D.: Petri net sagas. In: Köhler-Bußmeier, M., Kindler, E., Rölke, H. (eds.) Proceedings of the International Workshop on Petri Nets and Software Engineering 2021 co-located with the 42nd International Conference on Application and Theory of Petri Nets and Concurrency (PETRI NETS 2021), Paris, France, 25 June 2021 (due to COVID-19: virtual conference). CEUR Workshop Proceedings, vol. 2907, pp. 65–84. CEUR-WS.org (2021). http:// ceur-ws.org/Vol-2907

51. Röwekamp, J.H., Feldmann, M., Moldt, D., Simon, M.: Simulating place/transition nets by a distributed, web based, stateless service. In: Moldt, D., Kindler, E., Wimmer, M. (eds.) Petri Nets and Software Engineering. International Workshop, PNSE'19, Aachen, Germany, 24 June 2019. Proceedings. CEUR Workshop Proceedings, vol. 2424, pp. 163–164. CEUR-WS.org (2019). http://CEUR-WS.org/Vol-2424

52. Röwekamp, J.H., Moldt, D.: RENEWKUBE: reference net simulation scaling with RENEW and kubernetes. In: Donatelli, S., Haar, S. (eds.) PETRI NETS 2019. LNCS, vol. 11522, pp. 69–79. Springer, Cham (2019). https://doi.org/10.1007/978-3-030-21571-2_4

53. Röwekamp, J.H., Moldt, D., Feldmann, M.: Investigation of containerizing distributed Petri net simulations. In: Moldt, D., Kindler, E., Rölke, H. (eds.) Petri Nets and Software Engineering. International Workshop, PNSE 2018, Bratislava, Slovakia, 25–26 June 2018. Proceedings. CEUR Workshop Proceedings, vol. 2138, pp. 133–142. CEUR-WS.org (2018). http://ceur-ws.org/Vol-2138/

54. Röwekamp, J.H., Taube, M., Mohr, P., Moldt, D.: Cloud native simulation of reference nets. In: Köhler-Bußmeier, M., Kindler, E., Rölke, H. (eds.) Proceedings of the International Workshop on Petri Nets and Software Engineering 2021 co-located with the 42nd International Conference on Application and Theory of Petri Nets and Concurrency (PETRI NETS 2021), Paris, France, June 25th, 2021 (due to COVID-19: virtual conference). CEUR Workshop Proceedings, vol. 2907, pp. 85–104. CEUR-WS.org (2021). http://ceur-ws.org/Vol-2907

55. Schmidt, A.: Integration von ActiveMQ in eine Petrinetz-basierte Agentenumgebung. Bachelor thesis, University of Hamburg, Department of Informatics, Vogt-Kölln Str. 30, D-22527 Hamburg (2016)

56. Simon, M.: Concept and Implementation of Distributed Simulations in RENEW. Bachelor thesis, University of Hamburg, Department of Informatics, Vogt-Kölln Str. 30, D-22527 Hamburg (2014)

57. Simon, M., Moldt, D.: Extending Renew's algorithms for distributed simulation. In: Cabac, L., Kristensen, L.M., Rölke, H. (eds.) Petri Nets and Software Engineering. International Workshop, PNSE 2016, Toruń, Poland, 20–21 June 2016. Proceedings. CEUR Workshop Proceedings, vol. 1591, pp. 173–192. CEUR-WS.org (2016). http://CEUR-WS.org/Vol-1591

58. Stahl, Y.: Prototypische Umsetzung eines Kommunikationsprotokolls für die verteilte Simulation von P/T Netzen innerhalb des Petrinetz Simulators Renew. Bachelor thesis, University of Hamburg, Department of Informatics, Vogt-Kölln Str. 30, D-22527 Hamburg (2024)

59. Valk, R.: Modelling of task flow in systems of functional units. Technical report FBI-HH-B-124/87, University of Hamburg, Department of Computer Science, Vogt-Kölln Str. 30, D-22527 Hamburg (1987)

60. Valk, R.: Petri nets as token objects. In: Desel, J., Silva, M. (eds.) ICATPN 1998. LNCS, vol. 1420, pp. 1–24. Springer, Heidelberg (1998). https://doi.org/10.1007/3-540-69108-1_1

61. Voß, L.: Development of a Formalism for P/T Nets with Synchronous Channels and their Analysis using Siphons and Traps in Renew. Bachelor thesis, University of Hamburg, Department of Informatics, Vogt-Kölln Str. 30, D-22527 Hamburg (2022)

62. Voß, L., Willrodt, S., Moldt, D., Haustermann, M.: Between expressiveness and verifiability: P/T-nets with synchronous channels and modular structure. In: Köhler-Bußmeier, M., Moldt, D., Rölke, H. (eds.) Proceedings of the International Work-

shop on Petri Nets and Software Engineering 2022 co-located with the 43rd International Conference on Application and Theory of Petri Nets and Concurrency (PETRI NETS 2022), Bergen, Norway, June 20th, 2022. CEUR Workshop Proceedings, vol. 3170, pp. 40–59. CEUR-WS.org (2022). https://ceur-ws.org/Vol-3170

63. Wilde, T., Hess, T.: Forschungsmethoden der Wirtschaftsinformatik. Wirtschaftsinformatik **4**(49), 280–287 (2007)
64. Wollrath, A., Riggs, R., Waldo, J.: A distributed object model for the Java^ T^ M system. Comput. Syst. **9**, 265–290 (1996)

Persistent Permutations, Fairness, Asymmetric Choice Petri Nets, and Ochmański's Conjecture

Eike Best[1]([✉]) and Raymond Devillers[2]

[1] Department of Computing Science, Carl von Ossietzky Universität Oldenburg,
26111 Oldenburg, Germany
eike.best@informatik.uni-oldenburg.de
[2] Département d'Informatique, Université Libre de Bruxelles, Bruxelles, Belgium
raymond.devillers@ulb.be

Abstract. Ochmański's conjecture asserts that in a plain, pure and safe Petri net, and in the context of fairness, the existence of persistent permutations of finite computations allows one to deduce the existence of persistent permutations of infinite computations. In this paper, the conjecture is established for the class of pure and safe asymmetric choice Petri nets.

Keywords: Asymmetric choice · fairness · permutations · persistence · Petri nets

1 Introduction

In two documents [13,14], which are unfortunately not very easily available, Edward Ochmański formulated a hypothesis about deducing the existence of certain (fair) infinite firing sequences of a safe Petri net from the existence of finite ones. More precisely, if every finite firing sequence has a persistent equivalent (in terms of available permutations) then, it is claimed, the same is true for every fair infinite firing sequence. Up to this date, this conjecture appears to have resisted proofs as well as refutations. The present paper has been written with two aims in mind: (i) to reveal what the authors believe are the difficulties in proving or disproving the conjecture, and (ii) to delineate some specific Petri net classes for which it is provably true.

Ochmański's conjecture has a straightforward theoretical appeal, but its practical interest is perhaps less clear. On the one hand, one can guess that the existence of persistently equivalent executions of a Petri net might give rise to feasible concurrent scheduling strategies that minimise arbitrary choices. On the other hand, if the sequential behaviour of a Petri net is limited to persistent sequences, then it even becomes Turing powerful and zero tests could be simulated by such strategies [3].

© The Author(s), under exclusive license to Springer Nature Switzerland AG 2025
E. Amparore and Ł. Mikulski (Eds.): PETRI NETS 2025, LNCS 15714, pp. 155–173, 2025.
https://doi.org/10.1007/978-3-031-94634-9_8

The organisation of the paper is as follows. Section 2 sets the scene by defining basic notions and notations relating to Petri nets and to transition systems. Section 3 explains what is meant by the existence of (finite or infinite) persistent permutations of firing sequences, thus allowing Ochmański's conjecture to be cast in precise terms for general Petri nets (not just for safe ones as in [13,14]). Section 4 contains the main result and its proof. This result confirms Ochmański's conjecture for the class of pure, plain and safe asymmetric choice Petri nets and for safe free-choice nets (all of these notions are defined in Sect. 2). Asymmetric choice Petri nets are a proper superclass of the more widely known class of free-choice nets [7]. Section 5 describes an example which shows that the conjecture is false for 2-bounded (rather than safe) Petri nets, proving that safeness is an essential precondition for its truth. In Sect. 6, different concepts of fairness, as well as some permutation equivalence notions which are related to the ones discussed in the main part of the paper, are investigated. Section 7 concludes the paper.

2   Labelled Transition Systems and Petri Nets

The basic objects of our study are edge-labelled transition systems and unlabelled Petri nets, where the edge labels of a transition system and the transitions of a net correspond to each other, as in the reachability graph of a net.

Definition 1. LABELLED TRANSITION SYSTEMS
A labelled transition system with initial state, abbreviated lts, is a quadruple $TS = (S, \rightarrow, T, s_0)$ where S is a set of *states*, T is a set of *labels* with $S \cap T = \emptyset$, $\rightarrow \subseteq (S \times T \times S)$ is the *transition relation*, and $s_0 \in S$ is an *initial state*.
A label t is *enabled* in a state s, denoted by $s \xrightarrow{t}$, if there is some state $s' \in S$ such that $(s, t, s') \in \rightarrow$. For $t \in T$, $s \xrightarrow{t} s'$ iff $(s, t, s') \in \rightarrow$, meaning that s' is *reachable* from s through the execution of t. The definitions of enabledness and of the reachability relation are extended to sequences $\sigma \in T^*$:[1]

$s \xrightarrow{\varepsilon}$ and $s \xrightarrow{\varepsilon} s$ are always true, by definition, and
$s \xrightarrow{\sigma t} (s \xrightarrow{\sigma t} s')$ if $\exists q \in S$ with $s \xrightarrow{\sigma} q \xrightarrow{t} (s \xrightarrow{\sigma} q \xrightarrow{t} s'$, respectively)

For a state $s \in S$, $[s\rangle$ will denote the set of states s' reachable from s, i.e., such that $s \xrightarrow{\sigma} s'$ for some $\sigma \in T^*$. Enabledness may be extended to infinite sequences $\sigma \in T^\infty$: $s \xrightarrow{\sigma}$ iff $s \xrightarrow{\sigma'}$ for each finite prefix σ' of σ.
Two lts with the same label set, $(S, \rightarrow, T, s_0)$ and $(S', \rightarrow', T, s_0')$, will be called *isomorphic* if there is a bijection $\beta \colon S \rightarrow S'$ such that $s_0' = \beta(s_0)$ and $(r, t, s) \in \rightarrow$ iff $(\beta(r), t, \beta(s)) \in \rightarrow'$.
For a finite sequence $\sigma \in T^*$ of labels, the *Parikh vector* $\Psi(\sigma)$ is a T-vector (i.e., a vector of natural numbers with index set T), where $\Psi(\sigma)(t)$ denotes the

---

[1] We use the shorthands T^* and T^∞ to denote the set of finite and infinite sequences of transitions, respectively.

number of occurrences of t in σ. This may be extended to infinite sequences if we allow vectors of natural or infinite numbers: if $\sigma \in T^{\infty}$, $\Psi(\sigma)(t)$ is ∞ if t occurs infinitely often in σ; otherwise, it is the number of occurrences of t in σ.   □ 1

Definition 2. BASIC PROPERTIES OF LABELLED TRANSITION SYSTEMS
A labelled transition system $(S, \rightarrow, T, s_0)$ is called *finite* if S and T (hence also $\rightarrow$) are finite sets; *totally reachable* if $S = [s_0\rangle$; *singly live* if each transition $t \in T$ is enabled at some reachable state;[2] *deterministic* if for any states $s, s', s'' \in [s_0\rangle$ and sequences $\sigma, \tau \in T^*$ with $\Psi(\sigma) = \Psi(\tau)$: $(s \xrightarrow{\sigma} s' \wedge s \xrightarrow{\tau} s'') \Rightarrow s' = s''$ as well as $(s' \xrightarrow{\sigma} s \wedge s'' \xrightarrow{\tau} s) \Rightarrow s' = s''$ (i.e., from any one state, Parikh-equivalent sequences may not lead to two different successor states, nor come from two different predecessor states);[3] *persistent* [11] if for all reachable states s, s', s'', and labels t, u, if $s \xrightarrow{t} s'$ and $s \xrightarrow{u} s''$ with $t \neq u$, then there is some (reachable) state $r \in S$ such that both $s' \xrightarrow{u} r$ and $s'' \xrightarrow{t} r$ (i.e., once two different labels are both enabled at some state, none of them can disable the other, and executing both, in any order, leads to the same state).   □ 2

Definition 3. PETRI NETS
A (finite, initially marked, place-transition, arc-weighted) Petri net is a quadruple $N = (P, T, F, M_0)$ such that P is a finite set of *places*, T is a finite set of *transitions*, with $P \cap T = \emptyset$, F is a *flow* function $F : ((P \times T) \cup (T \times P)) \rightarrow \mathbb{N}$, and M_0 is the *initial marking*, where a *marking* is a mapping $M : P \rightarrow \mathbb{N}$ (indicating a number of *tokens* in each place). A transition $t \in T$ is *enabled by* a marking M, denoted by $M \xrightarrow{t}$, if for all places $p \in P$, $M(p) \geq F(p, t)$. If t is enabled at M, then t can *occur* (or *fire*) in M, leading to the marking M' defined by $M'(p) = M(p) - F(p, t) + F(t, p)$ (denoted by $M \xrightarrow{t} M'$). The set of markings reachable from M is denoted $[M\rangle$. The *reachability graph of* N, $RG(N)$, is the labelled transition system with the set of vertices $[M_0\rangle$ and set of edges $\{(M, t, M') \mid M, M' \in [M_0\rangle \wedge M \xrightarrow{t} M'\}$.   □ 3

Note on Definitions: All notions defined for transition systems (such as total reachability, single liveness, determinism, and persistence) are carried over to Petri nets through the fact that reachability graphs are transition systems.

Notes on Notation: We shall use letters $a, b, c, \ldots \in T$ (but also $t \in T$) for the labels of a transition system (or, respectively, for the corresponding transitions of a Petri net); $\alpha, \beta, \sigma, \tau \in T^*$ for sequences of transitions; p, q for the places of a net; and M, K, L for the markings of a net.

Definition 4. BASIC STRUCTURAL PROPERTIES OF PETRI NETS
For a place p and a transition t of a Petri net $N = (P, T, F, M_0)$, let ${}^{\bullet}t = \{p \in P \mid F(p, t) > 0\}$ be the *preset of* t containing *pre-places*, $t^{\bullet} = \{p \in P \mid F(t, p) > 0\}$ its *postset* containing *post-places*, ${}^{\bullet}p = \{t \in T \mid F(t, p) > 0\}$ the *preset of* p,

---

[2] This is sometimes also called weakly live.
[3] Often, determinism is defined only w.r.t. successor states, i.e., in forward direction.

and $p^\bullet = \{t \in T \mid F(p,t) > 0\}$ its *postset*.[4] N is called *plain* if $cod(F) \subseteq \{0,1\}$ (i.e., there are no multiple arcs); *pure* or *side-condition free* if $p^\bullet \cap {}^\bullet p = \emptyset$ for all places $p \in P$; *choice-free* (usually denoted by CF) [6] if $|p^\bullet| \leq 1$ for each place $p \in P$; *free-choice* (usually denoted by FC) [7,9] if it is plain and for all $t, t' \in T$, ${}^\bullet t \cap {}^\bullet t' \neq \emptyset$ entails ${}^\bullet t = {}^\bullet t'$;[5] and *asymmetric choice* (AC) [4,9] if N is plain and for all $t, t' \in T$, ${}^\bullet t \cap {}^\bullet t' \neq \emptyset$ entails ${}^\bullet t \subseteq {}^\bullet t' \vee {}^\bullet t' \subseteq {}^\bullet t$.[6]   □ 4

Definition 5. BASIC BEHAVIOURAL PROPERTIES OF PETRI NETS
A Petri net $N = (P, T, F, M_0)$ is *k-bounded*, for some $k \in \mathbb{N}$, if $\forall M \in [M_0)\colon \forall p \in P\colon M(p) \leq k$ (i.e., the number of tokens on any place never exceeds k); *safe* if it is 1-bounded; *bounded* if $\exists k \in \mathbb{N}\colon N$ is k-bounded; and *pps* if it is plain, pure, and safe. A marking M is a *deadlock* if no transition is enabled at M, and a finite firing sequence $M_0 \xrightarrow{\sigma} M$ is called a *deadlocking sequence* (or *maximal*, or *non-extendable*) if M is a deadlock marking.   □ 5

The class of pps Petri nets is closely related to *elementary nets* [18]. Elementary Petri nets have a strengthened firing rule: namely, t can occur if all its pre-places have exactly one token *and* all its post-places have exactly zero tokens. Every elementary net with this strengthened firing rule can be turned into an equivalent pps net with the usual firing rule, by adding appropriate complement places.[7] Conversely, for pps nets, the two firing rules coincide.

Proposition 1. PROPERTIES OF PETRI NET REACHABILITY GRAPHS *[16]*
Any Petri net N is totally reachable and deterministic. N is bounded iff $RG(N)$ is finite.   □ 1

Notes on Pictures: A transition system is represented as a directed graph whose nodes are states and whose edges are labelled with labels from the set T. As an example, consider the labelled transition system TS_0 shown on the left-hand side of Fig. 1. It has eight states, ten edges, and an initial state M_0. A Petri net, on the other hand, is represented by circles for places, tokens inside places to represent markings, squares for transitions, and directed arrows, inscribed by their weights, for arcs. For simplicity, arcs with weight zero are omitted altogether, and arcs with weight one are drawn without any explicit inscription.[8] As an example, the net N_0 shown on the right-hand side of Fig. 1 has five places, four transitions, eight arcs with weight 1, and an initial marking M_0 comprising three tokens.

---

[4] The notions of preset and postset extend immediately to sets of places or transitions.
[5] This is a generalisation of the original definition [9] which requires that, for any $t \in T$, either $|{}^\bullet t| = 1$ or $|({}^\bullet t)^\bullet| = 1$.
[6] This is a variant of the original definition [9] which requires that every transition has at most one shared pre-place, i.e., $\forall p_1, p_2 \in S\colon (p_1{}^\bullet \cap p_2{}^\bullet = \emptyset) \vee |p_1{}^\bullet| \leq 1 \vee |p_2{}^\bullet| \leq 1$.
[7] I.e., for each place p there is a complement place $\tilde{p}$ such that for each reachable marking M, $M(p) + M(\tilde{p}) = 1$.
[8] In this paper, no arc weights greater than 1 will be considered.

Definition 6. SOLVABILITY
A Petri net N *solves* (or *generates*) a transition system TS if $RG(N)$ and TS are isomorphic.   □ 6

As an example, the "token game" reveals instantly that N_0 solves TS_0.

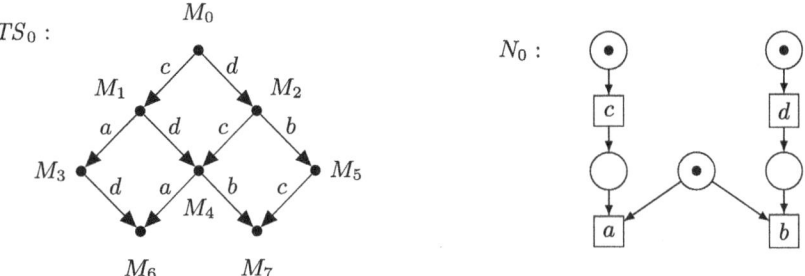

Fig. 1. An lts TS_0 and a generating Petri net N_0. Note that N_0 is plain, pure, safe, and singly live, but not asymmetric choice and, a fortiori, not free-choice.

3   Properties SPE and FPE

Ochmański's conjecture will be formulated in terms of Petri nets as in Definition 3, rather than elementary nets as in [13].[9] In order to be able to state the conjecture, we need to define persistent sequences, permutations, and fair sequences. In the following, let $N = (P, T, F, M_0)$ be a Petri net with some (initial) marking M_0.

Definition 7. PERSISTENCE OF NETS AND OF FIRING SEQUENCES [13]
A net (N, M_0) is called *persistent* if for each marking $M \in [M_0\rangle$ and for all transitions t, u, $(t \neq u \wedge M \xrightarrow{t} \wedge M \xrightarrow{u})$ implies $(M \xrightarrow{tu} \wedge M \xrightarrow{ut})$.[10] A finite or infinite firing sequence $M_0 \xrightarrow{a_1} M_1 \xrightarrow{a_2} \ldots$ is called *persistent* if for every $i > 0$ and $t \neq a_i$, if $M_{i-1} \xrightarrow{t}$ then also $M_i \xrightarrow{t}$.   □ 7

The persistence of a sequence $M_0 \xrightarrow{a_1} M_1 \xrightarrow{a_2} \ldots$ means that no single step $M_{i-1} \xrightarrow{a_i} M_i$ may switch some transition $t \neq a_i$ from enabled at M_{i-1} to disabled at M_i. This is similar to the persistence of a net, but restricted to a single (firing) sequence. The relationship between choice-freeness, persistence, and persistent firing sequences is clarified by the next two propositions and some concomitant examples.

---

[9] This is a slight strengthening of the conjecture, since if it turns out to be true for Petri nets, then it is also true for elementary nets (with an immediate proof).

[10] This is consistent with Definition 2; see the note after Definition 3.

Proposition 2. CHOICE-FREE NETS ARE PERSISTENT
Any choice-free net is persistent, whatever its initial marking.

Proof: Let t, u with $t \neq u$ be two distinct transitions and let M, M', M'' be reachable markings such that $M \xrightarrow{t} M'$ and $M \xrightarrow{u} M''$. By $t \neq u$ and choice-freeness, ${}^\bullet t \cap {}^\bullet u = \emptyset$. Thus firing t does not disable u and vice versa, so that both $M \xrightarrow{tu}$ and $M \xrightarrow{ut}$ are firable and lead to the same state, by determinism.   □ 2

Proposition 2 cannot be reversed. Figure 2 shows a persistent transition system TS_1 and a persistent pps Petri net solution N_1 which is not choice-free.[11]

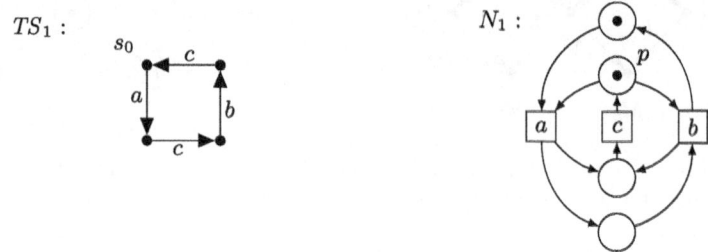

Fig. 2. A transition system TS_1 and a Petri net N_1 solving it. N_1 is (singly) live, plain, pure, safe, and persistent, but not choice-free (since a and b share an input place p). N_1 is not even asymmetric choice.

Proposition 3. PERSISTENT NETS HAVE PERSISTENT FIRING SEQUENCES
Let (N, M_0) be a persistent Petri net and let $M_0 \xrightarrow{\sigma}$ be a (finite or infinite) firing sequence of (N, M_0). Then $M_0 \xrightarrow{\sigma}$ is persistent.

Proof: By contraposition, deriving the non-persistence of (N, M_0) from the non-persistence of $M_0 \xrightarrow{\sigma}$. So, let $M_0 \xrightarrow{\sigma}$ be a non-persistent firing sequence in (N, M_0). Then there are (by definition) transitions a_i and t with $a_i \neq t$ and reachable markings M_{i-1} and M_i such that $M_{i-1} \xrightarrow{a_i} M_i$, $M_{i-1} \xrightarrow{t}$ and $\neg(M_{i-1} \xrightarrow{a_i t})$. Thus (N, M_0) is not persistent.   □ 3

The (implicit) implication in Proposition 3 cannot be reversed: there may be persistent sequences generated by a non-persistent net. For instance, in Fig. 1, consider the sequence $M_0 \xrightarrow{cad}$; more fully:

$$M_0 \xrightarrow{c} M_1 \xrightarrow{a} M_3 \xrightarrow{d} M_6$$

---

[11] Note, in passing, that the shared input place p of a and b describes a potential conflict which is, however, never effective, a phenomenon that has sometimes been called "behavioural choice-freeness" and corresponds to "persistence". In fact, TS_1 has no choice-free pps solution at all. It does have choice-free, non-pps solutions, but there are other examples of persistent transition systems whose solutions necessarily contain choices, no matter whether they are pps or not [4].

This sequence is persistent, even though the net (N_0, M_0) is not persistent. Indeed, the net also has a non-persistent sequence $M_0 \xrightarrow{cda}$ as follows:

$$M_0 \xrightarrow{c} M_1 \xrightarrow{d} M_4 \xrightarrow{a} M_6$$

The crucial point is that M_4 – as opposed to M_3 – enables b while M_6 does not.

We are interested in the case that firing sequences are permutation-equivalent to persistent firing sequences. Permutation equivalence is described by a relation $\equiv$ between pairs of firing sequences starting from the same state, which is normally the initial state M_0. The definition of permutation equivalence $\equiv$ is based on two auxiliary relations $\equiv_0$ (meaning a single permutation) and $\equiv_0^*$ (meaning a finite concatenation of single permutations, including the identity as a special case):

Definition 8. PERMUTATION-EQUIVALENCE OF FIRING SEQUENCES [5]
Whenever $M_0 \xrightarrow{\alpha t_i t_{i+1} \beta}$ and $M_0 \xrightarrow{\alpha t_{i+1} t_i \beta}$ for $\alpha \in T^*$ and $\beta \in T^* \cup T^\infty$, then

$$M_0 \xrightarrow{\alpha t_i t_{i+1} \beta} \equiv_0 M_0 \xrightarrow{\alpha t_{i+1} t_i \beta}$$

by definition; moreover, $\equiv_0^*$ is the reflexive and transitive closure of $\equiv_0$, and $\equiv \, = \, \equiv_0^*$ for finite firing sequences, by definition.

If $\sigma_1, \sigma_2 \in T^\infty$ are infinite firable sequences, then $M_0 \xrightarrow{\sigma_1} \equiv M_0 \xrightarrow{\sigma_2}$ if

- either there are "finitely many permutations", i.e.: $M_0 \xrightarrow{\sigma_1} \equiv_0^* M_0 \xrightarrow{\sigma_2}$
- or there are "infinitely many permutations", defined as follows:

  for every $n \geq 0$, there are $M_0 \xrightarrow{\sigma_1'}$ and $M_0 \xrightarrow{\sigma_2'}$ such that

  $M_0 \xrightarrow{\sigma_1} \equiv_0^* M_0 \xrightarrow{\sigma_1'}$ and σ_1', σ_2 agree on their prefixes of length n

  $M_0 \xrightarrow{\sigma_2} \equiv_0^* M_0 \xrightarrow{\sigma_2'}$ and σ_2', σ_1 agree on their prefixes of length n. 8

□ 8

For example, in the above case (Fig. 1), the two sequences $M_0 \xrightarrow{cad}$ and $M_0 \xrightarrow{cda}$ are permutation-equivalent ($M_0 \xrightarrow{cad} \equiv_0 M_0 \xrightarrow{cda}$, hence $M_0 \xrightarrow{cad} \equiv M_0 \xrightarrow{cda}$), with only one permutation $ad \rightsquigarrow da$ between them.

Note on Terminology: In the remainder of this paper, the words "equivalence" or "equivalent" always refer to the permutation equivalence just defined.

Proposition 4. EQUIVALENCE IMPLIES SAME PARIKH VECTOR
Let $\sigma_1, \sigma_2 \in T^* \cup T^\infty$. Then $(M_0 \xrightarrow{\sigma_1} \equiv M_0 \xrightarrow{\sigma_2}) \Rightarrow (\Psi(\sigma_1) = \Psi(\sigma_2))$.

Proof: Clearly, $(M_0 \xrightarrow{\sigma_1} \equiv_0 M_0 \xrightarrow{\sigma_2}) \Rightarrow (\Psi(\sigma_1) = \Psi(\sigma_2))$, hence also $(M_0 \xrightarrow{\sigma_1} \equiv_0^* M_0 \xrightarrow{\sigma_2}) \Rightarrow (\Psi(\sigma_1) = \Psi(\sigma_2))$ and the property is true for finite firing sequences.

If $\sigma_1, \sigma_2 \in T^\infty$, $M_0 \xrightarrow{\sigma_1} \equiv M_0 \xrightarrow{\sigma_2}$ and $t \in T$, then

- if t occurs infinitely often in σ_1, let $n \in \mathbb{N}$ and n_1 be the position of the nth occurrence of t in σ_1. By definition, there is σ_1' such that $M_0 \xrightarrow{\sigma_1} \equiv_0^* M_0 \xrightarrow{\sigma_1'}$ and σ_1', σ_2 agree on their prefixes of length n_1, meaning that $\Psi(\sigma_2)(t) \geq n$, hence that $\Psi(\sigma_2)(t) = \infty = \Psi(\sigma_1)(t)$ (and similarly if t occurs infinitely often in σ_2); we thus only have to consider the case where t occurs finitely often, both in σ_1 and σ_2;
- if $\Psi(\sigma_1)(t) = n \in \mathbb{N}$ and n_1 is the position of the nth occurrence of t in σ_1, by definition there is σ_1' such that $M_0 \xrightarrow{\sigma_1} \equiv_0^* M_0 \xrightarrow{\sigma_1'}$ and σ_1', σ_2 agree on their prefixes of length n_1, meaning that $\Psi(\sigma_2)(t) \geq n = \Psi(\sigma_1)(t)$: similarly, $\Psi(\sigma_1)(t) \geq \Psi(\sigma_2)(t)$, so that $\Psi(\sigma_1)(t) = \Psi(\sigma_2)(t)$.   □ 4

By contrast, having the same Parikh vector does not imply equivalence, even if the two sequences are firable from the same marking, as illustrated by Fig. 3.

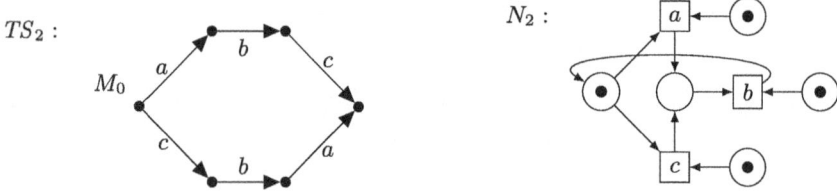

Fig. 3. A pps net on the right and its reachability graph on the left: we have $\Psi(abc) = (1,1,1) = \Psi(cba)$ but not $M_0 \xrightarrow{abc} \equiv M_0 \xrightarrow{cba}$, since neither $M_0 \xrightarrow{bac}$ nor $M_0 \xrightarrow{acb}$.

Ochmański's conjecture depends on a fairness property. The fairness notion he employs is very much standard and is called *strong fairness*, or just *fairness*, in the literature.[12] More precisely, (strong) fairness is a universally agreed concept for *infinite firing sequences*. Finite firing sequences usually play a minor role; we might either just call every finite sequence fair by default, or we might pick a subset of finite sequences which are called fair while others are not. Indeed, the literature suggests a reasonable choice between two alternative possibilities:

- The *finite-is-fair principle* (e.g., [12]) simply considers all finite sequences to be fair. Intuitively, this principle asserts that unfairness occurs only when some activity is neglected infinitely often.
- By contrast, the *maximal-finite-is-fair principle* (e.g., [8]) requires that a finite firing sequence is *deadlocking* (i.e., *not extendable*, hence *maximal*) in order to be fair. Intuitively, this principle asserts that unfairness also occurs in case some execution is "dragging its feet", so to speak, that is, delays indefinitely some possible activity.

---

[12] Other forms of fairness will be discussed below in Sect. 6.

Since in [14], Ochmański adopts the latter principle, we will do the same (see the first sentence of the next definition).[13]

As formal tools to capture fairness, we use the modified existential and universal quantifiers $\exists_i^\infty$ ("there are infinitely many i with ...") and $\forall_i^\infty$ ("for all but finitely many i, ...").

Definition 9. (STRONG) FAIRNESS
A finite firing sequence $M_0 \xrightarrow{\sigma} M$ is (strongly) fair with respect to $t \in T$ if $\neg M \xrightarrow{t}$, and it is *fair* if it is fair with respect to every transition $t \in T$, meaning that M is a deadlock.

An infinite firing sequence $M_0 \xrightarrow{t_1} M_1 \xrightarrow{t_2} M_2 \xrightarrow{t_3} M_3 \xrightarrow{t_4} \ldots$ is *fair with respect to $t \in T$* if

$$(\exists_i^\infty : (t = t_i)) \ \lor \ (\forall_i^\infty : \neg M_i \xrightarrow{t})$$

and it is *fair* if it is fair with respect to every transition $t \in T$.   □ 9

Definition 10. PERSISTENT PERMUTATION EQUIVALENTS [13]
The marked net (N, M_0) is called SPE (S for "short") if every finite firing sequence starting from M_0 has a persistent equivalent, and FPE (F for "fair") if every fair firing sequence starting from M_0 has a persistent equivalent.   □ 10

The pps net (N_0, M_0) in Fig. 1 satisfies SPE. For example, the non-persistent sequence $M_0 \xrightarrow{cda}$ has the persistent equivalent $M_0 \xrightarrow{cad}$. It also satisfies FPE, since there are no infinite firing sequences (so that all fair sequences are deadlocking and hence (of course) finite), and because it satisfies SPE. Ochmański's conjecture generalises this observation to infinite fair sequences:

Conjecture: (OCHMAŃSKI'S CONJECTURE)
Let (N, M_0) be a pps Petri net which satisfies SPE. Then it also satisfies FPE.

□

Note that by SPE, every fair finite sequence has a persistent equivalent, simply by its finiteness. The key problem is to investigate whether SPE $\Rightarrow$ FPE also for infinite firing sequences. This problem is dealt with in the next two sections.

Figure 2 illustrates a case where FPE is satisfied: there is a single infinite firing sequence $(acbc)^\infty$ which is both fair and persistent; hence FPE and Ochmański's conjecture are satisfied. As a more involved example, we may consider the system shown in Fig. 4, which is a slight variant of the one shown in Fig. 1. Its infinite firing sequences are of the form[14] $(cade|cdae|dcae|dbcf|dcbf|cdbf)^\infty$. They are fair unless they contain infinitely many *cdae* or *dcae* but only finitely many *dbcf*, *dcbf* and *cdbf*, since then transition b is enabled infinitely often but finitely often executed, or if they contain infinitely many *dbcf* or *cdbf* but only finitely many

---

[13] The results, proofs and (counter-) examples in the main part of this paper, Sects. 3 to 5, do not depend on this choice. This will be discussed more fully in Sect. 6.

[14] Where | means a choice between several patterns, instead of—and, as before, $(...)^\infty$ denotes an infinite repetition of finite patterns.

cade, cdae and *dcae*, since then transition a is enabled infinitely often but finitely often executed. They are non-persistent if they contain at least one *cdae, dcae, dcbf* or *cdbf*, since no path traversing M_4 is persistent. However all these infinite firing sequences are equivalent to a persistent sequence in $(cade|dbcf)^\infty$; indeed any *cdae* may be transformed into *cade* with one permutation, any *dcae* may be transformed into *cade* with two permutations, any *dcbf* may be transformed into *dbcf* with one permutation, and any *cdbf* may be transformed into *dbcf* with two permutations. Hence, N_0' (which inherits the pps property from N_0) satisfies FPE, and thus Ochmański's conjecture.

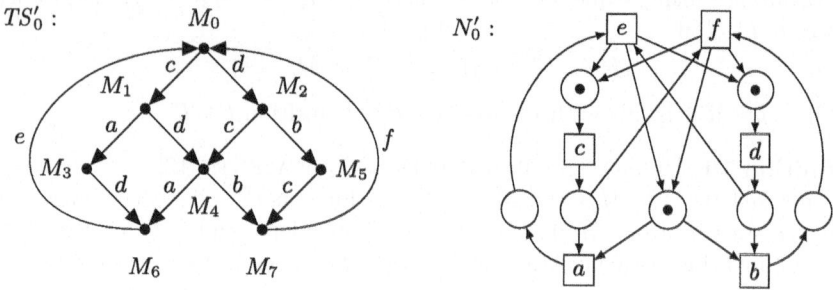

Fig. 4. A variant of Fig. 1 generating infinite sequences. The additional transitions e and f lead back, respectively, from the final states M_6 and M_7 to the initial state M_0 of TS_0'. The augmented net N_0' is still plain, pure and safe, and is a solution of TS_0'.

Figure 5 demonstrates that fairness is an essential ingredient of the claim, in the sense that not *every* infinite firing sequence needs to have a persistent equivalent. The net is safe and satisfies SPE, since every finite firing sequence can be permuted in such a way as to avoid enabling a and b (the only two transitions which are in a conflict due to their shared input place) simultaneously; for instance, the sequence $y(xac)^n$ may be permuted into $(xac)^n y$. However, there is an infinite firing sequence

$$M_0 \xrightarrow{y(xac)^\infty}$$

which has no equivalent persistent permutation. But this sequence is not fair since b is enabled infinitely often but never executed (in fact, it so happens that no infinite sequence whatsoever is fair). For this reason, this example does *not* violate the claim of the conjecture. Later, in Sect. 5, we shall see that the "safeness" part of the pps premise is also an essential ingredient of the claim.

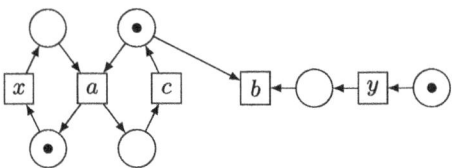

Fig. 5. This pps net satisfies SPE (essentially, the conflict between a and b can be delayed by executing y as late as necessary). It also has an infinite (unfair) firing sequence $M_0 \xrightarrow{y(xac)^\infty}$ which cannot be permuted into a persistent one (essentially, y cannot be delayed forever without violating permutation equivalence).

Remark 1. NON-INHERITANCE OF THE SPE PROPERTY
If a net N is fixed, a marking M of it will be said to satisfy SPE if so does (N, M). It may be observed that if M satisfies SPE, it may happen that this is not true for some of its successor markings. For example, in Fig. 1, M_0 satisfies SPE for N_0; in particular, the sequence $M_0 \xrightarrow{c} M_1 \xrightarrow{db}$ has a persistent equivalent, namely $M_0 \xrightarrow{dbc}$; but SPE is not valid for M_1 since the sequence $M_1 \xrightarrow{db}$ does not have a persistent equivalent. The same phenomenon occurs in Fig. 5 where, after $M_0 \xrightarrow{x} M_1$, the sequence $M_1 \xrightarrow{yb}$ is enabled but does not have a persistent equivalent. This lack of SPE inheritance complicates the analysis of the relationship between SPE and FPE in the general case.   □ 1

4   Two Petri Net Classes Satisfying SPE $\Rightarrow$ FPE

In this section, we prove Ochmański's conjecture for pps asymmetric choice nets and for safe free-choice nets. Along the way, we show that conflict-freeness is implied as well. First, we clarify the role of the pps property.

The safeness of a net almost entails its plainness, in the sense that any transition bordering on a non-plain arc is necessarily dead (i.e., non singly live):

Lemma 1. SAFENESS (ALMOST) IMPLIES PLAINNESS
A singly live, safe net is plain.

Proof: Let p be a place and t be a transition. If (p, t) is a non-plain arc, then any occurrence of t needs two or more tokens on p. If (t, p) is a non-plain arc, then any occurrence of t puts two or more tokens on p.   □ 1

The impact of pureness and safeness on persistence is captured by the following:

Lemma 2. SIMULTANEOUS ENABLING OF TRANSITIONS
Let M be a reachable marking of a pure and safe net and let a_1, a_2 be two different transitions with $M \xrightarrow{a_1} K$ and $M \xrightarrow{a_2} L$. Then $(K \xrightarrow{a_2})$ if and only if $(L \xrightarrow{a_1})$.

Proof: Suppose that $M \xrightarrow{a_1} K$ and $M \xrightarrow{a_2} L$ and $\neg(M \xrightarrow{a_1} K \xrightarrow{a_2})$. By the latter and by $M \xrightarrow{a_2} L$, there is a place $p \in {}^\bullet a_1 \cap {}^\bullet a_2$ such that $M(p) = 1$ (by safeness and because $M \xrightarrow{a_1} K$) and $K(p) = 0$ (by $\neg(K \xrightarrow{a_2})$). By pureness, $p \notin a_2^\bullet$, and hence also $L(p) = 0$ and $\neg(M \xrightarrow{a_2} L \xrightarrow{a_1})$. By symmetry, the claim follows. (Note that this is independent of whether $K = L$ or not.[15])   □ 2

The main result is proved next.

Theorem 1. SPE $\Rightarrow$ FPE FOR PURE AND SAFE AC NETS
Let (N, M_0) be a Petri net which is pure, safe, and asymmetric choice. Suppose that (N, M_0) satisfies SPE. Then N satisfies FPE.

Proof: Let (N, M_0) be a Petri net which is (plain,) pure, safe, asymmetric choice, and satisfies SPE. The proof is done in two steps.

In the first step, we show that (N, M_0) is persistent. This will be done by contradiction, assuming that (N, M_0) is not persistent and deriving a contradiction.

So, assume that (N, M_0) is not persistent. Then there is a reachable marking M violating the persistence property. That is, $M \in [M_0\rangle$ and $M \xrightarrow{a_1} K$ and $M \xrightarrow{a_2} L$ for some transitions a_1, a_2 with $a_1 \neq a_2$, and $\neg(K \xrightarrow{a_2})$ (possibly $K = L$). By Lemma 2, also $\neg(L \xrightarrow{a_1})$.

As in the proof of that lemma, there is a place $p \in {}^\bullet a_1 \cap {}^\bullet a_2$ which carries a token at M ($M(p) = 1$) and can lose its token either through the occurrence of a_1 ($K(p) = 0$) or through the occurrence of a_2 ($L(p) = 0$). By the asymmetric choice property, we may assume, without loss of generality, that ${}^\bullet a_1 \supseteq {}^\bullet a_2 \neq \emptyset$, so that if a_1 is enabled at some (reachable) marking, so is a_2.

Hence, any firing sequence including a_1 is not persistent, since a_2 is always enabled along with a_1, but disabled after the occurrence of a_1 (because of the existence of p). Any permutation of such a sequence contains a_1, and hence, cannot be persistent, either. This contradicts the assumed property SPE, since the sequence $M_0 \longrightarrow \ldots \longrightarrow M \xrightarrow{a_1} K$, which is not persistent, cannot have an equivalent persistent permutation (as it contains a_1 by permutation equivalence).

Thus, an asymmetric choice pps net satisfying SPE is persistent. This finishes the first step of the proof.

In a second step, we prove that the FPE property holds. From the first step, it is known that (N, M_0) is persistent, and from Proposition 3, we deduce that all of its firing sequences (whether finite or infinite) are persistent. Thus the identity permutation suffices in order to find a persistent equivalent to any infinite firing sequence. Note that fairness is not needed in this case.   □ 1

Figure 5 highlights the importance of the AC property in the first step of this proof. Indeed, the net shown there is pps and satisfies SPE, but it is neither asymmetric choice (as exhibited by transitions a and b) nor persistent (since we have the firing sequences $M_0 \xrightarrow{xya}$ and $M_0 \xrightarrow{xyb}$, but not $M_0 \xrightarrow{xyab}$); note that

---

[15] It is even independent of whether $a_1 = a_2$ or not; if so, the claim reduces to the fact that a transition cannot occur twice in a row in a pure and safe net.

SPE is satisfied since, for instance, the sequence $M_0 \xrightarrow{xya}$ is non-persistent (b is enabled after xy but not after xya), but it is equivalent to $M_0 \xrightarrow{xay}$ (where b is no longer enabled before the end).

By contrast, Fig. 2 shows that a non-asymmetric-choice pps net may nevertheless be persistent, while Fig. 3 exhibits a non-asymmetric-choice pps net which is neither persistent nor SPE (since in N_2, we have $M_0 \xrightarrow{a}$ and $M_0 \xrightarrow{c}$, and none of them has an equivalent persistent firing sequence).

When single liveness is added, we can derive global choice-freeness:

Corollary 1. SPE $\Rightarrow$ CF FOR SINGLY LIVE, PURE AND SAFE AC NETS
Let (N, M_0) be a singly live, pure, safe, asymmetric choice Petri net satisfying SPE. Then N is choice-free.

Proof: By contradiction. Suppose that $p \in {}^\bullet a_1 \cap {}^\bullet a_2$ for two transitions a_1, a_2 with $a_1 \neq a_2$. By single liveness, some firing sequence contains a_1. By pureness, $p \notin a_1{}^\bullet \wedge p \notin a_2{}^\bullet$. Thus firing a_1 always disables a_2 and firing a_2 always disables a_1. As in the previous proof, this contradicts the SPE property when a firing sequence containing a_1 (or a_2, for that matter) is considered.   □ 1

Finally, we show that when the class of nets is narrowed to free-choice ones, the pureness premise can be dispensed with (but then, of course, the equivalence with elementary nets is lost). Lemma 2 is then no longer useful, but the above proof can be adapted as follows:

Theorem 2. SPE $\Rightarrow$ FPE FOR SAFE FC NETS
Let (N, M_0) be a safe free-choice Petri net. Suppose that (N, M_0) satisfies SPE. Then N satisfies FPE.

Proof: Consider a non-persistent situation with $M_0 \xrightarrow{\alpha} M$, $M \xrightarrow{a_1} K$, $M \xrightarrow{a_2} L$ and $\neg(K \xrightarrow{a_2})$, for two transitions a_1, a_2 with $a_1 \neq a_2$. (We may have $L \xrightarrow{a_1}$ since pureness is not assumed.) Then ${}^\bullet a_1 \cap {}^\bullet a_2 \neq \emptyset$ and, by the free-choice property, ${}^\bullet a_1 = {}^\bullet a_2$, and thus, a_2 is enabled whenever a_1 is. This implies that no (finite) firing sequence containing a_1 can be persistent, contrary to the SPE premise. Again, fairness is not used in this case.   □ 2

5 A Plain and Pure Counterexample to SPE $\Rightarrow$ FPE

Figure 6 shows that the implication SPE $\Rightarrow$ FPE is false in general. Consider the labelled transition system shown on the left-hand side. In this lts, any finite firing sequence can be permuted equivalently in such a way that the last M' in it (if any) is reached via y rather than via a_1, and M'' (which can occur at most once) is reached via x rather than via b. Hence SPE is satisfied. However, consider the infinite firing sequence

$$M_0 \xrightarrow{y(xa_1a_2bc)^\infty} \qquad (1)$$

This sequence is fair but has no persistent equivalent, since, for taking account of the fact that the label y must occur at some point or other, the marking M

cannot be avoided, and the firing $M \xrightarrow{a_1} M'$ is non-persistent (as M enables b but M' does not). Hence FPE is not satisfied. The problem is that no $\equiv$-permutation of (1) can shift the label y "to the very end" without losing it altogether, and yet, no transition is treated unfairly.

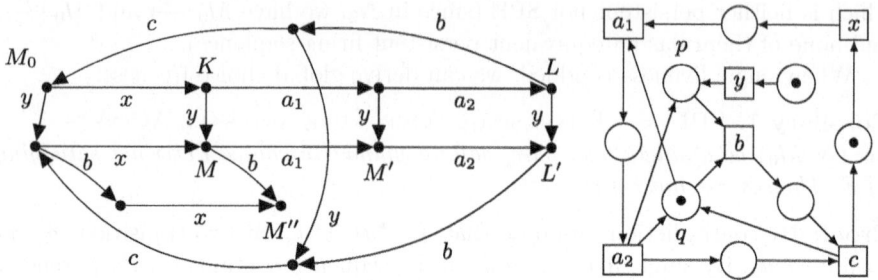

Fig. 6. An lts satisfying SPE but not FPE (l.h.s.) and a generating Petri net (r.h.s.).

The Petri net shown on right-hand side of Fig. 6 solves this transition system. This net is not safe; indeed, at L', place p carries two tokens. Note that p is the only unsafe place (with bound 2), and that L' is the only non-safe reachable marking. In this sense, the example demonstrates that the conjecture is sharp.[16]

The paper [2] develops a formal argument why there is no pps Petri net solving the transition system shown on the left-hand side of Fig. 6. Intuitively, observe that at L, we have both $L \xrightarrow{y}$ and $L \xrightarrow{b}$, but at K (and also at M_0), only yb, but not by, is enabled. Such a situation cannot occur in a pps net. In particular, all known constructions for turning a 2-bounded place p into two or more safe places (for example, pipelining the tokens on p, or duplicating p) introduce causal dependencies between (instances of) y and (instances of) a_2, by which the essence of the counterexample is necessarily destroyed.

This example demonstrates that safeness is absolutely instrumental for the conjecture. By comparison, plainness is less important, since it follows (almost) from safeness by Lemma 1, and the exact role of pureness is not fully clear, except that it is used in the first of our results. Nevertheless, we stress that the definitions of SPE and FPE do not depend on the pps premise. With this wider – as compared with [13,14] – context in mind, the authors are not aware of any (bounded but not necessarily safe) AC net for which SPE fails to imply FPE. This gives rise to a second open question: Is there an asymmetric choice Petri net which satisfies SPE but not FPE?

6   Fairness of Finite Firing Sequences, APE and JPE

The finite-is-fair principle has been adopted, for example, in [1,4,12,15]. The maximal-finite-is-fair principle is used (and justified, in various contexts) in [8,

---

[16] The authors also believe that it is a smallest example with this property.

10,14,17]. In the previous part of the paper, the latter was acceded to. Now, let us consider what happens if we adopt the former instead, i.e., if the first sentence in Definition 9 is replaced by:

Any finite firing sequence $M_0 \xrightarrow{\sigma} M$ is (strongly) fair.

Let the resulting permutation equivalence notions be called SPE^f and FPE^f, respectively. Of course, SPE^f equals SPE, since fairness plays no role there. But FPE^f differs from FPE; it is actually stronger. Still, Ochmański's conjecture remains unaffected, since under the premise SPE (or SPE^f), FPE is the same as FPE^f. However, for its *converse*, we have the following:

Lemma 3. THE CONVERSE OF OCHMAŃSKI'S CONJECTURE

(i) $\text{FPE}^f \Rightarrow \text{SPE}^f$ *(for any Petri net)*
(ii) $\neg\,(\,\text{FPE} \Rightarrow \text{SPE}\,)$

Proof: (i): By the finite-is-fair regime, all finite sequences are fair. Hence FPE^f reduces to SPE, thus to SPE^f, for finite sequences.

(ii): A counterexample, which is even pps, is shown in Fig. 7, as taken from [14]. The net N_3 satisfies FPE: Any maximal firing sequence must perform x, y, z and either a and c, or a and d, or b and d. The sequence $xyzac$ is not persistent, but the equivalent sequence $xaycz$ is. Similarly, the sequences $xazdy$ and $zdybx$ are persistent. But the finite sequences yb and yc have no persistent equivalents, so that SPE is not satisfied.   □ 3

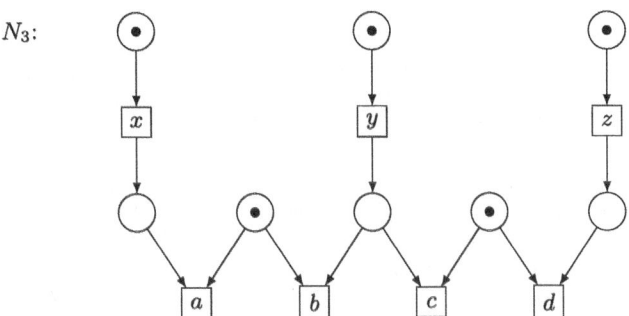

Fig. 7. A net N_3 which is FPE but not SPE.

Another distinction between the finite-is-fair and maximal-finite-is-fair principles is that the latter – but not, as shown in Lemma 4(i), the former – satisfies a compositionality property which might be interesting in checking the fairness of sequences: in a disjoint composition of nets, a sequence is fair if and only if its projections on the components are both fair (Lemma 4(ii)).

Lemma 4. COMPOSITIONALITY IN THE MAXIMAL-FINITE-IS-FAIR REGIME
Let N be composed of two disjoint nets N_1 and N_2.

(i) *Using the finite-is-fair regime, i.e., Definition 9 as modified above, it may be the case that some infinite sequence of N is unfair while its projections onto N_1 and N_2 are both fair.*
(ii) *Using Definition 9, every (finite or infinite) firing sequence of N is fair if and only if its projections onto N_1 and N_2 are both fair.*

Proof: (i): Consider the net N_4 shown on the left-hand side of Fig. 8. The infinite firing sequence $M_{4,0} \xrightarrow{a^\infty}$ in N_4 is not fair since b is constantly enabled and never occurs.[17] However, its (empty) projection $M_{4,0}^2 \xrightarrow{\varepsilon} M_{4,0}^2$ on N_4^2 is fair by virtue of being finite, and its projection $M_{4,0}^1 \xrightarrow{a^\infty}$ on N_4^1 is fair, too, since a occurs infinitely often (and b is not part of the scene). This does not happen with the maximal-finite-is-fair principle, since $M_{4,0}^2$ is not a deadlock in N_4^2, so that the empty firing sequence is not fair in this case.[18]
(ii): This follows from the fact that a sequence of N is deadlocking if and only if the same is true for its two projections onto N_1 and N_2.   □ 4

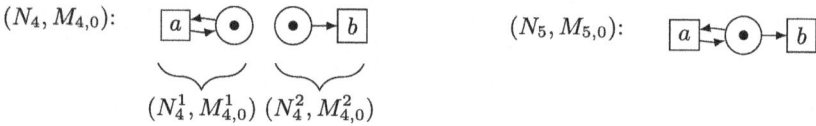

$(N_4, M_{4,0})$: $(N_5, M_{5,0})$:

$(N_4^1, M_{4,0}^1)$ $(N_4^2, M_{4,0}^2)$

Fig. 8. The net N_4 with initial marking $M_{4,0}$ is the disjoint sum of two component nets, N_4^1 with initial marking $M_{4,0}^1$, and N_4^2 with initial marking $M_{4,0}^2$. In N_4, the sequence $M_{4,0} \xrightarrow{a^\infty}$ is unfair and fails to satisfy progress. In N_5, with initial marking $M_{5,0}$, the sequence $M_{5,0} \xrightarrow{a^\infty}$ is also (weakly and strongly) unfair, but it satisfies progress.

Both [8] and [10] have remarked that the maximal-finite-is-fair principle is a merge of two concepts: that of strong fairness for infinite sequences and that of *progress* (as defined in [15], for instance) for finite sequences. Call a transition t *continuously enabled* along some computation if all markings enable it, and *constantly enabled* if, in addition, its enabling input tokens are not taken away (and immediately put back) by other transitions. (Thus, in N_4 of Fig. 8, the sequence a^∞ continuously and constantly enables b, while in N_5, the same sequence continuously enables b but does not constantly enable b.) An infinite sequence is *weakly fair* (or, in the parlance of [14], *just*) if no transition is ever continuously enabled without being executed, and it *satisfies progress* if no transition is ever constantly enabled without being executed. For finite sequences, progress is

---

[17] Constant enabling can be defined precisely; see the remarks after this proof.
[18] N_4 is not pure and thus not pps, but it can easily be modified into a pps example by splitting transition a into a beginning and an end.

nothing but maximality, so that there are no choices of definitions; but for weak fairness, we again have two possibilities: all finite sequences are weakly fair, or only the maximal ones are weakly fair. In [14], we also find two other properties besides SPE and FPE:

- APE, which requires *all* (finite or infinite) sequences to have persistent equivalents;
- JPE, which requires all *just* (weakly fair) sequences to have persistent equivalents.

When these are taken into account, together with their finite-is-fair variants, we get $APE^f = APE$, and moreover, the following implications:

Lemma 5. RELATIONSHIPS BETWEEN APE, JPE, SPE, AND FPE

(i) $APE^f \Rightarrow JPE^f \Rightarrow FPE^f \Rightarrow SPE^f$ *(for any Petri net)*
(ii) $\neg (FPE^f \Rightarrow JPE^f)$ *and* $\neg (JPE^f \Rightarrow APE^f)$
(iii) $APE \Rightarrow JPE$, $APE \Rightarrow SPE$, *and* $JPE \Rightarrow FPE$ *(for any Petri net)*
(iv) *The implications in* (iii) *cannot be reversed; SPE and JPE are incomparable; and FPE does not imply SPE.*

Proof: (i): Every fair sequence is just and every just sequence is just a sequence, together with Lemma 3(i).
(ii): Figure 5 disproves $FPE^f \Rightarrow JPE^f$ and Fig. 7 disproves $JPE^f \Rightarrow APE^f$.
(iii): Same argument as before.
(iv): Figure 5 disproves $FPE \Rightarrow JPE$, $SPE \Rightarrow APE$, and $SPE \Rightarrow JPE$, and Fig. 7 disproves $JPE \Rightarrow APE$, $JPE \Rightarrow SPE$ and $FPE \Rightarrow SPE$.   □ 5

Figure 9 provides a summary of the relationships proved, disproved, and conjectured so far.

Fig. 9. The implications without question mark are true for all nets (Lemmata 3(i) and 5(i,iii)). The implication(s) provided with a question mark are true for pps asymmetric choice nets and for safe free-choice nets (Sect. 4), while they are wrong for general nets (Sect. 5) and are conjectured to be true for pps nets in general (Sect. 3) and for (bounded) AC nets (Sect. 5). All other possible (bilateral) implications, except those arising from reflexivity and transitivity, are wrong (Figs. 5, 7, and Lemmata 3(ii), 5(ii,iv)). Also, $APE \Leftrightarrow APE^f$ and $SPE \Leftrightarrow SPE^f$ and $SPE \Rightarrow (FPE \Leftrightarrow FPE^f)$.

7   Concluding Remarks

In this paper, we have proved the correctness of Ochmański's conjecture for the class of pps asymmetric choice Petri nets and for the class of safe free-choice nets. The full conjecture, i.e., the claim that SPE $\Rightarrow$ FPE holds for all pps nets (not just for asymmetric choice ones), remains open.

In future work, we plan to take further steps towards proving the critical implication SPE $\overset{?}{\Rightarrow}$ FPE for the full class of pps nets. Along the way, it might be interesting to know whether a counterexample such as that given in Sect. 5 can be found within the class of (non-safe) asymmetric choice nets. Also, the complexity status of the properties APE, JPE, FPE, SPE remains to be investigated.

Acknowledgments. The first author was made aware of this line of research by Edward Ochmański during an enjoyable visit to Toruń in 2014. The authors gratefully acknowledge the reviewers' remarks.

References

1. Apt, K.R., Francez, N., Katz, S.: Appraising fairness in languages for distributed programming. Distrib. Comput. **2**(4), 226–241 (1988). https://doi.org/10.1007/BF01872848
2. Barylska, K., Best, E., Schlachter, U., Spreckels, V.: Properties of plain, pure, and safe Petri nets. In: Koutny, M., Kleijn, J., Penczek, W. (eds.) Transactions on Petri Nets and Other Models of Concurrency XII. LNCS, vol. 10470, pp. 1–18. Springer, Heidelberg (2017). https://doi.org/10.1007/978-3-662-55862-1_1
3. Barylska, K., Mikulski, L., Ochmanski, E.: On persistent reachability in Petri nets. Inf. Comput. **223**, 67–77 (2013). https://doi.org/10.1016/J.IC.2012.11.004
4. Best, E., Devillers, R.: Petri Net Primer - A Compendium on the Core Model, Analysis, and Synthesis. Springer, Cham (2024). https://doi.org/10.1007/978-3-031-48278-6. ISBN 978-3- 031-48277-9
5. Best, E., Devillers, R.: Sequential and concurrent behaviour in Petri net theory. Theor. Comput. Sci. **55**(1), 87–136 (1987). https://doi.org/10.1016/0304-3975(87)90090-9
6. Crespi-Reghizzi, S., Mandrioli, D.: A decidability theorem for a class of vector-addition systems. Inf. Process. Lett. **3**(3), 78–80 (1975). https://doi.org/10.1016/0020-0190(75)90020-4
7. Desel, J., Esparza, J.: Free Choice Petri Nets, vol. 40. Cambridge Tracts in Theoretical Computer Science. Cambridge University Press (1995)
8. van Glabbeek, R., Höfner, P.: Progress, justness, and fairness. ACM Comput. Surv. **52**(4), 69:1–69:38 (2019). https://doi.org/10.1145/3329125
9. Hack, M.H.Th.: Analysis of production schemata by Petri nets. Technical report, Massachussetts Institute of Technology, MAC TR-94, 1974 (based on his MSc thesis, 1972)
10. Kindler, E., van der Aalst, W.M.P.: Liveness, fairness, and recurrence in Petri nets. Inf. Process. Lett. **70**(6), 269–27 (1999). https://doi.org/10.1016/S0020-0190(99)00074-5

11. Landweber, L.H., Robertson, E.L.: Properties of conflict- free and persistent Petri nets. J. ACM **25**(3), 352–364 (1978). https://doi.org/10.1145/322077.322079
12. Lehmann, D., Pnueli, A., Stavi, J.: Impartiality, justice and fairness: the ethics of concurrent termination. In: Even, S., Kariv, O. (eds.) ICALP 1981. LNCS, vol. 115, pp. 264–277. Springer, Heidelberg (1981). https://doi.org/10.1007/3-540-10843-2_22
13. Ochmański, E.: On conflict-free executions of elementary nets. Syst. Sci. Wydawca Oficyna Wydawnicza Politechniki Wroc- lawskiej **27**(2), 89–105 (2001)
14. Ochmański, E.: Persistent runs in elementary nets. FoLCo (Formal Languages and Concurrency) Memorandum, pp. 2–6, University of Toruń (2014)
15. Reisig, W.: Modelling and verification of distributed algorithms. In: CONCUR 1996, Concurrency Theory, 7th International Conference, Pisa, Italy, 26–29 August 1996, Proceedings, pp. 579–595 (1996). https://doi.org/10.1007/3-540-61604-7_77
16. Reisig, W.: Petri Nets: An Introduction, vol. 4. EATCS Monographs on Theoretical Computer Science. Springer, Cham (1985). https://doi.org/10.1007/978-3-642-69968-9. ISBN 3-540-13723-8
17. Romijn, J., Vaandrager, F.W.: A note on fairness in I/O automata. Inf. Process. Lett. **59**(5), 245–250 (1996). https://doi.org/10.1016/0020-0190(9)00122-6
18. Rozenberg, G., Engelfriet, J.: Elementary net systems. In: Reisig, W., Rozenberg, G. (eds.) ACPN 1996. LNCS, vol. 1491, pp. 12–121. Springer, Heidelberg (1998). https://doi.org/10.1007/3-540-65306-6_14

Statistical Model Checking of Stochastic Timed-Arc Petri Nets

Tanguy Dubois[1], Kim G. Larsen[2], and Jiří Srba[2(✉)]

[1] Nantes Université, École Centrale Nantes, CNRS, LS2N, UMR 6004,
44000 Nantes, France
tanguy.dubois@ec-nantes.fr
[2] Aalborg University, Aalborg, Denmark
{kgl,srba}@cs.aau.dk

Abstract. Timed-Arc Petri nets (TAPN) are a timed extension of Petri nets where tokens have their age and each arc is associated with a time interval restricting the ages of tokens available for transition firing. Additionally, a TAPN can also contain place invariants constraining the ages of tokens in places, inhibitor arcs preventing a transition from firing and transport arcs that preserve a token's age upon firing. This set of features, as much as it allows us to model complex systems, also often makes verification problems computationally hard or even undecidable. Moreover, in order to model real-life examples, additional stochastic aspects are often necessary to capture the desired behaviour. We suggest (to the best of our knowledge) the first stochastic semantics for TAPNs and design and implement the Statistical Model Checking (SMC) algorithms in the model checker TAPAAL. We argue for the semantic choices we made in the stochastic semantics and prove that the semantics is well-behaving. On a number of case studies we demonstrate the practical applicability of our modelling formalism and its SMC implementation.

1 Introduction

In many complex systems, exact deterministic models fail to capture the inherent uncertainty and variability present in real-world processes. Introducing stochastic elements to formal models allows us to better represent these unpredictable dynamics, providing a more realistic framework for analysis [19]. However, the exact verification of these models often becomes infeasible due to the sheer computational complexity involved. Instead, statistical model checking [30] offers a practical alternative: it provides approximate but computationally efficient results. While statistical methods may not guarantee absolute precision, they offer a valuable trade-off between accuracy and performance, enabling meaningful insights into systems that are otherwise too complex to analyze.

Research on introducing stochastic semantics to Petri nets focuses on incorporating randomness into the modelling of systems with uncertain or probabilistic behaviour. Stochastic Petri nets [4,18] extend traditional Petri nets by assigning probabilistic firing rates to transitions, enabling the modelling of systems with random delays, such as in queuing or communication networks. This

E. Amparore and Ł. Mikulski (Eds.): PETRI NETS 2025, LNCS 15714, pp. 174–196, 2025.
https://doi.org/10.1007/978-3-031-94634-9_9

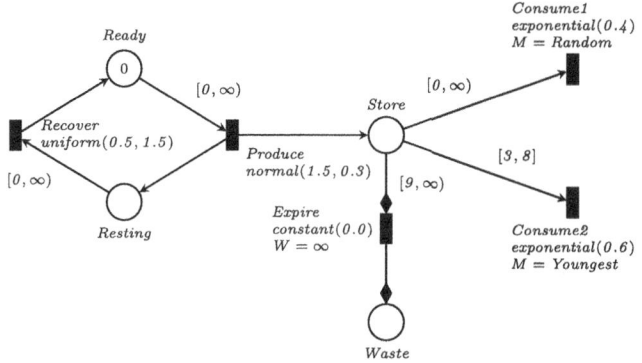

Fig. 1. Stochastic TAPN modelling a produced with two consumers

model is further extended to generalized stochastic Petri nets [3] which include also immediate transitions and further extensions with e.g. colored tokens [12]. A complex timing behaviour can be modelled by stochastic time Petri nets [14] where apart from the stochastic aspects every transition also contains a firing interval that defines the earliest and latest firing date of every transition. As there are only finitely many transitions (and hence we need to use only a fixed number of clocks), the SMC engine of UPPAAL is used in [14] to perform statistical model checking. On the other hand, the model of timed-arc Petri nets (TAPN) introduced in [7,21] (see also an overview paper [23]), requires a new clock for each token in the net and the total number of clocks cannot be a priori fixed. TAPNs equipped with transport arcs are more expressive than timed Petri nets and timed automata [9,37], however, a stochastic semantics for this model has not been suggested yet. In this paper, we consider the classical TAPN model extended with additional features like transport arcs, age invariants and inhibitor arcs as used in the model checker TAPAAL [15]. We suggest the first stochastic semantics to this extended model, implement a new statistical model checking engine as a part of the TAPAAL suite and evaluate it on a number of realistic case studies.

We finish the introduction by giving an intuition of our stochastic extension of TAPNs by describing a simple producer/consumer system depicted in Fig. 1. The net in the figure models a producer that is either in the place *Ready* or *Resting*. As soon as the transition *Produce* is enabled (e.g. in the initial marking that contains one token of age 0 in the place *Ready*), we sample its firing date $d \in \mathbb{R}_{\geq 0}$ from normal distribution with mean 1.5 and standard deviation 0.3. The system will then delay for d time units followed by firing the transition *Produce* which deposits new tokens of age 0 into the places *Resting* and *Store*. At this moment, two transitions become newly enabled, namely *Recover* and *Consume1*, and we sample their firing dates from uniform distribution between 0.5 and 1.5, and exponential distribution with rate 0.4, respectively. The system will now delay until the next interesting event occurs, meaning that either one of the two enabled transitions reaches its firing time and will fire, or a new transition becomes enabled. In the latter case, after three time units the transition *Consume2* becomes enabled as the token's age now fits into the interval [3, 8] and we sample its firing date from the exponential distribution with rate 0.6.

Now three transitions are enabled and the one with the earliest scheduled firing time will fire, unless another 5 time units pass and the transition *Consumer2* gets disabled and its scheduled firing time is deleted. Should no transition be scheduled for an additional time unit then *Expire* becomes enabled. Its firing date is sampled from the constant distribution with mean 0, meaning that it is scheduled to fire immediately and it will move the product from *Store* to *Waste* while preserving its age 9 (the special arrow types are the transport arcs which move tokens without changing their ages). The transition *Expire* has weight ∞, meaning that in case that at the same time other transitions are also enabled, *Expire* will get the priority in firing (assuming that the default weight of all other transitions is 1).

Finally, as the place *Store* can possibly contain multiple tokens of different ages, the firing modes of transitions *Consume1* and *Consume2* specify which tokens will be consumed during the firing. For *Consume1* we shall use randomly selected token that enables the transition, however, *Consume2* will always use the youngest available token in the interval [3, 8].

We may now ask what is the probability that there appears a product in the place *Waste* within the first 20 time units of execution of the system, which we can, using our SMC tool that executed 461 121 runs, estimate to have the probability 0.045 ± 0.002 with 95% confidence. This probability will approach the value 1 as the time horizon gets longer.

2   Related Work

There are two main approaches for the verification of probabilistic timed systems. Symbolic probabilistic verification methods [28] aim to produce the exact probability of observing a given property by computing or approximating the probabilistic measure. The tool PRISM [29] is known to implement such methods. On the other hand statistical model-checking (SMC) [30] generates random runs to produce an estimation of the evaluation of a property. While the symbolic method can be very costly by exploring the stochastic state space, the latter is a lot faster at the cost of precision. However, SMC is known to become costly and imprecise when it comes to the verification of rare events, and methods such as importance splitting [24] and importance sampling [25] have been introduced to solve this issue.

A version of probabilistic semantics for one-clock timed automata has been proposed in [2]. Although they cannot contain multiple clocks, their extension to networks of timed automata [17] can model systems implementing several concurrent timed probabilistic components. These are the semantics implemented in UPPAAL's SMC engine [16].

The first and the simplest stochastic extension of Petri nets (sPN) is detailed in [18]. Every transition becomes a stochastic component and samples its firing

Semantics	Time Constraints	Type
sPN (ORIS) [4]	No	Strong
DSPN (TimeNet) [39]	No	Strong
HPnGs (HYPEG) [35]	No	Strong
sTPN (UPPAAL) [14]	Yes	Strong
sTAPN (TAPAAL)	Yes	Weak

Fig. 2. Stochastic timed semantics for Petri nets

date according to a given distribution, resulting in a race where the transition with the earliest firing date wins and fires. This semantics is simple to understand but only handles transition delays without any timing constraints and every sampled date is considered valid. Verification of sPN can be done e.g. using the tool ORIS [33]. The basic stochastic Petri nets were extended with immediate transitions and transitions with deterministic delays into the Deterministic and Stochastic Petri nets (DSPN) [31], which are implemented in tools like TimeNet 4.0 [39]. Hybrid Petri nets with multiple general transitions (HPnGs) [35] also support these features, but also allow for continuous transitions. The verification of HPnGs can be done using the tool HYPEG (formerly libhpng) [34]. Both essentially use the same method for sampling dates, and thus have the same issues with missing time constraints, however, by mixing stochastic transitions with deterministic timed transitions and continuous transitions, this type of stochastic nets can model more complex timed behaviour.

There exists a stochastic extension of the Time Petri net (TPN) semantics, which achieves even more expressive power by associating a time firing interval to each transition [14]. These nets implement the *strong* semantics [8] that guarantess that a transition must be fired within its firing interval and the distributions used for the transitions are essentially scaled to match the firing intervals. An implementation using the UPPAAL SMC engine was initially used to perform SMC on these semantics [14]. In stochastic TPN time constraints are supported but are limited to the transitions in the net, which cannot be dynamically changed. On contrary, timed-arc Petri nets (TAPN) [7,21], studied in this paper, associate the timing information to potentially unbounded number of tokens in the net and input arcs to transition restrict the ages of tokens that can be consumed by transition firing. Traditionally, TAPNs rely on the *weak* semantics meaning that it is allowed to perform time delays that can disable currently enabled transitions. Further extensions with urgent transitions or age invariants [23] are needed to enforce urgent behaviour. Moreover, an extension of the timed-arc Petri net model with only transport arcs already makes the model more expressive than other types of timed nets and timed automata [9,37]. No stochastic semantics for TAPNs or other Petri net models that includes the weak (nonurgent) semantics has been given yet. This is, including an efficient implementation of the framework, the main contribution of this paper. An overview table of selected semantics and respective tools is given in Fig. 2.

3   Timed-Arc Petri Nets

We shall first introduce the semantics of timed-arcs Petri nets without the stochastic attributes. Let $\mathcal{I}$ be the set of *timed intervals* of the from $[a, b]$ where $a \in \mathbb{N}^0$, $b \in \mathbb{N}^0 \cup \{\infty\}$ and $a \leq b$. Let *time invariants* $\mathcal{I}_{inv} \subseteq \mathcal{I}$ be a subset of timed intervals of the form above where $a = 0$.

Definition 1 (Timed-Arc Petri net). *A timed-arc Petri net (TAPN) is a 7-tuple $\mathcal{N} = (P, T, IA, OA, \mathsf{Transport}, \mathsf{Inhib}, \mathsf{Inv})$ where*

- *P is a finite set of places,*
- *T is a finite set of transitions such that $P \cap T = \emptyset$,*
- *$IA \subseteq P \times \mathcal{I} \times \mathbb{N} \times T$ is the set of input arcs that connect a place to a transition and are annotated by a time interval and an arc weight, such that if $(p, I, w, t) \in IA$ and $(p, I', w', t) \in IA$ then $I = I'$ and $w = w'$,*
- *$OA \subseteq T \times \mathbb{N} \times P$ is the set of weighted output arcs that connect transitions to places such that if $(t, w, p) \in OA$ and $(t, w', p) \in OA$ then $w = w'$,*
- *$\mathsf{Transport} \subseteq IA \times OA$ is the set of transport arcs such that whenever $((p, I, w, t), (t', w', p')) \in \mathsf{Transport}$ then $t = t'$ and $w = w'$, and if $(\alpha, \beta), (\alpha, \beta') \in \mathsf{Transport}$ then $\beta = \beta'$ and symmetrically if $(\alpha, \beta), (\alpha', \beta) \in \mathsf{Transport}$ then $\alpha = \alpha'$,*
- *$\mathsf{Inhib} \subseteq P \times \mathbb{N} \times T$ is the set of weighted inhibitor arcs, and*
- *$\mathsf{Inv} : P \to \mathcal{I}_{inv}$ is the function assigning age invariants to places.*

We note that the definition implies that a given place and a transition cannot be connected by both the normal and transport arc at the same time. For a transition t, we denote by $\mathsf{Pre}(t) = \{p \in P \mid (p, I, w, t) \in IA\}$ the set of input places, and by $\mathsf{Post}(t) = \{p' \in P \mid (t, w', p') \in OA\}$ the set of output places. A marking represents the distribution of tokens together with their ages across the places in the net.

Definition 2 (Marking). *A marking M is a multiset over $P \times \mathbb{R}_{\geq 0}$ such that (p, x) represents a token of age x in the place p. A marking M is valid if, for each place $p \in P$, whenever $(p, x) \in M$ then $x \in \mathsf{Inv}(p)$. We denote the set of all valid markings by $\mathcal{M}$.*

For a marking $M \in \mathcal{M}$, we use the notation $M(p) = \{x \mid (p, x) \in M\}$ to denote the multiset of token ages in the place p.

Definition 3 (Enabled transition). *A transition $t \in T$ is enabled in a marking M if there exist multisets of tokens $In \subseteq \mathsf{Pre}(t) \times \mathbb{R}_{\geq 0}$ and $Out \subseteq \mathsf{Post}(t) \times \mathbb{R}_{\geq 0}$ such that*

- *$In \subseteq M$, i.e. the In set contains only tokens present in the marking M,*
- *for every $(p, I, w, t) \in IA$ and every $x \in In(p)$ we have $x \in I$ and $|In(p)| = w$, i.e. each input arc has exactly w tokens in its input place, all of them satisfying its time guard,*

- $|M(p)| < w$ for every $(p, w, t) \in$ Inhib, i.e. none of the inhibitor arcs has enough tokens in its input place to inhibit the firing of t,
- for every $((p, I, w, t), (t, w, p')) \in$ Transport and every $x \in In(p)$ we have $x \in$ Inv(p') and $x \in Out(p')$ where $|Out(p')| = w$, i.e. all w tokens in p can be moved to p' and still satisfy the age invariant of p', and
- for any $(t, w', p') \in OA$ that does not appear in Transport we have $|Out(p')| = w'$ and $x = 0$ for every $x \in Out(p')$, i.e. every output arc that is not a transport arc must create exactly w' new tokens of age 0.

We denote the set of transitions enabled in a marking M by en(M).

Definition 4 (Transition firing). If t is enabled in a marking M by the multisets of tokens In and Out then t can fire and produce a marking

$$M' = (M \backslash In) \cup Out$$

where $\backslash$ and $\cup$ are operations on multisets, and we write $M \xrightarrow{t} M'$.

Definition 5 (Delay). Let M be a marking and let $d \in \mathbb{R}_{\geq 0}$ be a real number. We can delay d time units from M if $x + d \in$ Inv(p) for all $(p, x) \in M$, and write $M \xrightarrow{d}$ to indicate that a delay is possible. If the delay is possible, we write $M \xrightarrow{\delta} M[d]$, where $M[d]$ is a marking defined by $M[d] = \{(p, x+d) \mid (p, x) \in M\}$.

A *run* of a TAPN from an initial marking M_0 is an alternating sequence of time delays and transition firings such that

$$M_0 \xrightarrow{d_0} M_0[d_0] \xrightarrow{t_0} M_1 \xrightarrow{d_1} M_1[d_1] \xrightarrow{t_1} M_2 \xrightarrow{d_2} M_2[d_2] \xrightarrow{t_2} \ldots$$

A run is *maximal* if it is either infinite or ends in a deadlock, i.e. in a marking M such that for any possible delay d we have en$(M[d]) = \emptyset$. By runs$(\mathcal{N})$ we denote the set of all maximal runs of $\mathcal{N}$.

3.1 A Logic for Reasoning About Runs

We shall now define a logic to argue about marking properties. A *marking property* φ is a Boolean combination of atomic propositions of the form $p \bowtie n$ where $p \in P$, $n \in \mathbb{N}^0$ and $\bowtie \in \{<, \leq, =, \geq, >\}$ with the obvious semantics that a marking M satisfies an atomic property $p \bowtie n$ if and only if $|M(p)| \bowtie n$, i.e. the number of tokens in a place p satisfies the constraint imposed by the atomic proposition. This is naturally extended to Boolean connectives and we write $M \models \varphi$ if a marking M satisfies the marking property φ. Let Φ be the set of all marking properties. Note that properties that involve the age of the tokens can be encoded by introducing "monitoring" transitions with appropriate time intervals and checking their enabledness.

A run of a TAPN satisfies the (reachability) formula $F\varphi$ if it contains a marking M such that $M \models \varphi$, and it satisfies the formulate $G\varphi$ if every marking M on the run satisfies $M \models \varphi$.

4   Stochastic Timed-Arc Petri Nets

Let $\mathcal{N} = (P, T, \mathsf{IA}, \mathsf{OA}, \mathsf{Transport}, \mathsf{Inhib}, \mathsf{Inv})$ be a TAPN as defined in the previous section. We shall now extend it with stochastic features by adding three additional functions[1] assigning (i) density function to transitions (for sampling firing delays), (ii) weights to transition (for the resolution for firing conflicts) and (iii) firing mode to transitions (for deciding which tokens are used in transition firing).

The function $\mathsf{density} : T \to \mathcal{F}(\mathbb{R}, \mathbb{R}_{\geq 0})$ assigns a probability density function[2] for each transition. It is used to sample firing dates of newly enabled transitions. When we sample a delay for a transition $t \in T$ according to $\mathsf{density}(t)$, we assume that we choose a random value according to the distribution, except for the case when the sampled value is negative, in which case we return the delay of 0.

The function $\mathsf{weight} : T \to \mathbb{R}_{\geq 0} \cup \{\infty\}$ assigns a weight (priority mass) to each transition. The weights are used in the event of a firing date collision in case several transitions sample the same firing date. Let $T' \subseteq T$ be the set of transitions that chose the same firing date then the probability of firing $t \in T'$ is given by a weighted uniform choice

$$\frac{\mathsf{weight}(t)}{\sum_{t' \in T'} \mathsf{weight}(t')} .$$

If one of the competing transitions has an infinite weight then it will always win the competition, except for the situation where there are several transitions with infinite weight—in this case we choose uniformly between them. On the other hand, if one of the competing transitions has a zero weight, then it will never be chosen, unless every other competing transition also has a zero weight—in that case we choose again uniformly among the zero-weight transitions.

Finally, the firing mode function $\mathsf{mode} : T \to \{\mathit{Youngest}, \mathit{Oldest}, \mathit{Random}\}$ is a function that determines for each transition which tokens are consumed when the transition is fired in a given marking. Assume that $t \in T$ is enabled in M by the multisets of tokens $In_1, \ldots, In_n$ (enumerating all the possiblities). If $\mathsf{mode}(t) = \mathit{Youngest}$ then the transition t will be fired using the set In_i, $1 \leq i \leq n$, that minimizes the sum of ages of all tokens in In_i and similarly if $\mathsf{mode}(t) = \mathit{Oldest}$ then the sum will be maximized. In case that there are several such sets that minimize/maximize the sum, we uniformly choose one such set. In case $\mathsf{mode}(t) = \mathit{Random}$, we select In_i with uniform probability i/n.

Stochastic TAPNs will use a *single-server policy* [5], meaning that once a transition has fired, it starts its firing process again (if enabled); and an *enabling memory policy* [5], meaning that firing dates are stored as long as their transition is enabled, and then get forgotten as soon as it gets disabled. Transition firings that both consume and produce a token to some place do not change enabledness of other transitions that also consume from such a place.

---

[1] These functions are defined over the domain of real numbers but in the actual implementation are represented as doubles.

[2] Where for each $t \in T$, the area under the function is one: $\int_{\mathbb{R}} \mathsf{density}(t)(\theta)d\theta = 1$.

4.1   Algorithm for Generating Random Runs

The heart of an SMC algorithm is the generation of random runs from an initial marking M_0, until we reach a deadlock, a given time or step (number of transition firings) bound, or a marking that satisfies a given marking property. The generation of random runs is executed according to the following strategy.

- Newly enabled transitions randomly sample their firing date according to the corresponding density function.
- The net then delays to the next interesting moment where either one of the enabled transitions is scheduled to fire or the enabledness of transitions changes and their scheduled firing dates get updated (if a transition becomes disabled, it is unscheduled and if it becomes newly enabled, we sample according to its density function).
- If several transitions are scheduled to fire at the same date, we select the winner according to the function weight and fire the transition using the tokens selected by the function mode. The winner then resets its firing date.
- By firing the winning transition, other transitions may become enabled or disabled. We update their scheduled firing dates accordingly and repeat the whole process until a marking property is satisfied (in which case we return true) or until we exceed the given number of steps (transition firings) or the given time horizon. In this case we return false.

The random run generation is formally described in Algorithm 1. An explanation of this algorithm follows.

1. While the accumulated delay and steps are under the specified bounds, we repeat until reaching a marking satisfying the given marking property, or a deadlock:
   (a) We look for the next smallest interesting delay, which is a date where the enabledness of some transition changes (line 7). In particular the condition (ii) expresses the fact that when an age of a token reaches the upper bound of a time interval on some arc, an arbitrarily small delay can disable a currently enabled transition.
   (b) If we cannot find such delay, and there are no scheduled transition firings, we reached a deadlock and terminate (line 8).
   (c) Otherwise, we delay up to the minimum between the smallest interesting delay and the earliest scheduled firing date (line 11).
   (d) If more than one transition is scheduled to fire at this date, we randomly choose a winner t using the weight function, and we fire it by selecting a multiset of consumed tokens according to mode(t) (lines 15 to 21).
   (e) We update the scheduled firing dates of each transition: disabled transitions are scheduled at ∞, the firing date of scheduled transitions is shifted according to the delay and newly enabled transitions are sampled according to their density functions (lines 22 to 28). The check at line 27 truncates negative sampled values to zero.
   (f) We check if any scheduled transition will get disabled after an arbitrary small delay and unschedule such a transition (line 29).

Algorithm 1. RandRun$_\mathcal{N}(M_0, \varphi, c, s)$

1: **Input:** Marking M_0, property φ, time bound $c \geq 0$, and steps bound $s \geq 0$
2: **Output:** Boolean indicating if a random run generated from M_0 contains a marking satisfying φ before c time units passed and k transitions were fired

3: **for all** $t \in T$: scheduled$(t) \leftarrow$ sample from density(t) if $t \in$ en(M_0), else ∞
4: AccumulatedDelay $\leftarrow 0$; AccumulatedSteps $\leftarrow 0$; $M \leftarrow M_0$
5: **while** AccumulatedDelay $\leq c$ and AccumulatedSteps $\leq s$ **do**
6:     **if** $M \models \varphi$ **then return** true
7:     select the smallest positive delay d, $0 < d \in \mathbb{R}_{\geq 0}$, satisfying
        (i) en$(M) \neq$ en$(M[d])$, or
        (ii) $\exists \delta > 0. \forall \varepsilon \in (0, \delta].$ en$(M) \neq$ en$(M[d + \varepsilon])$.
8:     **if** no such d exists and scheduled$(t) = \infty$ for all $t \in T$ **then**
9:         **return** false   ▷ We are in a deadlock
10:     **else**
11:         $d \leftarrow \min\{ d, \min_{t \in T}$ scheduled$(t) \}$ ▷ Select an earliest interesting delay
12:     **end if**
13:     $M \leftarrow M[d]$ ▷ Advance to the next interesting time point
14:     AccumulatedDelay $\leftarrow$ AccumulatedDelay $+ d$
15:     candidates $\leftarrow \{t \in T \mid$ scheduled$(t) - d = 0\}$ ▷ Transitions that can fire
16:     **if** candidates is not empty **then**
17:         randomly select $t_{\text{fired}} \in$ candidates using the weight function weights
18:         let $M \xrightarrow{t_{\text{fired}}} M'$ according to mode(t_{fired}); $M \leftarrow M'$
19:         scheduled$(t_{\text{fired}}) \leftarrow \infty$ ▷ The executed transition is unscheduled
20:         AccumulatedSteps $\leftarrow$ AccumulatedSteps $+ 1$
21:     **end if**
22:     **for** $t \in T$ **do**   ▷ Update the scheduled firing times of each $t \in T$
23:         **if** $t \notin$ en(M) **then**
24:             scheduled$(t) \leftarrow \infty$
25:         **else**
26:             scheduled$(t) \leftarrow \begin{cases} \text{scheduled}(t) - d & \text{if scheduled}(t) \neq \infty \\ \text{sample from density}(t) & \text{otherwise} \end{cases}$
27:             scheduled$(t) \leftarrow 0$ if scheduled$(t) < 0$
28:         **end if**
29:         **if** scheduled$(t) > 0$ and $\exists \delta > 0. \forall \varepsilon \in (0, \delta]. t \notin$ en$(M[\varepsilon])$ **then**
30:             scheduled$(t) \leftarrow \infty$ ▷ Unschedule t as it will get disabled
31:         **end if**
32:     **end for**
33: **end while**
34: **return** false

We shall first point out that we do not need urgent transitions (once enabled time cannot elapse) in the stochastic semantics, as urgency can be simulated by Dirac density function which always samples the value 0 and hence time cannot elapse as long as at least one such urgent transition is enabled. Contrary to non-stochastic semantics, age invariants in places cannot be used to enforce urgency.

Algorithm 2. ProbabilityEstimation$_\mathcal{N}(M_0, \varphi, c, s, \rho, \epsilon)$

1: **Input:** Marking M_0, property φ, time bound $c \geq 0$, steps bound $s \geq 0$, confidence
   $0 < \rho < 1$, precision $0 < \epsilon < 1$
2: **Output:** The probability $\pm\epsilon$ that φ is satisfied with confidence ρ in no more than
   c time units and s steps
3: $N \leftarrow \frac{\ln(2/(1-\rho))}{2\epsilon^2}$ ▷ The number of runs to execute
4: $x \leftarrow 0$
5: **for** $i = 1$ to N **do**
6:     $x \leftarrow x + \mathsf{RandRun}_\mathcal{N}(M_0, \varphi, c, s)$ ▷ Consider true as 1 and false as 0
7: **end for**
8: **return** $\frac{x}{N}$

Once a token age in a place reaches the invariant upper bound, the execution of the random run will end in a deadlock (unless some transition was actually scheduled to fire at that time).

The statistical model checking algorithm presented in Algorithm 2 now uses the random run generator to perform Monte-Carlo simulations for quantitative probability estimation [17]. It uses the Chernoff-Hoeffding bound to compute, for a given precision ϵ and a confidence level ρ, the number of runs N to be executed in the SMC algorithm. Now, a confidence-interval estimating process—that executes N runs and returns the proportion of φ-satisfying runs—will have probability larger than ρ of returning a value that is no more than ϵ away from the real (unknown) probability of φ being satisfied.

4.2 Examples of Random Runs

In order to better understand the details of the random run generation algorithm, we shall now exemplify it on a number of examples presented in Fig. 3. In these example, the default $[0, \infty)$ time intervals as well as the default weight 1 and the default random firing mode are omitted.

Let us first consider the net from Fig. 3a. Initially, t_0 is not enabled so we will delay to the next interesting event, which is after 3 time units where t_0 becomes enabled. At this moment, we sample a delay d uniformly from the interval $[0, 5]$. If $3 + d \leq 5$ then the next interesting delay is $3 + d$ where we fire t_0 and reach a deadlock. If $3 + d > 5$ then the next interesting delay is 2 time units and the token in p_0 reaches the age 5. As t_0 is not yet scheduled to fire at this point and after arbitrarily small delay t_0 becomes disabled, we unschedule the firing of t_0 as the check at line 29 of the algorithm succeeds. The run then deadlocks.

A race behaviour between two transitions is shown in Fig. 3b. Initially, t_0 is enabled and it chooses a firing date d_0 uniformly from the interval $[0, 30]$. If $d_0 \leq 10$ then t_0 will fire as the next interesting delay is d_0 where t_0 is the only enabled transition. If $d_0 > 10$ then at 10 units the transition t_1 will also sample its firing delay d_1 from the interval $[0, 30]$. Now if $d_0 - 10 < d_1$ and $d_0 \leq 15$ then the transition t_0 wins the race. Should $d_0 - 10 > d_1$ then t_1 wins the race and fires. If none of the transitions fired until 15 time units, t_0 gets unscheduled.

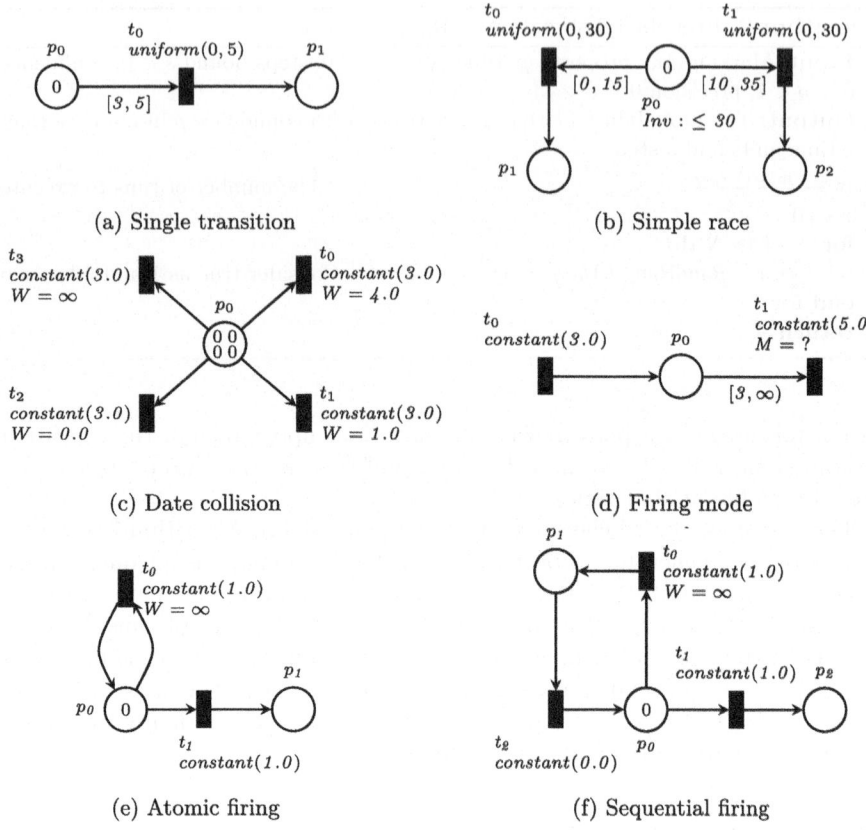

Fig. 3. Examples of stochastic timed-arc Petri nets

Depending on the sampled delay d_1, either t_1 fires provided that $d_1 + 10 \leq 30$, or the whole net deadlocks if $d_1 + 10 > 30$ due to the invariant on p_0 that blocks any time progress once the token in p_0 reaches the age 30. Note that the firing date collision of t_0 and t_1 is unlikely as the probability that the transition delays are sampled so that $d_0 = d_1 + 10$ is zero.

However, in case of constant distributions, collision of firing dates is possible as demonstrated in Fig. 3c. As there are four tokens of age 0 in the initial marking, each of the four transitions can fire. The scheduled firing data of each transition will be 3 and we need to resolve the probability in which order they fire. As the weight of t_3 is infinity, it will be always the first transition to fire. After firing t_3, three transitions remain enabled. Transition t_0 will fire next with the probability 4/5 and transition t_1 will fire with the probability 1/5. After both t_0 and t_1 fired, t_2 will fire last as its weight is 0 and it could not compete with t_0 and t_1.

The example in Fig. 3d shows the influence of the firing mode (depicted by $M = mode$ where $mode \in \{Random, Youngest, Oldest\}$) on executions. The transition t_0 produces a token every 3 time units. Once the token in p_0 reaches the age 3, also t_1 becomes enabled and will fire after 5 time units. At that point, there will be tokens of ages 8, 5, and 2 in p_0, and t_1 needs to choose which one of the token ages satisfying the time interval $[3, \infty)$ it consumes during the firing. If mode$(t_1) = $ *Youngest*, then t_1 select the token of age 5. If mode$(t_1) = $ *Oldest*, it select the token of age 8. If mode$(t_1) = $ *Random*, then t_1 uniformly chooses between the tokens of ages 5 and 8.

Finally, Figs. 3e and 3f demonstrate a difference when a transition gets unscheduled during the random run generation. In Fig. 3e the transitions t_0 and t_1 get scheduled at time 1 as they both sample from the constant distribution which always returns 1. As the weight of t_0 is ∞ and the default weight of t_1 is 1, the transition t_0 fires first, resets the age of the token in p_0 to 0, and samples its next firing date (which is again after 1 time unit). However, after the firing t_0, the transition t_1 is still enabled so its scheduled firing date does not change and it fires (without any further delay) immediately after t_0. The reader may wonder that during the firing of t_0, the transition t_1 got temporarily disabled and it should be resampled too. If such behaviour is desirable then it can be modelled as shown in Fig. 3f. In this case the probability that the transition t_1 fires is zero, as it gets always resampled after any firing of t_0 (which always wins the race as its weight is ∞).

4.3   Induced Probabilistic Semantics

The random run algorithm for a stochastic TAPN $\mathcal{N}$ defines an uncountable set of possible runs and induces a probabilistic measure $\mathbb{P}_\mathcal{N}$ on infinite sets of runs of the σ-algebra generated by cylinders of runs of the form

$$\pi = I_0.t_0.\ldots.I_n.t_n$$

where $I_i \in \mathcal{I}$ is a time interval and $t_i \in T$ is a transition for all i, $0 \leq i \leq n$. Given a run $\rho = M_0 \xrightarrow{d_0} M_0[d_0] \xrightarrow{t_0} \rho' \in \text{runs}(\mathcal{N})$ and a cylinder $\pi = I_0.t_0'.\pi'$, we write $\rho \in \pi$ if $d_0 \in I_0$, $t_0 = t_0'$ and $\rho' \in \pi'$; for the base-case we postulate that $\rho \in \varepsilon$ for any run ρ. In other words, the prefix of the concrete run must agree on the transition firings with the cylinder and the concrete delays must belong to the intervals in the cylinder.

For the sake of simplicity, we assume that every density function is continuous, hence no firing date collision may occur. This means that the scheduling function scheduled from Algorithm 1 satisfies that at most one transition may be fireable at the minimum scheduled delay. In the same way, we assume that the firing mode for each transition is deterministic, meaning that given an origin marking and a transition t_f, there is a unique marking reachable by firing t_f. At any point during the algorithm, the next possible delay d is deterministically determined as the minimum between scheduled firing date or the time points where transition enabledness changes. There are three cases, (i) either the next delay is a transition firing date and then a firing is triggered, or (ii) the next delay is an enabledness change and no transition fires, or (iii) there is no next delay and we reach a deadlock.

Therefore, given a set of persistent transitions pers (enabled before and after the delay and firing) and a set of newly-enabled transitions newen (disabled before, but enabled after the delay and firing), a previous scheduling function scheduled, and a delay $d \in \mathbb{R}_{\geq 0}$, we define the set of next scheduling functions SCH(d) as:

$$\mathsf{ns} \in \mathsf{SCH}(d) \iff \begin{cases} \mathsf{ns}(t_i) = \mathsf{scheduled}(t_i) - d & \text{if } t_i \in \mathsf{pers} \\ \mathsf{ns}(t_i) \in \mathbb{R}_{\geq 0} & \text{if } t_i \in \mathsf{newen} \\ \mathsf{ns}(t_i) = +\infty & \text{otherwise .} \end{cases} \tag{1}$$

By D we denote the density function on SCH(d) which assigns $D(\mathsf{ns}) \in \mathbb{R}_{\geq 0}$ to every scheduling function $\mathsf{ns} \in \mathsf{SCH}(d)$ given the set of newly-enabled transitions newen as follows:

$$D(\mathsf{ns}) = \prod_{t_i \in \mathsf{newen}} \mathsf{density}(t_i)(\mathsf{ns}(t_i)) . \tag{2}$$

Finally, let $\mathbb{1}_{cond}$ be an indicator variable equal to 1 if condition $cond$ is true, else 0. The probabilistic measure is now defined on cylinders using one of three formulas. In the main case (i), assuming that in a marking M the given function scheduled determines to fire the transition t_f after a delay d and reach a new marking M', the measure is defined inductively as:

$$\mathbb{P}_{\mathcal{N}}((M, \mathsf{scheduled}), I_0.t_0.\pi') = \mathbb{1}_{d \in I_0} \cdot \mathbb{1}_{t_f = t_0} \cdot \left(\int_{s \in \mathsf{SCH}(d)} D(s) \cdot \mathbb{P}_{\mathcal{N}}((M', s), \pi') \, \mathrm{d}s \right) .$$

The other two cases (ii) and (iii) are defined analogously. Observing that $\mathbb{P}_{\mathcal{N}}((M, \mathsf{scheduled}), \pi)$ decreases when extending the cylinder π, it follows from classical concepts of probability theory (see e.g. [1]) that $\mathbb{P}_{\mathcal{N}}$ extends uniquely to the smallest σ-algebra, $\mathfrak{S}^{\mathcal{N}}$, generated by the above cylinders.

Theorem 1. *Given a stochastic TAPN $\mathcal{N}$, $\mathbb{P}_{\mathcal{N}}$ is a probability measure over the σ-algebra $\mathfrak{S}^{\mathcal{N}}$ generated by the set of cylinders of the form $(\mathcal{I}.T)^*$.*

Most importantly $F\varphi$ and $G\varphi$ properties define measurable sets of runs, thus having well-defined probabilities as postulated by the next theorem.

Theorem 2. *Given a marking property $\varphi \in \Phi$, the formulae $F\varphi$ and $G\varphi$ describe sets of runs belonging to the σ-algebra $\mathfrak{S}^{\mathcal{N}}$, thus having well-defined probabilities $\mathbb{P}_{\mathcal{N}}(F\varphi)$ and $\mathbb{P}_{\mathcal{N}}(G\varphi)$.*

The last theorem proves that the computed probabilities indeed make sense, by exhibiting an equivalence with classical TAPNs.

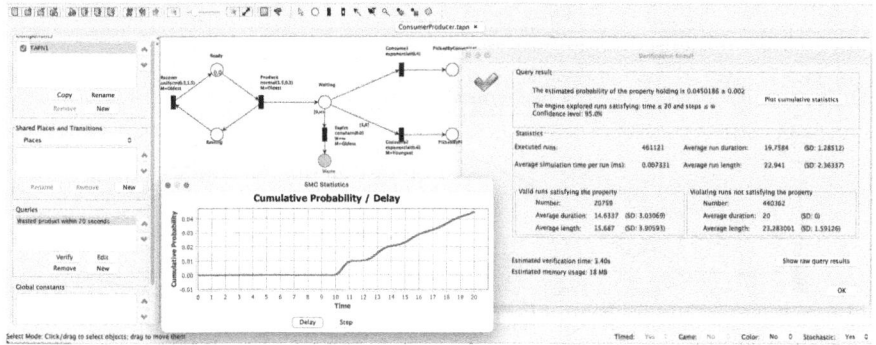

Fig. 4. TAPAAL's GUI with a running example and its SMC verification output

Theorem 3. *Let $\mathcal{N}$ be a stochastic TAPN and let $\mathcal{N}_d$ be a normal TAPN net obtained by removing all stochastic features in $\mathcal{N}$. Let φ be a marking property.*

(i) If there is no reachable marking in $\mathcal{N}_d$ that satisfies φ then $\mathbb{P}_{\mathcal{N}}(F\varphi) = 0$.
(ii) If all reachable markings in $\mathcal{N}_d$ satisfy φ then $\mathbb{P}_{\mathcal{N}}(G\varphi) = 1$.

5   Implementation in TAPAAL and Case Studies

The SMC run generation is implemented in C++ as part of the `verifydtapn` engine [26] and accepts a PNML description of stochastic TAPN together with the query and other SMC parameters and it executes Algorithm 2 and returns the computed probability as well as other useful statistics like average run length and duration, number of satisfying and violating runs and data for plotting cumulative probabilities and other plots that allow to visualize the average/minimum/maximum number of tokens in places during the simulations. The existing TAPAAL GUI [15] is extended to allow for the editing of stochastic TAPNs as well as queries, it communicates with the SMC engine and visualizes the results of SMC verification. A screenshot using our running producer/consumer example with the window showing the result with the statistics as well as the cumulative probability is shown in Fig. 4.

We currently support the uniform and geometric discrete distributions and dirac delta, uniform, exponential, normal, gamma, triangular and log normal continuous distributions. For the given set of distribution parameters, the GUI also visualizes the density function. We support three types of verification: quantitative probability estimation, using a Monte-Carlo algorithm; qualitative probability test, using a sequential probability ratio test; and a mode to generate and simulate traces, which can be either any traces, traces satisfying a property, or traces violating a property. The SMC engine has an option to use multi-threading and, if enabled, it executes one run generator per available core and achieves almost a linear speedup with increasing number of CPU cores. Moreover, we support the estimation of the total running time of the SMC algorithm from the

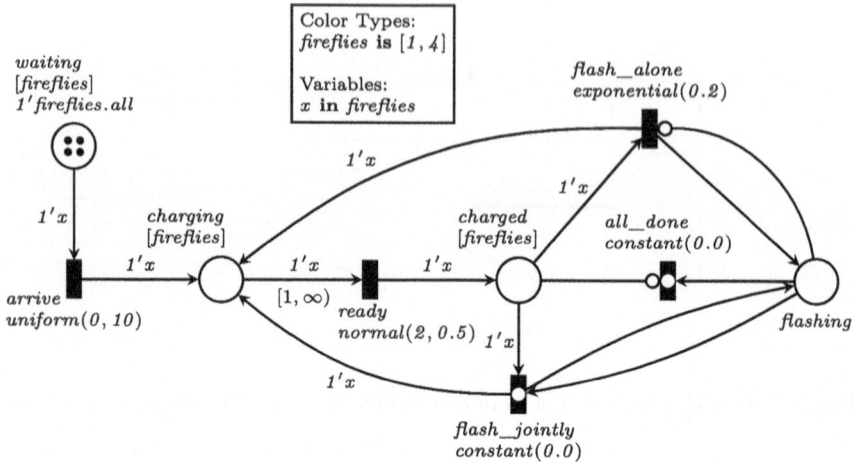

Fig. 5. Colored stochastic Petri net model of fireflies

given confidence and precision parameters, or alternatively suggest a precision to match a given running time and confidence level.

We shall now present three case studies demonstrating the capabilities of our tool. A release of TAPAAL SMC can be downloaded from https://www.tapaal.net/download/. The models used in the case studies can be found in other downloads at the bottom of the page.

5.1 Fireflies

Firefly is a beetle that is capable of emitting light flashes in periodic intervals. In nature, some firefly species like Photinus carolinus [38] show a remarkable synchronization capabilities so that a population of fireflies starts, after a while, to flash in synchrony. All this without any leader and in a completely distributed manner. This behaviour attracted the attention among scientists, trying to explain and model such behaviour (see e.g. [32,36]). In a simplified version, each firefly requires some amount of time in order to charge and be able to flash. The discharge (flash) can then occur after some random delay, unless the firefly notices a flash from other firefly, in which case it will join the flash (assuming that it is also charged).

In Fig. 5 we present a stochastic timed-arc Petri net model of the fireflies flashing behaviour. We use here the well-known colored extension of Petri nets where tokens can carry additional information (in our case the id of the firefly). Initially, there are four fireflies (tokens) in the place *waiting*, having ids one to four and initially of age 0. With uniform distribution between 0 to 10 s, the fireflies independently arrive to the place *charging*. Here each firefly must be charged before the transition *ready* moves it to the charged location. The charging takes at least one second due to the time interval $[1, \infty)$ on the arc. After the transition *ready* is enabled, an additional delay is sampled from a normal distribution

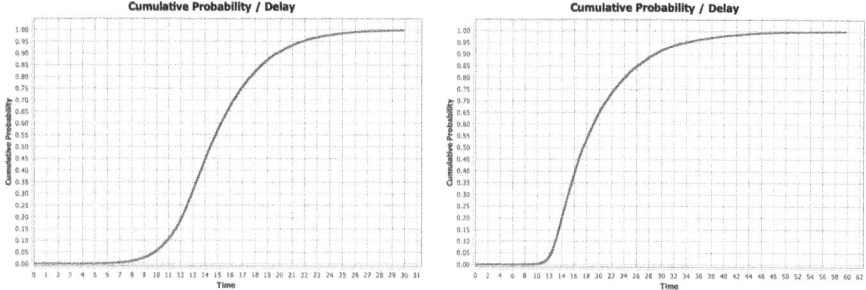

Fig. 6. Cumulative synchronization probability for 4 (left) and 10 (right) fireflies

with mean value of 2 s and a standard deviation of 0.5. Once a firefly arrives to the place *charged*, there are two situations. Either the place *flashing* has no token, meaning that there is currently no one emitting a flash. In this case the firefly waits an amount of time sampled from the exponential distribution of rate 0.2 (corresponding to the mean value of 5 s) and once it is schedule to fire the transition *flash_ alone*, it returns to the charging location. As a side-effect, a token (uncolored) is placed to the place *flashing*, which disables other fireflies from flashing alone. A token in the place *flashing* indicates that other fireflies in the place *charged* should join the flash immediately. This is done by firing the transition *flash_ jointly* which samples from the constant distribution which always returns the value 0. In other words, the transition *firing* is scheduled immediately and the transition hence becomes urgent (indicated by the circle in the middle of the transition). As soon as there are no further fireflies in the place *charged*, the urgent transition *all_ done* becomes enabled and removes the token from the place *flashing*.

In our tool, a colored Petri net is unfolded [6,13] to the standard timed-arc Petri net without colors by creating a copy of each place for each color and a copy of each transition for each binding of the variable x to different fireflies ids. The probability distributions of each transition are then simply overtaken by the unfolded copies of the transition.

We can now ask about the probability that all fireflies synchronize within 30 s. This can be formulated by the query

$$F(\text{charging} = 1 \text{ and } \text{flashing} = 1 \text{ and } \text{waiting} = 0) \qquad (3)$$

requiring that we reach a situation where there are no fireflies in the place *waiting*, exactly one firefly in the place *charging* and a token in *flashing*, indicating that a single firefly initiated a flash and all other fireflies (that are in the place *charged*) are joining in a synchronous flash. Our tool executes 18456 runs to estimate that the probability of synchronization within 30 s is 0.997 with 95% confidence and precision 0.01. If we increase the number of fireflies to 10, the probability of synchronous flashing within 30 s drops to 0.917 and we need at

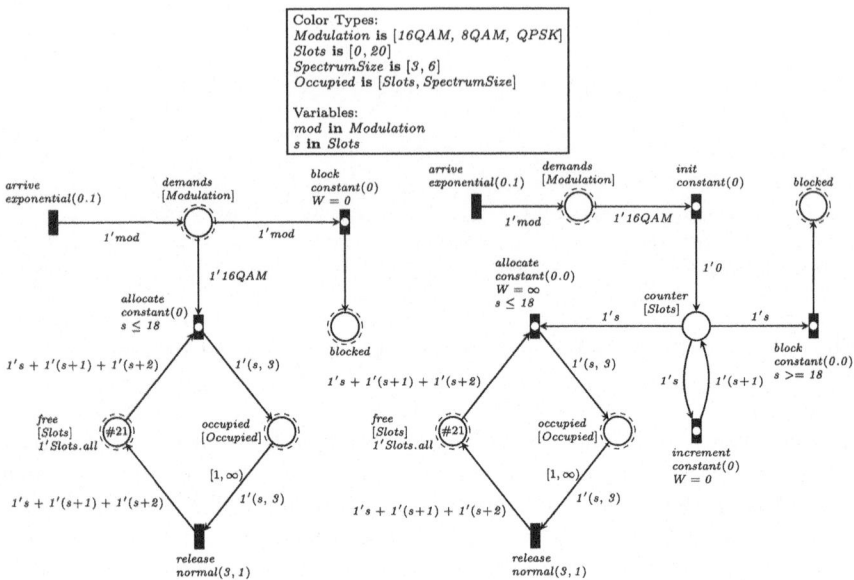

Fig. 7. Random fit (left) and first fit (right) spectrum allocation for 16QAM

least 60 s to achieve the synchronization probability 0.998. Cumulative probability plots for 4 and 10 fireflies, produced by our tool, are depicted in Fig. 6.

5.2 Frequency Spectrum Assignment in Elastic Optical Networks

Elastic all-optical networks [11] allow for fine-grained resource allocation technologies in order to schedule network traffic demands on different light frequencies inside a single optical fiber. When a new demand arrives, the spectrum allocation problem is to find for each optical fiber a sequence of available frequency slots that will carry the demand. The frequency slots must be contiguous [20] (follow each other) and the number of required slots depends on the modulation scheme [27]. After some time, a demand can be released, making the frequency slots allocated for the demand available again. This allocation/deallocation process can create a fragmentation in the allocated frequency slots, possibly resulting in a situation when a newly arrived demand is blocked (there are not enough available consecutive frequency slots to accommodate the demand).

In our stochastic timed-arc Petri net model depicted in Fig. 7, we study two spectrum allocation strategies: a *random fit* that randomly chooses from the available slots, trying to uniformly distribute the demands across the whole spectrum, and a strategy called *first fit* [10] that always uses the lowest possible available frequency slot. We consider three common modulation schemes 16QAM, 8QAM and QPSK that require 3, 4 and 6 frequency slots [27], respectively. In Fig. 7 we present the Petri net model for 16QAM only but the other two modulations are modelled anologously, just requiring a higher number of slots.

We use again the colored extension in order to allow for a more compact model description. In our model, we use the variable *mod* that ranges over the color type *Modulation*, the variable *s* ranging over the 21 frequency slots declared in the color type *Slots* and we also use the color type *Occupied* which is a Cartesian product of the *Slots* color type and the available spectrum size with values 3 to 6. The dashed circles around places denote the so-called shared places that appear also in the nets for the other modulations (not shown in our Petri net model). The shared place called *free* contains tokens with colors (frequency slots) that are currently available for allocation of a new demand. Initially, there are 21 tokens in the place *free* with color values from 0 to 20. On the other hand, the shared place *occupied* contains tokens with colors being pairs of integers (*frequency, size*) such that e.g. the value 10, 4 means that the frequency slots 10, 11, 12 and 13 are occupied by a demand that requires 4 consecutive frequency slots.

The random fit slot allocation (Fig. 7 left) gathers the arrived demands (assuming an exponential distribution of arrivals with rate 0.1) in the place called *demands*. Immediately after a demand arrival, the urgent transition *block* becomes enabled but it has weight 0, meaning that if the transition *allocate* is also enabled, it will have a priority. The transition *allocate* binds *s* to some initial frequency slot (no more than 18 as the demand requires three slots) and checks whether the frequency slots s, $s + 1$ and $s + 2$ are available (i.e. there are three tokens with these colors in the the place *free*). If this is the case, these three tokens are removed from the place *free* and a new token with the color $(s, 3)$ is added to *occupied*. After one second holding time, enforced by the interval $[1, \infty)$, the transition *release* becomes enabled and samples its firing date from the normal distribution with mean 3 and standard deviation 1 and releases these three slots once it fires. In case the transition *allocate* is enabled for several bindings of the variable *s* to different frequency slots, the used binding is selected using a uniform distribution.

In Fig. 7 (right) we describe the first fit spectrum allocation algorithm that always uses the lowest available frequency slots. This is achieved by placing a token with the color 0 (the lowest frequency slot) to the place *counter* immediately after any demand arrival. By firing the urgent transition *increment*, we then repeatedly increase the color value of the token by 1 until we find an available slot (using the same modelling approach as in the random assignment) or until we reach the frequency slot 18 where the transition *block* becomes enabled. The weight of the transition *allocate* is infinity, meaning that it has a priority over the transition *block* but in case the transition is still not enabled, the transition *block* will fire as the transition increment has weight 0.

We can now ask our tool to compare the blocking probability of the two allocation algorithms by issuing the SMC query `F blocked >= 1`, checking wheater at least one demand is blocked within the first 30 s. With 95% confidence and precision 0.01, our tool executes 18488 runs and estimates the blocking probability to 0.075 for the first fit slot allocation strategy and 0.167 for the random fit strat-

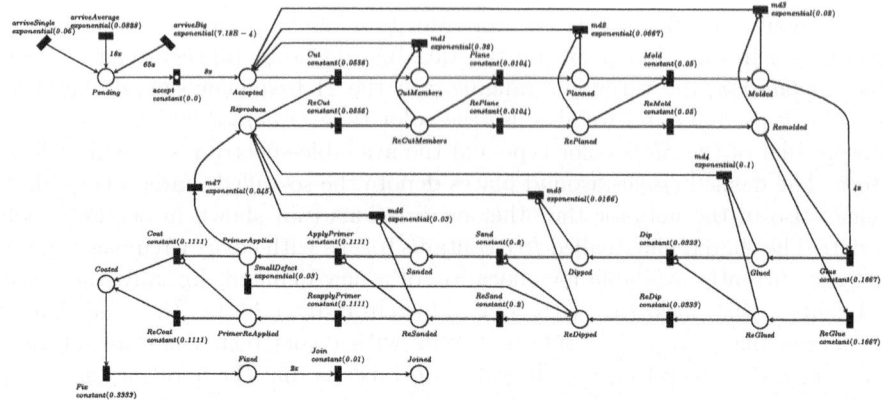

Fig. 8. Manufacturing process of wooden windows [22]

egy, showing an improved performance of the former strategy. This behaviour is also conformed by experiments in computer networking literature [10].

5.3 Manufacturing Process of Wooden Windows

The workflow of a Hungarian wooden window manufacturing company Holz-Team Ltd details a systematic production process, from preparing individual frame components to assembling and finishing complete windows. Orders vary in size, and the process includes measures to address material defects, with early-stage issues being easier to resolve than those found later. The Petri net model is given in Fig. 8 and contains timing probabilities derived from real measurements performed by the company as described in [22]. The workflow uses two types of transitions: exponentially delayed transitions for unpredictable events like order arrivals and material defects, and deterministic transitions for fixed-time production phases such as cutting, molding, and coating. For a detailed description of the process consult [22].

The production process faced a bottleneck where coated windows that required a hardware fix accumulated over time in the place called *Coated*. Indeed, using TAPAAL SMC engine we estimated that with probability 0.61 ± 0.01 and 95% confidence, the weekly production process generates more than 20 unprocessed coated windows. A solution to the bottleneck was suggested in [22] and required the hiring of an additional worker to reduce the hardware fixing operation from 20 min to 13 min. After this adjustment, the probability of storing 4 or more unprocessed windows dropped to 0.011. In Fig. 9 (left) we depict, based on 50 simulations of the workflow for a period of one year (2008 h), the number of accepted orders and the average number of unprocessed coated windows for the situation before the workflow adjustment. After the adjustment, the number of unprocessed coated windows stays on average slightly above 1 with a maximum between 3 to 5 windows, as shown in Fig. 9 (right). This allowed the company

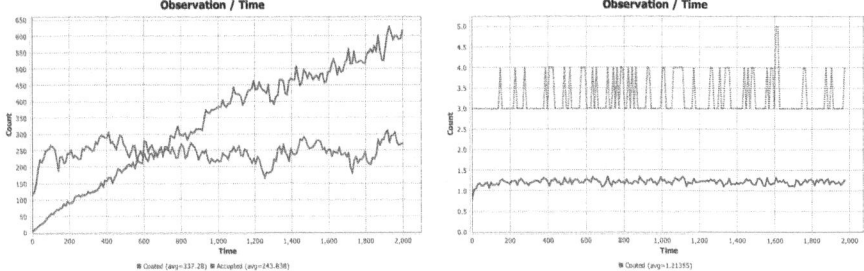

Fig. 9. Average number of orders and unprocessed coated windows over a year

to produce on average about 260 additional windows per year, increasing the productivity by almost 10%.

6   Conclusion

We proposed, to the best of our knowledge, the first stochastic semantics to the popular timed-arc Petri net model. The semantics is weak and race-based, implying that an enabled transition may not necessarily fire within its firing interval but can instead delay out of its enabledness zone, based on the firing date sampled from any probabilistic distribution (we currently support 7 continuous distributions and two discrete ones, but the approach is not limited only to these). This allows us to model a wide range of stochastic systems, as demonstrated in our case studies.

We discussed the design and implementation of the statistical model checking algorithm and argued that it is well-defined. All algorithms are implemented in the open-source model checker TAPAAL, including numerous performance optimizations (e.g. parallel execution) as well as a support for the visualization of the statistical model checking results. The initial experiments indicate that the formalism is applicable to a broad range of problems and allows us to reason about complex stochastic behaviours in an intuitive way.

In the future work, we would like to extend the SMC reachability formulae to a more powerful logic (like e.g. LTL) that will allow us to reason about probability of runs that satisfy additional requirements. An optimization of our SMC algorithm for handling rare events is another line of research.

Acknowledgements. We thank Mikkel Tygesen for his help with TAPAAL GUI and in particular the visualization of the SMC results. This paper was funded by the Villum Investigator project S4OS and the ANR project BisoUS ANR-22-CE48-0012.

References

1. Ash, R.B., Doleans-Dade, C.A.: Probability and Measure Theory. Academic Press (2000)

2. Baier, C., Bertrand, N., Bouyer, P., Brihaye, T., Größer, M.: Probabilistic and topological semantics for timed automata. In: Arvind, V., Prasad, S. (eds.) FSTTCS 2007. LNCS, vol. 4855, pp. 179–191. Springer, Heidelberg (2007). https://doi.org/10.1007/978-3-540-77050-3_15

3. Balbo, G.: Introduction to Generalized Stochastic Petri Nets, pp. 83–131. Springer, Cham (2007). https://doi.org/10.1007/978-3-540-72522-0_3

4. Bause, F., Kritzinger, P.: Stochastic Petri Nets: An Introduction to the Theory. AMS (2013)

5. Bérard, B., Cassez, F., Haddad, S., Lime, D., Roux, O.H.: Comparison of different semantics for time Petri nets. In: Peled, D.A., Tsay, Y.K. (eds.) Automated Technology for Verification and Analysis, pp. 293–307. Springer, Heidelberg (2005)

6. Bilgram, A., Jensen, P., Pedersen, T., Srba, J., Taankvist, P.: Methods for efficient unfolding of colored Petri nets. Fund. Inform. 189(3–4), 297–320 (2023)

7. Bolognesi, T., Lucidi, F., Trigila, S.: From timed petri nets to timed LOTOS. In: Proceedings of the IFIP WG6.1 Tenth International Symposium on Protocol, pp. 395–408 (1990)

8. Boyer, M., Roux, O.H.: Comparison of the expressiveness of arc, place and transition time petri nets. In: Kleijn, J., Yakovlev, A. (eds.) ICATPN 2007. LNCS, vol. 4546, pp. 63–82. Springer, Heidelberg (2007). https://doi.org/10.1007/978-3-540-73094-1_7

9. Boyer, M., Roux, O.H.: On the compared expressiveness of arc, place and transition time Petri nets. Fundam. Inf. 88(3), 225–249 (2008)

10. Chatterjee, B.C., Oki, E.: Performance evaluation of spectrum allocation policies for elastic optical networks. In: 2015 17th International Conference on Transparent Optical Networks (ICTON), pp. 1–4 (2015). https://doi.org/10.1109/ICTON.2015.7193485

11. Chatterjee, S., Pawlowski, S.: All-optical networks. Commun. ACM 42(6), 74–83 (1999). https://doi.org/10.1145/303849.303865

12. Chiola, G., Dutheillet, C., Franceschinis, G., Haddad, S.: Stochastic well-formed colored nets and symmetric modeling applications. IEEE Trans. Comput. 42(11), 1343–1360 (1993). https://doi.org/10.1109/12.247838

13. Christesen, N., Glavind, M., Schmid, S., Srba, J.: Latte: Improving the latency of transiently consistent network update schedules. In: Proceedings of 38th International Symposium on Computer Performance, Modeling, Measurements and Evaluation (PERFORMANCE 2020). ACM SIGMETRICS Performance Evaluation Review, vol. 48, no. 3, pp. 14–26. ACM (2020). https://doi.org/10.1145/3453953.3453957

14. Cicirelli, F., Nigro, C., Nigro, L.: Qualitative and quantitative evaluation of stochastic Time Petri Nets. In: 2015 Federated Conference on Computer Science and Information Systems (FedCSIS), pp. 763–772 (2015). https://doi.org/10.15439/2015F69, https://ieeexplore.ieee.org/abstract/document/7321519

15. David, A., Jacobsen, L., Jacobsen, M., Jørgensen, K.Y., Møller, M.H., Srba, J.: TAPAAL 2.0: integrated development environment for timed-arc petri nets. In: Flanagan, C., König, B. (eds.) TACAS 2012. LNCS, vol. 7214, pp. 492–497. Springer, Heidelberg (2012). https://doi.org/10.1007/978-3-642-28756-5_36

16. David, A., et al.: Statistical model checking for stochastic hybrid systems. Electron. Proc. Theor. Comput. Sci. 92, 122–136 (2012). https://doi.org/10.4204/EPTCS.92.9, http://arxiv.org/abs/1208.3856, arXiv:1208.3856 [cs]

17. David, A., et al.: Stochastic semantics and statistical model checking for networks of priced timed automata (2014). https://doi.org/10.48550/arXiv.1106.3961, http://arxiv.org/abs/1106.3961, arXiv:1106.3961 [cs] version: 2

18. Florin, G., Fraize, C., Natkin, S.: Stochastic petri nets: properties, applications and tools. Microelectron. Reliab. **31**(4), 669–697 (1991). https://doi.org/10.1016/0026-2714(91)90009-V
19. Gardiner, C.: Stochastic Methods: A Handbook for the Natural and Social Sciences. Springer Series in Synergetics, Springer, Heidelberg (2010)
20. Hai, D.T., Morvan, M., Gravey, P.: Combining heuristic and exact approaches for solving the routing and spectrum assignment problem. IET Optoelectron. **12**(2), 65–72 (2018). https://doi.org/10.1049/iet-opt.2017.0013
21. Hanisch, H.M.: Analysis of place/transition nets with timed arcs and its application to batch process control. In: Application and Theory of Petri Nets 1993, pp. 282–299. Springer, Cham (1993)
22. Horváth, Á.: Usability of deterministic and stochastic Petri nets in the wood industry: a case study. In: Advanced Computational Methods for Knowledge Engineering, pp. 119–127. Springer, Cham (2014)
23. Jacobsen, L., Jacobsen, M., Møller, M.H., Srba, J.: Verification of timed-arc petri nets. In: Černá, I., et al. (eds.) SOFSEM 2011. LNCS, vol. 6543, pp. 46–72. Springer, Heidelberg (2011). https://doi.org/10.1007/978-3-642-18381-2_4
24. Jegourel, C., Legay, A., Sedwards, S.: Importance splitting for statistical model checking rare properties, p. 576 (2013). https://doi.org/10.1007/978-3-642-39799-8_38, https://inria.hal.science/hal-01087826
25. Jegourel, C., Legay, A., Sedwards, S.: Command-based importance sampling for statistical model checking. Theor. Comput. Sci. **649**, 1–24 (2016). https://doi.org/10.1016/j.tcs.2016.08.009, https://www.sciencedirect.com/science/article/pii/S0304397516303966
26. Jensen, P., Larsen, K., Srba, J., Sørensen, M., Taankvist, J.: Memory efficient data structures for explicit verification of timed systems. In: Proceedings of the 6th NASA Formal Methods Symposium (NFM 2014). LNCS, vol. 8430, pp. 307–312. Springer, Cham (2014). https://doi.org/10.1007/978-3-319-06200-6_26
27. Kawaguchi, K., Seki, Y., Tanigawa, Y., Hirota, Y., Tode, H.: Proactive modulation format allocation method with selective downgrading to enhance inter-core crosstalk tolerance in sdm-eons. In: 2024 15th International Conference on Network of the Future (NoF), pp. 159–163 (2024). https://doi.org/10.1109/NoF62948.2024.10741410
28. Kwiatkowska, M., Norman, G., Parker, D.: Probabilistic model checking: advances and applications. In: Formal System Verification, pp. 73–121. Springer (2017)
29. Kwiatkowska, M., Norman, G., Parker, D.: PRISM 4.0: verification of probabilistic real-time systems. In: Gopalakrishnan, G., Qadeer, S. (eds.) Computer Aided Verification, pp. 585–591. Springer, Heidelberg (2011). https://doi.org/10.1007/978-3-642-22110-1_47
30. Legay, A., Lukina, A., Traonouez, L.M., Yang, J., Smolka, S.A., Grosu, R.: Statistical Model Checking, pp. 478–504. Springer, Cham (2019). https://doi.org/10.1007/978-3-319-91908-9_23
31. Marsan, M.A., Chiola, G.: On Petri nets with deterministic and exponentially distributed firing times. In: Rozenberg, G. (ed.) Advances in Petri Nets 1987, pp. 132–145. Springer, Heidelberg (1987).https://doi.org/10.1007/3-540-18086-9_23
32. McCrea Madeline, E.B., E., R.J.: A model for the collective synchronization of flashing in photinus carolinus. J. of Royal Soc. Interface (19) (2022)
33. Paolieri, M., Biagi, M., Carnevali, L., Vicario, E.: The ORIS tool: quantitative evaluation of non-Markovian systems. IEEE Trans. Softw. Eng. **47**(6), 1211–1225 (2021). https://doi.org/10.1109/TSE.2019.2917202, https://ieeexplore.ieee.

org/document/8719961, conference Name: IEEE Transactions on Software Engineering

34. Pilch, C., Edenfeld, F., Remke, A.: HYPEG: statistical model checking for hybrid petri nets: tool paper. In: Proceedings of the 11th EAI International Conference on Performance Evaluation Methodologies and Tools. VALUETOOLS 2017, pp. 186–191. Association for Computing Machinery, New York (2017). https://doi.org/10.1145/3150928.3150956

35. Pilch, C., Remke, A.: Statistical model checking for hybrid petri nets with multiple general transitions. In: 2017 47th Annual IEEE/IFIP International Conference on Dependable Systems and Networks (DSN), pp. 475–486 (2017). https://doi.org/10.1109/DSN.2017.41, https://ieeexplore.ieee.org/abstract/document/8023146, iSSN: 2158-3927

36. Ramírez-Ávila, G.M., Kurths, J., Depickère, S., Deneubourg, J.L.: Modeling Fireflies Synchronization, pp. 131–156. Springer, Cham (2019). https://doi.org/10.1007/978-3-319-78512-7_8

37. Srba, J.: Comparing the expressiveness of timed automata and timed extensions of petri nets. In: Cassez, F., Jard, C. (eds.) FORMATS 2008. LNCS, vol. 5215, pp. 15–32. Springer, Heidelberg (2008). https://doi.org/10.1007/978-3-540-85778-5_3

38. Wikipedia: Photinus carolinus — Wikipedia, the free encyclopedia (2025). http://en.wikipedia.org/w/index.php?title=Photinus%20carolinus&oldid=1237034011. Accessed 09 Jan 2025

39. Zimmermann, A., Knoke, M.: Timenet 4.0: a software tool for the performability evaluation with stochastic and colored petri nets. In: User manual (2007)

Energy Transfer in Timed Cyclic Networks

Luca Paparazzo⬤, Loïc Hélouët$^{(\boxtimes)}$⬤, and Nicolas Markey⬤

University of Rennes, Inria, CNRS, IRISA, Rennes, France
{luca.paparazzo,loic.helouet}@inria.fr, nicolas.markey@irisa.fr

Abstract. This paper considers a timed model tailored to study energy savings in transport networks. It focuses on regenerative braking, a situation where the kinetic energy of a vehicle is transformed in electrical energy and sent back in the electrical network. We first describe transport networks, and formalize their semantics as timed runs of an equivalent network of timed automata (NTA). We then consider three problems. The transfer existence problem checks whether a network has the ability to save energy. The ratio maximization problem aims at computing the maximal ratio of energy saved by time unit in the long run, and the threshold problem consists in verifying the existence of a strategy allowing the saving of more energy than a fixed minima. We show that the transfer problem is a reachability question in the region automaton of the NTA, which can be solved in PSPACE. The ratio problem can be solved using Karp's algorithm on a corner point abstraction for the NTA, yielding an EXPTIME complexity. Finally, the threshold problem requires to address properties of elementary cycles and finite paths of the corner-point automaton, yielding a PSPACE complexity.

Keywords: Transport networks · Timed automata · Energy

1 Introduction

Energy is an actual concern in urban transports, and considering the sizes, weights, and power used by equipments such as metros, the smallest improvement can save huge amounts of energy. Regenerative braking is a way to save energy: the kinetic energy of a braking vehicle is transformed in electrical energy that is either stored or transferred in the power network where it can be used to feed motors of other vehicles. Of course, these transformations induce losses at each conversion and transfer. However, considering the mass of a metro and the power needed to move such a mass, regenerative braking results in huge savings. The easiest way to transfer energy from a braking vehicle to a moving one is to synchronize brakings and departures of close trains (the Joule effect prevents energy transfer on long distances). This means that to save as much energy as possible, one may have to slightly delay departures of trains to benefit from the additional power sent back to the power network. However, saving energy is not

ⓒ The Author(s), under exclusive license to Springer Nature Switzerland AG 2025
E. Amparore and L. Mikulski (Eds.): PETRI NETS 2025, LNCS 15714, pp. 197–218, 2025.
https://doi.org/10.1007/978-3-031-94634-9_10

the primary concern of a metro: trains have to be operated to meet an expected quality of service (QoS), measured for instance as the number of kilometers ran per hour by passengers in the network, the number of trains leaving a station per hour, etc. Hence, the most appropriate strategy is not necessarily to synchronize pairs of trains to favor regenerative braking: such solution may bring situations where the first train in a pair is crowded, while the second is almost empty. Energy saving strategies are hence non-trivial. Due to the cost of metro operation, one cannot make too many real-life tests. Hence, a sound approach is to build formal models of metro networks, synthesize strategies and test them in sillico before considering an implementation in situ.

Metro networks have rather simple topologies that can be safely represented as cyclic graphs. We propose a model called *cyclic transport networks*, where energy transfer is measured as a function of time and distances between trains. In this architecture, metro positions at a given instant can be captured by a position on arcs of this graph. Metros move according to a predetermined speed profile, that specify the duration of the acceleration phase when a train leaves a station, of the cruise phase where a train maintains a constant speed, and of a braking phase when approaching the next station. The dwell durations of trains in station can be chosen in a predetermined range by an operator. QoS constraints are set station by station, and indicate the maximal time allowed between train stops at the considered station. We show that the semantics of this model can be captured by a network of priced timed automata.

We then consider three problems on these models. The first one is called the *transfer problem*: starting from a given configuration of the network, how to reach a situation in which energy transfer via regenerative braking is feasible? The answer to the transfer problem is not always positive: to use regenerative braking, one has to reduce distances between two consecutive trains. This can be enforced by choosing dwell times, but may induce behaviors that violate QoS constraints. The second problem is called the *ratio problem*, and aims at determining the maximal amount of energy that can be saved per time unit in average. The third problem is called the *threshold problem*, and consists in deciding if some run saves more energy than a given threshold λ.

We show that the transfer problem can be solved by finding a path of positive cost in the region automaton, and is in PSPACE. The ratio problem is more involved. Regions completely abstract time, so the cost of cyclic behaviors cannot be computed on a region automaton. We hence use a richer abstraction of priced timed automata, namely a corner point automaton. Then, the ratio problem can be solved by finding the cycles with optimal ratio in this corner point automaton. This is achieved by first translating the corner point automaton into a weighted automaton, and then using an adaptation of Karp's algorithm [10] to find optimal ratios. Last the threshold problem amounts to showing that either a cycle with strictly positive ratio exists, or that for some finite run, the saved energy exceeds the given threshold. Due to lack of space, some proofs are only sketched, but can be found in an extended version of this work [11].

Several works have considered energy or ratios in a timed setting. [9] considers multiweighted energy games on graphs with lower/upper level constraints and their extension to timed automata. Unsurprisingly, many questions on this model are undecidable. [7] considers one-clock priced timed automata where prices can grow linearly or exponentially. Existence of controllers to avoid negative rewards is shown in EXPTIME. [6] considers ratio questions for double priced automata: one price represents time, the other one a reward. The objective is to find cycles with minimal ratios, where a ratio is the time needed to get one unit of reward. [6] shows how to use a corner point abstraction to compute ratios. Though the ratio we want to maximize is a quantity of energy saved per time unit, the results of [6] can be adapted to our setting.

2 Preliminaries

In the rest of the paper, we will denote by $\mathbb{R}, \mathbb{Q}, \mathbb{N}$ the sets of reals, rationals, and non-negative integers. To simplify notations, for a pair of integers a, b chosen from a set $\{0, \ldots n\}$ we will write $a \oplus b$ instead of $a + b \mod n$ when n is clear from the context. We will mainly use real valued variables. For a given real or rational a, we will denote by $\lfloor a \rfloor$ its integral part, and by $frac(a)$ its fractional part. We hence have $a = \lfloor a \rfloor + frac(a)$. Let $X = \{x_1, \ldots, x_k\}$ be a set of real valued variables. A *valuation* for X is a map $\nu : X \to \mathbb{R}$ that associates a real value to every variable in X. The valuation $\nu + d$ assigns value $\nu(x) + d$ to every variable $x \in X$. Similarly, for $Y \subseteq X$, $\nu_{[Y:=0]}$ is the valuation that assigns $\nu(x)$ to each variable $x \in X \backslash Y$, and 0 to variables in Y. A *linear constraint* over X is an inequality of the form $a_1.x_1 + a_2.x_2 + \ldots a_n.x_n \leq b$, where b is a rational value and $a_1, \cdots, a_n$ are rational coefficients (which can have value 0). In the rest of the paper, we will only need simple constraints of the form $x_i - x_j \equiv b_{i,j}$, where $\equiv \in \{<, >, \leq, \geq\}$. A *guard* is a conjunction of simple constraints, and we will denote by Ψ_X the set of guards on X. We will assume that constants $b_{i,j}$ are integers. However, assuming rational values does not change the results of this paper. We will say that a valuation ν *satisfies* a guard $g \in \Psi_X$, denoted $\nu \models g$, iff replacing every variable x_i in g by value $\nu(x_i)$ yields a tautology.

Definition 1. *A* timed automaton *is a tuple* $\mathcal{A} = (S, s_0, Act, X, T, I)$ *where S is a set of states, $s_0 \in S$ is the initial state, Act is an alphabet of actions, X is a set of real valued variables called* clocks, $T \subseteq S \times Act \times \Psi_X \times 2^X \times S$ *is a set of transitions, and $I : S \to \Psi_X$ associates a guard called* invariant *to states of S.*

Let $\mathcal{A} = (S, s_0, Act, X, T, I)$ be a timed automaton. We will write $s \xrightarrow{a,g,R} s'$ when $(s, a, g, R, s') \in T$. A *configuration* of $\mathcal{A}$ is a pair (s, ν), where $s \in S$ and ν is a valuation for X. The semantics of timed automata is defined in terms of timed and discrete moves. A *timed move* of duration δ from a configuration $C = (s, \nu)$ is allowed if elapsing δ time units in configuration C does not violate the time invariant of state s, i.e. if $\nu + \delta \models I(s)$. Timed moves are denoted $C = (s, \nu) \xrightarrow{\delta} C' = (s, \nu + \delta)$. A *discrete move* from a configuration $C = (s, \nu)$

with a transition $t = (s, a, g, R, s')$ is allowed if $\nu \models g$, and $\nu_{[R:=0]} \models I(s')$. This discrete move transforms configuration C into a new configuration $C' = (s', \nu_{[R:=0]})$. A discrete move is denoted $C = (s, \nu) \xrightarrow{a} C' = (s', \nu_{[R:=0]})$. A *run* of a timed automaton is a sequence $\rho = C_0 \xrightarrow{e_1} C_1 \xrightarrow{e_2} C_2 \ldots$ where each e_i is either a timed or discrete move. We denote by $Runs(\mathcal{A})$ the set of runs of a timed automaton $\mathcal{A}$. We will sometimes distinguish between finite runs (denoted $Runs^*(\mathcal{A})$) and infinite ones (denoted $Runs^\infty(\mathcal{A})$).

An automaton $\mathcal{A}$ is *strongly non-Zeno* iff there exists $K \in \mathbb{N}$ such that for every run $\rho = C_0 \xrightarrow{e_1} C_1 \xrightarrow{e_2} C_2 \ldots$ of length k, $k \geq K$ implies $\sum_{e_i \in \mathbb{R}} \delta_i \geq 1$. Non-Zenoness is a standard assumption when considering timed models, and Zeno specifications are considered as ill-formed [5].

As one can see from the semantics, the space of configurations is infinite. A frequent approach to reason on timed automata is to build an abstraction for its state space, called the region abstraction [2,3]. We will define regions more precisely in Sect. 4.

Definition 2. *Let $\mathcal{A}_1 = (S_1, s_0^1, Act_1, X_1, T_1, I_1)$ and $\mathcal{A}_2 = (S_2, s_0^2, Act_2, X_2, T_2, I_2)$ be two timed automata. The* synchronous product *of $\mathcal{A}_1$ and $\mathcal{A}_2$ is a timed automaton $\mathcal{A}_1 \otimes \mathcal{A}_2 = (S_1 \times S_2, (s_0^1, s_0^2), Act_2 \cup Act_2, X_{1,2}, T_{1,2}, I_{1,2})$ where:*

- $(s_1, s_2) \xrightarrow{a,g,R} (s_1', s_2)$ *iff* $(s_1, a, g, R, s_1') \in T_1$ *and* $a \in Act_1 \backslash Act_2$
- $(s_1, s_2) \xrightarrow{a,g,R} (s_1, s_2')$ *iff* $(s_2, a, g, R, s_2') \in T_2$ *and* $a \in Act_2 \backslash Act_1$
- $(s_1, s_2) \xrightarrow{a,g,R} (s_1', s_2')$ *iff* $\exists (s_1, a, g_1, R_1, s_1') \in T_1, (s_2, a, g_2, R_2, s_2') \in T_2$ *and* $g = g_1 \wedge g_2, R = R_1 \cup R_2$
- $X_{1,2} = X_1 \cup X_2$ *and* $I_{1,2}(s_1, s_2) = I(s_1) \wedge I(s_2)$

The synchronous product of two timed automata is a timed automaton. The notion of synchronous product extends to an arbitrary number of automata in the obvious way: automata that share a common action a have to take jointly transitions labeled by a that meet all their respective guards, while other components do not move. In the rest of the paper, we will define products where at most two components are concerned by discrete moves.

Definition 3. *A priced timed automaton is a pair $(\mathcal{A}, Cost)$ where $\mathcal{A}$ is a timed automaton and $Cost : S \cup T \to \mathbb{Q}$ assigns a cost to every transition and a growth rate to every state of $\mathcal{A}$.*

Let $\rho = C_0 \xrightarrow{e_1} C_1 \ldots$ be a run of $\mathcal{A}$. The *cost* of a timed move $(s, \nu) \xrightarrow{\delta} (s, \nu + \delta)$ is $Cost(s) \times \delta$. The cost of a discrete move $(s, \nu) \xrightarrow{a} (s', \nu')$ using a transition $t = (s, a, g, R, s')$ is $Cost(t)$. Slightly abusing our notation, we will write $Cost(C \xrightarrow{e} C')$ to denote the cost of a timed or discrete move. The cost of a run ρ is the value $cost(\rho) = \sum_{i \in 1..|\rho|-1} Cost(C_i \xrightarrow{e_i} C_{i+1})$. Products of priced timed automata are built similarly to products of timed automata, with a particular cost function: the cost of a composite state $(s_1, s_2, \ldots s_k)$ of

k components $(\mathcal{A}_i, Cost_i)$ is $Cost(s_1, \ldots, s_k) = Cost_1(s_1) + Cost_2(s_2) + \cdots + Cost_k(s_k)$. In a product automaton, the cost of a transition $t = (S, a, g, R, S')$ involving a pair of transitions t_i, t_j from two components $\mathcal{A}_i, \mathcal{A}_j$ is $Cost(t) = Cost_i(t_i) + Cost_j(t_j)$.

Definition 4. *A weighted graph is a tuple $\mathcal{G} = (S, A, w)$ where S is a set of states, $A \subseteq S \times S$ is a set of arcs, and $w : A \to \mathbb{Q}$ is a weight function assigning a rational value to each arc.*

A *path* in a weighted graph is a sequence of states $\rho = s_1.s_2 \ldots s_k$ such that $(s_i, s_{i+1}) \in A$. Path ρ is a cycle iff $s_1 = s_k$. We denote by $Paths(\mathcal{G})$ the set of paths of a weighted graph $\mathcal{G}$ and by $Cycles(\mathcal{G})$ the set of cycles of a weighted graph $\mathcal{G}$. We will write $s \leadsto_\rho s'$ when ρ is a path from s to s' and $s \leadsto s'$ when a path from s to s' exists in $\mathcal{G}$. The *cost* of a path $\rho = s_1.s_2 \ldots s_k$ is the sum $w(\rho) = \sum_{i \in 1..k-1} w(s_i, s_{i+1})$. The length of ρ is denoted $|\rho|$ and is the number of arcs in ρ. The *ratio* of a cycle $\rho_c \in Cycles(\mathcal{G})$ is the value $rat(\rho_c) = \frac{w(\rho_c)}{|\rho_c|}$ and represents the average weight of a transition in cycle ρ_c. The minimal ratio of $\mathcal{G}$ is the value $rat_\mathcal{G}^{min} = \min_{\rho_c \in Cycles(\mathcal{G})} rat(\rho_c)$ and the maximal ratio $rat_\mathcal{G}^{max} = \max_{\rho_c \in Cycles(\mathcal{G})} rat(\rho_c)$.

The problem of the minimal ratio (i.e., the question of finding the minimal ratio in a weighted graph) has been studied by Karp [10]. The method proposed by Karp is to evaluate the minimal mean cycle value by iteratively computing, for a given state s, the minimal weight at fixed distance. In the rest of the paper, we will compute optimal energy saving strategies, and we are hence interested in maximal ratios. However, we show below that Karp's algorithm can be easily adapted to compute maximal ratios. We first assume, for simplicity, that all the graphs considered are strongly connected. We will then generalize the approach to all graphs, by focusing on their strongly connected components.

Definition 5 (maximal weight at fixed distance). *Let $\mathcal{G} = (S, A, w)$ be a weighted graph, $k \in \mathbb{N}$, and $v \in S$. We define by $P_{\max}^\mathcal{G}(k, v)$ the maximal weight at distance k from v as: $P_{\max}^\mathcal{G}(k, v) = \max\{w(\rho) \mid s \leadsto_\rho v \wedge |\rho| = k\}$.*

By convention, if there is no path of size k from s to v, then we set $P_{\max}^\mathcal{G}(k, v) = -\infty$. As $\mathcal{G}$ is strongly connected, then there exists an integer k such that $P_{\max}^\mathcal{G}(k, v) > -\infty$. To simplify notations, we will simply write $P_{\max}(k, v)$ instead of $P_{\max}^\mathcal{G}(k, v)$ when $\mathcal{G}$ is clear from the context.

Theorem 1. *Let $\mathcal{G} = (S, A, w)$ be a strongly connected weighted graph with $|S| = g$, and $\mu_\mathcal{G}^*$ be the maximal ratio for cycles of $\mathcal{G}$. Then,*

$$\mu_\mathcal{G}^* = \max_{v \in S} \min_{k \in [\![0, g-1]\!]} \frac{P_{\max}(g, v) - P_{\max}(k, v)}{g - k} \qquad (1)$$

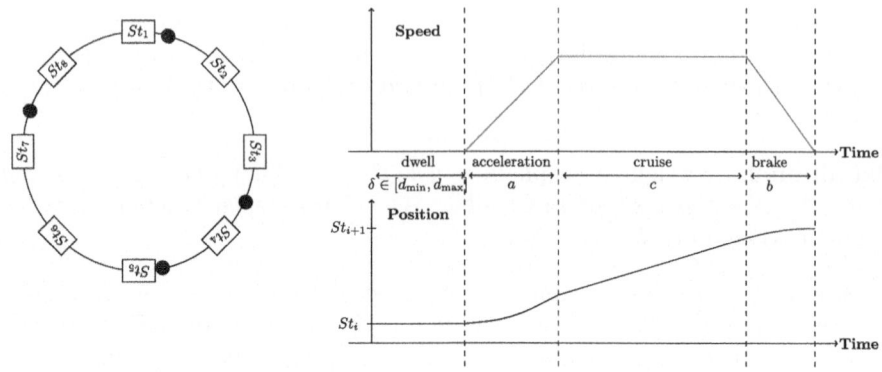

Fig. 1. A cyclic transport Network with 8 stations and 4 vehicles (left) and the four phases of a trip from a station St_i to a station St_{i+1} (right)

A complete proof for this theorem is given in [11]. Following Theorem 1, we can compute $\mu_{\mathcal{G}}^*(s)$ for a strongly connected graph $\mathcal{G}$ by computing $P_{\max}(k, v)$ for every $k \in [\![0, g-1]\!]$ and every $v \in S$. We then get $rat_{\mathcal{G}}^{max} = \mu_{\mathcal{G}}^*$. The computation of $rat_{\mathcal{G}}^{max}$ easily extends to graphs that are not strongly connected. Let $SCC_{\mathcal{G}}$ denote the set of strongly connected components of $\mathcal{G}$, and for a chosen component $C \in SCC_{\mathcal{G}}$, let $\mathcal{G}_C$ be the restriction of $\mathcal{G}$ to vertices of C. Then, we have that $rat_{\mathcal{G}}^{max} = \max\limits_{\rho_c \in Cycles(\mathcal{G})} rat(\rho_c) = \max\limits_{C \in SCC_{\mathcal{G}}} \max\limits_{\rho_c \in Cycles(\mathcal{G}_C)} rat(\rho_c) = \max\limits_{C \in SCC_{\mathcal{G}}} rat_{\mathcal{G}_C}^{max}$.

Now, the complexity to find connected components of $\mathcal{G}$ is linear w.r.t the number of edges in $\mathcal{G}$ [12], and for a given vertex $s \in S$ computing $\mu_{\mathcal{G}}^*(s)$ can be done by iteratively computing $P_{\max}^{\mathcal{G}}(k, v)$ for each v from the values of $P_{\max}^{\mathcal{G}}(k-1, v')$ for each $v' \in S$ computed at previous step $k-1$. This step can be done in $O(|S|^2)$ and is repeated at most g times. As in [10], we can then compute $rat_{\mathcal{G}}^{max}$ in $O(g.|A|)$.

3   Cyclic Transport Networks

A metro network can be seen as a ring connecting stations, with trains moving on this ring. The speed of trains is predetermined: trains are stopped in a station for a duration fixed by metro operators. Then they accelerate, maintain their cruise speed for a while, and brake to stop at the next station. Figure 1 represents a network, the speed variation for a train moving from a station St_i to a station St_{i+1} and its position between St_i and St_{i+1}.

Definition 6. *A Cyclic transport network (CTN for short) is a tuple $\mathcal{N} = (ST, P, E, freq)$, where $ST = \{St_1, \dots St_n\}$ is a set of stations, $P \subseteq (\mathbb{Q}^+)^3$ is a set of speed profiles, and $E : ST \times ST \to P$ is a partial map associating a speed profile to every pair of consecutive stations. We furthermore require that* *the support of E is a cycle. The map $freq : ST \to \mathbb{Q}$ is a partial map that adds a frequency constraint to a set $ST_C \subseteq ST$.*

Intuitively, a CTN models a cyclic physical network where vehicles move from one station to the next one, following the speed profile for this area of the network. A profile is a triple of positive values $(a, c, b) \in (\mathbb{Q}^+)^3$ describing an acceleration duration a, a duration c for a part of the trip where the vehicle is at cruise speed, and a braking duration b. $freq(St_i) = d_i$ indicates that a vehicle has to stop in station St_i at least every d_i time units to meet the network's service level. We do not make speeds nor distances explicit, as in the simple energy model that we propose later, the quantity that matters is the duration of overlapping braking and acceleration phases of vehicles around stations.

A *configuration* of a CTN with n stations and p vehicles is a triple $C_{CTN} = (pos, phase, trip)$, where $pos : 1..p \to ST$ indicates the last station visited by a vehicle, $phase : 1..p \to \{Stp, Ac, Cr, Br\}$ indicates the status of each vehicle's trip (stopped, accelerating, in cruise speed, braking) and $trip : 1..p \to \mathbb{R}^{\geq 0}$ indicates the time elapsed since a vehicle started a new phase of its trip. The semantics of a CTN is given in terms of timed and discrete moves. We define four types of possible events: a vehicle stops in a station, leaves a station, enters a cruise speed period or starts braking. To allow passenger transfers, the system ensures that a vehicle cannot leave less than $\delta_{min} > 0$ time units after its arrival in station. We also impose that the time spent in station is upper bounded by a value δ_{max}. Choosing an appropriate dwell duration in $[\delta_{min}, \delta_{max}]$ is a way to perform traffic management, for instance to recover from an unexpected delay, or in our case to favor energy transfer and meet frequency constraints. We can now define discrete and timed moves in a CTN.

- $(pos, phase, trip) \xrightarrow{leave\ St_i} (pos, phase', trip')$ iff
$$\exists k, pos(k) = St_i \wedge phase(k) = St \wedge trip(k) \in [\delta_{min}, \delta_{max}].$$
Then $phase'(k) = Ac, trip'(k) = 0$.
- $(pos, phase, trip) \xrightarrow{cruise\ St_i} (pos, phase', trip')$ iff
$$\exists k, pos(k) = St_i \wedge phase(k) = Ac \wedge trip(k) = a.$$
Then $phase'(k) = Cr, trip'(k) = 0$.
- $(pos, phase, trip) \xrightarrow{brake\ St_i} (pos, phase', trip')$ iff
$$\exists k, pos(k) = St_i \wedge phase(k) = Cr \wedge trip(k) = c.$$
Then $phase'(k) = Br, trip'(k) = 0$.
- $(pos, phase, trip) \xrightarrow{stop\ St_{i\oplus1}} (pos', phase', trip')$ iff
$$\exists k, pos(k) = St_i \wedge phase(k) = Br \wedge trip(k) = b.$$
Then $pos'(k) = St_{i\oplus1}, phase'(k) = Stp, trip'(k) = 0$
- A duration $\delta \in \mathbb{R}$ can elapse in configuration C_{CTN} if, letting this duration elapse does not violate the upper bound for the maximal duration of a trip phase for some vehicle. Formally, δ time units can elapse if:
$$\forall k, phase(k) = St \Rightarrow trip(k) + \delta \leq \delta_{max} \wedge phase(k) = Ac \Rightarrow trip(k) + \delta \leq a$$
$$\wedge phase(k) = Cr \Rightarrow trip(k) + \delta \leq c \wedge phase(k) = Br \Rightarrow trip(k) + \delta \leq a.$$
Elapsing δ time units in C_{CTN} then results in a configuration where each trip of a vehicle has progressed, leading to a new configuration:

$C_{CTN} + \delta = (pos, phase, trip + \delta)$, where $[trip + \delta](k) = trip(k) + \delta$.

Timed move are denoted by $C_{CTN} \xrightarrow{\delta} C_{CTN} + \delta$.

A *run* of a CTN from a configuration C_0 is a (possibly infinite) sequence of moves $C_0 \xrightarrow{e_1} C_1 \xrightarrow{e_2} \ldots$ where each e_i is either a discrete event of a duration. Given a run ρ, we denote by $\rho_{[i,j]}$ its restriction to steps from configuration C_i to configuration C_j. The duration of a finite run ρ is the value $T(\rho) = \sum_{i \in 1..|\rho|, e_i = \delta_i \in \mathbb{R}} \delta_i$. A run ρ of a CTN $\mathcal{N}$ meets its frequency constraints if, for every $St_i \in ST_C$, for every pair of consecutive stop events at station St_i at respective indexes i, j we have $T(\rho_{[i,j]}) \leq freq(St_i)$. We denote by $Runs(\mathcal{N})$ the set of runs of a transport network $\mathcal{N}$.

We adopt a simple energy model. Instead of measuring the total energy consumption along runs and trying to minimize this power demand, we assume that energy is saved mainly when trains can transfer energy via regenerative braking. We assume that energy is transferred from a braking train to an accelerating one around each station, but we do not measure how much energy is passed from one train to the next one. More realistic models should handle properties of the power network, trains characteristics and distances between trains. Instead, we take a simplified energy model and we say that there is an energy transfer at station St_k in a configuration $C = (pos, phase, trip)$ iff there exists a pair of vehicles i,j such that $pos(j) = pos(i) \oplus 1 = St_k$ and $phase(i) = Br$, $phase(j) = Ac$. For the whole time spent in configuration C there is an energy transfer between the train braking to stop in St_k and the train leaving St_k. In a configuration C, we define by $\mathcal{E}_C$ the number of stations where an energy exchange occurs. Formally, $\mathcal{E}_C = |\{St_i | \exists k, k' \in 1..p, pos(k) = i \wedge pos(k') = i \oplus 1 \wedge phase(k) = Br \wedge phase(k') = Ac\}|$. With this definition in mind, the energy transferred in a run ρ is $\mathcal{E}(\rho) = \sum_{C_i \xrightarrow{\delta_i} C_i + \delta_i \in \rho} \delta_i . \mathcal{E}_{C_i}$.

We are now ready to define problems addressing energy management in cyclic transport networks. Let $\mathcal{N} = (ST, P, E, freq)$ and C_0 be its initial configuration:

- **Transfer** : Is there a run ρ starting from C_0 such that ρ satisfies constraints in $freq$ and $\mathcal{E}(\rho) > 0$?
- **Ratio** : Given a CTN, what is the optimal ratio $r = \lim_{n \to \infty} \frac{\mathcal{E}(\rho_{[0,n]})}{T(\rho_{[0,n]})}$ for an infinite run ρ starting from C_0 that satisfies constraints in $freq$?
- **Threshold** : Given a value $\lambda \in \mathbb{Q}$, is there a finite run ρ starting from C_0 such that ρ satisfies constraints in $freq$ and $\mathcal{E}(\rho) > \lambda$?

We will show in the next sections that these three problems can be answered on an equivalent timed automaton, computed as a product of automata depicting the behaviors of trains and the status of stations, and that generates the same timed runs as the original CTN.

4   The Transfer Problem

Lemma 1. *Let* $\mathcal{N} = (ST, P, E, freq)$ *be a cyclic transport network with* n *stations and* p *vehicles. Then, there exists a timed automaton* $\mathcal{A}_\mathcal{N}$ *of size* $O((4.n)^p \cdot 4^n)$ *such that* $Runs(\mathcal{N})$ *and* $Runs(\mathcal{A}_\mathcal{N})$ *are isomorphic.*

Proof (sketch). For every vehicle in the system, we build a timed automaton $\overline{\mathcal{M}_i}$ that describes the position and status of the vehicle with $4.n$ states and one clock x_i to measure the time elapsed in each phase. For this automaton, discrete transitions use the same alphabet as the CTN, and are triggered in obvious way when $x_i = a, x_i = c, x_i = b$ to initiate beginning of cruising, or braking, or stop, or when $\delta_{min} \leq x_i \leq \delta_{max}$ to trigger a departure after an allowed dwell time. States memorize the *pos* and *phase* maps of the CTN. For each station St_i , we build an automaton $\overline{T_i}$ with 4 states that memorizes whether a train is leaving St_i, on its cruise from St_i to $St_{i\oplus 1}$, stopped in station, or braking to stopping in station St_i). Transitions are set the obvious way to memorize respective status of 0, 1 or 2 trains. In addition to these states, when $St_i \in ST_C$ automaton $\overline{T_i}$ has a clock y_i that measures the time elapsed between two stops and can be used to enforce the frequency constraint $freq(St_i)$. For each station automaton, we distinguish a particular state T_i, reached when a train is braking before the station and another one is leaving the station. As in CTNs, these are states where energy transfer can occur. Then, the overall automaton is obtained as a product $\overline{\mathcal{M}_1} \otimes \cdots \otimes \overline{\mathcal{M}_p} \otimes \overline{T_1} \otimes \cdots \otimes \overline{T_n}$. The size of this automaton is in $O((4.n)^p \cdot 4^n)$. It is easy to see that the timed transition systems underlying the two models are isomorphic. The full construction of these automata is detailed in [11].   □

The shape of states in $\mathcal{A}_\mathcal{N}$ is of the form $S = (s_{\overline{\mathcal{M}_1}}, \ldots, s_{\overline{\mathcal{M}_p}}, s_{\overline{T_1}}, \ldots, s_{\overline{T_n}})$, where each $s_{\overline{T_i}} \in \{N_i, A_i, B_i, T_i\}$ depicts the vehicles around station ST_i. N_i means that no vehicle is around the station, A_i that a single vehicle has left the station and is currently accelerating, B_i that a single vehicle is currently braking and will stop at the station, and T_i that there is both an accelerating and a braking vehicle around the station, i.e., a transfer. We can hence assign a cost function to $\mathcal{A}_\mathcal{N}$ as follows: for each composite state $S = (s_{\overline{\mathcal{M}_1}}, \ldots, s_{\overline{\mathcal{M}_p}}, s_{\overline{T_1}}, \ldots, s_{\overline{T_n}})$, we set $Cost(S) = |\{i \in 1..n \mid s_{\overline{T_i}} = T_i\}|$.

Translating a CTN into a timed automaton is not sufficient to get an algorithm that decides whether an energy transfer can occur or not. CTNs and their timed automata translations have an infinite set of configurations. Fortunately, we show below that we can rely on the standard region construction to build a finite abstraction and get a decision procedure.

Definition 7 (Regions [3]). *Let* $\mathcal{A}$ *be a timed automaton, and let* $C_{\max}$ *be the largest value that appears in guards and invariants of* $\mathcal{A}$*. Then, we can define an equivalence relation* $\equiv$ *on clock valuations as follows:* $\forall \nu, \nu' \in \mathbb{R}^X$, $\nu \equiv \nu'$ *iff*

- $\forall x \in X$, *either* $\lfloor \nu(x) \rfloor = \lfloor \nu'(x) \rfloor$, *or* $\lfloor \nu(x) \rfloor > C_{\max}$ *and* $\lfloor \nu'(x) \rfloor > C_{\max}$: *both valuations agree on the integral parts of their clocks.*

- $\forall x, y \in X$, such that $\nu(x) \leq C_{\max}$ and $\nu(y) \leq C_{\max}$ we have $frac(\nu(x)) \leq frac(\nu(y))$ if and only if $frac(\nu'(x)) \leq frac(\nu'(y))$: both valuations agree on the ordering of fractional parts of their clocks.
- $\forall x \in X$, such that $\nu(x) \leq C_{\max}, frac(\nu(x)) = 0$ if and only if $frac(\nu'(x)) = 0$

The *regions* of a timed automaton $\mathcal{A}$ are the equivalence classes of $\equiv$, and we will denote by $\mathcal{R}$ the set of regions. For a given valuation ν we will denote by $[\nu]$ the region that contains ν. Regions can have several forms. Consider a clock $x \in X$. As clocks take a positive value, we can split the domain of x into a subset of atomic constraints $C_x = \{\nu(x) = k \mid k \in 0..C_{max}\} \cup \{k - 1 < \nu(x) < k \mid k \in 1..C_{max}\} \cup \{\nu(x) > C_{max}\}$. A region can be represented by choosing for each $x \in X$ an atomic constraint in C_x, and for each pair of clocks x, y such that $k - 1 < \nu(x) < k$ and $k' - 1 < \nu(y) < k'$, choose whether $frac(\nu(x)) = frac(\nu(y))$, $frac(\nu(x)) < frac(\nu(y))$ or $frac(\nu(x)) > frac(\nu(y))$. This means that regions can be represented by selecting an atomic constraint in C_x for each clock $x \in X$ and an ordering on fractional parts of clocks valuations. The ordering on fractional parts can be described by a partition $X_0.X_1 \ldots X_q$ on X, such that $x, y \in X_i$ iff $frac(\nu(x)) = frac(\nu(y))$, and $x \in X_i, y \in X_j$ with $i < j$ iff $frac(\nu(x)) < frac(\nu(y))$. It is well known that the truth value of a guard is invariant in a region, and we will write $r \models g$ iff $\nu \models g$ for every valuation $\nu \in r$.

We can define a successor relation on regions, by saying that r' is a successor of r if and only if there exists $\nu \in r$ and $\nu' \in r'$ and a duration $d > 0$ such that $\nu' = \nu + d$. The set of successors of a region r can be efficiently computed from the representation with atoms and ordering on clocks (see [2] for details). For a region r where some clock x has a constant value (the chosen atomic constraint on x is of the form $x = k$), letting a duration $d > 0$ elapse means that for every $\nu \in r$ we have that $[\nu + d]$ is a successor of r, but $[\nu + d] \neq r$. For a region r where no clock has a constraint of the form $x = k$, then for every valuation $\nu \in r$, there exists a duration $d > 0$ such that $\nu + d \in r$. In this type of region, some time can elapse without changing the equivalence class, and hence without changing the guards or invariants that are satisfied in r. Regions with constraints of the form $k - 1 < frac(\nu(x)) < k$ only will be called *open regions*, and other regions will be called *closed region*.

Proposition 1 (Properties of regions [3]). $\forall r, r' \in \mathcal{R}$,

- $\forall g \in \Psi_X, \forall \nu, \nu' \in r$, we have $\nu \models g$ iff $\nu' \models g$: the valuations of a region satisfy the same guards and invariants. Consequently, it is sufficient to know the region of a valuation to decide which guards and invariants are satisfied, and hence which transitions are firable.
- $\exists \nu \in r, d \in \mathbb{R}$ such that $\nu + d \in r' \implies \forall \nu \in r, \exists d \in \mathbb{R}$ such that $\nu + d \in r'$: If one can let some time elapse from a valuation ν in r and reach a region r', then one can let some time elapse from any other valuation ν' of r and reach a valuation in r' (these durations do not need to be the same).
- $\forall Y \subseteq X, r_{[Y:=0]} \cap r' \neq \emptyset \implies r_{[Y:=0]} \subseteq r'$: After resetting clocks, all valuations in r are transformed into valuations of a single region r'.

Proposition 2 (finiteness of the region set [3]). *Let $\mathcal{A} = (S, s_0, Act, X, T, I)$ be a timed automaton. Then, the number of regions is in $O(|X|! \cdot 2^{|X|} \cdot \Pi_{x \in X}(2c_x + 2))$, where c_x is the maximal constant to which clock x is compared.*

Definition 8 (Region automaton [3]). *Let $\mathcal{A} = (S, s_0, Act, X, T, I)$ be a timed automaton, $\mathcal{R}$ be its set of regions, and let ν_0 be an initial valuation of clocks in X. We can define the region automaton $\mathcal{A}_\mathcal{R} = (S_\mathcal{R}, s_{0\mathcal{R}}, Act_\mathcal{R}, T_\mathcal{R})$ as follows:*

- *$S_\mathcal{R} \subseteq S \times \mathcal{R}$*
- *$s_{0\mathcal{R}} = (s_0, r_0)$ where r_0 is the region of ν_0*
- *$Act_\mathcal{R} = Act \cup \{\perp\}$*
- *$T_\mathcal{R} \subseteq S_\mathcal{R} \times Act_\mathcal{R} \times S_\mathcal{R}$ is defined by:*

  - *for every state s and every region r, there is a transition $(s, r) \xrightarrow{\perp} (s, r')$ if r' is a successor of r and $r, r' \models I(s)$.*
  - *For every discrete transition $t = s \xrightarrow{g, \alpha, Y} s' \in T$, there is a transition $(s, r) \xrightarrow{\alpha} (s', r')$ iff $r \models g \wedge I(s), r_{[Y:=0]} \subseteq r'$ and $r_{[Y:=0]} \models I(s')$.*

Usually, the whole set $S \times \mathcal{R} \times S$ is not reachable, and one can build $S_\mathcal{R}$ inductively starting from $s_{0\mathcal{R}}$. The automaton $\mathcal{A}_\mathcal{R}$ is hence a finite automaton over $\Sigma \cup \{\perp\}$. A run of the region automaton is a sequence of steps of the form

$$\hat{\rho} = (s_0, r_0) \xrightarrow{\sigma_1} (s_1, r_1) \xrightarrow{\sigma_2} \dots$$

Runs of the region automaton are said *progressive* iff for every clock x there exists an infinite number of indexes i such that $r_i \models x = 0 \vee x > C_{max}$. To distinguish between runs of $\mathcal{A}$ and runs of $\mathcal{A}_\mathcal{R}$ we will say that runs of $\mathcal{A}$ are *concrete runs* and runs of $\mathcal{A}_\mathcal{R}$ are *abstract runs*. The *region projection* of a concrete run $\rho = (s_0, \nu_0) \xrightarrow{d_0} (s_1, \nu_1) \xrightarrow{a_1} (s_2, \nu_2) \dots$ is an abstract run $proj_R(\rho) = (s_0, [\nu_0]) \xrightarrow{\perp} (s_1, [\nu_1]) \xrightarrow{a_1} (s_2, [\nu_2]) \dots$

Lemma 2 ([2]). *$\forall \rho$ run of $\mathcal{A}$, $proj_R(\rho)$ is a progressive abstract run of $\mathcal{A}_\mathcal{R}$.*

Lemma 3 ([2], Lemma 4.13)). *For every progressive abstract run γ of $\mathcal{A}_\mathcal{R}$ there exists a run $\mathcal{A}$ such that $\gamma = proj_R(\rho)$.*

One can tweak the region automaton of definition 8 to make it progressive [2]. Slightly abusing our constructions, we will assume that our region automaton is already progressive. Now let us consider an abstract state (s, r) of $\mathcal{A}_\mathcal{R}$ reachable from (s_0, r_0). This means that there exists a run ρ that ends in a state (s, ν) with $[\nu] = r$. If r is a region and there exists a transition $(s, r) \xrightarrow{\perp} (s, r')$ in $\mathcal{A}_\mathcal{R}$, then for every valuation ν' of r, one can let some duration $d > 0$ elapse, and reach valuation $\nu' + d \in r'$. This is in particular true for ν. Further, as $(s, r) \xrightarrow{\perp} (s, r')$ we have $r' \models I(s)$. Hence, ρ can always be extended with a timed move, i.e. there is a run of $\mathcal{A}$ that spends some time in state s. The cost of states is extended the obvious way in the region automaton and its runs, i.e., $Cost((s, r)) = Cost(s)$. A direct consequence is that finding a run where energy transfer occurs amounts to finding a run of $\mathcal{A}_R$ that contains a transition of the form $(s, r) \xrightarrow{\perp} (s, r')$ with $Cost(s) > 0$. This immediately gives us the following result:

Lemma 4. *A concrete run ρ of A has a positive cost iff $proj_R(\rho)$ is a run of $\mathcal{A}_R$ that contains a transition of the form $(s, r) \xrightarrow{\perp} (s, r')$ such that $Cost(s) > 0$.*

Theorem 2. *The transfer problem for CTNs is in PSPACE.*

Proof. The answer to the transfer problem for a CTN $\mathcal{N}$ is positive iff there exists a run ρ of $\mathcal{N}$ that meets all frequency constraints and such that $Cost(\rho) > 0$. We have shown in Lemma 1 that we can build an automaton $\mathcal{A}_\mathcal{N}$ with at most $n + p$ clocks that has the same set of runs as $\mathcal{N}$, and the same cost function. Hence, the answer to the transfer problem is positive if there exists a run $\rho = (s_0, \nu_0) \xrightarrow{e_0} \ldots \xrightarrow{e_k} (s, \nu) \xrightarrow{\delta} \ldots$ of $\mathcal{A}_\mathcal{N}$ that does not violate frequency constraints, visits a configuration (s, ν) with $Cost(s) > 0$ and lets a duration $\delta > 0$ elapse in (s, ν).

As shown by Lemma 4, this amounts to finding an accepting run in $\mathcal{A}_{\mathcal{N},\mathcal{R}}$ (the region automaton of $\mathcal{A}_\mathcal{N}$) that contains a transition of the form $(s, r) \xrightarrow{\perp} (s, r')$. Finding such a run π boils down to finding an accepting path of length $< |\mathcal{A}_{\mathcal{N},\mathcal{R}}|$ from $(s_0, [\nu_0])$ that contains a transition of the form $(s, r) \xrightarrow{\perp} (s, r')$ with $Cost(s) > 0$. Finding such an accepting path can be done in NLOGSPACE w.r.t the size of the region automaton, by a non-deterministic exploration of $\mathcal{A}_{\mathcal{N},\mathcal{R}}$, and memorizing whether a transition is of the form $(s, r) \xrightarrow{\perp} (s, r')$, which needs an additional bit of information. The number of states in $\mathcal{A}_\mathcal{N}$ is bounded by $(4.n)^p \cdot 4^n$, and the number of regions smaller than $(2 \cdot C_{max} + 2)^{n+p} \cdot (n+p)! \cdot 2^{n+p}$, that is the size of $\mathcal{A}_{\mathcal{N},\mathcal{R}}$ is exponential in the number of clocks, and accessibility can be checked in NPSPACE, by exploring non-deterministically successors of $(s_0, [\nu_0])$ along paths of length at most $K = (4.n)^p \cdot 4^n \cdot (2 \cdot C_{max} + 2)^{n+p} \cdot (n + p)! \cdot 2^{n+p}$. As NPSPACE=PSPACE, we get the PSPACE membership. □

5 The Ratio Problem

We can now address the ratio question for CTNs: For a given CTN, what is the maximal amount of energy saved by the network through transfer per time unit in average? Intuitively, an optimal strategy to save energy would be to repeat cyclic behaviors where the amount of energy saved by time unit is maximal. However, it is not sufficient to consider cycles of the timed automaton $\mathcal{A}_\mathcal{N}$ nor of its region automaton to get an optimal strategy. Consider for instance the example of Fig. 2, borrowed from [8]. In this example, Cassez et al. have shown that there exists a strategy to avoid the *Bad* state, namely playing $a.b.c^\omega$. The duration of a cycle is always 1. However, the sojourn stay in state ℓ_2 must be shorter at each cycle to preserve invariant of ℓ_1 and avoid *Bad*. We omit the details and refer the interested reader to [8]. Now, assume that the cost of ℓ_0, ℓ_1, Bad is 0 and the cost of ℓ_2 some positive value c_2. That is, the gain in each occurrence of the cycle $a.b.c$ decreases at each iteration. This cannot be observed in regions, where most of information on time is abstracted away. Interestingly, [8] shows that there is no effective machine (i.e. with decisions occurring at a fixed rate) that can control this model to avoid the Bad state, as the allowed sojourn time in ℓ_2 decreases. Another interesting point is that this issue cannot be observed on the region automaton. In the region automaton attached to this example, one will only see that a cycle occurs, that some time can be spent in ℓ_2 (witnessing a transfer), but as time is abstracted, one cannot decide whether the ratio is a positive value or converges to 0. An immediate conclusion it that the ratio question cannot be addressed directly as a property of the region automaton of a CTN. However, we will show hereafter that another abstract model called the *corner point abstraction* can be used to compute ratios.

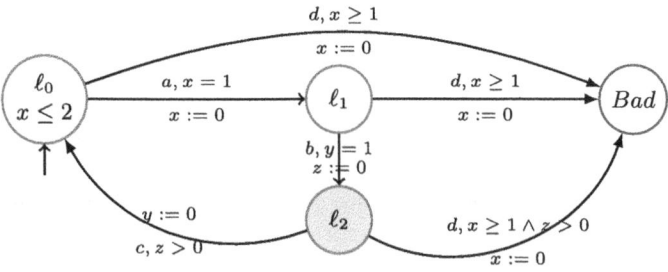

Fig. 2. A gadget with decreasing amount of saved energy at each cycle iteration.

Bouyer, et al. [6] consider a question similar to the ratio problem. They study double priced timed automata with two quantities: a reward, and a cost, and search for an infimum for the ratio $\frac{cost}{reward}$. The idea is to find a strategy giving the minimal cost to produce one unit of reward. The proposed solution only has to consider ratios of cycles, and the study is restricted to automata with non-negative costs and that are strongly reward diverging, i.e. such that for every infinite run ρ, $Reward(\rho) = +\infty$. Then, the authors build a corner-point abstraction of their initial automaton, and use Karp's algorithm (see [10] and Sect. 2) to compute optimal ratios. The solution of [6] does not immediately answer our ratio problem. First, we are interested in a supremum rather than in an infimum over the set of infinite runs. However, we have seen in Sect. 2 that Karp's algorithm can be adapted easily to compute maximal ratios. Second, our setting measures the amount of energy that can be transferred per unit of time. Hence, we can reuse the setting of [6] using time as "reward" and saved energy as "cost" only if time diverges in infinite runs, in other words, if the considered timed automaton is non-Zeno. Fortunately, this is always the case for CTNs.

Proposition 3. *Let $\mathcal{N}$ be a cyclic transport network, and $\mathcal{A_N}$ be the associated timed automaton. Then, $\mathcal{A}_{CTN}$ is non-Zeno.*

Proof. Let us first show that every execution of a cycle in $\mathcal{A}_\mathcal{N}$ has a strictly positive duration. Let C be a cycle of $\mathcal{A}_\mathcal{N}$ and let ρ_C be an execution that visits C. For each transition of C, there is a vehicle which changes state. Let m be the vehicle which state changes in the first transition of ρ_C. Let C_m be the projection of C on states of $\overline{\mathcal{M}_m}$ (obtained by keeping only component m of each tuple composing states of C). Then, C_m is also a cyclic path, and contains at least all states of $\overline{\mathcal{M}_m}$. Consider a pair of consecutive stations $St_i, St_{i\oplus1}$ and the profile $P(St_i, St_{i\oplus1}) = (a_i, c_i, b_i)$ Then, the minimal duration of an execution that visits a cyclic path of $\overline{\mathcal{M}_m}$ is $m_c = \sum\limits_{\substack{St_i \in St \\ P(St_i, St_{i\oplus1}) = (a_i, c_i, b_i)}} a_i + c_i + b_i + \delta_{min}$,

and $\mathcal{T}(\rho_C) \geq m_c > 0$.

Now, let ρ be an infinite run of $\mathcal{A}_\mathcal{N}$, and let $\overline{\rho_k}$ be the suffix of ρ starting after a prefix of size k. As the automaton is finite, there exists a finite set of cyclic paths $\mathcal{F}_C$ such that, after some index K, $\overline{\rho_K}$ only visits paths of $\mathcal{F}_C$, of strictly positive duration. We can then conclude that $\mathcal{T}(\rho) \geq \mathcal{T}(\rho_K) = +\infty$. □

A *corner point* is an integral valuation of clocks. A corner point ν is associated with a region r if it is in the closure of r. Intuitively, ν can be seen as a point (with integral coordinates) of the polyhedron that contains r and its borders. If a region partitions its fractional parts into q subsets (as proposed in the encoding $X_1. \ldots X_q$ in Sect. 4), then it has $q+1$ corner points. We are now ready to define a new abstraction, namely the *Corner-Point automaton*, with states of the form (s, r, ν), where s is a state of $\mathcal{A}$, r is one of its regions, and $\nu \in (\mathbb{N} \cup \{\top\})^X$ is a corner point of r. Intuitively, in a state of the form (s, r, ν), ν will indicate that valuations considered in the state are "close" to point ν. This additional information will be useful to differentiate regions where time can elapse from regions in which elapsing time changes the region. For convenience, we use value $\top$ to mean that the value of clock x has exceeded $C_{\max}$ and hence is in a region not bounded according to the coordinate x.

Definition 9 (Corner-point automaton [6]). *Let $\mathcal{A} = (S, s_0, Act, X, T, I)$ be a timed automaton. The* Corner-Point *automaton of $\mathcal{A}$ is denoted by $\mathcal{A}_{CP}$. The automaton $\mathcal{A}_{CP} = (S_{CP}, s_{0CP}, Act_{CP}, T_{CP})$ is defined as follows:*

- $S_{CP} = S \times \mathcal{R} \times CP$ *where $\mathcal{R}$ are regions of $\mathcal{A}$ and $CP = \{0, 1, \ldots, C_{\max}, \top\}^X$ are the possible corner points of $\mathcal{A}$.*
- $s_{0CP} = (s_0, r_0, \nu_0)$
- $Act_{CP} = Act \cup \{\bot\}$
- T_{CP} *is built from the transition relation $T_\mathcal{R}$ in the region automaton:*
  - *inter-region waiting transitions: For every waiting transition of the form $(s, r) \xrightarrow{\bot} (s, r') \in T_\mathcal{R}$, T_{CP} contains a transition of the form $(s, r, \nu) \xrightarrow{\bot} (s, r', \nu)$, where ν is a vertex of both r and r'.*
  - *discrete transitions: for every discrete transition $(s, r) \xrightarrow{\alpha} (s', r') \in T_\mathcal{R}$, T_{CP} contains transitions of the form $(s, r, \nu) \xrightarrow{\alpha} (s', r', \nu')$ such that, if $Y \subseteq X$ is the set of clocks reset by the transition, then $\nu' = \nu_{[Y:=0]}$.*

- *intra-region waiting transitions: for each pair* $(s, r) \in S \times \mathcal{R}$, *if* ν *and* $\nu + 1$ *are two vertices of* r, *then* $(s, r, \nu) \xrightarrow{\perp} (s, r, \nu + 1) \in T_{CP}$. *This type of transition represents a delay elapsing in a region. We add, by convention, that if* $\nu(x) \in \{C_{\max}, \top\}$ *then* $(\nu + 1)(x) = \top$.

Corner point automata do not yet contain information on energy, but we can easily decorate $\mathcal{A}_{CP}$ with a function $\mathcal{E}_{CP}$ giving the energy saved for each transition.

- $\forall t \in T_{CP}$, if t is a discrete transition, or an inter-region transition via the same corner point, then $\mathcal{E}_{CP}(t) = 0$
- If t is an intra-region waiting transition, i.e., if $t = (s, r, \nu) \xrightarrow{\perp} (s, r', \nu') \in T_{CP}$, then $\mathcal{E}_{CP}(t) = \mathcal{E}(s)$. Intuitively, the energy gain $\mathcal{E}(s)$ is the energy that can be saved in one time unit in state s, and t symbolizes elapsing of one time unit.

As for region automata, corner point automata defines abstract runs of the form $\pi = (s_0, r_0, \nu_0) \xrightarrow{a_1} (s_1, r_1, \nu_1) \xrightarrow{a_2} (s_2, r_2, \nu_2) \cdots$.

Consider a finite run $\pi = (s_0, r_0, \nu_0) \xrightarrow{a_1} \ldots \xrightarrow{a_k} (s_k, r_k, \nu_k)$, or equivalently a prefix of an infinite run of size k, i.e., composed of k consecutive transitions. Then we can define $\mathcal{E}_{CP}(\pi)$ as $\mathcal{E}_{CP}(\pi) = \sum_{i \in 1..k} \mathcal{E}_{CP}(t_i)$. We can also recover from an abstract run of a corner point automaton a duration: $\mathcal{T}(\pi) = |\{t_i \mid t_i \text{ is an intra-region transition }\}|$. When $\mathcal{T}(\pi) > 0$ then the ratio $rat(\pi) = \frac{\mathcal{E}(\pi)}{\mathcal{T}(\pi)}$ is defined. Last, the ratio of an infinite run π is: $rat(\pi) = \limsup_{k \to \infty} rat(\pi_{[0,k]})$ when it is defined. We can now define the optimal ratio in a corner point automaton as $rat^*_{CP} = \sup_{\pi \in \text{Runs}^\infty(\mathcal{A}_{CP})} rat(\pi)$.

Proposition 4. *Let* $\mathcal{N}$ *be a cyclic timed network, and* $(\mathcal{A}_{\mathcal{N}}, Cost)$ *the associated priced timed automaton. Then* rat^* *and* rat^*_{CP} *are always defined.*

Proof. The ratio is computed on infinite runs, i.e. runs that contain a cycle of $\mathcal{A}$. As shown in Proposition 3, $\mathcal{A}_{\mathcal{N}}$ is always non-Zeno, so for every infinite run ρ of $\mathcal{A}$, $\mathcal{E}(\rho) > 0$. Similarly, an infinite run of $\mathcal{A}_{CP}$ exists. Hence it visits a particular abstract state (s, r, ν) from which a cycle exists. As $\mathcal{A}$ is non-Zeno, every cycle of $\mathcal{A}$ contains at least one transition where time progresses, and hence the time progress of each cycle of $\mathcal{A}_{CP}$ can be of at least one time unit, i.e. there is a transition of the form (s', r', ν') to $(s', r', \nu' + 1)$ in the cycle. So, the duration of an abstract cycle in $\mathcal{A}_{CP}$ is strictly positive. $\square$

Proposition 5. $rat^*_{CP} \leq rat^*$

Proof. The proof was shown in [6]. If guards of $\mathcal{A}$ use non-strict inequalities (which is the case for $\mathcal{A}_{\mathcal{N}}$), then for every execution π of $\mathcal{A}_{CP}$ there exists a concrete execution with the same valuations as the corner points of π, and with the same ratio. So, if π is an optimal run of $\mathcal{A}_{CP}$, then $rat(\pi)$ is the ratio of a real execution of $\mathcal{A}$ that is not necessarily optimal, and hence $rat^*_{CP} \leq rat^*$. $\square$

Proposition 6. $rat^*_{CP} \geq rat^*$

Proof. The proof also comes from [6]. Let $\gamma = (l_0, \nu_0) \longrightarrow (l_1, \nu_1) \ldots$ be a sequence of configurations in $\mathcal{A}$. Then the set $proj_{CP}(\gamma)$ is the set of sequences of states that are paths of $\mathcal{A}_{CP}$ of the form:

$\pi = (l_0, [\nu_0], \alpha_{0,0}) \longrightarrow (l_0, [\nu_0], \alpha_{0,1}) \ldots (l_0, [\nu_0], \alpha_{0,p_0})(l_1, [\nu_1], \alpha_{1,0}) \ldots$ where each $\alpha_{i,j}$ is a corner point of region $r_i = [\nu_i]$. [6] also shows that ratios are affine functions which min and max are reached on integral coordinates. Then, for a non-Zeno timed automaton $\mathcal{A}$, for every run γ of $\mathcal{A}$, there exists a run π of $proj_{CP}(\gamma)$ such that $rat(\pi) \geq rat(\gamma)$ that is built by choosing corner points with the coordinates that maximize ratios in each r_i. In other words, for every run γ of $\mathcal{A}$, there exists a run π of $\mathcal{A}_{CP}$ with greater ratio. □

A direct consequence of Propositions 5 and 6 is that the optimal ratio computed on the corner point automaton $\mathcal{A}_{CP}$ is exactly the optimal ratio of $\mathcal{A}_\mathcal{N}$, i.e. we have $rat^* = rat^*_{CP}$. It now remains to show that rat^*_{CP} can be effectively computed. We show below that one can build a weighted graph $\mathcal{G}_\mathcal{N}$ such that, for every path ρ of $\mathcal{A}_{CP}$ there exists a path ρ_W of $\mathcal{G}_\mathcal{N}$ in which $Cost(\rho_W) = \mathcal{E}(\rho)$, and $\mathcal{T}(\rho) = |\rho_W|$.

We start from a corner-point automaton with a cost function $\mathcal{E}_{CP} : S \to \mathbb{Q}$ describing the energy attached to each state $s \in S$ of $\mathcal{A}$ (and hence to each state (s, r, ν) of $\mathcal{A}_{CP}$). Recall that $\mathcal{A}$ is non-Zeno. A consequence is that, in every cyclic behavior, there exists a minimal amount of time that must elapse. Recall also that transitions of $\mathcal{A}_{CP}$ are either discrete transitions, or inter-region transitions that represent moves of duration 0, or intra-region transitions that represent timed moves of duration up to 1. We denote respectively these sets of transitions $T_{d,CP}, T_{r,CP}, T_{i,CP}$. A consequence of propositions 5 and 6 and of the fact that guards use non-strict inequalities is that, as soon as some time can elapse in a region, then there is an execution visiting the same regions where the time spent in the region is 1 (otherwise ratios would be different in $\mathcal{A}$ and $\mathcal{A}_{CP}$). Further, using Proposition 4, we know that ratios are always defined for non-Zeno automata and in particular for $\mathcal{A}$. So, exploring paths of $\mathcal{A}_{CP}$ starting from a state (s, r, ν), it is guaranteed that in less than than $|\mathcal{A}_{CP}|$ steps, one will either find a deadlock state or a cycle with at least one intra-region timed transition. For a given state (s, r, ν) we define by $Seq(s, r, \nu)$ the set of sequences of transitions $\pi = t_0.t_1 \cdots t_k$ that start at (s, r, ν), and contain a single intra-region transition *i.e.*, paths of duration 1 time unit. In sequence π this implies among transitions of the form $t_i = (s_{i-1}, r_{i-1}, \nu_{i-1}) \xrightarrow{a_i} (s_i, r_i, \nu_i)$, a single transition t_j belongs to $T_{i,CP}$ and $a_j = \bot$, while other transitions t_i with $i \neq j$ belong to $T_{d,CP} \cup T_{r,CP}$ and we have $a_i = 0$.

$$Seq(s, r, \nu) = \left\{ \pi = t_0.t_1 \cdots t_k \in Runs(\mathcal{A}_{CP}) \mid \begin{array}{c} \pi \text{ starts from } (s, r, \nu) \\ \wedge| \bigcup_{i \in 0..k} t_i \cap T_{i,CP}| = 1 \end{array} \right\}$$

We are now ready to define the weighted automaton $\mathcal{G}_\mathcal{N} = (S_{\mathcal{G}_\mathcal{N}}, A, w)$:

- $S_{\mathcal{G}_\mathcal{N}} = \{(s, r, \nu) \in S_{CP} \text{ accessible from } (s_0, r_0, \nu_0)\}$,
- $A = \{((s, r, \nu), (s', r', \nu')) \in S_{\mathcal{G}_\mathcal{N}}^2 \mid \exists t_0.t_1 \cdots t_k \in Seq(s, r, \nu), \ t_0 = (s, r, \nu) \xrightarrow{a_0}$
  $(s_1, r_1, \nu_1) \wedge t_k = (s_{k-1}, r_{k-1}, \nu_{k-1}) \xrightarrow{a_k} (s', r', \nu')\}$
- For every arc $a \in A$, with $a = ((s, r, \nu), (s', r', \nu'))$ we set
  $$w(a) = \max_{\pi \in Seq(s,r,\nu) \pi \text{ ends in } (s',r',\nu')} \mathcal{E}_{CP}(\pi)$$

Proposition 7. *The maximal ratio rat_{CP}^* is equal to the maximal ratio of $\mathcal{G}_\mathcal{N}$.*

Proof. Let $C = a_1 \ldots a_k$ be the cycle of $\mathcal{G}_\mathcal{N}$ that achieves ratio μ^*. Then C is accessible from (s_0, r_0, ν_0). So, there exists a cycle $\pi_1 = \pi_{1,1}.\pi_{1,2} \cdots \pi_{1,k}$ in $\mathcal{A}_{CP}$ such that $\pi_{1,i}$ starts in (s_i, r_i, ν_i), ends in (s'_i, r'_i, ν'_i), $\pi_{1,i} \in Seq(s_i, r_i, \nu_i)$ and $(s'_i, r'_i, \nu'_i) = (s_{i+1}, r_{i+1}, \nu_{i+1})$ and maximizes $\mathcal{E}$. Similarly, there exists a sequence of arcs π_0 from (s_0, r_0, ν_0) to (s_1, r_1, ν_1). The ratio of C is $\frac{w(a_1) + \ldots w(a_k)}{k}$, and it is also the ratio of π_1. So, for every cycle in $\mathcal{G}_\mathcal{N}$, there exists a cycle of $\mathcal{A}_{CP}$ with the same ratio. This is in particular true for the cycle with the best ratio, so $rat_{CP}^* = \mu^*$. Conversely, every path π can be decomposed into finite sequences of transition that stop immediately after a transition of $T_{i,CP}$. These sequences give immediately a sequence of arcs in $\mathcal{G}_\mathcal{N}$ with the same ratio. $\qquad\square$

Proposition 7 immediately gives an algorithm to compute the optimal ratio of transferred energy per time unit in a $\mathcal{N}$. The first step is to compute the timed automaton $\mathcal{A}_\mathcal{N}$ by composition of the behaviors of the vehicles in the network, and the time constrained observers of the stations. The second step is to build $\mathcal{A}_{\mathcal{N},CP}$, the corner-point automaton of $\mathcal{A}_\mathcal{N}$. The third step is to compute the weighted graph $\mathcal{G}_\mathcal{N}$. The fourth step is to find the optimal ratio in connected components of $\mathcal{G}_\mathcal{N}$. This can be done in polynomial time using first Tarjan's algorithm [12] to find connected components and then our modification of Karp's algorithm shown in Sect. 2 to compute optimal ratios for all connected components. Notice that the fourth step is polynomial in the number of states in $\mathcal{A}_{CP}$, which is equal to the number of states in $\mathcal{A}_\mathcal{R}$ multiplied by the number of corner points, that is in $O((C_{max})^{n+p})$, where C_{max} is the largest constant appearing in $\mathcal{A}_\mathcal{N}$. We hence get the following result:

Theorem 3. *The optimal ratio of a CTN can be computed in exponential time.*

6   The Threshold Problem

Let us now consider the threshold problem: given a CTN $\mathcal{N}$ and an energy threshold λ we want to decide whether there exists a run of $\mathcal{N}$ for which the saved energy exceeds λ. The main idea behind this question is to ensure that there is a strategy for which the saved energy is significant. This problem differs from the transfer problem, as existence of a transfer does not guarantee that the transfer exceeds the lower bound λ. It also differs from the ratio problem, as one does not seek optimality. However, both problems are not disjoint. Consider a $\mathcal{N}$ for which a cyclic path C_r of the corner-point automaton of $\mathcal{N}$ has a (non-necessarily optimal) ratio $r > 0$. Then, for any threshold λ, one can build a run that contains more than $\frac{\lambda}{r \cdot \mathcal{T}(C_r)}$ occurrences of C_r and saves more than λ energy units. However, even if all cycles of the corner point automaton of $\mathcal{N}$ have ratio zero, one may still find acyclic paths which energy level is greater than λ.

Proposition 8. *Let $\mathcal{N}$ be a cyclic transport network, and $\lambda \in \mathbb{Q}$. Then, $\mathcal{N}$ has runs in which the saved energy exceeds λ iff $\mathcal{A}_{\mathcal{N},CP}$ has a reachable cycle of strictly positive ratio, or an acyclic path π such that $\mathcal{E}(\pi) > \lambda$.*

Proof. Recall from the proofs of Prop 6 that there is a correspondence (both is terms of visited transitions, regions and energy saved) between runs of $\mathcal{A}$ and paths of $\mathcal{A}_{\mathcal{N},CP}$. Indeed, for a run γ of $\mathcal{A}$, $proj_{CP}(\gamma)$ allows to find a run π with a greater ratio. Further, for every execution π of $\mathcal{A}_{CP}$ there exists a concrete execution with the same valuations as the corner points of π, and with the same ratio. The set of paths in $\mathcal{A}_{\mathcal{N},CP}$ is composed of paths that contain cycles, and acyclic paths. Assume that for every cycle C of $\mathcal{A}_{\mathcal{N},CP}$, we have $\frac{\mathcal{E}(C)}{\mathcal{T}(C)} = 0$. As $\mathcal{A}_{\mathcal{N}}$ is non-Zeno, this means in particular that $\mathcal{E}(C) = 0$. Let π be a path of the form $\pi = \rho_1.\beta_1.\rho_2\ldots$ where ρ_i's are acyclic paths and $\beta'_j s$ are cycles. Then, the amount of energy saved in π is the same as the amount of energy saved in a path $\pi' = \rho_1.\rho_2\ldots$ obtained by removing cycles. If ρ_i and ρ_j visit a common state, then π' still contains a cycle, and can again be reduced to a finite path $\pi'' = \rho'_1.\rho'_2\ldots\rho'_k$ where every $\rho'_i \neq \rho'_j$ have disjoint sets of states, and with the same energy level. Hence, the maximal energy level of a path in $\mathcal{A}_{\mathcal{N},CP}$ is achieved in an acyclic path. Now assume that for every acyclic path π of $\mathcal{A}_{\mathcal{N},CP}$, the energy level $\mathcal{E}(\pi)$ is smaller than λ. Then, there exists no path of $\mathcal{A}_{\mathcal{N},CP}$ and hence no run of $\mathcal{N}$ which allows to save more than λ units of energy.

On the other hand, if $\mathcal{A}_{\mathcal{N},CP}$ contains an accessible cycle C of positive ratio, then there exists a path $\pi_1 = \rho \cdot C$ where ρ is acyclic, and for every $k \in \mathbb{N}$, $\pi_k = \rho.C^k$ is also a path of $\mathcal{A}_{\mathcal{N},CP}$. The energy saved in cycle C is $\mathcal{E}(C) > 0$. So, the energy saved in π_k is $\mathcal{E}(\pi_k) = \mathcal{E}(\rho) + k.\mathcal{E}(c)$, and hence, for $k \geq \lceil \frac{\lambda}{\mathcal{E}(C)} \rceil$, $\mathcal{E}(\pi_k) \geq \lambda$. As π_k is a path of $\mathcal{A}_{\mathcal{N},CP}$, then there exists a concrete run ρ of $\mathcal{N}$ which has π_k as corner-point abstraction, and such that $\mathcal{E}(\rho) > \lambda$.

Conversely, if $\mathcal{A}_{\mathcal{N},CP}$ contains a run (involving cycles or not) π with $\mathcal{E}(\pi) \geq \lambda$, then there exists a run ρ of $\mathcal{A}_{\mathcal{N}}$ (and hence of $\mathcal{N}$) with $\mathcal{E}(\rho) = \mathcal{E}(\pi) \geq \lambda$. $\qquad\square$

Now notice that if there exists a cycle C of $\mathcal{A}_{\mathcal{N},CP}$ such that $\mathcal{E}(C) > 0$, then there exists an elementary cycle C' of $\mathcal{A}_{\mathcal{N},CP}$ (and of size smaller than $|\mathcal{A}_{\mathcal{N},CP}|$) that contains at least one of the transitions of C. With this property and using proposition 8, we can deduce that the threshold problem can be solved by exploring only acyclic paths of $\mathcal{A}_{\mathcal{N},CP}$. We hence have:

Theorem 4. *The threshold problem for CTNs is in PSPACE.*

Proof. (Sketch). The algorithm to check existence of a run π of $\mathcal{A}_{\mathcal{N},CP}$ with $\mathcal{E}(\pi) \geq \lambda$ is a standard non-deterministic exploration of paths of size at most $|\mathcal{A}_{\mathcal{N},CP}|$, that can be done in PSPACE (a complete proof is given in [11]).

7   Conclusion and Discussion

In this paper, we have shown that one can check if a transport network modeled as a cyclic graph with tokens moving at predetermined speeds can be controlled in order to save energy via regenerative braking. The paper voluntarily keeps the model simple: all vehicles have the same speed profile (acceleration, cruise speed, braking, ...) between two stations, and similar ranges of possible dwell times at each station. The only possible way to transfer energy is to synchronize braking of a train and acceleration of another one around the same station. So, the only way to control energy transfer is to choose appropriate dwell times. This setting can be easily extended: we can consider different speed profiles for each train, and let a controller choose a particular profile in addition to the dwell time. Overall, these improvements of the model are harmless for the techniques proposed, as they still allow to build a timed automaton, its region/corner point abstractions, and to check transfer, ratio and threshold properties with a slight increase in the number of states in $\mathcal{A}_{\mathcal{N}}$. Considering complex network architectures (e.g. with forks and joins) also seems to preserve the formal setting of this paper, even if construction of $\mathcal{A}_{\mathcal{N}}$ could be more involved than a simple product. Another desirable extension of the model is to make the energy consumptions and transfers more accurate. If several speed profiles are allowed for vehicles, as far as we are concerned by energy saving, there is a tradeoff to find between the energy saved when synchronizing trains and the energy used to travel at higher speed. We also consider a rough approximation of transfer: when a train brakes near a station St_i and its predecessor leaves, our model counts one unit of energy transferred per time unit around St_i. This is a coarse model, as the amount of energy that can be effectively transferred depends on the distance between vehicles. However, refining the model to handle variation of energy w.r.t. distances would call for more continuous models such as hybrid systems and the ratio or threshold problems may become undecidable, as reachability is already undecidable for hybrid automata [1].

Even for a simple model such as CTNs, several questions remain open. The first one is the lower bounds for transfer, optimal ratio, and threshold problems. We have shown that transfer and threshold problems are in PSPACE, and the calculus of an optimal ratio requires exponential time. The PSPACE complexity comes from the translation of CTNs into equivalent timed automata, and from the reduction of transfer and threshold to reachability questions. The complexity of ratio calculus comes from the need to explore all cycles of the corner-point automaton. Now the remaining question is whether these bounds are sharp. One can first notice that the translation from CTNs to timed automata leads to a subclass where each component has its own clock, that is reset at each transition. Yet, even restricted classes of timed automata such as event-clock automata [4] have PSPACE complexities for standard emptiness questions. The transfer problem, at least, seems of a very simple nature, and some instances of the problem have a trivial answer.

Consider for instance the CTN described in Fig. 3-left, with 4 stations and 2 vehicles. Assume the length of the network is 10 km, that the average speed of trains is 30 km/h, that dwell times range from 1 to 2 min, and that the only constraint in the network is that one train leaves station St_1 at least every 30 mn. Then, trivially, one can guide in a bounded amount of time the network to a situation represented in Fig. 3-right where the two vehicles are close enough for energy transfer, without violating the constraint on station St_1. For the same example, if one imposes a train every 11 min at station St_1, then one can only meet this constraint if vehicles are (up to distances covered during the dwell of the other vehicle) at opposite positions in the network. Obviously, this constraint and the existence of an energy transfer cannot be guaranteed with only two trains in the network, and this can be easily verified on the model.

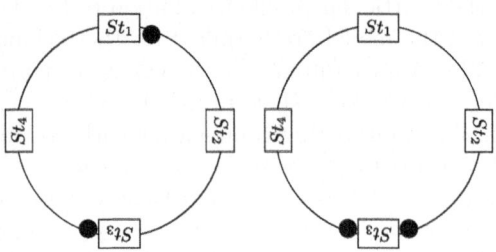

Fig. 3. A simple CTN with two trains. Initial configuration (left) and configuration allowing energy transfer (right).

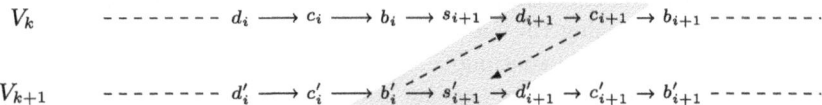

Fig. 4. Processes and transfer patterns for a pair of vehicles in a CTN

Another remark is that transfer occurs as soon as a pair of trains can synchronize their acceleration and braking phases. The behaviors of vehicles in a network such as a metro is very regular, and is a succession of events such as departures, beginning of cruise phase, beginning of braking phase, stop. These systems have a lot of concurrency, as the only constraint among vehicles is that they cannot occupy the same position at a given instant (or too close positions). The way vehicles moves are orchestrated can be seen as a partial order, where a single vehicle's behavior is a total order, ordering sequence events d_i, c_i, b_i, s_i denoting respectively departure from station st_i, beginning of cruise phase, beginning of braking phase, and stop at station st_i. Consider the example of Fig. 4. It depicts the mission of two consecutive trains V_k and V_{k+1}. In addition to the total ordering of events related to each train, we have an ordering $b'_i < d_{i+1} < c_{i+1} < s'_{i+1}$, meaning that train $k+1$ starts to brake before train k reaches its cruise speed.

These dependencies are symbolized by dashed arrow, and the whole pattern highlighted by a grey zone in the Figure. This representation of runs is similar to processes of a net, and there is no doubt that, one could build inductively this type of process for CTNs.

Let $\Delta(e)$ denote a realization date for event e. Then, there is an energy transfer if we can find consistent dates such that $\Delta(b'_i) \leq \Delta(d_{i+1}) \leq \Delta(c_{i+1}) \leq \Delta(s'_{i+1})$ and $[\Delta(d_{i+1}); \Delta(c_{i+1})] \cap [\Delta(b'_i); \Delta(s'_{i+1})]$ is not a singleton. So, the question of transfer amounts to finding appropriate patterns on a process while satisfying some constraints on its event dates. A natural question is whether a process view of CTN runs can reduce the complexity of energy problems. A second question is whether one can focus on a chosen pair of vehicles and abstract their environment. The orders generated by processes of CTNs seem to be regular (i.e. recognizable by some graph grammar). Now if the constraints attached to the times of generated processes also exhibit some form of regularity, that can be captured by Difference Bound Matrices, such a concurrent model could provide ways to solve efficiently energy transfer questions.

Acknowledgements. The authors would like to thank the anonymous reviewers of a preliminary version of this work for their precise reading and useful comments.

References

1. Alur, R., et al.: The algorithmic analysis of hybrid systems. Theor. Comput. Sci. **138**(1), 3–34 (1995)
2. Alur, R., Dill, D.: The theory of timed automata. In: Real-Time: Theory in Practice, REX Workshop. LNCS, vol. 600, pp. 45–73 (1991)
3. Alur, R., Dill, D.: A theory of timed automata. Theor. Comput. Sci. **126**(2), 183–235 (1994)
4. Alur, R., Fix, L., Henzinger, T.: Event-clock automata: a determinizable class of timed automata. Theor. Comput. Sci. **211**(1–2), 253–273 (1999)
5. Asarin, E., Maler, O., Pnueli, A., Sifakis, J.: Controller synthesis for timed automata. In: IFAC System Structure and Control, pp. 447–452 (1998)
6. Bouyer, P., Brinksma, E., Larsen, K.: Optimal infinite scheduling for multi-priced timed automata. Formal Methods Syst. Des. **32**(1), 3–23 (2008)
7. Bouyer, P., Fahrenberg, U., Larsen, K., Markey, N.: Timed automata with observers under energy constraints. In: Hybrid Systems: Computation and Control, HSCC 2010, pp. 61–70 (2010)
8. Cassez, F., Henzinger, T.A., Raskin, J.-F.: A comparison of control problems for timed and hybrid systems. In: Tomlin, C.J., Greenstreet, M.R. (eds.) HSCC 2002. LNCS, vol. 2289, pp. 134–148. Springer, Heidelberg (2002). https://doi.org/10.1007/3-540-45873-5_13
9. Fahrenberg, U., Juhl, L., Larsen, K.G., Srba, J.: Energy games in multiweighted automata. In: Cerone, A., Pihlajasaari, P. (eds.) ICTAC 2011. LNCS, vol. 6916, pp. 95–115. Springer, Heidelberg (2011). https://doi.org/10.1007/978-3-642-23283-1_9
10. Karp, R.: A characterization of the minimum cycle mean in a digraph. Discret. Math. **23**(3), 309–311 (1978)

11. Paparazzo, L., Hélouët, L., Markey, N.: Energy transfer in timed cyclic networks (extended version). Technicsl report, INRIA, CNRS, Univ. rennes (2025). https://inria.hal.science/hal-05006900
12. Tarjan, R.: Depth-first search and linear graph algorithms. SIAM J. Comput. $1(2)$, 146–160 (1972)

Leveraging Petri Nets for Workflow Anomaly Detection in Microservice Architectures

Priyanka Kamboj[1] , Cyrille Artho[1] , Roberto Guanciale[1 (✉)] ,
Reyhaneh Jabbarvand[2] , and Brighten Godfrey[2]

[1] KTH Royal Institute of Technology University, Stockholm, Sweden
{kamboj,artho,robertog}@kth.se
[2] University of Illinois Urbana-Champaign, Chicago, USA
{reyhaneh,pbg}@illinois.edu

Abstract. Modern microservice architectures pose challenges in understanding and managing the complex workflows within these decentralized services. In particular, it is difficult to identify anomalous behavior that could indicate a bug or attack. We use traces of microservice application activity (requests and responses) to infer a model of the application's normal behavior. Our approach mines Petri nets to formally represent concurrent operations and their temporal dependencies with a targeted delay injection approach that accurately and efficiently learns these dependencies. The models produced are both explainable and easy to inspect, which offers more transparency and control. Our evaluation shows that injecting delays during model training allows us to achieve perfect model and log fitness (Move-Model and Move-Log fitness of 1) with just 29 traces. In contrast, a straightforward approach requires over 10,000 traces to achieve similar accuracy. Our models successfully identify anomalies in various experiments, such as traces with one missing or multiple missing activities, and reordered sequences to simulate issues in real-world scenarios. Our approach outperforms the state-of-the-art method, demonstrating higher accuracy.

Keywords: Alignments · Cost function · Delay · Microservices · Process mining · Process discovery · Conformance Checking

1 Introduction

Microservices offer scalability and flexibility to meet evolving demands [13,35]. Unlike a monolithic architecture, a microservice-based architecture enables independent development, deployment, and updating of individual services. However, the diverse interaction patterns of application behavior and communication styles between microservices introduce complexity, making it challenging to ensure quality and security [15].

Our focus in this paper is to characterize the normal behavior of a microservice application automatically and to find anomalies in the running application

© The Author(s), under exclusive license to Springer Nature Switzerland AG 2025
E. Amparore and Ł. Mikulski (Eds.): PETRI NETS 2025, LNCS 15714, pp. 219–241, 2025.
https://doi.org/10.1007/978-3-031-94634-9_11

behavior that may indicate a bug or attack. Microservice systems employ tracing mechanisms such as Jaeger [2] and Zipkin [40] that observe behavior at the level of requests and responses in individual microservice instances. However, interpreting traces of such events to extract higher-level insights and patterns is difficult. In particular, non-deterministic timing and ordering of events are typical due to network latency, varying processing times, and system delays [24].

In our work, we apply process mining techniques to address the challenges in microservice architectures by providing insights into workflows within these systems. Process mining extracts process-related insights from business processes [33]. It analyzes the sequence of activities or events to help organizations understand and improve their complex business processes [6,39]. Through process mining, visual process models are generated to represent the sequence of activities, tasks, and service interactions by analyzing event logs produced by microservices. We utilize Petri nets to provide a precise and formalized representation of concurrent operations and their dependencies within microservice applications [23,34]. Petri nets are well-suited for this purpose as they capture key behaviors such as parallelism and dependencies between services, including partial-ordering or temporal ordering of activities. This makes them the best choice for representing complex interdependencies in a microservice architecture, where accurately modeling parallel and sequential behaviors is essential for optimizing performance and reliability.

The conformance-checking technique in process mining compares and matches the observed behavior in the event log to the behavior in the learned process model to verify that the event log conforms to the model [11,14]. In a microservice architecture, adopting conformance checking plays a crucial role in identifying discrepancies by ensuring the alignment of microservices execution with the designed process models.

To capture parallelism in microservices, we propose a novel methodology for learning the process model and detecting anomalies in the microservices. Our approach is organized into three stages: (1) event log generation, (2) process discovery, which involves (a) constructing Petri nets using raw trace data and (b) optionally applying a delay injection learning approach, and (3) conformance checking, where deviations from the process model or anomalies are detected and reported. The contributions of our work are as follows:

1. We design an approach for event log generation that accurately captures end-to-end processes in microservices, despite their decentralized and heterogeneous nature. This enables precise process discovery and provides a comprehensive view of system workflows.
2. We propose a delay injection algorithm to learn a process model faster as a Petri net from the distributed tracing data of microservice applications.
3. We implement a conformance checking mechanism that identifies deviations and anomalies in real-time.
4. We evaluate our approach through various experiments on the Social Network and Media Service from the DeathStarBench microservice benchmarking platform [15]. Our evaluation indicates that incorporating delay injection during

model training achieves perfect Move-Model and Move-Log fitness values of 1 with only 29 traces, an accuracy that a straightforward approach fails to reach even after 10,000 traces. Furthermore, the experimental results demonstrate that our approach outperforms the LogDP method [36] by effectively identifying anomalies and achieving higher precision.

The paper is structured as follows. Section 2 provides the background, followed by Sect. 3, which reviews related work. Sections 4 and 5 detail the proposed framework. Sections 6 and 7 present the experimental setup, results, and discussion. Finally, Sect. 8 concludes with the key findings and future directions.

2 Preliminaries and Background

In this section, we introduce key definitions essential to our work, including Petri nets for process modeling, legal moves for conformance checking, and cost functions to measure deviations.

Microservices decouple applications into loosely coupled, deployable services that communicate through well-defined protocols [21,32]. These services can be developed and deployed independently, with each service responsible for specific business functions. The distributed nature of microservices makes them more challenging to monitor and manage compared to monolithic applications. Mechanisms such as OpenTelemetry-based distributed tracing [28] have been introduced to capture traces for monitoring microservices.

Definition 1 (Trace and Event log). *A trace* $\sigma = (e_1, e_2, ..., e_n)$ *is a sequence of events representing a single instance of a process execution, where each* $e_i = (t_i, x_{z1}, x_{z2}, ...x_{zm})$ *is an event in the trace,* t_i *is its timestamp associated and* $x_{z1}, x_{z2}, ...x_{zm} \in \Sigma_L$ *are its attributes. An event log* $L = \sigma_1, \sigma_2, .., \sigma_k$ *is a collection of traces where each trace represents a different instance of a process.*

Definition 2 (Petri Net). *A Petri net is a tuple* $M = (P, T, F)$ *where* P *is the (finite) set of places,* T *is the (finite) set of transitions with* $P \cap T = \emptyset$ *, and* $F \subseteq (P \times T) \cup (T \times P)$ *is the flow relation that describes the set of directed edges between places and transitions.*

The state of a Petri net is a marking that associates (multiple) tokens to places. When a transition is fired (or triggered), the Petri net consumes a token from each input place and produces a token for each output place. A sequence of transitions from the initial to the final marking is called a *firing sequence*. The language of a Petri net denoted $\mathcal{L}_M$ comprises all possible firing sequences that it allows from the initial marking.

We use the Petri net of Fig. 1, modeling a social network microservice application, as a running example for this paper. Squares represent the activities (or transitions) in the process, while circles represent places denoting conditions or states in the system. The black box in the diagram denotes a silent (or invisible) transition in a Petri net that occurs without an explicit or observable event.

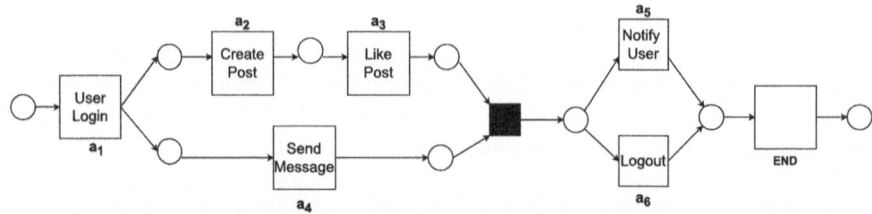

Fig. 1. Running example of a Petri net describing a Social Network microservice application process

Definition 3 (Legal move). *Let (P, T, F) be a Petri net and L a log. We define the synchronous moves $\mathcal{M}_{sync} = \Sigma_L \times T$, the moves on log $\mathcal{M}_{log} = \Sigma_L \times \{\tau\}$, the moves on model $\mathcal{M}_{model} = \{\tau\} \times T$, and the set of all legal moves as $\mathcal{M} = \mathcal{M}_{sync} \cup \mathcal{M}_{model} \cup \mathcal{M}_{log}$. An alignment γ is a sequence of legal moves in $\mathcal{M}^*$.*

An alignment γ is *valid* for a trace σ and a Petri net M if (1) $(\gamma|_{\mathcal{M}_{sync} \cup \mathcal{M}_{log}})_1 = \sigma$ and (2) $(\gamma|_{\mathcal{M}_{sync} \cup \mathcal{M}_{model}})_2 \in \mathcal{L}_M$, where $\mathcal{L}_M$ denotes the language of the Petri net. For a given sequence $\gamma \in \mathcal{M}^*$: The projection onto $\mathcal{M}_{sync} \cup \mathcal{M}_{log}$, denoted as $\gamma|_{\mathcal{M}_{sync} \cup \mathcal{M}_{log}}$, retains only moves that belong to these sets while preserving their order. Similarly, the projection onto $\mathcal{M}_{sync} \cup \mathcal{M}_{model}$, denoted as $\gamma|_{\mathcal{M}_{sync} \cup \mathcal{M}_{model}}$, retains only moves from these sets.

We define a projection function $\pi : \Sigma_L \to T$ that maps event attributes to transitions in the Petri net. This function extends to map log traces of the Petri net's firing sequences. The total cost of an alignment γ is the summation of the individual costs of all the moves in the alignment.

Definition 4 (Cost function). *The cost function $C(l, m)$ for an alignment pair (l, m) is defined as follows:*

$$C(l, m) = \begin{cases} 0, & \text{if } l \in \Sigma_L \text{ and } m \in T \text{ and } \pi(l) = m \text{ (synchronous move),} \\ 1, & \text{if } l \in \Sigma_L \text{ and } m = \tau \text{ (log move),} \\ 1, & \text{if } l = \tau \text{ and } m \in T \text{ (model move),} \end{cases} \quad (1)$$

We will denote the process of computing the cost function as *replaying* a trace against a model. In the cost function, we penalize log and model moves by assigning a cost because it indicates a deviation between the log and the process model. The synchronous move is the best case and incurs no cost. A small cost is assigned to silent moves to prevent the model from entering infinite loops. An alignment γ with minimum cost is considered *optimal* under the cost function C iff $\nexists \gamma' : C(\gamma') < C(\gamma)$, i.e., there does not exist another alignment γ' with a smaller cost.

Definition 5 (Fitness). *The Move-Model (f_{model}) and Move-Log (f_{log}) fitness of alignment γ are defined as:*

$$f_{model} = 1 - \frac{c_{model}}{c_{model} + c_{log}} \quad (2)$$

$$f_{log} = 1 - \frac{c_{log}}{c_{model} + c_{log}} \tag{3}$$

Here, c_{model} is the total cost of moves on model and c_{log} is the total cost of moves on log on alignment γ.

Fitness is a value between 0 and 1, where 0 indicates poor fitness and 1 indicates perfect fitness. A model achieves perfect fitness if all traces in the log can be replayed from start to finish without any deviations. A fitness value of less than 1 indicates discrepancies or deviations between the trace log and model.

In the example shown in Fig. 1, during regular operation, activities like a_2 (Create Post) and a_3 (Like Post) occur sequentially, while a_4 (Send Message) runs in parallel. An anomaly can occur if an activity is missing, such as a user creating a post without logging in, or if activities are reordered, for example, sending a notification to the user before creating a post. These deviations can indicate defects, system misconfigurations, or even an ongoing exploit that aims to manipulate the system workflow.

An anomaly refers to a system exhibiting a behavior that is not in the set of behaviors permitted by a model. In our setting, we apply model learning to a training set to detect anomalies in a test set.

Definition 6 (Anomaly). *An event e_i that is not stipulated by the model M and occurs in a trace $\sigma = (e_1, e_2, \ldots, e_n)$ from an event log L is considered anomalous. We define $\mathcal{E}_M(\sigma)$ as the set of valid events in σ, such that:*

$$\mathcal{E}_M(\sigma) = \{e_i \mid e_i \text{ corresponds to a transition } t \in T$$
$$\text{that is enabled at marking during the replay of } \sigma \text{ on } M\}.$$

An event e_i is defined as anomalous if $e_i \notin \mathcal{E}_M(\sigma)$.

Definition 7 (Precision and Recall). *The Precision and Recall are defined as:*

$$Precision = \frac{\sum_{\sigma \in L} |R_p(\sigma, M)| |M_{sync}|}{\sum_{\sigma \in L} |R_p(\sigma, M)| |M_{sync}| + \sum_{\sigma \in L} |R_p(\sigma, M)| |M_{model}|} \tag{4}$$

$$Recall = \frac{\sum_{\sigma \in L} |R_p(\sigma, M)| |M_{sync}|}{\sum_{\sigma \in L} |R_p(\sigma, M)| |M_{sync}| + \sum_{\sigma \in L} |R_p(\sigma, M)| |M_{log}|} \tag{5}$$

where $R_p(\sigma, M)$ represents optimal alignment mapping a trace σ to the model M, minimizing the alignment cost. The projections onto synchronous moves, model moves, and log moves are denoted as $|M_{sync}|$, $|M_{model}|$, and $|M_{log}|$, respectively.

Definition 8 (F$_1$-score). *The F$_1$-Score is defined as:* $F_1 = 2 \times \frac{Precision \times Recall}{Precision + Recall}$

3   Related Work

Several approaches exist to model microservices, analyze the structure, and visualize dependencies between services. Mystery machine [13] and GMAT [25] focus on constructing directed graphs to represent the relationship between services. The mystery machine uses data from *UberTrace* to determine causal relations among events, while GMAT [25] generates a Service Dependency Graph (SDG) to map the relationship between services. However, these approaches help to understand the architecture but do not fully capture the dynamic behavior of microservices and the concurrent behavior of microservice applications. In contrast, some works have focused on modeling microservices applications as process models, especially Petri nets. The authors in [10] proposed a method that utilizes microservice Orchestrator, an open-source framework offered by Netflix, to manage workflows. The method formalizes and analyzes the properties of microservices applications using Time Petri nets to define temporal constraints. Liu *et al.* [24] proposed a method to reduce the impact of failures and improve the reliability of microservices-based cloud applications. The authors utilized Petri nets to model the dependencies between microservice instances and verify properties such as state reachability.

One general approach that could be applied to characterize microservice application behavior involves using large and complex statistical models, like hidden Markov models (HMMs) [31] and latent Dirichlet allocation (LDA) [9]. However, these models lack transparency, making it difficult for engineers to interpret their results and understand the underlying process flows. They also fail to model choices or concurrency explicitly. To overcome these challenges, there is a need for more efficient and explainable models that can better capture dependencies within microservice architecture and provide better insights that are more accurate and easily understandable by engineers.

Existing anomaly detection methods like LogDP [36], LogFlash [19], Nezha [38], and LogSed [20] focus on sequence-based or dependency-based modeling, which fail to capture the parallelism and temporal dependencies inherent in microservice architectures. Consequently, anomaly detection in microservice architectures must account for deviations, such as missing, duplicate, or extra activities, in order to maintain the security and reliability of the system.

Various methods have been developed to check discrepancies or deviations between the process model and behavior reported in event logs. Carmona *et al.* [11] and van der Aalst *et al.* [33] outline two main approaches for conformance checking – log replay algorithms and trace alignment algorithms. An example of a token-based replay technique was effectively presented by Rozinat *et al.* [30], who introduced a token-based approach to verify the conformance between the Petri net and an event log. The method computes conformance by assessing the number of missing and added tokens after replaying traces against the Petri net. The token-based approach relied on heuristic techniques, which may sometimes lead to incorrect or incomplete diagnostic information.

Alignments have been proposed to match the observed behavior (i.e., event logs) with the intended behavior using a process model to identify deviations

[14,26]. Thus, alignments have been used for conformance checking to provide more precise and accurate diagnostic insights [16]. Bloemen *et al.* [8] studied how cost functions affect the resulting alignments. The authors presented a novel algorithm for computing alignments that leverages the cost function to maximize the number of correct matches. Grohs *et al.* [17] proposed an approach to identify process-level deviations by applying alignment-based conformance checking to provide insights into discrepancies between process execution and expected process models. However, their approach does not consider the challenges posed by microservice architectures, such as temporal misalignments.

Our delay injection approach is inspired by fault injection [18], which simulates faults in hardware or software, and combinatorial testing [27], which tests systematically how multiple parameters or faults in a system might interact. Unlike in fault injection, we do not inject hard faults (or complete failures of a network connection) but delays. We apply this technique systematically to single or pairs of activities, corresponding to combinatorial combinations of strengths 1 and 2 [27]. Unlike related work injecting delays in networked applications via proxies [37], we inject them directly into the application code. This is due to the higher complexity of the network structure of microservice applications, making it easier to perform the delay injection with minimal code changes rather than extensive reconfiguration of all microservices.

The state-of-the-art techniques focus on modeling microservices and visualizing their dependencies using approaches like directed graphs [13,25]. These methods provide a high-level understanding of the architecture and help identify bottlenecks. However, they fail to capture the concurrent behavior of microservices or detect anomalies. On the contrary, our proposed approach bridges this research gap by introducing a method for faster learning a process model from the distributed tracing data of microservices. Furthermore, our model detects anomalies, such as missing activities, reorder sequences of activities, and conflict activities, which existing methods overlook.

4   Proposed Framework

This section presents the proposed framework for process discovery and conformance checking to detect anomalies in a microservice architecture. Figure 2 illustrates the proposed framework, which is structured into three main stages—(1) event log generation, (2) process discovery, which involves (a) constructing Petri nets using raw trace data approach, and (b) optionally applying a delay injection learning approach and (3) conformance checking, where deviations from the process model or anomalies are detected and reported.

4.1   Event Log Generation

We use OpenTelemetry-based distributed tracing data to generate event logs, capturing the sequence of events across multiple traces. Each trace is composed

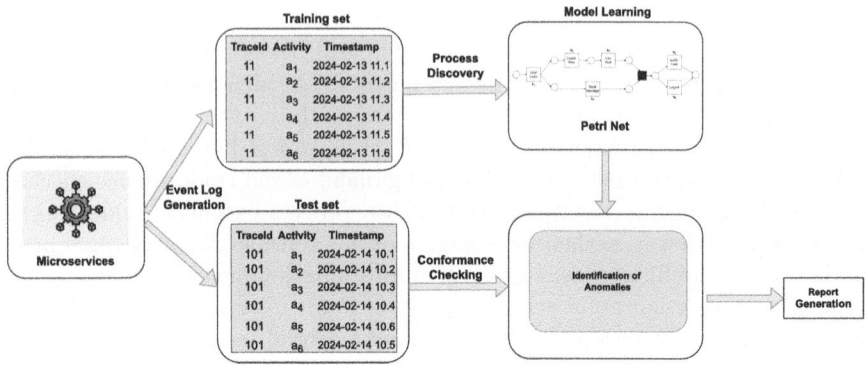

Fig. 2. Proposed Framework

of a series of events according to their parent-child relationships, forming a partial order that captures the logical dependencies between events. This structure ensures that the parent events precede child events, which reflects the hierarchical nature of execution in microservices. To evaluate the process models, we divided the generated trace data into training and test sets. The training set was used for process discovery to learn a model, while the test set was used for conformance checking to validate the model.

During this process, we encountered instances where the timestamps recorded from different processes were causally inconsistent. To address this challenge, we focused on the relationships between events as defined by their partial order, illustrating how events have occurred during their execution. We constructed a hierarchy of events by analyzing the structure of each trace in the event log and focusing on the TraceId fields generated using OpenTelemetry. Each child event is linked to its parent in this hierarchy, representing their causal relationship.

Within a single sub-process, all events are recorded inside a single *span*. Thus, within each span, events are totally ordered. We initially sort the events according to their timestamps to reflect the chronological sequence within each trace. This allows us to assess causal relationships while preserving the recorded sequence. As a result, our method facilitates a more precise analysis of workflows within microservice applications.

4.2   Process Discovery to Construct Petri Nets

As the distributed nature of network traffic necessitates a model that can capture parallelism, we use Petri nets. However, many Petri nets can potentially approximate a limited set of traces in a training set.

A significant challenge in process discovery is to find the "best" model that records the actual behavior of the real-world business process. A model that does not fit well cannot accurately reproduce all the behaviors documented in the event log. Another crucial quality criterion is *soundness*, ensuring that every process step can be executed and there is always a reachable satisfactory end state, represented by the final marking. For example, the Petri net is unsound if it experiences a deadlock, preventing the attainment of the final marking with only a single token in the ultimate position.

The researchers have proposed various process discovery algorithms to learn process models from event logs, such as *Alpha Miner* [7], *Heuristics Miner* [7]. One of the drawbacks of Alpha Miner and other process discovery algorithms is that it does not guarantee that the model will be sound and free of deadlocks. While experimenting with these algorithms, we encountered several issues.

For example, Alpha Miner produced an incomplete model that introduced loops in the model to capture certain patterns and only reflected direct activity relationships. In contrast, Inductive miner [22] employs an inductive approach, allowing it to capture more complex process patterns, including concurrency and choices. This flexibility makes it a robust choice for our analysis.

The initial part of the resulting Petri net from the Social Network dataset is shown in Fig. 3, demonstrating how our approach effectively captures concurrency patterns within the workflow. The Petri net structure represents both sequential and parallel activities, making it easier for engineers to validate the model's correctness. By comparing the Petri net with the event log, they can assess whether the learned model accurately reflects the underlying process.

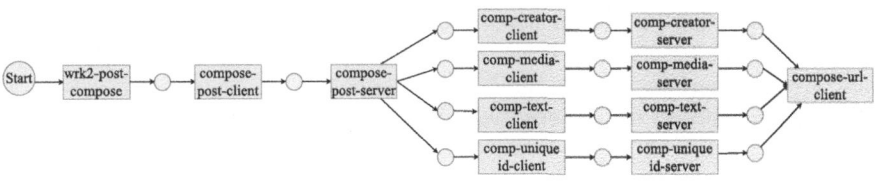

Fig. 3. Petri net representation of the Social Network dataset highlighting concurrency

In our work, we used the most popular Inductive Miner [22] algorithm from process discovery that builds a process tree from the event log. This approach uses a divide and conquer strategy, which consists of splitting the trace log, constructing a process tree, and, finally, splitting parts of the log are handled separately. The process tree often introduces silent or hidden activities to capture the model's behavior but avoids duplicating any activity. Further, the Inductive miner guarantees that the discovered Petri net from the process tree is sound [22].

4.3   Conformance Checking Using Trace Log and Process Model

We use alignment-based conformance checking to identify deviations by replaying events in the trace log against the process model. The approach adopts an A* algorithm to explore the state space to capture all the alignments [12]. The algorithm aims to find the best or all the alignments for each trace in the test log

(*TL*) and the learned model (*M*). For example, it helps pinpoint specific deviations, such as missing activity in the process model from the actual execution in the trace log. We consider three types of anomalies: a) missing activity—when an activity is skipped in the trace flow, b) reordered activity—when an activity sequence is interchanged, and c) conflict activity—when an activity conflicts with another in the trace flow. Our approach identifies deviations, including missing, unexpected, or reordering activities that could indicate potential anomalies. This information is generated as a report and later used to improve or enhance the process model. It also helps to take specific actions during process execution, like implementing security measures to analyze or monitor the malicious traffic in the network. This alignment-based technique serves as a cornerstone for maintaining the consistency and reliability of the entire system.

5   Proposed Delay Injection Learning Approach

When testing an application that consists of concurrent (parallel) microservices, each observation captures a certain interleaving of start and end time for these parallel activities. In cases where two parallel activities take a different amount of time, it is likely that the "faster" one almost always concludes earlier. Only in cases where the node that processes the faster request is under heavy computational load, scheduling, memory access, or network delays may result in "slower" services completing sooner. A new execution of a microservice, therefore, only results in a new observation if we see a different interleaving compared to the ones observed earlier. This is unlikely to happen if we execute services as is, as execution times of other activities typically do not vary much under a light testing load.

When multiple microservices execute concurrently, the number of possible interleavings of their start and end times grows exponentially. Each additional parallel activity introduces new potential sequences, which increases the complexity of testing and makes it nearly impossible to capture all possible interleavings.

To address this challenge, we propose a delay injection approach to speed up the learning process of behavior models. In this approach, we inject controlled delays in the system application. Our aim in introducing delays is that our model can better capture and learn the concurrent behavior of the system application. This incorporation of delay helps to effectively learn the temporal dependencies between the events in the system. When considering a delay like a "fault", this idea corresponds to using combinatorial testing [29], in the sense that injection delays in activities allow us to observe more diverse behaviors triggered by delays in different parts of the entire microservice ensemble.

For example, the Petri net in Fig. 1 starts with activity a_1, sequentially by a_2 and a_3. In Fig. 4, we illustrate the delay injection approach by considering four activities $\{a_1, a_2, ...a_4\}$ with initial timestamps 1, 2, 3, 2 (in ms), respectively. Additionally, activity a_4 runs concurrently with activity a_2. This figure depicts the delay injection approach and illustrates how introducing delays can modify the timing of these activities.

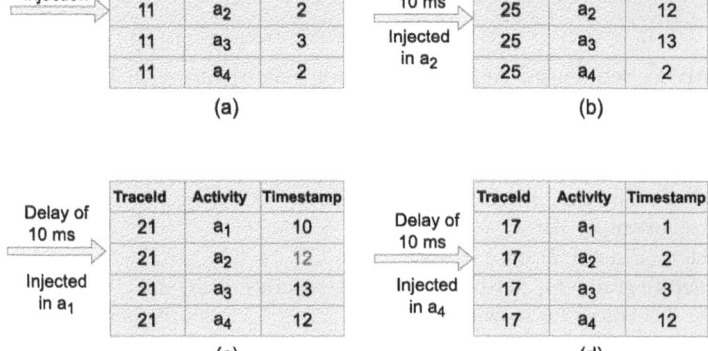

Fig. 4. Illustration of delay injection approach in application code

- Initial State without Delay: In the initial scenario, as shown in Fig. 4(a), no delay is injected into any activity. Therefore, the activities maintain their original timestamps: a_1 occurs at 1 ms, a_2 and a_4 both occur at 2 ms, and a_3 occurs at 3 ms.
- Delay Injection into a_2: In this case, as shown in Fig. 4(b), a delay of 10 ms is introduced into activity a_2. Despite the delay, the timestamp of a_1 and a_4 remains unchanged at 1 and 2 ms, respectively. However, delay in a_2 impacts the subsequent activities in the process flow. The updated timestamps are: activity a_2 now is delayed to 12 ms, and activity a_3 is shifted to 13 ms due to delay in a_2. This demonstrates how the delay in one activity (a_2) can affect the timing of subsequent activities.
- Delay Injection into a_1: In this case, as shown in Fig. 4(c), a delay of 10 ms is introduced into activity a_1. This delay shifts the timestamps of activities, changing to: activity a_1 now occurs at 11 ms, a_2 and a_4 occur at 12 ms, and a_3 now occurs at 13 ms.
- Delay Injection into a_4: In this case, as shown in Fig. 4(d), a delay of 10 ms is introduced into activity a_4. Despite the delay, the initial timestamp of activities a_1, a_2, and a_3 remains unchanged at 1, 2, and 3, respectively. However, the delay shifts the timestamp of a_4, causing it to now occur at 12 ms. The delay in a_4 does not affect the other activities, as a_4 runs concurrently with $\langle a_2, a_3 \rangle$. This parallel execution demonstrates that a_4 can be delayed such that it finishes after the parallel activities a_2 and a_3. With the delay for a_4, we can confirm the parallelism between $\langle a_2, a_3 \rangle$, and a_4 by observation.

Algorithm 1 outlines the delay injection approach for the system application. The sequential activity selection algorithm is designed to generate a comprehensive set of traces for all activities within the application by incorporating specified delays for each activity. In this approach, we sequentially inject a delay d_i into each activity a_i based on the total execution time of the process. Thus, our approach injects a delay of 1 ms in each activity in the event log and generates a combined trace log represented as T.

Algorithm 1. Sequential Activity Selection Algorithm with Delay Injection

> **Input** A: A set of activities in the application, $A = \{a_1, a_2, \ldots, a_n\}$.
> D: A corresponding set of delays for each activity, $D = \{d_1, d_2, \ldots, d_n\}$, where each d_i is the delay associated with activity a_i.
> **Output** T: The combined traces with delays.
> 1: Initialize an empty list $T \leftarrow [\ ]$.
> 2: **for** each a_i, d_i in A, D **do**
> 3:   Generate traces for activity a_i that incorporate delay d_i.
> 4: $T \leftarrow T \cup$ traces for a_i with delay d_i. $\triangleright$ Append new traces to T
> 5: **end for**
> 6: **return** T.

By introducing delays into both individual and multiple activities, we can observe the effect on the overall execution timings and examine how it affects the execution process of sequential and concurrent activities. The choice of a constant 1ms delay, rather than a randomized delay, allows us to maintain consistency across different traces and experiments. This consistency is crucial for isolating and analyzing the impact of delayed injection on the learning process without introducing additional variability. We chose a 1ms delay because it represents the worst-case execution time (WCET) of the entire process.

Our approach is useful for modeling the temporal dependencies between the activities. Therefore, the primary goal of our proposed algorithm is to fasten the learning process of temporal dependencies in the system application by injecting delays in the generated traces, with the delays being determined relative to the overall execution time of the process.

6   Implementation and Evaluation

In this section, we describe the experimental setup, the datasets used, the evaluation metrics, and the results of the experiments.

We carried out experiments in the ProM 6.3 framework to test the performance and validation of our proposed approach [4]. We ran all experiments on a Dell machine with an Intel Core i7 CPU (3.60 GHz), 32 GB RAM, and 1 GB BaseT network.

To implement delay injection, we modified the specific parts of the microservices code. For example, in the ComposeMedia function in the Media Service of Social Network dataset, we introduced a constant delay of 1 ms before executing the activity using *std::this_thread::sleep_for* function in C++. This delay simulates network latency or processing delays in real-world microservice systems. We inject a delay inside the start of each *Future* that connects to a child microservice and waits for its result. This way, we simulate a delay in the parallel activity of the microservice itself. Delays are injected into all activities using Algorithm 1. The implementation of this approach is shown in Listing 1.1. By systematically injecting delays, we can analyze their impact on process execution and evaluate how effectively our approach captures temporal dependencies.

Listing 1.1. Code snippet illustrating media composition with delay injection

```
1  void MediaHandler::ComposeMedia(
2      std::vector<Media> &_return, int64_t req_id,
3      const std::vector<std::string> &media_types,
4      const std::vector<int64_t> &media_ids,
5      const std::map<std::string, std::string> &carrier){
6      // Introduce a small delay to simulate network latency
7      std::this_thread::sleep_for(std::chrono::milliseconds(1));
8      // Extracting tracing information from the carrier
9      TextMapReader reader(carrier);
10     std::map<std::string, std::string> writer_text_map;
11     TextMapWriter writer(writer_text_map);
12     auto parent_span = opentracing::Tracer::Global()->Extract(reader);
13     // Create a new tracing span for the media composition operation
14     auto span = opentracing::Tracer::Global()->StartSpan(
15         "compose_media_server", {opentracing::ChildOf(parent_span->get())});
16         opentracing::Tracer::Global()->Inject(span->context(), writer);
17     }
```

Table 1. Characteristics and code composition of each microservice application

Service	Total New LoC	LoC for RPC handwritten	Autogen.	Unique Microservices
Social Network	15,198	9,286	52,283	36
Media Service	12,155	9,853	48,001	38

6.1 Case Studies

To evaluate our proposed approach, we generated datasets from two open-source microservice benchmark applications – *Social Network* [5] and *Media Service* [3] in DeathStarBench [15]. In the *Social Network* use case, the users send requests to *http*, implemented using a frontend generator, i.e., *Nginx*. Different services are responsible for composing posts, displaying user information, and the number of followers users follow or unfollow. In contrast, in the *Media Service* use case, users can search for information about their photos, videos, and movies and review information or insert new reviews into the application. The *Social Network* use case captures **parallel** behavior, while *Media Service* use case involves both **parallel** activities and **choice** between activities. All the services are packaged and deployed in Docker [1] containers, which helps to automate, deploy, and manage these containerized services. Table 1 summarizes the characteristics and code composition of each microservice application, including total new lines of code, manually written RPC-related code, auto-generated code, and the number of unique microservices. The languages used in Social Networks include C,

C++, Java, Node.js, Python, Scala, PHP, JavaScript, and Go, while the Media Service uses C, C++, Java, PHP, Scala, Node.js, Python, and JavaScript. The application traces are collected using Jaeger [2], which is a distributed tracing tool. The dataset of the work is publicly available at GitHub[1].

6.2 Experiments

In a microservice architecture, parallelism occurs when there is sufficient variance in activity durations. When execution time variance is limited, particularly in distributed and dynamic environments, delay injection is essential for identifying the concurrent behavior of the system.

Our experiments are designed to answer the following research questions:

- **RQ1** How does model learning using Petri nets capture concurrency?
- **RQ2** How can targeted delays improve the observability of parallelism in model learning?
- **RQ3** What is the anomaly detection accuracy of our approach?

To evaluate our approach, we measured key metrics: Move-Model and Move-Log fitness, $|\mathcal{L}_{TL}|$, Skipped Event, and Inserted Event, Precision, Recall, and F_1-score. Move-Model fitness evaluates how well a given Petri net model aligns with the behaviors captured in an event log. Move-Log fitness measures how well the event log corresponds to the expected behaviors defined by the Petri net model. The language of a log L, denoted as $\mathcal{L}_L$, refers to the set of all possible traces that can be derived from that log L.

$$\mathcal{L}_L = \{\pi(\sigma) \mid \sigma \text{ is a trace in } L\},$$

whereas language of the training log is denoted as $(\mathcal{L}_{Trl})$.

Furthermore, $|\mathcal{L}_{TL} \setminus \mathcal{L}_M|$ quantifies the difference between the language of the test log $(\mathcal{L}_{TL})$ and the language of the model $(\mathcal{L}_M)$. A *Skipped Event* corresponds to moves in $\mathcal{M}_{log}$, representing activities that are expected to be performed by the model but do not appear in the event log. On the other hand, an *Inserted Event* corresponds to moves in $\mathcal{M}_{model}$, which are the activities that appear in the event log but not part of the model's behavior. These are deviations from the expected behavior as per the model. We computed the dataset's average and standard deviation values over 30 runs. Additionally, we compare our proposed approach with the benchmark scheme – LogDP [36], which leverages the dependency relationships between log events to identify anomalies in large, unlabeled log data.

Experiment 1: Learning Approach Without Delay Injection. In Experiment 1, we evaluated the impact of increasing the number of traces on learning of the Petri net model. As we provided more traces, the model better captured and represented concurrent processes. We generated varying quantities of traces,

---

[1] https://github.com/prinkskam/Cyber-Safety-Cage-for-Networks.

Table 2. Results from model training without delays with an increasing number of traces on Social Networks dataset

# tr.	Move-Model Fitness		Move-Log Fitness		$\lvert\mathcal{L}_{TL}\rvert$		$\lvert\mathcal{L}_{TL}\setminus\mathcal{L}_{M}\rvert$		Skipped Event		Inserted Event	
	Avg	St. Dev	Avg	St. Dev	Avg	St. Dev	Avg	St. Dev	Avg	St. Dev	Avg	St. Dev
10	0.989630	0.00459	0.989986	0.00475	9.433	0.935	3531.166	1.085	9954.566	4409.164	9954.566	4409.164
100	0.999267	0.00055	0.999267	0.00055	75.433	5.487	3471.533	4.431	702.466	532.872	702.466	532.872
1000	0.999940	4.8E-05	0.999940	4.8E-05	462.800	53.192	3160.033	27.419	56.633	47.072	56.633	47.072
10000	0.999999	2.2E-06	0.999999	2.2E-06	2104.633	222.787	2178.233	86.523	0.500	2.013	0.500	2.013

Table 3. Results from model training without delays with an increasing number of traces on Media Service dataset

# tr.	Move-Model Fitness		Move-Log Fitness		$\lvert\mathcal{L}_{TL}\rvert$		$\lvert\mathcal{L}_{TL}\setminus\mathcal{L}_{M}\rvert$		Skipped Event		Inserted Event	
	Avg	St. Dev	Avg	St. Dev	Avg	St. Dev	Avg	St. Dev	Avg	St. Dev	Avg	St. Dev
10	0.982897	0.0128	0.98230	0.01200	10.000	0.00	11663.130	2.32	12804.270	9610.480	15307.066	9686.35
100	0.999193	0.00048	0.99821	0.00081	95.767	2.44	11606.700	4.77	630.366	380.930	1469.466	682.44
1000	0.999989	1.0E-05	0.99968	0.00019	811.466	27.37	11204.766	23.52	8.967	7.810	268.800	165.14
10000	0.999999	9.3E-07	0.99989	6.9E-05	5633.633	262.30	9401.400	68.54	0.300	0.466	94.266	61.52

ranging from 10 to 10,000 for each application, and used a test set of 30,000 traces to assess the model's performance through replay.

The Move-Model and Move-Log fitness metrics compute the alignment between the test log and the process model. From Tables 2 and 3, we observed that with 10 traces, the Move-Model fitness achieves average values of 0.989630 for the Social Network dataset and 0.982897 for the Media Service dataset. As the number of traces increased to 500, there was a substantial increase in fitness values, reaching 0.999869 and 0.999960 for the Social Network and Media Service datasets, respectively. When traces increased from 1000 to 10,000, the fitness values reached nearly 1 for both datasets. This illustrates that as the number of traces used to train the model increases, both Move-Model and Move-Log fitness show improvement, indicating that the learning approach performs better with more extensive training traces. Due to space limitations, we have reduced the size of the tables to focus on an exponential range of traces. In our analysis of the Social Network dataset, we observed an average of 14.6 skipped and inserted events across 5,000 traces. In contrast, the Media Service dataset had significantly higher averages of 2.1 skipped events and 104.133 inserted events. We noted some exceptions where the fitness values increased even as the number of skipped events decreased, indicating that fewer skipped events can improve alignment between observed data and the model.

Table 4. Model training with single delay injection on Social Network dataset

	# tr.	Move-Model Fitness	Move-Log Fitness	$\lvert\mathcal{L}_{TL}\rvert$	$\lvert\mathcal{L}_{TL}\setminus\mathcal{L}_{M}\rvert$	Skipped Event	Inserted Event
Average	a_1–a_{30}	0.999979	0.999979	30	3513.9	20	20
Standard Deviation	a_1–a_{30}	0	0	0	73.718	0	0

Table 5. Model training with single delay injection on Media Service Dataset

	# tr.	Move-Model Fitness	Move-Log Fitness	$|\mathcal{L}_{TL}|$	$|\mathcal{L}_{TL} \setminus \mathcal{L}_M|$	Skipped Event	Inserted Event
Average	a_1–a_{29} 1		0.999880	29	11666.76	0	90.5
Standard Deviation	a_1–a_{29} 0		0.000223	0	1.250	0	162.09

Experiment 2: Learning Approach with Delay. In Experiment 2, we injected delays to enable the model to recognize parallel behavior more quickly compared to traces without delays. The experiment introduced delay traces progressively, starting with a single delay trace from each activity (e.g., a_1, a_2, a_3, etc.). We sequentially added delay traces, starting from a_1, followed by a_1+a_2, $a_1+a_2+a_3$, and so on, up to a_1-a_n.

Table 4 and 5 represent the results computed for single delay traces for Social Network and Media Service datasets. From Table 4, we observed that with just 30 traces with delay, the Move-Model and Move-Log fitness achieves values of 0.999979 for the Social Network dataset. Similarly, from Table 4, with 29 delay traces, the Move-Model fitness achieves values of 1 and Move-Log fitness to 0.999880 for the Media Service dataset. Additionally, there are 20 skipped and inserted events in the Social Network dataset, whereas 0 skipped and 90.5 inserted events in the Media Service dataset.

We further experimented with 25 delay traces for each activity. By comparing Table 6 with Table 2, we observed that with 750 traces, the Move-Model and Move-Log fitness values reached 0.999979. In contrast, Experiment 1 (Table 2) required 10,000 traces for the learning approach without delay injection to reach comparable fitness levels. Additionally, with 750 traces, we observed an average of 9.33 skipped events, notably less than 14.6 skipped events observed for 5000 traces.

Table 6. Model training with 25 delay injected traces on Social Network dataset

	# tr.	Move-Model Fitness	Move-Log Fitness	$|\mathcal{L}_{TL}|$	$|\mathcal{L}_{TL} \setminus \mathcal{L}_M|$	Skipped Event	Inserted Event
Average	750	0.999990	0.999990	641.133	3316.730	9.333	9.333
Standard Deviation	750	1.07E-05	1.07E-05	25.306	35.772	10.148	10.148

By comparing Tables 2 and 3, with Tables 4 and 5, we examined that our proposed approach shows significant improvement in learning the behavior of Petri net model. Notably, our proposed approach achieves the Model-Move fitness value as 1 with just 29 traces compared to the learning approach without delay, which requires 10,000 traces to reach the fitness value 0.999999. These results demonstrate that our proposed delay injection approach significantly improves over the learning approach without delay injection.

Table 7. Model training with pairwise delay injection on Social network dataset

# tr.	Move-Model Fitness	Move-Log Fitness	$|\mathcal{L}_{TL}|$	$|\mathcal{L}_{TL} \setminus \mathcal{L}_M|$	Skipped Event	Inserted Event
25	0.995692	0.995692	25	3531	20	20
50	0.999979	0.999979	50	3523	20	20
		...				
435	0.999979	0.999979	415	3436	20	20

Experiment 3: Learning Approach with Pairwise Delay Injection. In Experiment 3, we introduced delays in two activities at a time by generating delay traces for each pairwise combination of activities. For each activity $a_i(1 \leq i \leq 30)$, we generated one trace with delays in each pair of activities. The different combinations of delay injection are $(a_1, a_2), (a_1, a_3), \ldots, (a_{29}, a_{30})$.

Table 7 represents the results for pairwise combinations of delay traces in the Social Network dataset. Due to space limitations, we have reduced the size of the table, focusing on the two varying columns:$|\mathcal{L}_{TL}|$ and $|\mathcal{L}_{TL} \setminus \mathcal{L}_M|$. By comparing Table 7 and Table 2, we noticed that with just 435 delay traces, the Model-Move and Model-Log fitness values reached 0.999979 when using pairwise delay injection. In contrast, Experiment 1 (Table 2) required 10,000 traces for the learning approach without delay injection to reach similar fitness levels. This comparison demonstrates that our proposed approach substantially accelerates the learning of the concurrent behavior of the application, achieving higher fitness values with fewer traces. The result highlights that the pairwise delay injection improves efficiency but also outperforms the learning approach without delay injection in terms of performance.

Experiment 4: Identification of Anomalies. Experiment 4 involved generating test log traces to identify deviations and anomalies within the system. These anomalies include traces with one missing activity, multiple missing activities, reordered sequences of activities, and conflict relations, all designed to simulate deviations from our learned model with delay injection. The anomalies were injected by manually altering the test log data, where the activities were randomly removed or reordered within the traces to simulate common system faults or potential security exploits. We introduced conflict relations by modifying the application code to create logical inconsistencies between dependent activities within the traces. The results, as shown in Tables 8 and 9, summarize the calculated results for the anomalies identified in the Social Network and Media Service datasets using our proposed approach.

From Table 8, we observe that for 1000 traces generated with a missing activity, the Move-Model fitness was 0.96875 on average, while Move-Log fitness reached a perfect score of 1.0 in the Social Network dataset. Similarly, in Table 9, the Move-Model fitness achieved an average of 0.962860, and Move-Log

Table 8. Anomaly identification in Social Network Dataset Logs and Models

# tr.	Move-Model Fitness	Move-Log Fitness	Skipped Event	Inserted Event
1000 (1 missing activity)	0.968750	1	1000	0
1000 (≥ 1 missing activity)	0.953749	1	1480	0
100 (Reorder activities)	0.969375	0.969375	98	98
200 (Reorder activities)	0.967187	0.967187	210	210

Table 9. Anomaly identification in Media Service Dataset Logs and Models

# tr.	Move-Model Fitness	Move-Log Fitness	Skipped Event	Inserted Event
1000 (1 missing activity)	0.962860	1	966	0
1000 (≥ 1 missing activity)	0.944014	1	1455	0
100 (Reorder activities)	0.975384	0.975384	64	64
200 (Reorder activities)	0.985961	0.985961	73	73
1000 (Conflict)	1	0.963079	0	997

fitness again reached 1.0 for the Media Service dataset. When 1000 traces contained multiple missing activities, the Move-Model fitness dropped to 0.953749 for the Social Network dataset and 0.944014 for the Media Service dataset, while the Move-Log fitness remained 1 for both cases. Additionally, for 200 traces with reordered sequences, Move-Model and Move-Log fitness values are 0.967187 and 0.985961 for the Social Network and Media Service datasets, respectively.

Table 10. Comparison of Anomaly Identification between LogDP and Proposed Work in Social Network Dataset

# tr.	LogDP			Proposed		
	Precision	Recall	F_1-score	Precision	Recall	F_1-score
1000 (1 missing activity)	0.847500	0.751267	0.796567	1	0.968750	0.984127
1000 (≥ 1 missing activity)	0.818433	0.706667	0.758300	0.953750	1	0.976328
100 (Reorder activities)	0.795167	0.840633	0.811800	0.969375	0.969375	0.969375
200 (Reorder activities)	0.774633	0.833833	0.802867	0.967187	0.967187	0.967187

Further, we analyzed and assessed the model's ability to handle conflict activities in the Media Service dataset, which include choices. Move-Model fitness remained at 1, demonstrating the model's capability to address such anomalies effectively. Since the Social Network application does not involve choices, we did

Table 11. Comparison of Anomaly identification between LogDP and Proposed Work in Media Microservice Dataset

# tr.	LogDP			Proposed		
	Precision	Recall	F_1-score	Precision	Recall	F_1-score
1000 (1 missing activity)	0.824567	0.739400	0.779933	1	0.962815	0.981055
1000 ($\geq$ 1 missing activity)	0.804233	0.745233	0.774367	1	0.943957	0.971170
100 (Reorder activities)	0.789700	0.824100	0.807267	0.975385	0.975385	0.975385
200 (Reorder activities)	0.778167	0.838533	0.808667	0.985964	0.985964	0.985964
1000 (Conflict)	0.767233	0.843533	0.804367	0.963084	1	0.981195

not generate traces with conflicts. Additionally, we verified the application code (ground truth) to confirm that our delay injection approach correctly learned the model for the Media Service dataset. These results confirm that the models learned from both datasets are robust and can effectively handle various anomalies, including missing activities, reordered sequences, and conflict activities.

To further validate the effectiveness of the proposed method, Table 10 and Table 11 presented a comparison between the LogDP method and the proposed approach to identify anomalies in Social Network and Media Service datasets. From both tables, we observed that our proposed approach outperforms the LogDP method in all scenarios. It achieved higher Precision, Recall, and F_1-score across all anomalies, including missing and reordered activities. Our approach performed better as it captures complex event dependencies, parallel activities, and temporal dependencies inherent in microservices that dependency-based methods like LogDP failed to address adequately. These results demonstrated that our approach effectively identifies such anomalies, improving anomaly detection in microservice architectures.

6.3 Summary

In our experiments, we found that learning without delay injection requires a substantial number of traces during the learning phase to achieve satisfactory model fitness. As the number of traces increased beyond 10,000, the Move-Model and Move-Log fitness values improved, approaching near-optimal levels. However, the model trained without delay injection does not effectively capture concurrency patterns. In contrast, when delays were incorporated into the training phase, we observed enhanced learning and significantly greater model accuracy. This demonstrates the critical role of delay injection in identifying concurrent patterns in microservice applications.

7   Discussion

Lessons Learned. As demonstrated in Sect. 6, incorporating delays through our delay injection approach allows for the accurate and efficient learning of the temporal dependencies between the events occurring within the microservice application. The models are designed to be explainable, human-readable, relatively simple, and fast, emphasizing transparency to make them accessible and understandable to users.

Our delay injection effectively detects deviations between the learned process model and the observed test logs in our experiments. The resulting models can accurately capture concurrency patterns and detect discrepancies such as missing, duplicated, or extra activities. The recovered models are precise and concise, making them easy for engineers to validate. We recommend incorporating delay injection in individual activities among the three training approaches tested, as it effectively captures concurrency patterns. This approach is helpful in environments with limited execution time variance and provides more significant insights into system behavior compared to methods without delay injection.

Threats to Validity. The setup and tooling used in this work may influence the results. For example, using Jaeger for tracing may induce overhead or delays, which could accumulate with our injected delays and impact performance measurements. Similarly, network capacity or memory limitations in the testing environment could impact trace accuracy and event timing. We conducted experiments on two open-source microservice applications with sequential and parallel activities with different durations. While commercial applications may be larger, the patterns exhibited in our experiments still reflect all possible cases of partial or total order among activities. Our current implementation supports C++ and Python environments, but our approach can also be extended to other platforms.

8   Conclusion

In this work, we have applied process mining to microservice-based applications to provide insights into their dynamic behavior and workflows. We evaluated our proposed methodologies on benchmark microservice applications—Social Network and Media Service—to optimize workflows and enhance the performance of decentralized systems. Our delay injection approach speeds up the learning process of behavior models as a Petri net from the application's trace data. By introducing delays, the model can better capture and learn the concurrent behavior of the system application. We conducted a series of experiments, progressively adding delay traces, and our approach demonstrated significant improvements. For example, in the Media Service use case, our model achieved perfect Move-Model and Move-Log fitness values of 1 with 29 delay traces, compared to 0.999999 after 10,000 traces without delay injection. These results demonstrate the impact of delay injection in improving the accuracy and efficiency of model learning. Moreover, our method significantly outperforms the LogDP method in anomaly detection experiments by achieving higher Precision and F_1-score. These findings highlight the ability of our approach to capture event dependencies and parallelism inherent in microservice architecture, enabling more accurate and reliable anomaly detection.

In this work, delay injection was implemented through manual code modifications to maintain flexibility and control over injection points. As future work, we want to automate the code changes needed for delay injection and investigate the resilience of microservices against connection loss. We will also investigate the many possible uses of our approach, ranging from security and reliability analysis to performance optimization of microservices.

References

1. Docker (2024). https://www.docker.com/
2. Jaeger (2024). https://www.jaegertracing.io/
3. MediaMicroservice (2020). https://github.com/delimitrou/DeathStarBench/tree/master/mediaMicroservices
4. ProMtool (2024). https://promtools.org/
5. SocialNetwork (2020). https://github.com/delimitrou/DeathStarBench/tree/master/socialNetwork
6. Agostinelli, S., Chiariello, F., Maggi, F.M., Marrella, A., Patrizi, F.: Process mining meets model learning: discovering deterministic finite state automata from event logs for business process analysis. Inf. Syst. **114**, 102180 (2023)
7. Augusto, A., et al.: Automated discovery of process models from event logs: review and benchmark. IEEE Trans. Knowl. Data Eng. **31**(4), 686–705 (2019). https://doi.org/10.1109/TKDE.2018.2841877
8. Bloemen, V., van Zelst, S.J., van der Aalst, W., van Dongen, B.F., van de Pol, J.: Maximizing synchronization for aligning observed and modelled behaviour. In: Weske, M., Montali, M., Weber, I., vom Brocke, J. (eds.) BPM 2018. LNCS, vol. 11080, pp. 233–249. Springer, Cham (2018). https://doi.org/10.1007/978-3-319-98648-7_14
9. Brito, M., Cunha, J., Saraiva, J.: Identification of microservices from monolithic applications through topic modelling. In: Proceedings of the 36th Annual ACM Symposium on Applied Computing, pp. 1409–1418 (2021)
10. Camilli, M., Bellettini, C., Capra, L., Monga, M.: A formal framework for specifying and verifying microservices based process flows. In: Software Engineering and Formal Methods: SEFM 2017 Collocated Workshops: DataMod, FAACS, MSE, CoSim-CPS, and FOCLASA, Trento, Italy, 4–5 September 2017, Revised Selected Papers 15, pp. 187–202. Springer (2018)
11. Carmona, J., van Dongen, B., Solti, A., Weidlich, M.: Conformance checking. Switzerland: Springer.[Google Scholar] **56**, 12 (2018)
12. Casas-Ramos, J., Mucientes, M., Lama, M.: Reach: researching efficient alignment-based conformance checking. Expert Syst. Appl. **241**, 122467 (2024)
13. Chow, M., Meisner, D., Flinn, J., Peek, D., Wenisch, T.F.: The mystery machine: end-to-end performance analysis of large-scale internet services. In: 11th USENIX Symposium on Operating Systems Design and Implementation, pp. 217–231 (2014)
14. Dunzer, S., Stierle, M., Matzner, M., Baier, S.: Conformance checking: a state-of-the-art literature review. In: Proceedings of the 11th International Conference on Subject-Oriented Business Process Management, pp. 1–10 (2019)

15. Gan, Y., et al.: An open-source benchmark suite for microservices and their hardware-software implications for cloud & edge systems. In: Proceedings of the Twenty-Fourth International Conference on Architectural Support for Programming Languages and Operating Systems, pp. 3–18 (2019)
16. Goulart Rocha, E., van Zelst, S.J., van der Aalst, W.M.: Mining behavioral patterns for conformance diagnostics. In: International Conference on Business Process Management, pp. 291–308. Springer (2024)
17. Grohs, M., van der Aa, H., Rehse, J.R.: Beyond log and model moves in conformance checking: discovering process-level deviation patterns. In: International Conference on Business Process Management, pp. 381–399. Springer (2024)
18. Hsueh, M., Tsai, T., Iyer, R.: Fault injection techniques and tools. IEEE Comput. 30(4), 75–82 (1997)
19. Jia, T., Wu, Y., Hou, C., Li, Y.: Logflash: real-time streaming anomaly detection and diagnosis from system logs for large-scale software systems. In: 32nd International Symposium on Software Reliability Engineering (ISSRE), pp. 80–90. IEEE (2021)
20. Jia, T., Yang, L., Chen, P., Li, Y., Meng, F., Xu, J.: Logsed: anomaly diagnosis through mining time-weighted control flow graph in logs. In: 10th International Conference on Cloud Computing (CLOUD), pp. 447–455. IEEE (2017)
21. Karabey Aksakalli, I., Çelik, T., Can, A.B., Tekinerdoğan, B.: Deployment and communication patterns in microservice architectures: a systematic literature review. J. Syst. Softw. 180, 111014 (2021). https://doi.org/10.1016/j.jss.2021.111014
22. Leemans, S.J., Fahland, D., Van Der Aalst, W.M.: Discovering block-structured process models from event logs-a constructive approach. In: Application and Theory of Petri Nets and Concurrency: 34th International Conference, PETRI NETS 2013, Milan, Italy, 24–28 June 2013. Proceedings 34, pp. 311–329. Springer (2013)
23. Liu, C., Zeng, Q., Cheng, L., Duan, H., Cheng, J.: Measuring similarity for data-aware business processes. IEEE Trans. Autom. Sci. Eng. 19(2), 1070–1082 (2022). https://doi.org/10.1109/TASE.2021.3049772
24. Liu, Z., Fan, G., Yu, H., Chen, L.: Modelling and analysing the reliability for microservice-based cloud application based on predicate petri net. Expert. Syst. 39(6), e12924 (2022)
25. Ma, S.P., Fan, C.Y., Chuang, Y., Lee, W.T., Lee, S.J., Hsueh, N.L.: Using service dependency graph to analyze and test microservices. In: 42nd Annual Computer Software and Applications Conference, vol. 2, pp. 81–86. IEEE (2018)
26. Mannel, L.L., van der Aalst, W.M.: Discovering process models with long-term dependencies while providing guarantees and handling infrequent behavior. In: International Conference on Applications and Theory of Petri Nets and Concurrency, pp. 303–324. Springer (2022)
27. Nie, C., Leung, H.: A survey of combinatorial testing. ACM Comput. Surv. (CSUR) 43(2), 1–29 (2011)
28. Otero, M., Garcia, J.M., Fernandez, P.: Towards a lightweight distributed telemetry for microservices. In: IEEE 44th International Conference on Distributed Computing Systems Workshops (ICDCSW), pp. 75–82 (2024). https://doi.org/10.1109/ICDCSW63686.2024.00018
29. Petke, J., Cohen, M.B., Harman, M., Yoo, S.: Practical combinatorial interaction testing: empirical findings on efficiency and early fault detection. IEEE Trans. Software Eng. 41(9), 901–924 (2015)
30. Rozinat, A., Van der Aalst, W.M.: Conformance checking of processes based on monitoring real behavior. Inf. Syst. 33(1), 64–95 (2008)

31. Samir, A., Pahl, C.: DLA: detecting and localizing anomalies in containerized microservice architectures using Markov models. In: 2019 7th International Conference on Future Internet of Things and Cloud (FiCloud), pp. 205–213 (2019). https://doi.org/10.1109/FiCloud.2019.00036
32. Thönes, J.: Microservices. IEEE Softw. **32**(1), 116 (2015)
33. Van Der Aalst, W.: Process mining: overview and opportunities. ACM Trans. Manag. Inf. Syst. (TMIS) **3**(2), 1–17 (2012)
34. Wen, L., Wang, J., van der Aalst, W.M., Huang, B., Sun, J.: A novel approach for process mining based on event types. J. Intell. Inf. Syst. **32**, 163–190 (2009)
35. Xie, R., Wang, L., Song, C.: Towards minimum latency in cloud-native applications via service-characteristic- aware microservice deployment. In: IEEE International Conference on Software Analysis, Evolution and Reengineering (SANER), pp. 35–46 (2024). https://doi.org/10.1109/SANER60148.2024.00010
36. Xie, Y., Zhang, H., Zhang, B., Babar, M.A., Lu, S.: LogDP: combining dependency and proximity for log-based anomaly detection. In: Hacid, H., Kao, O., Mecella, M., Moha, N., Paik, H. (eds.) ICSOC 2021. LNCS, vol. 13121, pp. 708–716. Springer, Cham (2021). https://doi.org/10.1007/978-3-030-91431-8_47
37. Yoneyama, J., Artho, C., Tanabe, Y., Hagiya, M.: Model-based network fault injection for IoT protocols. In: Proceedings of 14th International Conference on Evaluation of Novel Approaches to Software Engineering, (ENASE), Heraklion, Greece, pp. 201–209 (2019)
38. Yu, G., Chen, P., Li, Y., Chen, H., Li, X., Zheng, Z.: Nezha: interpretable fine-grained root causes analysis for microservices on multi-modal observability data. In: Proceedings of the 31st ACM Joint European Software Engineering Conference and Symposium on the Foundations of Software Engineering, pp. 553–565 (2023)
39. Zhang, Y., van der Aalst, W.M.: Explorative process discovery using activity projections. In: International Conference on Applications and Theory of Petri Nets and Concurrency, pp. 229–239. Springer (2023)
40. Zipkin: Zipkin. https://zipkin.io/

Translating Workflow Nets into the Partially Ordered Workflow Language

Humam Kourani[1,2]([⊠]) [ID], Gyunam Park[1] [ID], and Wil M. P. van der Aalst[1,2] [ID]

[1] Fraunhofer Institute for Applied Information Technology FIT,
Schloss Birlinghoven, 53757 Sankt Augustin, Germany
{humam.kourani,gyunam.park,wil.van.der.aalst}@fit.fraunhofer.de
[2] RWTH Aachen University, Ahornstraße 55, 52074 Aachen, Germany

Abstract. The Partially Ordered Workflow Language (POWL) has recently emerged as a process modeling notation, offering strong quality guarantees and high expressiveness. However, its adoption in practice is hindered by the prevalence of standard notations like workflow nets (WF-nets) and BPMN. This paper presents a novel algorithm for transforming safe and sound WF-net into equivalent POWL models. The algorithm recursively identifies structural patterns within the WF-net and translates them into their POWL representation. We formally prove the correctness of our approach, showing that the generated POWL model preserves the language of the input WF-net. Furthermore, we demonstrate the high scalability of our algorithm, and we show its completeness on a subclass of WF-nets that encompasses equivalent representations for all POWL models. This work bridges the gap between the theoretical advantages of POWL and the practical need for compatibility with established notations, paving the way for broader adoption of POWL in process analysis and improvement applications.

Keywords: Workflow Net · Process Modeling · Model Transformation

1 Introduction

The field of process modeling and analysis relies heavily on formal notations to represent and reason about the behavior of complex systems. While standard notations like Petri nets [26], and specifically Workflow Nets (WF-nets) [22], and Business Process Model and Notation (BPMN) [28] have gained widespread adoption, they suffer from limitations in terms of their ability to guarantee desirable quality properties such as *soundness* (i.e., the absence of deadlocks and other anomalies).

The recently introduced Partially Ordered Workflow Language (POWL) [15] addresses these limitations by providing a powerful yet formally sound framework for process modeling. POWL is a hierarchical modeling language that allows for the construction of complex process models by combining smaller submodels using a set of well-defined operators. These operators include exclusive choice

© The Author(s), under exclusive license to Springer Nature Switzerland AG 2025
E. Amparore and Ł. Mikulski (Eds.): PETRI NETS 2025, LNCS 15714, pp. 242–264, 2025.
https://doi.org/10.1007/978-3-031-94634-9_12

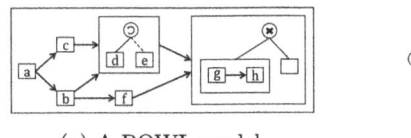   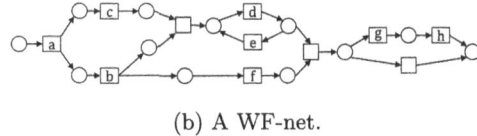

(a) A POWL model.   (b) A WF-net.

Fig. 1. Example process models.

(XOR), loop, and partial order. POWL can be viewed as a generalization of process trees [19], a notation widely used in various process mining techniques due to its quality guarantees. POWL preserves these desirable properties while increasing expressiveness through the partial order operator. Partial orders allow for the representation of activities that can be executed concurrently, but may have some ordering restrictions. Figure 1a illustrates an example POWL model, and Fig. 1b shows a WF-net that captures the same behavior.

The hierarchical nature of POWL offers several advantages. The structured representation can significantly enhance the understandability of complex process models for humans, making it easier to grasp the overall control flow and identify potential areas for improvement. Furthermore, POWL opens up opportunities for developing faster and more efficient techniques for different process mining tasks, similar to how specialized algorithms have been optimized for process trees. Previous work has demonstrated the advantages of POWL in process discovery from data [14,16] and process modeling from text [12].

Despite its demonstrated benefits, the practical adoption of POWL is challenged by the prevalence of standard notations in existing tools and workflows. This motivates the need for a robust conversion from standard notations to POWL, allowing us to harness these benefits without requiring the adjustment of existing tools and practices. While the conversion from POWL to a sound WF-net is relatively straightforward, the inverse transformation presents a more significant challenge. This asymmetry arises from the fact that POWL represents a subclass of sound WF-nets.

This paper proposes an algorithm that translates sound WF-nets into POWL. The proposed algorithm recursively decomposes the input WF-net into its smaller parts, identifies structural patterns corresponding to POWL's constructs, and assembles them into an equivalent POWL model. We formally prove the correctness of this transformation, demonstrating that the generated POWL model captures the language of the original WF-net. Furthermore, we define a subclass of WF-nets that encompasses equivalent representations for all POWL models, and we show the completeness of our approach on this subclass. We implement our algorithm and apply it on large WF-nets to demonstrates its high scalability. This work represents a significant step towards enabling the development of a process analysis and improvement techniques that utilize POWL internally while seamlessly accepting inputs in widely used formats.

The remainder of the paper is structured as follows. After discussing related work in Sect. 2, we introduce necessary preliminaries in Sect. 3. In Sect. 4, we detail our algorithm for translating WF-nets into POWL. Section 5 formally proves the algorithm's correctness and completeness guarantees. Finally, we assess the scalabilty of our approach in Sect. 6, and we conclude the paper in Sect. 7.

2   Related Work

Transformations between process modeling notations have been explored in various contexts. Some research focuses on transformations between different Petri net classes, such as the work on unfolding Colored Petri Nets into standard Place/Transition nets in [4] and the work on reducing free-choice Petri nets to either T-nets (also called marked graphs) or P-nets (also called state machines) in [24]. Other research addresses transformations between different types of process models. For example, [6] proposes an approach for converting BPMN models into Petri nets, [8] discusses translating UML diagrams into BPEL, and [17] explores the mapping of Event-driven Process Chains (EPCs) into colored Petri nets. An overview on different approaches for the translation between workflow graphs and free-choice workflow nets is provided in [7].

Transformations from graph-based formalisms like Petri nets into block-structured languages such as BPEL or process trees have been widely studied. The work on translating WF-nets to BPEL in [18,25] employs a bottom-up strategy, iteratively identifying patterns corresponding to BPEL fragments and substituting each identified pattern with a single transition to continue the recursion. This approach aims to maximize the size of the detected components in each iteration. The approach presented in [27] for translating WF-nets into process trees, while also employing a bottom-up strategy, restricts the search space to patterns of size two. This approach cannot be adapted to POWL due to the presence of advanced partial order constructs that cannot be decomposed into components of size two. The fundamental difference between our algorithm and the aforementioned approaches, besides using different modeling languages, is that our approach employs a top-down strategy to ensure high scalability.

3   Preliminaries

This section introduces fundamental preliminaries and notations.

3.1   Basic Notations

A *multi-set* generalizes the notion of a set by tracking the frequencies of its elements. A multi-set over a set X is expressed as $M = [x_1^{c_1}, ..., x_n^{c_n}]$ where $x_1, ..., x_n \in X$ are the elements of M (denoted as $x_i \in M$ for $1 \leq i \leq n$) and $M(x_i) = c_i \geq 1$ is the frequency of x_i for $1 \leq i \leq n$.

A sequence of length $n \geq 0$ over a set X is defined as function $\sigma \colon \{1, ..., n\} \to X$, and we express it as $\sigma = \langle \sigma(1), ..., \sigma(n) \rangle$. The set of all sequences over X is denoted by X^*. The concatenation of two sequences σ_1 and σ_2 is expressed as $\sigma_1 \cdot \sigma_2$, e.g., $\langle x_1 \rangle \cdot \langle x_2, x_1 \rangle = \langle x_1, x_2, x_1 \rangle$. For two sets of sequences L_1 and L_2, we write $L_1 \cdot L_2 = \{\sigma_1 \cdot \sigma_2 \mid \sigma_1 \in L_1 \ \wedge \ \sigma_2 \in L_2\}$.

Let $\prec \subseteq X \times X$ be a binary relation over a set X. We use $x_1 \prec x_2$ to denote $(x_1, x_2) \in \prec$ and $x_1 \not\prec x_2$ to denote $(x_1, x_2) \notin \prec$. We define the *transitive closure* of $\prec$ as $\prec^+ = \{(x, y) \mid \exists_{x_1, \ldots, x_n \in X} \ x = x_1 \ \wedge \ y = x_n \ \wedge \forall_{1 \leq i < n}, x_i \prec x_{i+1}\}$.

A *strict partial order* (*partial order* for short) over a set X is a binary relation that is *irreflexive* ($x \not\prec x$ for all $x \in X$) and *transitive* ($x_1 \prec x_2 \wedge x_2 \prec x_3 \Rightarrow x_1 \prec x_3$). Irreflexivity and transitivity imply *asymmetry* ($x_1 \prec x_2 \Rightarrow x_2 \not\prec x_1$). For $n \geq 2$, we use $\mathcal{O}^n$ to denote the set of all partial orders over $\{1, \ldots, n\}$. Let $X = \{x_1, \ldots, x_n\}$ be a set of size $n \geq 2$ and $\prec \in \mathcal{O}^n$. Then we write $\prec(x_1, \ldots, x_n)$ to denote the partial order $\prec'$ defined over X as follows: $i \prec j \Leftrightarrow x_i \prec' x_j$ for all $i, j \in \{1, \ldots, n\}$.

Let $\sigma_1, \ldots, \sigma_n \in X^*$ be $n \geq 2$ sequences over a set X and $\prec \in \mathcal{O}^n$. The *order-preserving shuffle operator* $\sqcup\kern-0.5em\sqcup_\prec$ generates the set of sequences resulting from interleaving $\sigma_1, \ldots, \sigma_n$ while preserving the partial order $\prec$ of the sequences and the sequential order within each sequence. For example, let $\sigma_1 = \langle a, b \rangle$, $\sigma_2 = \langle c \rangle$, $\sigma_3 = \langle d, e \rangle$, and $\prec = \{(1, 2), (1, 3)\} \in \mathcal{O}^3$. Then, $\sqcup\kern-0.5em\sqcup_\prec(\sigma_1, \sigma_2, \sigma_3) = \{\langle a, b, c, d, e \rangle, \langle a, b, d, c, e \rangle, \langle a, b, d, e, c \rangle\}$.

For a set X, a *partition* of X of size $n \geq 1$ is a set of subsets $P = \{X_1, \ldots, X_n\}$ such that $X = X_1 \cup \ldots \cup X_n$, $X_i \neq \emptyset$, and $X_i \cap X_j = \emptyset$ for $1 \leq i < j \leq n$. For any $x \in X$, we write P_x to denote the subset of the partition (also called *part*) that contains x, i.e., $P_x \in \{X_1, \ldots, X_n\}$ and $x \in P_x$. For example, let $P = \{\{a, b\}, \{c\}\}$ be a partition of $\{a, b, c\}$ of size 2. Then, $P_a = P_b = \{a, b\}$ and $P_c = \{c\}$.

3.2 Workflow Nets

We use Σ to denote the set of all activities. We use $\tau \notin \Sigma$ to denote the *silent activity*, which is used to model a choice between executing or skipping a path in process model, for example. To enable creating process models with duplicated activities, we introduce the notion of *transitions*, and we use $\mathcal{T}$ to denote the set of all transitions. Each transition is mapped to an activity, denoted as the *label* of the transition. We use $l: \mathcal{T} \to \Sigma \cup \{\tau\}$ to denote the labeling function.

A *Petri net* is a directed bipartite graph consisting of two types of nodes: *places* and *transitions*. Transitions represent instances of activities, while places are used to model dependencies between transitions.

Definition 1 (Petri Net). *A Petri net is a triple* $N = (P, T, F)$, *where* $T \subset \mathcal{T}$ *is a finite set of transitions,* P *is a finite set of places such that* $T \cap P = \emptyset$, *and* $F \subseteq (P \times T) \cup (T \times P)$ *is the flow relation.*

Let $N = (P, T, F)$ be a Petri net. We define the following notations:

- For $x \in P \cup T$, $\bullet x = \{y \mid (y, x) \in F\}$ is the *pre-set* of x, and $x \bullet = \{y \mid (x, y) \in F\}$ is the *post-set* of x.
- For $T' \subseteq T$, we define the *projection* of P on T' as $P|_{T'} = \{p \in P \mid (\bullet p \cup p \bullet) \cap T' \neq \emptyset\}$.
- For $P' \subseteq P$ and $T' \subseteq T$, we define the *projection* of F on P and T as $F|_{P', T'} = F \cap ((P' \times T') \cup (T' \times P'))$.
- For $T' \subseteq T$, two places $p, p' \in P$ are *equivalent with respect to* T', denoted as $p \approx_{T'} p'$, iff $(\bullet p \cap T' = \bullet p' \cap T') \wedge (p \bullet \cap T' = p' \bullet \cap T')$.

Places hold *tokens*, and a transition is considered *enabled* if each of its preceding places has at least one token. *Firing* an enabled transition consumes one token from each of its preceding places and produces a token in each of its succeeding places. A *marking* is a multi-set of places indicating the number of tokens in each place. A Petri net is called *safe* if each place in the net cannot hold more than one token. Next, we define three subclasses of Petri nets. More details on these classes can be found in [5, 21, 22].

Definition 2 (Marked Graph, Free-Choiceness, Workflow Net). *Let $N = (P, T, F)$ be a Petri net. Then, the following holds:*

- N *is a marked graph iff for any $p \in P$: $|\bullet p| \leq 1 \wedge |p\bullet| \leq 1$.*
- N *is free-choice iff for any $t_1, t_2 \in T$: $(\bullet t_1 \cap \bullet t_2 \neq \emptyset) \Rightarrow (\bullet t_1 = \bullet t_2)$.*
- N *is a workflow net (WF-net) iff places $N_{source}, N_{sink} \in P$ exist such that:*
    - **Unique source:** $\{N_{source}\} = \{p \in P \mid \bullet p = \emptyset\}$.
    - **Unique sink:** $\{N_{sink}\} = \{p \in P \mid p\bullet = \emptyset\}$.
    - **Connectivity:** *each node is on a path from N_{source} to N_{sink}.*

Let $N = (P, T, F)$ be a WF-net. We use N_{source} to denote the unique source place and N_{sink} to denote the unique sink place. Furthermore, we define:

- For $T' \subseteq T$, $\triangleright T' = \{p \in P \mid T' \cap p\bullet \neq \emptyset \wedge (p = N_{source} \vee (T \backslash T') \cap \bullet p \neq \emptyset)\}$ is the set of *entry points* of T'.
- For $T' \subseteq T$, $T'\triangleright = \{p \in P \mid T' \cap \bullet p \neq \emptyset \wedge (p = N_{sink} \vee (T \backslash T') \cap p\bullet \neq \emptyset)\}$ is the set of *exit points* of T'.
- For a partition $G = \{T_1, \ldots, T_n\}$ of T, the *execution order* of G within N is the binary relation $order(N, G) = \{(i, j) \mid i, j \in \{1, \ldots, n\} \wedge (T_i \triangleright \cap \triangleright T_j) \neq \emptyset\}$.

WF-nets may suffer from quality anomalies (e.g., transitions that can never be enabled). WF-nets without such undesirable properties are called *sound*.

Definition 3 (Soundness). *Let $N = (P, T, F)$ be a WF-net. N is sound iff the following conditions hold:*

- **No dead transitions:** *for each transition $t \in T$, there exists a marking M reachable from $[N_{source}]$ that enables t.*
- **Option to complete:** *for every marking M reachable from $[N_{source}]$, there exists a firing sequence leading from M to $[N_{sink}]$.*
- **Proper completion:** *$[N_{sink}]$ is the only marking reachable from $[N_{source}]$ with at least one token in N_{sink}.*

The WF-net shown in Fig. 1b is sound. Note that the option to complete implies proper completion.

3.3   POWL Language

A POWL model is constructed recursively from a set of activities, combined either as partial orders or using the control flow operators $\times$ and $\circlearrowleft$. The operator

$\times$ models an exclusive choice between submodels, while $\circlearrowleft$ model cyclic behavior between two submodels: the *do-part* is executed first, and each time the *redo-part* is executed, it is followed by another execution of the do-part. In a partial order, all submodels are executed, while respecting the given execution order.

POWL models are defined in [15] over activities. We redefine POWL models over transitions to allow for models with multiple instances of the same activity.

Definition 4 (POWL Model). *POWL models are defined as follows:*

- *Any transition $t \in T$ is a POWL model.*
- *Let $\psi_1, ..., \psi_n$ be $n \geq 2$ POWL models.*
  - *$\times(\psi_1, ..., \psi_n)$ is a POWL model.*
  - *$\circlearrowleft(\psi_1, \psi_2)$ is a POWL model.*
  - *For any partial order $\prec \in \mathcal{O}^n$, $\prec(\psi_1, ..., \psi_n)$ is a POWL model.*

Figure 1a shows an example POWL model. The language a POWL model is defined recursively based on the semantics of its operators.

Definition 5 (POWL Semantics). *The language of a POWL model ψ is recursively defined as follows:*

- *$\mathcal{L}(t) = \{\langle a \rangle\}$ for $t \in T$ with $l(t) = a \in \Sigma$.*
- *$\mathcal{L}(t) = \{\langle \rangle\}$ for $t \in T$ with $l(t) = \tau$.*
- *Let $\psi_1, ..., \psi_n$ be $n \geq 2$ POWL models with $L_i = \mathcal{L}(\psi_i)$ for $1 \leq i \leq n$.*
  - *$\mathcal{L}(\times(\psi_1, ..., \psi_n)) = \bigcup\limits_{1 \leq i \leq n} L_i.$*
  - *$\mathcal{L}(\circlearrowleft(\psi_1, \psi_2)) = L_1 \cdot (L_2 \cdot L_1)^*.$*
  - *For $\prec \in \mathcal{O}^n$, $\mathcal{L}(\prec(\psi_1, ..., \psi_n)) = \{\sigma \in \sqcup\sqcup_\prec(\sigma_1, ..., \sigma_n) \mid \forall_{1 \leq i \leq n} \sigma_i \in L_i\}.$*

3.4 Semi-block-Structured WF-Nets

POWL is less expressive than WF-nets, meaning that not all WF-nets have equivalent POWL models. To clearly define the scope of our WF-net to POWL translation algorithm, we extend the concept of *block-structured* workflow nets defined in [19]. A block-structured WF-net is a sound WF-net that can be divided recursively into parts having single entry and exit points. In other words, there must be a unique mapping between every place/transition with multiple outgoing arcs and every place/transition with multiple incoming arcs to mark the start and end of a block.

POWL can represent all block-structured WF-nets, and the introduction of the partial order operator in POWL extends its expressiveness beyond this class. We can relax the block-structure requirements since the partial order operator allows for the concurrent executions of transitions without imposing block structure on them. In other words, the block structure is only required for decision points (i.e., places) due to the usage of the process tree operators $\times$ and $\circlearrowleft$.

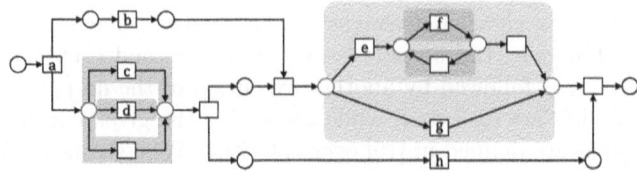

Fig. 2. A semi-block-structured WF-net with three blocks, highlighted in different colors.

Definition 6 (Semi-Block-Structured WF-nets). *Let $N = (P, T, F)$ be a WF-net. N is semi-block-structured* **iff** *it satisfies the following conditions:*

- *N is sound and safe.*
- ***Explicit decision points:*** *For each $(x, y) \in F : |\bullet y| = 1 \vee |x \bullet| = 1$.*
- ***Unique mapping between split and join decision points:*** *Let $P_{aligned} = \{p \in P \mid |p\bullet| > 1\}$ and $P_{join} = \{p \in P \mid |\bullet p| > 1\}$. There exists a bijective mapping $\mathcal{B} : P_{aligned} \to P_{join}$ such that for each $(p, p') \in \mathcal{B}$: $|p\bullet| = |\bullet p'|$.*
- ***Disjoint subnets between decision points:*** *For each pair $(p, p') \in \mathcal{B}$, let $k = |p\bullet| = |\bullet p'|$. There exist k disjoint, non-empty sets of transitions $T_1, \ldots, T_k \subseteq T$ such that for each i $(1 \leq i \leq k)$, $P_i = P|_{T_i}$, and $F_i = F|_{P_i, T_i}$:*
  - *$P_i \cap P|_{T \setminus T_i} = \{p, p'\}$.*
  - *$N_i = (P_i, T_i, F_i)$ is a workflow net with $\{N_{i_{source}}, N_{i_{sink}}\} = \{p, p'\}$.*

The WF-net in Fig. 2 is semi-block-structured WF-net but not block structured. Note that a semi-block-structured WF-net with no blocks is a marked graph. Furthermore, a semi-block-structured WF-net is free-choice due to the explicit decision points requirement. Therefore, we can conclude that semi-block-structured WF-nets are a subset of sound free-choice WF-nets, a superset of block-structured WF-nets, and superset of sound marked graph WF-nets.

It is trivial to observe that, for any POWL model, there exists at least one equivalent semi-block-structured WF-net. Furthermore, after introducing our conversion algorithm, we will show in Sect. 5.4 its completeness on this class of WF-nets; i.e., we will show the our algorithm successfully converts any semi-block-structured WF-net into a POWL model that captures the same language.

4   Transforming Workflow Nets into POWL

This section presents a recursive algorithm for transforming safe and sound WF-nets into equivalent POWL models. First, the WF-net can be preprocessed by applying a set of reduction rules. Then, the algorithm checks whether the WF-net forms a structural pattern that corresponds to a POWL component (i.e., choice, loop, or partial order). The identified pattern is then translated into its corresponding POWL representation, and the WF-net is projected onto subsets of transitions to continue the recursion.

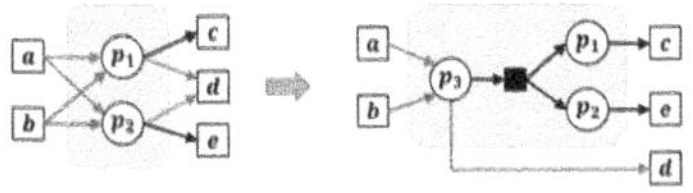

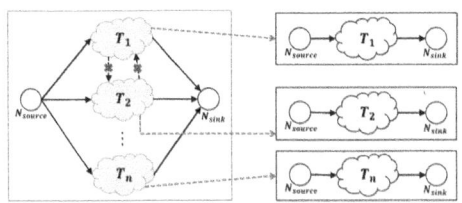

(a) Introducing explicit XOR split places. 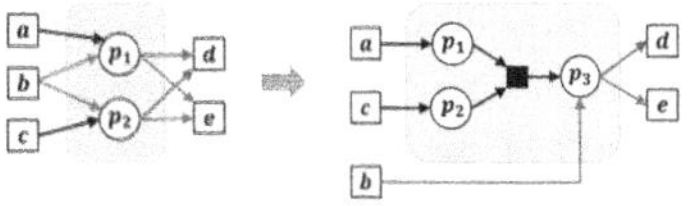   (b) Introducing explicit XOR join places.

Fig. 3. Reduction rules illustrated with examples.

Fig. 4. XOR pattern and projection.

4.1 Preprocessing

To expand the applicability of our algorithm, the input WF-net can be preprocessed by applying a series of reduction rules. This is an optional step, aiming
at bringing the WF-net closer to the structure expected in the subsequent steps
of the algorithm. Numerous reduction rules have been proposed in the literature, such as those presented in [1,2,5,20]. Any reduction rule can be applied
as long as it preserves the essential structural properties of the WF-net, namely
its language, safeness, and soundness. In Fig. 3, we illustrate reduction rules for
introducing explicit places for decision points. Additional reduction rules are
introduced in Sect. 4.3 to enable the detection of special loop structures.

4.2 Identifying Choices

This section addresses partitioning a WF-net into smaller subnets such that the
given WF-net corresponds to an XOR structure over the identified parts.

   We first introduce the concept of *transition reachability*, capturing the notion
of one transition being reachable from another through a path in the WF-net.

Definition 7 (Transition Reachability). *Let $N = (P, T, F)$ be a Petri net.
The transition reachability relation $\leadsto \subseteq T \times T$ is defined as follows for $t, t' \in T$:*

$$t \leadsto t' \Leftrightarrow (t, t') \in F^+.$$

   An *XOR pattern*, as illustrated in Fig. 4, represents a WF-net where transitions can be partitioned such that exactly one part can be executed in a single
process instance. Intuitively, transitions belonging to different parts cannot be
reachable from each other.

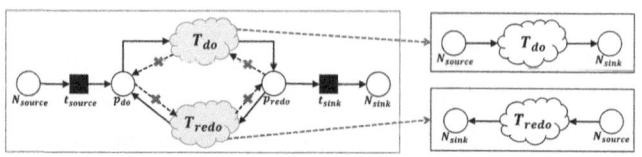

Fig. 5. Loop pattern and projection.

Definition 8 (XOR Pattern). *Let $N = (P, T, F)$ be a safe and sound WF-net. Let $G = \{T_1, \ldots, T_n\}$ be a partition of transitions of size $n \geq 2$. The tuple (N, G) is called an XOR pattern iff for all $(t, t') \in T \times T$:*

$$\text{if } t \rightsquigarrow t', \text{ then } G_t = G_{t'}.$$

For a WF-net N, we use $Partition_\times(N)$ to denote the partition generated by iteratively grouping transitions based on their reachability relation, aligning with Definition 8. Note that $(N, Partition_\times(N))$ is an XOR pattern if $|Partition_\times(N)| \geq 2$.

After identifying an XOR pattern, the WF-net is projected onto the different parts, creating several subnets for the recursive application of the algorithm. The projection is achieved by selecting the relevant places and flow relations, isolating the chosen part.

Definition 9 (XOR Projection). *Let (N, G) be an XOR pattern with $N = (P, T, F)$ and $T' \in G$ be a part. The XOR projection of N on T' is defined as $Project_\times(N, T') = (P', T', F')$ where $P' = P|_{T'}$ and $F' = F|_{P', T'}$.*

4.3　Identifying Loops

This section addresses partitioning a WF-net into subnets such that the given WF-net corresponds to a loop structure over the identified do- and redo-parts.

We first define the concept of *in-between places reachability* to identify the transitions that lie on paths between two specific places in a WF-net.

Definition 10 (In-Between Places Reachability). *Let $N = (P, T, F)$ be a Petri net. The set of reachable transitions between two places $p, p' \in P$, denoted as $R_{PP}(p, p')$, is defined as follows:*

$$t \in R_{PP}(p, p') \Leftrightarrow \exists t_1, \ldots, t_n \in T \text{ and } p_1, \ldots, p_{n+1} \in P$$

such that

$$p_1 = p \ \wedge \ p_{n+1} = p' \ \wedge \ t \in \{t_1, \ldots, t_n\},$$

and for each i $(1 \leq i \leq n)$:

$$p_i \neq p' \ \wedge \ (p_i, t_i) \in F \ \wedge \ (t_i, p_{i+1}) \in F.$$

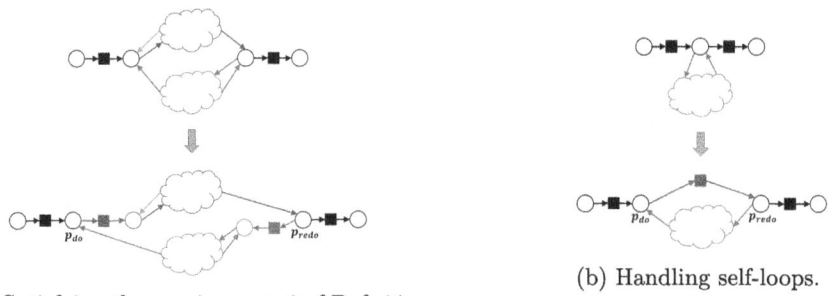

(a) Satisfying the requirement *6* of Def. 11.

(b) Handling self-loops.

Fig. 6. Reduction rules for enabling loop pattern detection.

An *loop pattern*, as illustrated in Fig. 5, represents a WF-net where transitions can be partitioned into three parts: do-part, redo-part, and silent transitions for loop entry and exit. A loop pattern must also include two special places, p_{do} and p_{redo}, which mark the entry and exit points of the loop. The do-part consists of transitions reachable from p_{do} to p_{redo}, while the redo-part consists of transitions reachable from p_{redo} to p_{do}.

Definition 11 (Loop Pattern). *Let $N = (P, T, F)$ be a safe and sound WF-net. Let $G = \{T_{do}, T_{redo}, T_{\tau}\}$ be a partition of transitions of size 3. The tuple (N, G) is called a loop pattern iff places $p_{do}, p_{redo} \in P$ exist such that:*

1. *Two transitions t_{source} and t_{sink} exist such that $t_{source} \neq t_{sink}$, $T_{\tau} = \{t_{source}, t_{sink}\}$, and $\mathcal{L}(t_{source}) = \mathcal{L}(t_{sink}) = \tau$.*
2. *$N_{source}\bullet = \{t_{source}\}$ and $\bullet N_{sink} = \{t_{sink}\}$.*
3. *$\bullet t_{source} = \{N_{source}\}$ and $t_{source}\bullet = \{p_{do}\}$.*
4. *$\bullet t_{sink} = \{p_{redo}\}$ and $t_{sink}\bullet = \{N_{sink}\}$.*
5. *$T_{do} = R_{PP}(p_{do}, p_{redo})$ and $T_{redo} = R_{PP}(p_{redo}, p_{do})$.*
6. *$\bullet p_{do} \cap T_{do} = \emptyset$ and $\bullet p_{redo} \cap T_{redo} = \emptyset$.*
7. *$p_{redo}\bullet \cap T_{do} = \emptyset$ and $p_{do}\bullet \cap T_{redo} = \emptyset$.*

Note that a WF-net satisfying the requirements *1–5* of Definition 11 can be preprocessed, as illustrated in Fig. 6a, to satisfy *6* without affecting its language, soundness, or safeness. Furthermore, the requirement *7* is implied by *1–5*.

Identifying Self-Loops with Missing Do-Part. A WF-net may represent a self-loop with a single place marking both the start and end of the loop. In such cases, all activities belong to the redo-part, which can be skipped or repeated. To enable the detection of these loops, we restructure the WF-net, as illustrated in Fig. 6b, by adding a silent transition to represent the do-part.

After detecting a loop pattern, the WF-net is projected into the do-part and redo-parts, creating two subnets, and the recursion continues on the created subnets. As illustrated in Fig. 5, the loop projection is done by selecting the appropriate subset of transitions and adjusting the flow relation to correctly connect the subnet to the source and sink place.

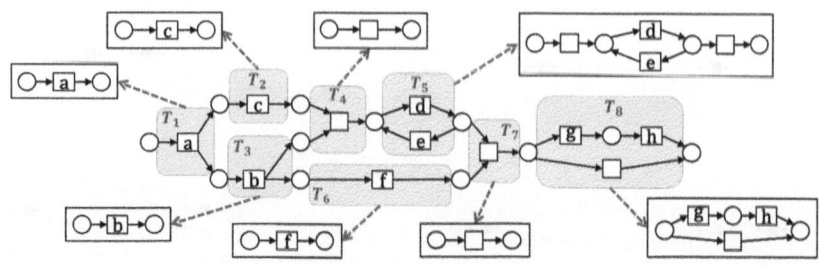

Fig. 7. An example illustrating a partial order pattern $(N, \{T_1, \ldots, T_8\})$ and the partial order projection of the WF-net N on the identified parts.

Definition 12 (Loop Projection). *Let (N, G) be an loop pattern with $N = (P, T, F)$; $T_{do}, T_{redo} \in G$; and $p_{do}, p_{redo} \in P$ as defined in Definition 11. The loop projection of N on $T' \in \{T_{do}, T_{redo}\}$ is $Project_{\circlearrowleft}(N, T') = (P', T', F')$ where*

$$P' = (P|_{T'} \setminus \{p_{do}, p_{redo}\}) \cup \{N_{source}, N_{sink}\}$$

and

$$F' = F|_{P', T'} \cup \{(N_{source}, t) \mid t \in T' \wedge (p_{start}, t) \in F\}$$
$$\cup \{(t, N_{sink}) \mid t \in T' \wedge (t, p_{end}) \in F\}$$

with

$$(p_{start}, p_{end}) = \begin{cases} (p_{do}, p_{redo}) & \text{if } T' = T_{do}, \\ (p_{redo}, p_{do}) & \text{if } T' = T_{redo}. \end{cases}$$

4.4 Identifying Partial Orders

This section addresses partitioning a WF-net into smaller subnets such that the given WF-net corresponds to a partial order over the identified parts.

A *partial order pattern* represents a WF-net with a partition of transitions such that all parts are executed and the execution order forms a partial order.

Definition 13 (Partial Order Pattern). *Let $N = (P, T, F)$ be a safe and sound WF-net. Let $G = \{T_1, \ldots, T_n\}$ be a partition of transitions of size $n \geq 2$. The tuple (N, G) is called a partial order pattern iff the following conditions hold:*

1. *For all $t, t' \in T$ and $p \in P$:*

$$\text{if } t, t' \in \{t \in T \mid \exists_{t_1, t_2 \in p\bullet} \ t_1 \rightsquigarrow t \wedge t_2 \not\rightsquigarrow t\}, \text{ then } G_t = G_{t'}.$$

2. $\prec = order(N, G)^+$ *is a partial order.*
3. **Unique local start:** *for all $i \in \{1, \ldots, n\}$ and $p, p' \in \triangleright T_i$: $p \approx_{T_i} p'$.*
4. **Unique local end:** *for all $i \in \{1, \ldots, n\}$ and $p, p' \in T_i \triangleright$: $p \approx_{T_i} p'$.*

The first condition of Definition 13 states that if a place has outgoing flows leading to different transitions, then all such transitions must fall into the same

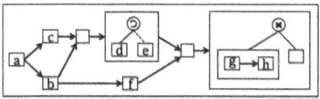

Fig. 8. The POWL model generated by Algorithm 1 for the WF-net from Fig. 1b.

part, as the place represents a decision point (i.e., there is a potential choice or loop within this part). The second condition requires that the transitive closure of the execution order between the parts of the partition forms a partial order. The third and fourth conditions ensure that the identified parts form cleanly separable components that can be executed independently with well-defined entry and exit points. Figure 7 shows a partial order pattern detected for the WF-net from Fig. 1b.

For a WF-net N, we use $Partition_{order}(N)$ to denote the partition generated by iteratively grouping transitions based on their reachability relations with respect to decision points, aligning with the first condition of Definition 13. Note that $(N, Partition_{order}(N))$ is a partial order pattern if $|Partition_{order}(N)| \geq 2$ and the remaining three requirements of Definition 13 are met.

Normalization. Before defining the projection for partial order patterns, we introduce the concept of normalization. Let N be a Petri net with known unique start and end places $p_s \in P$ and $p_e \in P$, respectively. Then P is normalized into a new Petri net $Normalize(N, p_s, p_e)$ by (i) inserting a new start place and connecting it to p_s through a silent transition in case $\bullet p_s \neq \emptyset$ and (ii) inserting a new end place and connecting it to p_e through a silent transition in case $p_e \bullet \neq \emptyset$. Normalization aims at ensuring conformance with the requirements of WF-nets (c.f. Definition 2) by adding new source and sink places if needed.

After detecting a partial order pattern, the WF-net is projected on the identified parts as illustrated in the example shown in Fig. 7. This projection is done by selecting the appropriate subset of transitions, adding unique start and end places to represent the entry and exit points of the part, adjusting the flow relation accordingly, and applying normalization if needed.

Definition 14 (Partial Order Projection). *Let (N, G) be a partial order pattern with $N = (P, T, F)$. Let $T' \in G$ be a part. The partial order projection of N on T' is $Project_{order}(N, T') = Normalize(N', p_s, p_e)$ where $p_s, p_e \notin P$ are two fresh places and $N' = (P', T', F')$ is constructed as follows:*

- $P' = (P|_{T'} \setminus (\triangleright T' \cup T' \triangleright)) \cup \{p_s, p_e\}$.
- $F' = F|_{P', T'}$
  $\cup \{(p_s, t) \mid (p, t) \in F \text{ for } p \in \triangleright T'\} \ \cup \ \{(t, p_s) \mid (t, p) \in F \text{ for } p \in \triangleright T'\}$
  $\cup \{(p_e, t) \mid (p, t) \in F \text{ for } p \in T' \triangleright\} \ \cup \ \{(t, p_e) \mid (t, p) \in F \text{ for } p \in T' \triangleright\}$.

4.5   WF-Net to POWL Converter

Algorithm 1 converts a safe and sound WF-net into an equivalent POWL model. First, the algorithm checks whether the WF-net consists of a single transition

Input: A safe and sound WF-net $N = (P, T, F)$.
Output: A POWL model or *null* if no translation is possible.

1 **Function** ConvertNetToPOWL(N):
2   **if** $|T| = 1$ *with* $T = \{t\}$, $|P| = 2$, *and* $F = \{(N_{source}, t), (t, N_{sink})\}$ **then**
3   **return** t
4     $G \leftarrow Partition_{\times}(N)$
5   **if** (N, G) *is an XOR pattern* **then**
6   **for** $T_i \in G = \{T_1, \ldots, T_n\}$ **do**
7             $\psi_i \leftarrow$ ConvertNetToPOWL($Project_{\times}(N, T_i)$)
8   **return** $\times(\psi_1, \ldots, \psi_n)$
9   **if** *a loop pattern exists with* $p_{do}, p_{redo} \in P$ *as described in Definition 11* **then**
10         $\psi_{do} \leftarrow$ ConvertNetToPOWL($Project_{\circlearrowleft}(N, R_{PP}(p_{do}, p_{redo}))$)
11         $\psi_{redo} \leftarrow$ ConvertNetToPOWL($Project_{\circlearrowleft}(N, R_{PP}(p_{redo}, p_{do}))$)
12   **return** $\circlearrowleft(\psi_{do}, \psi_{redo})$
13     $G \leftarrow Partition_{order}(N)$
14   **if** (N, G) *is a partial order pattern* **then**
15         $\prec \leftarrow order(N, G)^+$
16   **for** $T_i \in G = \{T_1, \ldots, T_n\}$ **do**
17             $\psi_i \leftarrow$ ConvertNetToPOWL($Project_{order}(N, T_i)$)
18   **return** $\prec(\psi_1, \ldots, \psi_n)$
19   **return** *null*

Algorithm 1: Conversion of a WF-Net into a POWL Model.

(base case). If a base case is not detected, the algorithm attempts to identify an XOR, loop, or partial order pattern. If a pattern is found, the algorithm projects the WF-net on the identified parts, recursively converts the created subnets into POWL models, and combines them using the appropriate POWL operator. If no pattern is detected, the algorithm returns *null*, indicating that the WF-net cannot be converted into a POWL model. Note that the order in which the algorithm searches for the different patterns is irrelevant, as at most one pattern can match the WF-net's structure at a time.

Example Applications: Figure 8 shows the POWL model generated by applying Algorithm 1 on the WF-net from Fig. 1b. The generated POWL model, while structurally different from the manually crafted POWL model in Fig. 1a, is semantically equivalent. Figure 9 illustrates three examples where Algorithm 1 returns null:

- The first WF-net (Fig. 9b), while free-choice, its decision points are not arranged in a block structure, and no equivalent POWL model exists for its language. The algorithm attempts to find XOR or partial order patterns but ultimately produces partitions of size 1, leading to the return of null.
- The second WF-net (Fig. 9b) exhibits a choice between activities a and b, followed by a non-free-choice between d and e that is influenced by the preceding choice. This long-term dependency choice cannot be represented in POWL. Algorithm 1 attempts to identify a partial order pattern, resulting in a partition that violates the requirements of Definition 13, e.g., no unique local end in the first part $T_1 = \{a, b\}$ since $\bullet p_3 \cap T_1 = \{a\} \neq \{b\} = \bullet p_4 \cap T_1$.
- In the third WF-net (Fig. 9c), the places p_2 and p_3 model a choice between d or executing both b and c concurrently. When attempting to identify a partial

(a) Free-choice WF-net with non-block-structured decision points.

(b) Non-free-choice WF-net with a long-term dependency between choices.

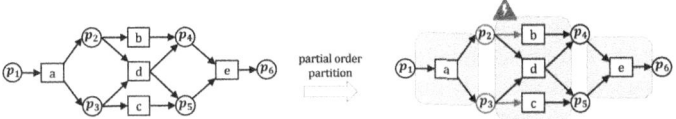

(c) Non-free-choice WF-net where decision points combine choice with concurrency.

Fig. 9. Examples where Algorithm 1 returns null.

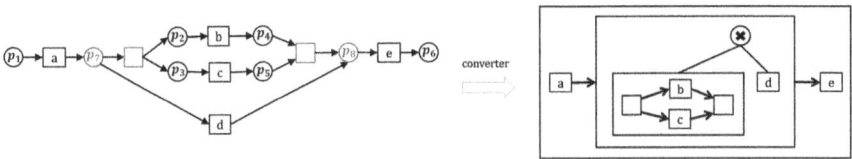

Fig. 10. Result of applying the reduction rules (c.f. Fig. 3) to the WF-net from Fig. 9c, followed by a successful conversion into POWL using Algorithm 1.

order pattern, the requirements are violated, e.g., no unique local start in the second part $T_2 = \{b, c, d\}$ since $p_2 \bullet \cap T_2 = \{b, d\} \neq \{c, d\} = p_3 \bullet \cap T_2$. However, applying the reduction rules from Fig. 3 before applying Algorithm 1 enables the successful conversion into POWL, as illustrated in Fig. 10.

5   Correctness and Completeness Guarantees

In this section, we prove the correctness and completeness guarantees of Algorithm 1. Our proof strategy for correctness relies on structural induction on the WF-net. We show that for each pattern (XOR, loop, and partial order), the projection operation preserves safeness and soundness, allowing for the recursive application of the algorithm on the created subnets (Sect. 5.1). Then, we demonstrate that combining the languages of these subnets using the respective POWL operators accurately reflects the language of the original WF-net (Sect. 5.2). We combine these findings to prove the overall correctness of the algorithm (Sect. 5.3). Finally, we show the completeness of our algorithm on semi-block-structured WF-nets (Sect. 5.4).

5.1 Projection Structural Guarantees

This section proves that the XOR, loop, and partial order projections, when applied to safe and sound WF-nets, produce safe and sound WF-nets. This ensures that the recursive calls in Algorithm 1 are always applied to valid inputs.

Lemma 1 (XOR Projection Structural Guarantees). *Let (N, G) be an XOR pattern with $N = (P, T, F)$ and $G = \{T_1, \ldots, T_n\}$. Let $N_i = (P_i, T_i, F_i) = Project_\times(N, T_i)$ for each i ($1 \leq i \leq n$). Then N_i is a safe and sound WF-net.*

Proof. **(1)** N_i **is a WF-net:** By construction, no place $p \in P_i \backslash \{N_{source}, N_{sink}\}$ is connecting transitions from both T_i and $T \backslash T_i$ in N. Therefore, all transitions and places in N_i remain connected on a path from N_{source} to N_{sink}.

(2) We prove that N_i **is safe and sound.** N_i can be replaced by a silent transition $t^+ \notin T$ in N generating another WF-net $N' = (P', T', F')$ as follows:

- $T' = (T \backslash T_i) \cup \{t^+\}$.
- $P' = P|_{T \backslash T_i}$.
- $F' = F|_{P', T \backslash T_i} \cup \{(N_{source}, t^+), (t^+, N_{sink})\}$.

N can be reconstructed from N' as described in [23, Theorem 3.4] by replacing t^+ by N_i if $P_i \cap P' = \{N_{source}, N_{sink}\}$. This condition is satisfied since no place connecting transitions from both T_i and $T \backslash T_i$ exists in N. Therefore, N_i is safe and sound by [23, Theorem 3.4].

Lemma 2 (Loop Projection Structural Guarantees). *Let (N, G) be an loop pattern with $N = (P, T, F)$; $T_{do}, T_{redo} \in G$; and $p_{do}, p_{redo} \in P$ as defined in Definition 11. Let $N_{do} = (P_{do}, T_{do}, F_{do}) = Project_\circlearrowright(N, T_{do})$ and $N_{redo} = (P_{redo}, T_{redo}, F_{redo}) = Project_\circlearrowright(N, T_{redo})$. Then N_{do} and N_{redo} are safe and sound WF-nets.*

Proof. **(1) We show that** N_{do} **is a WF-net (analogous proof for** N_{redo}**).**
(1.1) Unique source and sink: Assume for contradiction that there exists a place $p \in P_{do} \backslash \{N_{source}, N_{sink}\}$ such that $\bullet p = \emptyset$ or $p \bullet = \emptyset$ in N_{do}. This place must be connecting transitions from both T_{do} and T_{redo} since $\bullet p \neq \emptyset$ and $p \bullet \neq \emptyset$ in N. Without loss of generality, assume that there exist transitions $t \in T_{do}$ and $t' \in T_{redo}$ where $(t, p) \in F$ and $(p, t') \in F$. Since $t \in T_{do} = R_{PP}(p_{do}, p_{redo})$, there exists a path starting from p_{do} to t, without passing through p_{redo} in-between. From t, the path can continue through p to reach t'. From t', we know that we can eventually reach p_{redo} (we reach p_{do} first and then take any path from p_{do} to p_{redo}). Combining all of these sequences ($p_{do} \rightarrow \ldots \rightarrow t \rightarrow p \rightarrow t' \rightarrow \ldots \rightarrow p_{redo}$) implies that $t' \in T_{do}$. This contradicts our assumption that $t' \in T_{redo}$.
(1.2) Connectivity: We showed that no place is connecting transitions from both T_{do} and T_{redo} except the loop entry and exit places. Therefore, all transitions and places in N_{do} remain connected on a path from N_{source} to N_{sink}.
(2) Safeness and soundness: The proof that N_{do} and N_{redo} are safe and sound is analogous to the safeness and soundness proof of Theorem 1.

Lemma 3 (Partial Order Projection Structural Guarantees). *Let* (N, G) *be a partial order pattern with* $N = (P, T, F)$ *and* $G = \{T_1, \ldots, T_n\}$. *Let* $N_i = (P_i, T_i, F_i) = Project_{order}(N, T_i)$ *for each* i $(1 \leq i \leq n)$. *Then* N_i *is a safe and sound WF-net.*

Proof. **(1)** N_i **is a WF-net:** By construction, every node in N_i is on a path from p_s and p_e. The applied normalization ensures that additional source and/or sink places are inserted in case $\bullet p_s \neq \emptyset$ and/or $p_e \bullet \neq \emptyset$, respectively.

(2) We prove that N_i **is sound.**

(2.1) No dead transitions: Consider any transition $t \in T_i$. Since N is sound, there exists a reachable marking M in N that enables t. Consider the marking M_i in N_i defined as follows for $p \in P_i$:

$$M_i(p) = \begin{cases} M(p) & \text{if } p \in P, \\ 1 & \text{if } p = p_s \text{ and } M(p) = 1 \text{ for all } p \in \triangleright T_i, \\ 1 & \text{if } p = p_e \text{ and } M(p) = 1 \text{ for all } p \in T_i \triangleright, \\ 0 & \text{otherwise (for places added by normalization).} \end{cases}$$

The projection operation only removes places and transitions that are not in T_i and replaces the connections to $\triangleright T_i$ and $T_i \triangleright$ with p_s and p_e, respectively. Thus, any firing sequence leading to M in N can be transformed into a firing sequence leading to M_i in N_i by removing transitions not in T_i (and potentially firing the additional silent transition added by normalization). Therefore, M_i is reachable from $[N_{isource}]$ in N_i. The unique local start and end properties (c.f. Definition 13) ensure that if t needs to consume a token from a place $p \in \triangleright T_i$ (or $p \in T_i \triangleright$) in N, then all other places in $\triangleright T_i$ (or in $T_i \triangleright$, respectively) must be included in M as well. This implies that M and M_i agree on all places that are needed to enable t, including start and end places. Therefore, t is enabled in M_i.

(2.2) Option to complete: Consider any marking M_i reachable from $[N_{isource}]$ in N_i. Due to the unique local start and end properties (c.f. Definition 13), there must exist a reachable marking M in N that enables the transitions in T_i in the same way as M_i does in N_i, i.e., M is defined for $p \in P|_{T_i}$ as follows:

$$M(p) = \begin{cases} M_i(p_s) & \text{if } p \in \triangleright T_i, \\ M_i(p_e) & \text{if } p \in T_i \triangleright, \\ M_i(p) & \text{otherwise.} \end{cases}$$

Since N is sound, there exists a firing sequence σ from M to $[N_{sink}]$ in N. We can construct a corresponding firing sequence σ_i from M_i to $[N_{isink}]$ in N_i by taking only the transitions in σ that belong to T_i (and potentially firing the additional silent transition added by normalization).

(3) N_i **is safe:** Assume, for the sake of contradiction, that there exist a reachable marking M_i in N_i and a place $p \in P_i$ such that $M_i(p) \geq 2$. There must exist a reachable marking M in N that enables the transitions in T_i in the same way as M_i does in N_i (c.f. the proof of "option to complete"). Then there exists $p' \in P|_{T_i}$ such that $M(p') = M_i(p) \geq 2$. This violates the safeness of N.

5.2   Pattern Language Preservation Guarantees

This section establishes the language preservation guarantees of the identified patterns. We prove that the XOR, loop, and partial order patterns, when translated into their corresponding POWL representations, result in POWL models that have the same language as the original WF-net.

Lemma 4 (XOR Pattern Language Preservation). *Let (N, G) be an XOR pattern and $G = \{T_1, \ldots, T_n\}$. Let $\psi_1, \ldots, \psi_n$ be POWL models such that $\mathcal{L}(\psi_i) = \mathcal{L}(Project_\times(N, T_i))$ for each i $(1 \leq i \leq n)$. Then, $\mathcal{L}(\times(\psi_1, \ldots, \psi_n)) = \mathcal{L}(N)$.*

Proof. Since (N, G) forms an XOR Pattern, N enforces a choice of transitions from exactly one part $T_i \in G$ to be executed. $Project_\times(N, T_i)$ captures exactly the executions involving the transitions in T_i. Therefore, we conclude:

$$\mathcal{L}(N) = \bigcup_{i=1}^{n} \mathcal{L}(Project_\times(N, T_i)).$$

By combining this equality with Definition 5 and the assumption that $\mathcal{L}(\psi_i) = \mathcal{L}(Project_\times(N, T_i))$ for each $1 \leq i \leq n$, we get: $\mathcal{L}(N) = \mathcal{L}(\times(\psi_1, \ldots, \psi_n))$.

Lemma 5 (Loop Pattern Language Preservation). *Let (N, G) be a loop pattern with $N = (P, T, F)$; $T_{do}, T_{redo} \in G$; and $p_{do}, p_{redo} \in P$ as defined in Definition 11. Let ψ_{do} and ψ_{redo} be POWL models such that $\mathcal{L}(Project_\circlearrowleft(N, T_{do})) = \mathcal{L}(\psi_{do})$ and $\mathcal{L}(Project_\circlearrowleft(N, T_{redo})) = \mathcal{L}(\psi_{redo})$. Then, $\mathcal{L}(\circlearrowleft(\psi_{do}, \psi_{redo})) = \mathcal{L}(N)$.*

Proof. Let $N_{do} = Project_\circlearrowleft(N, T_{do})$ and $N_{redo} = Project_\circlearrowleft(N, T_{redo})$. Due to the soundness of N_{do} and N_{redo} and the absence of places connecting transitions from both the do- and redo-parts except p_{do} and p_{redo} (c.f., the proof of Lemma 2), we conclude that the language of N consists of sequences that can be segmented into subsequences of complete executions of N_{do} and N_{redo}, interleaved as follows:

$$\mathcal{L}(N) = \mathcal{L}(N_{do}) \cdot (\mathcal{L}(N_{redo}) \cdot \mathcal{L}(N_{do}))^*.$$

By combining this equality with Definition 5 and the assumption that $\mathcal{L}(\psi_{do}) = \mathcal{L}(N_{do}) \wedge \mathcal{L}(\psi_{redo}) = \mathcal{L}(N_{redo})$, we get: $\mathcal{L}(N) = \mathcal{L}(\circlearrowleft(\psi_{do}, \psi_{redo}))$.

Lemma 6 (Partial Order Pattern Language Preservation). *Let (N, G) be a partial order pattern with $N = (P, T, F)$, $G = \{T_1, \ldots, T_n\}$, and $\prec \in \mathcal{O}^n$ as defined in Definition 13. Let $\psi_1, \ldots, \psi_n$ be POWL models such that $\mathcal{L}(Project_{order}(N, T_i)) = \mathcal{L}(\psi_i)$ for each i $(1 \leq i \leq n)$. Then, $\mathcal{L}(\prec(\psi_1, \ldots, \psi_n)) = \mathcal{L}(N)$.*

Proof. Let $N_i = Project_{order}(N, T_i)$ for each i $(1 \leq i \leq n)$. By combining the semantics of partial orders (c.f. Definition 5) with the assumption that $\mathcal{L}(\psi_i) = \mathcal{L}(N_i)$ for each i $(1 \leq i \leq n)$, we can write:

$$\mathcal{L}(\prec(\psi_1, \ldots, \psi_n)) = \{\sigma \in \sqcup\!\sqcup_\prec(\sigma_1, \ldots, \sigma_n) \mid \forall_{1 \leq i \leq n} \sigma_i \in \mathcal{L}(N_i)\}.$$

(1) Proof for $\mathcal{L}(N) \subseteq \{\sigma \in \sqcup\!\sqcup_{\prec}(\sigma_1, ..., \sigma_n) \mid \forall_{1 \leq i \leq n} \sigma_i \in \mathcal{L}(N_i)\}$ **:** Let $\sigma \in \mathcal{L}(N)$ be any firing sequence of N. We construct subsequences $\sigma_1, ..., \sigma_n$ by projecting σ onto T_i for each i ($1 \leq i \leq n$). We need to show that σ can be expressed as a shuffle of $\sigma_1, ..., \sigma_n$, respecting the partial order $\prec$. This can be derived by proving the following three key points:

-   All parts $T_i \in G$ are present within σ.
-   Each σ_i is a firing sequence from N_i (i.e., $\sigma_i \in \mathcal{L}(N_i)$).
-   The partial order between the subsequences is preserved in σ (i.e., $\sigma \in \sqcup\!\sqcup_{\prec}(\sigma_1, ..., \sigma_n)$).

(1.1) All parts $T_i \in G$ **are present within** σ: Assume, for the sake of contradiction, that there exists a part $T_i \in G$ not present within σ, i.e., $\sigma_i = \langle\rangle$. Since N is a WF-net, this implies that there must be a *decision point* $p \in P$ where some outgoing paths from p eventually lead to transitions in T_i and other outgoing paths from p do not. This violates the first condition of Definition 13, which states that if a place has outgoing flows leading to different transitions, then all such transitions must fall into the same part in G.

(1.2) Each σ_i **is a firing sequence from** N_i: The unique local start and end properties (c.f. Definition 13) ensure each subnet is executed independently from start to end within N. Therefore, σ_i must be a firing sequence from N_i.

(1.3) The partial order $\prec$ **is preserved in** σ: Assume $i \prec j$ for any $i, j \in \{1, ..., n\}$. Since $\prec = order(N, G)^+$, transitions from T_i are executed first, producing tokens that are needed to eventually enable T_j. Suppose, for the sake of contradiction, that after the execution of transitions from T_j, transitions from T_i are re-enabled (i.e., tokens are produced in $\triangleright T_i$). Then we have two possible scenarios:

-   (i) The re-enabling of T_i does not depend on the completion of T_j (i.e., it does not require the consumption of tokens from $T_j \triangleright$): This means that we can perform a full execution of the subnet of T_i and reach the subnet of T_j again before its completion, violating safeness.
-   (ii) The re-enabling of T_i depends on the completion of T_j (i.e., it requires the consumption of tokens from $T_j \triangleright$): This implies the existence of a sequence of dependencies in the execution order $j \prec ... \prec i$. By transitivity, $j \prec i$ holds. This violates the asymmetry requirement of partial orders since $i \prec j$.

(2) Proof for $\{\sigma \in \sqcup\!\sqcup_{\prec}(\sigma_1, ..., \sigma_n) \mid \forall_{1 \leq i \leq n} \sigma_i \in \mathcal{L}(N_i)\} \subseteq \mathcal{L}(N)$ **:** Consider any sequence $\sigma \in \sqcup\!\sqcup_{\prec}(\sigma_1, ..., \sigma_n)$ where $\sigma_i \in \mathcal{L}(N_i)$ for $1 \leq i \leq n$. We showed that all parts $T_i \in G$ must be visited in N (c.f. the proof of 1.1). Due to the unique local start and end properties in N (c.f. Definition 13), each subnet can be executed independently in N, without violating the execution order $\prec = order(N, G)^+$. Therefore, the interleaved sequence σ constitutes a valid firing sequence in N.

5.3 Overall Correctness Guarantee

In this section, we prove the correctness of Algorithm 1. Specifically, we show that the algorithm, if successfully producing a POWL model, then the POWL model has the same language as the input WF-net.

Theorem 1 (Correctness). *Let $N = (P, T, F)$ be a safe and sound WF-net. If Algorithm 1 successfully converts N into a POWL model ψ, then $\mathcal{L}(N) = \mathcal{L}(\psi)$.*

Proof. We prove the theorem by induction on the number of transitions in N.

- **Base case:** The theorem trivially holds for a WF-net that contains a single transition.
- **Inductive hypothesis:** For $n > 1$, assume the theorem holds for all safe and sound WF-net with fewer transitions than n (i.e., $|T| < n$).
- **Inductive step ($|T| = n$):** We consider the different cases in the algorithm:
  - **XOR pattern:** Suppose an XOR pattern (N, G) with $G = \{T_1, \ldots, T_n\}$ is identified. For each $T_i \in G$, N is projected onto T_i to obtain $N_i = Project_\times(N, T_i)$. By Lemma 1, each N_i is a safe and sound WF-net with fewer transitions than n. By the inductive hypothesis, the POWL model ψ_i obtained from N_i satisfies $\mathcal{L}(\psi_i) = \mathcal{L}(N_i)$. The algorithm returns $\psi = \times(\psi_1, \ldots, \psi_n)$. By Lemma 4, $\mathcal{L}(\psi) = \mathcal{L}(N)$.
  - **Loop or partial order pattern:** The proof is analogous to the XOR case, using the appropriate lemmas for structural guarantees (Lemma 2 or Lemma 3) and language equivalence (Lemma 5 or Lemma 6).
  - **No pattern is detected:** If no pattern is detected, then the algorithm returns *null*, which does not contradict the theorem.

By induction, the theorem holds for all safe and sound WF-nets successfully converted by Algorithm 1.

5.4 Completeness Guarantee on Semi-block-Structured WF-Nets

In this section, we show that Algorithm 1 is complete when applied to semi-block-structured WF-nets (c.f. Definition 6).

Theorem 2 (Completeness). *Let N be a semi-block-structured WF-net. Algorithm 1 successfully converts N into a POWL model that has the same language as N.*

Proof. We distinguish between three cases:

- **Case 1: N has a single transition:** This matches the base case of the algorithm.
- **Case 2: N corresponds to an XOR or loop pattern:** Projecting the net on each part yields another semi-block-structured WF-net.

- **Case 3: N does not correspond to an XOR or loop pattern:** The algorithm creates a partition $G = Partition_{order}(N)$ where transitions within the same block are grouped into the same part of the partition. The transitive closure of the execution order $order(N, G)^+$ must form a partial order because (i) substituting each part with a single transition turn the net into a marked graph and (ii) soundness implies acyclicity for marked graph WF-nets. Since the top-level blocks have unique entry and exit points, the unique local start and end requirements of Definition 13 are also met. Therefore, a partial order pattern is detected. Projecting N on each part yields either (i) a base case for single transitions or (ii) a semi-block-structured WF-net that corresponds to an XOR or loop pattern for the blocks.

After projection, sub-nets are recursively handled in the same manner. Thus, the algorithm successfully produces a POWL model. The language equivalence follows by Theorem 1.

6   Implementation and Scalability Assessment

To assess the scalability of Algorithm 1, we implemented it, incorporating the reduction rules illustrated in Fig. 3 and Fig. 6. We then performed two experiments (code and data are available at https://github.com/humam-kourani/WF-net-to-POWL). In the first experiment, we utilized the process tree generator from [9,10] to generate 1000 process trees. These trees were then translated into WF-nets using PM4Py [3], resulting in a diverse set of WF-nets varying in size from 21 to 370 transitions and 15 to 305 places. In the second experiment, we used the ground truth WF-nets of the 20 processes from [11], which were originally derived from POWL models. For comparison, we also applied the WF-net to process tree converter from [27] in both experiments.

Results. In the first experiment, Algorithm 1 successfully generated POWL models for all 1000 WF-nets, which was expected since process trees represent a subclass of POWL. The experiment demonstrated the high scalability of our approach, as illustrated in Fig. 11. Conversion times for our approach ranged from 0.002 to 2.48 s, whereas the tree-based converter from [27] required between 0.013 and 126 s. These results highlight the efficiency of our algorithm's top-down decomposition strategy over the bottom-up approach used by the process tree converter. In the second experiment, while our algorithm was successful on all WF-nets, the tree-based converter was only able to convert 17 out of the 20 WF-nets into process trees.

   In summary, our experiments demonstrate the advantage of our approach in both supporting a broader range of structures and its superior performance compared to the process tree-based converter. Note that the implemented algorithm is also available in ProMoAI [13] (https://promoai.streamlit.app/), powering the redesign feature for improving existing process models via large language models.

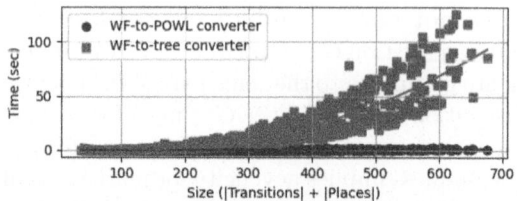

Fig. 11. Comparison of conversion times between Algorithm 1 and the process tree converter from [27] on the 1000 WF-nets dataset.

7   Conclusion

This paper introduced a novel algorithm for translating safe and sound Workflow Nets (WF-nets) into the Partially Ordered Workflow Language (POWL). The algorithm leverages the hierarchical structure of POWL by recursively identifying patterns within the WF-net that correspond to POWL's operators. We formally proved the correctness of our approach, showing that the resulting POWL model preserves the language of the original WF-net. Furthermore, we demonstrated the high scalability of the proposed algorithm and showed its completeness on semi-block-structured WF-nets, a subclass that contains equivalent workflow-nets for any POWL model.

This work paves the way for broader adoption of POWL in different process mining applications. The main avenue for future work is the development of optimized process mining techniques that leverage the structural properties of POWL, such as efficient algorithms for conformance checking. Furthermore, we aim to provide a platform that allows users to visualize process models in the POWL language and implements new methods for interactive process analysis and improvement.

References

1. Berthelot, G.: Transformations and decompositions of nets. In: Brauer, W., Reisig, W., Rozenberg, G. (eds.) ACPN 1986. LNCS, vol. 254, pp. 359–376. Springer, Heidelberg (1987). https://doi.org/10.1007/978-3-540-47919-2_13
2. Berthelot, G., Lri-Iie.: Checking properties of nets using transformations. In: Rozenberg, G., (ed.) APN 1985, vol. 222, pp. 19–40. Springer, Heidelberg (1986). https://doi.org/10.1007/BFb0016204
3. Berti, A., van Zelst, S.J., Schuster, D.: PM4Py: a process mining library for Python. Softw. Impacts **17**, 100556 (2023)
4. Dal Zilio, S.: MCC: a tool for unfolding colored petri nets in PNML format. In: Janicki, R., Sidorova, N., Chatain, T. (eds.) PETRI NETS 2020. LNCS, vol. 12152, pp. 426–435. Springer, Cham (2020). https://doi.org/10.1007/978-3-030-51831-8_23
5. Desel, J., Esparza, J.: Free Choice Petri Nets, vol. 40. Cambridge University Press, Cambridge (1995)
6. Dijkman, R.M., Dumas, M., Ouyang, C.: Semantics and analysis of business process models in BPMN. Inf. Softw. Technol. **50**(12), 1281–1294 (2008)

7. Favre, C., Fahland, D., Völzer, H.: The relationship between workflow graphs and free-choice workflow nets. Inf. Syst. **47**, 197–219 (2015)
8. Gardner, T.: UML modelling of automated business processes with a mapping to BPEL4WS. Orientation Web Serv. **30** (2003)
9. Jouck, T., Depaire, B.: PTandLogGenerator: a generator for artificial event data. In: Azevedo, L., Cabanillas, C., (eds.) Proceedings of the BPM Demo Track 2016 Co-located with the 14th International Conference on Business Process Management (BPM 2016). CEUR Workshop Proceedings. Rio de Janeiro, Brazil, September 21, 2016, vol. 1789, pp. 23–27. CEUR-WS.org (2016)
10. Jouck, T., Depaire, B.: Generating artificial data for empirical analysis of control-flow discovery algorithms - a process tree and log generator. Bus. Inf. Syst. Eng. **61**(6), 695–712 (2019)
11. Kourani, H., Berti, A., Schuster, D., van der Aalst, WM.P.: Evaluating large language models on business process modeling: framework, benchmark, and self-improvement analysis. *CoRR*, abs/2412.00023 (2024)
12. Kourani, H., Berti, A., Schuster, D., van der Aalst, W.P.: Process modeling with large language models. In: van der Aa, H., Bork, D., Schmidt, R., Sturm, A. (eds.) BPMDS EMMSAD 2024 2024. LNBIP, vol. 511, pp. 229–244. Springer, Cham (2024). https://doi.org/10.1007/978-3-031-61007-3_18
13. Kourani, H., Berti, A., Schuster, D., van der Aalst, W.M.P.: ProMoAI: process modeling with generative AI. In: Proceedings of the Thirty-Third International Joint Conference on Artificial Intelligence, IJCAI 2024, Jeju, South Korea, August 3-9, 2024, pp. 8708–8712. ijcai.org (2024)
14. Kourani, H., Schuster, D., van der Aalst, W.M.P.: Scalable discovery of partially ordered workflow models with formal guarantees. In: 5th International Conference on Process Mining, ICPM 2023, Rome, Italy, October 23-27, 2023, pp. 89–96. IEEE (2023)
15. Kourani, H., van Zelst, S.J.: POWL: partially ordered workflow language. In: Di Francescomarino, C., Burattin, A., Janiesch, C., Sadiq, S. (eds.) BPM 2023. LNCS, vol. 14159, pp. 92–108. Springer, Cham (2023). https://doi.org/10.1007/978-3-031-41620-0_6
16. Kourani, H., van Zelst, S.J., Schuster, D., van der Aalst, W.: Discovering partially ordered workflow models. Inf. Syst. **128**, 102493 (2025)
17. Langner, P., Schneider, C., Wehler, J.: Petri net based certification of event-driven process chains. In: Desel, J., Silva, M. (eds.) ICATPN 1998. LNCS, vol. 1420, pp. 286–305. Springer, Heidelberg (1998). https://doi.org/10.1007/3-540-69108-1_16
18. Lassen, K.B., van der Aalst, W.: WorkflowNet2BPEL4WS: a tool for translating unstructured workflow processes to readable BPEL. In: Meersman, R., Tari, Z. (eds.) OTM 2006. LNCS, vol. 4275, pp. 127–144. Springer, Heidelberg (2006). https://doi.org/10.1007/11914853_9
19. Leemans, S.J.J.: Robust Process Mining with Guarantees - Process Discovery, Conformance Checking and Enhancement. LNBIP, vol. 440. Springer, Cham (2022)
20. Murata, T.: Petri nets: properties, analysis and applications. Proc. IEEE **77**(4), 541–580 (1989)
21. Reisig, W., Rozenberg, G.: Lectures on Petri Nets I: Basic Models, Advances in Petri Nets, the volumes are based on the Advanced Course on Petri Nets. LNCS vol. 1491. Springer, Cham (1998)
22. Salimifard, K., Wright, M.: Petri net-based modelling of workflow systems: an overview. Eur. J. Oper. Res. **134**(3), 664–676 (2001)

23. Aalst, W.: Workflow verification: finding control-flow errors using petri-net-based techniques. In: van der Aalst, W., Desel, J., Oberweis, A. (eds.) Business Process Management. LNCS, vol. 1806, pp. 161–183. Springer, Heidelberg (2000). https://doi.org/10.1007/3-540-45594-9_11

24. Aalst, W.: Reduction using induced subnets to systematically prove properties for free-choice nets. In: Buchs, D., Carmona, J. (eds.) PETRI NETS 2021. LNCS, vol. 12734, pp. 208–229. Springer, Cham (2021). https://doi.org/10.1007/978-3-030-76983-3_11

25. van der Aalst, W., Lassen, K.B.: Translating unstructured workflow processes to readable BPEL: theory and implementation. Inf. Softw. Technol. **50**(3), 131–159 (2008)

26. van Hee, K.M., Sidorova, N., van der Werf, J.: Business process modeling using petri nets. Trans. Petri Nets Other Model. Concurr. **7**, 116–161 (2013)

27. van Zelst, S.J., Leemans, S.: Translating workflow nets to process trees: an algorithmic approach. Algorithms **13**(11), 279 (2020)

28. von Rosing, M., White, S., Cummins, F., de Man, H.: Business process model and notation - BPMN. In: von Rosing, M., von Scheel, H., Scheer, A-W., (eds.), The Complete Business Process Handbook: Body of Knowledge from Process Modeling to BPM, vol. I, pp. 429–453. Morgan Kaufmann/Elsevier (2015)

Distributed Places and Safe Net Reduction

Victor Khomenko[1,2], Maciej Koutny[2(✉)], and Alex Yakovlev[3]

[1] Renesas Electronics, Delta 200 Office Park, Welton Rd, Swindon SN5 7XB, UK
[2] School of Computing, Newcastle University, 1 Science Square,
Newcastle upon Tyne NE4 5TG, UK
{victor.khomenko,maciej.koutny}@ncl.ac.uk
[3] School of Engineering, Newcastle University, Merz Court,
Newcastle upon Tyne NE1 7RU, UK
alex.yakovlev@ncl.ac.uk

Abstract. Being able to find small Petri nets with the same behaviour as formal specifications of concurrent systems benefits both effective verification and practical implementation of such systems. This paper considers specifications given in the form of safe nets and process expressions.

The paper introduces a novel concept of 'distributed place' which abstracts the behaviour of an individual net place. It is shown that if distributed places cover a safe Petri net, then it is possible to delete some places without changing the behaviour. Crucially, the reduction is carried out both statically and locally, making it computationally feasible in practice.

The resulting reduction technique is then considered for an algebra of safe Petri nets (boxes) derived from process (box) expressions. Though the original derivation can yield exponentially large boxes, the recent research demonstrated that if a box expression does not involve cyclic behaviours, the exponential number of places can be reduced down to polynomial (quadratic). In this paper, using distributed places, it is demonstrated that similar optimisation can also be achieved in the cyclic case.

Keywords: safe net · distributed place · static reduction · local reduction · box expression · control flow · composition · connection graph · cograph · edge clique cover

1 Introduction

Petri nets are a formal modelling technique designed specifically to deal with concurrent and distributed computing systems. They support a simple yet expressive semantics, intuitive graphical notation, and the possibility of capturing behaviours concisely without making subsequent formal verification or synthesis undecidable. There are several software tools for Petri nets, and they have been extensively used as a modelling formalism and as an intermediate representation to which designs initially expressed in other formalisms are translated. In fact, developing translations from, *e.g.* process algebras, concurrent programming languages and other formalisms to Petri nets has been extensively pursued for the past four decades, see *e.g.* [4,9,14,17].

© The Author(s), under exclusive license to Springer Nature Switzerland AG 2025
E. Amparore and Ł. Mikulski (Eds.): PETRI NETS 2025, LNCS 15714, pp. 265–286, 2025.
https://doi.org/10.1007/978-3-031-94634-9_13

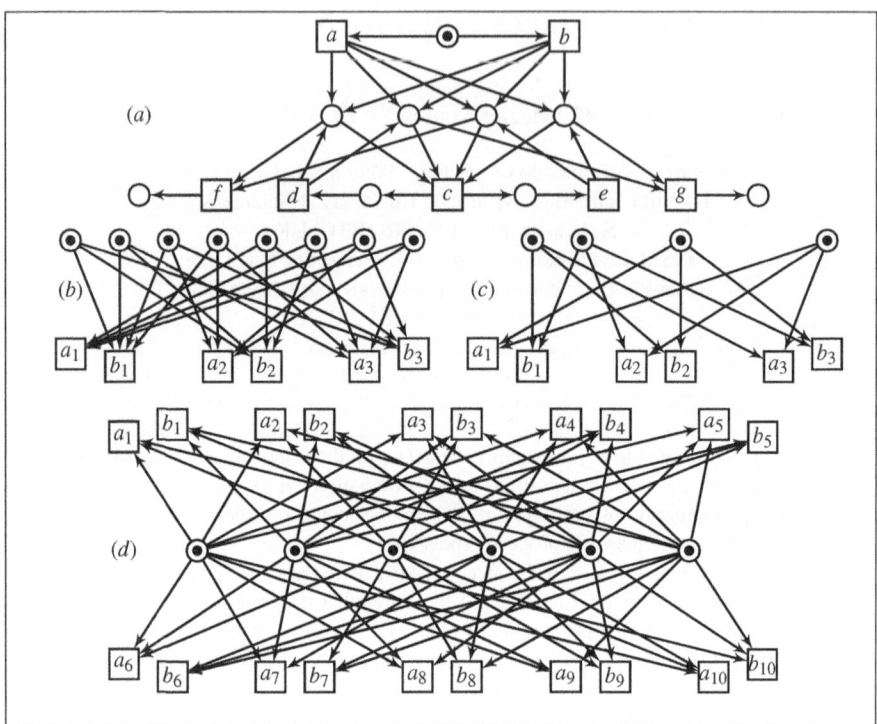

Fig. 1. (a) A net translated from box expression $[a \square b * c ; (d \| e) * f \| g]$. (b) A net expressing choice between three pairs of concurrent actions, a_i and b_i, and (c) its reduced version exhibiting the same behviour. (d) A reduced version of a marked net expressing choice between 10 pairs of concurrent actions, a_i and b_i.

Box Algebra [4,5] provides a generic process-algebraic framework for Petri nets called *(Petri) boxes*. Compositionally defined boxes can also be derived from *box expressions* which are a process algebra in the usual language-theoretic sense. For example, Fig. 1(a) shows a box derived from box expression $[a \square b * c ; (d \| e) * f \| g]$. Such an expression specifies a concurrent system which starts with the execution of either a or b, then executes (possibly zero times) a loop 'c followed by d and e concurrently', and terminates by executing e and f concurrently. Box Algebra has several concrete incarnations, including CCS [16] and TCSP [11]. In general, being able to find small Petri nets with the same behaviour as formal specifications of concurrent systems benefits both effective verification and practical implementation of such systems. This paper considers specifications given in the form of safe Petri nets and box expressions.

The possibility to create concise system models is often the key advantage of Petri nets over simpler formalisms like Finite State Machines (FSMs), where one often encounters the exponential *state space explosion* [19] already during the modelling stage. Unfortunately, as we observed in [12], translating even simple control flows to

Petri nets may lead to an exponential explosion in the Petri net size. A typical example where such a situation tends to occur is when groups ('bursts') of two or more concurrent and simultaneously enabled actions are put in a mutually exclusive choice. Consider, for instance the net in Fig. 1(b) which expressed a choice between concurrent executions of a_1 and b_1, of a_2 and b_2, and of a_3 and b_3 (this corresponds to a box expression like $(a_1 \| b_1) \square (a_2 \| b_2) \square (a_3 \| b_3)$). The standard construction (based on cross-product) in Fig. 1(b) uses $2^3 = 8$ places, but the same behaviour can be realised using only 4 places, as shown in Fig. 1(c). In general, to express choice between n such pairs of actions, the standard construction would use 2^n places. However, an optimised construction could only require a small fraction of these places. For instance, Fig. 1(d) shows a reduced solution for $n = 10$ which uses 6 rather than $2^{10} = 1024$ places.

Previous Work. As a motivating example, [12] considers *Burst Automata* [6] used in the area of asynchronous circuits design. They are like FSMs with arcs labelled by sets of concurrently executed actions (bursts). In [6] a language-preserving linear size translation is proposed, that prefixes each burst with a silent 'fork' transition and then uses another silent 'join' transition after the burst to detect completion. Unfortunately, there are situations when silent transitions are unacceptable [7]. First of all, such silent transitions turn a deterministic model into a non-deterministic one, which is often undesirable (*e.g.* non-determinism cannot be directly implemented physically, say in an asynchronous logic circuit [7]). Second, language equivalence may be too weak (*e.g.* it does not preserve branching time temporal properties or even deadlocks), and prefixing bursts with silent transitions breaks not only strong bisimulation but also weak bisimulation.

To preserve strong bisimulation, the *cross-product* is traditionally used, see *e.g.* [3, 4, 8, 9]. To express a choice between several bursts (*i.e.* sets of concurrent transitions) $B_1, B_2, \ldots, B_n$, this construction would create a set of places corresponding to tuples in $B_1 \times B_2 \times \cdots \times B_n$, and so the Petri net size is exponential in the number of bursts. For example, [6] developed translations from Burst Automata to Petri nets based on cross-product preserving weak or strong bisimulation.

In [12] we proposed an alternative to cross-product, that uses at most quadratic (in the total size of all bursts) number of places to express a choice between bursts, thereby reducing the size of Burst Automata to Petri net translation from exponential [6] down to polynomial. Furthermore, in some cases a logarithmic number of places is sufficient, yielding a double-exponential reduction w.r.t. cross-product. The latter case is illustrated in Fig. 1(b, c, d), as for a choice between n binary bursts we get (asymptotically) double-exponential reduction from 2^n to $O(\log n)$. The technique was based on showing the equivalence between the modelling problem of expressing a choice between bursts of concurrent events and the problem of finding an edge clique cover of a complete multipartite graph.

For specifications of concurrent systems in the form of process expressions, in [13], we generalised the above technique to box expressions involving choice, concurrency, and sequencing operators (but not iteration). More precisely, we proposed a polynomial translation of such box expressions that preserves strong bisimulation (in fact, it guarantees the isomorphism of reachability graphs, which is a stronger equivalence).

The developed translation is compositional—this is ensured by augmenting Box Algebra [4] with the notion of *interface graphs* in a way that allowed importing many results from Box Algebra.

The construction based on finding an *edge clique cover* of a certain *complement-reducible graph (cograph* [15]), where some of the edges may already be considered as 'covered', yields a translation with the number of created places corresponding to the number of cliques in the cover. It is then easy to see that at most polynomial (quadratic) number of cliques are always sufficient (because the relevant cograph has a linear number of vertices), which yields a polynomial Petri net. Hence, the results of [12, 13] demonstrated that if the box expression does not involve cyclic behaviours, the exponential number of places created by the cross-product construction can be reduced from exponential down to polynomial (quadratic) even in the worst case, and to logarithmic in the best (non-degraded) case.

In this paper, we extend the results of [13] to the case of process expressions with iteration. To this end, we introduce a novel notion of *distributed place* which provides a modelling abstraction for the individual places of Petri nets.

The Approach Followed in This Paper. The general problem addressed in this paper can be formulated in the following way:

> *Given a formal specification of a concurrent system expressed in some formalism, find a small Petri net with the same behaviour.*

Two particular instances of such a broad problem concern specifications formulated using Petri nets and process algebras:

> *Given a safe Petri net $\mathcal{N}$, find a small safe Petri net $\mathcal{N}'$ with the same behaviour as $\mathcal{N}$ (e.g. with an isomorphic reachability graph).*   **(Prob I)**

> *Given a process algebra expression, find possibly smallest safe Petri net with the same behaviour.*   **(Prob II)**

Aiming to solve (Prob I), a naive way of finding $\mathcal{N}'$ could be to derive the sequential reachability graph $reach_{\mathcal{N}}$ of $\mathcal{N}$ and then apply Petri net synthesis techniques for $reach_{\mathcal{N}}$ specifically aimed at generating a minimal net solution like those in [1, 2, 18]. Though theoretically sound, such a method is hardly practical as the size of $reach_{\mathcal{N}}$ is too often exponential in the size of $\mathcal{N}$.

With a rather bleak outlook at solving (Prob I) in full generality, one might seek limited but still practically relevant solutions. In particular, one can aim at developing a compositional solution, along the following lines:

> *Given a safe Petri net $\mathcal{N}$ constructed from components $\mathscr{C}_1, \ldots, \mathscr{C}_k$, find possibly smallest behaviourally equivalent safe Petri net $\mathcal{N}'$ constructed from optimised versions of (some of) the components. In addition, all the decisions should be made as 'locally' as posible.*

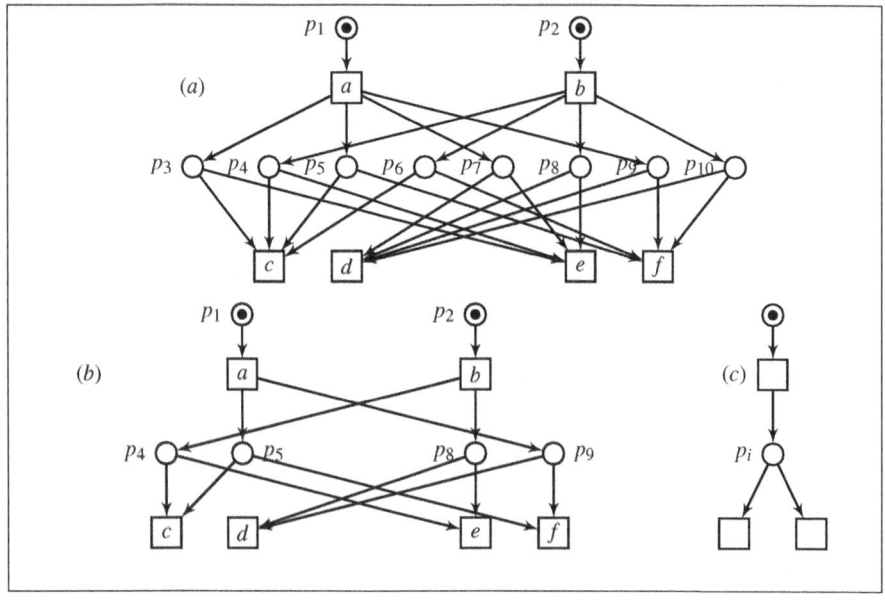

Fig. 2. (*a*) A safe Petri net $\mathcal{N}_0$, (*b*) its reduced behaviourally equivalent net $\mathcal{N}_0'$, and (*c*) a sequential component $\mathcal{S}_i$ of $\mathcal{N}_0$ ($i = 3, 4, \ldots, 10$).

As an example of such an approach consider $\mathcal{N}$ derived from sequential Petri nets $\mathcal{S}_1, \ldots, \mathcal{S}_k$ composed by gluing (synchronising) on common transitions. In such a case, deleting from each $\mathcal{S}_i$ some places leading to $\mathcal{S}_i'$ generating the same firing sequences and/or deleting some $\mathcal{S}_i$'s would lead to $\mathcal{N}'$ with the same behaviour as $\mathcal{N}$.

Consider, for example, the Petri net $\mathcal{N}_0$ shown in Fig. 2(*a*) which specifies a system such that it first concurrently executes transitions a and b, and then either c and d (concurrently) or e and f (also concurrently). $\mathcal{N}_0$ can be decomposed uniquely onto eight sequential subnets $\mathcal{S}_3, \mathcal{S}_4, \ldots, \mathcal{S}_{10}$ as depicted in Fig. 2(*c*). None of these can be reduced in the way described above. However, one can remove four of these sequential nets leaving $\mathcal{S}_4, \mathcal{S}_5, \mathcal{S}_8, \mathcal{S}_9$ in the composition. The result is $\mathcal{N}_0'$ with an isomorphic reachability graph depicted in Fig. 2(*b*). Notice that removing $\mathcal{S}_3, \mathcal{S}_6, \mathcal{S}_7, \mathcal{S}_9$ could not be done locally as one needs to take into account other component sequential subnets.

In this paper, we will show that there is an alternative compositional solution to the problem at hand. To this end, we will consider a covering of the places P of a safe Petri net by *distributed places* $\mathscr{P}_1, \ldots, \mathscr{P}_k$. Each distributed place $\mathscr{P}_i$ is a set of places satisfying properties which depend purely on the arcs between $\mathscr{P}_i$ and the transitions in ${}^\bullet\mathscr{P}_i \cup \mathscr{P}_i^\bullet$. What is important is that other arcs incident to these transitions are irrelevant, and so the definition of a distributed place is purely *local*. We then demonstrate that it is possible to delete from the $\mathscr{P}_i$'s some of the places leading to $\mathscr{P}_1', \ldots, \mathscr{P}_k'$ forming places of a behaviourally equivalent safe Petri net. And, what is of utmost importance, the reduction from $\mathscr{P}_i$ to $\mathscr{P}_i'$ is both *static* and *local*.

As an example, consider again $\mathcal{N}_0$ shown in Fig. 2(a). Its places can be partitioned into two distributed places: $\mathscr{P}_1 = \{p_1, p_2\}$ and $\mathscr{P}_2 = \{p_3, \ldots, p_{10}\}$. In this case, $\mathscr{P}_1$ cannot be reduced and so $\mathscr{P}_1' = \mathscr{P}_1$, but $\mathscr{P}_2$ can be reduced to $\mathscr{P}_2' = \{p_4, p_5, p_8, p_9\}$. The result is a behaviourally equivalent net $\mathcal{N}_0'$ shown in Fig. 2(b) but now the reduction is done locally.

In this paper, we introduce a novel notion of 'distributed place' which intuitively provides an abstraction for the individual places of Petri nets. We also formulate and prove correct the above procedure based on the reduction of distributed places.

Moving on to (Prob II), we will consider boxes (safe Petri nets) derived from process expressions supporting sequential ($E\,;F$), choice ($E\,\square\,F$), parallel ($E\,\|\,F$), and iteration ($[E*F*G]$) compositions. For each such expression E, there is a standard translation yielding a safe Petri net $box(E)$, called a box. For compositionally derived boxes, distributed places emerge 'by construction'. What is more, to reduce $box(E)$ one does not need to explicitly derive $box(E)$ that can be exponential in the size of E. Instead, we construct an undirected 'connection graph' (a cograph [15]) $cg(E)$ which is linear in the size of E, and derive the $\mathscr{P}_i'$'s directly from an edge clique covering of $cg(E)$. Note that that edge clique cover on cographs is in NP and is suspected to be NP-complete, but there are polynomial heuristic algorithms, and even the trivial cover that covers each edge by a separate clique yields at most quadratic solution, which is much better than the exponential solution resulting from the cross-product construction.

Structure of this Paper. This paper has three parts. Following a preliminary Sect. 2, Part I (Sect. 3) introduces distributed places. In particular, Theorem 2 provides a formal justification of the suitability of distributed places as a means of behaviour-preserving reduction of safe Petri nets. Part II (Sects. 4 and 5) first identifies two operations for composing distributed placed, and proves some basic properties of distributed places. It then demonstrates how these operations are used in the standard translation of box expressions to boxes. Part III (Sects. 6 and 7) presents basic facts concerning connection graphs and shows how they can be employed in an optimised translation of box expressions to behaviourally equivalent boxes. The proofs are omitted due to the page limit.

2   Basic Definitions

We start by recalling basic notions about safe Petri nets and then introducing specific notions and notations used throughout the paper.

Nets. A net is a tuple (P, T, Fl), where P is a finite set of $places$, T is a disjoint finite set of $transitions$, $Fl \subseteq T \times P \cup P \times T$ is the $flow\ relation$.

For every $x \in P \cup T$, $^{\bullet}x = \{y \mid yFx\}$, $x^{\bullet} = \{y \mid xFy\}$, and $^{\bullet}x^{\bullet} = {}^{\bullet}x \cup x^{\bullet}$. These notations extend in the usual way to sets of places and transitions. We assume that each place p has at least one $input$ transition ($^{\bullet}p \neq \varnothing$) or at least one $output$ transition, ($p^{\bullet} \neq \varnothing$), and p is not a $side$-$condition$ ($p^{\bullet} \cap {}^{\bullet}p = \varnothing$). Moreover, each transition t has at least one pre-$place$, ($^{\bullet}t \neq \varnothing$).

Marked Nets. A *marked net* is a tuple $\mathcal{N} = (P, T, Fl, M_{init})$ such that (P, T, Fl) is a net from which $\mathcal{N}$ inherits the notions introduced above, and $M_{init} \subseteq P$ is the *initial marking*. In general, any set of places is a *marking*.

Firing Sequences. A transition $t \in T$ is *fireable* at a marking M if $^{\bullet}t \subseteq M$. Such a transition can be *fired* leading to marking $M' = (M \setminus {}^{\bullet}t) \cup t^{\bullet}$. We denote this by $t \in$ *fireable*$_{\mathcal{N}}(M)$ and $M[t\rangle_{\mathcal{N}} M'$, respectively.

A *firing sequence* of $\mathcal{N}$ is a sequence of transitions $\sigma = t_1 \ldots t_k \in T^*$ such that there are markings $M_0, \ldots, M_k$ satisfying $M_{init} = M_0$ and $M_{i-1}[t_i\rangle_{\mathcal{N}} M_i$, for every $1 \leq i \leq k$. Moreover, each marking M_i is *reachable* in $\mathcal{N}$. The sets of all firing sequences and all reachable markings are denoted by *fseq*$(\mathcal{N})$ and *reach*$(\mathcal{N})$, respectively.

A marking M' is *reachable* from marking M if $M[t_1\rangle_{\mathcal{N}} M_1[t_2\rangle_{\mathcal{N}} M_2 \ldots M_{k-1}[t_k\rangle_{\mathcal{N}} M'$, for some $M_1, \ldots, M_{k-1}$ and $t_1, \ldots, t_k$. We also denote $M[t_1 \ldots t_k\rangle_{\mathcal{N}} M'$.

Reachability Graph. The triple *reachgraph*$_{\mathcal{N}} = (reach(\mathcal{N}), A, M_{init})$, where *reach*$(\mathcal{N})$ are the nodes, $A = \{(M, t, M') \mid M \in reach(\mathcal{N}) \wedge M[t\rangle_{\mathcal{N}} M'\}$ are the directed arcs, and M_{init} is the initial node, is called the *reachability graph* of $\mathcal{N}$.

Two reachability graphs, *reachgraph*$_{\mathcal{N}} = (reach(\mathcal{N}), A, M_{init})$ and *reachgraph*$_{\mathcal{N}'} = (reach(\mathcal{N}'), A', M'_{init})$, are *isomorphic* if there is a bijection $\iota :$ *reach*$(\mathcal{N}) \rightarrow reach(\mathcal{N}')$ such that $\iota(M_{init}) = M'_{init}$ and $(M, t, M') \in A \iff (\iota(M), t, \iota(M')) \in A'$, for all $M, M' \in reach(\mathcal{N})$.

Safe Nets. A marked net $\mathcal{N}$ is *safe* if $t^{\bullet} \cap M = \varnothing$, for all reachable markings M and transitions $t \in$ *fireable*$_{\mathcal{N}}(M)$.

Net Size. The total number of tokens in the initial marking cannot exceed the number of places, so one can define $\mathcal{N}$'s size as the total number of places, transitions, and arcs. Hence the size of $\mathcal{N}$ is in practice dominated by its arcs. When translating a process expression to nets, the set of transitions will be given and the objective will be to use small numbers of places and arcs.

Sequences of Transitions. Let σ be a finite sequence of transitions and $V \subseteq T$. Then: (i) *pref*(σ) is the set of all the prefixes of σ; (ii) $\sigma|_V$ is obtained from σ by deleting all the transitions outside V; and (iii) $\sigma \circ \sigma'$ is the concatenation of σ and a sequence of transitions σ'. The empty sequence is denoted by λ.

3   Distributed Places

We now introduce a concept which is aimed at capturing a simple structural notion of a 'distributed place' (set of places) whose state (marking)—as the execution progresses— monotonically changes between being empty and being full. For example, $\mathcal{P} = \{p_3, p_4, \ldots, p_{10}\}$ is a distributed place of the net in Fig. 2(a). Initially $\mathcal{P}$ is empty, and firing a and b inserts tokens into its places. Note that while being filled, no transition

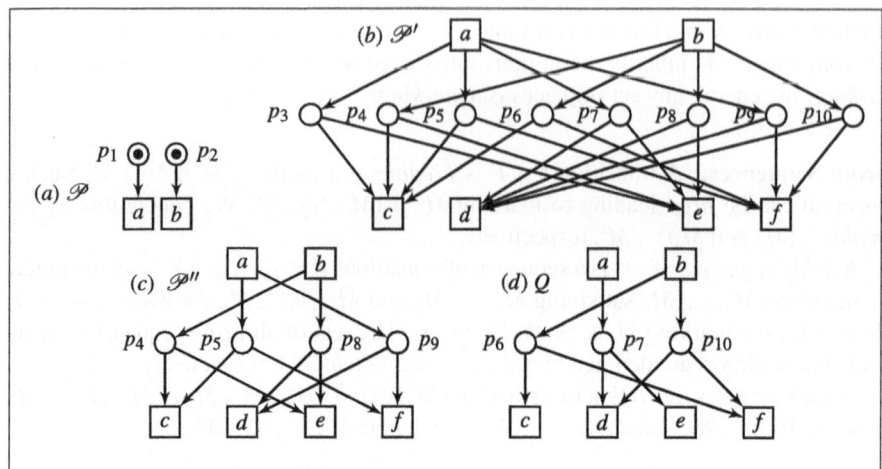

Fig. 3. (a, b, c) Three distributed places with surrounding transitions for the marked nets in Fig. 2(a, b), and (d) a set of places $Q = \{p_6, p_7, p_{10}\}$ with surrounding transitions which is not a distributed place for the marked net $\mathcal{N}_0$ in Fig. 2(a).

can remove tokens from $\mathcal{P}$. Then, *e.g.* transitions c and d can fire and remove all the tokens in $\mathcal{P}$ and, while being emptied, no transition can insert tokens into $\mathcal{P}$.

We start by specifying what is meant by a monotonic change of the marking of a set of places Q. Note that, in general, firing a single transition is not enough to fill with tokens an empty Q. Rather, we may need a sequence of transitions to achieve the desired effect. Similarly, firing a single transition is in general not enough to empty Q filled with tokens.

Definition 1 (in-sequence and out-sequence). *Let Q be a nonempty set of places of a net such that $Q^\bullet \cap {}^\bullet Q = \varnothing$.*

- *An* in-sequence *of Q is a nonempty sequence $\sigma = t_1 \ldots t_k$ of transitions in ${}^\bullet Q$ such that $Q \cap t_i^\bullet \cap \{t_1, \ldots, t_{i-1}\}^\bullet = \varnothing$ $(1 < i \leq k)$. Also, σ is* complete *if $Q \subseteq \{t_1, \ldots, t_k\}^\bullet$.*
- *An* out-sequence *of Q is a nonempty sequence $\sigma = t_1 \ldots t_k$ of transitions in $Q^\bullet$ such that $Q \cap {}^\bullet t_i \subseteq Q \setminus {}^\bullet \{t_1, \ldots, t_{i-1}\}$ $(1 < i \leq k)$. Also, σ is* complete *if $Q \subseteq {}^\bullet \{t_1, \ldots, t_k\}$.*

All in- and out-sequences are *inseq$_Q$* and *outseq$_Q$*, respectively. Moreover, all complete in- and out-sequences are *cinseq$_Q$* and *coutseq$_Q$*, respectively.

Intuitively, an in-sequence σ is a possible way of monotonically (due to $Q^\bullet \cap {}^\bullet Q = \varnothing$) and safely (due to $Q \cap t_i^\bullet \cap \{t_1, \ldots, t_{i-1}\}^\bullet = \varnothing$) inserting tokens into an empty Q, and if this results in full Q then σ it is complete. Similarly, an out-sequence σ is a possible way of monotonically removing tokens from a fully marked Q, and if this results in empty Q then σ is a complete. It is important to emphasize that the in- and out-sequences are defined statically and are not necessarily executable when Q is part of a larger net.

Definition 2 (distributed place). *A* distributed place *of a net is a nonempty set of places $\mathscr{P}$ such that:*

1. $\mathscr{P}^{\bullet} \cap {}^{\bullet}\mathscr{P} = \varnothing$.
2. $t^{\bullet} \cap {}^{\bullet}u \cap \mathscr{P} \neq \varnothing$, *for all* $t \in {}^{\bullet}\mathscr{P}$ *and* $u \in \mathscr{P}^{\bullet}$.
3. *inseq$_{\mathscr{P}}$ = pref(cinseq$_{\mathscr{P}}$) \ $\{\lambda\}$ and outseq$_{\mathscr{P}}$ = pref(coutseq$_{\mathscr{P}}$) \ $\{\lambda\}$.*

Note that the above definition is local as it only depends on the arcs with incident to places in the distributed place.

Distributed places of a marked net are the distributed places of its underlying net. Moreover, two distributed places, $\mathscr{P}$ and $\mathscr{P}'$, are *separated* if ${}^{\bullet}\mathscr{P}^{\bullet} \cap {}^{\bullet}(\mathscr{P}')^{\bullet} = \varnothing$.

By Definition 2(1), there can be two kinds of transitions adjacent to a distributed place $\mathscr{P}$: (i) transitions in ${}^{\bullet}\mathscr{P}$ inserting tokens into $\mathscr{P}$ but not removing any; and (ii) transitions in $\mathscr{P}^{\bullet}$ removing tokens from $\mathscr{P}$ but not inserting any. (Intuitively, Q is not a 'distributed side-condition' to any transition.)

The following two crucial properties of a distributed place $\mathscr{P}$ manifest themselves when $\mathscr{P}$ is a part of a marked net $\mathcal{N}$: (i) if $\mathscr{P}$ is empty at some reachable marking, then it is not possible to fire a transition $u \in \mathscr{P}^{\bullet}$ after an incomplete (projected) in-sequence σ; and (ii) if $\mathscr{P}$ is full at some reachable marking, then it is not possible to fire a transition $u \in {}^{\bullet}\mathscr{P}$ after an incomplete (projected) out-sequence σ.

Indeed, suppose that $u \in \mathscr{P}^{\bullet}$ and $\sigma \circ \omega$ is a firing sequence of $\mathcal{N}$ such that: (i) σ leads to a marking at which $\mathscr{P}$ is empty; and (ii) $\omega|_{{}^{\bullet}\mathscr{P}^{\bullet}} = t_1 \ldots t_k$ is an incomplete in-sequence of $\mathscr{P}$. Then, by Definition 2(3), there is $t \in {}^{\bullet}\mathscr{P}$ such that $t_1 \ldots t_k t$ is an in-sequence of $\mathscr{P}$. Hence, by Definition 2(2), there is $q \in t^{\bullet} \cap {}^{\bullet}u \cap \mathscr{P}$. Moreover, since $\mathscr{P} \cap t^{\bullet} \cap \{t_1, \ldots, t_k\}^{\bullet} = \varnothing$, we have $q \notin \{t_1, \ldots, t_k\}^{\bullet}$, and so u is not fireable in $\mathcal{N}$ after $\sigma \circ \omega$.

Example 1. Figure 3(a,b) depicts two distributed places together with the surrounding transitions for the safe marked net $\mathcal{N}_0$ shown in Fig. 2(a): $\mathscr{P} = \{p_1, p_2\}$ and $\mathscr{P}' = \{p_3, p_4, p_5, p_6, p_7, p_8, p_9, p_{10}\}$. Also, their complete in- and out-sequences are as follows:

$$cinseq_{\mathscr{P}} = \varnothing \qquad coutseq_{\mathscr{P}} = \{ab, ba\}$$
$$cinseq_{\mathscr{P}'} = \{ab, ba\} \qquad coutseq_{\mathscr{P}'} = \{cd, dc, ef, fe\}.$$

We then observe that $\mathscr{P}'' = \{p_4, p_5, p_8, p_9\}$ in Fig. 2(c) is a distributed place of $\mathcal{N}_0$ with the same (complete) in- and out-sequences as $\mathscr{P}'$. We also observe that the set of places $Q = \{p_6, p_7, p_{10}\}$ (shown in Fig. 3(d)) is not a distributed place as $a^{\bullet} \cap {}^{\bullet}c \cap Q = \varnothing$. ◇

It is easy to observe that all permutations of in-sequences (out-sequences) are in-sequences (out-sequences). Such an observation leads to a static characterisations of distributed places. A result of this kind could help in finding distributed places in marked nets before applying Theorem 2 which underpins the proposed net reduction.

Proposition 1. *A nonempty set of places $\mathscr{P}$ of a net is a distributed place iff we have:*

1. $\mathscr{P}^{\bullet} \cap {}^{\bullet}\mathscr{P} = \varnothing$.

2. $t^\bullet \cap {}^\bullet u \cap \mathscr{P} \neq \varnothing$, for all $t \in {}^\bullet \mathscr{P}$ and $u \in \mathscr{P}^\bullet$.
3. If $U \subseteq {}^\bullet \mathscr{P}$ is a maximal nonempty set of transition such that $t^\bullet \cap u^\bullet \cap \mathscr{P} = \varnothing$, for all $t \neq u \in U$, then $\mathscr{P} \subseteq U^\bullet$.
4. If $U \subseteq \mathscr{P}^\bullet$ is a maximal nonempty set of transitions such that ${}^\bullet t \cap {}^\bullet u \cap \mathscr{P} = \varnothing$, for all $t \neq u \in U$, then $\mathscr{P} \subseteq {}^\bullet U$.

We also have the following immediate result as it was assumed that there are no side-conditions.

Proposition 2. $\mathscr{S}_p = \{p\}$ is a distributed place, for every place p of a marked net.

Hence, the places P of a marked net $\mathscr{N}$ have a cover by trivial singleton distributed places $\mathscr{S}_p$. And an immediate corollary is that having a covering of P by distributed places does not guarantee the safeness of $\mathscr{N}$ (in contrast, having a cover by sequential components implies that $\mathscr{N}$ is safe).

Intuitively, a distributed place is an implementation of a local state (variable) which can be cyclically and monotonically filled and emptied. This is evident in the next result which asserts that the behaviour of each distributed place of safe marked net can be understood as execution of alternating in-sequences and out-sequences, provided that the initial marking of the distributed place is set to 'empty' or to 'filled'.

Theorem 1. Let $\mathscr{P}$ be a distributed place of a safe $\mathscr{N} = (P, T, Fl, M_{init})$. Then:

$$fseq(\mathscr{N})|_{{}^\bullet \mathscr{P}^\bullet} \subseteq \begin{cases} pref((cinseq_{\mathscr{P}} \circ coutseq_{\mathscr{P}})^+) & \text{if } \mathscr{P} \cap M_{init} = \varnothing \\ pref((coutseq_{\mathscr{P}} \circ cinseq_{\mathscr{P}})^+) & \text{if } \mathscr{P} \subseteq M_{init}. \end{cases}$$

The next result captures an essential property of distributed places, namely that in safe marked nets with places covered by distributed places, one can apply reductions at local level (*i.e.* within individual distributed places) without changing the semantics.

Theorem 2. Let $\mathscr{N} = (P, T, Fl, M_{init})$ be a safe marked net and, for every place $p \in P$, let $\mathscr{P}_p$ be a distributed place in $\mathscr{N}$ comprising p such that $\mathscr{P}_p \cap M_{init} = \varnothing$ or $\mathscr{P}_p \subseteq M_{init}$. Moreover, let $P' \subseteq P$ be a set of places such that

$$\mathscr{N}' = (P', T, Fl', M'_{init}) = (P', T, Fl|_{T \times P' \cup P' \times T}, M_{init} \cap P')$$

is a marked net satisfying, for all $p \in P$, $t \neq u \in {}^\bullet \mathscr{P}_p^\bullet$ and $\mathscr{P}'_p = \mathscr{P}_p \cap P'$:

$${}^\bullet(\mathscr{P}'_p)^\bullet = {}^\bullet(\mathscr{P}_p)^\bullet \tag{1a}$$

$$t \in {}^\bullet \mathscr{P}_p \wedge u \in \mathscr{P}_p^\bullet \implies t^\bullet \cap {}^\bullet u \cap \mathscr{P}'_p \neq \varnothing \tag{1b}$$

$${}^\bullet t \cap {}^\bullet u \cap \mathscr{P}_p \neq \varnothing \implies {}^\bullet t \cap {}^\bullet u \cap \mathscr{P}'_p \neq \varnothing. \tag{1c}$$

Then the following hold:

1. $fseq(\mathscr{N}') = fseq(\mathscr{N})$.
2. $\mathscr{N}'$ is a safe marked net.

3. $reachgraph_{\mathcal{N}} \cong reachgraph_{\mathcal{N}'}$ *provided that, for each* $p \in P$ *and each reachable marking* M *of* $\mathcal{N}$, *there exists a marking* M' *reachable from* M *such that* $\mathscr{P}_p \cap M' = \varnothing$ *or* $\mathscr{P}_p \subseteq M'$.

The above result provides a static condition for a successful reduction of distributed places in safe marked nets. Also, as shown below, none of the key assumptions in Theorem 2 can be dropped.

Example 2. We first note that in the formulation of Theorem 2 one cannot remove the assumption that $\mathcal{N}$ is safe. Indeed, otherwise we could take any non-safe $\mathcal{N}$ together with the trivial singleton distributed places and wrongly conclude that $\mathcal{N}' = \mathcal{N}$ is safe.

We cannot drop Eq. (1b) either. Indeed, consider $\mathcal{N}_1$ in Fig. 4(a) with three distributed places: $\mathscr{P}_1 = \{q_1, q_2\}$, $\mathscr{P}_2 = \{q_3, q_4, q_5, q_6\}$, and $\mathscr{P}_3 = \{q_7, q_8\}$. We then can take $P' = \{q_1, q_2, q_3, q_6, q_7, q_8\}$, and the resulting reduced net $\mathcal{N}_1'$ in Fig. 4(b) generates more firing sequences than $\mathcal{N}_1$ (*e.g. acbd*) even though assumptions other than Eq. (1b) are satisfied (in this case $\mathscr{P}_2 \cap P' = \{q_3, q_6\}$ and Eq. (1b) does not hold for $t = b$ and $u = d$).

Finally, one cannot drop Eq. (1c). Indeed, consider $\mathcal{N}_2$ in Fig. 4(c) with two distributed places: $\mathscr{P}_1 = \{r_1, r_2, r_3, r_4\}$ and $\mathscr{P}_2 = \{r_5, r_6, r_7, r_8\}$. We can take $P' = \{r_1, r_4, r_5, r_6, r_7, r_8\}$, and the resulting $\mathcal{N}_2'$ in Fig. 4(d) generates more firing sequences than $\mathcal{N}_2$ (*e.g. ad*) even though assumptions other than Eq. (1b) are satisfied (in this case $\mathscr{P}_1 \cap P' = \{r_1, r_4\}$ and Eq. (1c) does not hold for $t = a$ and $u = d$).   ◇

In view of Proposition 2, one could always choose the covering by singleton distributed places $\mathscr{S}_p$. But then, due to Eq. (1a), there would be no reduction at all. Having said that, the choice of suitable covering by distributed places and P' is not unique, leaving a scope for possibly substantial improvement.

It is also worth noting that to apply Theorem 2 in cases when only a partial cover by distributed places is provided, one can always complete the partial cover with trivial singleton distributed places.

4   Composing Distributed Places

It is possible to derive distributed places in a systematic (compositional) way. Since any marked net $\mathcal{N} = (P, T, Fl, M_{init})$ dealt with in the rest of this paper is such that no two different places have the same input transitions and the same output transitions (*i.e.* $^\bullet p = {}^\bullet r \wedge p^\bullet = r^\bullet$ implies $p = r$), we will adopt a convention that a place p is identified by the sets $^\bullet p$ and $p^\bullet$, and so p will be denoted as π_Z, where Z is a set of indexed input and output transitions of p: $Z = \{t^{out} \mid (p,t) \in Fl\} \cup \{t^{in} \mid (t,p) \in Fl\}$.

With such a notation, transitions and arcs of $\mathcal{N}$ become implicit, as we have:

$$T = \{t \mid \exists \pi_Z \in P: t^{in} \in Z \vee t^{out} \in Z\}$$
$$Fl = \{(t, \pi_Z) \in T \times P \mid t^{in} \in Z\} \cup \{(\pi_Z, t) \in T \times P \mid t^{out} \in Z\}. \tag{2}$$

Hence the whole static graph structure of $\mathcal{N}$ can be represented by its set of places. We also denote $\mathscr{I}_t = \mathscr{S}_{\pi_{\{t^{in}\}}} = \{\pi_{\{t^{in}\}}\}$ and $\mathscr{O}_t = \mathscr{S}_{\pi_{\{t^{out}\}}} = \{\pi_{\{t^{out}\}}\}$, for every transition t. By Proposition 2, $\mathscr{I}_t$ and $\mathscr{O}_t$ are singleton distributed places.

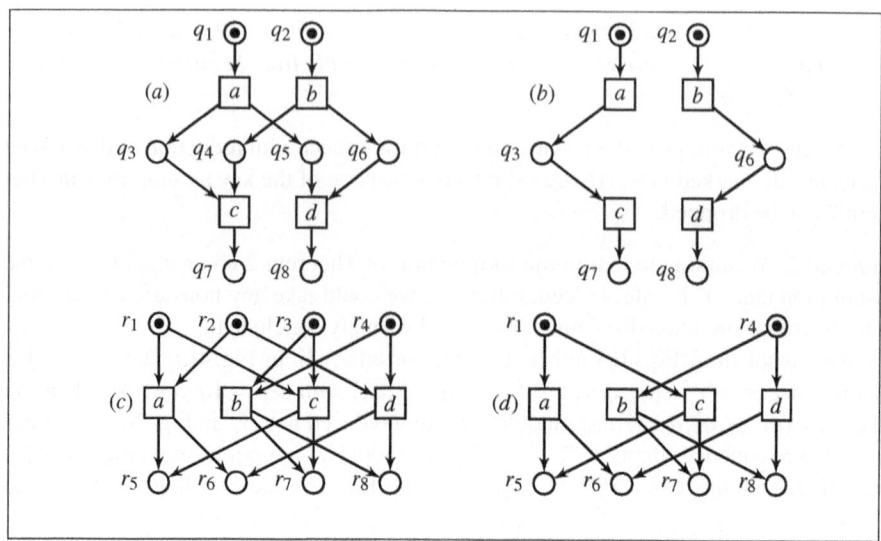

Fig. 4. (a) A net $\mathscr{N}_1 = box((a \parallel b);(c \parallel d))$, and (b) its unsuccessful reduction $\mathscr{N}_1'$. (c) A net $\mathscr{N}_2 = box((a \parallel b) \square (c \parallel d))$, and (d) its unsuccessful reduction $\mathscr{N}_2'$.

What we just introduced is a mere notational convenience allowing simple additions of new and removal of existing places without having to modify the flow relation explicitly. As a result, the resulting formulae are more concise.

Example 3. Consider $\mathscr{N}_0$ in Fig. 2(a). By the above convention, $p_1 = \pi_{a^{out}}$ and $p_3 = \pi_{a^{in}c^{out}e^{out}}$. And the transitions and arcs of $\mathscr{N}_0$ can be represented by:

$$\pi_{a^{out}} \quad \pi_{a^{in}c^{out}e^{out}} \quad \pi_{b^{in}c^{out}e^{out}} \quad \pi_{a^{in}c^{out}f^{out}} \quad \pi_{b^{in}c^{out}f^{out}}$$
$$\pi_{b^{out}} \quad \pi_{a^{in}d^{out}e^{out}} \quad \pi_{b^{in}d^{out}e^{out}} \quad \pi_{a^{in}d^{out}f^{out}} \quad \pi_{b^{in}d^{out}f^{out}}$$

For example, $a \in {}^\bullet p_3$ since $p_3 = \pi_Z$ and $a^{in} \in Z$.   ◇

When composing two distributed places, $\mathscr{P}$ and $\mathscr{P}'$, we will assume that they are *separated* which means that they are connected to disjoint sets of transitions, *i.e.* ${}^\bullet\mathscr{P}^\bullet \cap {}^\bullet(\mathscr{P}')^\bullet = \varnothing$.

The first way of constructing new distributed places is to apply set union.

Proposition 3. *Let $\mathscr{P}$ and $\mathscr{P}'$ be two separated distributed places such that ${}^\bullet\mathscr{P} = {}^\bullet(\mathscr{P}') = \varnothing$ or $\mathscr{P}^\bullet = (\mathscr{P}')^\bullet = \varnothing$. Then $\mathscr{P} \oplus \mathscr{P}' = \mathscr{P} \cup \mathscr{P}'$ is a distributed place.*

In some frameworks aimed at supporting step-wise construction (or hierarchical analysis) of distributed systems, a central role is played by some notion of synchronisation between components exhibiting concurrency. In terms of net construction, a suitable synchronisation between different process threads can be achieved through the following cross-product of sets of places.

Definition 3 (cross-product). *The* cross-product *of two nonempty sets of places,* $\mathscr{P}$ *and* $\mathscr{P}'$, *satisfying* $^{\bullet}\mathscr{P}^{\bullet} \cap {}^{\bullet}(\mathscr{P}')^{\bullet} = \varnothing$ *is the following set of places:* $\mathscr{P} \otimes \mathscr{P}' = \{\pi_{Z \cup Z'} \mid \pi_Z \in \mathscr{P} \wedge \pi_{Z'} \in \mathscr{P}'\}$.

Example 4. Consider the distributed place $\mathscr{P}'$ in Fig. 3(*b*). We observe that

$$
\begin{aligned}
\mathscr{P}' &= \{\pi_{a^{in}c^{out}e^{out}}, \pi_{b^{in}c^{out}e^{out}}, \pi_{a^{in}c^{out}f^{out}}, \pi_{b^{in}c^{out}f^{out}}, \\
&\qquad \pi_{a^{in}d^{out}e^{out}}, \pi_{b^{in}d^{out}e^{out}}, \pi_{a^{in}d^{out}f^{out}}, \pi_{b^{in}d^{out}f^{out}}\} \\
&= \{\pi_{a^{in}}, \pi_{b^{in}}\} \otimes \{\pi_{c^{out}e^{out}}, \pi_{d^{out}e^{out}}, \pi_{c^{out}f^{out}}, \pi_{d^{out}f^{out}}\} \\
&= \{\pi_{a^{in}}, \pi_{b^{in}}\} \otimes (\{\pi_{c^{out}}, \pi_{d^{out}}\} \otimes \{\pi_{e^{out}}, \pi_{f^{out}}\}) \\
&= (\mathscr{I}_a \oplus \mathscr{I}_b) \otimes ((\mathscr{O}_c \oplus \mathscr{O}_d) \otimes (\mathscr{O}_e \oplus \mathscr{O}_f)) \, .
\end{aligned}
$$

Similar decompositions exist for the distributed place $\mathscr{P}_1 = \{q_3, q_4, q_5, q_6\}$ of $\mathscr{N}_1$ in Fig. 4(*a*), and the distributed place $\mathscr{P}_2 = \{r_1, r_2, r_3, r_4\}$ of $\mathscr{N}_2$ in Fig. 4(*c*):

$$
\begin{aligned}
\mathscr{P}_1 &= \{\pi_{a^{in}c^{out}}, \pi_{b^{in}c^{out}}, \pi_{a^{in}d^{out}}, \pi_{b^{in}d^{out}}\} = \{\pi_{a^{in}}, \pi_{b^{in}}\} \otimes \{\pi_{c^{out}}, \pi_{d^{out}}\} \\
&\qquad\qquad\qquad\qquad\qquad\qquad\qquad\qquad\quad = (\mathscr{I}_a \oplus \mathscr{I}_b) \otimes (\mathscr{O}_c \oplus \mathscr{O}_d) \\
\mathscr{P}_2 &= \{\pi_{a^{in}c^{in}}, \pi_{b^{in}c^{in}}, \pi_{a^{in}d^{in}}, \pi_{b^{in}d^{in}}\} = \{\pi_{a^{in}}, \pi_{b^{in}}\} \otimes \{\pi_{c^{in}}, \pi_{d^{in}}\} \\
&\qquad\qquad\qquad\qquad\qquad\qquad\qquad\qquad\quad = (\mathscr{I}_a \oplus \mathscr{I}_b) \otimes (\mathscr{I}_c \oplus \mathscr{I}_d) \, .
\end{aligned}
$$

Cross-product can be used to construct new distributed places.

Proposition 4. *If* $\mathscr{P}$ *and* $\mathscr{P}'$ *are separated distributed places then* $\mathscr{P}'' = \mathscr{P} \otimes \mathscr{P}'$ *is also a distributed place such that*

$$
\begin{array}{ll}
inseq_{\mathscr{P}''} = inseq_{\mathscr{P}} \cup inseq_{\mathscr{P}'} & cinseq_{\mathscr{P}''} = cinseq_{\mathscr{P}} \cup cinseq_{\mathscr{P}'} \\
outseq_{\mathscr{P}''} = outseq_{\mathscr{P}} \cup outseq_{\mathscr{P}'} & coutseq_{\mathscr{P}''} = coutseq_{\mathscr{P}} \cup coutseq_{\mathscr{P}'} \, .
\end{array}
$$

Intuitively, the pair of distributed places in Proposition 4 is such that a process thread entering $\mathscr{P} \otimes \mathscr{P}'$ through an in-sequence of $\mathscr{P}$ or an in-sequence of $\mathscr{P}'$ can leave $\mathscr{P} \otimes \mathscr{P}'$ either by an out-sequence of $\mathscr{P}$ or an out-sequence of $\mathscr{P}'$.

Example 5. Coming back to Example 4, the decompositions of sets of places $\mathscr{P}'$, $\mathscr{P}_1$, and $\mathscr{P}_2$, together with Propositions 2, 3, and 4, imply that these three sets of places are in fact distributed places. ◇

5 Box Algebra and Distributed Places

We consider a fragment of Box Algebra [4] focusing on nets derived from four control flow operators. Omitting communication and synchronisation operators does not diminish the importance of the results obtained below as such operators generate new transitions rather than places.

In this paper, process expressions (or *box expressions*), are derived as follows:

$$
E ::= a \mid E; E \mid E \square E \mid E \| E \mid [E * E * E] \tag{3}
$$

where a is an atomic action. Intuitively, a denotes a process which can execute atomic action a and terminate, $E; F$ denotes sequential composition of two processes, $E \square F$

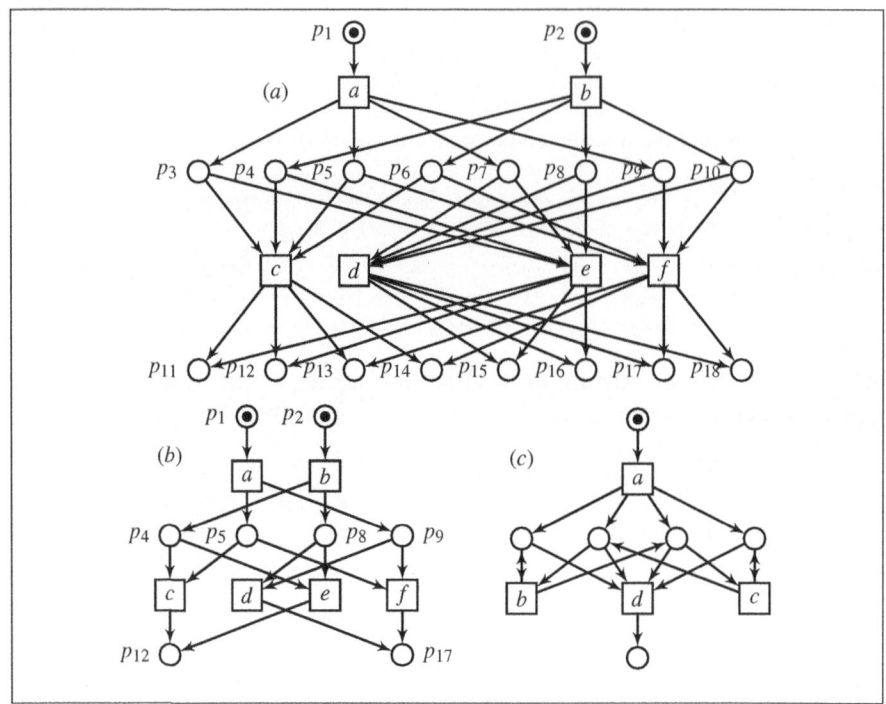

Fig. 5. (a) $\mathcal{N}_3 = box((a \| b); ((c \| d) \,\square\, (e \| f)))$ and (b) its reduced version $\mathcal{N}_3'$, and (c) a non-safe net corresponding to the box expression $[a * (b \| c) * d]$.

denotes choice composition, $E \| F$ denotes parallel composition, and $[E * F * G]$ denotes iteration which starts by executing E, then executes F any number of times (including zero), and terminates the whole iteration by executing G.

Since we deal only with control flow and actions do not have any special semantics, we assume that no action occurs more than once in a box expression.

The syntax in Eq. (3) admits box expressions which would result in non-safe nets (e.g. the box expression $[a * (b \| c) * d]$ corresponds to the non-safe net shown in Fig. 5(c)). However, there is a simple way of avoiding this by modifying the syntax:

$$
\begin{aligned}
E &::= a \mid E \| E \mid E ; E \mid E \,\square\, E \mid [E * H * E] \\
H &::= \qquad\quad E ; E \mid H \,\square\, H \mid [E * H * E]
\end{aligned}
\tag{4}
$$

The standard translation from box expressions to boxes adapted for the notation used in this paper is carried out in two steps. The translation first associates with each box expression E a triple of places, $\Pi_E = (\Pi_E^e, \Pi_E^x, \Pi_E^i)$. Then the marked net associated with E is given by:

$$
box(E) = (\Pi_E^e \cup \Pi_E^x \cup \Pi_E^i, T, Fl, \Pi_E^e),
$$

where T are the actions occurring in E, and Fl are the arcs derived as in Eq. (2). The set Π_E^e comprises the *entry places* (initially marked), Π_E^x comprises the *exit places* (initially empty), and the Π_E^i comprises the *internal places* (initially empty). The entry and exit places play an active part in the compositional derivation of boxes, and the internal places are carried over without any changes to the resulting boxes.

The syntax-driven derivation of Π_E proceeds as follows:

$$
\begin{aligned}
\pi_a &= (\mathcal{O}_a, \mathcal{I}_a, \varnothing) \\
\pi_{E\parallel F} &= (\Pi_E^e \cup \Pi_F^e, \Pi_E^x \cup \Pi_F^x, \Pi_E^i \cup \Pi_F^i) \\
\pi_{E;F} &= (\Pi_E^e, \Pi_F^x, \Pi_E^i \cup \Pi_F^i \cup (\Pi_E^x \otimes \Pi_F^e)) \\
\pi_{E\square F} &= (\Pi_E^e \otimes \Pi_F^e, \Pi_E^x \otimes \Pi_F^x, \Pi_E^i \cup \Pi_F^i) \\
\Pi_{[E*F*G]} &= (\Pi_E^e, \Pi_G^x, \Pi_E^i \cup \Pi_F^i \cup \Pi_G^i \cup ((\Pi_E^x \otimes \Pi_F^x) \otimes (\Pi_F^e \otimes \Pi_G^e)))
\end{aligned}
\tag{5}
$$

Example 6. For the box expression $(a \parallel b);(c \parallel d)$ the translation proceeds as follows:

$$
\begin{aligned}
\Pi_{(a\parallel b);(c\parallel d)} &= (\Pi_{a\parallel b}^e, \Pi_{c\parallel d}^x, \pi_{a\parallel b}^i \cup \pi_{c\parallel d}^i \cup (\Pi_{a\parallel b}^x \otimes \Pi_{c\parallel d}^e)) \\
&= (\Pi_a^e \cup \Pi_b^e, \Pi_c^x \cup \Pi_d^x, \Pi_a^i \cup \Pi_b^i \cup \Pi_c^i \cup \Pi_d^i \cup ((\Pi_a^x \cup \Pi_b^x) \otimes (\Pi_c^e \cup \Pi_d^e))) \\
&= (\mathcal{O}_a \cup \mathcal{O}_b, \mathcal{I}_c \cup \mathcal{I}_d, \varnothing \cup \varnothing \cup \varnothing \cup \varnothing \cup ((\mathcal{I}_a \cup \mathcal{I}_b) \otimes (\mathcal{O}_c \cup \mathcal{O}_d))) \\
&= (\{\pi_{a^{out}}, \pi_{b^{out}}\}, \{\pi_{c^{in}}, \pi_{d^{in}}\}, \{\pi_{a^{in}c^{out}}, \pi_{b^{in}c^{out}}, \pi_{a^{in}d^{out}}, \pi_{b^{in}d^{out}}\}) \, .
\end{aligned}
$$

Hence the places of $box((a\parallel b);(c\parallel d))$ are as follows: $\pi_{a^{out}}$, $\pi_{b^{out}}$, $\pi_{c^{in}}$, $\pi_{d^{in}}$, $\pi_{a^{in}c^{out}}$, $\pi_{b^{in}c^{out}}$, $\pi_{a^{in}d^{out}}$, and $\pi_{b^{in}d^{out}}$. Moreover, the initial marking is $M_{init} = \{\pi_{a^{out}}, \pi_{b^{out}}\}$. As a result, $box((a\parallel b);(c\parallel d))$ is the marked net $\mathcal{N}_1$ of Fig. 4(a).

For the box expression $(a\parallel b)\square(c\parallel d)$ the translation proceeds as follows:

$$
\begin{aligned}
\Pi_{(a\parallel b)\square(c\parallel d)} &= (\Pi_{a\parallel b}^e \otimes \pi_{c\parallel d}^e, \Pi_{a\parallel b}^x \otimes \Pi_{c\parallel d}^x, \pi_{a\parallel b}^i \cup \pi_{c\parallel d}^i) \\
&= ((\mathcal{O}_a \oplus \mathcal{O}_b) \otimes (\mathcal{O}_c \oplus \mathcal{O}_d), (\mathcal{I}_a \oplus \mathcal{I}_b) \otimes (\mathcal{I}_c \oplus \mathcal{I}_d), \varnothing) \\
&= (\{\pi_{a^{out}c^{out}}, \pi_{a^{out}d^{out}}, \pi_{b^{out}c^{out}}, \pi_{b^{out}d^{out}}\}, \{\pi_{a^{in}c^{in}}, \pi_{a^{in}d^{in}}, \pi_{b^{in}c^{in}}, \pi_{b^{in}d^{in}}\}, \varnothing) \, .
\end{aligned}
$$

Hence the places of $box((a\parallel b)\square(c\parallel d))$ are as follows: $\pi_{a^{out}c^{out}}$, $\pi_{a^{out}d^{out}}$, $\pi_{b^{out}c^{out}}$, $\pi_{b^{out}d^{out}}$, $\pi_{a^{in}c^{in}}$, $\pi_{a^{in}d^{in}}$, $\pi_{b^{in}c^{in}}$, and $\pi_{b^{in}d^{in}}$. Moreover, the initial marking is given by $M_{init} = \{\pi_{a^{out}c^{out}}, \pi_{a^{out}d^{out}}, \pi_{b^{out}c^{out}}, \pi_{b^{out}d^{out}}\}$. As a result, $box((a\parallel b)\square(c\parallel d))$ is the marked net $\mathcal{N}_2$ of Fig. 4(c).

Two more examples of the translation are shown in Figs. 1(a) and 5(a).   ◇

The translation from a box expression E to box $box(E)$ generates a marked net where one can identify distributed places $dp(E)$ on which the reduction procedure captured in Theorem 2 can be based. To this end, we define a triple of sets of distributed places $\Delta_E = (\Delta_E^e, \Delta_E^x, \Delta_E^i)$ such that Δ_E^e is a cover of the entry places, Δ_E^x is a cover of the exit places, and Δ_E^i is a cover of the internal places.

The syntax-driven derivation of Δ_E proceeds as follows:

$$
\begin{aligned}
\Delta_a &= (\{\mathcal{O}_a\}, \{\mathcal{I}_a\}, \varnothing) \\
\Delta_{E\parallel F} &= (\{\Pi_E^e \cup \Pi_F^e\}, \{\Pi_E^x \cup \Pi_F^x\}, \Delta_E^i \cup \Delta_F^i) \\
\Delta_{E;F} &= (\Delta_E^e, \Delta_F^x, \Delta_E^i \cup \Delta_F^i \cup \{\Pi_E^x \otimes \Pi_F^e\}) \\
\Delta_{E\square F} &= (\{\Pi_E^e \otimes \Pi_F^e\}, \{\Pi_E^x \otimes \Pi_F^x\}, \Delta_E^i \cup \Delta_F^i) \\
\Delta_{[E*F*G]} &= (\Delta_E^e, \Delta_G^x, \Delta_E^i \cup \Delta_F^i \cup \Delta_G^i \cup \{(\Pi_E^x \otimes \Pi_F^x) \otimes (\Pi_F^e \otimes \Pi_G^e)\}) \, .
\end{aligned}
\tag{6}
$$

We then define the *distributed places* of E as $dp(E) = \Delta_E^e \cup \Delta_E^x \cup \Delta_E^i$.

Example 7. Referring to Fig. 4(a, c), we have

$$dp((a \parallel b);(c \parallel d)) = \{\{q_1, q_2\}, \{q_3, q_4, q_5, q_6\}, \{q_7, q_8\}\}$$
$$= \{\{\pi_{a^{out}}, \pi_{b^{out}}\}, \{\pi_{a^{in}c^{out}}, \pi_{b^{in}c^{out}}, \pi_{a^{in}d^{out}}, \pi_{b^{in}d^{out}}\}, \{\pi_{c^{in}}, \pi_{d^{in}}\}\}$$
$$dp((a \parallel b) \square (c \parallel d)) = \{\{r_1, r_2, r_3, r_4\}, \{r_5, r_6, r_7, r_8\}\}$$
$$= \{\{\pi_{a^{out}c^{out}}, \pi_{a^{out}d^{out}}, \pi_{b^{out}c^{out}}, \pi_{b^{out}d^{out}}\},$$
$$\{\pi_{a^{in}c^{in}}, \pi_{a^{in}d^{in}}, \pi_{b^{in}c^{in}}, \pi_{b^{in}d^{in}}\}\}.$$

Moreover, for the marked net in Fig. 5(a), we have:

$$dp((a \parallel b);((c \parallel d) \square (e \parallel f))) = \{\{p_1, p_2\}, \{p_3, \ldots, p_{10}\}, \{p_{12}, \ldots, p_{18}\}\}.$$

We then obtain some basic properties of boxes and their distributed places.

Proposition 5. *Let E be a box expression defined by the syntax in Eq. (4), $box(E) = (P, T, Fl, M_{init})$, and M be a reachable marking of $box(E)$.*

1. *$box(E)$ is a safe marked net.*
2. *$M_{init} = \Pi_E^e = \{p \in P \mid {}^\bullet p = \varnothing\}$ and $M_{fin} = \Pi_E^x = \{p \in P \mid p^\bullet = \varnothing\}$.*
3. *M_{fin} can be reached from any reachable marking of $box(E)$.*
4. *$M_{init} \subseteq M$ implies $M = M_{init}$, and $M_{fin} \subseteq M$ implies $M = M_{fin}$.*
5. *$\Delta_E^e = \{\Pi_E^e\}$ and $\Delta_E^x = \{\Pi_E^x\}$.*
6. *$dp(E)$ is a set of mutually disjoint distributed places covering the places of $box(E)$.*
7. *M_{init} and M_{fin} are separated distributed places for E generated by the second line of the syntax in Eq. (4).*
8. *If $\mathscr{P} \neq \mathscr{P}' \in cg(E)$ then ${}^\bullet \mathscr{P} \cap {}^\bullet(\mathscr{P}') = \mathscr{P}^\bullet \cap (\mathscr{P}')^\bullet = \varnothing$.*

The above proposition will be used in the proof that reductions captured by Theorem 2 can be applied to compositionally defined boxes.

Note that, in view of Proposition 5(4), Eq. (6) can be regarded as an alternative translation from box expression E to $box(E)$ carried out using distributed places rather than ordinary places. An additional information conveyed by $dp(E)$ is a useful structuring of the places of $box(E)$ into distributed places.

6   Cographs and Distributed Places

To carry out effective reduction of separated pure distributed places Δ_E partitioning the set of places of $box(E)$, we will employ undirected graphs without self-loops $G = (V, A)$, where V is a finite set of vertices and A is a set of edges. A graph $(\{v\}, \varnothing)$ comprising a single vertex v will be denoted by v.

The *complementation* of G (comprising exactly all the edges which do not belong to G) is denoted $\overline{G}$, and the *disjoint union* of of two graphs, G and G', is denoted by $G \uplus G'$. The *join* of two graphs, G and G', is the graph $G - G'$ constructed from their disjoint union by adding an edge between every vertex of G and every vertex of G'.

A *clique* in G is a nonempty set of vertices clq which are pairwise connected by edges (note that a single vertex graph is a clique). A clique clq is *maximal* (or maxclique) if it is not a subset of any other clique in G. We denote this by $clq \in maxClq(G)$.

A set of cliques in a graph forms an *edge clique cover* (ECC) if, for every edge, there is at least one clique that contains both endpoints of this edge (in our case, we will also include singleton cliques covering isolated vertices). The *edge clique cover problem* (ECCP) consists in finding an ECC using the smallest number of maximal cliques. A variant of ECCP is a *partial ECCP* (PECCP), where some of the edges are assumed to be already covered. Note that for our purposes heuristic algorithms solving (P)ECCP can be used as result will still be polynomial (even the trivial ECC covering each edge separately results in at most a quadratic (in the number of vertices) number of cliques).

There is a strong and direct connection between distributed places in boxes and the ECCP for a class of graphs defined next.

Definition 4 ([15]). *The* complement-reducible *graphs* (cographs) *are recursively defined as follows: (i) a single vertex graph is a cograph and the empty graph is a coraph; (ii) the complement of a cograph is a cograph; and (iii) the disjoint union of cographs is a cograph.*

As the join operation can be expressed via disjoint union and complementation (note that $G - G' = \overline{\overline{G} \uplus \overline{G'}}$), any graph derived by repeatedly applying the join and disjoint union operations (starting from the single-vertex graphs) is a cograph.

Representing sets of places, and generating sets of places through cross-product can be done implicitly using graphs where vertices are indexed actions.

Definition 5. *A* connection graph *is a graph* $\Gamma = (V, A)$ *such that each vertex is of the form* a^{in} *or* a^{out}, *where a is an action used in box expressions. Moreover,* $\Pi_\Gamma = \{\pi_{clq} \mid clq \in maxClq(\Gamma)\}$ *are the* places generated by Γ.
Note: Each $clq = \{a_1^{in}, \ldots, a_k^{in}, b_1^{out}, \ldots, b_l^{out}\}$ *generates* $\pi_{clq} = \pi_{a_1^{in} \ldots a_k^{in} b_1^{out} \ldots b_l^{out}}$.

The next result captures a fundamental link between two operations on disjoint places, $\oplus$ and $\otimes$, and the union and join operations on connection graphs.

Proposition 6. *Let* Γ *and* Γ' *be two connection graphs such that* Π_Γ *and* $\Pi_{\Gamma'}$ *are separated distributed places. Then:*

1. $\Pi_{\Gamma \uplus \Gamma'}$ *is a distributed place satisfying* $\Pi_{\Gamma \uplus \Gamma'} = \Pi_\Gamma \oplus \Pi_{\Gamma'}$ *provided that* ${}^\bullet(\Pi_\Gamma) = {}^\bullet(\Pi_{\Gamma'}) = \varnothing$ *or* $(\Pi_\Gamma)^\bullet = (\Pi_{\Gamma'})^\bullet = \varnothing$.
2. $\Pi_{\Gamma - \Gamma'}$ *is a distributed place satisfying* $\Pi_{\Gamma - \Gamma'} = \Pi_\Gamma \otimes \Pi_{\Gamma'}$.

Hence, the distributed places in Δ_E (obtained by repeated applications of the cross-product and union) can be generated by suitable connection graphs.

This is done by first associating with each box expression E a triple of cographs $\Gamma_E = (\Gamma_E^e, \Gamma_E^x, \Gamma_E^i)$, where the meaning of Γ_E^e, Γ_E^x, and Γ_E^i is that each cograph generates places in the corresponding component in Π_E, e.g. $\Pi_{\Gamma_E^i} = \Pi_E^i$.

The syntax-driven derivation of Γ_E proceeds as follows:

$$
\begin{aligned}
\Gamma_a &= (a^{out}, a^{in}, \varnothing) \\
\Gamma_{E \parallel F} &= (\Gamma_E^e \uplus \Gamma_F^e, \Gamma_E^x \uplus \Gamma_F^x, \Gamma_E^i \uplus \Gamma_F^i) \\
\Gamma_{E;F} &= (\Gamma_E^e, \Gamma_F^x, \Gamma_E^i \uplus \Gamma_F^i \uplus (\Gamma_E^x - \Gamma_F^e)) \\
\Gamma_{E \square F} &= (\Gamma_E^e - \Gamma_F^e, \Gamma_E^x - \Gamma_F^x, \Gamma_E^i \uplus \Gamma_F^i) \\
\Gamma_{[E*F*G]} &= (\Gamma_E^e, \Gamma_G^x, \Gamma_E^i \uplus \Gamma_F^i \uplus \Gamma_G^i \uplus ((\Gamma_E^x - \Gamma_F^x) - (\Gamma_F^e - \Gamma_G^e))) .
\end{aligned}
\tag{7}
$$

We then define the *connection graph* of E as $cg(E) = \Gamma_E^e \uplus \Gamma_E^x \uplus \Gamma_E^i$.

Example 8. For the box expression $(a \parallel b) ; (c \parallel d)$ we have:

$$\Gamma_{(a\parallel b);(c\parallel d)} = (\Gamma_{a\parallel b}^e, \Gamma_{c\parallel d}^x, \Gamma_{a\parallel b}^i \uplus \Gamma_{c\parallel d}^i \uplus (\Gamma_{a\parallel b}^x - \Gamma_{c\parallel d}^e))$$
$$= (a^{out} \uplus b^{out}, c^{in} \uplus d^{in}, (a^{in} \uplus b^{in}) - (c^{out} \uplus d^{out})) \,.$$

Hence $cg((a \parallel b) ; (c \parallel d)) = a^{out} \uplus b^{out} \uplus c^{in} \uplus d^{in} \uplus ((a^{in} \uplus b^{in}) - (c^{out} \uplus d^{out}))$ as shown in Fig. 6(a).   ◇

Proposition 7. *Let E be a box expression defined by the syntax in Eq. (4).*

1. Γ_E^e, Γ_E^x, Γ_E^i *are disjoint connection graphs such that* $\Pi_{\Gamma_E^e} = \Pi_E^e$, $\Pi_{\Gamma_E^x} = \Pi_E^x$, *and* $\Pi_{\Gamma_E^i} = \Pi_E^i$.
2. $cg(E)$ *is a connection graph such that* $\Pi_{cg(E)}$ *is the set of places of* $box(E)$.

Note that, in view of Proposition 7(1), Eq. (7) can be regarded as an alternative translation from box expression E to $box(E)$ carried out using connection graphs rather than ordinary places. An additional information conveyed by $cg(E)$ is, *e.g.* a compact representation of places of $box(E)$.

7   Cograph-Based Reduction

We now can devise a three-stage procedure for constructing a reduced behaviourally equivalent version of $box(E)$ for a box expression E defined by the syntax in Eq. (4).

– **Stage I.** Using the derivation rules Eq. (7), construct the connection graph $cg(E)$.
– **Stage II.** Apply any algorithm for finding solution $\mathscr{C}$ of ECCP for $cg(E)$.
– **Stage III.** Using the derivation rules Eq. (2), construct a marked net $\mathscr{N}_{\mathscr{C}}$ with the set of places $\Pi_{\mathscr{C}}$ and the initial marking $\{p \in \Pi_{\mathscr{C}} \mid {}^\bullet p = \varnothing\}$.

As the number of vertices of $cg(E)$ is $2n$, where n is the number of actions occurring in E, one needs at most $2n(2n-1)/2 = n(2n-1)$ cliques to form a suitable $\mathscr{C}$, and so there are at most $n(2n-1)$ places in the resulting net. Sometimes, the number of places can be much smaller, *e.g.* $O(\log n)$, as shown by the example in Fig. 1(d). This clearly compares favourably with the result of the original translation which can yield $box(E)$ of exponential size.

The validity of the above procedure is established in the next result.

Theorem 3. $\mathscr{N}_{\mathscr{C}}$ *is a safe marked net generating the same reachability graph as* $box(E)$.

It is also possible to apply the following two optimisations which do not invalidate the last result.

– **Optimisation I.** At the end of Stage I, mark all the edges of the form (a^{in}, b^{in}) as already covered. Then, in Stage II, use any algorithm for solving PECCP.
– **Optimisation II.** At the end of Stage III, delete all the places in $\{p \in \Pi_{\mathscr{C}} \mid p^\bullet = \varnothing\}$.

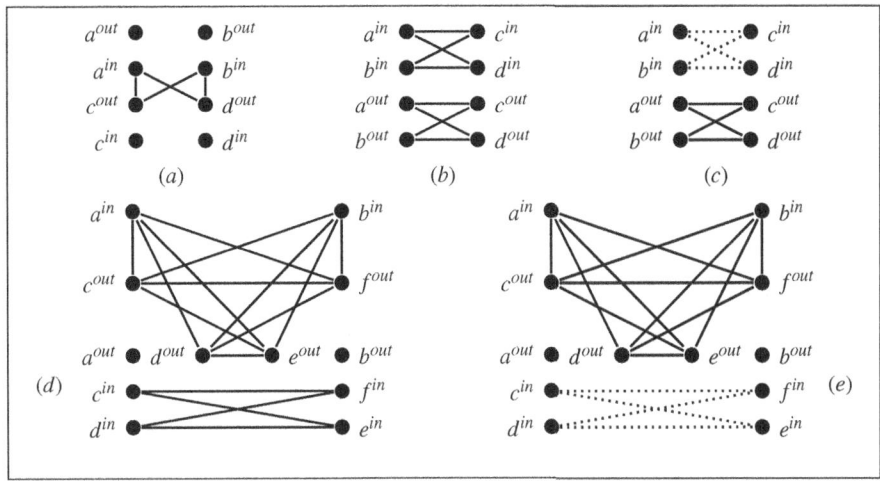

Fig. 6. Connection graphs of: (a) of $(a \| b);(c \| d)$, (b) of $(a \| b) \square (c \| d)$, (c) of $(a \| b) \square (c \| d)$ with dotted edges already covered, (d) of $(a \| b);((c \| d) \square (e \| f))$, and (e) of $(a \| b);((c \| d) \square (e \| f))$ with dotted edges already covered.

Note that PECCP is trivially in NP and many existing ECCP algorithms can be adapted to solve PECCP, *e.g.* [10]. Moreover, since ECCP is NP-complete for general graphs and is a special case of PECCP, it follows that PECCP is NP-complete for general graphs. However, for connection graphs derived for box expressions—a very restricted subclass of graphs—many NP-complete problems become polynomial when restricted to this class. Having said that, to our knowledge, the question whether ECCP or PECCP is NP-complete on cographs is still open. Note also that for modelling control flows the optimality is not required, so fast heuristic algorithms computing small but not necessarily smallest covers would be sufficient. In fact, the trivial ECC covering each edge by a separate clique, which is then arbitrarily extended to a maximal one, already avoids the exponential explosion resulting from the cross-product construction.

Example 9. Figure 6(b) shows the connection graph for $(a \| b) \square (c \| d)$. Moreover, Fig. 6(c) indicates the same connection graph after applying the first optimisation. A possible outcome of solving PECC yields the following cover: $\{a^{out}, c^{out}\}$, $\{a^{out}, d^{out}\}$, $\{b^{out}, c^{out}\}$, $\{b^{out}, d^{out}\}$, $\{a^{in}, c^{in}\}$, and $\{b^{in}, d^{in}\}$. Hence, the solution obtained is $\mathcal{N}_1$ in Fig. 4(c) with the places r_5 and r_7 removed.   ◇

Example 10. Figure 6(d) shows the connection graph for $E = (a \| b);((c \| d) \square (e \| f))$ (note that $box(E)$ is shown in Fig. 5(a)). Figure 6(e) shows the same connection graph after designating some edges as already covered (Optimisation I). A possible outcome of solving PECC yields the following cover: $\{a^{out}\}$, $\{c^{in}, e^{in}\}$, $\{a^{in}, c^{out}, f^{out}\}$, $\{a^{in}, d^{out}, f^{out}\}$, $\{c^{out}\}$, $\{d^{in}, f^{in}\}$, $\{b^{in}, c^{out}, e^{out}\}$, and $\{b^{in}, d^{out}, e^{out}\}$. Hence, the solution obtained is $\mathcal{N}_3'$ of Fig. 5(b). Moreover, after applying Optimisation II, we obtain $\mathcal{N}_1'$ shown in Fig. 2(b).   ◇

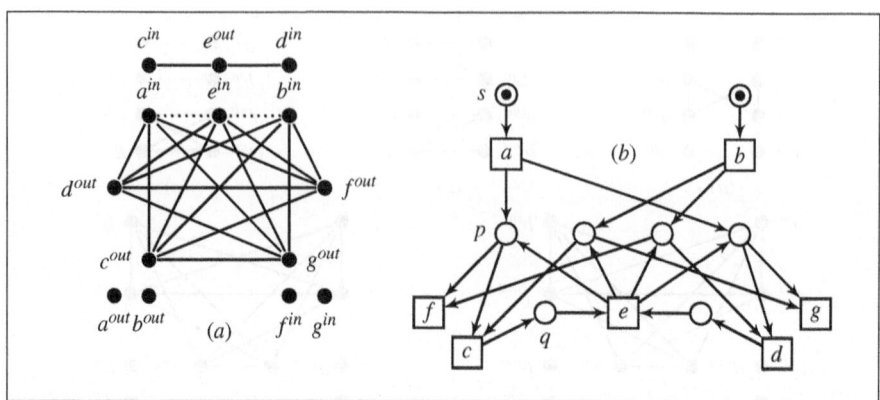

Fig. 7. (a) Connection graph of box expression $[a \| b * (c \| d); e * f \| g]$, and (b) its optimised translation. For example, $s = \{a^{out}\}$, $p = \{a^{in}, e^{in}, f^{out}, c^{out}\}$, and $q = \{c^{in}, e^{out}\}$.

Example 11. For the box expression $[a \| b * (c \| d); e * f \| g]$ we have the following after noting that $\Gamma_{w \| z} = (w^{out} \uplus z^{out}, w^{in} \uplus z^{in}, \varnothing)$, for all actions w and z:

$$
\begin{aligned}
&\Gamma_{[a \| b * (c \| d); e * f \| g]} \\
&= (\Gamma^e_{a \| b}, \Gamma^x_{f \| g}, \Gamma^i_{a \| b} \uplus \Gamma^i_{(c \| d); e} \uplus \Gamma^i_{f \| g} \uplus ((\Gamma^x_{a \| b} - \Gamma^x_{(c \| d); e}) - (\Gamma^e_{(c \| d); e} - \Gamma^e_{f \| g}))) \\
&= (a^{out} \uplus b^{out}, f^{in} \uplus g^{in}, \\
&\quad ((c^{in} \uplus d^{in}) - e^{out}) \uplus ((a^{in} \uplus b^{in}) - e^{in}) - ((c^{out} \uplus d^{out}) - (f^{out} \uplus g^{out})))) \,.
\end{aligned}
$$

Hence $cg([a \| b * (c \| d); e * f \| g])$ is as shown in Fig. 7(a) with dotted lines indicating edges already covered. This cograph has 14 max-cliques, but the following 10 provide a minimal ECC: $\{a^{out}\}$, $\{b^{out}\}$, $\{a^{in}, e^{in}, f^{out}, c^{out}\}$, $\{a^{in}, e^{in}, d^{out}, g^{out}\}$, $\{b^{in}, e^{in}, g^{out}, c^{out}\}$, $\{b^{in}, e^{in}, d^{out}, f^{out}\}$, $\{f^{in}\}$, $\{g^{in}\}$, $\{c^{in}, e^{out}\}$, $\{d^{in}, e^{out}\}$. Then after Optimisation II, we obtain the net in Fig. 7(b).   ◇

8   Concluding Remarks

The way of deleting places from the boxes has been demonstrated in [13] to satisfy *local optimality* for all box expressions derived from the syntax in Eq.(3) without iteration (*i.e.* deleting any remaining place changes the behaviour). However, the same does not hold for *global optimality* [13].

In the future work we plan to extend the results obtained in this paper to box expressions employing other operators. Moreover, we are interested in developing efficient algorithms for identifying distributed places in general safe marked nets.

Acknowledgments. This research was supported by the Leverhulme Trust funded grant RPG-2022-025, and the EPSRC funded grant EP/X036006/1 (SONNETS). The authors are grateful to the three anonymous referees, whose comments contributed to the final version of this paper.

Disclosure of Interests. The authors have no competing interests to declare that are relevant to the content of this article.

References

1. Badouel, É., Bernardinello, L., Darondeau, P.: The synthesis problem for elementary net systems is NP-complete. Theor. Comput. Sci. **186**(1–2), 107–134 (1997)
2. Badouel, É., Bernardinello, L., Darondeau, P.: Petri Net Synthesis. Texts in Theoretical Computer Science. An EATCS Series. Springer, Heidelberg (2015). https://doi.org/10.1007/978-3-662-47967-4
3. Best, E., Devillers, R., Hall, J.G.: The box calculus: a new causal algebra with multi-label communication. In: Rozenberg, G. (ed.) Advances in Petri Nets 1992. LNCS, vol. 609, pp. 21–69. Springer, Heidelberg (1992). https://doi.org/10.1007/3-540-55610-9_167
4. Best, E., Devillers, R.R., Koutny, M.: Petri net algebra. Monographs in Theoretical Computer Science. An EATCS Series. Springer, Heidelberg (2001). https://doi.org/10.1007/978-3-662-04457-5
5. Best, E., Devillers, R.R., Koutny, M.: The box algebra = Petri nets + process expressions. Inf. Comput. **178**(1), 44–100 (2002)
6. Chan, A., Sokolov, D., Khomenko, V., Lloyd, D., Yakovlev, A.: Burst automaton: framework for speed-independent synthesis using burst-mode specifications. IEEE Trans. Comput. Aided Des. Integr. Circuits Syst. **42**(5), 1560–1573 (2023)
7. Cortadella, J., Kishinevsky, M., Kondratyev, A., Lavagno, L., Yakovlev, A.: Logic Synthesis for Asynchronous Controllers and Interfaces. Springer, Heidelberg (2002). https://doi.org/10.1007/978-3-642-55989-1
8. van Glabbeek, R., Goltz, U.: Refinement of actions in causality based models. In: de Bakker, J.W., de Roever, W.-P., Rozenberg, G. (eds.) REX 1989. LNCS, vol. 430, pp. 267–300. Springer, Heidelberg (1990). https://doi.org/10.1007/3-540-52559-9_68
9. Goltz, U., Mycroft, A.: On the relationship of CCS and petri nets. In: Paredaens, J. (ed.) ICALP 1984. LNCS, vol. 172, pp. 196–208. Springer, Heidelberg (1984). https://doi.org/10.1007/3-540-13345-3_18
10. Gramm, J., Guo, J., Hüffner, F., Niedermeier, R.: Data reduction and exact algorithms for clique cover. ACM J. Exp. Algor. **13** (2009)
11. Hoare, C.: Communicating Sequential Processes. Prentice-Hall, Upper Saddle River (1985)
12. Khomenko, V., Koutny, M., Yakovlev, A.: Avoiding exponential explosion in petri net models of control flows. In: Bernardinello, L., Petrucci, L. (eds.) Application and Theory of Petri Nets and Concurrency - 43rd International Conference, PETRI NETS 2022, Bergen, Norway, 19–24 June 2022, Proceedings. Lecture Notes in Computer Science, vol. 13288, pp. 261–277. Springer, Heidelberg (2022). https://doi.org/10.1007/978-3-031-06653-5_14
13. Khomenko, V., Koutny, M., Yakovlev, A.: Slimming down Petri boxes: compact Petri net models of control flows. In: Klin, B., Lasota, S., Muscholl, A. (eds.) 33rd International Conference on Concurrency Theory, CONCUR 2022, Warsaw, Poland, 12–16 September 2022. LIPIcs, vol. 243, pp. 8:1–8:16. Schloss Dagstuhl - Leibniz-Zentrum für Informatik (2022)
14. Khomenko, V., Meyer, R., Hüchting, R.: A polynomial translation of pi-calculus FCPs to safe Petri nets. Logical Methods Comput. Sci. **9**(3) (2013)
15. Lerchs, H.: On cliques and kernels. Technical Report, Dept. of Comp. Sci., Univ. of Toronto (1971)
16. Milner, R.: A Calculus of Communicating Systems. Springer, Heidelberg (1980). https://doi.org/10.1007/3-540-10235-3_11
17. Olderog, E.-R.: Operational Petri net semantics for CCSP. In: Rozenberg, G. (ed.) APN 1986. LNCS, vol. 266, pp. 196–223. Springer, Heidelberg (1987). https://doi.org/10.1007/3-540-18086-9_27

18. Pietkiewicz-Koutny, M.: Synthesis of ENI-systems using minimal regions. In: Sangiorgi, D., de Simone, R. (eds.) CONCUR 1998. LNCS, vol. 1466, pp. 565–580. Springer, Heidelberg (1998). https://doi.org/10.1007/BFb0055648
19. Valmari, A.: The state explosion problem. In: Reisig, W., Rozenberg, G. (eds.) ACPN 1996. LNCS, vol. 1491, pp. 429–528. Springer, Heidelberg (1998). https://doi.org/10.1007/3-540-65306-6_21

Analysing Probabilistic Hornets

Michael Köhler-Bußmeier[1]([⊠])[ID] and Lorenzo Capra[2][ID]

[1] University of Applied Science Hamburg, Berliner Tor 7, 20099 Hamburg, Germany
michael.koehler-bussmeier@haw-hamburg.de
[2] Dipartimento di Informatica, Università degli Studi di Milano, Via Celoria 18,
Milan, Italy
capra@di.unimi.it

Abstract. In this paper, we study HORNETS extended with firing probabilities. HORNETS are a Nets-within-Nets formalism, that is, a Petri net formalism where the tokens are Petri nets again. Each of these net-tokens has its own firing rate that is independent from the rates of other net-tokens. HORNETS provide algebraic operations to modify net-tokens during the firing. For our probabilistic extension these operators could also modify the net-token's firing rate individually.

We use our model to analyse self-modifying systems quantitatively. HORNETS are very well suited to model self-adaptive systems performing a MAPE-like loop (monitor-analyse-plan-execute). Here, the system net describes the feedback loop, and the net-tokens describe the adapted model elements.

We introduce a sub-class of HORNETS that can be translated into Algebraic Nets. Therefore, we can exploit more tools to generate state spaces with probabilities, i.e., in our stochastic setting: discrete Markov chains.

Keywords: Hornets · Nets-within-Nets · Discrete Markov Chains · Maude · Adaptive Systems · MAPE-K-Loop · Self-Modification at Run-Time

1 Introduction

HORNETS are a well-suited formalism to specify self-modification as it has built-in constructs to support structural modifications of Petri nets as an effect transitions firing. We use HORNETS to model the self-adaption processes in multi-agent systems (cf. [17,30]) specified in SONAR [18,19] – forming a so-called MAPE-loop (short for: monitor-analyse-plan-execute, cf. [32]). In this application area the adaptation also includes structural modifications of the Petri net-token, at run-time; it is the system's architecture that is dynamic.

HORNETS [12] follow the *Nets-within-Nets* approach [31], i.e., we have Petri nets that have *nets as tokens* and we have algebraic operations on the net-tokens. A net-token is a pair $[N, M]$, where N is the object-net defining the topology and M is the current marking of the net-token. Firing transforms the net-token's marking M and the algebraic operators modify the net topology N.

© The Author(s), under exclusive license to Springer Nature Switzerland AG 2025
E. Amparore and Ł. Mikulski (Eds.): PETRI NETS 2025, LNCS 15714, pp. 287–309, 2025.
https://doi.org/10.1007/978-3-031-94634-9_14

In this paper, we would like to extend our approach by quantitative information on the relative frequencies of self-modifications. Our main contribution is the definition of *Probabilistic* HORNETS, where object-nets have the form N^Λ; the mapping Λ assigns firing rates to the object net's transitions (cf. our example in Sect. 5). Rates induces probabilities to conflicting events – we do not consider timed transitions here.

We would like to mention two special features: Firstly, we have an independent rate for each net-token $[N^\Lambda, M]$. Secondly, for HORNETS, we could use the operators to *modify* the firing rates Λ. We exploit this feature during the monitoring phase of our self-adaptation loop. From the two specialties, we find that the state space is harder to analyse when compared to e.g. Stochastic Nets [23].

Related Work. Probabilistic choices in process models go back to the work on probabilistic automata [27]. For Petri nets, probability is heavily used to model the firing time distribution, like for Stochastic Nets [23]; so, the concept is more related to 'time' than to 'alternatives'. However, probabilistic choices are introduced by immediate transitions and their rates for Generalised Stochastic Petri Nets (GSPN) [24]. Probabilistic choices for Process Algebra are studied in [3]. These formalisms are also used for adaptive systems, e.g., [4] argues that self-adaptive software needs verification at run-time using stochastic model-checking [20]. However, these formalism use 'flat' models; here we consider nested structures, where execution is embedded into a context, like done by the Ambient Calculus [7] and the π-calculus [25] (the process algebra perspective) or nets-with-nets [31] formalisms, like nested nets [22], elementary object systems (EOS) [14], and HORNETS [12].

Object Nets can be seen as the Petri net perspective on mobility, in contrast to the Ambient Calculus [7] or the π-calculus [25], which form the process algebra perspective. While probabilistic extensions exist for context-oriented process algebras (cf. [21] and [26]), to the best of our knowledge there are no such extensions for nets-within-nets.

Structure. The paper has the following structure. In Sect. 2 we recall the definition of HORNETS as given in previous publications. In Sect. 3 we present the main contribution of this paper, namely Probabilistic EHORNETS. In Sect. 4 we show how we use this formalism to model and analyse self-adaptive systems and how the firing rates of the system-net are connected to the transformation complexity. We present an example for a MAPE-like adaption in Sect. 5, where we study the well-known *battle-of-sexes* coordination game. In Sect. 6 we will investigate a sub-class of our EHORNETS, called GSM-EHORNETS, which is reducible to Algebraic Nets [28]; this will simplify the algorithmic construction of discrete Markov chains. The work ends with a conclusion and an outlook.

2   Algebraic Nets-Within-Nets: Hornets

We have defined HORNETS in [12] as a generalisation of our object nets [15], which follow the *nets-within-nets* paradigm as proposed by Valk [31].

In the following we will present the simplified model of *Elementary Hornets* (EHORNETS) from [13], where the nesting structure is restricted to two levels, while HORNETS [12] allow for an arbitrarily nested structure. This is done in analogy to the class of *elementary object net systems* (EOS) [15], which are the two-level specialisation of general object nets [15].

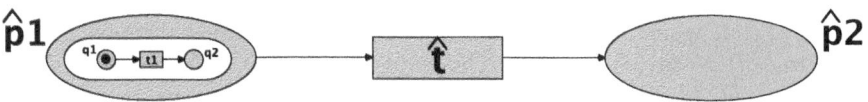

Fig. 1. Nets within Nets: Nets as Tokens

Example 1. With HORNETS we study Petri nets where the tokens are nets again, i.e., we have a nested marking. Assume that we have the object net N with places $P = \{q_1, q_2\}$ and transitions $T = \{t_1\}$. The marking of the HORNET of Fig. 1 is denoted by the nested multiset: $\widehat{p}_1[N, q_1]$. Events are also nested. We have three different kinds of events – as illustrated by the example given in Fig. 1:

1. System-autonomous: The system net transition $\widehat{t}$ fires autonomously, which moves the net-token from $\widehat{p}_1$ to $\widehat{p}_2$ without changing its marking.

$$\widehat{p}_1[N, q_1] \quad \rightarrow \quad \widehat{p}_2[N, q_1]$$

2. object-autonomous: The object-net fires transition t_1 "moving" the black token from q_1 to q_2. The object-net remains at its location $\widehat{p}_1$.

$$\widehat{p}_1[N, q_1] \quad \rightarrow \quad \widehat{p}_1[N, q_2]$$

3. Synchronisation: Whenever we add matching synchronisation inscriptions (using communication channels) at the system net transition $\widehat{t}$ and the object net transition t_1, then both must fire synchronously: The object-net is moved to $\widehat{p}_2$ and the black token moves from q_1 to q_2 inside. Whenever synchronisation is specified, autonomous actions are disabled.

For HORNETS we extend object-nets with algebraic concepts that allow to modify the structure of the net-tokens as a result of a firing transition. This is a generalisation of the approach of algebraic nets [28], where algebraic data types replace the anonymous black tokens. The general HORNET formalism is Turing-complete: In [12] we have proven that there are several possibilities to simulate counter programs: One could use the nesting to encode counters. Another possibility is to encode counters in the algebraic structure of the net operators.

In the following we recall the definition of EHORNETS from [13]. First, we recall notations for p/t nets; then we will define the algebraic structure of the net-token and the logic used for guards. We introduce nested multisets as the marking structure. Finally, we define the firing rule, that, in general, involves a synchronisation of system net transitions with transitions of the net-tokens, i.e., we define nested events. The reader familiar with the firing rule of EHORNETS (Definition 2) can safely skip the remainder of this section.

Multisets and P/T Nets. A multiset $\mathbf{m}$ on the set D is a mapping $\mathbf{m} : D \to \mathbb{N}$. Multisets can also be represented as a formal sum in the form $\mathbf{m} = \sum_{i=1}^{n} x_i$, where $x_i \in D$. Multiset addition is defined component-wise: $(\mathbf{m}_1 + \mathbf{m}_2)(d) := \mathbf{m}_1(d) + \mathbf{m}_2(d)$. The empty multiset $\mathbf{0}$ is defined as $\mathbf{0}(d) = 0$ for all $d \in D$. Multiset-difference $\mathbf{m}_1 - \mathbf{m}_2$ is defined by $(\mathbf{m}_1 - \mathbf{m}_2)(d) := \max(\mathbf{m}_1(d) - \mathbf{m}_2(d), 0)$. The cardinality of a multiset is $|\mathbf{m}| := \sum_{d \in D} \mathbf{m}(d)$. A multiset $\mathbf{m}$ is finite if $|\mathbf{m}| < \infty$. The set of all finite multisets over the set D is denoted $MS(D)$. The *domain* of a multiset is $dom(\mathbf{m}) := \{d \in D \mid \mathbf{m}(d) > 0\}$.

Any mapping $f : D \to D'$ extends to a multiset-homomorphism $f^\sharp : MS(D) \to MS(D')$ by $f^\sharp(\sum_{i=1}^{n} x_i) = \sum_{i=1}^{n} f(x_i)$.

A p/t net N is a tuple $N = (P, T, \mathbf{pre}, \mathbf{post})$, such that P is a set of places, T is a set of transitions, with $P \cap T = \emptyset$, and $\mathbf{pre}, \mathbf{post} : T \to MS(P)$ are the pre- and post-condition functions. A marking of N is a multiset of places: $\mathbf{m} \in MS(P)$. We denote the enabling of t in marking $\mathbf{m}$ by $\mathbf{m} \xrightarrow{t}$. Firing of t is denoted by $\mathbf{m} \xrightarrow{t} \mathbf{m}'$.

Net-Signatures. We define the algebraic structure of object-nets. For a general introduction of algebraic specifications cf. [10]. Let K be a set of net-types (kinds). A (many-sorted) *specification* (Σ, X, E) consists of a signature Σ, a family of variables $X = (X_k)_{k \in K}$, and a family of axioms $E = (E_k)_{k \in K}$.

A *signature* is a disjoint family $\Sigma = (\Sigma_{k_1 \cdots k_n, k})_{k_1, \cdots, k_n, k \in K}$ of operators. The set of terms of type k over a signature Σ and variables X is denoted $\mathbb{T}_\Sigma^k(X)$.

We use (many-sorted) predicate logic, where the terms are generated by a signature Σ and formulae are defined by a family of predicates Ψ. The set of formulae is denoted PL_Γ, where $\Gamma = (\Sigma, X, E, \Psi)$ is the *logic structure*.

Object Nets and Net-Algebras. Let Σ be a signature over K. A *net-algebra* assigns to each type $k \in K$ a set $\mathcal{U}_k$ of object-nets – the net universe. Each object $N \in \mathcal{U}_k, k \in K$ net is a p/t net $N = (P_N, T_N, \mathbf{pre}_N, \mathbf{post}_N)$. We assume the family $\mathcal{U} = (\mathcal{U}_k)_{k \in K}$ to be disjoint.

The places of the object-nets in $\mathcal{U}_k$ are not disjoint, since the firing rule allows tokens to be moved between net-tokens within the same set $\mathcal{U}_k$. Such a transfer is possible, if we assume that all nets $N \in \mathcal{U}_k$ have the same set of places P_k. P_k is the place universe for all object-nets of kind k. However, object-nets of a given kind behave independently from each other (i.e., they have their markings) except when they synchronize with the upper level.

The family of object-nets $\mathcal{U}$ is the universe of the algebra. A *net-algebra* $(\mathcal{U}, \mathcal{I})$ assigns to each constant $\sigma \in \Sigma_{\lambda, k}$ an object-net $\sigma^{\mathcal{I}} \in \mathcal{U}_k$ and to each operator $\sigma \in \Sigma_{k_1 \cdots k_n, k}$ with $n > 0$ a mapping $\sigma^{\mathcal{I}} : (\mathcal{U}_{k_1} \times \cdots \times \mathcal{U}_{k_n}) \to \mathcal{U}_k$.

A variable assignment $\alpha = (\alpha_k : X_k \to \mathcal{U}_k)_{k \in K}$ maps each variable onto an element of the algebra. For a variable assignment α the evaluation of a term $t \in \mathbb{T}^k_\Sigma(X)$ is uniquely defined and will be denoted as $\alpha(t)$.

A net-algebra, such that all axioms of (Σ, X, E) are valid, is called *net-theory*.

Nested Markings. A marking of an EHORNET assigns to each system net place one or many net-tokens. The places of the system net are typed by the function $k : \widehat{P} \to K$, meaning that a place $\widehat{p}$ contains net-tokens of kind $k(\widehat{p})$. Since the net-tokens are instances of object-nets, a *marking* is a *nested* multiset:

$$\mu = \sum_{i=1}^{n} \widehat{p}_i[N_i, M_i] \quad \text{where} \quad \widehat{p}_i \in \widehat{P}, N_i \in \mathcal{U}_{k(\widehat{p}_i)}, M_i \in MS(P_{N_i}), n \in \mathbb{N}$$

Each addend $\widehat{p}_i[N_i, M_i]$ denotes a net-token on the place $\widehat{p}_i$ that has the structure of the object-net N_i and the marking $M_i \in MS(P_{N_i})$. The set of all nested multisets is denoted as $\mathcal{M}_H$. We define the partial order $\sqsubseteq$ on nested multisets by setting $\mu_1 \sqsubseteq \mu_2$ iff $\exists \mu : \mu_2 = \mu_1 + \mu$.

The projection $\Pi^1_N(\mu)$ is the multiset of system-net places containing the object-net N (where $\mathbf{1}(\phi) = 1$ if ϕ holds and $= 0$ otherwise):

$$\Pi^1_N \left(\sum\nolimits_{i=1}^{n} \widehat{p}_i[N_i, M_i] \right) := \sum\nolimits_{i=1}^{n} \mathbf{1}(N_i = N) \cdot \widehat{p}_i \tag{1}$$

Analogously, the projection $\Pi^2_N(\mu)$ is the multiset of all net-tokens' markings (that belong to the object-net N):

$$\Pi^2_N \left(\sum\nolimits_{i=1}^{n} \widehat{p}_i[N_i, M_i] \right) := \sum\nolimits_{i=1}^{n} \mathbf{1}(N_i = N) \cdot M_i \tag{2}$$

The projection $\Pi^2_k(\mu)$ is the sum of all net-tokens' markings belonging to the same type $k \in K$:

$$\Pi^2_k (\mu) := \sum\nolimits_{N \in \mathcal{U}_k} \Pi^2_N (\mu) \tag{3}$$

2.1 Elementary HORNETS (EHORNETS)

Assume a fixed logic $\Gamma = (\Sigma, X, E, \Psi)$ and a net-theory $(\mathcal{U}, \mathcal{I})$. An *elementary higher-order object net* (EHORNET) is composed of a system net $\widehat{N}$ and the set of object-nets $\mathcal{U}$. W.l.o.g. we assume $\widehat{N} \notin \mathcal{U}$.

The system net is a net $\widehat{N} = (\widehat{P}, \widehat{T}, \mathbf{pre}, \mathbf{post}, \widehat{G})$, where each arc is labelled with a multiset of terms: $\mathbf{pre}, \mathbf{post} : \widehat{T} \to (\widehat{P} \to MS(\mathbb{T}_\Sigma(X)))$. Each transition is labelled by a guard predicate $\widehat{G} : \widehat{T} \to PL_\Gamma$. The places of the system net are typed by the function $k : \widehat{P} \to K$. As a typing constraint we have that each arc inscription has to be a multiset of terms that are all of the kind that is assigned to the arc's place: $\mathbf{pre}(\widehat{t})(\widehat{p}), \mathbf{post}(\widehat{t})(\widehat{p}) \in MS(\mathbb{T}^{k(\widehat{p})}_\Sigma(X))$.

For each variable binding α we obtain the evaluated functions $\mathbf{pre}_\alpha, \mathbf{post}_\alpha : \widehat{T} \to (\widehat{P} \to MS(\mathcal{U}))$ in the obvious way. Let $\mathcal{U}_k(\widehat{t})$ be the set of object nets occurring in the arc expressions given the binding α:

$$\mathcal{U}_k(\widehat{t}) := \bigcup_{\widehat{p} \in \widehat{P} : k(\widehat{p}) = k} dom\big(\mathbf{pre}_\alpha(\widehat{t})(\widehat{p})\big) \cup dom\big(\mathbf{post}_\alpha(\widehat{t})(\widehat{p})\big) \tag{4}$$

Synchronisation. The transitions in a HORNET are labelled with synchronisation inscriptions. We assume a fixed set of channels $C = (C_k)_{k \in K}$.

- For each kind $k \in K$ and each binding α a system net transition $\hat{t}$ is labelled with a multiset of channels: $\hat{l}^k_\alpha(\hat{t}) \in MS(C_k)$. The intention is that $\hat{t}$ fires synchronously with a multiset of object-net transitions with the same multiset of labels.
- For each $N \in \mathcal{U}_k$ the function l_N assigns to each transition $t \in T_N$ either a channel $c \in C_k$ or $\perp_k$, whenever t fires without synchronisation, i.e., autonomously.

Definition 1. *An eHORNET is a tuple* $EH = (\widehat{N}, \mathcal{U}, \mathcal{I}, k, l, \mu_0)$ *such that:*

1. $\widehat{N}$ *is an algebraic net, called the* system net.
2. $(\mathcal{U}, \mathcal{I})$ *is a finite net-theory for the logic* Γ.
3. $k : \widehat{P} \to K$ *is the typing of the system net places.*
4. $l = (\widehat{l}, l_N)_{N \in \mathcal{U}}$ *is the labelling.*
5. $\mu_0 \in \mathcal{M}_H$ *is the initial marking.*

Example 2. We will illustrate Definition 1 with the example given in Fig. 2. We assume that we have one net type: $K = \{\text{WFN}\}$. We have only one operator $\|$ for parallel composition: $\| \in \Sigma_{\text{WFN}^2, \text{WFN}}$. The operator is interpreted by $\mathcal{I}$ as the usual AND operation on workflow nets. We have the universe with three object nets: $\mathcal{U}_{\text{WFN}} = \{N_1, N_2, N_3\}$. All places of the system net have the same type, i.e., $k(\widehat{p}) = k(\widehat{q}) = k(\widehat{r}) = \text{WFN}$. The structure of the system net $\widehat{N}$ and the object nets is given the usual way as shown in Fig. 2 (For readability, we do not draw isolated places here. Net-tokens are considered equal modulo isolated places). This eHORNET uses the channel ch for synchronisation: $\hat{l}^{WFN}_\alpha(\hat{t}) = \{ch\}$ and $l_N(e) = ch$ and $= \perp_k$ for all $t \neq e$. In the initial marking we consider a eHORNET with two nets N_1 and N_2 as tokens (as shown on the left): $\mu_0 = \widehat{p}[N_1, v] + \widehat{q}[N_2, s]$. The firing rule is explained below.

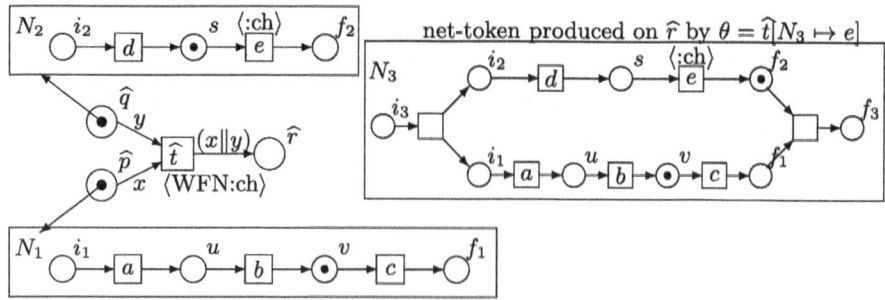

Fig. 2. Modification of the Net-Token's Structure.

2.2 Events and Firing Rule

The synchronisation labelling generates the set of system events Θ. We have three kinds of events:

1. Synchronised firing: There is at least one object-net that has to be synchronised, i.e., there is a kind k such that $\widehat{l}_\alpha^k(\widehat{t})$ is not empty.
   Such an event is a pair $\theta = \widehat{t}^\alpha[\vartheta_\alpha]$, where $\widehat{t}$ is a system net transition, α is a variable binding, and ϑ_α is a function that maps each object-net to a multiset of its transitions, i.e., $\vartheta_\alpha(N) \in MS(T_N)$. It is required that $\widehat{t}$ and $\vartheta_\alpha(N)$ have matching multisets of labels, i.e., $\widehat{l}_\alpha^k(\widehat{t}) = \sum_{N \in \mathcal{U}_k(\widehat{t})} l_N^\sharp(\vartheta_\alpha(N))$, i.e., $\widehat{t}$ fires synchronously with all the object-net transitions $\vartheta_\alpha(N)$.
2. System-autonomous firing: The transition $\widehat{t}$ of the system net fires autonomously, whenever $\widehat{l}(\widehat{t})$ is the empty multiset $\mathbf{0}$. We consider system-autonomous firing as a special case of synchronised firing generated by the function ϑ_{id}, defined as $\vartheta_{id}(N) = \mathbf{0}$ for all $N \in \mathcal{U}$.
3. Object-autonomous firing: An object-net transition t in N fires autonomously within the place $\widehat{p}$ whenever $l_N(t) = \bot_k$. For the sake of uniformity we denote object-autonomous events as $id_{\widehat{p},N}[\vartheta_t]$, where $\vartheta_t(N') = \{t\}$ if $N = N'$ and $\mathbf{0}$ otherwise. We define $\widehat{G}(id_{\widehat{p},N}) :=$ TRUE and:

$$\mathbf{pre}_\alpha(id_{\widehat{p},N})(\widehat{p}')(N') = \mathbf{post}_\alpha(id_{\widehat{p},N})(\widehat{p}')(N') = \mathbf{1}(\widehat{p}' = \widehat{p} \wedge N' = N)$$

The set of all *events* generated by the labelling l is $\Theta_l := \Theta_1 \cup \Theta_2$, where Θ_1 contains synchronous events (including system-autonomous events as a special case) and Θ_2 contains the object-autonomous events:

$$\begin{aligned} \Theta_1 &:= \left\{\widehat{\tau}^\alpha[\vartheta_\alpha] \mid \forall k \in K : \widehat{l}_\alpha^k(\widehat{\tau}) = \sum_{N \in \mathcal{U}_k(\widehat{t})} l_N^\sharp(\vartheta_\alpha(N))\right\} \\ \Theta_2 &:= \left\{id_{\widehat{p},N}[\vartheta_t] \mid \widehat{p} \in \widehat{P}, N \in \mathcal{U}_{k(\widehat{p})}, t \in T_N\right\} \end{aligned} \quad (5)$$

Firing Rule. A system event $\theta = \widehat{\tau}^\alpha[\vartheta_\alpha]$ removes net-tokens together with their individual internal markings. Firing the event replaces a nested multiset $\lambda \in \mathcal{M}_H$ that is part of the current marking μ, i.e., $\lambda \sqsubseteq \mu$, by the nested multiset ρ. The enabling condition is expressed by the *enabling predicate* $\phi_{EH}(\widehat{\tau}^\alpha[\vartheta_\alpha], \lambda, \rho)$:

$$\begin{aligned} \forall k \in K : \forall \widehat{p} \in k^{-1}(k) : \forall N \in \mathcal{U}_k : \Pi_N^1(\lambda)(\widehat{p}) &= \mathbf{pre}_\alpha(\widehat{\tau})(\widehat{p})(N) \wedge \\ \forall \widehat{p} \in k^{-1}(k) : \forall N \in \mathcal{U}_k : \Pi_N^1(\rho)(\widehat{p}) &= \mathbf{post}_\alpha(\widehat{\tau})(\widehat{p})(N) \wedge \\ \Pi_k^2(\lambda) &\geq \sum_{N \in \mathcal{U}_k(\widehat{t})} \mathbf{pre}_N^\sharp(\vartheta_\alpha(N)) \wedge \\ \Pi_k^2(\rho) = \Pi_k^2(\lambda) - \sum_{N \in \mathcal{U}_k(\widehat{t})} \mathbf{pre}_N^\sharp(\vartheta_\alpha(N)) &+ \sum_{N \in \mathcal{U}_k(\widehat{t})} \mathbf{post}_N^\sharp(\vartheta_\alpha(N)) \end{aligned} \quad (6)$$

- Conjunct (1) states that the removed submarking λ contains in $\widehat{p}$ the right number of net tokens that are removed by $\widehat{\tau}$.
- Conjunct (2) states that generated sub-marking ρ contains on $\widehat{p}$ the right number of net-tokens, that are generated by $\widehat{\tau}$.

- Conjunct (3) states that the sub-marking λ enables all synchronised transitions $\vartheta_\alpha(N)$ in the object N.
- Conjunct (4) states that the marking of each object net N is changed according to the firing of the synchronised transitions $\vartheta_\alpha(N)$.

Note that conjuncts (1) and (2) ensure that only the relevant net tokens for firing are included in λ and ρ. Conditions (3) and (4) allow for additional tokens in the net-tokens.

For system-autonomous events $\widehat{t}^\alpha[\vartheta_{id}]$ the enabling predicate ϕ_{EH} can be simplified further: Conjunct (3) is always true since $\mathbf{pre}_N(\vartheta_{id}(N)) = \mathbf{0}$. Conjunct (4) simplifies to $\Pi_k^2(\rho) = \Pi_k^2(\lambda)$, which means that no token of the object nets get lost when a system-autonomous events fires.

Analogously, for an object-autonomous event $\widehat{\tau}[\vartheta_t]$ we have an idle-transition $\widehat{\tau} = id_{\widehat{p},N}$ and $\vartheta = \vartheta_t$ for some t. Conjuncts (1) and (2) simplify to $\Pi^1_{N'}(\lambda) = \widehat{p} = \Pi^1_{N'}(\rho)$ for $N' = N$ and to $\Pi^1_{N'}(\lambda) = \mathbf{0} = \Pi^1_{N'}(\rho)$ otherwise. This means that $\lambda = \widehat{p}[M]$, M enables t, and $\rho = \widehat{p}[M - \mathbf{pre}_N(t) + \mathbf{post}_N(t)]$.

Definition 2 (Firing Rule). *Let $\mu, \mu' \in \mathcal{M}_H$ be markings.*

- *The event $\widehat{\tau}^\alpha[\vartheta_\alpha]$ is enabled in μ for (λ, ρ) iff $\lambda \sqsubseteq \mu$, $\phi_{EH}(\widehat{\tau}[\vartheta_\alpha], \lambda, \rho)$ holds and the guard $\widehat{G}(\widehat{t})$ follows from the axioms, i.e., $E \models^\alpha_\mathcal{I} \widehat{G}(\widehat{\tau})$.*
- *An event $\widehat{\tau}^\alpha[\vartheta_\alpha]$ that is enabled in μ can fire – denoted $\mu \xrightarrow[EH]{\widehat{\tau}^\alpha[\vartheta_\alpha](\lambda,\rho)} \mu'$.*
- *The resulting successor marking is defined as $\mu' = \mu - \lambda + \rho$.*

Note that the firing rule has no a-priori decision how to distribute the marking on the generated net-tokens. Therefore we need the mode (λ, ρ) to formulate the firing of $\widehat{\tau}^\alpha[\vartheta_\alpha]$ in a functional way.

Example 3. We will illustrate the firing rule considering the HORNET from Fig. 2 again. To model a run-time adaption, we combine N_1 and N_2 resulting in the net $N_3 = (N_1 \| N_2)$. This modification is modelled by system net transition $\widehat{t}$ of the HORNET. Consider the event $\theta_1 = \widehat{t}^\alpha[\vartheta_\alpha]$, where $\vartheta_\alpha(N_1) = \vartheta_\alpha(N_2) = \mathbf{0}$ and $\vartheta_\alpha(N_3) = \{e\}$, which formalises that $\widehat{t}$ fires synchronously with e. In a binding α with $x \mapsto N_1$ and $y \mapsto N_2$ the event is enabled. Let $(x\|y)$ evaluate to $(N_1\|N_2) = N_3$ for α, i.e., the net-token generated on $\widehat{r}$ has the structure of N_3. The event removes the two net-tokens from $\widehat{p}$ and $\widehat{q}$. As $\sum_{N \in \mathcal{U}_k(\widehat{t})} \mathbf{pre}^\sharp_N(\vartheta_\alpha(N)) = \mathbf{pre}^\sharp_{N_1}(\vartheta_\alpha(N_1)) + \mathbf{pre}^\sharp_{N_2}(\vartheta_\alpha(N_2)) + \mathbf{pre}^\sharp_{N_3}(\vartheta_\alpha(N_3)) = \mathbf{pre}^\sharp_{N_1}(\mathbf{0}) + \mathbf{pre}^\sharp_{N_2}(\mathbf{0}) + \mathbf{pre}^\sharp_{N_3}(e) = \{s\}$ and $\sum_{N \in \mathcal{U}_k(\widehat{t})} \mathbf{post}^\sharp_N(\vartheta_\alpha(N)) = \{f_2\}$ the marking of net-token generated on $\widehat{r}$ is the sum of the two incoming markings combined with the object-net transition $e : \{s\} \to \{f_2\}$:

$$\widehat{p}[N_1, v] + \widehat{q}[N_2, s] \xrightarrow{\theta_1} \widehat{r}\big[(N_1\|N_2), (v + f_2)\big]$$

This transfer of markings is possible since all the places of N_1 and N_2 are also places in N_3 and tokens can be transferred as an injection. It is also possible that the net-tokens fire object autonomously. E.g., the net-token on place $\widehat{p}$ enables the object net transition c in μ_0, i.e., the event $\theta_2 = id_{\widehat{p},N_1}[\vartheta_c]$:

$$\widehat{p}[N_1, v] + \widehat{q}[N_2, s] \xrightarrow{\theta_2} \widehat{p}[N_1, f_1] + \widehat{q}[N_2, s]$$

Note, that the net-token on $\widehat{q}$ does not enable the transition e in an autonomous way due to the channel inscription.

3   Probabilistic Extensions of EHornets

Let $EH = (\widehat{N}, \mathcal{U}, \mathcal{I}, k, \Theta, \mu_0)$ be an EHORNET. The *reachability graph* $RG(EH) = (V, E, \mu_0)$ contains all reachable nested markings as vertices/nodes $V = RS(EH)$, $E = \{(\mu, \theta, \mu') \mid \mu \xrightarrow{\theta} \mu'\}$ as edges, and the initial marking μ_0 as a distinguished node. We equip the EHORNET model with a function $\Lambda : \Theta \to \mathbb{R}^{>0}$ that assigns rates (i.e., firing weights) to events $\theta \in \Theta$. Note that rates are dependent on the binding α and the involved net-tokens N^Λ.

For stochastic Petri nets (SPN) [23] the usual way to derive probabilities from these rates is to normalise over all transitions enabled in a given marking. In the following we extend this idea to nested events of EHORNETS. Let $En(\mu) := \{\theta \in \Theta \mid \mu \xrightarrow{\theta}\}$ be the set of all events enabled in the nested marking μ. Then, the probability of firing $\theta \in En(\mu)$ is proportional to its rate $\Lambda(\theta) \in \mathbb{R}^{>0}$. For EHORNETS we define the firing probability as:

$$Pr_\mu(\theta) := \frac{\Lambda(\theta)}{\sum_{\theta' \in En(\mu)} \Lambda(\theta')} \tag{7}$$

For an arc (μ, θ, μ') in the reachability graph $RG(EH) = (V, E, \mu_0)$ we define

$$Pr((\mu, \theta, \mu')) := Pr_\mu(\theta) \tag{8}$$

For a state-based stochastic analysis purposes, the events are usually ignored and we will use the stochastic matrix defined as: $\mathbf{P}(\mu, \mu') := \sum_{\theta \in \Theta} Pr(\mu, \theta, \mu')$ if $En(\mu) \neq \emptyset$, 0 otherwise. Conversely, $\mathbf{P}(\mu, \mu) := 1$ if $En(\mu) = \emptyset$, 0 otherwise.

Definition 3. *A Probabilistic EHORNET $PEH = (EH, \Lambda)$ is given by an EHORNET EH and a rate (weight) function $\Lambda : \Theta \to \mathbb{R}^{>0}$. The induced discrete-time Markov chain (DTMC) is given by the matrix $\mathbf{P}$.*

Nested Firing Rates. The induced Markov chain is defined in a 'global' fashion since it considers rates $\Lambda(\theta)$ of events that may involve synchronisation of transitions from both system- and object-net. Therefore, it seems coherent to derive these rates from the transitions involved in the event. We can calculate the rates of an event by assigning rates to the system net and the object nets' transitions: $\Lambda(\widehat{t})$ and $\Lambda_N(t)$, respectively. (For the special transitions $\widehat{\tau} = id_{\widehat{p},N}$ we define $\Lambda(id_{\widehat{p},N}) = 1$.)

The event $\theta = \widehat{\tau}^{\alpha}[\vartheta_{\alpha}]$ synchronizes the system net transition $\widehat{\tau}$ with a multiset of transitions $\vartheta_{\alpha}(N)$ for each object net N. The rate of the event θ is generated from the rates of these transitions using the product rule of probability theory. Therefore, rates must be within the interval $[0;1]$. To ensure this, we apply a normalising function σ to rates, e.g. $\sigma(x) = 1 - \exp(-a \cdot x)$, $a > 0$.

$$\Lambda_{pr}(\theta) := (\sigma \circ \Lambda)(\widehat{\tau}) \cdot \prod_{k \in K} \prod_{N \in \mathcal{U}_k} \prod_{t \in T_N} \left((\sigma \circ \Lambda_N)(t) \right)^{|\vartheta_{\alpha}(N)(t)|} \tag{9}$$

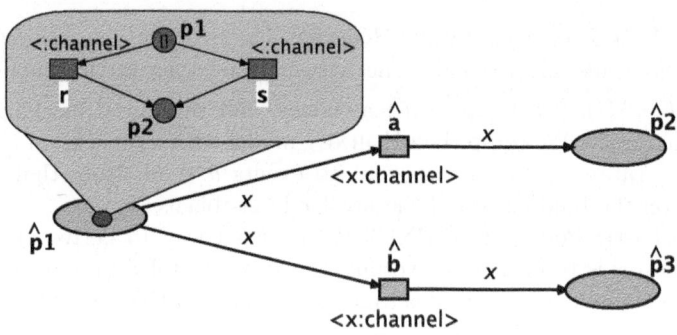

Fig. 3. A Probabilistic eHORNET

Example 4. In Fig. 3 we have two system-net transitions $\widehat{a}$ and $\widehat{b}$ in conflict. They have to synchronise either with the object net transition r or s. From the perspective of synchronisation we have a symmetric situation here. We have the events: $\widehat{a}[N \mapsto r]$, $\widehat{a}[N \mapsto s]$, $\widehat{b}[N \mapsto r]$, and $\widehat{b}[N \mapsto s]$.

Assume the rates $\Lambda' := (\sigma \circ \Lambda)$ are $\widehat{a} \mapsto 0.2$, $\widehat{b} \mapsto 0.3$, $r \mapsto 0.5$, and $s \mapsto 0.7$. From (7) and (9) we obtain the firing probability for e.g. $\widehat{a}[N \mapsto r]$ as

$$Pr\left(\widehat{a}[N \mapsto r]\right) = \frac{\Lambda'(\widehat{a}) \cdot \Lambda'(r)}{\Lambda'(\widehat{a}) \cdot \Lambda'(r) + \Lambda'(\widehat{a}) \cdot \Lambda'(s) + \Lambda'(\widehat{b}) \cdot \Lambda'(r) + \Lambda'(\widehat{b}) \cdot \Lambda'(s)}$$

$$= \frac{0.2 \cdot 0.5}{0.2 \cdot 0.5 + 0.2 \cdot 0.7 + 0.3 \cdot 0.5 + 0.3 \cdot 0.7} = 0,166.$$

The symmetry in the synchronisation leads to a symmetry in probabilities; e.g., the probability of firing the system-net transition $\widehat{a}$ with any possible object-net transition is:

$$\frac{\Lambda'(\widehat{a}) \cdot \Lambda'(r) + \Lambda'(\widehat{a}) \cdot \Lambda'(s)}{\Lambda'(\widehat{a}) \cdot \Lambda'(r) + \Lambda'(\widehat{a}) \cdot \Lambda'(s) + \Lambda'(\widehat{b}) \cdot \Lambda'(r) + \Lambda'(\widehat{b}) \cdot \Lambda'(s)} = \frac{\Lambda'(\widehat{a})}{\Lambda'(\widehat{a}) + \Lambda'(\widehat{b})}$$

This is the firing probability one would expect when considering the system-net alone. We have an analogous situation for the object. Let us consider the probability of firing an event containing the object-net transition r:

$$\frac{\Lambda'(\widehat{a}) \cdot \Lambda'(r) + \Lambda'(\widehat{b}) \cdot \Lambda'(r)}{\Lambda'(\widehat{a}) \cdot \Lambda'(r) + \Lambda'(\widehat{a}) \cdot \Lambda'(s) + \Lambda'(\widehat{b}) \cdot \Lambda'(r) + \Lambda'(\widehat{b}) \cdot \Lambda'(s)} = \frac{\Lambda'(r)}{\Lambda'(r) + \Lambda'(s)}$$

This is the firing probability when considering the object-net alone.

Note that this effect is due to the symmetry in synchronisation; whenever the system-net transitions would have different synchronisation partners the probabilities would be different. Assume that, e.g., $\widehat{b}$ has much more synchronisation partners than $\widehat{a}$, then the probability of events containing $\widehat{a}$ would have been dominated by the events containing $\widehat{b}$.

In general, the stochastic behaviour of a model depends on conflicting events. To enable a fair definition of rates, in ongoing work, we study the possibility of extending the structural conflict relation for p/t nets (which provides a necessary condition for conflicts to occur) at θ.

4 Modeling Rates in Self-adaptive Systems

Usually, EHORNETS are used to specify self-modifying systems. With EHORNETS we can easily integrate adaptable firing rates into the structure of the net-token: Object Nets are pairs of net topology N and the firing rate Λ, denoted N^Λ. The mapping Λ assigns firing rates to transitions. Therefore, we have an independent rate for each net-token $[N^\Lambda, M]$ and the system net may modify these rates.

In the following, we consider a very common scenario, where the system-net describes the MAPE-loop of adaption (monitor-analyse-plan-execute [32]), while the object-nets are workflow nets [1] composed by basic process algebraic operations, like sequence, and-composition, and xor-choices. For these scenarios we like to show how the firing rates can be naturally derived in an automated way. For the object-nets this is straightforward: For workflow nets as net-tokens, we have natural candidates for the rates $\Lambda(t)$ in the object-nets – we usually derive them from execution logs, i.e. by monitoring data. Here, rates are used to describe probabilities of xor-branches.

For the system-net the natural candidates for rates are less obvious: When we have self-modifying systems, a transition $\widehat{t}$ in the system-net describes a transformation of the workflow net-tokens. The rate $\Lambda(\widehat{t})$ should describe the probability of executing the transformation modelled by $\widehat{t}$ during the MAPE-loop. When considering self-adaptive systems, we like to express that there is a relationship between the complexity of an adaption and its probability. The motivation for this is that transformations have to be evaluated during some kind of planning process and transformations that are less complex are usually considered more frequently during this planning. For EHORNETS we have natural candidates that describe the *transformation complexity*: the guard $\widehat{G}(\widehat{t})$ and the arc inscriptions $\mathbf{pre}(\widehat{t})(\widehat{p})$ and $\mathbf{post}(\widehat{t})(\widehat{p})$. The transformation complexity of a system-net transition $\widehat{t}$ equals the number of operations $\|\cdot\|$ that occur in its arc inscriptions and guards:

$$TC(\widehat{t}) := \|\widehat{G}(\widehat{t})\| + \sum_{\widehat{p}\in\widehat{P}} \|\mathbf{pre}(\widehat{t})(\widehat{p}))\| + \sum_{\widehat{p}\in\widehat{P}} \|\mathbf{post}(\widehat{t})(\widehat{p}))\| \qquad (10)$$

The idea to derive the firing rate from $TC(\hat{t})$ is straightforward: Since the search space for planning grows exponentially in the number of transformation operators, we define that the rates drop exponentially, too:

$$\Lambda_{mape}(\hat{t}) := \gamma^{TC(\hat{t})}, \qquad \gamma \in [0;1] \tag{11}$$

Here, γ is a meta-parameter that specifies the discount for the planning horizon. Note, in this MAPE-loop setting the firing rates are directly derived from the model without any need for extra modeling effort. Thus, (11) turns an qualitative EHORNET into a quantitative probabilistic model – and this for free.

5   Example: Modelling Adaption in a Coordination Game

The following example for a self-adaptive system is based on the battle-of-sexes scenario, which is well-known in game theory. Two agents, named 0 and 1, must choose between two actions, labelled as a_i and b_i, $i = 0, 1$. They receive a positive reward if they choose the same action (i.e., coordinate) and zero otherwise.

In this game, the first agent prefers action a, while the second prefers b. If we assume that the reward for the preferred outcome is three times higher than for the other, then the game is specified by the following payoff matrix.

	a_1	b_1
a_0	$(3,1)$	$(0,0)$
b_0	$(0,0)$	$(1,3)$

Let $(a^{\langle x \rangle} \oplus^{\langle y \rangle} b)$ describe the probabilistic xor-choice between action a and b where $\Lambda(a) = x$ and $\Lambda(b) = y$. The object net that models this game is shown as a net-token in Fig. 4; it is a parallel composition (denoted by $\_\|\_$) of two choices (for some initial values of $x_0, y_0, x_1,$ and y_1):

$$N_1^\Lambda = (a_0^{\langle x_0 \rangle} \oplus^{\langle y_0 \rangle} b_0) \parallel (a_1^{\langle x_1 \rangle} \oplus^{\langle y_1 \rangle} b_1) \tag{12}$$

The system net observes the decision history and adapts by modifying the rates (cf. the EHORNET in Fig. 4). We have four transitions named play game on the right side corresponding to the four different ways of choosing the actions. We give the payoff as a reward signal to the agents. (There might be more appropriate ways of adapting, but for this simple example we do not care about the efficiency of the learning process.) For example, when the agents play (a_0, a_1) then we update the rates in the the workflow by the payoff $(3, 1)$ and we obtain:

$$N_2^\Lambda = (a_0^{\langle x_0+3 \rangle} \oplus^{\langle y_0 \rangle} b_0) \parallel (a_1^{\langle x_1+1 \rangle} \oplus^{\langle y_1 \rangle} b_1) \tag{13}$$

We have another source of adaption in the system net: Choices that are chosen quite regularly over a longer time period are converted into fixed structures without choice by the two transitions named adapt XOR on the left-hand side. In this example the transformation is allowed whenever a_0 is chosen in more than 80% of the time. This is expressed by the transition guard $\frac{x_0}{(x_0+y_0)} > 0.8$. Then, we obtain $N_3^\Lambda = a_0 \parallel (a_1^{\langle x_1 \rangle} \oplus^{\langle y_1 \rangle} b_1)$ as the modified net structure. Analogously, whenever b_0 dominates. (For simplicity we omit modifications in the model whenever the second agent has a dominating option.)

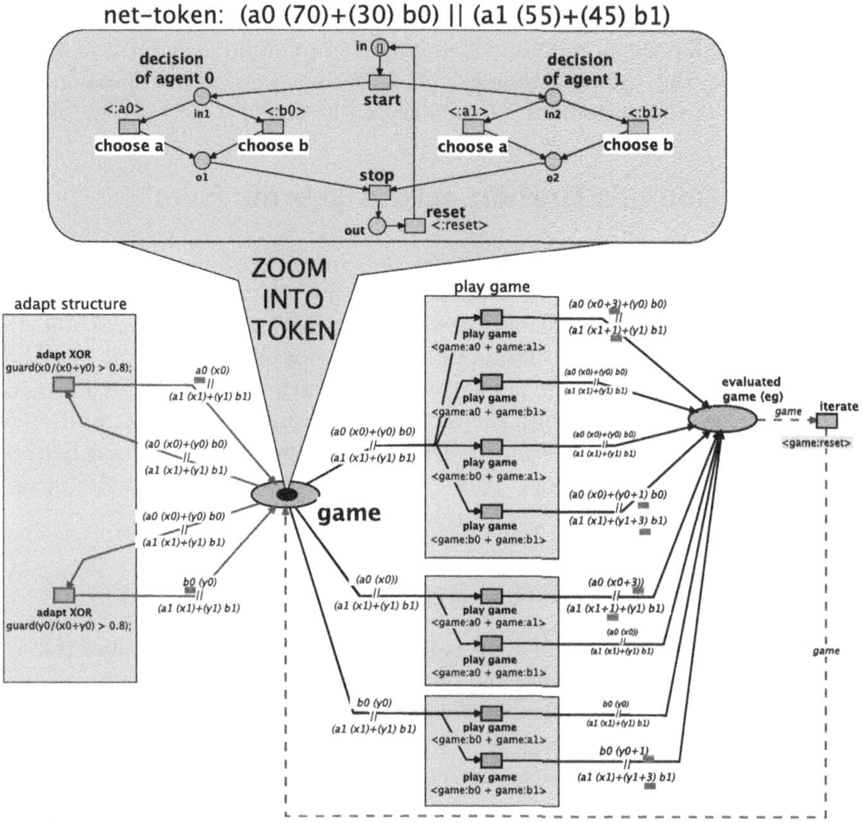

net–token: (a0 (70)+(30) b0) || (a1 (55)+(45) b1)

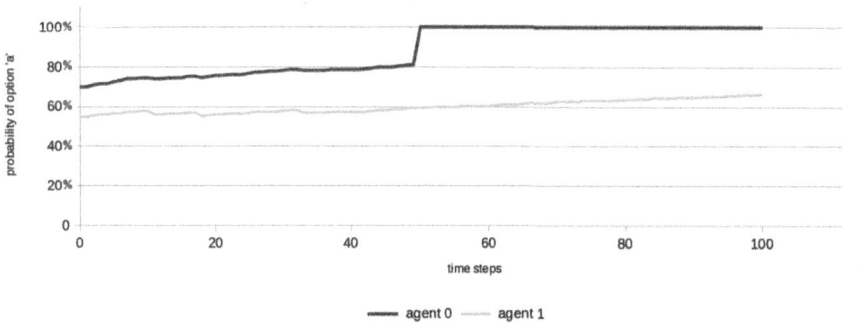

Fig. 4. The Probabilistic EHORNET: System-Net containing the Battle-of-Sexes Interaction (right) and the Structural Adaption Logic (left)

Fig. 5. A Sample Run: The Dynamics of the Probabilities of Options a_0 and a_1

For an example run (with initial rates $x_0 = 70$, $y_0 = 30$, $x_1 = 55$, and $y_1 = 45$) the probabilities of choosing option a is given in Fig. 5. We choose $\gamma = 0.5$ to balance the frequencies of structural modifications and rate updates. One can clearly observe that the update rule increases the probabilities in favor of options a_0 and a_1. Note, the 'jump' at time $t \approx 50$: Here, a structural modification takes place, which sets the probability of choosing a_0 to 100%.

6   Translation of EHornets into Algebraic Nets

In recent application studies we observed that our models have the interesting structural property that the system net does not combine or distribute net-tokens of the same net type (no forks or joins). For this sub-class of (probabilistic) EHORNETS we generate a translation into a *flat* probabilistic algebraic Petri net, which would make it possible to use conventional Petri net tools (e.g., CPN-Tools [11] or GreatSPN [2] to generate the model's state space and export a discrete-time Markov chain (DTMC) from it. Due to the different abstraction between algebraic nets and coloured nets, however, we decided to use Maude [9] as a modelling framework.

6.1   Generalised State Machines (GSM-eHornet)

In [14] we defined a sub-class of EOS, called Generalised State Machines (GSM), which have exactly one incoming and one outgoing arc for each net type. Consequently, the number of net-tokens for a given type never changes. Since a GSM has exactly one net-token of each type, all tokens of object net places are located inside the same net-token, i.e., whenever we know the system net place $\widehat{p}$ where the unique net-tokens is located, we know the location of all the object-nets's tokens. For EHORNETS we can define a similar concept: For each kind k each event of a GSM-EHORNET removes one unique net-token for the object net N_k in the preset and one net-token for N'_k in the postset – or the kind does not occur in the event – and the initial marking contains at most one net-token of kind k.

Definition 4. *An* EHORNET *is a* GSM-EHORNET *whenever for each event* $\theta = \widehat{\tau}^\alpha[\vartheta]$ *we have for all* $k \in K$: $|\Pi_k^2(\mu_0)| \leq 1$ *and*

$$\sum_{N \in \mathcal{U}_k} \sum_{\widehat{p} \in \widehat{P}} \mathbf{pre}_\alpha(\widehat{\tau})(\widehat{p})(N) = \sum_{N \in \mathcal{U}_k} \sum_{\widehat{p} \in \widehat{P}} \mathbf{post}_\alpha(\widehat{\tau})(\widehat{p})(N) \in \{0,1\}$$

The definition of a GSM-EHORNET implies that there is at most one place $\widehat{p}_k$ for each kind $k \in K$ in the preset of a transition (analogously for the postset). Additionally, for this place, the algebraic inscription $\mathbf{pre}(\widehat{\tau})(\widehat{p}_k)$ must also be a singleton multiset, i.e., we have $\mathsf{N}_k := \mathbf{pre}(\widehat{\tau})(\widehat{p}_k)$. Analogously, we have the term $\mathsf{N}'_k := \mathbf{post}(\widehat{\tau})(\widehat{p}'_k)$ for the unique place $\widehat{p}'_k$ in the postset.

For each channel c_i we have to select one transition t_i that is able to synchronise on c_i: $l_N(t_i) = c_i$. Concretely, $Y(k)$ maps k onto the multiset of synchronised object net transitions: $Y(k) = t_1 + \cdots + t_{|\widehat{l}^k(\widehat{t})|}$. Note, that we cannot be sure that the selected transitions t_i are present in the net-tokens, since we haven't bound the variables N_k and N'_k to object-nets, yet. Therefore, we call $\widehat{\tau}[Y]$ a *pseudo-synchronisation* and check the condition that the synchronised net-tokens do really *contain* the synchronised transitions as a firing guard:

$$\psi_{alg}(\widehat{\tau}[Y]) := \forall k \in K : dom(Y(k)) \subseteq \left(T_{\alpha(N_k)} \cup T_{\alpha(N'_k)} \right) \tag{14}$$

The set of all such *pseudo-synchronisation* $\widehat{\tau}[Y]$ is denoted Θ_{alg}.

Proposition 1. *Let EH be a GSM-*EHORNET*. The firing rule for the $\widehat{\tau}[Y]$ can be expressed as the following conditional multiset rewrite rule:*

$$\xrightarrow[EH]{\widehat{\tau}[Y]} \frac{\sum_{k \in K(\tau)} \widehat{p}_k \left[N_k, M_k^0 + \mathbf{pre}_k^\sharp(Y(k)) \right]}{\sum_{k \in K(\tau)} \widehat{p}'_k \left[N'_k, M_k^0 + \mathbf{post}_k^\sharp(Y(k)) \right]} \quad if \quad \widehat{G}(\widehat{\tau}) \wedge \psi_{alg}(\widehat{\tau}[Y]) \tag{15}$$

The firing rates for this rule is obtained by evaluating the expressions N_k with the variable binding α to obtain the concrete rate function $\Lambda_{\alpha(N_k)}$:

$$\Lambda_{pr}^\alpha(\widehat{\tau}[Y]) := (\sigma \circ \Lambda)(\widehat{\tau}) \cdot \prod_{k \in K(\tau)} \prod_{t \in Y(k)} \left((\sigma \circ \Lambda_{\alpha(N_k)})(t) \right)^{|Y(k)(t)|} \tag{16}$$

Note that due to the nesting we cannot drop the object net context M_k^0 in (15).

6.2   Flattening the GSM-eHornets

Analogously to our "flattening" construction for EOS used in [14, Thm. 6.2] we can define the "flat" version of an EHORNET, i.e., a P/T net, called the *reference net*. We obtain the flat, un-nested structure by removing the nesting of multisets:

$$\text{R}_\text{N} : \quad \widehat{p}\,[N, M] \quad \mapsto \quad \widehat{p}(N) + M \tag{17}$$

Here, we use the coloured token $\widehat{p}(N)$ to denote that $\widehat{p}$ contains a net-token of the object net N. This information is needed especially for object-autonomous events $\widehat{\tau}^\alpha[\vartheta]$, where $\tau = id_{\widehat{p}_0, N}$ to check that the object net transition that is fired is really contained in the object net N.

In general, this flattening construction does not preserve the behaviour: Each firing of an EOS can be simulated by the reference net, but not vice versa [14, Thm. 6.1]. However, for the sub-class of GSM, which are fork/join-free EOS, the reference net bisimulates the GSM [14, Thm. 6.2].

Instead of unfolding the typed system net place of the EHORNET into an uncoloured version in the EOS we generate an *algebraic* Petri net. From Prop. 1 we obtain the flattened rewrite rule:

$$\xrightarrow{\text{R}_\text{N}(\widehat{\tau}[Y])} \frac{\left(\sum_{k \in K(\tau)} \widehat{p}_k(N_k) + M_k^0 + \mathbf{pre}_k^\sharp(Y(k)) \right)}{\left(\sum_{k \in K(\tau)} \widehat{p}'_k(N'_k) + M_k^0 + \mathbf{post}_k^\sharp(Y(k)) \right)} \quad if \widehat{G}(\widehat{\tau}) \wedge \psi_{alg}(\widehat{\tau}[Y]) \tag{18}$$

In the following, we omit the object net context M_k^0, which can be done without changing the behaviour, since M_k^0 is un-constrained.

Example 5. The rule for synchronous firing from (18) also subsumes autonomous firing as special cases:

1. For system-autonomous transitions we have no synchronisations, i.e., $Y(k) = \mathbf{0}$:

$$\left(\sum_{k \in K(\tau)} \widehat{p}_k(\mathrm{N}_k) \right) \xrightarrow{\frac{\mathrm{RN}(\widehat{\tau}[Y])}{\mathrm{RN}(EH)}} \left(\sum_{k \in K(\tau)} \widehat{p}_k'(\mathrm{N}_k') \right) \text{ if } \widehat{G}(\widehat{\tau})$$

2. Similarly, object-autonomous is expressed with $\tau = id_{\widehat{p}_0, \mathrm{N}_k}$ and $Y(k) = t$ for $k = k(\widehat{p}_0)$ and $Y(k) = \mathbf{0}$ otherwise:

$$\left(\widehat{p}_0(\mathrm{N}_k) + \mathbf{pre}_k^\#(t) \right) \xrightarrow{\frac{\mathrm{RN}(id_{\widehat{p}_0, \mathrm{N}_k}[Y])}{\mathrm{RN}(EH)}} \left(\widehat{p}_0(\mathrm{N}_k) + \mathbf{post}_k^\#(t) \right) \text{ if } t \in T_{\alpha(\mathrm{N}_k)}$$

Note, that the side condition $\widehat{p}_0(\mathrm{N}_k)$ is needed to check whether the object net really contains the object net transition t.

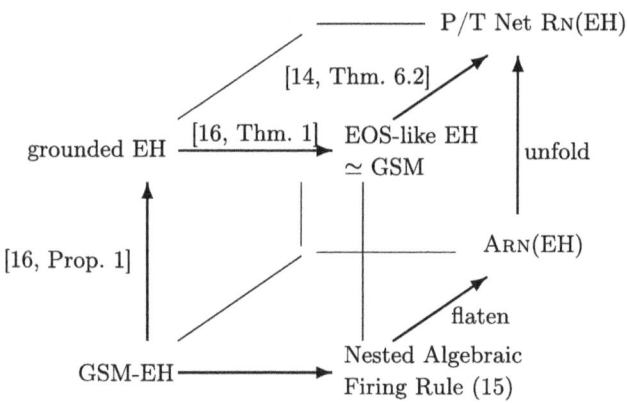

Fig. 6. Translation Paths of EHORNETS into un-nested Petri Nets

Since the rule in (18) is un-nested we finally obtain an *algebraic reference net*, which is simply the collection of all the conditional multiset rewrite rules interpreted as a Petri net. We consider the union of the places of the system-net and the places of all the object-nets. Here, places of the system-net have the same type as before and the places from the objects nets carry the typed `black-token`, which is interpreted by the universe $\mathcal{U}_{\texttt{black-token}}$ of nets with an empty place set: $P_{\texttt{black-token}} = \emptyset$. So, all net-tokens must have $\mathbf{0}$ as its marking. Each transition $\mathrm{RN}(\widehat{\tau}[Y])$ is labeled with the rate expression $\Lambda_{pr}^\alpha(\mathrm{RN}(\widehat{\tau}[Y]))$.

Definition 5. *Let* $EH = (\widehat{N}, \mathcal{U}, \mathcal{I}, k, l, \mu_0)$ *be a GSM-EHORNET. We define the algebraic reference net* $\mathrm{ARN}(EH) = (P_A, T_A, d_A, \mathbf{pre}_A, \mathbf{post}_A, G_A, m_A)$, *where* $P_A = \widehat{P} \cup \bigcup_{k \in K} P_k$, $T_A = \Theta_{alg}$, $d_A(p) = k(p)$ *whenever* $p \in \widehat{P}$ *and* $d_A(p) =$

black-token otherwise, and $m_A = \mathrm{Rn}^\#(\mu_0)$. *The arc inscriptions* $\mathbf{pre}_A$, $\mathbf{post}_A$, *and guards* G_A *are given by the flattened firing rule (18).*

For a probabilistic EHORNET $PEH = (EH, \Lambda)$ be a we use (16) to obtain the probabilistic algebraic reference net $\mathrm{ARN}(PEH) := (\mathrm{ARN}(EH), \Lambda^\alpha_{pr})$.

In analogy to the result for GSM, where the reference net bisimulates the GSM [14, Thm. 6.2], we obtain the same result for GSM-EHORNET.

Proposition 2. *Let EH be a* GSM-EHORNET. *Then the algebraic reference net* $\mathrm{ARN}(EH)$ *is bisimilar to EH.*

Proof (Sketch). Follows directly from the construction of the flattened firing rule (18) and the fact that for a GSM-EHORNET we have a one-to-one correspondence of kind $k \in K$ and the net-tokens of this kind for each reachable marking. □

Our construction preserves the algebraic structure, but translates into an un-nested (i.e., flat) model. Figure 6 summarises the relationship of the current translation in a three-dimensional cube: Going 'up' unfolds the bindings (i.e., a variable substitution, a grounding); going to the 'right' translates an EHORNET into an equivalent (bisimilar) structure (here: either an Eos-like EHORNET or algebraic rewrite rules); and going to the 'rear' removes the nesting structure, i.e., it translates a nets-within-nets formalism into a flat net. Note, that grounding and the translation into an Eos-like EHORNET as defined in [16, Thm. 1] is possible for any EHORNET, not only for GSM-EHORNETS. Additionally, it is always possible to flaten an Eos or an EHORNET, but the resulting p/t net is not bismilar in general – this is guaranteed for GSM only. Therefore, we start the constructions only for GSM-EHORNETS. In this case the construction of the Eos-like EHORNET in fact generates a GSM. In [16] we worked out the "upper" path (up-right-rear), while this presentation proposes to take the "lower" path (right-rear-up). Both construction paths will generate the so-called reference net (i.e., a p/t net), which is bisimilar to the EHORNET, whenever we restrict ourselves to GSM-EHORNETS. However, as the lower, "algebraic" path of the diagram preserves the variables in the inscriptions it allows algebraic reduction techniques, e.g., techniques based on symmetries, like [8].

6.3 Construction of a DTMC for a GSM-eHornet

The EHORNET PEH in Fig. 4, which models the coordination game, is a GSM-EHORNET. The transformation into a probabilistic algebraic net $\mathrm{ARN}(PEH)$ is shown in Fig. 7. The huge number of arcs is mainly due to the synchronization, which leads to a combinatoric explosion of events, i.e., we have a small number of transitions, but exponentially many events.

The state space of our algebraic net and the firing rates generate a DTMC. In principle, all states are very simple, since we only have two system net places and we have exactly one net-token in each marking – each reachable marking has the form $\mu = \widehat{p}[N^\Lambda, M]$ where $\widehat{p} = $ game or $\widehat{p} = $ evaluated game (eg); we have four different markings M within each net-token $[N^\Lambda, M]$; finally, the structure N^A of the net-tokens has the form $N^A = (N_0 \| N_1)^A$, where $N_0 \in \{(a_0{}^{\langle x_0 \rangle} \oplus^{\langle y_0 \rangle} b_0), a_0, b_0\}$ and $N_1 = (a_1{}^{\langle x_1 \rangle} \oplus^{\langle y_1 \rangle} b_1)$. However, note that we have an infinite state space as the rates $x_i, y_i, i = 0, 1$ are unbounded.

We use Maude [9] to generate all possible transition bindings and a reachability graph (RG) for a GSM-eHORNET that includes all the information to derive the DTMC stochastic matrix. Conceptually, translating an algebraic net into Maude is quite simple, as rewriting logic subsumes multiset rewriting, which includes Petri nets as a special case [29]. However, to automatically generate the RG we must pay attention to the Maude pattern-matching mechanism. Translating the eHORNETS transitions into Maude rewrite rules requires some syntactical manipulation of transition inscriptions to adhere to this mechanism perfectly. A special encoding of markings using *canonical forms* [5] allows us to operate modulo object-net isomorphism. The resulting specification[1] of eHORNETS is a compromise between readability on one side and efficiency on the other. In this contribution we do not investigate the details of our Maude specification. Instead we like to demonstrate the possibility to generate the RG using the Maude search feature.

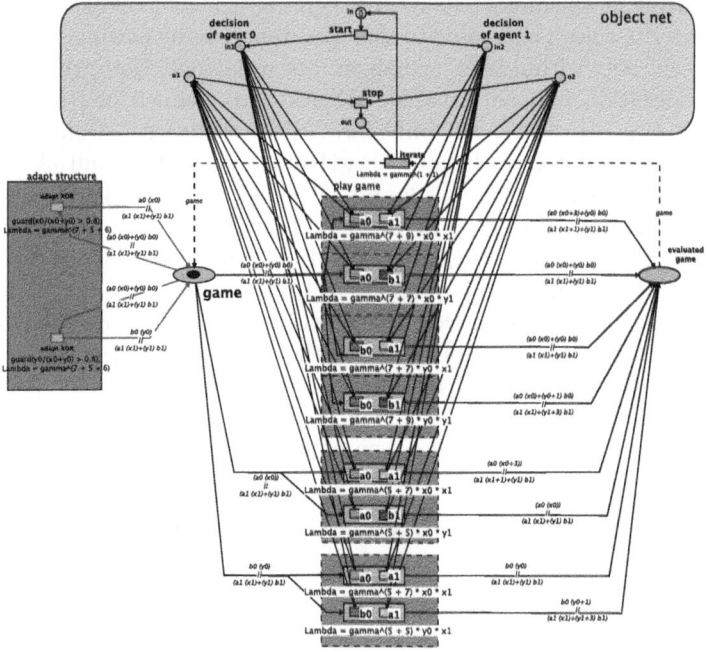

Fig. 7. The eHORNET of Fig. 4 translated into a Probabilistic Algebraic Net

An initial part of the reachability graph for the coordination game is shown in Fig. 8. For simplicity, we show the rates $(x_0, y_0; x_1, y_1)$ for each state block.

---

[1] Source: https://github.com/lgcapra/rewpt/tree/main/algPT/EHornets.

Since rates are never decreased we have a branching 'back-bone' with some loops. Branching occurs whenever the game has received a non-zero payoff, i.e., agents coordinate; otherwise we reproduce the state and obtain a loop. The state space generates the following stochastic matrix $\mathbf{P}(\mu_i, \mu_j)$:

$\mathbf{P}(\mu_i, \mu_j)$	0	1	2	3	4	5	6	7	$\ldots$	x_0	y_0	x_1	y_1
0		1							$\ldots$	70	30	55	45
1			.385	.480	.135				$\ldots$	70	30	55	45
2						1			$\ldots$	**73**	30	**56**	45
$\vdots$	$\vdots$	$\vdots$	$\vdots$	$\vdots$	$\vdots$	$\vdots$	$\vdots$	$\vdots$	$\ddots$	$\vdots$	$\vdots$	$\vdots$	$\vdots$

In this paper, we are mainly concerned with the principle representation of states and calculation of successors. Our Maude specification generates a bounded expansion of the state space as we do not expand states whenever the rates are greater than max_rate. We will postpone the question of how to analyse an infinite process – using either a bounded approximation of the state space or a symbolic representation – to future research.

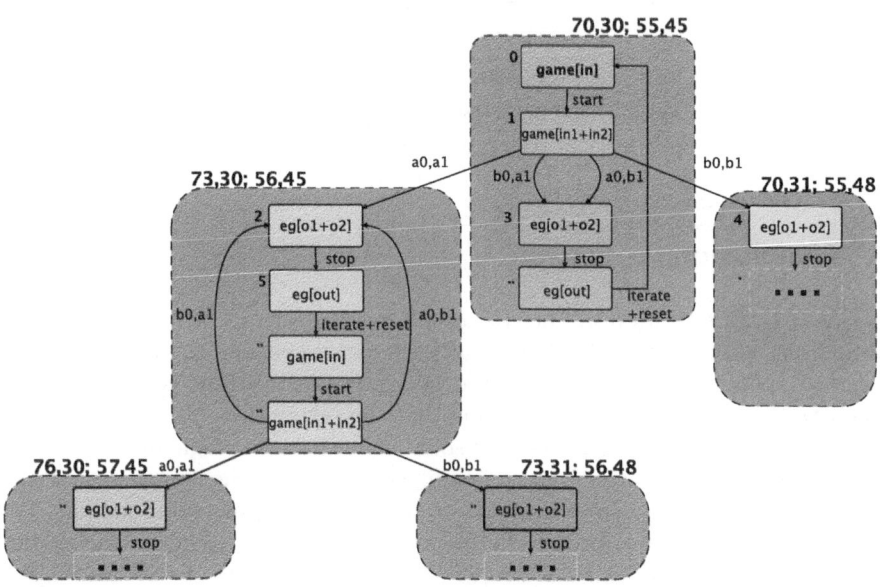

Fig. 8. Initial Part of the Reachability Graph

From Fig. 8 one might expect an exponential number of states after g (successful) games. However, since any permutation of games leads to the same state, we obtain a much smaller number. Each 'block' of four states is described by (i, j), where i and j denote the number of games coordinating on a and b, respectively. Since $i + j \leq g$ we have $O(g^2)$ states. As the average payoff is 2 we expect $g \approx$ max_rate$/2$.

In Table 1 we report the state space size as the max_rate parameter varies, along with the time taken by the Maude (v3.5) search command. The results match our theoretical expectations. Using the Maude facilities, we have model-checked that the DTMC contains several closed communicating classes (each built of four states and with its own unique stationary distribution) that are entered after an initial transient preamble whose length depends on max_rate.

In ongoing work we analyse the generated DTMC with external tools. A preliminary analysis allowed us to check e.g. that the probability to enter a communicating class in at most $n = 100$ steps is 0.77 for and max_rate $= 125$ and 0.81 for max_rate $= 250$ – which is consistent with the example run of Fig. 5. Currently, we evaluate the Python library PyDTMC[2] for analysis purposes.

Table 1. State Space Sizes and Generation Times (intel core i5 12th gen, 32 GB RAM)

max_rate	#states	#edges ($\simeq$)	time (s)
125	2,060	3,700	0.5
250	19,784	36,000	7
500	111,872	200,000	40
1000	527,696	980,000	159
2000	2,297,480	4,300,000	887
4000	9,622,024	18,200,000	6932

7   Conclusion

In this contribution we introduced *Probabilistic* EHORNETS, a Nets-within-Nets formalism, where the system net and each net-token are equipped with firing rates. Our formalism is well-suited to express quantitative aspects for self-adaptive systems. It is a typical setting that the system-net describes the adaption loop and the net-tokens describe some kind of process logic (e.g., protocols or workflow nets) that are subject to modification at run-time. For these and similar scenarios the firing rates arise naturally from the application domain: The workflows contain rates for xor-choices and for adaption events in the system net we argued that the rate is proportional to the inverse of the transformation complexity $TC(\hat{t})$. For the special class of GSM-EHORNETS we obtained a translation into probabilistic algebraic nets and their reachability state space is nothing different than a DTMC. Here, we used a Maude implementation of EHORNETS to generate the states and probabilities of the Markov chain.

---

[2] https://pypi.org/project/PyDTMC/.

In ongoing work we will deepen the aspect of formal analysis for these stochastic EHORNETS. Even for bounded state spaces the state space of an EHORNET grows much faster than that of a p/t net: The reachability problem requires exponential space even for safe EHORNETS [13,16], while it is well-known that it is in PSpace for safe p/t nets. Therefore, we like to exploit symmetry in the structure of an EHORNET using the concept of automorphism (cf. our previous definition of an automorphism for EOS in [6]). Another approach is to use a symbolic representation of the state space itself, e.g., we could use a BDD-representation as used in the probabilistic model checker PRISM [20].

We are also investigating approximation techniques: A major source of complexity is the interaction of the system net with the object nets and the huge size of possible object nets – even for small numbers of places. We already know that EHORNET firing is preserved by projection onto the system net, i.e., for each event $\mu \xrightarrow{\widehat{t}[\vartheta_\alpha]} \mu'$ the system net transition $\widehat{t} = \Pi^1(\widehat{t}[\vartheta_\alpha])$ would be enabled in the system net $\widehat{N}$ for the marking $\Pi^1(\mu)$; the contrary does not hold in general. Therefore, we like to study to which extend we can approximate our model when considering the – much smaller – projection of the state space onto the system net alone.

References

1. Aalst, W.: Verification of workflow nets. In: Azéma, P., Balbo, G. (eds.) ICATPN 1997. LNCS, vol. 1248, pp. 407–426. Springer, Heidelberg (1997). https://doi.org/10.1007/3-540-63139-9_48

2. Amparore, E.G., Balbo, G., Beccuti, M., Donatelli, S., Franceschinis, G.: 30 years of GreatSPN. In: Fiondella, L., Puliafito, A. (eds.) Principles of Performance and Reliability Modeling and Evaluation. SSRE, pp. 227–254. Springer, Cham (2016). https://doi.org/10.1007/978-3-319-30599-8_9

3. Baeten, J., Bergstra, J.A., Smolka, S.A.: Axiomatizing probabilistic processes: ACP with generative probabilities. In: Cleaveland, W.R. (ed.) CONCUR 1992. LNCS, vol. 630, pp. 472–485. Springer, Heidelberg (1992). https://doi.org/10.1007/BFb0084810

4. Calinescu, R., Ghezzi, C., Kwiatkowska, M., Mirandola, R.: Self-adaptive software needs quantitative verification at runtime. Commun. ACM 55(9), 69–77 (2012). https://doi.org/10.1145/2330667.2330686

5. Capra, L.: Canonization of reconfigurable pt nets in Maude. In: Lin, A.W., Zetzsche, G., Potapov, I. (eds.) Reachability Problems, pp. 160–177. Springer, Cham (2022). https://doi.org/10.1007/978-3-031-19135-0_11

6. Capra, L., Köhler-Bußmeier, M.: A "symbolic" representation of object-nets. In: Bramas, Q., et al. (eds.) Distributed Computing and Intelligent Technology - 21th International Conference, ICDCIT25. vol. 15507, pp. 68–74. Springer, Cham (2025). https://doi.org/10.1007/978-3-031-81404-4_6. https://arxiv.org/abs/2411.00149

7. Cardelli, L., Ghelli, G., Gordon, A.D.: Mobility types for mobile ambients. In: Wiedermann, J., van Emde Boas, P., Nielsen, M. (eds.) ICALP 1999. LNCS, vol. 1644, pp. 230–239. Springer, Heidelberg (1999). https://doi.org/10.1007/3-540-48523-6_20

8. Chiola, G., Dutheillet, C., Franceschinis, G., Haddad, S.: On well-formed coloured nets and their symbolic reachability graph. In: Rozenberg, G. (ed.) Proceedings of the 11th International Conference on Application and Theory of Petri Nets. Lecture Notes in Computer Science, vol. 524, pp. 387–410. Springer, Heidelberg (1990). https://doi.org/10.1007/978-3-642-84524-6_13
9. Clavel, M., et al. (eds.): All About Maude - A High-Performance Logical Framework, How to Specify, Program and Verify Systems in Rewriting Logic, Lecture Notes in Computer Science, vol. 4350. Springer, Heidelberg (2007). https://doi.org/10.1007/978-3-540-71999-1
10. Ehrig, H., Mahr, B.: Fundamentals of algebraic Specification. EATCS Monographs on TCS, Springer, Heidelberg (1985)
11. Jensen, K., Kristensen, L.M.: Coloured Petri Nets: Modelling and Validation of Concurrent Systems. Springer, Heidelberg (2009). https://doi.org/10.1007/b95112
12. Köhler-Bußmeier, M.: Hornets: nets within nets combined with net algebra. In: Franceschinis, G., Wolf, K. (eds.) PETRI NETS 2009. LNCS, vol. 5606, pp. 243–262. Springer, Heidelberg (2009). https://doi.org/10.1007/978-3-642-02424-5_15
13. Köhler-Bußmeier, M.: On the complexity of the reachability problem for safe, elementary Hornets. Fundamenta Informaticae **129**, 101–116 (2014)
14. Köhler-Bußmeier, M.: A survey on decidability results for elementary object systems. Fund. Inf. **130**(1), 99–123 (2014)
15. Köhler-Bußmeier, M., Heitmann, F.: On the expressiveness of communication channels for object nets. Fund. Inf. **93**(1–3), 205–219 (2009)
16. Köhler-Bußmeier, M., Heitmann, F.: Complexity results for elementary hornets. In: Colom, J.-M., Desel, J. (eds.) PETRI NETS 2013. LNCS, vol. 7927, pp. 150–169. Springer, Heidelberg (2013). https://doi.org/10.1007/978-3-642-38697-8_9
17. Köhler-Bußmeier, M., Sudeikat, J.: Studying the micro-macro-dynamics in MAPE-like adaption processes. In: Intelligent Distributed Computing XVI. IDC 2023. Studies in Computational Intelligence, vol. 1138. Springer, Heidelberg (2024).https://doi.org/10.1007/978-3-031-60023-4_24
18. Köhler-Bußmeier, M., Wester-Ebbinghaus, M.: Sonar*: a multi-agent infrastructure for active application architectures and inter-organisational information systems. In: Braubach, L., van der Hoek, W., Petta, P., Pokahr, A. (eds.) Conference on Multi-Agent System Technologies, MATES 2009. Lecture Notes in Artificial Intelligence, vol. 5774, pp. 248–257 (2009). https://doi.org/10.1007/978-3-642-04143-3_27
19. Köhler-Bußmeier, M., Wester-Ebbinghaus, M., Moldt, D.: A formal model for organisational structures behind process-aware information systems. In: Transactions on Petri Nets and Other Models of Concurrency. Special Issue on Concurrency in Process-Aware Information Systems, vol. 5460, pp. 98–114 (2009). https://doi.org/10.1007/978-3-642-00899-3_6
20. Kwiatkowska, M., Norman, G., Parker, D.: PRISM 4.0: verification of probabilistic real-time systems. In: Gopalakrishnan, G., Qadeer, S. (eds.) CAV 2011. LNCS, vol. 6806, pp. 585–591. Springer, Heidelberg (2011). https://doi.org/10.1007/978-3-642-22110-1_47
21. Kwiatkowska, M., Norman, G., Parker, D., Vigliotti, M.G.: Probabilistic mobile ambients. Theor. Comput. Sci. **410**(12), 1272–1303 (2009). https://doi.org/10.1016/j.tcs.2008.12.058
22. Lomazova, I.A.: Nested Petri nets - a formalism for specification of multi-agent distributed systems. Fund. Inf. **43**(1–4), 195–214 (2000)
23. Marsan, M.A.: Stochastic Petri Nets: An Elementary Introduction, pp. 1–29. Springer, Heidelberg (1990)

24. Marsan, M.A., Balbo, G., Conte, G., Donatelli, S., Franceschinis, G.: Modelling with Generalized Stochastic Petri Nets. Wiley Series in Parallel Computing, John Wiley and Sons, Hoboken (1995)
25. Milner, R., Parrow, J., Walker, D.: A calculus of mobile processes, parts 1–2. Inf. Comput. **100**(1), 1–77 (1992)
26. Pradalier, S., Palamidessi, C.: Expressiveness of probabilistic π-calculus. Electron. Notes Theor. Comput. Sci. **164**(3), 119–136 (2006). https://doi.org/10.1016/j.entcs.2006.07.015
27. Rabin, M.O.: Probabilistic automata. Inf. Control **6**(3), 230–245 (1963). https://doi.org/10.1016/S0019-9958(63)90290-0
28. Reisig, W.: Petri nets and algebraic specifications. Theor. Comput. Sci. **80**, 1–34 (1991)
29. Stehr, M.-O., Meseguer, J., Ölveczky, P.C.: Rewriting logic as a unifying framework for petri nets. In: Ehrig, H., Padberg, J., Juhás, G., Rozenberg, G. (eds.) Unifying Petri Nets. LNCS, vol. 2128, pp. 250–303. Springer, Heidelberg (2001). https://doi.org/10.1007/3-540-45541-8_9
30. Sudeikat, J., Köhler-Bußmeier, M.: On combining domain modeling and organizational modeling for developing adaptive cyber-physical systems. In: Proceedings of the 14th International Conference on Agents and Artificial Intelligence, pp. 330–336. INSTICC, SciTePress (2022). https://doi.org/10.5220/0010881200003116
31. Valk, R.: Object petri nets. In: Desel, J., Reisig, W., Rozenberg, G. (eds.) ACPN 2003. LNCS, vol. 3098, pp. 819–848. Springer, Heidelberg (2004). https://doi.org/10.1007/978-3-540-27755-2_23
32. Weyns, D.: An Introduction to Self-Adaptive Systems: A Contemporary Software Engineering Perspective. John Wiley & Sons Ltd., Hoboken (2020)

Enjoy the Silence, Part II: Probability-Based Queries on Stochastic Labelled Petri Nets

Sander J. J. Leemans[1,2]([✉]), Marco Montali[3], Timo Gersing[1],
Felix Engelhardt[1], and Natalia Sidorova[4]

[1] RWTH Aachen University, Aachen, Germany
s.leemans@bpm.rwth-aachen.de
[2] Fraunhofer, Aachen, Germany
[3] Free University of Bozen-Bolzano, Bolzano, Italy
[4] Eindhoven University of Technology, Eindhoven, The Netherlands

Abstract. A stochastic process model combines control flow and stochasticity in a single representation. Answering queries on the behaviour and probabilities of such a model is essential not only for analysis and verification, but also towards stochastic process mining, in which the frequency of events and traces is explicitly taken into account. In this paper, we focus on probability-based queries on stochastic process models, dealing with questions like "what are the 10 most likely traces?" and "what are the traces with a probability higher than 1%?", and "what are the most likely traces that together cover 80% of the behaviour in the model?". We formalise these queries in the setting where the model is represented as a stochastic labelled Petri net with repeated labels and silent transitions. We provide a representation of the stochastic deterministic state space induced by the net suitable for answering probability-based queries, and introduce an algorithm to do so. Finally, we implement our approach and evaluate its applicability and feasibility on real-life event logs.

Keywords: stochastic labelled Petri net · stochastic process mining · probability queries

1 Introduction

Capturing dynamic systems by combining behaviour and stochasticity has a long tradition in the theory and applications of Petri nets [4,24]. Interest in these models has recently resurged within process science, in order to capture business processes using stochasticity to quantitatively tackle uncertainty on the duration of activities and/or the way deferred choices are resolved.

In modelling and mining, dealing with stochastic processes is central. Process mining aims to provide business analysts with techniques to analyse business processes, based on traces (sequences of process steps) of process instances recorded in information systems. A process model, such as a Petri net, describes the process control-flow and can be obtained from an event log through automated process discovery, from domain knowledge or from industry process benchmarks [1,8].

© The Author(s), under exclusive license to Springer Nature Switzerland AG 2025
E. Amparore and L. Mikulski (Eds.): PETRI NETS 2025, LNCS 15714, pp. 310–332, 2025.
https://doi.org/10.1007/978-3-031-94634-9_15

While trace frequencies have always been central when analysing processes, e.g., to distinguish and compare common and infrequent behaviour and to single out outliers, traditional process mining techniques do not convey this information at the model level. In particular, a decision point in the process is described as a non-deterministic choice in the Petri net, while trace frequencies in the log may be used to provide more refined, quantitative information.

This missing link has recently been tackled in *stochastic process mining*, where non-deterministic choices in process models are refined into stochastic choices that quantify the relative probability of selecting a route from the available options, moment by moment. Consequently, different challenges emerge with frequencies, stochastic choices and other forms of uncertainty [5].

When dealing with stochastic process modelling and mining, stochastic Petri nets constitute a natural candidate formalism, with three essential features that are not readily supported by traditional approaches [4,24]:

- Transitions are optionally labelled with activity names, and different transitions may have the same labels.
- Unlabelled (*silent*) transitions represent internal steps in the process, which are not logged. Silent transitions are omnipresent in process mining [1] to represent optional behaviour and process gateways (i.e. routing points).
- Process executions consist of unbounded, yet finitely many, steps, hence nets need to have a clear notion of termination, with a probability of termination attached. We solve this by treating every deadlock as a final marking.

These three features are tackled in so-called *stochastic labelled Petri nets* (SLPNs) [19], which have been extensively targeted in process mining to tackle discovery [3,14,18,27], conformance checking [15,19,20] and performance analysis [14]. However, models with unbounded state spaces and livelocks, i.e., progression is made but a final state cannot be reached, are typically not supported.

Conducting model analysis and verification in this spectrum, as well as dealing with process mining tasks, calls for analytic techniques of *querying* SLPNs and retrieving combined information on probabilities and model behaviour. Additionally, queries can be used to inspect a process model and extract useful insights, and constitute an essential building block to compute conformance and performance indicators. There are three main classes of queries: *state queries* - returning the probability that the process reaches a given marking; *behavioural queries* - returning the probability that the process generates a given trace or a trace that satisfies a given temporal specification; and *probability queries* - returning traces that meet a given condition on probabilities.

In particular, [19] has shown that state and behavioural queries can be homogeneously answered over *bounded* SLPNs by combining techniques from Markov chain analysis and formal verification. These queries are in turn useful to compute several conformance metrics that can otherwise only be approximated [22,23]. Such queries are particularly difficult to answer due to the presence of silent transitions, duplicated activities, and livelocks. The queries supported in [19] all take as input a given state or behaviour and return a probability mass. This leaves an important question, which we tackle in this paper: *how to answer probability-based queries*, that is, queries whose input refers to a probability, and whose output is a set of most-likely behaviour that fills that probability mass. Importantly, these ideas are *independent of the adopted stochastic modelling formalism* (e.g., SLPNs), but touch the core of the modelled stochastic behaviour. Examples of such queries are: (1) What are the 10 most-likely traces in the SLPN? (2) What are the traces of the SLPN that have a probability higher than 1%? (3) What are the most-likely traces that together cover 80% of the model? (4) After executing a trace prefix, in what markings can the system be with what probability? All these queries can be generalised to one problem: *Problem: answering probability queries.* Given a stochastic process model and a stopping criterion, find all traces in decreasing order of probability such that the criterion is met.

Probability queries enable process mining tasks related to: (1) *model inspection*, to obtain information on models that are too complex to be analysed manually; (2) *model validation* based on logs, such as sanity checks on the generalisation of stochastic process discovery techniques, for example to check whether the most frequent traces in the log remain so in the model; (3) *prediction*, querying the most likely outcomes given a prefix of a trace; and (4) *conformance checking*, since trace probabilities can be computed linearly on top of the structures we build to answer probability-based queries.

In this paper, we provide algorithmic techniques to answer probability queries over SLPNs, in a more general setting than that of [19] as we deal with *unbounded* nets with livelocks (Sect. 3). Reasoning directly over traces requires to on-the-fly removal of silent moves and merge distinct moves with identical labels, which is not supported by previous techniques. We provide a representation of the stochastic state space induced by the SLPNs that natively addresses these two aspects (Sect. 4). We then introduce an algorithm that iteratively explores such a state space to answer different types of probability queries (Sect. 5), guaranteeing termination in the large class of SLPNs without unbounded silent cycles under a conjecture – that is, loops of silent transitions that lead to unbounded markings (cf. Definition 10). The algorithm comes with an open-source implementation, which we use to evaluate applicability and feasibility of our approach on real-life event logs (Sect. 6). Finally, we relate our approach to existing literature (Sect. 7) and conclude the paper (Sect. 8).

2   Preliminaries

A weighted set is a function $M : E \to \mathbb{R}_{0+}$ from a set E of elements to the non-negative real numbers, mapping each element in E to its weight in the set. We denote with $\boldsymbol{M} \in \mathbb{R}_{0+}^{|E|}$ the incidence vector of a weighted set M. For weighted sets M and M' over E, we define multiset union $(M \uplus M')(x) = M(x) + M'(x)$ and difference $(M \setminus M')(x) = \max(0, M(x) - M'(x))$ for $x \in E$. We say that M is a subset of M', written $M \subseteq M'$, if $\forall_x M(x) \leq M'(x)$. In case all weights are non-negative integers, we refer to the weighted set as a *multiset*. The set of all multisets over E is denoted by $E^{\mathbb{N}}$.

A probability distribution (*q-set*) $Q\colon E \to \mathbb{R}_{0+}$ is a weighted set whose elements' weights sum to one: $\sum_{e \in E} Q(e) = 1$. We denote with $\widetilde{Q}$ the set $\{e \mid Q(e) > 0\}$ of the elements with a positive probability in Q.

Definition 1 (Stochastic language). *A* trace *is a sequence of* activities. *A* stochastic language *is a weighted set of traces, such that the sum of their probabilities is less than or equal to 1. A stochastic language is* complete *if it is a q-set, i.e., a probability distribution.*

A *Markov chain* is a directed graph whose edges are annotated with probabilities, that is, a tuple (S, p) where S is a possibly infinite set of states and $p\colon S \times S \to [0, 1]$ a function such that for every $s \in S$, $\sum_{s' \in S} p((s, s')) = 1$.

In this paper, we employ *absorbing* Markov chains, that is, Markov chains where states are partitioned into two sets: a set of *absorbing states* containing states without any successor (or, equivalently, having only a self-loop), and a set of *transient states* containing non-absorbing states that are directly or indirectly connected to at least one absorbing state.

3   Stochastic Labelled Petri Nets

We define stochastic labelled Petri nets (SLPNs) and their language. Furthermore, we explore several challenging aspects of SLPNs, which are not typically considered in process mining contexts [19,27]: non-positive transition weights (to accomodate dynamic weights in future work), livelocks and unboundedness.

Definition 2 (Stochastic labelled Petri net). *A* stochastic labelled Petri net *(SLPN) is a tuple* $(P, T, F, M_0, \Sigma, \lambda, w)$ *in which* P *is a finite set of places,* T *is a finite set of transitions such that* $P \cap T = \emptyset$, *and* $F \in ((P \times T) \cup (T \times P))^{\mathbb{N}}$ *is a flow relation. A marking* $M \in P^{\mathbb{N}}$ *is a multiset of places (indicating tokens on places), and we call* $M_0 \in P^{\mathbb{N}}$ *the initial marking.* Σ *is a finite alphabet of activities and* $\lambda\colon T \to \Sigma \cup \{\tau\}$ *is a labelling function mapping the transitions to the activities or to the silent transition* $\tau \notin \Sigma$. *Furthermore,* $w\colon T \to \mathbb{R}$ *is a weight function on the transitions.*

Fig. 1 shows an example of an SLPN.

We write ${}^\bullet t = [p \mid (p, t) \in F] \in P^{\mathbb{N}}$ and $t^\bullet = [p \mid (t, p) \in F] \in P^{\mathbb{N}}$ for the *preset* and *postset* of a transition $t \in T$. A transition $t \in T$ is *enabled* for a marking M if ${}^\bullet t \subseteq M \wedge w(t) > 0$. For an SLPN in a marking M, we write $M \xrightarrow{a} M'$ if any transition t with $\lambda(t) = a$ is enabled in M such that firing t leads to M', that is $M' = M \uplus t^\bullet \setminus {}^\bullet t$. Similarly, $M \xrightarrow{\tau} M'$ indicates that a silent transition can fire in M and lead to M'.

The probability that a transition t fires in a marking M is inversely proportional to the summed weight of the enabled transitions:

Definition 3 (Likelihood of transition firing). *Let* $(P, T, F, M_0, \Sigma, \lambda, w)$ *be an SLPN, let* $M \in P^{\mathbb{N}}$ *be a marking, and let* $t \in T$ *be a transition that is enabled in* M. *Then, the probability that* t *fires in* M *is* $p(M, t) = \dfrac{w(t)}{\sum_{t' \in T\colon {}^\bullet t' \subseteq M \wedge w(t') > 0} w(t')}$.

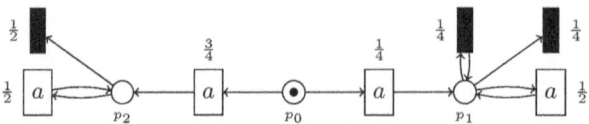

Fig. 1. An irregular SLPN. Silent transitions are black.

As usual, we call a marking M in which no transition is enabled a *deadlock*:

Definition 4 (Deadlock). *Let* $(P, T, F, M_0, \Sigma, \lambda, w)$ *be an SLPN, and let* M *be a marking. Then,* M *is a deadlock if and only if* $\nexists_{t \in T} {}^{\bullet}t \nsubseteq M \wedge w(t) > 0$.

A *partial path* $\langle t_1 \dots t_n \rangle$ of an SLPN is formed by transitions that sequentially bring the SLPN from its initial marking M_0 to the markings $M_1, \dots, M_n$ with $\forall_{1 \leq i \leq n} M_i = M_{i-1} \uplus t_i^{\bullet} \setminus {}^{\bullet}t_i$ such that each transition t_i is enabled by M_{i-1}. The probability of a path is the product of the probabilities of transitions it consists of. A *complete path* is a partial path that ends in a deadlock. A (partial) *trace* is a (partial/complete) path projected by λ to labels of Σ, dropping silent transitions and retaining only its visible steps. The q-set of traces of an SLPN, denoted with $\mathcal{L}$, is its stochastic language. We say that a marking M_j is *reachable* from a marking M_i if there exists a partial path from M_i to M_j.

3.1 SLPNs with Livelocks

Let M be a reachable marking from the initial marking M_0 of an SLPN; if it is impossible to reach a deadlock marking from M, then M is part of a *livelock*. An SLPN with a livelock has an incomplete stochastic language, witnessing that with some probability the modelled system keeps moving and can never complete.

Definition 5 (Livelock). *Let* $(P, T, F, M_0, \Sigma, \lambda, w)$ *be an SLPN, and let* M *be a marking that is reachable from* M_0. M *is part of a* livelock $(\bigcirc(M))$ *if and only if there is no marking* M' *that is a deadlock and that is reachable from* M.

The labelling of transitions and the stochastic perspective are not relevant for livelocks. Therefore, decidability results of the livelock-freedom of standard Petri nets carry over to SLPNs directly. In particular, livelock-freedom is decidable [9, thms 3.2.25 & 3.2.26], though practical decision algorithms are expensive.

A particular type of livelock is the silent livelock, in which the modelled process can only progress internally, that is, through silent transitions, but cannot fire a visible transition anymore, and cannot terminate:

Definition 6 (Silent livelock). *Let* $(P, T, F, M_0, \Sigma, \lambda, w)$ *be an SLPN, and let* M *be a marking that is reachable from* M_0 *and that is part of a livelock. Let* $\mathcal{M}$ *be the set of all markings reachable from* M. *If for all* $M' \in \mathcal{M}$ *it holds that only silent transitions are enabled in* M', *then* M *is part of a silent livelock.*

While it is decidable whether a marking is part of a livelock, this is not a cheap computation [9]. Therefore, we consider a special case of livelock that is

---

Algorithm 1. Non-decreasing livelock decision

1: **procedure** NON-DECREASING LIVELOCK(SLPN $(P, T, F, M_0, \Sigma, \lambda, w)$, marking M)
2:     $\mu \leftarrow \langle M \rangle$ ▷ sequence of markings
3:     **while** $t \leftarrow$ only transition enabled in M **do**
4:         $M \leftarrow M \uplus t^\bullet \setminus {}^\bullet t$ ▷ marking after firing t
5:         $\mu \leftarrow \mu \cdot \langle M \rangle$
6:         **if** $\mu = \langle \ldots, M_k, \ldots, M_\ell \rangle$ with $M_k \subseteq M_\ell$ **then**   ▷ we are in a cycle
7:             $\Delta \leftarrow M_\ell \setminus M_k$ ▷ difference by cycle
8:             **return** $\forall_{k \leq i < \ell} \{ t' \in T \mid w(t') > 0 \wedge {}^\bullet t' \subseteq M_i \uplus \biguplus_{j=1}^\infty \Delta \} = 1$
9:   ▷ cycling enables no other transition
10:        **end if**
11:    **end while**
12:    **return** false
13: **end procedure**

---

cheaper to compute: in a *non-decreasing* livelock, at any point in the livelock, only one transition is enabled, and thus the probability of firing that transition is 1. As such, in a non-decreasing livelock, the probablity never decreases:

Definition 7 (Non-decreasing livelock). *Let $\langle t_1, \ldots t_n \rangle$ be a partial path with $\langle M_0, \ldots M_n \rangle$ its markings, such that M_n is part of a livelock. If there is only one infinite partial path of which $\langle t_1, \ldots t_n \rangle$ is a prefix, then M_n is part of a non-decreasing livelock.*

An efficient algorithm to decide whether marking M is part of a non-decreasing livelock is shown in Algorithm 1. Intuitively, the algorithm walks along the cycle, and at every step only one transition can be enabled (line 3). At some point (see Lemma 8 below), we will encounter a marking that is larger or equal to a marking encountered before (line 6). Then, the algorithm can decide: if the cycle between these two markings would be repeated forever, it still would not enable any other transition at any point along the cycle (line 9). If that requirement hold, then M is part of a non-decreasing livelock.

Lemma 8. *Algorithm 1 decides after finitely many steps whether marking M is part of a non-decreasing livelock.*

Proof. Assume M is part of a non-decreasing livelock. Then, there exists one enabled transition t in line 3. This yields an infinite sequence of markings $\mu = \langle M_1, \ldots \rangle$. By Dickson's Lemma [7], there exist indices $k < \ell \in \mathbb{N}$ with $M_k \subseteq M_\ell$, which we detect in line 6. The sequence of transitions fired between M_k and M_ℓ can be fired again, starting in M_ℓ instead of M_k. This yields a cycle that is traversed infinitely often. Let $\Delta = M_\ell \setminus M_k$ be the increase in markings that we gain after traversing the cycle once. Then the markings that we visit after M_ℓ are all given by $M_i \uplus \biguplus_{j=1}^\alpha \Delta$, with $k \leq i < \ell$ and $\alpha \in \mathbb{N}$ being the number of times we traverse the cycle. Since we are in a non-decreasing livelock, there is only one marking enabled for all $k \leq i < \ell$ and $\alpha \in \mathbb{N}$, hence line 9 returns true.

Assume that M is not part of a non-decreasing livelock., and that line 6 is true at some point. Otherwise, the while loop terminates because there is not exactly one enabled transition, and the algorithm returns false. After detecting $k < \ell \in \mathbb{N}$ with $M_k \nsubseteq M_\ell$, the transitions fired between M_k and M_ℓ can be fired infinitely often. Since we are not in a non-decreasing livelock, there must be another enabled transition at some point. Such a transition is enabled in marking $M_i \uplus \biguplus_{j=1}^{\alpha} \Delta$ with $k \leq i < \ell$ and $\alpha \in \mathbb{N}$. Hence, $w(t) > 0 \wedge {}^\bullet t \nsubseteq M_i \uplus \biguplus_{j=1}^{\infty} \Delta$ for more than one transition $t \in T$, and the algorithm returns false in line 9.   □

3.2 Unbounded SLPN

We recall the standard notion of *boundedness*:

Definition 9 (Bounded SLPN). *Let* $(P, T, F, M_0, \Sigma, \lambda, w)$ *be an SLPN, and let* $\bar{M}$ *be the set of all reachable markings from* M_0. *If* $\bar{M}$ *is finite, then the SLPN is called* bounded.

Worded differently, a place is bounded if there exists a k such that every reachable marking has at most k tokens in that place. An SLPN is bounded if all of its places are bounded. An unbounded marking can only be "reached" through the execution of infinitely many transitions. Since T is finite, for every unbounded net, there must be a sequence of transitions that can be repeated infinitely often and that puts infinitely many tokens into a place:

Definition 10 (Unbounded (silent) cycle). *Let* $(P, T, F, M_0, \Sigma, \lambda, w)$ *be an SLPN, and let* $\bar{M}$ *be the set of all reachable markings from* M_0. *Let* $M_i \in \bar{M}$ *be a reachable marking and let* $\langle t_1 \ldots t_n \rangle$ *be a sequence of transitions such that* $M_i \xrightarrow{t_1} \ldots \xrightarrow{t_n} M_j$. *If* $\forall_{p \in P} M_i(p) \leq M_j(p)$ *and* $\exists_{p \in P} M_i(p) < M_j(p)$, *then* $(M_i, \langle t_1 \ldots t_n \rangle)$ *is an* unbounded cycle. *The cycle is an* unbounded silent cycle *if all of the transitions* $t_1 \ldots t_n$ *are silent.*

An SLPN with an unbounded cycle is unbounded: this cycle can be repeated, causing a place to have unboundedly many tokens. Conversely, every unbounded SLPN has at least one unbounded cycle. The proof is analogous to [10].

4   Q-State Space of an SLPN

We define a q-state space over the behaviour of an SLPN, introduce a method to compute it, and show how the q-state space can be explored to address our Problem. The q-state space yields a stochastic deterministic automaton, that is, it is fully deterministic, but may have an unbounded number of states.

4.1 Q-State Semantics

A *probabilistic state* (q-state) Q is a q-set of markings, describing the distribution over markings the net *could* be in. We use exponents to denote the marking probabilities within a q-state, e.g., $[[p_1]^{0.5}, [p_2]^{0.5}]$ indicates a 0.5 probability of being in either marking $[p_1]$ or $[p_2]$. The initial q-state of a net with initial marking M_0 is $Q_0 = [M_0^1]$. In our example, it is $[[p_0]]$.

SLPN transition t is *enabled* in q-state Q if Q has an enabling marking for t with positive probability: $\exists_{M \in \widetilde{Q}} {}^\bullet t \subseteqq M \wedge w(t) > 0$. A transition from a q-state Q to a q-state Q' with activity a, denoted with $Q \xrightarrow[p]{a} Q'$, indicates that from at least one of the markings in $\widetilde{Q}$, we can take zero or more silent transitions in the SLPN, followed by a single firing of a transition labelled with a. After the firing of the transition labelled with a, no more silent transitions must be executed. The resulting Q' is the distribution over markings we may end up in, and p is the probability that a is executed in Q.

While an SLPN can only deterministically terminate – in a deadlock –, a q-state space can terminate stochastically, if one of its markings is a deadlock. In a q-state Q, execution can terminate with probability $p_\perp(Q)$.

Finally, a q-state Q may partially or completely be a silent livelock. That is, for some of its markings, it may not be possible to eventually terminate or execute an activity. The sum probability of these markings in Q is called the *silent livelock probability* which we denote with $\bullet(Q)$. Obviously, it holds that $p_\perp(Q) + \bullet(Q) + \sum_{Q \xrightarrow[p]{a} Q'} p = 1$. The *livelock probability* $p_\circ(Q)$ is the probability that the SLPN is in a livelock in Q: $p_\circ(Q) = \sum_{M \in \widetilde{Q} \wedge \circ(M)} Q(M)$.

4.2   Q-State Transitions

We compute the outgoing edges of a q-state Q in four steps, detailed next. (1) We construct a Markov chain with its states representing all markings that can be reached from any marking of $\widetilde{Q}$ by executing any number of silent transitions, followed by one labelled transition. (2) We transform this Markov chain into an absorbing Markov chain. (3) We solve the Markov chain to obtain the likelihoods of each next marking. (4) From the solved Markov chain, we obtain the outgoing transitions of Q, as well as the successor q-states they lead to.

Constructing the Markov Chain. From a set of markings $\widetilde{Q}$ we create a Markov chain, where states represent the markings that can be reached in the SLPN, and transitions represent the transitions of the SLPN. Silent transitions of the SLPN are traversed without limitation, however if a labelled transition is encountered, it is taken but the reached marking is not explored further.

We define the states S of the Markov chain by using an auxiliary set S', which allows us to follow silent transitions until we have fired a labelled transition, and then stop. That is, let S and S' be the smallest sets such that: (1) all markings of $\widetilde{Q}$ are in S and S'; (2) all markings reachable from a state in S' by firing silent transitions are in S and S'; and (3) all markings reachable from a state in S' by firing a transition labelled with an activity $a \in \Sigma$ are in S (we annotate these markings with activity label a, and refer to them as *after-activity states*).

$$M \in \widetilde{Q} \Rightarrow (M \in S \wedge M \in S') \tag{1}$$

$$(M \in S' \wedge M \xrightarrow{\tau} M') \Rightarrow (M' \in S \wedge M' \in S') \tag{2}$$

$$(M \in S' \wedge M \xrightarrow{a} M') \Rightarrow M'_a \in S \qquad \textit{after-activity state} \tag{3}$$

The probability function p', denoting the edges of the Markov chain, is constructed according to the mentioned SLPN transitions, with the probabilities they have in the SLPN ((4) and (5)). Additionally, the after-activity states (6) and deadlock markings (7) become absorbing states:

$$\begin{pmatrix} M \in S \wedge \\ M \xrightarrow{\tau} M' \end{pmatrix} \Rightarrow p'(M, M') = \sum_{{}^\bullet t \subseteq M \wedge \lambda(t) = \tau \wedge M \uplus t^\bullet \setminus {}^\bullet t = M'} p(M, \text{t}) \tag{4}$$

$$\begin{pmatrix} M \in S \wedge \\ M \xrightarrow{a} M'_a \end{pmatrix} \Rightarrow p'(M, M'_a) = \sum_{{}^\bullet t \subseteq M \wedge \lambda(t) = a \wedge M \uplus t^\bullet \setminus {}^\bullet t = M'} p(M, \text{t}) \tag{5}$$

$$M_a \in S \Rightarrow p'(M_a, M_a) = 1 \tag{6}$$

$$\begin{pmatrix} M \in S \wedge \\ \forall_t {}^\bullet t \not\subseteq M \end{pmatrix} \Rightarrow p'(M, M) = 1 \tag{7}$$

For our example from Fig. 1 (the second q-state), the Markov chain with its transition matrix $\mathcal{T}$ is shown in Fig. 2.

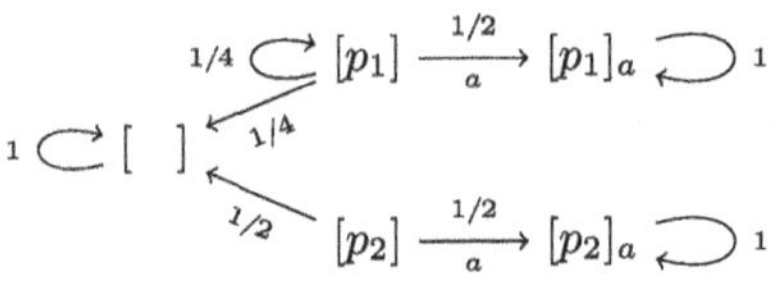

$$\begin{array}{c} \\ [p_1]_a \\ [\,] \\ [p_2]_a \\ [p_1] \\ [p_2] \end{array} \begin{array}{c} \begin{array}{ccccc} [p_1]_a & [\,] & [p_2]_a & [p_1] & [p_2] \end{array} \\ \begin{bmatrix} 1 & 0 & 0 & 0 & 0 \\ 0 & 1 & 0 & 0 & 0 \\ 0 & 0 & 1 & 0 & 0 \\ 1/2 & 1/4 & 0 & 1/4 & 0 \\ 0 & 1/2 & 1/2 & 0 & 0 \end{bmatrix} \end{array}$$

Fig. 2. Markov chain and its matrix of marking set $\widetilde{Q} = \{[p_1], [p_2]\}$ of our example from Fig. 1.

Note that in case an unbounded silent cycle (Definition 10) is encountered, Eq. (2) covers an unbounded number of states, and a practical computation would not terminate, as the Markov chain has infinitely many states. Even though the Markov chain is still well defined, we leave it for future work to establish whether such Markov chains can be solved.

Making the Markov Chain Absorbing. For the Markov chain to be solved, it needs to be absorbing. The after-activity states (6) and deadlock-marking states (7) are absorbing by construction. Furthermore, if the SLPN is free of silent livelocks (Definition 6), then from every state a deadlock-marking state (7) or an after-activity state (3) can be reached, thus the Markov chain is absorbing.

Algorithm 2. Make a Markov chain absorbing

1: **procedure** ABSORBISE(Markov chain (S, p))
2:     $X \leftarrow$ absorbing states of S
3:     **while** X changes **do**   ▷ search states that can reach an absorbing state
       $X \leftarrow X \cup \{s \in S \mid s' \in X \wedge p(s, s') > 0\}$ **end while**
4:     **for** $s \in S \setminus X$ **do**   ▷ make all other states absorbing
5:         mark s as a livelock state
6:         $p(s, s) \leftarrow 1$
7:         **for** $s' \in S \setminus \{s\}$ **do** $p(s, s') \leftarrow 0$ **end for**
8:     **end for**
9:     **return** (S, p)
10: **end procedure**

However, if a silent livelock can be reached, then in the Markov chain there is a state from which neither an after-activity state (3) nor a deadlock-marking state (7) can be reached. Algorithm 2 shows a procedure to identify and address these states. First, the absorbing states are identified through their self-loop probability of 1. Second, a backward search is performed from the absorbing states to identify all states that can reach an absorbing state (line 3). Each state that is not encountered in this search can thus not reach an absorbing state, which means that it is a state that is part of a livelock of silent transitions. As such, we mark it as a *silent-livelock state* for later use, and all its outgoing transitions are removed and replaced with a single self-transition, which makes the state absorbing (line 6). Intuitively, this changes the behaviour of the livelock, however this obviously does not change the behaviour represented by the Markov chain. Applying this procedure to all such states ensures that every state can reach an absorbing state, which makes the Markov chain absorbing.

Solving the Markov Chain. We solve the absorbing Markov chain using a standard procedure [12]. First, $\mathcal{T}$ is put in canonical form: all rows and columns of absorbing markings are listed first. Then, $\mathcal{T}$ has the following shape: $\begin{bmatrix} I & 0 \\ \mathcal{A} & \mathcal{B} \end{bmatrix}$, where I is an identity matrix and 0 is a matrix of zeroes. Second, the fundamental matrix $\mathcal{F} = (I - \mathcal{B})^{-1}$ is computed. As the Markov chain is absorbing, the existence of the matrix inverse is guaranteed [12]. Each entry $\mathcal{F}_{i,j}$ denotes the average number of times we visit marking j if we start in marking i before moving to an absorbing state. Then, $\mathcal{P} = \mathcal{F}\mathcal{A}$ is computed, which we substitute into $\begin{bmatrix} I & 0 \\ \mathcal{P} & 0 \end{bmatrix}$ to obtain the matrix $\mathcal{T}^\infty$. Given states of marking M, M', $\mathcal{T}^\infty_{M,M'}$ is the probability that the net ends up in M' when starting in M. For our example:

$$\mathcal{A} = \begin{array}{c} [p_1] \\ [p_2] \end{array} \begin{bmatrix} [p_1]_a & [\,] & [p_2]_a \\ 1/2 & 1/4 & 0 \\ 0 & 1/2 & 1/2 \end{bmatrix} \quad \mathcal{B} = \begin{array}{c} [p_1] \\ [p_2] \end{array} \begin{bmatrix} [p_1] & [p_2] \\ 1/4 & 0 \\ 0 & 0 \end{bmatrix} \quad \mathcal{F} = (I - \mathcal{B})^{-1} = \begin{array}{c} [p_1] \\ [p_2] \end{array} \begin{bmatrix} [p_1] & [p_2] \\ 4/3 & 0 \\ 0 & 1 \end{bmatrix}$$

$$\mathcal{P} = \mathcal{F}\mathcal{A} = \begin{array}{c} \\ [p_1] \\ [p_2] \end{array} \begin{array}{c} \begin{array}{ccc} [p_1]_a & [\,] & [p_2]_a \end{array} \\ \begin{bmatrix} 2/3 & 1/3 & 0 \\ 0 & 1/2 & 1/2 \end{bmatrix} \end{array} \qquad \mathcal{T}^\infty = \begin{array}{c} \\ [p_1]_a \\ [\,] \\ [p_2]_a \\ [p_1] \\ [p_2] \end{array} \begin{array}{c} \begin{array}{ccccc} [p_1]_a & [\,] & [p_2]_a & [p_1] & [p_2] \end{array} \\ \begin{bmatrix} 1 & 0 & 0 & 0 & 0 \\ 0 & 1 & 0 & 0 & 0 \\ 0 & 0 & 1 & 0 & 0 \\ 2/3 & 1/3 & 0 & 0 & 0 \\ 0 & 1/2 & 1/2 & 0 & 0 \end{bmatrix} \end{array}$$

Extracting the Next q-States. Next, we multiply the incidence vector $\boldsymbol{Q}$ of q-state Q, with added zeroes where necessary, with $\mathcal{T}^\infty$, to obtain a distribution over states $\mathcal{N}$ that we end up in if we proceed from Q. For our example:

$$\boldsymbol{Q} = \begin{array}{c} \begin{array}{ccccc} [p_1]_a & [\,] & [p_2]_a & [p_1] & [p_2] \end{array} \\ \begin{bmatrix} 0 & 0 & 0 & 1/4 & 3/4 \end{bmatrix} \end{array}$$

$$\mathcal{N} = \boldsymbol{Q}\mathcal{T}^\infty = \begin{array}{c} \begin{array}{ccccc} [p_1]_a & [\,] & [p_2]_a & [p_1] & [p_2] \end{array} \\ \begin{bmatrix} 1/6 & 11/24 & 3/8 & 0 & 0 \end{bmatrix} \end{array}$$

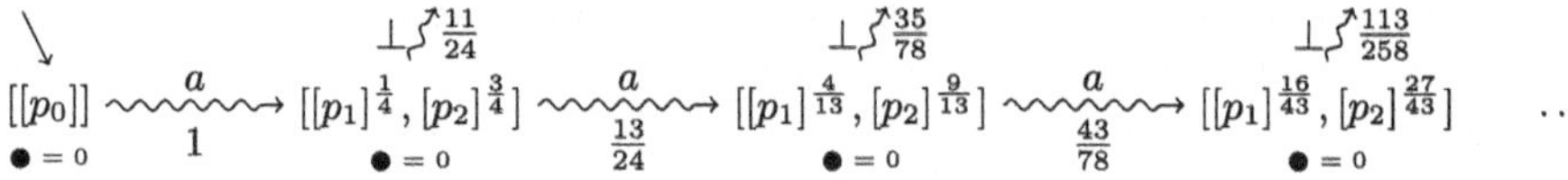

Fig. 3. Q-state space of our example SLPN from Fig. 1.

There are two after-activity states that indicate that an a was executed ($[p_1]_a$ and $[p_2]_a$). The probability that we execute an a in Q is the sum of these probabilities ($13/24$). If we execute an a, then by definition of the q-state space, we first fire any number of silent transitions in the SLPN, followed by a single firing of a transition with the label a. The markings we may then end up in are thus, in our example, $[[p_1]^{1/6}, [p_2]^{3/8}$. The sum of these markings is the probability that we execute an a; to obtain the next q-state Q', the last step is to normalise this into a distribution: in our example, $Q \overset{a}{\underset{13/24}{\rightsquigarrow}} [[p_1]^{\frac{4}{13}}, [p_2]^{\frac{9}{13}}]$.

Similarly, the probability to terminate from Q is the sum of probabilities of all markings that are deadlocks in $\mathcal{N}$. In our example, this is $p_\perp(Q) = 11/24$.

Finally, the probability to get stuck in a silent livelock in Q, denoted with $\bullet(Q)$, is the sum of probabilities of the states that were marked as silent-livelock states in Algorithm 2. In our example, this is 0.

The effect of firing activity a can be summarized removing unnecessary rows and columns from $\mathcal{P}$, obtaining a smaller matrix $\mathcal{P}'_a$. In our example, we get:

$$
\mathcal{P}'_a = \begin{array}{c} \\ [p_1] \\ [p_2] \end{array} \overset{\begin{array}{cc} [p_1]_a & [p_2]_a \end{array}}{\left[\begin{array}{cc} 2/3 & 0 \\ 0 & 1/2 \end{array} \right]}
$$

This method, for any SLPN, defines a state space of q-states (that is, an SDA), and provides a way to traverse this space by computing states incrementally. For our example, the first states of the q-state space are shown in Fig. 3. In Sect. 5, we use this method to answer probabilistic queries on SLPNs. But first, we discuss the computability of the q-state space.

4.3 Computability

Next, we prove the computability of the procedure to traverse the q-state space. Notice that due to the Markov chain construction, we can only do this for SLPNs without unbounded silent cycles.

Lemma 11. *Let N be an SLPN without unbounded silent cycles, and let Q be a q-state that is reachable from $[M_0]$. Then, the successors of Q can be computed.*

Proof. Prove by induction that $\widetilde{Q}$ is finite. Base case: the initial q-state contains one marking (M_0), thus is finite. Induction step: assume that $\widetilde{Q}_i$ is finite. Construction of the Markov chain (Sect. 4.2) follows a state space exploration of markings of the SLPN, and stops at every labelled transition (Eq. 3). As N contains no unbounded silent cycles, the Markov chain has a bounded number of states, and thus any successor $\widetilde{Q}_j$ of $\widetilde{Q}_i$ is finite. Through the method of this section, computing the successors of any reachable q-state $\widetilde{Q}$ terminates.   □

5 Answering Probabilistic Queries

We now use the q-state space to answer probabilistic queries in the sense of our Problem: given an SLPN and a stopping criterion $\varphi \colon (\Sigma^*)^{\mathbb{R}} \times \mathbb{R} \to \mathbb{R}$, we introduce an algorithm that adds the traces of the SLPN in descending order of likelihood, until either all traces have been added or φ returns true. Algorithm 3 shows the procedure. Intuitively, it performs a search of all prefixes supported by the model, in order of *potential probability*, that is, an upper bound on how much probability mass the prefix may contribute to the final result.

We define the potential probability of a prefix trace as $\rho(\sigma) = p_\sigma * (1 - p_\circ(\sigma))$, in which we with $p_\circ(\sigma)$ denote the probability that the last q-state after executing σ is part of a livelock. Moreover, p_σ is the probability that $\sigma = \langle a_1 \ldots a_k \rangle$ is executed: $p_\sigma = \prod_{a_i \in \sigma \wedge Q_{i-1} \overset{a_i}{\underset{p}{\to}} Q_i} p$. The prefix with the highest potential probability is expanded from its last q-state Q by:

Algorithm 3. Answer a probabilistic query

1: **procedure** QUERY(SLPN $(P, T, F, M_0, \Sigma, \lambda, w)$, stopping criterion φ)
2:     $R \leftarrow \{(\text{prefix}, \langle\,\rangle) \text{ with potential probability } 1\}$ ▷ queue
3:     $S \leftarrow []$ ▷ result
4:     **while** remove (Y, σ) with highest priority v from R **do**
5:         **if** $\varphi(S, v)$ **then return** S **end if**
6:         **if** $Y = \text{prefix}$ **then**   ▷ prefix at head of queue
7:             $Q \leftarrow$ last q-state of σ
8:             **if** $p_\perp(Q) > 0$ **then**   ▷ Q is final; trace found
9:                 add (trace, σ) to R with priority $p_\sigma * p_\perp(Q)$
10:            **end if**
11:            **for** $Q \xrightarrow{a} Q'$ with probability $p_a > 0$ **do**   ▷ using method of Section 4
12:                **if** $\exists_{M \in \tilde{Q}'} \neg \bigcirc(M)$ **then**   ▷ Q has a non-livelocked marking
13:                    add $(\text{prefix}, \sigma \cdot \langle a \rangle)$ to R with priority $p_\sigma * p_a * (1 - p_\circ(Q'))$
14:                **end if**
15:            **end for**
16:        **else**   ▷ completed trace at head of queue
17:            add (σ, v) to S
18:        **end if**
19:    **end while**
20:    **return** S
21: **end procedure**

- If the prefix can terminate with probability $p_\perp(Q)$, the corresponding trace is added to the queue with its probability as priority (line 8). This ensures that final traces are processed in order of their probability (Lemma 13).
- For each outgoing edge $Q \xrightarrow{a}_p Q'$ that does not lead to a certain livelock (line 12), the prefix is expanded to include a (line 13). The potential probability of $\sigma \cdot \langle a \rangle$ is the product of the probability to execute σ, the probability p of executing a in Q, and the probability that Q' is not in a livelock.

As soon as φ returns true or an error (lines 5), or all traces have been traversed (line 20), the procedure ends.

Next, we prove that traces are added in order of increasing probability (Lemma 13) and that the algorithm terminates (Theorem 16). Towards Lemma 13, we first prove that potentials are non-increasing and an upper bound on trace probabilities (Lemma 12). As a prefix of a trace uniquely identifies a q-state, we lift the functions $p_\circ$ and $p_\perp$ to prefixes of traces. Recall that $\mathcal{L}(\sigma) = p_\sigma * p_\perp(\sigma)$.

Lemma 12. *Let N be an SLPN and σ be a trace of N. Then, for all traces σ' in N of which σ is a prefix, it holds that $\mathcal{L}(\sigma') \leq \rho(\sigma') \leq \rho(\sigma)$.*

Proof. Note that $\rho(\sigma') = p_{\sigma'} * (1 - p_\circ(\sigma'))$. As "not being in a livelock" includes termination, $p_\perp(\sigma') \leq 1 - p_\circ(\sigma')$ holds, which proves the first inequality. For the second inequality, we show that $(p_{\sigma'}/p_\sigma) * (1 - p_\circ(\sigma')) \leq 1 - p_\circ(\sigma)$. Let $\sigma' = \sigma * \langle a_1 \dots a_k \rangle$ and note that $p(\sigma')/p(\sigma)$ is the probability of executing $a_1 \dots a_k$

after σ. The value $(p_{\sigma'}/p_\sigma) * (1 - p_o(\sigma'))$ is the probability to execute $a_1 \ldots a_k$ and not be in a livelock afterwards. This is only possible if we are not already in a livelock after σ. Hence this probability is included in $1 - p_o(\sigma)$, which proves the second inequality. □

Lemma 13 (Algorithm 3 returns traces in order of probability). *Let N be an SLPN. Let σ_1 be a trace that is added to S before a trace σ_2 in Algorithm 3. Then, $\mathcal{L}(\sigma_1) \geq \mathcal{L}(\sigma_2)$.*

Proof. When σ_1 is added to S, it has the highest priority in R, being $\mathcal{L}(\sigma_1)$.

- If σ_2 is already in the queue as a trace with priority $\mathcal{L}(\sigma_2)$, then by the priority queue it holds that $\mathcal{L}(\sigma_1) \geq \mathcal{L}(\sigma_2)$.
- If σ_2 is not in the queue yet, then there is some σ_3 in the queue that is a prefix of σ_2 with potential $\rho(\sigma_3) \leq \mathcal{L}(\sigma_1)$. By Lemma 12, $\mathcal{L}(\sigma_2) \leq \rho(\sigma_3)$, and thus $\mathcal{L}(\sigma_1) \geq \mathcal{L}(\sigma_2)$.
- If σ_2 is already in the queue as a prefix, the previous case applies as σ_2 is also a prefix of itself. □

Towards the proof of termination (Theorem 16), we use a conjecture that states that for every infinite sequence of prefixes, the potential converges to zero. The only only way the potential could not converge to zero, is if both the prefix probability p_σ does not converge to zero, and the livelock probability $p_o(\sigma)$ does not converge to one. However, if p_σ does not converge to zero, then the probability of firing the next activity converges to one. We conjecture that this is only possible if the probability of being in a livelock converges to one.

Conjecture 14. For all $i \in \mathbb{N}$, let $\sigma_i \in \Sigma^i$ be a prefix trace such that the potential $\rho(\sigma_i)$ is maximal among all prefix traces of length i. The sequence of potentials $(\rho(\sigma_i))_{i \in \mathbb{N}}$ converges to zero.

Given the conjecture, we prove that finitely many prefixes are added in front of any trace on the queue.

Lemma 15. *Let N be an SLPN and σ be a trace of N. Let φ always return false. If Conjecture 14 holds, then eventually σ gets added to S.*

Proof. According to Conjecture 14, there exists an $i \in \mathbb{N}$ with $\rho(\tilde{\sigma}) < \mathcal{L}(\sigma)$ for all prefix traces $\tilde{\sigma}$ of length at least i. Since Lemma 12 implies $\mathcal{L}(\sigma) \leq \rho(\sigma')$ for all prefixes σ' of σ, we consider every prefix of σ at the top of the priority queue before considering any prefix trace of length i or longer. In particular, σ will be added as a trace to the queue and will be at the top of the queue before any trace of length at least i is considered. Since there are only finitely many traces shorter than i, we add σ to S after finitely many steps. □

Theorem 16 (Algorithm 3 terminates). *Let $N = (P, T, F, M_0, \Sigma, \lambda, w)$ be an SLPN without unbounded silent cycles. Let φ be a stopping function that in context of Algorithm 3 returns true after a bounded number of calls. Then, if Conjecture 14 holds, Algorithm 3 applied to N terminates.*

Proof. As line 11 terminates by Lemma 11, Σ is bounded, and line 12 is decidable by [9, thms 3.2.25 & 3.2.26], all lines in isolation terminate. Next, towards contradiction, assume that Algorithm 3 does not terminate, which means that R never gets empty. Still, by our assumption on φ, a bounded number of traces gets added to S. Then, by line 12, some trace σ of N never reaches the head of the queue, which contradicts Lemma 15. Hence, Algorithm 3 terminates.   □

Next, we provide three examples of stopping conditions φ, each corresponding to a particular type of probabilistic query.

5.1 Most-Likely Traces

The first type of probability query asks for a given number (n) of most-likely traces. The corresponding stopping condition returns whether n has been reached:

$$\varphi_{\text{most likely}}(S, v) = |S| \geq n \tag{8}$$

Obviously, $\varphi_{\text{most likely}}$ returns true at a bounded number of traces (n). Hence by Theorem 16, if Conjecture 14 holds, Algorithm 3 with $\varphi_{\text{most likely}}$ terminates.

5.2 Probability Mass

The second type of probabilistic query is to find the most likely traces that together cover a probability mass larger than a given f. In two cases, it is not possible to obtain such a set S. First, if the probabilty that the SLPN ends up in a livelock from the initial marking is ψ, and f is larger than $1 - \psi$, then the query is not satisfiable (9a). Second, if the SLPN contains an unbounded number of traces (represented by ℓ) and the requested probability mass f is equal to the livelock probability, the query is unsatisfiable as well (9b).

Otherwise, the corresponding stopping function φ_{cover} returns whether the requested mass f has been gathered (9c):

$$\varphi_{\text{cover}}(S, v) = \begin{cases} \text{query not satisfiable} & \text{if } f > 1 - \psi & \text{(9a)} \\ \text{query not satisfiable} & \text{if } f = 1 - \psi \wedge \ell & \text{(9b)} \\ f \leq \sum_{(p,\sigma) \in S} p & \text{otherwise} & \text{(9c)} \end{cases}$$

The measure ψ (livelock probability) could, for bounded SLPNs, be computed using Markov chain techniques similar to [19]. The measure ℓ (loops) can be computed, for bounded SLPNs, by waiting for a trace to be added to S that has two q-states that contain the same marking. For both ψ and ℓ, detection on unbounded models remains future work.

As the SLPN has a total probability mass of $1 - \psi$, it holds that φ_{cover} returns true at a bounded number of traces. Thus, if ψ and ℓ are known, by Theorem 16, Algorithm 3 with φ_{cover} terminates.

5.3   Minimum Probability

The third type of probabilistic query is to obtain all traces that have a probability equal to or higher than a given f. If $f > 0$, the SLPN has at most $1/f$ traces with a probability $\geq f$, and the criterion returns true if $v < f$, that is, when the potential v of the last-visited prefix cannot lead to a trace with a probability $\geq f$ anymore. Otherwise, their combined probability would exceed 1 (10b). If the number of traces in the SLPN is unbounded, then there are an unbounded number of traces with a probability ≥ 0. Therefore, if $f = 0$ and ℓ hold, then the query has no answer (10a).

$$\varphi_{\text{minimum probability}}(S, v) = \begin{cases} \text{query not satisfiable} & \text{if } f = 0 \wedge \ell \quad \text{(10a)} \\ v < f & \text{otherwise} \quad \text{(10b)} \end{cases}$$

Consequently, if Conjecture 14 holds, Algorithm 3 with $\varphi_{\text{minimum probability}}$ terminates by Theorem 16. By Lemma 13, after termination S contains the traces of the SLPN with probability greater than f.

However, for $\varphi_{\text{minimum probability}}$, we can weaken the check for livelocks on line 12 to non-decreasing livelocks (Definition 7). The more times a decreasing livelock is executed, the lower the probability of the partial path, thereby steadily decreasing its probability, until eventually f is reached. Hence, we do not need to explicitly check for decreasing livelocks. This argument does not hold for non-decreasing livelocks, which do not decrease in probability, thus line 12 can be replaced with a (cheaper) check for non-decreasing livelocks, as in Algorithm 1. We refer to this adapted algorithm as Algorithm 3'.

Lemma 17 (Algorithm 3' terminates with $\varphi_{\text{minimum probability}}$). *Let N be an SLPN without unbounded silent cycles. Then, if Conjecture 14 holds, Algorithm 3' applied to N terminates.*

Proof. Algorithm 1 visits each transition at most once, thus the replaced line 12 in isolation terminates. With reference to the proof of Theorem 16, left to prove: eventually, p drops below f. Assume towards contradiction that N has no non-decreasing livelock, $f > 0$ and that p never drops below f. Then, there exist infinitely many prefixes of probability $\geq f$. A prefix can be expanded with either (i) termination, (ii) a choice between multiple transitions, (iii) a forced single enabled transition, and (iv) a silent livelock. Cases (i), (ii) and (iv) can only be applied a bounded number of times, so in order to have infinitely many prefixes, case (iii) must be applicable infinitely many times, which can only happen infinitely many times in a non-decreasing livelock, which is a contradiction. If $f = 0$ and $\neg\ell$, the proof is similar. Hence, Algorithm 3' terminates.   □

6   Evaluation

We now evaluate our approach for answering probabilistic queries on SLPNs: we discuss its implementation, and evaluate its feasibility and applicability.

6.1   Implementation

The method of Sect. 5 has been implemented in the open-source stochastic process mining framework Ebi (https://ebitools.org [17]). The three types probabilistic queries have been implemented: most likely traces ($\varphi_{\text{most likely}}$) and probability mass ($\varphi_{\text{cover}}$) have been implemented without livelock check but with non-decreasing livelock check, while minimum-probability ($\varphi_{\text{minimum probability}}$) has been fully implemented.

Given the sequential nature of the computation of q-states and the potentially low probabilities of traces (we encountered fractional numbers of 18 000 divided by 19 000 bits), IEEE double precision floating point numbers may not provide the necessary range and accuracy. Therefore, the implementation uses infinite-precision fractions throughout, though a user can disable this and revert back to double precision for run time considerations. The experiments reported hereafter were all performed with exact arithmetic.

Conceptually, the methods of Sect. 5 support application to any stochastic modelling formalism (with the exception of the livelock check). Subsequently, the implementation supports any modelling formalism with stochastic semantics, which currently includes SLPNs, SDFAs, finite stochastic languages and event logs (though direct, more efficient, special cases apply for the latter two).

6.2   Feasibility

To test the feasibility of our implementation, we apply it to all 522 XES logs listed by the IEEE Task Force on Process Mining (https://www.tf-pm.org/resources/logs, accessed 7-5-2024). To resemble real-life use cases, we first apply 4 process discovery techniques to obtain a variety of control-flow models for each log, after which each of these models is enriched with a stochastic perspective by 3 stochastic process discovery techniques. More details are shown in Table 1. To the resulting 6 253 stochastic process models[1], we apply the three types of probabilistic queries:

- We extract the traces that have a 0.1% or higher probability, and measure the time this takes with a timeout of 6 h;
- We extract the 10 most likely traces, and measure the time this takes with a timeout of 6 h;
- For coverage, we take a different approach as coverage is highly dependent on the model's number of traces. We extract the most-likely traces that together cover 99.999% of the probability mass, with a timeout of 1 min. We measure the time that this takes, and the number of traces reported in that 1 min.

All measures were taken on an AMD 5964wx CPU with 256GB of RAM available.

The results are shown in (Figs. 4 and 5). The minimum probability querying timed out in 151 cases, while the most-likely trace querying timed out in 116 cases.

---

[1] The ALI stochastic process discovery ran out of RAM for 11 models, thus these were not further considered.

Table 1. Details of the experimental set-up.

(a) Control-flow discovery.

Inductive Miner - infrequent (0.8) [16] transformed to a labelled Petri net	IMf
Directly Follows Model Miner (0.8) [21] transformed to a labelled Petri net	DFM
flower miner: allow for any behaviour	FM
trace model: a model with the language of the log	TM

(b) Stochastic discovery.

uniform: assign each transition a weight of 1	UNI
occurrence: assign each transition the occurrence of its label, and 1 to silent transitions [3]	OCC
alignments: perform alignments and assign each transition its occurrences [3]	ALI

(a) Minimum probability. (b) Most-likely traces.

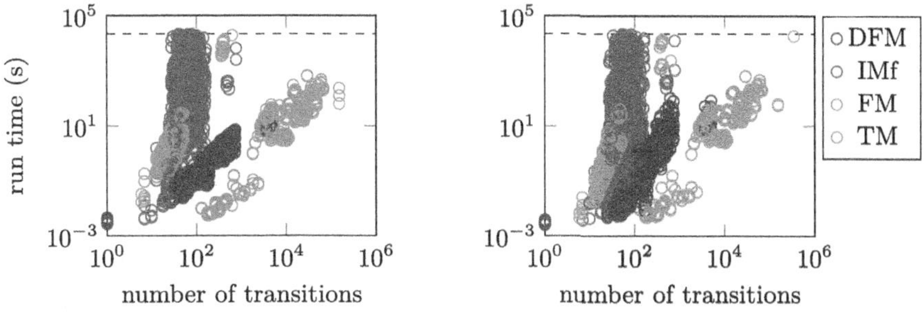

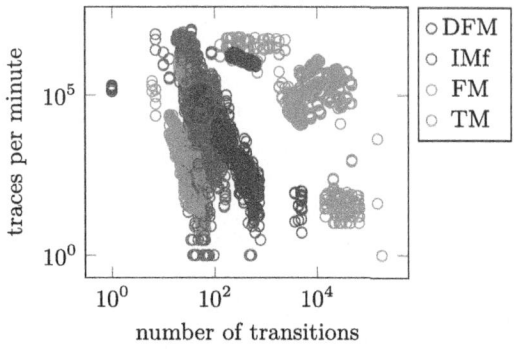

Fig. 4. Results of the feasibility experiment. Dashed lines indicate the 6h timeout.

The different discovery techniques have different characteristics: the cheapest to compute were DFM and FM, as their models have a small marking space, due to the models being deterministic and having a limited number of silent transitions. Hence, all q-states of DFM-models have only one marking, and the q-states of FM models have at most two markings.

TM models have very large q-states, as initially a transition is enabled for every trace of the event log. The q-states get smaller whenever a transition is fired, and as such, TM models utilise the q-state space extensively. For complex models with lots of activities, the q-states get smaller quite quickly, so for most large models, there is an initial expensive q-state, but subsequent q-states

Fig. 5. Results of the coverage feasibility experiment.

get small quickly; TM-models with 10^6 transitions could still be handled in a few minutes. However, the worst case is IMf, which tends to introduce silent transitions, which may involve high numbers of large q-states.

We conclude that our method is feasible in practical cases. Nevertheless, we observed run times of over a day for the most complex models, which involved over a million transitions and used more than 150GB of RAM, though these were clearly exceptions, as smaller models took a few minutes or seconds.

6.3   Applicability

To illustrate the applicability of our approach, we consider an analyst tasked with analysing a loan application process (BPIC2012). A process model was created using IMf [16], and annotated with the occurrence miner [3]. To get an idea of the quality of the model, we can compute the uEMSC [22] value, which with 0.015 is rather low. To investigate the quality of the model further, we apply our approach to obtain the 5 most-likely traces from this stochastic model:

Trace	probability in model	probability in log
⟨submitted, partlysubmitted, afhandelen leads⟩	0.139	0
⟨submitted, partlysubmitted⟩	0.099	0
⟨submitted, partlysubmitted, preaccepted ⟩	0.062	0
⟨submitted, partlysubmitted, afhandelen leads, afhandelen leads⟩	0.018	0
⟨submitted, partlysubmitted, completeren aanvraag⟩	0.015	0

The list confirms that the model does not describe the log well, as of the 5 most likely traces *none* appears in the event log *at all*. As such, the analyst should proceed with caution with this stochastic model and perhaps apply a different discovery approach.

7   Related Work

Stochastic labelled Petri nets (SLPNs) have been used before to describe stochastic business processes: they can be discovered [18] and their conformance to an event log or another stochastic model can be computed [15,19]. To compute some types of conformance, trace queries must be answered to obtain the probability that the LSPN will generate a given trace [22]. Silent steps and repeated labels make such queries non-trivial to answer. In [19], it is shown that trace queries can be answered analytically over bounded LSPNs, by extracting a Markov chain from the LSPN (considering transitions), and by defining a suitable notion of cross-product with a given temporal specification (expressed over labels). Unfortunately, the approach in [19] cannot be employed to answer probability queries as considered in this paper. Intuitively, this is because probabilities are measured over traces, and one needs to associate trace prefixes to probabilities, incrementally updating them towards full traces. This calls for a state space where silent transitions have been removed (retaining their probabilities), and where transitions with the same label are suitably merged, while at the same time retaining the information on which distinct states they lead to, each with each probability. This is what our q-state space captures (cf. Sect. 4).

Generalised stochastic Petri nets (GSPN [4,27]) extend SLPNs by distinguishing between immediate and timed transitions, where immediate transitions have priority over timed transitions. Notably, if one is interested only in traces and their probabilities, immediate and timed transitions behave homogeneously, due to the fact that in GSPNs timed transitions have exponentially distributed durations. Our results seamlessly carries over to GSPNs, which do not support silent transitions, following the same line of reasoning from [19].

The construction of the q-state space we provide resembles the construction of a (a possibly infinite-state) stochastic deterministic automaton. Interestingly, this tightly corresponds to the (deterministic versions of) automata models introduced in the seminal works [28,29]. A property discussed in [28] is that stochastic non-deterministic finite automata (SNFAs, called PFAs in [28]) are strictly more expressive than their deterministic counterpart, which corresponds to our q-state spaces being either finite or infinite. The addition of silent transitions to SNFAs (called λ-PFAs in [28]) does not increase their expressivity. Nevertheless, on λ-PFAs, many problems, including the probabilistic queries we address in this paper, become intractable and are NP-hard. This could be exploited to provide complexity results for the probability queries studied in this paper.

Hidden Markov models (HMMs) are equivalent to SNFAs, with the exception that in contrast to SNFAs, HMMs cannot express the empty trace [29]. Furthermore, HMMs are typically used to study state-based properties, whereas within process mining, we are interested in transitions and their sequencing. A more elaborate discussion on the relation between HMMs and SDFAs can be found in [29]. At the surface, Markov decision processes (MDPs) [25] appear closer to SDFAs, due to their focus on actions and decision making. However, they are substantially different: in MDPs transition choices are not stochastic but purely non-deterministic, while the state in which the systems evolves in after picking a transition is probabilistic. Due to the state-based nature of Markov models, the problem of identifying the most-likely trace (possibly involving silent transitions) is not a typical focus of Markov analysis [13], and to the best of our knowledge, has not been addressed, in particular as we consider infinite state spaces. Nevertheless, the methods described in this paper apply directly to Markov models with labelled transitions.

Another type of stochastic automaton (R-SDFA) is defined in [26]. This model is closer to MDPs, but differs from our q-state spaces. That is, R-SDFAs will either accept or reject a trace with a given probability, while our models will generate a trace with a certain probability. For R-SDFAs, the following problems are undecidable: i) is there a trace that has a larger probability than a threshold, and ii) do there exist traces whose probability is arbitrarily close to a given probability [2,11]. It remains to be studied if such undecidability results carry over to our setting.

8   Conclusion

Stochastic process mining calls for answering probabilistic queries on stochastic process models. These are queries that retrieve traces satisfying a criterion expressed through conditions on probability masses. This is instrumental to to validate the model, to inspect the model if it is too complex to understand in full, to formulate predictions, and to check conformance.

In this paper, we have tackled the problem of answering such probability queries. We defined a stochastic deterministic state space of activities, and showed how it can be computed incrementally for unbounded SLPNs with livelocks and silent transitions, with the exception of unbounded silent cycles. On this state space, we showed how to answer three types of probability queries, and proved that answering such queries is computable if Conjecture 14 holds and two facts are known: whether the model has an unbounded number of traces, and the total probability that the model ends up in a livelock. The approach was implemented in the Ebi framework [17] and is open source. We evaluated its feasibility on real-life and artificial logs, and applicability on a real-life log.

As for future work, we plan to build on the connection between the q-state space as presented here and the stochastic automata models discussed in Sect. 7, towards establishing (un)decidability and complexity results related to the exact computation of probability queries for different SLPN classes. Furthermore, we suggest to further explore implementation improvements and strategies to truncate the resulting state space when answering probability queries. Finally, it would be interesting to explore an adaptation of our approach for stochastic Time Petri Nets, employing the results from [6] on transient probability calculations in semi-Markov processes.

Acknowledgments. Marco Montali has been partially funded by the NextGenerationEU FAIR PE0000013 project MAIPM (CUP C63C22000770006) and by the PRIN MIUR project PINPOINT Prot. 2020FNEB27.

References

1. van der Aalst, W.: Process Mining - Data Science in Action, 2nd edn. Springer, Heidelberg (2016)
2. Blondel, V.D., Canterini, V.: Undecidable problems for probabilistic automata of fixed dimension. Theory Comput. Syst. **36**(3), 231–245 (2003)
3. Burke, A., Leemans, S., Wynn, M.T.: Discovering stochastic process models by reduction and abstraction. In: Buchs, D., Carmona, J. (eds.) PETRI NETS 2021. LNCS, vol. 12734, pp. 312–336. Springer, Cham (2021). https://doi.org/10.1007/978-3-030-76983-3_16
4. Chiola, G., Marsan, M.A., Balbo, G., Conte, G.: Generalized stochastic petri nets: a definition at the net level and its implications. IEEE Trans. Softw. Eng. **19**(2), 89–107 (1993)
5. Cohen, I., Gal, A.: Uncertain process data with probabilistic knowledge: problem characterization and challenges. In: PROBLEMS, vol. 2938. CEUR (2021)

6. Dengler, G., Carnevali, L., Budde, C.E., Vicario, E.: Transient evaluation of non-markovian models by stochastic state classes and simulation. In: QEST+FORMATS. Lecture Notes in Computer Science, vol. 14996, pp. 213–232. Springer, Heidelberg (2024). https://doi.org/10.1007/978-3-031-68416-6_13

7. Dickson, L.E.: Finiteness of the odd perfect and primitive abundant numbers with n distinct prime factors. Am. J. Math. **35**(4), 413–422 (1913)

8. Dumas, M., Rosa, M.L., Mendling, J., Reijers, H.A.: Fundamentals of Business Process Management, 2nd edn. Springer, Heidelberg (2018)

9. Esparza, J.: Petri nets lecture notes. PNSkript. pdf (2019)

10. Finkel, A.: The minimal coverability graph for Petri nets. In: Rozenberg, G. (ed.) ICATPN 1991. LNCS, vol. 674, pp. 210–243. Springer, Heidelberg (1993). https://doi.org/10.1007/3-540-56689-9_45

11. Gimbert, H., Oualhadj, Y.: Probabilistic automata on finite words: decidable and undecidable problems. In: Abramsky, S., Gavoille, C., Kirchner, C., Meyer auf der Heide, F., Spirakis, P.G. (eds.) ICALP 2010. LNCS, vol. 6199, pp. 527–538. Springer, Heidelberg (2010). https://doi.org/10.1007/978-3-642-14162-1_44

12. Grinstead, C.M., Snel, J.L.: Introduction to Probability, 2nd edn. American Mathematical Society (1997)

13. Hartmanns, A., Junges, S., Quatmann, T., Weininger, M.: A practitioner's guide to MDP model checking algorithms. In: TACAS (1). LNCS, vol. 13993. Springer, Heidelberg (2023). https://doi.org/10.1007/978-3-031-30823-9_24

14. Kalenkova, A.A., Mitchell, L., Roughan, M.: Performance analysis: discovering semi-markov models from event logs. CoRR arxiv:2206.14415 (2022)

15. Leemans, S., van der Aalst, W., Brockhoff, T., Polyvyanyy, A.: Stochastic process mining: earth movers' stochastic conformance. Inf. Syst. **102**, 101724 (2021)

16. Leemans, S., Fahland, D., van der Aalst, W.: Discovering block-structured process models from event logs containing infrequent behaviour. In: Lohmann, N., Song, M., Wohed, P. (eds.) BPM 2013. LNBIP, vol. 171, pp. 66–78. Springer, Cham (2014). https://doi.org/10.1007/978-3-319-06257-0_6

17. Leemans, S.J.J., Li, T., van Detten, J.N.: Ebi - a stochastic process mining framework. In: ICPM Doctoral Consortium/Demo. CEUR Workshop Proceedings, vol. 3783. CEUR-WS.org (2024)

18. Leemans, S.J.J., Li, T., Montali, M., Polyvyanyy, A.: Stochastic process discovery: can it be done optimally? In: CAiSE. LNCS, vol. 14663. Springer, Heidelberg (2024). https://doi.org/10.1007/978-3-031-61057-8_3

19. Leemans, S.J.J., Maggi, F.M., Montali, M.: Enjoy the silence: analysis of stochastic petri nets with silent transitions. Inf. Syst. (2024)

20. Leemans, S., Polyvyanyy, A.: Stochastic-aware precision and recall measures for conformance checking in process mining. Inf. Syst. **115**, 102197 (2023)

21. Leemans, S.J.J., Poppe, E., Wynn, M.T.: Directly follows-based process mining: exploration & a case study. In: ICPM, pp. 25–32. IEEE (2019)

22. Leemans, S.J.J., Syring, A.F., van der Aalst, W.M.P.: Earth movers' stochastic conformance checking. In: BPM Forum. LNBIP (2019)

23. Li, T., Leemans, S.J., Polyvyanyy, A.: The Jensen-Shannon distance for stochastic conformance checking. In: International Conference on Process Mining, pp. 70–83. Springer, Heidelberg (2024). https://doi.org/10.1007/978-3-031-82225-4_6

24. Marsan, M.A., Conte, G., Balbo, G.: A class of generalized stochastic petri nets for the performance evaluation of multiprocessor systems. ACM Trans. Comput. Syst. **2**(2), 93–122 (1984). https://doi.org/10.1145/190.191

25. Puterman, M.L.: Markov decision processes. In: Stochastic Models, Handbooks in Operations Research and Management Science, vol. 2, pp. 331–434. Elsevier (1990)

26. Rabin, M.O.: Probabilistic automata. Inf. Control **6**(3), 230–245 (1963)
27. Rogge-Solti, A., van der Aalst, W., Weske, M.: Discovering stochastic petri nets with arbitrary delay distributions from event logs. In: Lohmann, N., Song, M., Wohed, P. (eds.) BPM 2013. LNBIP, vol. 171, pp. 15–27. Springer, Cham (2014). https://doi.org/10.1007/978-3-319-06257-0_2
28. Vidal, E., Thollard, F., de la Higuera, C., Casacuberta, F., Carrasco, R.C.: Probabilistic finite-state machines-part I. IEEE Trans. Pattern Anal. Mach. Intell. **27**(7), 1013–1025 (2005)
29. Vidal, E., Thollard, F., de la Higuera, C., Casacuberta, F., Carrasco, R.C.: Probabilistic finite-state machines-part II. IEEE Trans. Pattern Anal. Mach. Intell. **27**(7), 1026–1039 (2005)

Decidability Problems for Weak Time Petri Nets with Read, Reset and Transfer Arcs

Didier Lime[ID], Rémi Parrot[(✉)][ID], and Olivier H. Roux[ID]

Nantes Université, École Centrale Nantes, CNRS, LS2N, UMR 6004, 44000 Nantes, France
{didier.lime,remi.parrot,olivier.roux}@ls2n.fr

Abstract. We address the class of Time Petri Nets (TPNs) where time intervals are associated with transitions and constrain when those transitions can be fired. For TPN with strong semantics, i.e. where time elapsing cannot disable transitions, reachability, coverability and boundedness problems are undecidable. They are however decidable with the weak semantics in which time elapsing can disable transitions.

We first propose an intermediate semantics allowing us to study the decidability border between weak and strong semantics. We prove that with only one transition in strong semantics, reachability, coverability and boundedness are undecidable.

We then consider the so-called read, reset and transfer arcs for which the coverability problem is decidable in the untimed context and we study their impact on TPN with a weak semantics (weak TPN). We prove that coverability becomes undecidable for weak TPN when we add either 2 read arcs, or 2 reset arcs, or 2 transfer arcs.

Lastly, considering bounded nets, we propose a state space computation algorithm for weak TPN with read, reset and transfer arcs that proves the decidability of the reachability problem.

Keywords: Time Petri nets · Weak semantics · Read arc · Reset arc · Transfer arc · Decidability · State Class Graph

1 Introduction

A number of extensions of Petri Nets with time have been proposed where time can be a duration or an interval that can be associated with arcs, places or transitions (see [4] for a survey). In these timed extensions of Petri nets, two types of semantics can be considered for time elapsing. In the *strong semantics*, time elapsing cannot disable transitions whereas in the *weak semantics*, all time delays are allowed.

In this paper, we address the class of Time Petri Nets [9] where time intervals are associated with transitions, which is one of the most commonly-used subclass of Petri nets with time. It will henceforth be referred to as TPN.

E. Amparore and L. Mikulski (Eds.): PETRI NETS 2025, LNCS 15714, pp. 333–353, 2025.
https://doi.org/10.1007/978-3-031-94634-9_16

For bounded TPN, reachability and coverability problems are decidable for both semantics [2,12]. But for unbounded nets these problems are undecidable with the strong semantics [7] and decidable with the weak one [12].

We propose an intermediate model between these two semantics with a cursor allowing us to study the decidability border between weak and strong semantics.

In the same spirit, we propose to study the classical read, reset and transfer arcs for which the coverability problem is decidable in the untimed case and to study their impact on TPNs in weak semantics.

Reset and transfer nets have been studied as particular cases of the so-called Generalized Self-Modifying nets [6]. Reachability is undecidable [1] and coverability is decidable for reset and transfer nets, whereas boundedness is decidable for transfer but not for reset nets.

Read arcs describe reading without consuming [13] and have been introduced as test arcs [5] or as particular case of Contextual nets [11]. Without taking true concurrency into account, a read arc can be simulated by both an input arc and an output arc between the read place and the reading transition. Hence reachability and coverability are decidable for Petri Nets with read arcs.

Contributions. We introduce a new formalism blending together the weak and strong semantics pushing the idea of [8] further. We call it Weakly-strong TPNs (WS-TPNs).

We prove that with TPNs with weak semantics, we are very close to the decidability border: if we allow a single transition to behave with strong semantics, then coverability is undecidable. Similarly, if we allow two read arcs, or two reset arcs, or two transfer arcs, we also have undecidability of coverability.

In light of these negative results, we then focus on bounded TPNs mixing strong and weak semantics and give and prove an algorithm to explicitly compute a finite astraction of its state-space, as an extension of the classical state class graph of Berthomieu [2].

Outline. In Sect. 2, we introduce the formalism of WS-TPNs and related definitions, as well as TPNs (with weak semantics) with read, reset, and transfer arcs. In Sect. 3, we prove the aforementioned undecidability results. In Sect. 4, we extend the state class graph construction to WS-TPNs. Finally, we conclude in Sect. 5.

2   Weakly-Strong Time Petri Nets

$\mathbb{N}$, $\mathbb{Z}$ and $\mathbb{R}_+$ are respectively the sets of natural integers, integers and non-negative real numbers. For a finite set X with of size $|X| = n$, an element of A^X can also be seen as a vector in A^n (with $A = \mathbb{N}, \mathbb{Z}, \mathbb{R}_+$), by fixing an arbitrary order on the elements of X. The usual operators $+, -, \times, <, \leq, >, \geq$ and $=$ are used on vectors of A^n and are the point-wise extensions of their counterparts in A. We call *valuation* an element $\nu \in A^X$. For such a valuation and for $d \in A$, $\nu + d$ denotes the vector $(\nu + d)(x) = \nu(x) + d$.

Let $\mathbb{I}_\mathbb{N}$ and $\mathbb{I}_{\mathbb{R}_+}$ be the set of intervals whose finite bounds (or end-points) are respectively in $\mathbb{N}$ and $\mathbb{R}_+$. Let $I \in \mathbb{I}_{\mathbb{R}_+}$, $\underline{I}$ is its lower bound and $\overline{I}$ is its upper bound if it exists, $+\infty$ otherwise. Let $d \in \mathbb{R}_+$, $I - d$ denotes the (possibly empty) interval defined by: $I - d = \{x - d \mid x \in I \text{ and } x \geq d\}$. The downward closure of I is denoted by $\downarrow I = \{x \in \mathbb{R}_+ \mid \exists y \in I \text{ and } x \leq y\}$. The closure of I is denoted by $\mathsf{cl}\,(I) = \{x \in \mathbb{R}_+ \mid \underline{I} \leq x \text{ and if } \overline{I} \neq +\infty, x \leq \overline{I}\}$.

2.1   Definition of Weakly-Strong Time Petri Nets (WS-TPNs)

Definition 1 (Petri nets). *A* Petri net *(PN) is a tuple* $N = \langle P, T, F, m_0 \rangle$ *where*

- *P is a finite set of places;*
- *T is a finite set of transitions with* $P \cap T = \emptyset$;
- $m_0 \in \mathbb{N}^P$ *is the initial marking;*
- $F : (P \times T) \cup (T \times P) \to \mathbb{N}$ *is a flow (and weight) function where* $F(x, y)$ *is the weight of the arc from x to y and is zero if there is no arc from x to y.*

A marking m (state) is a mapping in $\mathbb{N}^P$ and $m(p)$ is the number of tokens in place p.

$\forall t \in T$, we denote ${}^\bullet t$ and $t^\bullet$ the backward and forward vectors in $\mathbb{N}^{|P|}$ s.t. $\forall p \in P, {}^\bullet t(p) = F(p, t)$ and $t^\bullet(p) = F(t, p)$.

The state (marking) of a net evolves according to usual firing rules of transitions:

- A transition $t \in T$ is *enabled* by m if each input place p of t is marked with at least $F(p, t)$ tokens i.e. $m \geq^\bullet t$. Let $\mathsf{enab}\,(m) = \{t \in T \mid m \geq^\bullet t\}$ be the set of transitions enabled by m.
- An enabled transition $t \in T$ can fire leading to a new marking $m' = m -^\bullet t + t^\bullet$

A time Petri net extends the definition of Petri net, by considering that a transition requires some time to be ready to fire.

Definition 2. *A* Time Petri Net *(TPN) is the tuple* $\langle P, T, F, m_0, I_s \rangle$ *where* $\langle P, T, F, m_0 \rangle$ *is a Petri net and* $I_s : T \to \mathbb{I}_\mathbb{N}$ *is a map called the* static firing interval *function that associates a temporal (firing) interval with any transition.*

In a TPN, a transition $t \in T$ can fire if it has been enabled for a time belonging to its interval $I_s(t)$. There exists two semantics defining what happens at the upper bound of the firing interval. Intuitively,

- Under the *strong* semantics, time is not allowed to elapse anymore if the end of the interval has been reached. The transition has to fire, or has to be disabled by the firing of another transition.
- Under the *weak* semantics, time can be elapsed regardless of enabled transitions. However, if an enabled transition outreaches its firing interval, then it will not be able to fire anymore. It is considered *asleep*, until it is disabled by the firing of another transition.

We propose a new class of time Petri net allowing to express both semantics, and some in-betweens, in a unique framework.

Definition 3. *A Weakly-Strong Time Petri Net (WS-TPN) is the tuple* $N = \langle P, T, F, m_0, I_s, \Pi \rangle$ *where* $\langle P, T, F, m_0, I_s \rangle$ *is a TPN and* $\Pi \subseteq 2^T$ *is a set of subsets of* T *called* packs.

In WS-TPNs, time can be elapsed only if no pack $\pi \in \Pi$ sees *all* its enabled transitions $t \in \pi$ outreach their firing intervals (fall asleep). In other words, it is possible to outreach the firing interval of some transitions by elapsing time, but not *all* the enabled transitions of one pack.

A time Petri net is usually represented by a bipartite directed graph where places are circles, with tokens inside representing the marking, and transitions are rectangles with static firing intervals below. The packs are represented with dashed frames (in red).

Example 1. An example of WS-TPN is represented in Fig. 1.

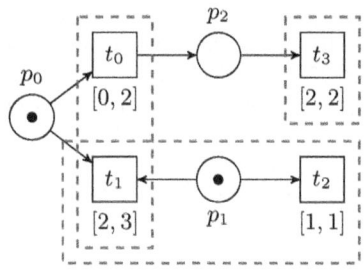

Fig. 1. The WS-TPN N_1 with initial marking $m_0 = (1, 1, 0)$ and packs $\Pi = \{\{t_0, t_1\}, \{t_1, t_2\}, \{t_3\}\}$ (Color figure online)

2.2   Semantics of WS-TPNs

A state of a WS-TPN is a pair (m, I) where m is a marking and $I : T \rightarrow \mathbb{I}_{\mathbb{R}_+}$ is a map called *dynamic firing interval* function such that for each enabled transition t in m, $I(t)$ denotes the remaining durations to wait before t can fire. An enabled transition t can fire when its dynamic interval contains zero: $0 \in I(t)$.

A transition t_k is *newly enabled* after firing t_i from marking m if "it is not enabled by $m -^\bullet t_i$ and it is enabled by $m' = m -^\bullet t_i + t_i^\bullet$". Let $\uparrow \mathsf{enab}\,(m, t_i)$ be the set of transitions *newly enabled* by firing t_i from m, formally:

$$\uparrow \mathsf{enab}\,(m, t_i) = \{t_k \mid t_k \in \mathsf{enab}\,(m -^\bullet t_i + t_i^\bullet)$$
$$\text{and } (t_k \notin \mathsf{enab}\,(m -^\bullet t_i) \text{ or } t_k = t_i)\}$$

We say that a transition is *awake* in a state (m, I), if it is enabled by m and can be fired immediately or in the future i.e. it has been enabled for less than its upper bound $(I(t) \neq \emptyset)$. Then a pack $\pi \in \Pi$ is *awake* in a state (m, I) if at least one transition t of π is awake.

Definition 4 (Awake transitions and packs in a state). *Let* (m, I) *be a state. The set of awake transitions in* (m, I) *is* $\mathsf{awake}((m, I)) = \{t \in T \mid t \in \mathsf{enab}(m) \text{ and } I(t) \neq \emptyset\}$. *The set of awake packs in* (m, I) *is* $\pi\text{-}\mathsf{awake}((m, I)) = \{\pi \in \Pi \mid \exists t \in \mathsf{awake}((m, I)) \cap \pi\}$.

In WS-TPNs, a pack that is awake in a given state should remain awake by elapsing time.

We need the definition of a timed transition system to define the semantics of WS-TPNs.

Definition 5 (Timed Transition System). *A* Timed Transition System *(TTS) over the set of actions* Σ *is a tuple* $S = \langle Q, q_0, \Sigma, \rightarrow \rangle$ *where* Q *is a set of states,* $q_0 \in Q$ *is the initial state,* Σ *is a finite set of actions disjoint from* $\mathbb{R}_+$, $\rightarrow \subseteq Q \times (\Sigma \cup \mathbb{R}_+) \times Q$ *is a set of edges. If* $(q, e, q') \in \rightarrow$, *we also write* $q \xrightarrow{e} q'$.

We now give the definition of the semantics of WS-TPNs.

Definition 6 (Semantics of WS-TPN). *The semantics of a WS-TPN* $N = \langle P, T, F, m_0, I_s, \Pi \rangle$ *is a timed transition system* $S_N = \langle Q, q_0, T, \rightarrow \rangle$ *where:* $Q = (\mathbb{N})^P \times (\mathbb{I}_{\mathbb{R}_+})^T$, $q_0 = (m_0, I_s)$, $\rightarrow \in Q \times (T \cup \mathbb{R}_+) \times Q$ *consists of the discrete and continuous transition relations:*

- *the discrete transition relation is defined for all* $t_i \in T$ *by* $(m, I) \xrightarrow{t_i} (m', I')$ *iff:*

$$\begin{cases} t_i \in \mathsf{enab}(m) \text{ and } m' = m - {}^\bullet t_i + t_i^\bullet \\ 0 \in I(t_i) \\ \forall t_k \in T, I'(t_k) = \begin{cases} I_s(t_k) & \text{if } t_k \in\uparrow \mathsf{enab}(m, t_i) \\ I(t_k) & \text{otherwise.} \end{cases} \end{cases}$$

- *the continuous transition relation is defined for all* $d \in \mathbb{R}_+$ *by* $(m, I) \xrightarrow{d} (m, I')$ *iff:*

$$\begin{cases} \forall t_k \in \mathsf{enab}(m), I'(t_k) = I(t_k) - d \\ \pi\text{-}\mathsf{awake}((m, I)) = \pi\text{-}\mathsf{awake}((m, I')) \end{cases}$$

A *run* of N is a possibly infinite sequence $q_0 a_1 q_1 a_2 q_2 \ldots$ such that $\forall i > 0$, $q_i \in Q$, $a_i \in (T \cup \mathbb{R}_+)$ and $q_{i-1} \xrightarrow{a_i} q_i$. We note $q \xrightarrow{t@d} q'$ for the sequence elapsing $d \in \mathbb{R}_+$ followed by the firing of t from q: $q \xrightarrow{d} q_d \xrightarrow{t} q'$. We write $q \xrightarrow{t} q'$ if there exists $d \in \mathbb{R}_+$ such that $q \xrightarrow{t@d} q'$.

Example 2. Three runs of N_1 (from Fig. 1) are represented in Fig. 2: $q_0 \xrightarrow{t_0@0} q_1 \xrightarrow{t_2@1} q_2 \xrightarrow{t_3@1} q_3$, $q_0 \xrightarrow{t_0@1.3} q_5 \xrightarrow{t_3@2} q_6$ and $q_0 \xrightarrow{t_1@2.5} q_4$. In the first one t_0 fires before t_2. In the second one, t_2 falls asleep at time 1 then t_0 fires at time 1.3. In the third, t_2 and t_0 fall asleep at time 1 and 2 (resp.), then t_1 fires at time 2.5.

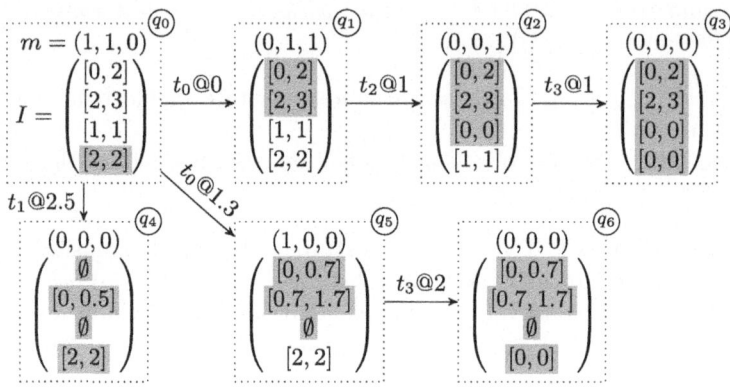

Fig. 2. Some runs of N_1 (from Fig. 1). The vectors m and I are represented in order of index of the places and the transitions (resp.). The dynamic intervals of disabled/asleep transitions are greyed out.

2.3   Particular Structures of Packs

In [8], a weakly-strong semantics on time Petri nets based on structural properties of the net (conflicts and bundles, which are extended conflicts) was presented. The framework of WS-TPN generalizes this notion to any possible packs, and therefore can be used to model these properties.

Weak and Strong Semantics. This framework is able to express both the *weak* and the *strong* semantics over TPNs. Let $N = \langle P, T, F, m_0, I_s, \Pi \rangle$ be a WS-TPN, and $N' = \langle P, T, F, m_0, I_s \rangle$ its corresponding TPN.

If N has no packs ($\Pi = \emptyset$), then there are necessarily no awake packs in all the states. Therefore, the condition in the semantics to elapse time is always true, which is equivalent to the weak semantics on N'.

If the packs of N are all the singletons of the power set of T ($\Pi = \bigcup_{t \in T} \{\{t\}\}$), then the awake packs in one state are exactly the ones with an enabled transition. Thus, it is possible to elapse time only if no transition falls asleep by doing so. This exactly expresses the strong semantics on N'.

2.4   Weak TPN with Read, Reset and Transfer Arcs

We conclude this section by introducing the classical read, reset, and transfer arcs. For simplicity we add them to TPNs and not to WS-TPNs as the following undecidability results using those arcs need only the weak semantics of TPNs. It would be straightforward to adapt this definition to WS-TPNs.

Reset arcs extend Petri Nets with the possibility of emptying places regardless of their previous contents. Transfer arcs allow to transfer the whole contents of some place to some other one. Theses arcs increase expressive power of Petri Nets making reachability undecidable [1,6]. Read arcs describe reading without consuming [13]. In the untimed setting, ignoring the issue of true concurrency, they do not increase expressive power of Petri Nets and reachability problem remains decidable.

Definition 7 (Read, Reset and Transfer nets). *A Petri net with read, reset and transfer arcs is a tuple* $N = \langle P, T, F, F_r, m_0 \rangle$ *where*

- *P is a finite set of places,*
- *T is a finite set of transitions*
- *m_0 is the initial marking*
- *$F : (P \times T) \cup (T \times P) \to \mathbb{N} \cup P$ is a flow function where $F(x,y) = p$ with $p \in P$ means that $F(x,y)$ is equal to the number of tokens in place p. Moreover, $\forall (p,t,p') \in (P \times T \times P)$, we have :*
  - *if $F(p,t) = p'$ then $p = p'$*
  - *if $F(t,p') = p$, then $F(p,t) = p$ and we say that (p,p') is a transfer arc*
  - *if $F(p,t) = p$ and $\nexists p'' \in P$ with $F(t,p'') = p$ then (p,t) is a reset arc*
- *$F_r : (P \times T) \to \mathbb{N}$ is the read function*

Reset and transfer arcs are not involved in the enabling of a transition since when $F(p,t) \notin \mathbb{N}$ then $F(p,t)$ is equal to the number of tokens in place p whatever that number may be, including zero. On the other hand, read arcs are involved in the enabling of transitions. For all $t \in T$ we denote by ${}^\bullet t$, the reading vector in $\mathbb{N}^{|P|}$ s.t. $\forall p \in P, {}^\bullet t(p) = F_r(p,t)$.

Moreover for a marking m, we have ${}^\bullet t(p) = m(p)$ when $F(p,t) = p \in P$ and ${}^\bullet t(p) = F(p,t)$ otherwise, and $t^\bullet(p') = m(p)$ when $F(t,p') = p \in P$ and $t^\bullet(p') = F(t,p')$ otherwise.

- Transition t is *enabled* by m if each input place p of t is marked with at least $F(p,t)$ and $F_r(p,t)$ tokens i.e. $m \geq^\bullet t + {}^\bullet t$
- An enabled transition $t \in T$ can fire leading to a new marking $m' = m - {}^\bullet t + t^\bullet$.

3   Undecidability Results

Definition 8 (Problems). *Let $N = \langle P, T, F, m_0 \rangle$ a Petri net and* reach(N) *the set of reachable markings.*

- *The* marking reachability *problem: given* $m \in \mathbb{N}^P$, *does* $m \in$ reach (N) ?
- *The* marking coverability *problem: given* $m \in \mathbb{N}^P$, *does there exist* $m' \in \mathbb{N}^P$ *such that* $m' \geq m$ *and* $m' \in$ reach (N) ?
- *The* boundedness *problem: does there exist* $b \in \mathbb{N}$ *such that for all* $m \in$ reach (N) *and for all* $p \in P$, $m(p) \leq b$?

These definitions extend to the other classes of Petri nets that we study in this section, and in particular to TPN.

3.1   Reachability and Coverability Problems for WS-TPN

The marking reachability, the marking coverability and the boundedness problems are decidable for TPN with weak semantics [12].

We show that the halting problem for 2-counter machines can be reduced to the reachability problem of weakly-strong TPN, with only one transition in strong semantics.

Definition 9 (2-counter Machine). *A deterministic 2-counter machine is a tuple* (c_1, c_2, L, I) *where:*

- c_1 *and* c_2 *are counters with non-negative integer values;*
- *L is a finite set of lines including an initial state* init *and a halting state* halt;
- *I is a finite set of instructions of the form* (l, c, l') *or* (l, c, l', l'') *with* $c \in \{c_1, c_2\}$, *representing respectively increment (we write also* inc (l, c, l') *for readability) and conditional decrement (*jzdec (l, c, l', l'')*) instructions.*
  Calling l the source line of the instruction in both cases, there can be at most one instruction in I for any given source line (be it an inc *or a* jzdec*).*

Definition 10 (Configuration of a 2-counter machine). *A configuration of a 2-counter machine* (c_1, c_2, L, I) *is a tuple* (v_1, v_2, s) *with* $v_1, v_2 \in \mathbb{N}$, *and* $l \in L$. *The machine can go from configuration* (v_1, v_2, l) *to configuration* (v'_1, v'_2, l') *if and only if:*

- *either there exists an instruction* inc (l, c_i, l') *such that* $v'_i = v_i + 1$ *and* $v'_{3-i} = v_{3-i}$
- *or there exists an instruction* jzdec (l, c_i, l_z, l_{nz}) *such that either* $l' = l_{nz}$, $v_i > 0$, $v'_i = v_i - 1$ *and* $v'_{3-i} = v_{3-i}$, *or* $l' = l_z$, $v_i = 0$, $v'_i = v_i$ *and* $v'_{3-i} = v_{3-i}$.

Theorem 1 (Undecidability of the halting problem [10]). *The problem of knowing whether a deterministic 2-counter machine can go from configuration* $(0, 0, \text{init})$ *to some configuration* (v_1, v_2, halt), *with* $v_1, v_2 \in \mathbb{N}$, *is undecidable.*

First recall that for TPNs with strong semantics, coverability is undecidable [7]. This result is obtained by reduction from the halting problem for 2-counter machines, with a TPN widget encoding increments (Fig. 3a) and one encoding conditional decrement (Fig. 3b).

We have one place for each counter, where the counter value is given by the number of tokens in the place, and one place for each line. By abusing the notations, we use the same names for those corresponding objects in the machine and in the TPN. Then the following lemma holds:

Lemma 1 ([7]). *The place* halt *can be marked in the TPN obtained by assembling the widget for all the instructions if and only if the 2-counter machine halts.*

Proof. The increment widget is quite straightforward. The conditional decrement works by using time to ensure a priority of transition nz_j over transition z_j. □

Now assume those widgets are TPNs with weak semantics. The increment widget still works as intended: if the machine halts we can still traverse the widget and increment the number of tokens in c_i and if the machine does not halt, the only new behaviors are those in which the inc_j transition falls asleep, but then, since the machine is deterministic that transition will never become newly enabled again and the whole net is blocked.

The situation is, as expected (since coverability is decidable), not the same for the jzdec instruction as now nothing prevents the net from firing the z_j transition even when c_i is marked, simply by letting nz_j fall asleep.

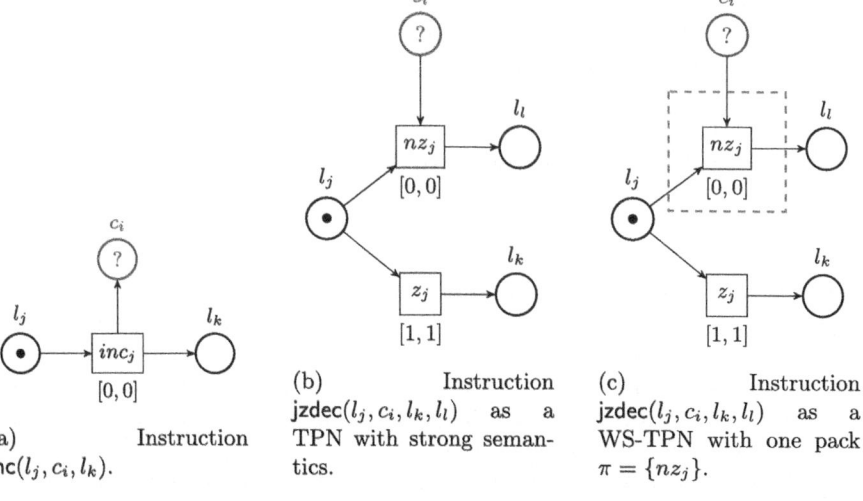

(a) Instruction inc(l_j, c_i, l_k).

(b) Instruction jzdec(l_j, c_i, l_k, l_l) as a TPN with strong semantics.

(c) Instruction jzdec(l_j, c_i, l_k, l_l) as a WS-TPN with one pack $\pi = \{nz_j\}$.

Fig. 3. Instructions inc and jzdec as a TPN with strong semantics and WS-TPN. The places in blue (c_i) are shared between all the instructions. (Color figure online)

Actually only the nz_j transition needs to have strong semantics, so the WS-TPN in Fig. 3c faithfully models the conditional decrement instruction.

Lemma 2. *The place* halt *can be marked in the WS-TPN obtained by assembling the widgets in Fig. 3a (treated as WS-TPN) for all* inc *instructions and the widgets in Fig. 3c for all* jzdec *instructions if and only if the 2-counter machine halts.*

Proof. If the machine halts then by correctness of the construction of [7], there is an execution in the WS-TPN that marks place halt.

If the machine does not halt, either no transition falls asleep, and the resulting execution would be an execution of the same net with strong semantics, and then by correctness of the construction of [7] that execution does not mark halt. Or there is exists a transition that falls asleep in one of the widget. Since nz_j is alone in its pack, it cannot fall asleep and there remains only two cases:

1. if that transition is an inc_j transition in a inc widget, as explained before, since the machine is deterministic the token enabling that transition (in the l_j place) can never be removed and the transition will stay asleep forever, blocking the execution that therefore cannot mark halt.
2. if it is a z_j transition in a jzdec widget, then it must be the case that c_i is empty otherwise nz_j would have fired before z_j could fall asleep. Then nz_j can never fire, and for the same reasons as in the previous case the execution is blocked without reaching halt. □

With this construction, we have as many transitions with strong semantics as there are jzdec instructions in the machine. We now show how this can be reduced to only one such transition.

Consider the new version of the jzdec widget as a WS-TPN shown in Fig. 4. The idea is to share the zero-test both between jzdec instruction on a given counter, and between the two counters. For the former, we decouple the shared widget and the control flow: when in line l_j, to do a jzdec, we send a token to shared place p_{test} and wait on transitions $jump_k$ and $jump_l$. The shared zero-test proceeds as before and a token arrives either in p_z if the tested counter is zero or in p_{nz} if it is not. In the former case, $jump_k$ becomes enabled and can fire and mark l_k, and in the latter case l_l will become marked (unless, as always transitions fall asleep), as expected. Note that when p_z is marked, only $jump_k$ is possible, and none of the other $jump_{k'}$, because only place p_{test_j} is marked and none of the other $p_{test_{j'}}$. The same holds symmetrically for p_{nz} and $jump_l$.

To share the widget between counters, we actually ask for both places c_1 and c_2 to be non-zero to mark p_{nz}, but we add one token in the counter that is not involved in the instruction before the test, so that it is for sure non-zero. That token will be removed when exiting either by $jump_k$ or $jump_l$. In that way, either the tested counter, say c_1, is non-zero. Then by construction so is c_2 and nz fires before z because of the strong semantics. Or c_1 is zero, then nz cannot fire, and z will eventually fire (if it does not fall asleep). The test therefore works as expected.

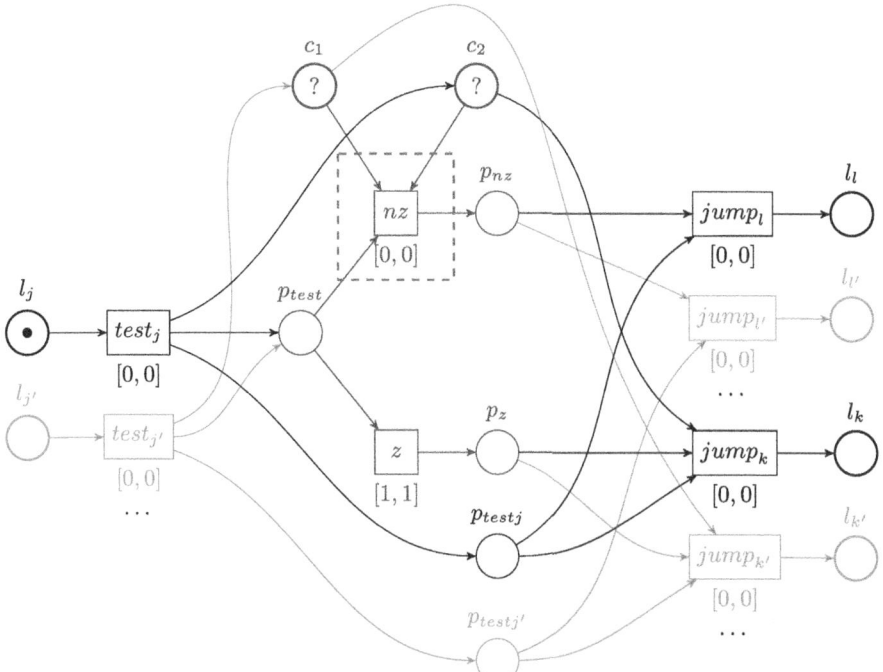

Fig. 4. Instructions $\mathsf{jzdec}(l_j, c_1, l_k, l_l)$ and $\mathsf{jzdec}(l'_j, c_2, l'_k, l'_l)$ as a WS-TPN with packs $\Pi = \{\{nz\}\}$. All the instructions jzdec share the blue part of the widget ($\{c_1, c_2, p_{test}, p_{nz}, p_z, nz, z\}$).

With these modification the usual lemma holds, and in addition the obtained WS-TPN has only a single pack, which contains a single transition, i.e., there is a single transition with strong semantics in the whole net.

Lemma 3. *The place* halt *can be marked in the WS-TPN obtained by assembling the widgets in Fig. 3a (treated as WS-TPN) for all* inc *instructions and the widgets in Fig. 4 for all* jzdec *instructions if and only if the 2-counter machine halts.*

Proof. With the above explanations, it is clear that if the machine halts, there is an execution in the net that mimics the execution of the machine and eventually marks the place halt.

When the machine does not halt, we just need to make sure that we have not added spurious executions that would still mark place halt. As we have seen above, when no transition ever falls asleep, the zero-test and then the jzdec instruction, are mimicked correctly. In both the zero and non-zero case, there is only a single possible execution that marks respectively l_k or l_l as expected.

If some transition falls asleep, we have additional cases: transitions $test_j$, $jump_k$, and $jump_l$.

- if transition $test_j$ falls asleep, as before, because of the determinism of the machine the token will stay in l_j forever and the WS-TPN is just blocked, and place halt is not marked;
- if transition $jump_k$ falls asleep (assuming the counter was zero), then place l_k will not become marked and the widget for the next instruction cannot execute so there is no way that place p_z is ever emptied by another instruction widget, and thus the WS-TPN is blocked without ever marking place halt;
- if transition $jump_l$ falls asleep, the case is symmetrical the $jump_k$ case. □

It follows that:

Theorem 2. *The marking coverability problem is undecidable for WS-TPNs with a single pack, which contains a single transition.*

In general, coverability is weaker than reachability. Indeed suppose one wants to test if a marking m is coverable using reachability, one can just add a transition with for all places p an input arc with weight $m(p)$ and an output arc to some common witness place. Also add for all places an additional transition with just an input arc to that place. Finally, ask if the marking with only one token in the witness place and none in the others is reachable.

Corollary 1. *The marking reachability problem is undecidable for WS-TPNs with a single pack, which contains a single transition.*

Lemma 4. *The WS-TPN obtained with the same construction as Lemma 3 and by adding a place and arcs going from all the transitions of the net to that place is bounded if and only if the 2-counter machine halts.*

Proof. The added place (called p_c) simply counts the number of transitions fired during the execution. The proof is similar to the previous one:

- If the machine halts, then in all possible executions of the net the counters and the place p_c are bounded. All the others places are either empty or contain one token (if we reached a deadlock).
- If the machine does not halt, then there is (at least one) infinite execution of the net. Therefore the marking of p_c is not bounded.

It follows that:

Theorem 3. *The boundedness problem is undecidable for WS-TPNs with a single pack, which contains a single transition.*

Corollary 2. *The marking coverability, the marking reachability and the boundedness problems are undecidable for TPNs with weak semantics augmented with a single transition with strong semantics.*

3.2   Coverability Problem for Weak TPN with Read, Reset or Transfer Arcs

In the untimed setting, reachability, coverability and boundedness problems are decidable for Petri Nets with read arcs. For reset and transfer nets, reachability is undecidable [1] and coverability is decidable but boundedness is decidable for transfer nets but not for reset nets [6].

We propose to study the decidability impact on these arcs on TPNs under weak semantics. Like in the previous section, we reduce the halting problem for 2-counter machine to the marking coverability problem for TPN with read, reset and transfer arcs.

The widget encoding inc instruction has been given in the previous section (Fig. 3a), then we focus on the jzdec instruction.

Read. The encoding is given in Fig. 5. The encoding is based on the transition *ion* (note that *i.o.n.* means *immediately or never*) whose time interval is $[0, 0]$ and which can either fire immediately after being enabled or never fire at all if it falls asleep.

Lemma 5. *The place* halt *can be marked in the weak TPN obtained by assembling the widgets in Fig. 3a for all* inc *instructions and the widgets in Fig. 5 for all* jzdec$(., c_i, ., .)$ *instructions if and only if the 2-counter machine halts.*

Proof. We start with a marking m where, apart from c_i, l_j is the only marked place with $m(l_j) = 1$.

Assume the value of c_i is equal to 0 (then $m(c_i) = 0$). The net can only fire the sequence of transitions $test_j.one.ion.z$ reaching a marking where the only marked place is l_k and the place c_i remains empty.

Assume now that the value of c_i is greater than 0 (i.e. $m(c_1) \geq 1$). $test_j$ is the only firable transition, then it fires and both transitions *one* and *ion* are enabled.

- If *ion* fires, the place *ongoing* becomes empty and the transition *one* is no more firable because of the read arc. We reach a deadlock.
- If we elapse 1 time unit, *ion* falls asleep and *one* can fire. Now, the only transition which can be fired is nz leading to a marking m' where (apart from c_1) the only marked place is l_l and the value of the counter is now $m'(c_i) = m(c_i) + 1 - 2 = m(c_i) - 1$.

If any other transition than *ion* falls asleep, the net reaches a deadlock.

Therefore, if the machine halts, there is an execution of the net that mimics the execution of the machine and eventually marks the place halt. If the machine does not halt, either the net mimics the execution of the machine or it reaches a deadlock, but in no case it marks the place halt.   □

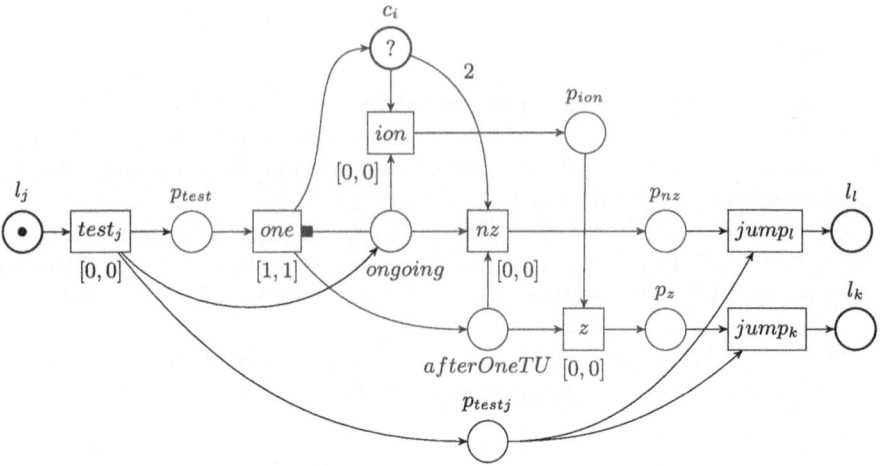

Fig. 5. The jzdec(l_j, c_1, l_k, l_l) for weak TPN with one read arc: $F_r(ongoing, one) = 1$. All the instructions jzdec($., c_i, ., .$) share the blue part of the widget ($\{c_i, p_{test}, ongoing, afterOneTU, p_{ion}, p_{nz}, p_z, one, ion, nz, z\}$). (Color figure online)

Reset. The encoding is given in Fig. 6. As for the read arc, the idea is to fire the transition *ion* (i.e. *immediately or never*) only when the counter is zero and after the firing of *one*. This time, when the counter is non-zero, the immediate firing of *ion* will lead to a deadlock, and its firing after 1 time unit must be made impossible. The transition *ion* must therefore remain continuously enabled after the firing of *test$_j$* and must not be newly enabled by the the firing of *one*. To do this, *test$_j$* produces 2 tokens in *ongoing*.

Lemma 6. *The place* halt *can be marked in the weak TPN obtained by assembling the widgets in Fig. 3a for all* inc *instructions and the widgets in Fig. 6 for all* jzdec($., c_i, ., .$) *instructions if and only if the 2-counter machine halts.*

Proof. We start with a marking m where, apart from c_i, l_j is the only marked place with $m(l_j) = 1$.

Assume the value of c_i is equal to 0 (then $m(c_i) = 0$). The net can only fire the sequence of transitions *test$_j$.one.ion.z* reaching a marking where the only marked place is l_k and the place c_i remains empty.

Assume now that the value of c_i is greater than 0 (i.e. $m(c_i) \geq 1$). *test$_j$* is the only firable transition, then it fires and both transitions *one* and *ion* are enabled.

-   If *ion* fires, it remains a token in the place *ongoing* and the place *noMoreThanOnce* has been reset by the reset arc. Then, the transition *one* is disabled. We are either in a deadlock if c_i is empty, or we can fire again the transition *ion* and we reach a deadlock.
-   If we elapse 1 time unit after the firing of *test$_j$*, *ion* falls asleep and we can fire the transition *one*. Since there are 2 tokens in *ongoing*, the transition *ion*

is not newly enabled by the firing of *one* and thus stays asleep. Now, the only transition which can fire is nz removing the 2 tokens of *ongoing* and leading to a marking m' where (apart from c_i) the only marked place is l_l. The value of the counter is now $m'(c_i) = m(c_i) + 1 - 2 = m(c_i) - 1$.

As in the proof of Lemma 5, if any other transition than *ion* falls asleep, the net reaches a deadlock.   □

Transfer. The model is the same as for the reset arc but we add a place *garbage* and we replace the reset arc by a transfer arc from $noMoreThanOnce$ to *garbage* as follows:

- $F(noMoreThanOnce, ion) = noMoreThanOnce$
- $F(ion, garbage) = noMoreThanOnce$

Lemma 7. *The place* halt *can be marked in the weak TPN obtained by assembling the widgets in Fig. 3a for all* inc *instructions and the widgets in Fig. 6 with the transfer arc describe previously for all* jzdec$(., c_i, ., .)$ *instructions if and only if the 2-counter machine halts.*

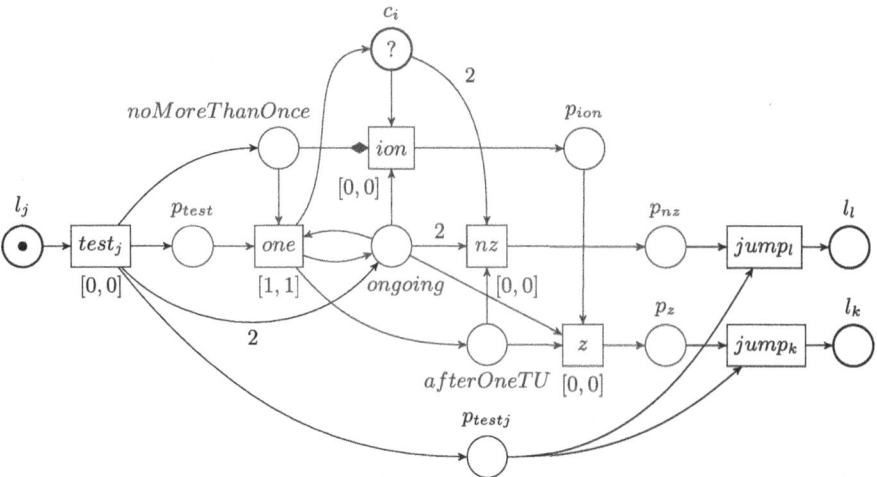

Fig. 6. The jzdec(l_j, c_i, l_k, l_l) for weak TPN with one reset arc $(noMoreThanOnce, ion)$. All the instructions jzdec$(., c_i, ., .)$ share the blue part of the widget $(\{c_i, p_{test}, noMoreThanOnce, ongoing, afterOneTU, p_{ion}, p_{nz}, p_z, one, ion, nz, z\})$. (Color figure online)

Proof. The proof is similar than the one of Lemma 6. The tokens will accumulate in the *garbage* place each time the instruction is executed. This is not a problem since we are interested in the coverability of places l.

Theorem 4. *The marking coverability, the marking reachability and the boundedness problems are undecidable for the following classes of TPN in weak semantics:*

- *TPN with 2 read arcs*
- *TPN with 2 reset arcs*
- *TPN with 2 transfer arcs*

Proof. The widgets used in lemmas 5, 6 and 7 require one special arc for all the instructions jzdec$(., c_i, ., .)$. Then we need 2 special arcs, one per counter.

To get back to a marking reachability or a boundedness problem one can use the same tricks as in the Sect. 3.1.   □

4   State Class Graph

As always with continuous time, the models have an infinity of states. Therefore we propose a new symbolic representation of the states space of WS-TPNs based on classes [2,3]. This representation is finite for bounded nets.

The idea is to group all the states that can be reached by a sequence of transitions $\sigma = t_1 \ldots t_n \in T^*$ from the initial state q_0. We propose here an algorithm to compute these sets $\{q \in Q \mid q_0 \xrightarrow{t_1} q_1 \ldots \xrightarrow{t_n} q\}$ called state classes.

4.1   Firing and Sleeping Domain and State Classes

Classically a *firing time* is a variable associated with each enabled transition of the net. For WS-TPNs, we have to consider the set $W = \{t_1, \ldots, t_k\}$ of awake transitions. A firing time θ_i is associated with each $t_i \in W$, and is compacted in a vector notation $\boldsymbol{\theta} = (\theta_1, \ldots, \theta_k) \in (\mathbb{R}_+)^W$. Additionally, a *sleeping time* is associated with each awake transition that may fall asleep. Let $W_s = \{t_i \in W \mid \overline{I_s}(t_i) \neq +\infty\}$. A sleeping time λ_i is associated with each $t_i \in W_s$, and is compacted in a vector notation $\boldsymbol{\lambda} = (\lambda_1, \ldots, \lambda_{k'}) \in (\mathbb{R}_+)^{W_s}$.

A *firing and sleeping domain* D is a set of firing and sleeping times $(\boldsymbol{\theta}, \boldsymbol{\lambda})$. It is a convex polyhedron of $(\mathbb{R}_+)^{|W|+|W_s|}$ described by a system of linear inequalities. These inequalities are of two types: rectangular $v_i \sim c$ and diagonal $v_i - v_j \sim c$, $v_i = \theta_i$ or $v_i = \lambda_i$, $\sim \in \{\leq, \geq, <, >\}$ and $c \in \mathbb{N}$. For the sake of simplicity, we use the same index for the variables θ_i and λ_i and the awake transitions t_i. A domain is empty $D = \emptyset$ when there is no solution to the system of linear inequalities.

Definition 11 (State class). *A state class is a tuple $K = \langle m, W, D \rangle$, where m is a marking, $W \subseteq T$ is a set of awake transitions, $W_s = \{t_i \in W \mid \overline{I_s}(t_i) \neq +\infty\}$ and $D \subseteq (\mathbb{R}_+)^{|W|+|W_s|}$ is a firing and sleeping domain.*

Let $q = (m, I)$ be a state of the net. The dynamic interval function I can be seen, like a firing domain, as a polyhedron of $(\mathbb{R}_+)^T$. We denote $I|_A$ the restriction of I to $A \subseteq T$. We say that a state belongs to a state class if its dynamic interval function is included in the firing domain, and the supremum of its dynamic interval function is in the sleeping domain.

Definition 12 (Membership of states in a class). *Let a state class* $K = \langle m, W, D \rangle$ *and a state* $q = (m', I)$. *We say* $q \in K$ *iff* $m = m'$, $\mathsf{awake}(q) = W$ *and* $\forall \boldsymbol{\theta} \in I|_W, \exists (\boldsymbol{\theta}, \boldsymbol{\lambda}) \in D$ *s.t.* $\forall t_i \in W_s, \lambda_i = \overline{I}(t_i)$.

4.2 Construction of the State Class Graph

In contrast to simple TPNs, for WS-TPNs we have to consider two types of events: firing a transition $t \in T$ and putting a set of transitions $S \subseteq T$ to sleep, i.e. exceeding their firing intervals, which will be denoted $\widehat{S}$. The set of actions consisting of putting a transition to sleep is denoted $\widehat{T} = \{\widehat{S} \mid S \subseteq T\}$.

The State Class Graph (SCG) of a WS-TPN $N = \langle P, T, F, m_0, I_s, \Pi \rangle$ is the set of classes $(K_\sigma)_{\sigma \in (T \cup \widehat{T})^*}$ obtained as follows:

- The initial class is $K_\epsilon = \langle m_0, W_0, D_0 \rangle$, where $W_0 = \mathsf{awake}((m_0, I_s))$ and D_0 is the domain s.t. $\forall (\boldsymbol{\theta}, \boldsymbol{\lambda}) \in D_0, \forall t_i \in W_0, \theta_i \in I_s(t_i)$ and if $\overline{I_s}(t_i) \neq +\infty$, $\lambda_i = \overline{I_s}(t_i)$.
- If the sequence σ is feasible, i.e., we can find delays so that transitions in σ fire in order with those delays putting the corresponding transitions in σ to sleep, and $K_\sigma = \langle m, W, D \rangle$, then we have to check if $t_f \in W$ can fire and what set of transitions can fall sleep before that. Let $W_s = \{t_i \in W \mid \overline{I_s}(t_i) \neq +\infty\}$, and let $S \subseteq W_s$.

  S can fall asleep and t_f can fire after that iff

$$\begin{cases} D \wedge \bigwedge_{t_i \in S} \{\theta_f \succ_i \lambda_i\} \wedge \bigwedge_{\substack{t_i \in W \setminus S \\ t_i \neq t_f}} \{\theta_i \geq \theta_f\} \neq \emptyset \\ \forall \pi \in \Pi, \pi \cap W \neq \emptyset \iff \pi \cap (W \setminus S) \neq \emptyset \end{cases}$$

  where $\succ_i => $ if $D \wedge \{\theta_i = \lambda_i\} \neq \emptyset$ and $\succ_i = \geq$ otherwise.
- If S can fall asleep and t_f can fire after that, then the sequence $\sigma.\widehat{S}.t_f$ is feasible, then $K_{\sigma.\widehat{S}.t_f} = \langle m', W', D' \rangle$ is computed from $K_\sigma = \langle m, W, D \rangle$:
  - $m' = m - {}^\bullet t_f + t_f{}^\bullet$
  - $W' = ((W \setminus S) \cap \mathsf{enab}(m')) \cup (\uparrow \mathsf{enab}(m, t_f))$
  - Let $W_s' = \{t_i \in W' \mid \overline{I_s}(t_i) \neq +\infty\}$. D' is described by inequalities over variables θ_i' (resp. λ_i') added for all $t_i \in W'$ (resp. $t_i \in W_s'$). Formally, $\forall (\boldsymbol{\theta}', \boldsymbol{\lambda}') \in D', \exists (\boldsymbol{\theta}, \boldsymbol{\lambda}) \in D$ s.t.:

$$\begin{cases} \forall t_i \in W' \cap (\uparrow \mathsf{enab}(m, t_f)), \theta_i' \in I_s(t_i) \\ \forall t_i \in W' \setminus (\uparrow \mathsf{enab}(m, t_f)), \theta_i' \geq 0 \text{ and } \theta_i' = \theta_i - \theta_f \\ \forall t_i \in W_s' \cap (\uparrow \mathsf{enab}(m, t_f)), \lambda_i' = \overline{I_s}(t_i) \\ \forall t_i \in W_s' \setminus (\uparrow \mathsf{enab}(m, t_f)), \lambda_i' \geq 0 \text{ and } \lambda_i' = \lambda_i - \theta_f \end{cases}$$

The construction of SCG is also valid for nets with reset, read and transfer arcs. Since they only affect the discrete behaviour, we just need to adapt the firing of transitions by following the semantics.

Remark 1. It is important to explicitly put transitions to sleep in the SCG. Otherwise it does not capture all the possible discrete executions of the net. It may happen that a transition t_s can fall asleep in a class but cannot do it after firing another transition t_f. Indeed, firing t_f may disable all the others transitions in the pack of t_s thus forbidding it to fall asleep.

In a real implementation of the algorithm constructing the SCG, the computation of the new domain D' after firing and putting transitions to sleep is done by adding the new variables (θ', λ') and constraints to D (by conjunction), and by eliminating the variables (θ, λ) with Fourier-Motzkin.

Example 3. The state class graph of N_1 (from Fig. 1) is represented in the Fig. 7.

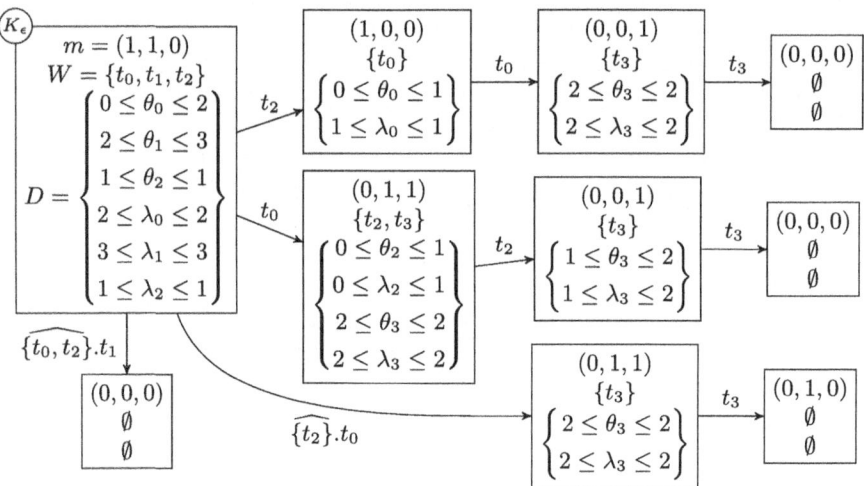

Fig. 7. SCG of N_1 (from Fig. 1).

4.3   Properties of the SCG

To construct the state class graph, as in [2,3], we just build the potentially infinite tree of state classes for all possible sequences σ and merge on-the-fly classes with the same marking, same awake transitions and same solutions to their firing and sleeping domains. For a bounded net, this procedure terminates:

Proposition 1. *Let N be a bounded WS-TPN, then its State Class Graph is finite.*

Proof. Compared to the construction of Berthomieu [2,3], the main difference is the addition of the λ variables. The same arguments as for the classical state class graph, then ensure that there can only be a finite number of state classes.

Let $\sigma \in \left(T \cup \widehat{T}\right)^*$ be a sequence of actions, namely firing and putting transitions to sleep. We note $\breve{\sigma} \in T^*$ the sequence of actions derived from σ where only the firings remain. Formally, if $\sigma = \widehat{S_1} t_1 \widehat{S_2} t_2 \ldots \widehat{S_n} t_n$ then $\breve{\sigma} = t_1 \ldots t_n$.

Proposition 2. *Let a WS-TPN $N = \langle P, T, F, m_0, I_s, \Pi \rangle$, its initial state $q_0 = (m_0, I_s)$ and its SCG $(K_\sigma)_{\sigma \in (T \cup \widehat{T})^*}$. Let $\tau = t_1 \ldots t_n \in T^*$ and $q \in Q$. There exists $q_1, \ldots, q_{n-1} \in Q$ s.t. $q_0 \xrightarrow{t_1} q_1 \ldots \xrightarrow{t_{n-1}} q_{n-1} \xrightarrow{t_n} q$ iff there exists $\sigma \in \left(T \cup \widehat{T}\right)^*$ s.t. $\breve{\sigma} = \tau$, K_σ is in the SCG and $q \in K_\sigma$.*

Proof. Let $H(n)$ be the induction hypothesis that the lemma holds for all sequences τ of size $n \geq 0$.

The initial case $H(0)$ is simply $q_0 \in K_\epsilon$, which is straightforward.

Let $n \geq 0$, we suppose $H(n)$ and prove $H(n+1)$. Let $\tau = t_1 \ldots t_n \in T^*$, $t_{n+1} \in T$ and $q \in Q$.

$\boxed{\Rightarrow}$ We suppose that there exists $q_1, \ldots, q_n \in Q$ s.t. $q_0 \xrightarrow{t_1} q_1 \ldots \xrightarrow{t_n} q_n \xrightarrow{t_{n+1}} q$. From $H(n)$ we know that there exists $\sigma \in \left(T \cup \widehat{T}\right)^*$ s.t. $\breve{\sigma} = \tau$, $K_\sigma = \langle m, W, D \rangle$ is in the SCG and $q_n = (m, I) \in K_\sigma$.

By definition, there exists $d \in \mathbb{R}_+$ s.t. $q_n \xrightarrow{d} q' \xrightarrow{t_{n+1}} q_{n+1}$. Let $S = $ awake$(q_n) \setminus$ awake(q') be the set of transitions that fall asleep by elapsing d from q_n. Let $t_f = t_{n+1}$. We first prove that $\sigma.\widehat{S}.t_f$ is feasible in the SCG.

Let $t_i \in S$. First, for all $\theta_i \in I(t_i)$, $\theta_i < d$. Thus, $\overline{I(t_i)} < d$ if cl $(\downarrow I(t_i)) = \downarrow I(t_i)$ and $\overline{I(t_i)} \leq d$ otherwise. Then, there exists $\boldsymbol{\theta} \in I|_W$ s.t. $d \leq \theta_f$. Moreover, since $q_n \in K_\sigma$, $\exists (\boldsymbol{\theta}, \boldsymbol{\lambda}) \in D$ s.t. $\lambda_i = \overline{I(t_i)}$. Therefore $\lambda_i < \theta_f$ if $D \wedge \{\theta_i = \lambda_i\} \neq \emptyset$ and $\lambda_i \leq \theta_f$ otherwise.

Let $t_i \in W \setminus (S \cup \{t_f\})$. First, there exists $\theta_i' \in I(t_i)$ s.t. $d \leq \theta_i'$. Moreover, since $0 \in I(t_f) - d$, there exists $\theta_f' \in I(t_f)$ s.t. $\theta_f' \leq d$. Therefore, $\exists \boldsymbol{\theta} \in I|_W$ s.t. $\theta_i = \theta_i'$ and $\theta_f = \theta_f'$. Then $\exists (\boldsymbol{\theta}, \boldsymbol{\lambda}) \in D$ s.t. $\theta_f \leq \theta_i$.

The condition on the packs follows exactly the semantics, and it then true. Therefore $\sigma.\widehat{S}.t_f$ is feasible in the SCG. Let $K_{\sigma.\widehat{S}.t_f} = \langle m', W', D' \rangle$. We prove now that $q_{n+1} \in K_{\sigma.\widehat{S}.t_f}$.

It is straightforward to show that $q_{n+1} = (m', I')$, for some I', and awake$(q_{n+1}) = W'$. Let $\boldsymbol{\theta}' \in I'|_{W'}$. For all $t_i \in W'$, $\theta_i' \in I_s(t_i)$ if $t_i \in \uparrow$ enab (m, t_f), else $\exists \theta_i \in I(t_i)$ s.t. $\theta_i' = \max(0, \theta_i - d)$. Since $0 \in I(t_f) - d$, it follows that $d \in I(t_f)$. Let $\theta_f = d$, then if $t_i \in \uparrow$ enab (m, t_f), $\theta_i' = \max(0, \theta_i - \theta_f)$. Since $q_n \in K_\sigma$, $\exists (\boldsymbol{\theta}, \boldsymbol{\lambda}) \in D$ s.t. $\forall t_i \in W_s$, $\lambda_i = \overline{I(t_i)}$. We define $\boldsymbol{\lambda}'$ s.t. $\forall t_i \in W_s'$, $\lambda_i = \overline{I_s(t_i)}$ if $t_i \in \uparrow$ enab (m, t_f), else $\lambda_i' = \lambda_i - \theta_f = \lambda_i - d = \overline{I'(t_i)}$. Hence $\exists (\boldsymbol{\theta}', \boldsymbol{\lambda}') \in D'$ s.t. $\forall t_j \in W'$, $\lambda_j = \overline{I'(t_j)}$, which terminates to prove that $q_{n+1} \in K_{\sigma.\widehat{S}.t_f}$.

$\Leftarrow$ We suppose that there exists $\sigma \in \left(T \cup \widehat{T}\right)^*$ s.t. $\breve{\sigma} = \tau.t_{n+1}$, K_σ is in the SCG and $q \in K_\sigma$. Let $\sigma = \sigma_n.\widehat{S}.t_{n+1}$ with $\sigma_n = \widehat{S_1}t_1 \ldots \widehat{S_n}t_n$. It is straightforward to show that $\exists d \in \mathbb{R}_+$, $q_n, q' \in Q$ s.t. $q_n = (m, I) \xrightarrow{d} q' \xrightarrow{t_{n+1}} q_{n+1}$ and $S = \mathsf{awake}\,(q_n) \setminus \mathsf{awake}\,(q')$. The predecessor is given by the definition of the domain, and d is one of the possible firing time $\theta_{n+1} \in I(t_{n+1})$. Then $H(n)$ terminates the proof. $\square$

The (discrete) langage of a WS-TPN N is the set of sequences that matches runs of the net. It is denoted $\mathcal{L}(N) = \{\sigma = t_1 \ldots t_n \mid \exists q_1, \ldots, q_n \in Q$ and $q_0 \xrightarrow{t_1} q_1 \ldots \xrightarrow{t_n} q_n\}$. Similarly the langage of a SCG $G = (K_\sigma)_{\sigma \in (T \cup \widehat{T})^*}$ is the set of sequences of firing derived from feasible sequences: $\mathcal{L}(G) = \{\sigma \in T^* \mid \exists \sigma' \in \left(T \cup \widehat{T}\right)^*$ s.t. $\breve{\sigma'} = \sigma$ and $K_{\sigma'}$ exists $\}$.

Corollary 3. *A WS-TPN and its State Class Graph have the same (discrete) langage.*

5 Conclusion

We have proposed a new class of time Petri nets, called WS-TPNs, that seamlessly integrate both the strong semantics, related to urgency, and the weak semantics of time Petri nets.

Using this class, we prove that while (unbounded) TPNs with weak semantics have nice decidability results (even reachability is decidable), coverability becomes undecidable if we allow even a single transition to behave along the strong semantics.

We complete this negative result by proving that the same holds when adding instead two read arcs, or two reset arcs, or two transfer arcs, while those arcs had no such impact on regular Petri nets, and [12] has shown that a TPN with weak semantics and the corresponding Petri net (obtained by removing time intervals) have the same set of reachable markings.

We finally show how to construct a finite abstraction of the state space for bounded WS-TPNs that capture the set of discrete sequences of the net, in a similar fashion to regular bounded TPNs with strong semantics [2,3].

Further work include investigating structural restrictions on the use of transitions following a strong semantics in WS-TPNs, to obtain decidable subclasses. Similarly we would also find relevant structural sufficient conditions, mostly on conflicts, that ensure that no transition ever falls asleep.

References

1. Araki, T., Kasami, T.: Some decision problems related to the reachability problem for Petri nets. Theoret. Comput. Sci. **3**(1), 85–104 (1976)

2. Berthomieu, B., Diaz, M.: Modeling and verification of time dependent systems using time petri nets. IEEE Trans. Softw. Eng. **17**(3), 259–273 (1991)
3. Berthomieu, B., Menasche, M.: An enumerative approach for analyzing time petri nets. In: Proceedings IFIP, pp. 41–46. Elsevier Science Publishers (1983)
4. Boyer, M., Roux, O.H.: On the compared expressiveness of arc, place and transition time Petri nets. Fund. Inform. **88**(3), 225–249 (2008)
5. Christensen, S., Hansen, N.D.: Coloured petri nets extended with place capacities, test arcs and inhibitor arcs. In: Ajmone Marsan, M. (ed.) ICATPN 1993. LNCS, vol. 691, pp. 186–205. Springer, Heidelberg (1993). https://doi.org/10.1007/3-540-56863-8_47
6. Dufourd, C., Finkel, A., Schnoebelen, P.: Reset nets between decidability and undecidability. In: Larsen, K.G., Skyum, S., Winskel, G. (eds.) ICALP 1998. LNCS, vol. 1443, pp. 103–115. Springer, Heidelberg (1998). https://doi.org/10.1007/BFb0055044
7. Jones, N.D., Landweber, L.H., Lien, Y.E.: Complexity of some problems in Petri nets. Theoret. Comput. Sci. **4**, 277–299 (1977)
8. Komenda, J., Lahaye, S., Parrot, R., Roux, O.H.: Weakly strong semantics of time Petri nets for performance evaluations. IFAC-PapersOnLine **58**(1), 66–71 (2024). 17th IFAC Workshop on discrete Event Systems WODES 2024
9. Merlin, P.M.: A study of the recoverability of computing systems. Ph.D. thesis, Dep. of Information and Computer Science, University of California, Irvine (1974)
10. Minsky, M.L.: Computation: finite and infinite machines. Prentice-Hall Inc, Upper Saddle River, NJ, USA (1967)
11. Montanari, U., Rossi, F.: Contextual nets. Acta Informatica **32**(6), 545–596 (1995)
12. Reynier, P.-A., Sangnier, A.: Weak time petri nets strike back! In: Bravetti, M., Zavattaro, G. (eds.) CONCUR 2009. LNCS, vol. 5710, pp. 557–571. Springer, Heidelberg (2009). https://doi.org/10.1007/978-3-642-04081-8_37
13. Vogler, W., Semenov, A., Yakovlev, A.: Unfolding and finite prefix for nets with read arcs. In: Sangiorgi, D., de Simone, R. (eds.) CONCUR 1998. LNCS, vol. 1466, pp. 501–516. Springer, Heidelberg (1998). https://doi.org/10.1007/BFb0055644

SkiNet: A User-Oriented Tool for Petri Net-Based Analysis of Robotic Skills

Baptiste Pelletier[1]($\boxtimes$), Charles Lesire[2], and Karen Godary-Dejean[3]

[1] DCAS, ISAE, 10 Av. Marc Pélegrin, 31055 Toulouse, France
`baptiste.pelletier@isae-supaero.fr`
[2] DTIS, ONERA, Univ. de Toulouse, 2 Av. Marc Pélegrin, 31000 Toulouse, France
`charles.lesire@onera.fr`
[3] LIRMM-CNRS, Univ. de Montpellier, 161 Rue Ada, 34095 Montpellier, France
`karen.godary-dejean@umontpellier.fr`

Abstract. Despite the rise in applications of formal methods for the design and analysis of robotic systems, their integration into existing frameworks remains a challenge. Part of the problem is due to the important changes in the software architecture of robotic design tools needed to accommodate for formal methods, but another challenge is the inherent difficulty for non-experts to use formal methods. In this work, we present SkiNet, a user-oriented tool which aims to assist in the design of skill-based robotic systems using a formal model based on Petri nets. The main idea for SkiNet is to interpret the user-written specifications and properties, into elements from the formal model, and vice versa. This allows the user to be more independent when diagnosing and validating the specified behavior, without having to directly manipulate the back-end formal tools. An example of the verification process of robotic skills specifications using SkiNet is given to showcase this paradigm.

Keywords: Formal methods · Robotics · Model-checking · Software engineering

1 Introduction

Recent trends in robotics and autonomous systems have shown a rise in the use of skill-based architectures to describe behaviors [4,16]. Indeed, the design of automation and decision-making functions is simplified by dividing the often complex behavior of a robot into a set of skills it can perform, i.e. modular elementary actions that integrate the sensing or acting capabilities of the robot. On top of this, these skills can run concurrently using abstract resources models for software and hardware components. Domain-specific languages (DSLs) such as SkiROS [15] or Robot-Language [1,2], offer frameworks to design robotic skills.

In parallel, the rise of formal methods applied to autonomous systems allowed an increase in robustness and safety of behaviors, with domains ranging from robotics to autonomous vehicles [13]. However, their integration into frameworks

E. Amparore and Ł. Mikulski (Eds.): PETRI NETS 2025, LNCS 15714, pp. 354–365, 2025.
https://doi.org/10.1007/978-3-031-94634-9_17

used for the design and analysis of robotic systems, especially for end-user oriented tools, is not as popular. Indeed, high-level skills and mission specification is still being done using semi-formal methods, such as behavior trees [8] or UML-based languages [20], or using DSLs such as Robot-Language [1]. In contrast, approaches relying on formal methods, such as DiNeROS [6], ASPiC [11] or the work of Ziparo [22], struggle to gain in popularity. The fact that these frameworks are supposed to be handled by end-users, who often have no experience with formal methods or robotic programming, makes them less appealing than their semi-formal counterparts, even if formal verification becomes impossible. On the other hand, toolchains like the Genom3/Fiacre/Hippo framework [5] are formally defined, but more expert-oriented. This leads to a need for user-oriented formal tools, especially when the behaviors become too complex and require a formal simulation and verification. Related works like VeriPlan [10] or Frama-C [14] have tried to tackle this issue in their respective domains. However, to the best of our knowledge, there are no tool who offer, with a user-oriented approach, an automatic verification of specifications of robotic behaviors, based on a formal model faithful to the behavior generated from these specifications, not just the verification of the specifications themselves.

In this work, we tackle robotic skills specified using Robot-Language (RL), a DSL aimed at users who are familiar with robotics programming and who wish to create skills for their robots [1,2]. This DSL allows to describe a skillset: a set of skills defining the elementary actions of the robot, a set of resources abstracting its software and hardware components, and a set of events, considered uncontrollable, that affect resources states. The RL skillset comes with a tool for generating a ROS-compliant C++ code that follows the specified behavior of the skills, with empty functions that need to be filled by the user with the functional code of the robot. In a previous work [17], we showed how RL skillsets can be translated into Petri net models and what interesting properties could be verified via model-checking of the Petri net. The translation and analysis were integrated in a tool called SkiNet, which we present in this paper, with an emphasis on user-friendliness.

As the tool started to be used by more and more users with various experience in robotic skills programming and formal methods, the discrepancy between the specification language and the PN elements and model-checking results turned out to be a challenge. This is a known issue, highlighted by Espiau et al. [7], called the "error diagnosis problem". SkiNet was therefore designed as a high-level tool, translating the skillset paradigm, manipulated by the user, and the PN environment, used for simulation and model-checking. The latter partly relies on Tina, a general PN analysis toolbox [3].

The paper is divided as follows: the SkiNet architecture, and how it interfaces between the user, the PN model, the RL parser tool and the Tina toolbox, are presented in Sect. 2. Then, Sect. 3.2 presents a case study of how users can rely on SkiNet to simulate and diagnose unwanted behaviors. Finally, we will conclude with our future goals for the tool. The SkiNet tool is available, with the example skillsets shown in this paper, as well as tutorials, on Gitlab: https://gitlab.com/onera-robot-skills/skinet-release.

2   SkiNet

In this section, we present in detail the SkiNet tool, how it interfaces with the user, the Robot-Language skillset specification file, the generated PN models and the back-end tools it relies on.

Two domains are tackled by the tool, as shown in Fig. 1: the user-space, i.e. what the user interacts with, and the Petri net (PN) space. The user-space is the Robot-Language (RL) skillset specification files (extension .rl), while the PN-space is the PN model and the interfaces with tools from the Tina toolbox. In the center is SkiNet, used either via a command-line interface or the instance of a Python class called `SkillsetNet`. The goal of SkiNet is to assist the user in analyzing its robotic skills, with the PN analysis running in the background, the user being never exposed to the PN-space unless requested.

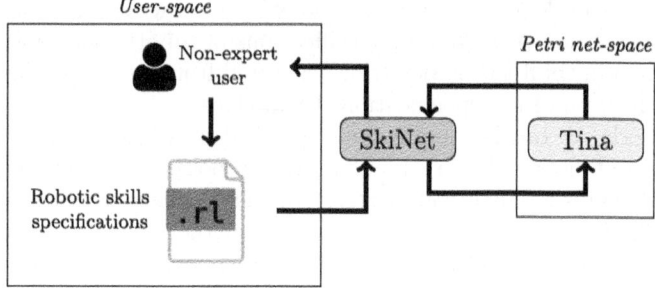

Fig. 1. SkiNet acts as an interpreter between the user-space and the Petri net-space for the user to analyze Robot-Language skills.

2.1   Robot-Language (User-Space)

A RL skillset [1] contains sets of resources $\mathcal{R}$, events $\mathcal{V}$ and skills $\mathcal{S}$. They have been applied to a variety of robotic systems, such as ground, aerial or underwater vehicles [2,17,18]. In this work, we present a use-case of a skillset used to model a Spot®robot from Boston Dynamics, mounted with a Jaco® manipulator arm from Kinova, as seen in Fig. 2.

Resources $r \in \mathcal{R}$ are state machines with states $S^r = \{S_0^r, S_1^r, ..., S_n^r\}$, with transitions between these states, which help modelling the functional layer. Events $\nu \in \mathcal{V}$ are used to model internal or external changes in the system that would impact the resources states, while skills $s \in \mathcal{S}$ model the elementary actions the robot can do. Both are refined with guards and effects on the resources to specify their execution, and the combination of a guard and an effect is called a skillset transition τ. An example of skill, `safe_poweroff`, taken from the `spot_and_arm.rl` skillset, used to turn off the motor power on the Spot®robot, is given in Listing 1.1. This skill has three preconditions and can only be started if these are met (l.2 to 6), and when started, the motors become busy (resource `control_mode` goes to the *Busy* state, l.7). Finally, it can end in a success (l.8 to 11), or a failure (l.12), switching `control_mode` back to *Idle*. In case of a success, the resource `power_status` goes to *PowerOff*.

Fig. 2. Spot®robot, controlled by a Robot-Language skillset to execute various autonomous missions.

```
1    skill safe_poweroff {
2        precondition {
3            is_sitting : spot_status == Sitting
4            can_move   : lease_status == AutoMode and control_mode == Idle
5            is_powered : power_status == PowerOn
6        }
7        start {control_mode -> Busy}
8        success is_poweredoff {
9            effect control_mode -> Idle
10           postcondition power_status == PowerOff
11       }
12       failure couldnot_poweroff {effect control_mode -> Idle}
13   }
```

Listing 1.1. Skill `safe_poweroff`, written in the Robot-Language DSL [1].

2.2 Petri Net Generation

The first step to analyze and simulate a RL skillset will be to generate a `SkillsetNet` class instance (`SkN`) from the skillset specification (`.rl` file). This `SkN` instance will contain the skillset elements and its PN model, as shown in Fig. 3. From the specification file, the RL parsing tool is used to obtain a `Skillset` instance, a class describing the skillset content as a tree-like structure, with the skills, resources and events. From this, the PN model of the skillset is generated, giving a `PetriNet` instance, as well as a `.net` file (textual PN extension used by the Tina toolbox). Finally, the `SkN` instance is populated using the places, transitions and arcs from the `PetriNet` instance, by associating them back to elements in the skillset.

In our previous work [17], we have formalized the automated generation of skillsets into PNs modelling their execution. We summarize the important elements here. Resources $r \in \mathcal{R}$ are each modelled with a set of places p_i^r for each of their states S_i^r. The transitions of the PN are generated from skillset transitions, generating a set of more than one PN transition T_τ for each τ, to take into account all possible previous and next states of resources when τ is triggered.

On top of this, skills $s \in \mathcal{S}$ have two states, idle or executing, modeled as PN places p_i^s and p_e^s respectively. As an example, the PN generated from the skill `safe_poweroff` in Lst. 1.1 is given Fig. 4.

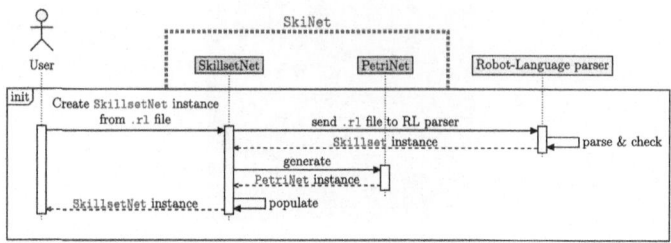

Fig. 3. SkiNet uses the RL parser with the input file to get a `Skillset` tree-like instance, before generating the PN model and finally the high-level `SkillsetNet`.

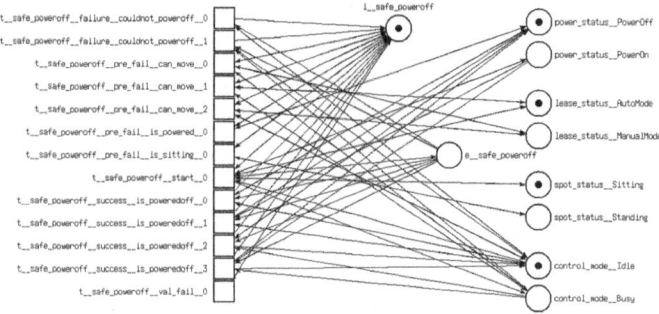

Fig. 4. PN generated from the specifications of the `safe_poweroff` skill.

The `SkN` class provides various high-level methods, the middle-level `PetriNet` class methods, as well as lower-level methods which call the tools from the Tina toolbox. The latter can be found either in the `PetriNet` class, if the action to be realized can be used for generic PNs, or the `SkN` class, if the action is specialized in the context of robotic skills analysis.

Manipulating the Generated Petri Net. While the generated PN is initially saved as a `.net`, users can use SkiNet to call the `ndrio` tool from Tina to convert it to another format. This allows expert users to use other tools than Tina and SkiNet, if they are already familiar with other PN analysis tools. The PN can also be exported in the graphical format `.ndr` of Tina using SkiNet. Figure 5 compares the graphical PN produced by `ndrio` and SkiNet, with the latter being able to position the elements between skills, resources and events. However, as we can see, despite the improved arrangement, the readability is still very difficult.

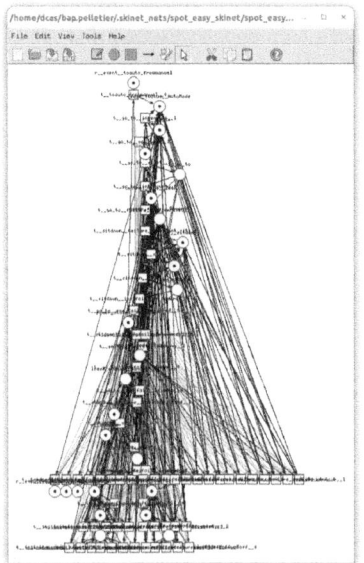

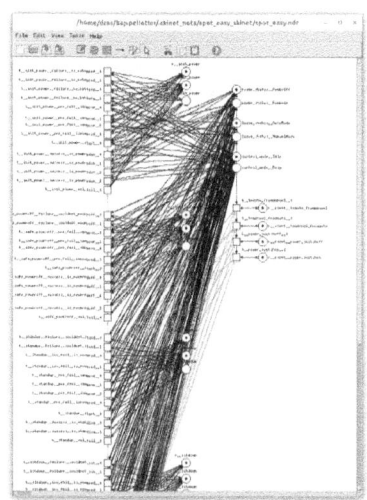

(a) Graphical PN model (.ndr) generated by ndrio tool of Tina using the *graphplace* placement filter.

(b) Graphical model rearranged by SkiNet, with skill transitions sorted on the left, skill state places in the center, and resource state places and event transitions on the right.

Fig. 5. Comparison of the PN graphical representation capabilities.

Examples. While graphical representation usually makes it easier to simulate the execution of models, it remains challenging to visually follow all simulation events and changes for big and complex models. Indeed, some of our skillsets would generate PN models too large for manual simulation, as seen in Tab. 1, with skillsets taken from the examples folder of the SkiNet Gitlab repository. We can see that some models can have a lot of PN transitions compared to the number of skillset transitions τ, making the PN difficult to manipulate using the former, this discrepancy being due to the concurrency in the resources states. Therefore, a textual simulation using skillset transitions instead of PN transitions can help the user, as we will see in the next section.

2.3 Simulation

Textual simulation using SkiNet was designed to help the user by hiding places that are not marked and transitions that are not enabled or firable, in order to only show the "active" elements of the PN. The textual simulation tool play of Tina can also be used, however the automated generation of the PN can lead to places and transitions names being hard to read or put back into context, as well as multiple places/transitions being generated from the same skillset element.

Table 1. Examples of RL skillsets used for various robot types. $|\mathcal{R}|$, $|\mathcal{S}|$ and nb.τ are the number of resources, skills and skillset transitions respectively, $|P|$ and $|T|$ are the number of places and transitions in the generated PN. The state-space size is also given (number of states).

Robot type	RL skillsets files	$|\mathcal{R}|$	$|\mathcal{S}|$	nb. τ	$|P|$	$|T|$	States
Wheeled	ugv_skillset.rl	3	5	44	24	60	37
Quadruped	spot.rl	6	10	77	48	164	400
Flying	uav_skillset.rl	6	10	116	54	297	1512
Arm (simple)	manipulator.rl	4	6	68	49	266	51340
Arm (detailed)	manipulator_6dof.rl	6	1	36	63	16089	16265

On top of this, to be really helpful for the user, the simulation process should remind essential keywords or variable names of the user-space (such as resource states, skill invariants and preconditions, or events). SkiNet will perform the translation of the PN elements back to the user-space.

This simulation is managed as illustrated in Fig. 6: first, the *state* of the skillset is determined from the PN marking, sorting active elements as skill states, resource state places and firable skillset transitions, and second, the PN is simulated using the high-level skillset transitions τ, rather than the transitions $t \in T_\tau$ generated from them.

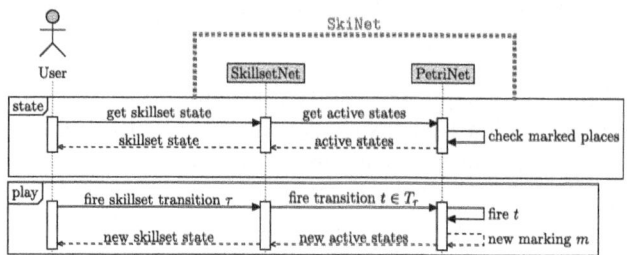

Fig. 6. Skillset state and skillset transition firing sequences.

Skillset State. The state of a skillset is the set of the active states of each skill (idle or running), and resources (states S_i^r). If the place p_i^s of a skill is marked, the skill is considered idle; otherwise, if p_e^s is active, the skill is executing. Similarly, the place p_i^r of a resource r being marked indicates that its state S_i^r is active. This will already help the textual simulation, as only the active states of skills and resources will be shown to the user.

Skillset Execution. In the user-space, the simulation of the skillset execution is done using the skillset transitions τ of skills or events. In the PN-space, the PN is simulated using the firing of the transitions $t \in T_\tau$. This, and the previous skillset state definitions, make for the textual simulation interface that will allow users to simulate their skillsets more intuitively. For instance, if the user decides to start a skill s, SkiNet will take the corresponding skillset transition and verify if a transition $t \in T_\tau$ is firable from the current marking of the PN, and fire it. If none are firable, the skillset transition τ is directly hidden to the user.

Figure 7 compares the textual simulation of a skillset with its PN model using Tina and SkiNet. The `play` tool from Tina, taking as input the generated `.net` or `.ndr` file, shows the marking of the PN and the enabled and firable transitions at the given marking. The Skinet interface gives the skillset state, with resources and skill states clearly indicated, as well as the executable skills and events.

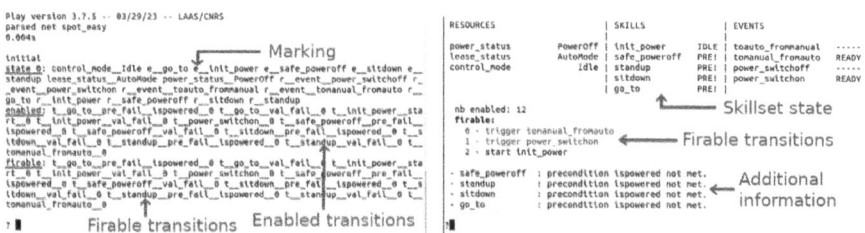

(a) Interface of `play`, the command line PN player from the Tina toolbox.

(b) Interface of the skillset simulation from SkiNet.

Fig. 7. Comparison of the PN simulation environments.

3 Analysis Using SkiNet

Finally, we will showcase how SkiNet can help the user to verify and validate its skillset model, with an example.

3.1 Verification Process

Because the skillset specified by the user in the `.rl` file may not match the desired behavior, SkiNet will help with the validation step, i.e. making sure that the skillset can accomplish the desired behaviors, but also with the verification, i.e. checking that the way it accomplishes these behaviors is correct. This analysis is done in two steps: first, the state-space of the skillset must be calculated to generate all the reachable states of the system, and then, model-checking can be performed to check the desired properties. Figure 8 illustrates how SkiNet serves as interface between the user and the analysis tools from Tina.

For state-space construction, the `sift` tool is used with the PN model to create a Kripke structure [12], stored as a `.ktz` binary file. This Kripke structure is used by the model-checking tools `selt` and `muse` of Tina, for linear temporal logic (LTL) and computational tree logic (CTL) respectively. For each of these tools, a child process is created in the `SkN` instance, namely `selt_child` and `muse_child`, and loaded with the Kripke structure. Once again, manipulating these tools is a challenge for non-expert users. Having to calculate the state-space is not a reflex for most users, so SkiNet will do it automatically for them. On top of this, temporal logics can be challenging to use, and the fact that the tools from Tina are in the PN-space makes them even more challenging, as the user would have to use the names of places and transitions as they appear in the generated .net file. SkiNet helps in two ways to assist the user, by both providing generic properties to be verified on the skillset, but also by serving as an intermediary for the user when verifying temporal logic formulae.

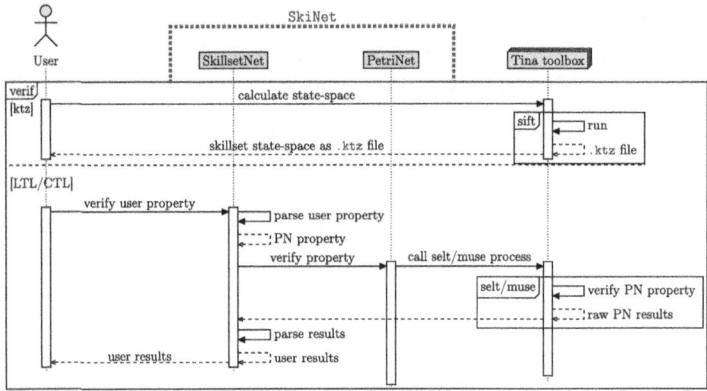

Fig. 8. State-space construction and temporal logic verification sequences.

Generic Properties. Generic properties, defined in our previous work [17], can be called using methods from the `SkN` instance or the command line interface of SkiNet, to verify skillset deadlocks, skill liveness, or the presence of useless elements, defined as skillset transitions τ where all transitions $t \in T_\tau$ are dead. These methods will call the child processes to verify predefined temporal logic formulae, get the results sent back by the processes, and parse them from the PN-space to the user-space, as seen in Fig. 8.

User Properties. Users can write, in the user-space, their own properties, using either CTL or LTL. For LTL, the `selt` tool will return a sequence of transitions from the initial state if the property has a counter example. For CTL, a set of states verifying the property is returned by `muse`, from which the `pathto` tool from Tina can be used to find the sequence of transitions leading to each state. Then, SkiNet will interpret each transition of these sequences in the user-space, so that the user can see what event or skill execution led to these states.

3.2   Example

We consider a user trying to validate a skillset written for the Spot®robot. The specified skillset is the `spot_and_arm.rl` example in the Gitlab repository. Using SkiNet, the user wants to verify a specific property: can the robotic arm be in an unknown state while the Spot®is turning off (skill `safe_poweroff`, Lst. 1.1)? The user writes the property in the user-space, using CTL, to find states that would violate the property, giving Eq. 1:

$$EG \ (safe\_poweroff.executing \ and \ arm\_status.Unknown) \qquad (1)$$

SkiNet will convert the property to the PN-space and return the set of states where the property is true, using `muse`, as shown in Fig. 9: four states are found, and SkiNet can return a sequence of transitions leading to each state. For instance, the sequence leading to state n.181 is: the skill `init_power` and standup of Spot®succeed, then the `goto_articular` skill of the robotic arm is started, during which the event `end_effector_blocked` is triggered, leading to a failure of this skill. After that, if the Spot®uses the `sitdown` skill, followed by the `safe_poweroff` skill, the property will be true. (postcondition of the success, l.10). The user can then modify the skills and events to fix this behavior.

```
Skillset: spot_and_arm
Target logic : CTL
Property       : EG ( safe_poweroff.executing and arm_status.Unknown )
EG ( e__safe_poweroff /\ arm_status__Unknown )
- Analysis -
Result (set): [181, 226, 230, 250]
Play a result[enter state number, q or press enter to stop, m to show/hide markings, f to show/hide full trace]:181
> 0 : init_power.start
> 1 : init_power.success.is_poweredon
> 2 : standup.start
> 3 : standup.success.is_standing
> 4 : goto_articular.start
> 5 : triggered end_effector_blocked
> 6 : goto_articular.inv_fail.is_ok
> 7 : sitdown.start
> 8 : sitdown.success.is_sitting
> 9 : safe_poweroff.start
Result (set): [181, 226, 230, 250]
Play a result[enter state number, q or press enter to stop, m to show/hide markings, f to show/hide full trace]:█
```

Fig. 9. Result for the verification of Eq. 1 using SkiNet, showing the property to verify, the same property written in the user-space, and the set of states verifying the property. The sequence leading to state n.181 is shown.

4   Conclusion

In this work, we have presented SkiNet, a user-oriented tool for the modeling and verification and validation of robotic skills specifications using Petri nets. We have demonstrated the ability of SkiNet to maintain a boundary between the "user-space", where the user only tackles elements from the specifications, and the "Petri net-space". This allows users who are unfamiliar with formal methods to formally simulate, verify and diagnose the skills behaviors.

SkiNet has received positive feedback from developers and users familiar with Robot-Language, familiar with formal methods and model-checking, or both, as well as graduate students who used the tools as part of supervised projects, but a proper case-study should be conducted. On top of this, SkiNet was used to verify skillsets for aerial, ground and underwater robots, with some examples available on the Gitlab repository. Further improvements on SkiNet involve:

- Improve model-checking to go beyond just translating transition sequences, shown Fig. 9, by grouping results to avoid redundant traces.
- Create an ecosystem for verifying robotic skills from other DSLs than RL, or using other formal methods, such as the work of Raïs et al. [19], as well as an interface for planning tools to query SkiNet when solving problems.
- Plugging the LTL/CTL property specification to user-oriented tools such as FRET [9] or LLM-based tools like NL2CTL [21] and VeriPlan [10].

Acknowledgement. This research was by the Occitanie Region (France) and the National Research Agency (ANR) grant ANR-22-CE92-0011-02.

Disclosure of Interests. The authors have no competing interests to declare that are relevant to the content of this article.

References

1. Albore, A., Doose, D., Grand, C., Guiochet, J., Lesire, C., Manecy, A.: Skill-based design of dependable robotic architectures. Robot. Auton. Syst. **160** (2023). https://doi.org/10.1016/j.robot.2022.104318
2. Albore, A., Doose, D., Grand, C., Lesire, C., Manecy, A.: Skill-based architecture development for online mission reconfiguration and failure management. In: International Workshop on Robotics Software Engineering (RoSE) (2021). https://doi.org/10.1109/RoSE52553.2021.00015
3. Berthomieu, B., Vernadat, F., dal Zilio, S.: The tina toolbox home page - time petri net analyzer - by laas/cnrs (2004)
4. Bøgh, S., Nielsen, O., Pedersen, M., Krüger, V., Madsen, O.: Does your robot have skills? In: International Symposium on Robotics (2012)
5. Dal Zilio, S., Hladik, P.E., Ingrand, F., Mallet, A.: A formal toolchain for offline and run-time verification of robotic systems. Robot. Auton. Syst. **159** (2023). https://doi.org/10.1016/j.robot.2022.104301, https://www.scopus.com/inward/record.uri?eid=2-s2.0-85141912955&doi=10.1016%2fj.robot.2022.104301&partnerID=40&md5=0fc8e276f09ce286efe7549f74207eb5, cited by: 3; All Open Access, Bronze Open Access, Green Open Access
6. Ebert, S., Mey, J., Schöne, R., Götz, S., Aßmann, U.: Dineros: a model-driven framework for verifiable ROS applications with petri nets. In: International Conference on Model Driven Engineering Languages and Systems Companion (MODELS-C) (2023). https://doi.org/10.1109/MODELS-C59198.2023.00127
7. Espiau, B., Kapellos, K., Jourdan, M.: Formal verification in robotics: why and how? Robot. Res. (1996)
8. Ghzouli, R., Berger, T., Johnsen, E.B., Dragule, S., Wąsowski, A.: Behavior trees in action: a study of robotics applications. In: International Conference on Software Language Engineering (2020). https://doi.org/10.1145/3426425.3426942

9. Giannakopoulou, D., Mavridou, A., Rhein, J., Pressburger, T., Schumann, J., Shi, N.: Formal requirements elicitation with fret. In: International Working Conference on Requirements Engineering: Foundation for Software Quality (REFSQ-2020). No. ARC-E-DAA-TN77785 (2020)

10. Lee, C., Porfirio, D., Wang, X.J., Zhao, K., Mutlu, B.: Veriplan: integrating formal verification and LLMs into end-user planning. arXiv preprint arXiv:2502.17898 (2025)

11. Lesire, C., Pommereau, F.: Aspic: an acting system based on skill petri net composition. In: International Conference on Intelligent Robots and Systems (IROS) (2018). https://doi.org/10.1109/IROS.2018.8594328

12. Liu, Z., Xing, Z.: Characterizing petri nets with the temporal logic CTL. In: Conference on Information Technology and Computer Science (2012). https://doi.org/10.2991/citcs.2012.97

13. Luckcuck, M., Farrell, M., Dennis, L.A., Dixon, C., Fisher, M.: Formal specification and verification of autonomous robotic systems: a survey. ACM Comput. Surv. **52**(5) (2019). https://doi.org/10.1145/3342355

14. Maroneze, A., Perrelle, V., Kirchner, F.: Advances in usability of formal methods for code verification with frama-c. Electron. Commun. EASST **77** (2019). https://doi.org/10.14279/tuj.eceasst.77.1108, https://eceasst.org/index.php/eceasst/article/view/2235

15. Mayr, M., Rovida, F., Krueger, V.: Skiros2: A skill-based robot control platform for ros (2023)

16. Pedersen, M.R., et al.: Robot skills for manufacturing: from concept to industrial deployment. Robot. Comput. Integr. Manuf. **37** (2016). https://doi.org/10.1016/j.rcim.2015.04.002

17. Pelletier, B., Lesire, C., Doose, D., Godary-Dejean, K., Dramé-Maigné, C.: Skinet, a petri net generation tool for the verification of skillset-based autonomous systems. In: International Workshop on Formal Methods for Autonomous Systems (FMAS) (2022). https://doi.org/10.4204/EPTCS.371.9

18. Pelletier, B., Lesire, C., Grand, C., Doose, D., Rognant, M.: Predictive runtime verification of skill-based robotic systems using petri nets. In: 2023 IEEE International Conference on Robotics and Automation (ICRA), pp. 10580–10586 (2023). https://doi.org/10.1109/ICRA48891.2023.10160434

19. Raïs, S., Brunel, J., Doose, D., Herbreteau, F.: Cross-layer formal verification of robotic systems. In: International Workshop on Formal Methods for Autonomous Systems (FMAS) (2024). https://doi.org/10.4204/eptcs.411.9

20. Thomas, U., Hirzinger, G., Rumpe, B., Schulze, C., Wortmann, A.: A new skill based robot programming language using UML/P statecharts. In: International Conference on Robotics and Automation (ICRA) (2013). https://doi.org/10.1109/ICRA.2013.6630615

21. Zhao, M., Tao, R., Huang, Y., Shi, J., Qin, S., Yang, Y.: Nl2ctl: automatic generation of formal requirements specifications via large language models. In: Ogata, K., Mery, D., Sun, M., Liu, S. (eds.) Formal Methods and Software Engineering, pp. 1–17. Springer, Singapore (2024)

22. Ziparo, V.A., Iocchi, L., Lima, P.U., Nardi, D., Palamara, P.F.: Petri net plans: a framework for collaboration and coordination in multi-robot systems. Auton. Agents Multi-Agent Syst. **23** (2011)

Deciding (Sub-Marking) Reachability in $O(P^2 + T^2)$ for Sound Acyclic Free-Choice Workflow Nets

Thomas M. Prinz[1]($\boxtimes$)[iD], Christopher T. Schwanen[2][iD], and Wil M. P. van der Aalst[2][iD]

[1] Course Evaluation Service, Friedrich Schiller University Jena, Jena, Germany
Thomas.Prinz@uni-jena.de
[2] Chair of Process and Data Science (PADS), RWTH Aachen University, Aachen, Germany
{schwanen,wvdaalst}@pads.rwth-aachen.de

Abstract. Reachability is a central decision problem in Petri net theory deciding whether a given marking can be reached from the initial marking. Sub-marking reachability (the covering problem) asks whether there is a reachable marking, which consists of at least the tokens in the given marking. The current state of the art describes the computational complexity of both problems as polynomial for live and bounded free-choice nets as well as for sound free-choice workflow nets. This paper refines this complexity on the class of sound acyclic (simple) free-choice workflow nets to $O(P^2 + T^2)$. The presented approach uses three new concepts: admissibility, maximum admissibility, and diverging transitions. *Admissibility* requires that all places in a given marking are pairwise concurrent. *Maximum admissibility* states that adding a place to an admissible marking would make it inadmissible. A *diverging transition* is a transition which originally "produces" the concurrent tokens that lead to a given marking. All three concepts can additionally provide explanations why a (sub-)marking is not reachable.

Keywords: Workflow Nets · Reachability · Covering problem · Soundness · Free-choice

1 Introduction

A central problem in Petri net theory is *reachability*: deciding if a given marking can be reached from the initial marking. Reachability of a given marking is crucial to show whether a given system, which shall be analyzed, fulfills a given property, or not. Such properties can be, for example, the absence of deadlocks, livelocks, or undesired states during conformance checking. As it forms the basis of many verification approaches, deciding reachability and deciding it computationally effective is important. The general reachability problem for Petri nets falls within the NONELEMENTARY complexity class [7]. Cheng et al. [4] give a PSPACE complexity class for general safe nets. Esparza [11] stated that the reachability problem in safe and live free-choice nets can be reduced to the *CNF-SAT* problem; thus, deciding reachability for this class is NP-complete. However, Desel and Esparza [8] already showed earlier that reachability in cyclic free-choice nets, a restricted subclass of live and bounded free-choice nets where the initial marking is a home marking, can be decided in polynomial time. Regarding sound

E. Amparore and Ł. Mikulski (Eds.): PETRI NETS 2025, LNCS 15714, pp. 366–387, 2025.
https://doi.org/10.1007/978-3-031-94634-9_18

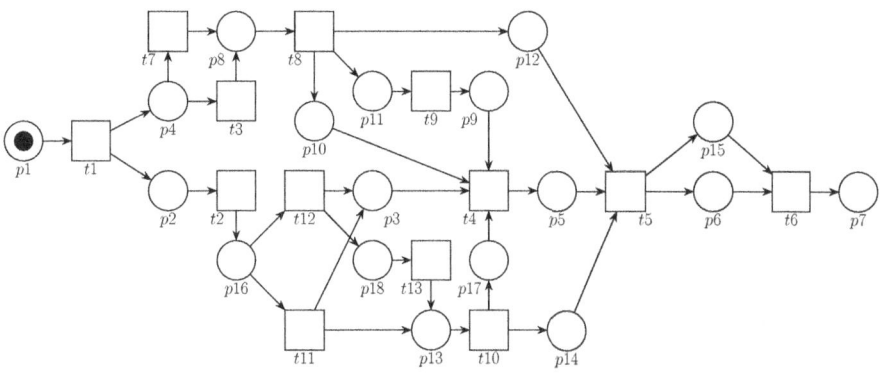

Fig. 1. A sound acyclic free-choice workflow net.

extended free-choice workflow nets, Yamaguchi [22] confirms a polynomial time complexity of deciding reachability, but the concrete polynomial is unknown.

In this paper, we show that reachability for *sound acyclic (simple) free-choice workflow nets* can be solved in quadratic time, $O(P^2 + T^2)$. We further show that *sub-marking reachability* [12] (the *covering problem*, i.e., if a partial marking is reachable) can be solved in the same computational complexity.

Figure 1 shows an example of a sound acyclic free-choice workflow net. A question could be: Is a marking reachable in which $p9$, $p12$, and $p16$ have tokens? To answer such questions, instead of focusing on *concrete occurrence sequences* (traces) to the desired marking, the presented algorithm demonstrates that most decisions on whether a marking is reachable can be decided by *concurrency*. Thereby, we introduce the concept of *admissible* and *maximum admissible* markings. A marking M is *admissible* if all pairs of marked places are in a concurrency relation. M is further *maximum admissible* if it is not possible to add a further token to the marking without destroying admissibility. We show that each reachable marking in a sound acyclic free-choice workflow net must be maximum admissible and that neither computing the concurrency relation [14,19] nor checking maximality requires knowledge of concrete occurrence sequences.

Although maximum admissibility is a necessary condition and good heuristics of deciding reachability, it is unfortunately not sufficient. There are nets with markings that are maximum admissible but *not* reachable. For this reason, we argue that concurrency is always introduced by diverging transitions. This manifests in occurrence nets of sound acyclic free-choice workflow nets as subgraphs which only diverge in transitions. Hence, the identification of these diverging transitions implies the existence of an occurrence net that leads to the (maximum) admissible marking. For this reason, deciding reachability can be achieved by checking the given marking for maximum admissibility and determining whether a diverging transition exists that leads to the marking—examining concrete occurrence sequences is not necessary. Similarly, a sub-marking is reachable if the marking is just admissible and there is such a diverging transition. The overall approach has a quadratic computational complexity in the worst case. As a further benefit, the approach can also explain why a marking is *not* reachable rather than just deciding reachability.

Of course, this paper examines a quite limited class of Petri nets regarding (sub-marking) reachability. There are two main reasons why investigating this class of nets is still important: (1) Industrial business process models strongly correlate with free-choice workflow nets [13]. Soundness is seen as an important quality criterion [10] of process models and can be checked in cubical computational time complexity with detailed diagnostic information [18]. Finally, many industrial process models are acyclic. For this reason, for an industrial setting, sound acyclic free-choice workflow nets are an interesting class of nets. (2) There is an interesting trend investigating *free-choice nets with a home cluster* [3], which showed that this class of nets correlates strongly with *perpetual nets* finally correlating with sound free-choice workflow nets. Mapping the new method of *loop decomposition* for industrial workflow graphs [17–19] to free-choice workflow nets should eventually fill the gap to investigate the class of free-choice nets with a home cluster with the class of sound acyclic free-choice workflow nets. For this reason, this paper can be interpreted as a further step to achieve a low computational polynomial time complexity to decide (sub-marking) reachability while providing diagnostic information at the same moment.

The remainder of this paper is structured as follows: Sect. 2 introduces basic concepts of Petri nets, markings, reachability, paths, and soundness. Maximum admissible markings and their application are discussed in Sect. 3 which are then used in Sect. 4 to finally decide reachability in sound acyclic free-choice workflow nets. Eventually, Sect. 5 concludes this paper.

2   Preliminaries

This paper uses standard notions for Petri nets, which are provided in the following. We also recall the *Path-to-End Theorem* for simple free-choice nets.

2.1   Multisets, Petri Nets, and Paths

$\mathscr{B}(A)$ is the set of all *multisets* over some set A. For a multiset $b \in \mathscr{B}(A)$, $b(a)$ denotes the number of times element $a \in A$ appears in b. For example, $b_1 = [\ ] = \varnothing$, $b_2 = [x,x,y]$, $b_3 = [x,y,z]$, $b_4 = [x,x,y,x,y,z]$, and $b_5 = [x^3,y^2,z]$ are multisets over the set $A = \{x,y,z\}$. b_1 is the empty multiset, b_2 and b_3 consist of three elements, and $b_4 = b_5$, i.e., the ordering of elements is irrelevant and b_5 uses a more compact notation for repeating elements. The standard set operators can be extended to multisets, e. g., $x \in b_2$, $b_2 \uplus b_3 = b_4$, $b_5 \setminus b_2 = b_3$, etc.

Definition 1 (Petri Nets). *A Petri net (or simply a net) N is a triple (P,T,F) with P and T are finite, disjoint sets of places and transitions, and $F \subseteq (P \times T) \cup (T \times P)$ is the flow relation.*   ⏌

$P \cup T$ can be interpreted as *nodes* and F as *edges* between those nodes. For $x \in P \cup T$, $\bullet x := \{p \mid (p,x) \in F\}$ is the *preset* of x (all directly preceding nodes) and $x \bullet := \{s \mid (x,s) \in F\}$ is the *postset* of x (all directly succeeding nodes). Each node in $\bullet x$ is an *input* of x and each node in $x\bullet$ is an *output* of x. The preset and postset of a set of nodes $X \subseteq P \cup T$ is defined as $\bullet X := \bigcup_{x \in X} \bullet x$ and $X\bullet := \bigcup_{x \in X} x\bullet$, respectively. N is *proper* iff $\forall t \in T$: $\bullet t \neq \emptyset \wedge t\bullet \neq \emptyset$. N is *(extended) free-choice* iff $\forall t_1, t_2 \in T$: $\bullet t_1 \cap \bullet t_2 \neq \emptyset \Rightarrow \bullet t_1 = \bullet t_2$. N is *simple* free-choice iff $\forall p \in P$: $|p \bullet| \geq 2 \Rightarrow \bullet(p\bullet) = \{p\}$, i.e., $\forall t_1, t_2 \in T$: $\bullet t_1 \cap \bullet t_2 \neq \emptyset \Rightarrow \{p\} = \bullet t_1 = \bullet t_2$ [9].

Without loss of generality, this paper focuses on simple free-choice nets as Murata [15] *presents a linear time transformation algorithm of extended to simple and simple to extended free-choice nets.*

A *path* $(n_1, \ldots, n_m)$ is a sequence of nodes $n_1, \ldots, n_m \in P \cup T$ with $m \geq 1$ and $\forall i \in \{1, \ldots, m-1\}$: $n_i \in \bullet n_{i+1}$. Note that places and transitions alternate on paths. $[(n_1, \ldots, n_m)]$ depicts the set of all nodes on the path. If all nodes of a path are pairwise different, the path is *acyclic*; otherwise, it is *cyclic*. $Paths(x, y)$ denotes the set of all paths between nodes x and y, where $x, y \in P \cup T$. N is *acyclic* if all its paths are acyclic. In the nets shown here, circles represent places, rectangles transitions, and directed edges represent flows as done in Fig. 1.

Definition 2 (Workflow Nets, FC-WF-Nets, and AFC-WF-Nets). *A* workflow net *$N = (P, T, F, i, o)$ is a net (P, T, F) with $i, o \in P$, $\bullet i = o\bullet = \emptyset$. i is the* source *and o is the* sink *of N. All nodes are on a path from i to o. If N is (simple) free-choice, then N is called a* FC-WF-net. *If N is acyclic (simple) free-choice, N is called an* AFC-WF-net. ⌐

This paper focuses on AFC-WF-nets.

2.2 Markings, Reachability, Properties, and Soundness

The behavior of nets is defined via *markings*, which describe the number of *tokens* on places in a specific state.

Definition 3 (Marking). *A* marking M *of a net $N = (P, T, F)$ is a multiset of places, $M \in \mathscr{B}(P)$. (N, M) is a marked net. $\langle M \rangle := \{x \in M\}$ depicts the set of places of M.* ⌐

Transitions whose input places all have tokens are *enabled* in a marking and can be fired, leading to the net's semantics:

Definition 4 (Enabledness, Firing, and Reachability). *Let (N, M) be a marked net $N = (P, T, F)$. A transition $t \in T$ is* enabled *in M, denoted as $(N, M)[t\rangle$, iff every place $\bullet t$ contains at least one token in M, $\bullet t \subseteq M$. $en(N, M) := \{t \in T \mid (N, M)[t\rangle\}$ is the set of enabled transitions in (N, M).*

If t is enabled in M, then t may fire, *which removes one token from each of t's input places and adds one token to each of t's output places. $M' = (M \setminus \bullet t) \uplus t\bullet$ is the marking resulting from firing t in (N, M). $(N, M)[t\rangle(N, M')$ denotes that t is enabled in (N, M) and firing t would result in (N, M').*

A sequence $\sigma = \langle t_1, t_2, \ldots, t_n \rangle \in T^*$, $n \geq 1$, is an occurrence sequence *if there are markings $M_1, M_2, \ldots, M_{n+1}$ such that $\forall 1 \leq i \leq n$: $(N, M_i)[t_i\rangle(N, M_{i+1})$. For applying such a sequence σ, we write $(N, M_1)[\sigma\rangle(N, M_{n+1})$. We say that $t_1, t_2, \ldots, t_n$ occur in σ.*

A marking M' is reachable *from M if there is an occurrence sequence σ such that $(N, M)[\sigma\rangle(N, M')$. $R(N, M) := \{M' \in \mathscr{B}(P) \mid \exists \sigma \in T^*: (N, M)[\sigma\rangle(N, M')\}$ denotes the set of all reachable markings of (N, M).*

If $N = (P,T,F,i,o)$ is a workflow net, $[i]$ is its initial marking *and $[o]$ is its* final
marking.   ⌐

Next, we list some behavioral properties important for analysis:

Definition 5 (Live, Bounded, Safe, and Dead). *Let (N,M) be a marked net $N =
(P,T,F)$. (N,M) is* live *iff for every reachable marking $M' \in R(N,M)$ and for every
transition $t \in T$, there is a reachable marking $M'' \in R(N,M')$, which enables t. (N,M)
is k-bounded iff for every reachable marking $M' \in R(N,M)$ and for every place $p \in
P$: $M'(p) \leq k$. (N,M) is* bounded *iff there is a k such that (N,M) is k-bounded. (N,M)
is* safe *iff (N,M) is 1-bounded.*

A place $p \in P$ is dead *in (N,M) iff $\forall M' \in R(N,M)$: $M'(p) = 0$. A transition $t \in T$ is*
dead *in (N,M) iff $\forall M' \in R(N,M)$: $t \notin en(N,M')$.*   ⌐

Workflow nets can be *sound* [1]:

Definition 6 (Soundness). *A workflow net $N = (P,T,F,i,o)$ with its initial marking $[i]$
and its final marking $[o]$ is* sound *iff*

(1) $\forall M \in R(N,[i])$: $[o] \in R(N,M)$,
(2) $\forall M \in R(N,[i])$: $(o \in M \Rightarrow M = [o])$, and
(3) there is no dead *transition in N: $\forall t \in T \; \exists M,M' \in R(N,[i])$: $(N,M)[t\rangle(N,M')$.*

*An equivalent definition of soundness is that the marked short-circuited net $(N',[i])$ of
N, where N' is defined as*

$$(P,T \cup \{t\}, F \cup \{(o,t),(t,i)\})$$

for a transition $t \notin T$, is live and bounded [1].   ⌐

Sound free-choice workflow nets are further *safe*:

Lemma 1 (Safeness). *Sound free-choice workflow nets are safe.*   ⌐

Proof (Lemma 1). See Verbeek et al. [21].   □

Sound simple free-choice workflow nets ensure that each path from an arbitrary
place to the sink place o contains at most one token [19]:

Theorem 1 *(Path-to-End Theorem). Let $W = (P,T,F,i,o)$ be a simple FC-WF-net. It
holds: On all paths ρ from any place p to o in any reachable marking $M \in R(W,[i])$, the
sum of all tokens on all places of ρ is at most 1, i.e.,*

$$\forall p \in P \; \forall \rho \in Paths(p,o) \; \forall M \in R(W,[i]): |[\rho] \cap M| \leq 1.$$ ⌐

Proof (Theorem 1). See Theorem 2 in Prinz et al. [19].

Note that this is a special case of Lemma 5.11 in Van der Aalst [2].   □

3   (Maximum) Admissible Markings

In the following, we describe, argue, and prove why reachability and *concurrency* strongly interact in sound AFC-WF-nets. We will show that a given marking can only be reachable if all its places are pairwise concurrent. Two places are concurrent if there is a reachable marking with tokens on both places:

Definition 7 (Concurrency). *Let $N = (P, T, F, i, o)$ be a sound AFC-WF-net. Two places $x, y \in P$ are concurrent in N, denoted $x \parallel y$, iff $\exists M_{\parallel} \in R(N, [i])$: $[x, y] \subseteq M_{\parallel}$. $\parallel(x) := \{y \in P \mid x \parallel y\}$ denotes the set of all places to which x is concurrent. Due to safeness of sound AFC-WF-nets, a place x is not concurrent to itself.*   ⌟

For example, in Fig. 1, $p9 \parallel p10$ and $p5 \parallel p12$ as well as $\parallel(p15) = \{p6\}$.

Sound AFC-WF-nets have benefits regarding their complexity during analysis. One of them is that two concurrent places must not have a path between them:

Lemma 2. *Let $N = (P, T, F, i, o)$ be a sound AFC-WF-net with two places $x, y \in P$, $x \neq y$.*

$$x \parallel y \quad \Longrightarrow \quad Paths(x, y) = \varnothing \wedge Paths(y, x) = \varnothing$$ ⌟

Proof (Lemma 2). See Prinz et al. [19] (Cor. 4). The interested reader can also use the Path-to-End Theorem 1 to confirm the path absence of concurrent nodes.   □

The interested reader can check that, e. g., $p9$ in Fig. 1 has no path to $p10$ and vice versa.

By Lemma 1, sound AFC-WF-nets are safe. For this reason, all concurrent places x and y, $x \parallel y$, must be joined by a transition on all pairs of paths from x and y to o:

Lemma 3 (Two Concurrent Places are Joined by Transitions). *Let $N = (P, T, F, i, o)$ be a sound AFC-WF-net. Furthermore, let $x, y \in P$. It holds:*

$$x \parallel y$$

$$\Longrightarrow$$

$$\forall \rho_x = (x_1, \ldots, x_n, o) \in Paths(x, o), n \geq 1$$
$$\forall \rho_y = (y_1, \ldots, y_m, o) \in Paths(y, o), m \geq 1$$
$$\exists i \in \{1, \ldots, n\} \, \exists j \in \{1, \ldots, m\}:$$
$$\{x_1, \ldots, x_i\} \cap \{y_1, \ldots, y_j\} = \{x_i\} = \{y_j\} \subseteq T$$ ⌟

Proof (Lemma 3). Constructive proof. Let $x, y \in P$ with $x \parallel y$. By Definition 7 of concurrency and from $x \parallel y$, it follows that $\exists M \in R(N, [i])$: $[x, y] \subseteq M$. By Lemma 1, N is safe and, thus, $x \neq y$.

By Definition 2 of workflow nets, there are at least two paths $\rho_x = (x_1, \ldots, x_n, o) \in Paths(x, o)$ and $\rho_y = (y_1, \ldots, y_m, o) \in Paths(y, o)$ from $x = x_1$ and $y = y_1$ to the sink o, respectively. It follows also: $[\rho_x] \cap [\rho_y] \neq \varnothing$. Therefore, ρ_x and ρ_y must have a *first*

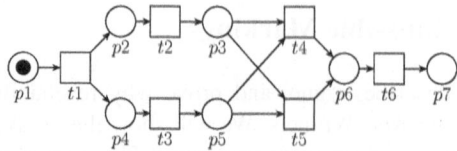

Fig. 2. In sound extended free-choice workflow nets, Lemma 3 does not hold without modifications as *p6* is *not* a joining transition.

common node $x_i = y_j$ with $x_i \in [\rho_x] \cap [\rho_y]$ and $\{x_1, \ldots, x_{i-1}\} \cap \{y_1, \ldots, y_{j-1}\} = \emptyset$ and $i \in \{1, \ldots, n\}, j \in \{1, \ldots, m\}$.

According to the proof of the Path-to-End Theorem 1 in Prinz et al. [19], p. 136, Eq. (4), it holds for each transition on a path ρ to the sink o in a sound (simple) AFC-WF-net because of simple free-choiceness:

$$\forall t \in ([\rho] \cap T): \quad |\bullet t \cap [\rho]| = 1 \quad \wedge \quad |t \bullet \cap [\rho]| \geq 1$$

For this reason, removing a token from ρ_x or ρ_y starting from M, respectively, can only be achieved by a place p with $|p \bullet| \geq 2$ (a decision). Since N is simple free-choice, $\bullet(p \bullet) = \{p\}$, any output transition of such p can be fired in each reachable marking $M' \in R(N, M)$, $p \in M'$—thus, also output transitions on ρ_x ($p \bullet \cap [\rho_x]$) and ρ_y ($p \bullet \cap [\rho_y]$). As a consequence, we can treat the tokens on both paths ρ_x and ρ_y starting from x and y, respectively, to "remain" on ρ_x and ρ_y, i.e., if a transition t_x on ρ_x fires, it puts a token back on ρ_x, and if a transition t_y on ρ_y fires, it puts a token back on ρ_y. Since N is sound, this treatment of tokens to "remain" on ρ_x and ρ_y cannot lead to a dead transition. Furthermore, since $x \parallel y$ and Lemma 2, there is no path from x to y and from y to x, i.e., no token can get from x to y and vice versa.

Now, there are two possibilities for the first common node $x_i = y_j$:

$x_i \in P$: Without loss of generality, once the token of ρ_x "reaches" x_i before the token of ρ_y in a reachable marking $M' \in R(N, M)$, then in M' are at least two tokens on path ρ_y to o. This contradicts the Path-to-End Theorem 1. ↯ Therefore, $x_i \notin P$.

$x_i \in T$: This is the only remaining possibility. ✓

Both cases state that all pairs of paths ρ_x and ρ_y have a transition as a first common node. For this reason, this lemma holds. □

Figure 2 shows an example of a sound *extended* free-choice workflow net, for which Lemma 3 does not hold. However, since this paper focuses on *simple* free-choiceness, such cases are not possible as *t4* and *t5* would be merged into a single transition.

The concurrency relation is symmetric by definition. The relationship is crucial for checking reachability since all places in a reachable marking must be in a concurrency relation by Definition 7. We call a marking where all places are pairwise concurrent *admissible*:

Definition 8 (Admissible Markings). *Let N be a sound AFC-WF-net. A marking $M_a \in \mathscr{B}(P)$ is admissible if all different places in M_a are pairwise concurrent:*

$$\forall x, y \in P, x \neq y: \quad [x, y] \subseteq M_s \implies x \parallel y$$

Each marking containing a single place is admissible. Following from Definition 8, admissibility of a marking is a necessary condition for its (sub-marking) reachability. If a marking is *not* admissible, we can quickly decline its reachability. In addition, admissibility of markings limits the "size" of reachable markings by the concurrency relation. There must be *maximum* admissible markings:

Definition 9 (Maximum Admissible Markings). *Let* $N = (P,T,F,i,o)$ *be a sound AFC-WF-net. A marking* $M \in \mathscr{B}(P)$ *is* maximum admissible *iff*

1. $\forall x,y \in M, x \neq y: x \parallel y$ *(M is admissible) and*
2. $\bigcap_{x \in M} \parallel(x) = \varnothing$ *(adding places of* $P \setminus \langle M \rangle$ *to M would make M inadmissible).*   ⌐

A marking M composed of concurrent places can only be *not* reachable by Definition 8 for three reasons: (1) M contains not enough places, (2) M contains too much places, or (3) M has a "correct" number of places but all places cannot be concurrent at the same time. Fortunately, we can show that each *reachable* marking in a sound AFC-WF-net is *maximum admissible*, i.e., removes reasons (1) and (2):

Theorem 2 (Reachable Markings are Maximum Admissible). *Let* N *be a sound AFC-WF-net. All reachable markings* M_r *from the initial marking* $[i]$ *are maximum admissible.*   ⌐

Proof (Theorem 2). Let $N = (P,T,F,i,o)$ be a sound AFC-WF-net. Furthermore, let $M_r = [p_1, \ldots, p_m]$, $m \geq 1$, be a reachable marking $M_r \in R(N,[i])$. The proof is done by contradiction: M_r is *not* maximum admissible. As M_r is admissible by Definition 8 but M_r is not *maximum* admissible, by Definition 9, there must be a place $p \in (P \setminus M_r)$ not in M_r being concurrent to all places in M_r:

$$\exists p \in (P \setminus M_r) \, \forall p' \in M_r: p' \parallel p \tag{1}$$

Let p be such a place in the following.

For each $p_i \in M_r$, $1 \leq i \leq m$, there is a path $\rho_i \in Paths(p_i, o)$ to the sink o by Definition 2 of workflow nets. In addition, let $\rho_p \in Paths(p, o)$ be a path to o, which contains p by Definition 2. By Theorem 1, for each ρ_i there is exact one token ρ_i on p_i in M_r (ρ_i is safe). There are exactly two cases for ρ_p:

Case 1: $\exists p' \in ([\rho_p] \cap P): p' \in M_r$ (there is a token on ρ_p in M_r). Thus, there is a path from p to p', $Paths(p, p') \neq \varnothing$. Since $p \parallel p'$ by (1), this contradicts Lemma 2 that concurrency requires the absence of paths. ⨍ This case does not hold.

Case 2: $\forall p' \in ([\rho_p] \cap P): p' \notin M_r$ (there is no token on ρ_p in M_r). The entire situation is abstractly illustrated in Fig. 3. By Lemma 3, all such two paths ρ_x and ρ_y, $\rho_x, \rho_y \in \{\rho_1, \ldots, \rho_m, \rho_p\}$, $\rho_x \neq \rho_y$, contain a transition t with at least two input places $in_x \in ([\rho_x] \cap \bullet t)$ and $in_y \in ([\rho_y] \cap \bullet t)$, $in_x \neq in_y$; even for ρ_p with any other path ρ_i, $1 \leq i \leq m$. For the moment, we say ρ_x and ρ_y "converge" in t. By this case, ρ_p is without any token in M_r, i.e., the transition(s) $t_1, \ldots, t_m$, in which ρ_p converges with any of the other paths ρ_i, $1 \leq i \leq m$, are dead when the token(s) on those paths reach any of $t_1, \ldots, t_k$ in a marking $M' \in R(N, M_r)$. This contradicts Definition 6 of soundness. ⨍ This case does not hold.

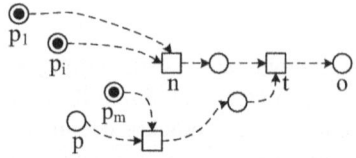

Fig. 3. For each p_i with a token in M_r, there is a path ρ_i to the sink o.

Since both cases do not hold, the contradiction does not hold. For this reason, the theorem holds that each reachable marking is maximum admissible.   □

For example, the marking $[p9, p10]$ in Fig. 1 is *not* maximum admissible since $p9$ and $p10$ have $p3$, $p17$, etc. as common concurrent places. The marking $[p3, p5]$ is *not* maximum admissible because it is not admissible at all: $p3 \nparallel p5$. However, the marking $[p5, p12, p14]$ is maximum admissible as all places, to which $p12$ is concurrent (e. g., $p9$, $p11$, $p18$, etc.), are not concurrent to $p5$ and $p14$.

It follows for checking if a marking is reachable in a sound AFC-WF-net to investigate first its maximum admissibility. By Definition 9, checking maximum admissibility of a marking depends on two simple rules. Following these two rules, the computation whether a given marking is (maximum) admissible or not, can be decided in $O(|P|^2 + |T|^2)$:

Theorem 3 (Computational Complexity of (Maximum) Admissibility). *Let $N = (P, T, F, i, o)$ be a sound AFC-WF-net. Deciding, whether a given marking M_r is maximum admissible in N or not, can be achieved in $O(|P|^2 + |T|^2)$.*   ⌋

Proof (Theorem 3). It is possible with the algorithm of Prinz et al. [19] to determine the concurrency relation by Definition 7 in $O(|P|^2 + |T|^2)$ for sound AFC-WF-nets.

The second rule of Definition 9 states:

$$\bigcap_{x \in M_r} \|(x) = \varnothing. \tag{2}$$

Since $x \in M_r$ is concurrent to all other places of M_r except to itself, x is only not in the intersection of all concurrency sets because x is missing in its own concurrency set $\|(x)$. For this reason, temporarily adding x to its own concurrency set leads to a combination of both rules:

$$\bigcap_{x \in M_r} \left(\|(x) \cup \{x\} \right) = M_r. \tag{3}$$

Once the concurrency relation is computed and the concurrency sets of all places are stored within a computationally efficient data structure for mathematical sets being able to compute the intersection in constant time, e. g., by a BitSet in Java, Eq. (3) can be checked in linear time, $O(|P|)$.

In summary, the overall computational complexity is dominated by the computation of the concurrency relation and can, therefore, be achieved in $O(|P|^2 + |T|^2)$. Furthermore, if the concurrency relation is stored, checking (maximum) admissibility can be achieved in linear time, $O(|P|)$.   □

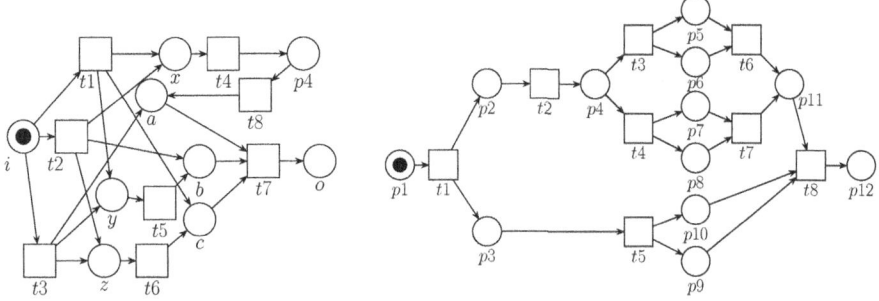

Fig. 4. A net with a maximal marking [x,y,z] that is not reachable.

Fig. 5. A sound AFC-WF-net.

Unfortunately, maximum admissibility is only a necessary condition for reachability but *not* sufficient since not each maximum admissible marking is also reachable as the net in Fig. 4 demonstrates. This net contains three places x, y, and z. They are pairwise concurrent, i.e., $x \parallel y$, $x \parallel z$, and $y \parallel z$. The resulting marking $[x,y,z]$ is maximum admissible. However, as the reader can confirm, this marking is not reachable since the places are not concurrent to each other at the same time. As a consequence, not every maximum admissible marking is reachable. We have to confirm maximum admissible markings, which are reachable as well.

4 Reachability

This section will show the sufficient condition of reachability in sound AFC-WF-nets. Besides the absence of paths between concurrent places, another benefit of sound AFC-WF-nets is that their *occurrence nets* [16] (i.e., a net, which represents how a corresponding net was executed) are simplified to *run nets* [20]. In contrast to occurrence nets, run nets are just subgraphs (sub *nets*) of sound AFC-WF-nets and, therefore, each node can only occur once. A run net is defined as follows [20]:

Definition 10 (Run Nets). *Let $N = (P,T,F,i,o)$ be a sound AFC-WF-net. An occurrence sequence $\sigma \in T^*$ is a run iff σ leads from the initial marking $[i]$ to the final marking $[o]$. A net $\pi = (P_\sigma, T_\sigma, F_\sigma)$ is a run net of a run σ of N iff*

$$P_\sigma = \left\{ p \in P \mid p \in \bigcup_{t \in \sigma}(\bullet t \cup t \bullet) \right\}$$

$$T_\sigma = \{t \in T \mid t \in \sigma\}$$

$$F_\sigma = F \cap \left((P_\sigma \cup T_\sigma) \times (P_\sigma \cup T_\sigma)\right).$$

$n \in \sigma$ occurs in π, depicted as $n \in \pi$. $\Pi(N)$ depicts the set of all run nets of N. ⌋

If a run net $\pi = (P_\sigma, T_\sigma, F_\sigma)$ of a sound AFC-WF-net $N = (P,T,F,i,o)$ contains a transition $t \in T_\sigma$, then it must also contain its input and output places, $\forall t \in T_\sigma: (\bullet t \cup t \bullet) \subseteq P_\sigma$.

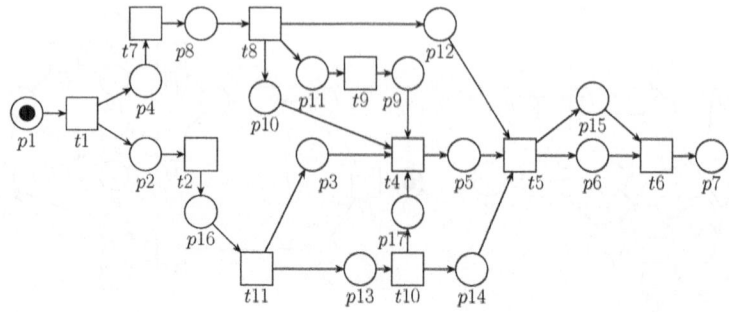

Fig. 6. Example of a run net of the example workflow net in Fig. 1.

If a run net contains a place $p \in P_\sigma$, $p \notin \{i, o\}$, it must also contain exactly one of its input and one of its output transitions, $\forall p \in (P_\sigma \setminus \{i, o\}): |p \bullet \cap T_\sigma| = 1$ and $|\bullet p \cap T_\sigma| = 1$. Figure 6 shows an example of a run net of our net in Fig. 1.

Let M_m be a marking for which we want to decide reachability in a given sound AFC-WF-net N. M_m must be maximum admissible by Theorem 2 and, thus, all places of M_m are pairwise concurrent. Regarding the run nets of N and Lemma 2 about the absence of paths between concurrent places, there cannot be paths between any possibly concurrent places in any run net of N. This fact was also confirmed in Theorem 2 in Prinz et al. [20]. For this reason, deciding whether a maximum admissible or admissible sub-marking M_m is reachable in N corresponds with the existence of a run net that contains all places of M_m:

Corollary 1. *Let $N = (P, T, F, i, o)$ be a sound AFC-WF-net and a marking M_m. It holds:*

$$M_m \in R(N, [i]) \iff M_m \text{ is maximum admissible } \land \tag{4}$$
$$\exists (P_\sigma, T_\sigma, F_\sigma) \in \Pi(N): \langle M_m \rangle \subseteq P_\sigma$$
$$\exists M_s \in R(N, [i]): M_m \subseteq M_s \iff M_m \text{ is admissible } \land \tag{5}$$
$$\exists (P_\sigma, T_\sigma, F_\sigma) \in \Pi(N): \langle M_m \rangle \subseteq P_\sigma$$

Proof (Corollary 1). Constructive proof by focusing on Eq. (4):

$\Rightarrow M_m \in R(N, [i])$. Since M_m is reachable from $[i]$, there must be a run net by Definition 10 of run nets, which contains all places of M_m. Furthermore, by Theorem 2, each reachable marking is maximum admissible, and, therefore, M_m. ✓

$\Leftarrow M_m$ is maximum admissible $\land \exists (P_\sigma, T_\sigma, F_\sigma) \in \Pi(N): \langle M_m \rangle \subseteq P_\sigma$. Since M_m is maximum admissible, all places of M_m are pairwise concurrent. Furthermore, there is a run net π, which contains all places of M_m. For this reason, and by Lemma 2, all places of M_m do not have paths to each other in N as well as in π. Thus, all places in M_m can have tokens in the same reachable marking from $[i]$ as they can appear in any order from each other. As a consequence, M_m is a reachable marking from $[i]$. ✓

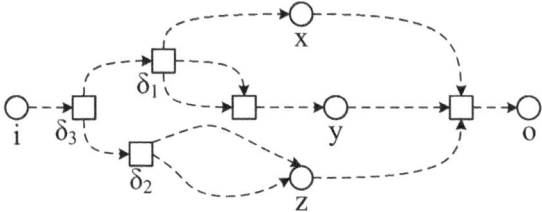

Fig. 7. For the marking $[x,y,z]$ there are diverging points δ_1 (for x and y) and δ_3 (for x and z as well as y and z). The source i is *no* diverging point of $[x,y,z]$ as all paths to them pass the same output transition of i. δ_2 is no diverging point of $[x,y,z]$ as it has only paths to z.

The proof of Eq. (5) can be done similarly.   □

Since we are able to decide the (maximum) admissibility of a marking in $O(|P|^2 + |T|^2)$, we need an approach to check if there is at least one run net, which contains all places of a given marking. Deriving such a run net is not trivial as decisions (places with multiple outgoing flows) can lead to an exponential growth of possible run nets. However, we do not have to derive a concrete run net—we only have to show that such a run net must exist, or not. For this reason, we will focus in the following on transitions, which diverge the control-flow and, therefore, "produce" concurrency.

If a sound AFC-WF-net has a reachable marking M_r with at least two places x and y, then there must be at least one transition t with multiple output places $\{x_t, y_t, \ldots\}$ (as the net has just one source) and x_t and y_t have disjoint paths to x and y so that the tokens are not joint before x and y. We call such t a *diverging point* of $\{x,y\}$ as well as a *diverging point* of the set $M_r = D$. Figure 7 illustrates the concept of diverging points abstractly for three places x, y, and z. δ_1 is a diverging point of $\{x,y,z\}$ although it only has disjoint paths to x and y. δ_2 is no diverging point of $\{x,y,z\}$ since it only has paths to z. The transition δ_3 is a diverging point of $\{x,y,z\}$ as it has disjoint paths to x and z as well as y and z. The source i is *no* diverging point of $\{x,y,z\}$ since each path from i to them goes through δ_3. The following definition describes that formally:

Definition 11 (Diverging Points). *Let $N = (P,T,F,i,o)$ be an AFC-WF-net. A node $\delta \in P \cup T$ is a diverging point of a set $D \subseteq P \cup T$, $|D| \geq 2$, of nodes, depicted $\delta \ll D$, if for at least two nodes $d_1, d_2 \in D$, $d_1 \neq d_2$, and for at least two output nodes $o_1, o_2 \in \delta\bullet$, there are two disjoint paths ρ_1 from o_1 to d_1 and ρ_2 from o_2 to d_2.*

$$\exists d_1, d_2 \in D \; \exists o_1, o_2 \in \delta \bullet \; \exists \rho_1 \in Paths(o_1, d_1) \; \exists \rho_2 \in Paths(o_2, d_2): \quad [\rho_1] \cap [\rho_2] = \varnothing$$

Let $\Delta(D) := \{\delta \in P \cup T \mid \delta \ll D\}$ be the set of all diverging points of D.

For a node $o_1 \in \delta\bullet$ and the set D, the set $\mathscr{R}_D(o_1) = \{d \in D \mid Paths(o_1, d) \neq \varnothing\}$ defines those nodes of $d \in D$ for which there is a path from o_1 to d.   ⌐

In the AFC-WF-net of Fig. 4, the set $D = \{x,y,z\} = \langle[x,y,z]\rangle$ of places has the diverging points $\Delta(\{x,y,z\}) = \{t1, t2, t3, i\}$. For example, there are disjoint paths from $t1$ to x and from $t1$ to y. For the outputs $\{x,y,c\}$ of $t1$, the subsets of nodes of $\{x,y,z\}$ they have paths to are $\mathscr{R}_{\{x,y,z\}}(x) = \{x\}$, $\mathscr{R}_{\{x,y,z\}}(y) = \{y\}$, and $\mathscr{R}_{\{x,y,z\}}(c) = \varnothing$. In the example

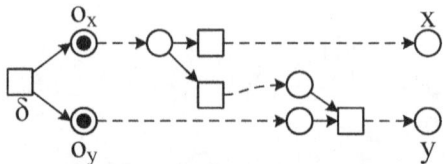

Fig. 8. A diverging transition δ for two places x and y has two output places o_x and o_y with two disjoint paths ρ_x from o_x to x and ρ_y from o_y to y, respectively. Through simple free-choiceness, each decision on ρ_x and ρ_y may lead to a remaining token on ρ_x and ρ_y, respectively. In addition, soundness ensures that transitions on ρ_x and ρ_y, respectively, with multiple input places will be fired. Furthermore, since $x \parallel y$ there are no paths between x and y and vice versa by Lemma 2, i.e., no transition on ρ_x and ρ_y may require the token of y and x, respectively.

AFC-WF-net in Fig. 5, the set $\{p7, p8, p9\}$ has $t4$ and $t1$ as diverging points although $t1$ has no disjoint paths for $p7$ and $p8$ (as it follows the same output).

Assume a transition δ is a diverging point of two concurrent places x and y, $\delta \lll \{x, y\}$. In this case, the tokens can "travel" from this transition via disjoint paths to x and y, i.e., from all markings in which δ is enabled, there is at least one reachable marking, in which x and y have tokens:

Lemma 4. *Let $N = (P, T, F, i, o)$ be a sound AFC-WF-net and $x, y \in P$. It holds:*

$$x \parallel y \quad \wedge \quad \delta \in \left(\Delta(\{x, y\}) \cap T \right)$$

$$\Longrightarrow$$

$$\forall M_\delta \in R(N, [i]), \ \delta \bullet \subseteq M_\delta \ \exists M_{xy} \in R(N, M_\delta) : [x, y] \subseteq M_{xy}$$

Proof (Lemma 4). Constructive proof. By $\delta \in \left(\Delta(\{x, y\}) \cap T \right)$ and Definition 11, it follows that there are two disjoint paths $\rho_x \in Paths(o_x, x)$ and $\rho_y \in Paths(o_y, y)$ from two output places $o_x, o_y \in \delta \bullet$ to x and y, $[\rho_x] \cap [\rho_y] = \varnothing$.

Since N is sound, there is at least one reachable marking $M_\delta \in R(N, [i])$ with $\delta \bullet \subseteq M_\delta$ (any marking after firing δ). Let M_δ be such a marking with $o_x, o_y \in M_\delta$. N is (simple) free-choice, i.e., for each transition t on ρ_x and ρ_y, $t \in \left(([\rho_x] \cup [\rho_x]) \cap T \right)$, with $|\bullet t| \geq 2$, it holds $\forall p \in \bullet t : |p \bullet| = 1$. Since $[\rho_x] \cap [\rho_y] = \varnothing$ and since $\forall p \in P : |p \bullet| \geq 2 \Rightarrow \bullet(p\bullet) = \{p\}$ (simple free-choiceness), we can force tokens on ρ_x and ρ_y (starting with o_x and o_y) to always put a token on a place on ρ_x and ρ_y, respectively (i.e., transitions on ρ_x and ρ_y fire regardless if they are outputs of diverging places, which could also "take another path"). Since N is sound, there is no dead transition (on ρ_x and ρ_y). Furthermore, since $x \parallel y$ and N is sound there are no paths between x and y and there are no paths between y and x by Lemma 2, i.e., no transition on ρ_x and ρ_y may "require" the token of y and x to fire, respectively. Figure 8 illustrates the situation. For this reason, at least until reaching x and y, it is possible that the number of tokens on ρ_x and ρ_y will not decrease. Eventually, tokens on ρ_x and ρ_y "arrive" at x and y in a marking M_{xy} that is reachable from M_δ, i.e., $\forall M_\delta \in R(N, [i]), \ \delta \bullet \subseteq M_\delta \ \exists M_{xy} \in R(N, M_\delta) : [x, y] \subseteq M_{xy}$. ✓ $\square$

Assume there is a maximum admissible marking M_m and it shall be checked if it is reachable from the initial marking. If all diverging points $\Delta(M_m)$ are transitions ($\Delta(M_m) \subseteq T$), then we can imply that there is a run net, which contains these $\Delta(M_m)$ transitions with all their output places and the paths between them. Thus, by Corollary 1, it directly follows that M_m is reachable. This is the simplest case. Unfortunately, places can also be diverging points as the above examples have already shown. This complicates the situation since it is not sure which output transition of a diverging place we should add to a run net to check the reachability of M_m. Trying each combination of output transition may lead again to an exponential growth of combinations. Fortunately, soundness and admissibility of M_m simplifies matters as with a sound AFC-WF-net and a admissible marking, there is always at least one output transition to take since the output transitions of such diverging places must always share the nodes to which they have paths to:

Lemma 5. *Let $N = (P,T,F,i,o)$ be a sound AFC-WF-net with a marking $M \in \mathscr{B}(P)$, a place as diverging point $\delta \in (\Delta(M) \cap P)$, and $\{o_x, o_y\} \subseteq \delta\bullet$. It holds:*

$$M \text{ is admissible} \implies \left(\mathscr{R}_{\langle M \rangle}(o_x) \subseteq \mathscr{R}_{\langle M \rangle}(o_y)\right) \vee \left(\mathscr{R}_{\langle M \rangle}(o_y) \subseteq \mathscr{R}_{\langle M \rangle}(o_x)\right). \quad \lrcorner$$

Proof. (Lemma 5). Proof by contradiction.

$$M \text{ is admissible} \wedge \mathscr{R}_{\langle M \rangle}(o_x) \not\subseteq \mathscr{R}_{\langle M \rangle}(o_y) \wedge \mathscr{R}_{\langle M \rangle}(o_x) \not\subseteq \mathscr{R}_{\langle M \rangle}(o_y) \quad (6)$$

$$\Rightarrow \quad \exists x \in \mathscr{R}_{\langle M \rangle}(o_x) \, \exists y \in \mathscr{R}_{\langle M \rangle}(o_y): \quad x \notin \mathscr{R}_{\langle M \rangle}(o_y) \wedge y \notin \mathscr{R}_{\langle M \rangle}(o_x) \quad (7)$$

Let $x, y \in M$ be such places. There are exactly two cases:

Case 1: $Paths(o_x, y) \neq \varnothing \vee Paths(o_y, x) \neq \varnothing$. Then, however, $y \in \mathscr{R}_{\langle M \rangle}(o_x)$ or $x \in \mathscr{R}_{\langle M \rangle}(o_y)$ by Definition 11 of diverging points. This contradicts (7) and this case does not hold. $\frac{\ell}{}$

Case 2: $Paths(o_x, y) = \varnothing \wedge Paths(o_y, x) = \varnothing$. Thus:

$$\forall \rho_x \in Paths(o_x, x) \, \forall \rho_y \in Paths(o_y, y): \quad [\rho_x] \cap [\rho_y] = \varnothing. \quad (8)$$

For this reason, concurrency of x and y cannot "start" from $[\delta]$. Figure 9 illustrates the situation in this case.

Let $\rho_x \in Paths(o_x, x)$ and $\rho_y \in Paths(o_y, y)$. By the admissibility of M in Eq. (6) and its Definition 8, it follows $x \parallel y$. Therefore, all paths from x and from y to the sink o join at first at converging transitions by Lemma 3. Let $c \in T$ be such a converging transition and $\rho_{cx} \in Paths(x, c_x)$ with $c_x \in \bullet c$ is a path without any other converging transition of x and y (i.e., c is the "first" of such transitions). There is also a disjoint path $\rho_{cy} \in Paths(y, c_y)$ with $c_y \in \bullet c$ by Lemma 3, $[\rho_{cy}] \cap [\rho_{cx}] = \varnothing$. As (a) there is no path from o_x to y, (b) there is no path from o_y to x, (c) c_x is the first converging transition on ρ_{cx}, and (d) N is sound and simple free-choice ($\bullet o_x = \bullet o_y = \{\delta\}$), whether o_x or o_y is fired is independent ("free") from other tokens in any reachable marking M of N. For this reason and N is sound, let $M_\delta \in R(N, [i])$ and $M_{c_y, \delta} \in R(N, M_\delta)$ be two reachable markings with $\delta \in M_\delta$ and $[c_y, \delta] \subseteq M_{c_y, \delta}$ so that c can

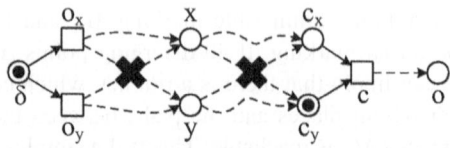

Fig. 9. A diverging place δ has two output transitions o_x and o_y. o_x has a path to x but not to y and o_y has a path to y but not to x. By $x \parallel y$ and Lemma 3, there is a "first" joining transition c. Thus, there is no path from y to c_x and there is no path from x to c_y. If δ decides for o_x, no token "after" δ can reach to c_y enabling c. Thus, since soundness is required, there must be a marking $M_{c_y,\delta}$ with $[c_y,\delta] \subseteq M_{c_y,\delta}$ to avoid a dead c. However, the path from δ via o_y, c_y, and c to o is not safe.

be fired if δ decides for o_x. However, in any $M_{c_y,\delta}$, there is an "unsafe" path from δ via c_y to the sink o by Theorem 1. This contradicts (6) that N is sound. ⨏ This case does not hold.

Both cases do not hold. Thus, contradiction (6) does not hold. ⨏ Thus, the lemma must hold.   □

The above Lemma 5 guarantees that the output transitions $\delta\bullet$ of a diverging place δ of an admissible marking M_m either have paths to the same subset of places of M_m or at least one output transition has paths to more places of M_m. To simplify what matters: there is always one output transition of a diverging place that has paths to the union of all subsets of M_m of all other output transitions. Since we are interested in the (sub-marking) reachability of all places of M_m, it is always the best option to take that output transition with the most reachable places of M_m. In summary, if we derive $\Delta(M_m)$ for an admissible marking M_m, we could take those output transitions of diverging places containing the greatest subset of places of M_m.

The remaining open question is if there is at least one transition in $\Delta(M_m)$, which has paths to all places in M_m. By Definition 11, such a transition $\delta \in \Delta(M_m)$ exists if $\bigcup_{o_\delta \in \delta\bullet} \mathscr{R}_{\langle M_m \rangle}(o_\delta) = M_m$. If such δ exists, then we can imply a partial run net starting by δ and adding all transitions and necessary paths to these transitions as well as to the places of M_m and the necessary paths to those places. Although this partial run net may not be a "full" run net, this partial run net implies the existence of a "full" run net.

Let us consider our example in Fig. 1. Assume, we want to check if the marking $M_m = [p3, p8, p14, p17]$ is reachable. All places are pairwise concurrent and it is not possible to add any further place (M_m is maximum admissible). For this reason, we have to check whether there is a diverging transition, which can cause this marking. The diverging points for $M_m = [p3, p8, p14, p17]$ cover the set $\{t10, t11, t12, p16, t1\}$ by Definition 11. $t10$'s output places $p14$ and $p17$ are part of M_m, thus, $\mathscr{R}_{\langle M_m \rangle}(p14) = \{p14\}$ and $\mathscr{R}_{\langle M_m \rangle}(p17) = \{p17\}$. For $t12$'s output places $p3$ and $p18$, it holds $\mathscr{R}_{\langle M_m \rangle}(p3) = \{p3\}$ and $\mathscr{R}_{\langle M_m \rangle}(p18) = \{p14, p17\}$. For $t11$, the information are similar $\mathscr{R}_{\langle M_m \rangle}(p3) = \{p3\}$ and $\mathscr{R}_{\langle M_m \rangle}(p13) = \{p14, p17\}$. The place $p16$ is a decision with two output transitions $t11$ and $t12$. By Lemma 5, both transitions either have the same information ($\mathscr{R}_{\langle M_m \rangle}(t11) = \mathscr{R}_{\langle M_m \rangle}(t12) = \{p3, p14, p17\}$) or one is a superset of the other. Eventually, transition $t1$ has the output places $p4$ with $\mathscr{R}_{\langle M_m \rangle}(p4) = \{p8\}$ and

$p2$ with $\mathscr{R}_{\langle M_m \rangle}(p2) = \{p3, p14, p17\}$. Now, we can induce a partial run net from this information. Figure 10 illustrates the induction steps. First, we can add the diverging point $t1$, which is a diverging point of all places of M_m. Furthermore, we add its output places $p2$ and $p4$ as well as a path from $p1$ (the source) to $t1$. We further add $p8$ as a place of M_m and a path to it as we know that $p4$ has a disjoint path to $p8$. The diverging place $p16$ with a path to it is added next (all paths from $t1$ to $t11$ or $t12$ pass $p16$). Furthermore, we add that output transition with the most information (i.e., that output transition t for which $\mathscr{R}_{\langle M \rangle}(t)$ is the greatest). In this case, it does not matter, which of $t11$ or $t12$ we add as they have the same information. In the next step, we add the diverging transition $t12$ (which we already have inserted) with its output places $p3$ and $p18$. Eventually, diverging transition $t10$ is added with a path to it as well as its output places. This final partial run net contains all places of $M_m = [p3, p8, p14, p17]$ and implies the existence of a "full" run net containing all places of M_m. Such steps are applied to finally deciding the (sub-marking) reachability in AFC-WF-nets:

Theorem 4 ((Sub-Marking) Reachability). *Let $N = (P, T, F, i, o)$ be a sound AFC-WF-net with a marking $M \in \mathscr{B}(P)$. It holds:*

$$M \in R(N, [i]) \iff M \text{ is maximum admissible} \quad \wedge \qquad (9)$$
$$\exists \delta \in (\Delta(M) \cap T): \bigcup_{o_\delta \in \delta \bullet} \mathscr{R}_{\langle M \rangle}(o_\delta) = M$$

$$\exists M_s \in R(N, [i]): M \subseteq M_s \iff M \text{ is admissible} \quad \wedge \qquad (10)$$
$$\exists \delta \in (\Delta(M) \cap T): \bigcup_{o_\delta \in \delta \bullet} \mathscr{R}_{\langle M \rangle}(o_\delta) = M \quad \lrcorner$$

Proof (Theorem 4). Constructive proof. By Corollary 1 it holds:

$$M \in R(N, [i]) \iff M \text{ is maximum admissible} \wedge \exists (P_\sigma, T_\sigma, F_\sigma) \in \Pi(N): \langle M \rangle \subseteq P_\sigma.$$

Furthermore, by Lemma 5, it follows for each diverging transition $\delta \in \Delta(M) \cap T$ and for each of its output places $o_\delta \in \delta \bullet$ that there is at least one sub-net of all $\mathscr{R}_{\langle M \rangle}(o_\delta)$ from o_δ to all $x \in \mathscr{R}_{\langle M \rangle}(o_\delta)$, which diverges only in transitions. As a consequence, for each diverging transition $\delta \in \Delta(M) \cap T$, the union $\bigcup_{o_\delta \in \delta \bullet} \mathscr{R}_{\langle M \rangle}(o_\delta)$ describes the subset $M' \subseteq M$ of M to which the transition δ has such a sub-net to, which only diverges in transitions. If there is no diverging transition $\delta \in \Delta(M) \cap T$ with $M = \bigcup_{o_\delta \in \delta \bullet} \mathscr{R}_{\langle M \rangle}(o_\delta)$, then there is no run net that can contain a sub-net, which only diverges at transitions to the places of M. Therefore, there is no run net that contains all places of M. If there is such a diverging transition δ with $M = \bigcup_{o_\delta \in \delta \bullet} \mathscr{R}_{\langle M \rangle}(o_\delta)$, then there must be at least one run net, which contains all places in M. $\checkmark$ The proof of sub-marking reachability by (10) follows the same argumentation. $\checkmark$ $\qquad \square$

With Theorem 4, we can decide if a (sub-)marking is reachable based on the structure of a sound AFC-WF-net rather than over its concrete behavior in terms of occurrence sequences, state space exploration, or similar techniques. In addition, maximum admissibility and diverging points help to understand why a given specific marking is *not*

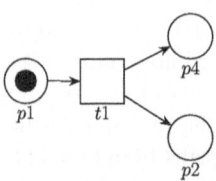

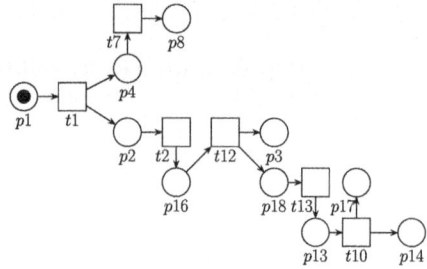

Adding diverging point $t1$ with its output places $p2$ and $p4$.

Adding a path from $p4$ to $p8$. Furthermore, adding diverging point $p16$ with its output transition $t12$. It does not matter, which of $t11$ or $t12$ is added as they have the same information ($\mathcal{R}_{\langle M_m \rangle}(t11) = \mathcal{R}_{\langle M_m \rangle}(t12)$).

Adding diverging point $t12$ with its output places $p3$ and $p18$.

Adding a path to the diverging point $t10$ with its output places $p14$ and $p17$. This net contains all places $[p3, p8, p14, p17]$ and ...

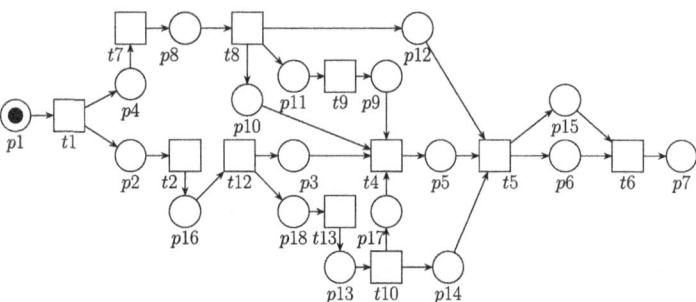

... implies the existence of a run net.

Fig. 10. Step-wisely inducing a partial run net for the exemplary AFC-WF-net in Fig. 1 out of the diverging point information for the marking $M_m = [p3, p8, p14, p17]$. The final partial run net implies the existence of a run net presented at the end.

reachable. For example, the marking $[p8, p5, p14]$ in Fig. 1 is *not* reachable from $[p1]$ since $p8$ is not concurrent to $p5$ (there is a path from $p8$ to $p5$ following Lemma 2). In Fig. 4, the marking $[x, y, z]$ is not reachable from $[i]$ as there is no diverging transition,

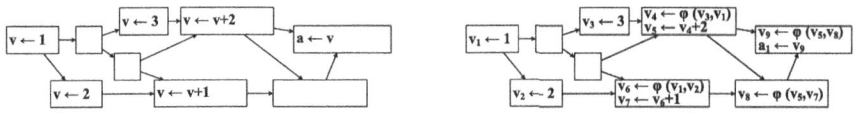

Fig. 11. Transforming a program with program variables v and a (left) into SSA form (right) by inserting new definitions of variables and ϕ-functions at converging nodes.

which leads to simultaneous tokens on x, y, and z—just on x and y, x and z, or y and z. This diagnostic information may support repairing improperly developed nets, in which a desired marking is not reachable or an undesired marking is reachable.

Algorithmic Derivation: There is a concept in compiler theory to identify diverging points: the placement of so-called ϕ-functions to derive the minimal *Static Single Assignment* (SSA) form [6]. The SSA form requires that each *program variable* in a computer program gets statically only once a value. If there are converging control-flows, a ϕ-function selects that program variable, which was used during execution and assigns it to a new program variable. Figure 11 illustrates an exemplary control-flow graph (left) being transferred into SSA form (right). Cytron et al. [6] derived algorithms to compute the *minimal* SSA form from normal programs. The minimal SSA form only inserts ϕ-functions if necessary. To achieve a SSA form out of a computer program, the transformation inserts a new program variable (which is called a *definition*) for each assignment of this variable. Subsequently, it inserts a ϕ-function of the form $d_n = \phi(d_1, \ldots, d_m)$, $m \geq 2$, for each program variable at each program point where different definitions $d_1, \ldots, d_m$ can occur. This placement of ϕ-functions in the minimal SSA assumes fairness, i.e., each path can be followed by a token. The positions of nodes where a ϕ-function is required for a program variable d are positions where two paths from two different definitions of the same variable first meet. As a consequence, for all $d_1, \ldots, d_m$ there are disjoint paths from the position of the ϕ-function to their assignment. Actually, this is the same definition as for diverging points, only in the reverse direction.

We can simply reuse the algorithm of placing ϕ-functions to identify all diverging points $\Delta(M)$ of a given admissible marking M and to compute the sets $\mathcal{R}_{\langle M \rangle}(o_\delta)$ for all $o_\delta \in \bigcup_{\delta \in \Delta(M)} \delta \bullet$. In doing so, we assume that on all places in M there is an assignment on the same single program variable v. Then, we perform the ϕ-function placements, however, on the *reverse* net, i.e., we change the direction of each flow. Thus, the SSA algorithm will replace the assignments on the program variable v with unique definitions of v. On all nodes, from where two different definitions have disjoint paths, the SSA approach has placed a ϕ-function with those definitions as arguments and a new definition where the result of the ϕ-function is assigned. It is important to note that the different unique definitions allow for collecting $\mathcal{R}_{\langle M \rangle}(o)$ for the outputs o of the diverging points and identifying disjoint paths easily.

Algorithm 1 summarizes the steps of checking (sub-marking) reachability of a given marking M_r in a sound AFC-WF-net N. At first, (maximum) admissibility of M_r is checked in line 2. If M_r is *admissible* or *maximum admissible*, then M_r may be (sub-marking) reachable. Otherwise, if M_r is *not admissible*, M_r is stated to be *not reachable*.

Algorithm 1. Checking (sub-marking) reachability of a given marking $M_r \in \mathcal{B}(P)$ for a sound AFC-WF-net N.

1: **function** ISREACHABLE(N, M_r)
2:       *admissibility* ← CHECKMAXIMUMADMISSIBILITY(N, M_r)
3:       **if** *admissibility* ∈ { admissible, maximum admissible } **then**
4:           Compute $\Delta(M_r)$ of N by the ϕ-placement algorithm.
5:           **for all** $\delta \in (\Delta(M_r) \cap T)$ **do**
6:               $M_\delta \leftarrow \bigcup_{o_\delta \in \delta \bullet} \mathcal{R}_{\langle M_r \rangle}(o_\delta)$
7:               **if** $M_r = M_\delta$ **then**
8:                   **if** *admissibility* = admissible **then**
9:                       **return** sub-marking reachable
10:                  **else**
11:                      **return** reachable
12:      **return** not reachable

Line 4 computes the diverging points of M_r with the ϕ-placement algorithm of building the minimal SSA form [6] in the reverse net of N. This placement includes for all outputs o of all diverging points those diverging points and places of M_r, $\mathcal{R}_{\langle M_r \rangle}(o)$, to which they have paths. Finally, all diverging transitions δ are investigated in lines 5–11. The set M_δ is computed in line 6 regarding Theorem 4. Line 7 checks if M_r is equal to M_δ. If this check holds, the marking is either *sub-marking reachable* (if M_r is *admissible*) or *reachable* (if M_r is maximum admissible). Otherwise, other diverging transitions are investigated until either there is one, for which the check holds, or the marking is stated as *not reachable*. This algorithm does not require checking reachability by investigating the state space, computing occurrence sequences, etc. This explains why the computational complexity is just quadratic in the worst case:

Theorem 5 (Computational Complexity). *Let $N = (P, T, F, i, o)$ be a sound AFC-WF-net and M_r a marking to check. Deciding reachability of M_r can be computed in $O(|P|^2 + |T|^2)$.*

Proof (Theorem 5). Constructive proof by line-by-line investigation of Algorithm 1.

Line 2: Checking admissibility and maximum admissibility can be achieved in quadratic computational time complexity, $O(|P|^2 + |T|^2)$, by Theorem 3.
Line 4: Following Cytron et al. [6], placing ϕ-functions requires to compute the "dominance frontier", which is feasible in quadratic time, $O(|P|^2 + |T|^2)$, by Cooper et al. after building the "tree of dominance" in the same complexity $O(|P|^2 + |T|^2)$ [5]. The placement of ϕ-functions for M_r can also be achieved in quadratic time $O(|P|^2 + |T|^2)$ for a single variable by Cytron et al. [6] as it is needed in our case. Thus, overall, this line takes $O(|P|^2 + |T|^2)$.
Lines 5–11: Lines 6–11 are investigated $O(|T|)$ times by line 5. Computing M_δ is linear $O(|P|)$ with an implementation requiring constant time for set unions (e. g., a BitSet in Java). The remaining lines 7–11 are at most $O(|P|)$. Thus, lines 5–11 can be achieved in $O(|T| \cdot |P|)$ in general.

In summary, checking the (sub-marking) reachability of a given marking is dominated by the term $O(|P|^2 + |T|^2)$. ✓

Termination: All steps in Algorithm 1 are either known to terminate (as placing ϕ-functions) or are performed over finite sets (as over the set of diverging points, M_r, etc.). For this reason, Algorithm 1 terminates.   □

5   Conclusion

This paper showed for the Petri net class of sound acyclic free-choice workflow nets that all places of a reachable (sub-)marking must be pairwise concurrent (the marking is admissible). Furthermore, reachable markings must be maximum admissible, i.e., they are admissible and adding a further place to them would destroy their admissibility. Since maximum admissibility is just a necessary condition of (sub-marking) reachability, this paper introduced diverging transitions as sufficient condition, whose existence manifests in subgraphs of sound acyclic free-choice workflow nets representing possible occurrence graphs and, therefore, enabling the reachability of (maximum) admissible markings. The overall approach presented in this paper has a $O(|P|^2 + |T|^2)$ computational complexity. Furthermore, the new concepts of admissibility, maximum admissibility, and diverging transitions allow for explaining *why* a given (sub-)marking is *not* reachable.

The Petri net and business process management communities get a new computationally efficient algorithm to check reachability, especially for process-like and industrial related nets. Since the reachability of markings is central to many other decision problems of Petri nets, the here presented approach empowers to speed up other algorithms. For example, business analysts could request their process models if undesired states of their models could happen to improve quality, check compliance, and to avoid cost-intensive failures.

In the future, the concepts of admissibility and maximum admissibility shall be generalized to the class of proper free-choice nets with a home cluster. Furthermore, the method of loop decomposition [18] shall be introduced for free-choice nets leading to an unfolding of loops increasing a model just quadratically in the worst case. For industrial settings, we want to provide an implementation of our approach, which allows for checking reachability for process models by simply selecting some nodes of the model. Furthermore, it is of interest whether the new concepts introduced in this paper may be applicable to solve other problems in Petri nets.

References

1. van der Aalst, W.M.P.: Verification of workflow nets. In: Azéma, P., Balbo, G. (eds.) ICATPN 1997. LNCS, vol. 1248, pp. 407–426. Springer, Heidelberg (1997). https://doi.org/10.1007/3-540-63139-9_48
2. van der Aalst, W.: Free-choice nets with home clusters are lucent. Fundam. Informaticae **181**(4), 273–302 (2021). https://doi.org/10.3233/FI-2021-2059
3. van der Aalst, W., et al.: Soundness of workflow nets: classification, decidability, and analysis. Formal Aspects Comput. **23**(3), 333–363 (2011). https://doi.org/10.1007/s00165-010-0161-4

4. Cheng, A., Esparza, J., Palsberg, J.: Complexity results for 1-safe nets. Theor. Comput. Sci. **147**(1&2), 117–136 (1995). https://doi.org/10.1016/0304-3975(94)00231-7
5. Cooper, K.D., Harvey, T.J., Kennedy, K.: A simple, fast dominance algorithm. Technical Report TR-06-33870, Institut für Informatik, Technische Universität München, Rice University (2001)
6. Cytron, R., Ferrante, J., Rosen, B.K., Wegman, M.N., Zadeck, F.K.: Efficiently computing static single assignment form and the control dependence graph. ACM Trans. Program. Lang. Syst. **13**(4), 451–490 (1991). https://doi.org/10.1145/115372.115320
7. Czerwinski, W., Lasota, S., Lazic, R., Leroux, J., Mazowiecki, F.: The reachability problem for Petri nets is not elementary. J. ACM **68**(1), 7:1–7:28 (2021)
8. Desel, J., Esparza, J.: Reachability in cyclic extended free-choice systems. Theor. Comput. Sci. **114**(1), 93–118 (1993). https://doi.org/10.1016/0304-3975(93)90154-L
9. Desel, J., Esparza, J.: Free Choice Petri Nets. Cambridge University Press, pbk version ed. edition edn. (2008)
10. van Dongen, B.F., Mendling, J., van der Aalst, W.M.P.: Structural patterns for soundness of business process models. In: Tenth IEEE International Enterprise Distributed Object Computing Conference (EDOC 2006), 16–20 October 2006, Hong Kong, China, pp. 116–128. IEEE Computer Society (2006). https://doi.org/10.1109/EDOC.2006.56
11. Esparza, J.: Reachability in live and safe free-choice Petri nets is NP-complete. Theor. Comput. Sci. **198**(1–2), 211–224 (1998). https://doi.org/10.1016/S0304-3975(97)00235-1
12. Esparza, J., Nielsen, M.: Decidability issues for Petri nets - a survey. CoRR abs/2411.01592 (2024). https://doi.org/10.48550/ARXIV.2411.01592
13. Favre, C., Fahland, D., Völzer, H.: The relationship between workflow graphs and free-choice workflow nets. Inf. Syst. **47**, 197–219 (2015)
14. Kovalyov, A., Esparza, J.: A Polynomial Algorithm to Compute the Concurrency Relation of Free-Choice Signal Transition Graphs. Sonderforschungsbereich 342: Methoden und Werkzeuge für die Nutzung paralleler Rechnerarchitekturen TUM-19528, SFB-Bericht Nr. 342/15/95 A, Institut für Informatik, Technische Universität München, München, Germany (1995)
15. Murata, T.: Petri nets: properties, analysis and applications. Proc. IEEE **77**(4), 541–580 (1989). https://doi.org/10.1109/5.24143
16. Polyvyanyy, A., Weidlich, M., Conforti, R., Rosa, M.L., ter Hofstede, A.H.M.: The 4C Spectrum of fundamental behavioral relations for concurrent systems. In: Ciardo, G., Kindler, E. (eds.) Application and Theory of Petri Nets and Concurrency - 35th International Conference, PETRI NETS 2014, Tunis, Tunisia, 23–27 June 2014. Proceedings. Lecture Notes in Computer Science, vol. 8489, pp. 210–232. Springer (2014)
17. Prinz, T.M., Choi, Y., Ha, N.L.: Understanding and decomposing control-flow loops in business process models. In: Ciccio, C.D., Dijkman, R.M., del-Río-Ortega, A., Rinderle-Ma, S. (eds.) Business Process Management - 20th International Conference, BPM 2022, Münster, Germany, 11–16 September 2022, Proceedings. Lecture Notes in Computer Science, vol. 13420, pp. 307–323. Springer (2022)
18. Prinz, T.M., Choi, Y., Ha, N.L.: Soundness unknotted: an efficient soundness checking algorithm for arbitrary cyclic process models by loosening loops. Inf. Syst. **128**, 102476 (2025). https://doi.org/10.1016/J.IS.2024.102476
19. Prinz, T.M., Klaus, J., van Beest, N.R.T.P.: Pushing the limits: concurrency detection in acyclic sound free-choice workflow nets in $O(P^2 + T^2)$. In: Köhler-Bussmeier, M., Moldt, D., Rölke, H. (eds.) Proceedings of the International Workshop on Petri Nets and Software Engineering 2024 co-located with the 45th International Conference on Application and Theory of Petri Nets and Concurrency (PETRI NETS 2024), 24–25 June 2024, Geneva, Switzerland. CEUR Workshop Proceedings, vol. 3730, pp. 132–154. CEUR-WS.org (2024). https://ceur-ws.org/Vol-3730/paper08.pdf

20. Prinz, T.M., Welsch, T., Ha, N.L.: Recognizing relationships: detecting the 4C Spectrum in $O(P^2 + T^2)$ for acyclic sound process models. In: Borbinha, J., Sales, T.P., da Silva, M.M., Proper, H.A., Schnellmann, M. (eds.) Enterprise Design, Operations, and Computing - 28th International Conference, EDOC 2024, Vienna, Austria, September 10-13, 2024, Revised Selected Papers. Lecture Notes in Computer Science, vol. 15409, pp. 281–299. Springer (2024). https://doi.org/10.1007/978-3-031-78338-8_15
21. Verbeek, H., Basten, T., van der Aalst, W.: Diagnosing workflow processes using Woflan. Comput. J. **44**(4), 246–279 (2001). https://doi.org/10.1093/comjnl/44.4.246
22. Yamaguchi, S.: Polynomial time verification of reachability in sound extended free-choice workflow nets. IEICE Trans. Fundam. Electron. Commun. Comput. Sci. **97-A**(2), 468–475 (2014). https://doi.org/10.1587/TRANSFUN.E97.A.468

Complexity of Alignments on Sound Free-Choice Workflow Nets

Christopher T. Schwanen[1]([envelope]) [iD], Wied Pakusa[2] [iD],
and Wil M. P. van der Aalst[1] [iD]

[1] Chair of Process and Data Science (PADS), RWTH Aachen University, Aachen, Germany
{schwanen,wvdaalst}@pads.rwth-aachen.de
[2] Koblenz University of Applied Sciences, Koblenz, Germany
pakusa@hs-koblenz.de

Abstract. An optimal alignment consists of a minimal number of edit operations (deletions and insertions) to fit an observed event trace with a process model. In conformance checking, alignments are used to quantify in how far reality deviates from the predefined business norm and constitute probably the most important tool. In practice, however, it has frequently been observed that finding optimal alignments is computationally expensive. In this paper, we extend the proof of the Shortest Sequence Theorem for live, bounded, free-choice Petri nets to make it also applicable to moves in alignments on this model class. This way, we are able to show that computing alignments on sound free-choice workflow nets is NP-complete. While this still rules out an efficient algorithm, our result opens the door for a new set of tools to attack the alignment problem which go beyond the standard reachability approach used in most implementations. Eventually, we will demonstrate that soundness alone is not a sufficient criterion by proving that computing alignments on general safe and sound workflow nets is PSPACE-complete and thus indeed incurring immense algorithmic costs on more general model classes.

Keywords: Process Mining · Conformance Checking · Alignments · Computational Complexity · Workflow Nets · Free-Choice Petri Nets

1 Introduction

The goal of *conformance checking* is to compare the observed behavior of a process against a reference model, typically given as Petri net or BPMN diagram. This allows the identification of inefficiencies, regulatory violations, and so on. As of today, *alignments* are considered to be the state-of-the-art technique in conformance checking; we refer to [8] for a thorough introduction to this field.

The system in Fig. 1 specifies a process with two observable events a and b in the form of a Petri net. We view a Petri net as both, a formal model defining a process, and as a *language acceptor*. In both cases, we agree on an *initial* and a *final* marking and consider runs of the model (so called *firing sequences*) from

E. Amparore and Ł. Mikulski (Eds.): PETRI NETS 2025, LNCS 15714, pp. 388–410, 2025.
https://doi.org/10.1007/978-3-031-94634-9_19

the initial to the final marking. For each run, the transition labels form a finite *word*, also called a *trace*. Transitions can also be unlabeled (those are called *silent*). The collection of generated words constitutes the *language* accepted by the Petri net. For the net in Fig. 1, the initial marking is highlighted (one single token in place p_{init}) and the intended final marking has a single token in p_{final}. Relative to these markings, the Petri net accepts the words *aabb* and *abab*, but not *abaa*. The accepted language is $(aab|aba)^+b$.

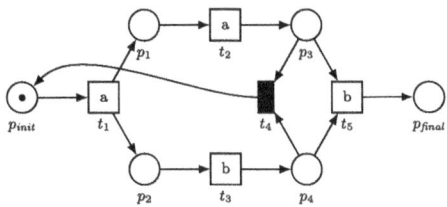

Fig. 1. Example of a Petri net (sound free-choice workflow net)

Dissimilarity is quantified by *"aligning"* a given trace against the model, which means that we *insert* and *delete* activities into and from the trace until it fits with the model. The goal is to find an *optimal alignment* which minimizes the number of required change operations. This is formalized as follows: we assume that the business process and the trace are executed and generated concurrently. While the model evolves from its initial to its final marking, and, concurrently, the trace from its first to its last letter, there are three different types of *moves*:

1. the system *and* the trace take one step *synchronously*, i.e., the system fires a transition t labeled with a and the trace moves on via its next letter a; this is called a *synchronous* move (a, t);
2. the system fires t, but the trace maintains its state; this corresponds to an *insert* operation where the label of t (which might be empty) is inserted into the trace; such move is called *model move* and is denoted by $(\gg, t)$ where $\gg$ is a distinguished "no-move" symbol;
3. dually, the model can stay in its current state while only the trace proceeds one letter further which corresponds to a *deletion* operation; such steps are called *log moves* and can be written as $(a, \gg)$.

Figure 2 shows two alignments for a trace $w = abaa$ and the system from Fig. 1. The first row indicates the progress in the trace, while the last two rows contain the labels and transitions fired by the model. For the first alignment, we have four insert operations (model moves) including a silent one (t_4 is unlabeled, indicated by τ). The trace generated by the model is *abaaabb*. The model trace *abab* in the second alignment results in one deletion (log move) and one insertion (model move), which is optimal regarding the number of change operations.

It has frequently been observed that finding optimal alignments is computationally hard. Given the pivotal role of alignments, this situation calls for a

$$\frac{a \quad b \gg \gg a \quad a \gg \gg}{\begin{matrix} a & b & a & \tau & a & a & b & b \\ t_1 & t_3 & t_2 & t_4 & t_1 & t_2 & t_3 & t_5 \end{matrix}}$$

$$\frac{a \quad b \quad a \gg a}{\begin{matrix} a & b & a & b \\ t_1 & t_3 & t_2 & t_5 \gg \end{matrix}}$$

(a) Four synchronous moves, three model moves (inserts), one silent model move

(b) Three synchronous moves, one model move (insert), one log move (deletion)

Fig. 2. Two possible alignments for $w = abaa$

thorough and systematic analysis of the complexity of the alignment problem. In particular, for practical applications it would be beneficial to know which parameters of the process model influence the complexity of the alignment problem and, thus, which algorithmic approaches are most promising on certain inputs. Surprisingly, such an analysis is, to date, still missing. In this work, we make several important steps towards such a complexity-theoretic classification of alignments.

In [26,27], we studied the complexity of alignments on process trees. We showed that the alignment problem is in NP for general process trees and in P for process trees with unique labels. We now take one step further by considering *live, bounded, free-choice Petri nets*, a model class strongly generalizing process trees, but still known for its good algorithmic properties. Our first main result is to show that alignments on this class can also be computed in NP (see Sect. 5). This is a significant improvement over the case of safe Petri nets where the alignment problem is PSPACE-complete (see Sect. 7). Our proof is based on an extension of the *Shortest Sequence Theorem* for live, bounded, free-choice Petri nets [14] by which we can bound the length of an optimal alignment polynomially. To complement the NP upper bound, we also show NP-hardness via a reduction from the membership problem for shuffle languages (see Sect. 5). This gives NP-completeness for alignments on important model classes studied in process mining, such as sound free-choice workflow nets or process trees (see Sect. 6). Interestingly, for both classes, the reachability problem is known to be in P, which means that the alignment problem is provably harder (assuming $P \neq NP$).

Our second main contribution is to show that the *free-choice* assumption is crucial. We prove that on safe and sound workflow nets, the alignment problem is PSPACE-complete (see Sect. 7). For this, we significantly extend a construction for simulating a PSPACE-computation by a live and safe Petri net. This shows that the alignment problem is indeed intractable on safe and sound workflow nets, the standard model class used in process mining.

2   Related Work

The groundwork for conformance checking was laid in [24] where metrics based on a technique known as *token-based replay* were proposed. Although token-based replay is efficient from an algorithmic point of view, it fails to provide accurate results in presence of common model constructs. This is why alignments, introduced by [3], have replaced token-based replay as "the gold standard" in conformance checking [9]. An overview of recent developments is given in [8,9].

It is ubiquitously mentioned that computing alignments (with respect to the standard algorithms based on A^*) suffers from the so-called *state explosion problem* [cf. 30]. In general, it has been recognized that the reachability problem for Petri nets is a lower bound for the complexity [8,9]. Most relevant for us are complexity bounds for classes of safe Petri nets (which means that places can hold at most one token). For safe Petri nets, it was shown that reachability is PSPACE-complete [19]. Later on, it turned out that almost all interesting computational problems on safe Petri nets are PSPACE-hard. A comprehensive overview of results is given in [10,17,18]. Because of the high complexity, more restricted classes of Petri nets with better algorithmic properties have been studied as well. Here, *free-choice* Petri nets form the most relevant example on which important computational problems become tractable, see [13].

Different algorithmic approaches were investigated to improve scalability of alignment computations. E.g., in [5] the authors use symbolic representations of the (exponential) state space to reduce the memory footprint of the alignment computation. Another common approach is to encode alignments into related problems for which well-adapted algorithms exists. This idea is investigated, for example, in [21] where alignment computations are represented as a planning problem. Another angle is to improve heuristics for the A^*-algorithm: in [15], for instance, Petri net theory and linear programming are combined to improve the runtime on several benchmarks significantly. Besides this, also approximative algorithms have been proposed, see, e.g., [28] for a scheme based on integer linear programming and [29] for a genetic method to compute optimal alignments.

However, these studies do not provide guarantees and we did hardly find any source that investigates the algorithmic complexity of alignments. A statement on specific model classes is provided in [6,7]. There, the authors showed that computing alignments is NP-complete on the class of safe Petri nets when the length of permissible alignments is limited. To the best of our knowledge, the approach in [27] was the first to truly break the PSPACE-barrier when computing alignments on the class of process trees. When further restricting process trees to only have unique labels, the complexity drops from NP even further to P [26]. Beyond that, we are not aware of any results studying the complexity of alignments over different process models.

Finally, we like to mention work on the related *error correction problem* for regular languages. A first efficient algorithm for this problem was given by Wagner in the 1970s [31], a complexity-theoretic analysis can be found in [23]. For context-free grammars (CFGs), a polynomial-time error-correction algorithm can be found in [4]. Our work might be interesting for the error correction problem as well, as it provides a new perspective on the complexity of the problem for other presentations of regular languages.

3   Preliminaries

Let $\mathbb{N} := \{0, 1, 2, \dots\}$ denote the natural numbers. For any tuple a, $\pi_i(a)$ denotes the *projection* on its ith element, i.e., $\pi_i \colon A_1 \times \cdots \times A_n \to A_i, (a_1, \dots, a_n) \mapsto a_i$.

Definition 1 (Multiset). A *multiset* M over a set A is a function $M \colon A \to \mathbb{N}$; thus, for any $a \in A$, $M(a)$ indicates how often a is contained in the multiset M. The set of all multisets over A is given by $\mathbb{N}^A$. We also use the notation $[a^{M(a)} | a \in A]$ for a multiset $M \in \mathbb{N}^A$. Any set A can also be considered a multiset by assigning 1 to each item. For multisets $M, M' \in \mathbb{N}^A$, we use the standard notation for functions, e.g., $M + M'$, $M \le M'$, etc. The *support* of a multiset $M \in \mathbb{N}^A$, denoted by $\langle M \rangle$, is the set of distinct elements contained in M, i.e., $\langle M \rangle := \{a \in A | M(a) > 0\}$. For multisets $M \in \mathbb{N}^A$ and $M' \in \mathbb{N}^B$ over two sets A and B, $M \oplus M'$ denotes their addition after extending them by 0.

Definition 2 (Sequence, Permutation). *Sequences* with index set I over a set A are denoted by $\sigma = \langle a_i \rangle_{i \in I} \in A^I$. The *length* of a sequence σ is written as $|\sigma|$ and the set of all finite sequences over A is denoted by A^*. For a sequence $\sigma = \langle a_i \rangle_{i \in I} \in A^I$, the notation $\sum \sigma$ is used as a shorthand for $\sum_{i \in I} a_i$. Given two sequences σ and σ', $\sigma \cdot \sigma'$ (or $\sigma \sigma'$ in short) denotes the concatenation of the two sequences. The restriction of a sequence $\sigma \in A^*$ to a set $B \subseteq A$ is the subsequence $\sigma|_B$ of σ consisting of all elements in B. A function $f \colon A \to B$ can be applied to a sequence $\sigma \in A^*$ given the recursive definition $f(\langle \rangle) := \langle \rangle$ and $f(\langle a \rangle \cdot \sigma) := \langle f(a) \rangle \cdot f(\sigma)$. The *multiset representation* of a sequence $\sigma \in A^*$, denoted by $\boldsymbol{\sigma}$, is defined by $\boldsymbol{\sigma}(a) := |\sigma|_{\{a\}}|$ for every $a \in A$, and $\langle \boldsymbol{\sigma} \rangle$ provides the *support* of σ. A sequence $\sigma' \in A^*$ is a *permutation* of σ if and only if $\boldsymbol{\sigma'} = \boldsymbol{\sigma}$.

Definition 3 (Alphabet). An *alphabet* Σ is a finite, non-empty set of *labels*.

Definition 4 (Petri Net). Let Σ be an alphabet. A *Petri net* N is a bipartite directed graph $N = (P, T, F, \ell)$ where P and T, $P \cap T = \emptyset$ are disjoint finite sets of vertices and $F \subseteq (P \times T) \cup (T \times P)$ is the set of arcs. In a Petri net, P is called the set of *places*, T the set of *transitions*, and F the *flow relation*. In addition, $\ell \colon T \to \Sigma \cup \{\tau\}$ is a *labeling function*. A transition $t \in T$ is *labeled* if $\ell(t) \in \Sigma$ and it is *silent* if $\ell(t) = \tau$. Given a vertex $v \in P \cup T$, its *pre-set* $\bullet v$ and *post-set* $v \bullet$ are defined by $\bullet v := \{u \in P \cup T | (u, v) \in F\}$ and $v \bullet := \{u \in P \cup T | (v, u) \in F\}$. With regard to a place (transition), its pre- and post-set are also called *input* and *output* transitions (places).

Definition 5 (Marking, System, Firing Rule). Given a Petri net $N = (P, T, F, \ell)$, a *marking* $M \in \mathbb{N}^P$ is a multiset where $M(p)$ is the number of *tokens* at place $p \in P$. A place $p \in P$ is *marked* at M if $M(p) > 0$. The pair (N, M) of a Petri net $N = (P, T, F, \ell)$ and a marking $M \in \mathbb{N}^P$ is called a *system*. A transition $t \in T$ is *enabled* in M, denoted by $(N, M)[t\rangle$, if and only if each of its input places $p \in \bullet t$ is marked, i.e., $\forall p \in \bullet t \colon M(p) > 0$. An enabled transition may *fire*, denoted by $(N, M)[t\rangle(N, M')$, and firing results in a new marking M':

$$M'(p) = \begin{cases} M(p) - 1 & \text{if } p \in \bullet t \wedge p \notin t \bullet, \\ M(p) + 1 & \text{if } p \notin \bullet t \wedge p \in t \bullet, \\ M(p) & \text{otherwise.} \end{cases}$$

A sequence of transitions $\sigma = \langle t_i \rangle_{i=1}^n \in T^*$ is called a *firing sequence* of (N, M) if for every transition t_i of the sequence holds that $(N, M_{i-1})[t_i\rangle$ and $(N, M_{i-1})[t_i\rangle(N, M_i)$ where $M_0 = M$ and $M_n = M'$. Firing such a sequence is denoted by $(N, M)[\sigma\rangle(N, M')$. The empty sequence $\langle\rangle$ is always enabled and firing the empty sequence leaves the marking unchanged, i.e., $(N, M)[\langle\rangle\rangle(N, M)$. A marking M' is *reachable* if a firing sequence $\sigma \in T^*$ exists such that M' is the resulting marking, i.e., $(N, M)[\sigma\rangle(N, M')$. The set of all reachable markings of (N, M) is denoted by $[N, M\rangle := \{M' \in \mathbb{N}^P | \exists \sigma \in T^* : (N, M)[\sigma\rangle(N, M')\}$.

Definition 6 (Boundedness, Safeness). Given a $k \in \mathbb{N}$, a system (N, M_0) with $N = (P, T, F, \ell)$ is *k-bounded* if k is a bound for any reachable marking, i.e., $\forall M \in [N, M_0\rangle : \forall p \in P : M(p) \leq k$. (N, M_0) is *safe* if it is 1-bounded.

Definition 7 (Accepting System). An *accepting system* (N, M_{init}, M_{final}) extends a system (N, M_{init}), $N = (P, T, F, \ell)$, with a *final marking* $M_{final} \in \mathbb{N}^P$.

Definition 8 (Complete Firing Sequence, Trace, Behavior and Language of a System). Let $S = (N, M_{init}, M_{final})$ be an accepting system with $N = (P, T, F, \ell)$ and $\ell : T \to \Sigma \cup \{\tau\}$ over Σ. A firing sequence $\sigma \in T^*$ is a *complete firing sequence* of S if $(N, M_{init})[\sigma\rangle(N, M_{final})$. The set of complete firing sequences $\phi(S) := \{\sigma \in T^* | (N, M_{init})[\sigma\rangle(N, M_{final})\}$ is the *behavior* of S. Considering the labels of a complete firing sequence, this is referred to as a *trace* $\sigma \in \Sigma^*$. The set of all traces $\mathcal{L}(S) := \{\ell(\sigma)|_\Sigma | \sigma \in \phi(S)\}$ is the *language* of S.

Definition 9 (Easy Soundness). An accepting system (N, M_{init}, M_{final}) is *easy sound* if and only if the final marking is reachable, i.e., $M_{final} \in [N, M_{init}\rangle$.

4   Alignments

Alignments juxtapose an observed trace with a complete firing sequence of the process model. For this, activities in the trace are compared in pairs with the transitions of the complete firing sequence. These pairs are called *moves*:

Definition 10 (Moves). Let Σ be an alphabet and $S = (N, M_{init}, M_{final})$ an accepting system with Petri net $N = (P, T, F, \ell)$ and labeling function $\ell : T \to \Sigma \cup \{\tau\}$. Furthermore, let $\gg$ be a distinguished *"no move"* symbol. A *move* is an ordered pair $(a, t) \in (\Sigma \cup \{\gg\}) \times (T \cup \{\gg\})$. We distinguish between three types of *legal* moves: The move (a, t) is a

- *synchronous move* if $a \in \Sigma$, $t \in T$, and $a = \ell(t)$,
- *log move* if $a \in \Sigma$ and $t = \gg$,
- *model move* if $a = \gg$ and $t \in T$.

All other moves are considered illegal. A model move $(\gg, t)$ is a *silent move* if $\ell(t) = \tau$. The set $LM_{\Sigma,S}$ denotes all legal moves between Σ and S:

$$LM_{\Sigma,S} := \{(a, t) \in \Sigma \times T | a = \ell(t)\} \cup (\Sigma \times \{\gg\}) \cup (\{\gg\} \times T).$$

An *alignment* is a sequence of legal moves whose first components form the observed trace and whose second components form a complete firing sequence of the process model (ignoring the $\gg$-symbol and τ-labels), formally:

Definition 11 (Alignment). Let Σ be an alphabet, $\sigma \in \Sigma^*$ a trace, and $S = (N, M_{init}, M_{final})$ an accepting system with $N = (P, T, F, \ell)$ and $\ell \colon T \to \Sigma \cup \{\tau\}$. An *alignment* $\gamma \in LM_{\Sigma,S}^*$ between σ and S is a sequence of moves such that $\sigma = \pi_1(\gamma)|_\Sigma$ and $\pi_2(\gamma)|_T \in \phi(S)$. $\Gamma_{\sigma,S}$ denotes all alignments between σ and S:

$$\Gamma_{\sigma,S} := \{\gamma \in LM_{\Sigma,S}^* \mid \pi_1(\gamma)|_\Sigma = \sigma \wedge \pi_2(\gamma)|_T \in \phi(S)\}.$$

In easy-sound systems, a trivial alignment always exists: first generate the input trace via log moves, and then generate a firing sequence from the initial to the final marking via model moves. This trivial alignment corresponds to the worst possible scenario: the model and the trace have nothing in common. However, we are really interested in *optimal* alignments which *maximize* the synchronization between the trace and the model. This is formalized by assigning costs to moves and then finding an alignment with minimal costs.

Definition 12 (Alignment Cost). Let $S = (N, M_{init}, M_{final})$ be an easy-sound system, i.e., $\phi(S) \neq \emptyset$, with $N = (P, T, F, \ell)$ and $\ell \colon T \to \Sigma \cup \{\tau\}$, and let $LM_{\Sigma,S}$ be the set of all legal moves between Σ and S. An *alignment cost function* is a function $c \colon LM_{\Sigma,S} \to \mathbb{Q}_{\geq 0}$. The cost of an alignment $\gamma \in LM_{\Sigma,S}^*$ is given by the sum of costs of each move in the sequence, i.e., $\sum c(\gamma)$.

The *standard cost function* $c \colon LM_{\Sigma,S} \to \mathbb{Q}_{\geq 0}$ is defined as

$$(a, t) \mapsto c(a, t) := \begin{cases} 0 & a \in \Sigma \wedge t \in T \wedge a = \ell(t), \quad \text{or} \quad a = \gg \wedge \ell(t) = \tau, \\ 1 & a \in \Sigma \wedge t = \gg, \qquad\qquad\qquad \text{or} \quad a = \gg \wedge \ell(t) \in \Sigma. \end{cases}$$

Definition 13 (Optimal Alignment). Let S be an easy-sound system, i.e., $\phi(S) \neq \emptyset$, and let $c \colon LM_{\Sigma,S} \to \mathbb{Q}_{\geq 0}$ be an alignment cost function. Given a trace $\sigma \in \Sigma^*$, an alignment $\gamma_{opt} \in \Gamma_{\sigma,S}$ is *optimal* if and only if no other alignment between σ and S has lower costs, i.e., $\sum c(\gamma_{opt}) = \min_{\gamma \in \Gamma_{\sigma,S}} \{\sum c(\gamma)\}$.

Of course, computing an optimal alignment is a *functional* optimization problem, the corresponding decision problem is the following:

*Problem 1 (Alignment (*ALIGN*)).*

Input: An alphabet Σ, an easy-sound system (N, M_{init}, M_{final}) with Petri net $N = (P, T, F, \ell)$ and labeling function $\ell \colon T \to \Sigma \cup \{\tau\}$, a trace $\sigma \in \Sigma^*$ over Σ, and a cost function $c \colon LM_{\Sigma,S} \to \mathbb{Q}_{\geq 0}$.

Question: Given $k \in \mathbb{Q}_{\geq 0}$, is there an alignment $\gamma \in \Gamma_{\sigma,S}$ with $\sum c(\gamma) \leq k$?

For our purposes the (binary) decision problem ALIGN is more adequate, since we are interested in classifying its computational complexity. However, it is also clear that an algorithm for the decision version can be transformed into an algorithm for the functional variant by performing a binary search on the cost threshold k. This only requires a polynomial number of calls to the decision algorithm and does not change the complexity class of the problem.

5   ALIGN on Live, Bounded, Free-Choice Systems

First, we turn our attention to *live, b-bounded, free-choice systems* (LBFC-systems, for short). These nets have been thoroughly studied in Petri net theory and enjoy nice structural properties [cf. 13]. In our main result (Theorem 4), we show that for each trace and each LBFC-system there exists an optimal alignment of polynomial length. From this, we obtain a simple guess-and-verify NP-algorithm for the alignment problem which reduces the complexity from PSPACE to NP.

Definition 14 (Live, b-Bounded, Free-Choice (LBFC) Systems). A system (N, M_0) over $N = (P, T, F, \ell)$ is a live, b-bounded, free-choice system (LBFC-system) if and only if it is

live, if:	$\forall M \in [(N, M_0)\rangle, \forall t \in T : \exists M' \in [(N, M)\rangle : (N, M')[t\rangle$
b-bounded, if:	$\forall M \in [(N, M_0)\rangle, \forall p \in P : M(p) \leq b$
free-choice, if:	$\forall (s, t) \in F : \bullet t \times s \bullet \subseteq F.$

In this paper, we consider LBFC-systems with respect to a *fixed* value of $b \in \mathbb{N}$, i.e., b is a constant which does not vary along different inputs. Implicitly, we think of $b = 1$ (safe systems), but any other fixed value for b is possible too.

Our NP-algorithm for LBFC-systems is based on the following key property: whenever a marking M' can be reached from M in an LBFC-system, then there *exists some* connecting firing sequence of polynomial length. This was first shown in [14], but we rely on the textbook by the same authors [13].

Theorem 1 (Shortest Sequence Theorem [13, Theorem 9.17]). *Let (N, M_0) be an LBFC-system with n transitions and let M be a reachable marking. Then, there is a firing sequence σ such that $(N, M_0)[\sigma\rangle(N, M)$ and*

$$|\sigma| \leq b \cdot \frac{n \cdot (n+1) \cdot (n+2)}{6}.$$

Unfortunately, the Shortest Sequence Theorem cannot be directly applied to alignments as it only guarantees the *existence* of a "short" firing sequence. In particular, when we start from an arbitrary firing sequence and then, using the above theorem, pass over to a short one, this new sequence can be completely different from the original sequence. This is problematic in context of alignments, since the "short" sequence could contain different transitions, potentially with much higher costs of moves. In a nutshell, a "shorter" sequence is not necessarily a "cheaper" sequence. Luckily, the result can be stated in a more general form.

Theorem 2 (Shortest Sequence Theorem – Generalized Form). *Let (N, M_0) with $N = (P, T, F, \ell)$ and bound b be an LBFC-system. Moreover, let $\sigma \in T^*$ and $M_1, M_2 \in [N, M_0\rangle$ such that $(N, M_1)[\sigma\rangle(N, M_2)$. Then, there exists a firing sequence $\sigma' \in T^*$ with $\sigma' \leq \sigma$, $(N, M_1)[\sigma'\rangle(N, M_2)$ and*

$$|\sigma'| \leq b \cdot \frac{|T| \cdot (|T|+1) \cdot (|T|+2)}{6}.$$

To prove Theorem 2, we go through the machinery of [13] while making necessary changes to results and proofs. For those parts that do not require any adaptation, we refer to [13] for details. The proof consists of two steps. First, we consider the case of *T-systems*, a subclass of free-choice systems where each place has at most one input and one output transition. In such systems, whenever a transition is enabled, it cannot be disabled by firing any other transition.

Definition 15 (T-Net, T-System). A Petri net $N = (P, T, F, \ell)$ is a *T-net* if each place has at most one input and one output transition, i.e., $\forall p \in P \colon |{\bullet}p| \leq 1, |p{\bullet}| \leq 1$. A system (N, M_0) is a *T-system* if N is a T-net.

The syntactic restrictions of T-nets allow us to rearrange firing sequences in such a way that the same sets of transitions are repeatedly fired until, eventually, more and more transitions die out and the set of occurring transitions becomes smaller and smaller (assuming b-boundedness).

Definition 16 (Biased Firing Sequence [cf. 13, Definition 3.22]**).** A firing sequence $\sigma \in T^*$ is *biased* if for all $t_1, t_2 \in \langle \sigma \rangle$, $t_1 \neq t_2$, it holds that ${\bullet}t_1 \cap {\bullet}t_2 = \emptyset$.

Note that firing sequences in T-systems are always biased. In biased sequences, we can move each occurring transition to an initial segment:

Lemma 1 ([13, Lemma 3.24 p. 56]**).** *Let (N, M_0) be a system with $N = (P, T, F, \ell)$ and let $\sigma \in T^*$ be a biased firing sequence. Then, there exists a permutation $\rho = \sigma_1 \sigma_2$ of σ with $\boldsymbol{\rho} = \boldsymbol{\sigma}$ such that $\exists M \in \mathbb{N}^P \colon (N, M_0)[\sigma\rangle(N, M) \wedge (N, M_0)[\sigma_1\sigma_2\rangle(N, M)$ and no transition occurs more than once in σ_1 and every transition that occurs in σ_2 occurs also in σ_1, i.e., $\langle \boldsymbol{\sigma_2} \rangle \subseteq \langle \boldsymbol{\sigma_1} \rangle$.*

For the following lemma we need to generalize the result from [13]:

Lemma 2 (Generalized Form of [13, Lemma 3.25]**).** *Let (N, M_0) with $N = (P, T, F, \ell)$ and bound b be an LBFC-system and let $(N, M_0)[\sigma\rangle(N, M)$ for a firing sequence $\sigma \in T^*$ and a marking M such that σ is biased and non-empty. Then, there exists $\rho = \sigma_1 \sigma_2$ with $\boldsymbol{\rho} \leq \boldsymbol{\sigma}$ such that*

- *$(N, M_0)[\sigma\rangle(N, M)$ and $(N, M_0)[\sigma_1\sigma_2\rangle(N, M)$,*
- *each transition occurs at most b times in σ_1, and*
- *$\langle \boldsymbol{\sigma_1} \rangle \supset \langle \boldsymbol{\sigma_2} \rangle$ (set of occurring transitions decreases).*

Proof. First, we repeatedly apply Lemma 1 in order to obtain a permutation of σ which is of the form $\rho_1 \rho_2 \cdots \rho_n$ with $\rho_i \neq \langle\rangle$ with the following properties:

- for all $i = 1, \ldots, n$, no transition occurs more than once in ρ_i, and
- for all $i < n$, $\langle \boldsymbol{\rho_i} \rangle \supseteq \langle \boldsymbol{\rho_{i+1}} \rangle$.

If $n \leq b$, we are done: simply choose $\sigma_2 = \langle\rangle$ and $\sigma_1 = \rho_1 \cdots \rho_n$. So, let us assume $n \geq b + 1$. If $\langle \boldsymbol{\rho_1} \rangle \supset \langle \boldsymbol{\rho_{b+1}} \rangle$, we can stop our argument at this point as well, since then $\sigma_1 = \rho_1 \cdots \rho_b$ and $\sigma_2 = \rho_{b+1} \cdots \rho_n$ would have the desired properties. So, let us also assume that $\langle \boldsymbol{\rho_1} \rangle = \langle \boldsymbol{\rho_{b+1}} \rangle$. Choose the maximal m such that

$\langle \rho_1 \rangle = \langle \rho_m \rangle$ (note that $b + 1 \leq m \leq n$). Let $X := \langle \rho_1 \rangle = \langle \rho_m \rangle \subseteq T$. Since each transition in X occurs precisely once in each of the ρ_i, we know that for each place the same number of tokens is added or removed after firing each of the subsequences ρ_i. Since each place can hold at most b tokens (since the net is b-bounded) this means that firing all transitions in X must leave the number of tokens in each place invariant. But this, in turn, implies that each of the ρ_i, $1 \leq i \leq m$ does not modify the initial marking M_0. Hence, we can simply choose $\sigma_1 = \rho_1$ and $\sigma_2 = \rho_{m+1} \cdots \rho_n$ which completes our proof. $\qquad\square$

This already implies the Shortest Sequence Theorem for T-systems. For later use, let us state the concrete implications in a generalized form of what the authors in [13] call the *Biased Sequence Lemma*:

Lemma 3 (Biased Sequence Lemma, Generalized Form of [13, Lemma 3.26]**).** *Let (N, M_0) with $N = (P, T, F, \ell)$ and bound b be an LBFC-system and let $(N, M_0)[\sigma\rangle(N, M)$ for a firing sequence $\sigma \in T^*$ and a marking M such that σ is biased and non-empty. Let k denote the number of (distinct) transitions in σ, i.e., $k := |\langle \sigma \rangle|$. Then, there exists a firing sequence ρ such that $\rho \leq \sigma$, $(N, M_0)[\rho\rangle(N, M)$, and*

$$|\rho| \leq b \cdot \frac{k \cdot (k+1)}{2}.$$

Proof. By repeatedly applying Lemma 2, we find $\rho = \rho_1 \rho_2 \cdots \rho_k$, $\rho \leq \sigma$ where

- $(N, M_0)[\rho_1 \rho_2 \cdots \rho_k\rangle(N, M)$,
- each transition occurs at most b times in ρ_i, and
- ρ_i contains at most $(k + 1 - i)$ many different transitions.

Hence, $|\rho| \leq b \cdot \sum_{i=1}^{k} i = b \cdot k \cdot (k+1)/2$ as claimed. $\qquad\square$

Next, we consider the case of general free-choice systems. To proceed, we need to introduce a couple of notions and definitions. In what follows, we implicitly refer to some LBFC-system (N, M_0) with $N = (P, T, F, \ell)$.

Definition 17 (Cluster [cf. 13, Definition 4.4]). For two transitions $t, t' \in T$ we let $t \sim t'$ if $\bullet t = \bullet t'$. This gives rise to an equivalence relation $\sim$ on T whose equivalence classes $[t]$ are called *clusters*.

Definition 18 (Conflict Order[cf. 13, Definition 9.8]). A *conflict order* $\preceq \subseteq T \times T$ is any partial order on T such that two transitions $t, t' \in T$ are comparable if and only if $[t] = [t']$, i.e., $\bullet t \cap \bullet t' \neq \emptyset$. The corresponding strict partial order is denoted by $\prec$, i.e., $t \prec t'$ if and only if $t \preceq t'$ and $t \neq t'$.

To put it differently, a conflict order is composed of separate linear orderings on the clusters of the LBFC-system. The key idea (and main challenge) in the proof of the *Shortest Sequence Theorem* is to show that each firing sequence can be rearranged in such a way that the permuted firing sequence *is ordered* with respect to *some* conflict order with which the firing sequence *agrees*:

Definition 19 (Ordered Firing Sequence [cf. 13, Definition 9.8]**).** A firing sequence $\sigma \in T^*$ *is ordered* with respect to a conflict order $\preceq$ if for all $t \prec t'$ there is no occurrence of t in σ after an occurrence of t'. Furthermore, σ *agrees with* $\preceq$ if for each cluster c either no transition of c occurs in σ or the last transition of c that occurs in σ is the maximal transition in c (according to $\preceq$).

The central result is the following Theorem from [13] which, conveniently, we can use without modification:

Theorem 3 ([13, Proposition 9.16]). *Let (N, M_0) with $N = (P, T, F, \ell)$ be an* LBFC-*system. Moreover, let $(N, M_0)[\sigma\rangle(N, M)$ for a firing sequence $\sigma \in T^*$ and let $\preceq$ be any conflict order which agrees with σ. Then, there exists a $\preceq$-ordered permutation ρ of σ such that $(N, M_0)[\rho\rangle(N, M)$.*

With Theorem 3 and the Biased Sequence Lemma (Lemma 3) we can finally prove our generalized version of the Shortest Sequence Theorem.

Proof (Theorem 2). First, note that for $M \in [N, M_0\rangle$ the system (N, M) is an LBFC-system as well. Hence, it suffices to shorten a firing sequence σ of the form $(N, M_0)[\sigma\rangle(N, M)$. Moreover, due to Theorem 3, we can assume that σ is $\preceq$-ordered for some conflict order $\preceq$ (if not, we can permute σ accordingly).

We iteratively split σ up into parts σ_i, $i = 1, \ldots, k$, i.e., $\sigma = \sigma_1 \sigma_2 \cdots \sigma_k$, with respect to the following property: σ_i is a maximal prefix of $\sigma_i \cdots \sigma_k$ such that σ_i is biased. We claim that the number of distinct transitions in σ_{i+1} is smaller than the number of distinct transitions in σ_i. This is because the first transition t' of σ_{i+1} must be in the same cluster $t' \in [t]$ of some transition $t \neq t'$ that occurs in σ_i (because of the maximality of σ_i). Since σ is ordered, $t \prec t'$ and t cannot occur in $\sigma_{i+1} \cdots \sigma_k$. Hence, σ_i contains at most $(|T| - i + 1)$ distinct transitions. In particular, $k \leq |T|$. We have $(N, M_{i-1})[\sigma_i\rangle(N, M_i)$ where $M_k = M$. By Lemma 3 we find firing sequences σ_i', $\boldsymbol{\sigma_i'} \leq \boldsymbol{\sigma_i}$ of length at most $b \cdot (|T| - i + 1) \cdot (|T| - i + 2)/2$ with $(N, M_{i-1})[\sigma_i'\rangle(N, M_i)$. The shortened sequence $\sigma' = \sigma_1' \sigma_2' \cdots \sigma_k'$, $\boldsymbol{\sigma'} \leq \boldsymbol{\sigma}$ has length at most

$$\frac{b}{2} \cdot \sum_{i=1}^{|T|} (|T| - i + 1) \cdot (|T| - i + 2) = \frac{b}{2} \cdot \frac{|T| \cdot (|T| + 1) \cdot (|T| + 2)}{3},$$

and satisfies $(N, M_0)[\sigma'\rangle(N, M)$, which completes our proof. $\square$

With this result, we can bound the length of sequences of consecutive model moves in alignments. This, in turn, allows us to show that LBFC-systems always have optimal alignments of polynomial length:

Theorem 4 (Alignments in LBFC-Systems). *Let $S = (N, M_{init}, M_{final})$ be an* LBFC-*system with $N = (P, T, F, \ell)$, bound b, and labeling function $\ell : T \to \Sigma \cup \{\tau\}$ and let $\sigma \in \Sigma^*$ be a trace over the alphabet Σ. Then, there exists an optimal alignment $\gamma \in \Gamma_{\sigma, S}$ between σ and S such that*

$$|\gamma| \leq (|\sigma| + 1) \cdot \left(b \cdot \frac{|T| \cdot (|T| + 1) \cdot (|T| + 2)}{6} + 1 \right).$$

Proof. Let $\gamma = \langle \gamma_i \rangle_{i=1}^{|\gamma|} \in \Gamma_{\sigma,S}$ be an optimal alignment of minimal length. We show that $|\gamma|$ satisfies the above inequality.

Let us denote the length of the trace σ by $q = |\sigma|$. Since $\pi_1(\gamma)|_\Sigma = \sigma$, we can find a subsequence $\langle \gamma_i \rangle_{i \in I}$ of γ with $I \subseteq \{1, \ldots, |\gamma|\}$, $I = \{i_1, i_2, \ldots, i_q\}$, $i_1 < i_2 < \cdots < i_q$ and such that $\pi_1(\langle \gamma_i \rangle_{i \in I}) = \sigma$. We use the positions $i_1, i_2, \ldots, i_q \in I$ in order to split γ up into $q + 1$ parts:

$$\delta_0 = \langle \gamma_1, \ldots, \gamma_{i_1} \rangle, \ \delta_1 = \langle \gamma_{i_1+1}, \ldots, \gamma_{i_2} \rangle, \ \ldots, \ \delta_q = \langle \gamma_{i_q+1}, \ldots, \gamma_{|\gamma|} \rangle.$$

For an illustration, see Fig. 3. Next, we show that for each $j \in \{0, \ldots, q\}$ we have $|\delta_j| \leq b \cdot |T| \cdot (|T| + 1) \cdot (|T| + 2)/6 + 1$. If we can verify this, the claim follows. Pick some $j \in \{0, \ldots, q\}$ and let

$$M_j \in \mathbb{N}^P: \ (N, M_{init})[\pi_2(\langle \gamma_1, \ldots, \gamma_{i_j} \rangle)|_T \rangle (N, M_j),$$
$$M_j' \in \mathbb{N}^P: \ (N, M_{init})[\pi_2(\langle \gamma_1, \ldots, \gamma_{i_{j+1}-1} \rangle)|_T \rangle (N, M_j'),$$

where for the cases $j = 0$ we let $M_0 := M_{init}$ and for $j = q$ we let $M_q' := M_{final}$. In words, M_j is the marking that we obtain from M_{init} by firing the transitions in γ of the first j parts, i.e., $\pi_2(\langle \gamma_1, \ldots, \gamma_{i_j} \rangle)|_T$, and M_j' is the marking that we get by firing the transitions in the first $j + 1$ parts, i.e., $\pi_2(\langle \gamma_1, \ldots, \gamma_{i_{j+1}-1} \rangle)|_T$, except for the very last transition from the move γ_{i_j}. The motivation for looking at these two markings is as follows:

- by definition, we can reach the marking M_j' from marking M_j by firing the intermediate sequence $\pi_2(\langle \gamma_{i_j+1}, \ldots, \gamma_{i_{j+1}-1} \rangle)$,
- each of the moves in this sequence $\langle \gamma_{i_j+1}, \ldots, \gamma_{i_{j+1}-1} \rangle$ is of the form $(\gg, t)$, i.e., we only move in the system, but not in the trace,
- the length of this sequence is $|\delta_j| - 1$.

Hence, it suffices to show that the sequence $\langle \gamma_{i_j+1}, \ldots, \gamma_{i_{j+1}-1} \rangle$ is of length at most $b \cdot |T| \cdot (|T| + 1) \cdot (|T| + 2)/6$. To see this, we make use of Theorem 2. In fact, by this result we know that from $\pi_2(\langle \gamma_{i_j+1}, \ldots, \gamma_{i_{j+1}-1} \rangle)$ we could construct, by deleting and rearranging transitions, a firing sequence which leads from M_j to M_j' in the underlying system of length at most $b \cdot |T| \cdot (|T|+1) \cdot (|T|+2)/6$. Since *all* moves in $\langle \gamma_{i_j+1}, \ldots, \gamma_{i_{j+1}-1} \rangle$ are of the form $(\gg, t)$, we could lift the necessary deletion and rearrangement steps to the level of γ without doing any harm to the alignment properties. Also note that since we only delete and rearrange moves, the costs of the alignment do not increase. Since γ was chosen to be an optimal alignment of minimal length, the claim follows. $\square$

Theorem 4 yields an NP-strategy for the alignment problem on LBFC-systems: on input S, σ, and k, where S is an accepting LBFC-system, σ is a trace as above, and $k \in \mathbb{Q}_{\geq 0}$ denotes a threshold, the problem is to decide whether some alignment $\gamma \in \Gamma_{\sigma,S}$ exists with costs $\sum c(\gamma) \leq k$. We make use of Theorem 4 and non-deterministically construct an alignment γ of length at most $(|\sigma| + 1) \cdot (b \cdot |T| \cdot (|T| + 1) \cdot (|T| + 2)/6 + 1)$ and verify that: (1) γ is a valid alignment between σ and S, and (2) its costs $\sum c(\gamma)$ do not exceed k. By Theorem 4, an optimal

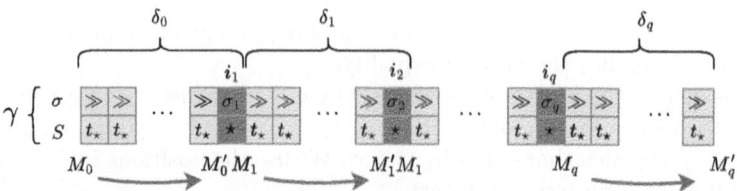

Fig. 3. Decomposing the alignment γ into parts δ_j at non-model move positions i_j. The proof idea is to shorten model move sequences (purple arrows) to connecting sequences of length at most $b \cdot |T| \cdot (|T| + 1) \cdot (|T| + 2)/6$. t_* is a placeholder for any transition in T, $\star$ for a transition in T or the no-move symbol $\gg$.

alignment between σ and S is among the potential candidates. Note that (since b is a constant) the length of γ is bounded polynomially in σ and S. Moreover, it is easy to verify that γ is a valid alignment (just simulate the system S and σ accordingly) and to check that its costs do not exceed k. Hence, we obtain:

Corollary 1. *On the class of* LBFC-*systems,* ALIGN *is in* NP.

6   ALIGN on Sound Free-Choice Workflow Nets

In this section, we focus on sound free-choice workflow nets. Workflow nets are characterized by a distinct source place and a distinct sink place and are thus tightly related to accepting systems. In particular, sound free-choice workflow nets are not only considered a subclass of LBFC-systems for which we have shown above that ALIGN $\in$ NP, but also of cyclic LBFC-systems (i.e., their initial marking is always reachable) for which reachability is in P [12]. This raises hope for an efficient algorithm for ALIGN—contrary to reachability on LBFC-systems, which is NP-complete [16]—; yet, we show that ALIGN remains NP-hard.

Definition 20 (Workflow Net). A Petri net $N = (P, T, F, \ell)$ with labeling function $\ell \colon T \to \Sigma \cup \{\tau\}$ is a *workflow net* if

- there is a single *source place* $p_{init} \in P$, i.e., $\{p_{init}\} = \{p \in P | \bullet p = \emptyset\}$,
- a single *sink place* $p_{final} \in P$, i.e., $\{p_{final}\} = \{p \in P | p \bullet = \emptyset\}$, and
- every vertex $v \in P \cup T$ of the Petri net is on a path from p_{init} to p_{final}.

Implicitly, in workflow nets, the initial marking is $M_{init} = [p_{init}]$, and the final marking is $M_{final} = [p_{final}]$. The *short-circuited net* $\bar{N} := (P, T \cup \{\bar{t}\}, F \cup \{(p_{final}, \bar{t}), (\bar{t}, p_{init})\}, \ell \cup \{(\bar{t}, \ell(\bar{t}) := \tau)\})$, $\bar{t} \notin T$, is an important tool to transfer the behavioral characteristics of LBFC-systems to workflow nets.

Definition 21 (Soundness [1, Theorem 11]). A workflow net is *sound* if and only if the *short-circuited* net is live and bounded.

Note that a sound free-choice workflow net is always safe [2, Lemma 1]. Using Definitions 20 and 21 and Corollary 1, we immediately obtain NP-membership:

Corollary 2. *On the class of sound free-choice workflow nets,* ALIGN *is in* NP.

We use a small detour via a closely related problem to show that NP is indeed also a lower bound. The *membership problem* determines whether a trace is part of the language of a given Petri net, i.e., whether it occurs as the labeling of a complete firing sequence of a given accepting system.

Problem 2 (Membership (MEMBER)).
    Input: An alphabet Σ, an accepting system $S = (N, M_{init}, M_{final})$ with $N = (P, T, F, \ell)$ and labeling function $\ell \colon T \to \Sigma \cup \{\tau\}$, and a trace $\sigma \in \Sigma^*$.
    Question: Is $\sigma \in \mathcal{L}(S)$?

It is easy to see that MEMBER is a special case of ALIGN where we look for a *perfect* alignment with costs 0.

Lemma 4. MEMBER *is polynomial-time reducible to* ALIGN.

Proof. Let Σ, $S = (N, M_{init}, M_{final})$, $N = (P, T, F, \ell)$, $\ell \colon T \to \Sigma \cup \{\tau\}$, and $\sigma \in \Sigma^*$ be an input of the MEMBER problem. For the reduction to ALIGN we make use of the standard cost function and set $k = 0$. In effect, we are looking for a *perfect* alignment between σ and S (i.e., with costs 0), which can only consist of synchronous moves and silent model moves. Thus, it requires that the trace σ can be obtained as the labeling of a firing sequence of the system S from M_{init} to M_{final}. This is precisely the decision problem MEMBER. There is, however, a small issue: the system S is not necessarily easy-sound, but this is required for inputs of ALIGN. To solve this, we add a new transition $t_{\blacktriangleright\!\blacktriangleright}$ with $\bullet t_{\blacktriangleright\!\blacktriangleright} = \langle M_{init} \rangle$, $t_{\blacktriangleright\!\blacktriangleright}\bullet = \langle M_{final} \rangle$ and a new label not present in σ. This ensures easy-soundness, but, if this transition is taken in some alignment, we neither have a synchronous nor silent move, and thus the costs are at least 1. □

To show NP-hardness of ALIGN, we use the fact that a live, safe T-system can emulate a *shuffle language* for which MEMBER is NP-complete [22,32]. Since T-systems are free-choice, we can combine Theorem 5 with Corollaries 1 and 2, to conclude the same for LBFC-systems and sound free-choice workflow nets.

Theorem 5. *On the class of live and safe T-systems,* ALIGN *is NP-hard. Even when this class is further restricted to acyclic systems,* ALIGN *remains NP-hard.*

Proof. Let $x \sqcup\!\sqcup y := \{v_1 w_1 \cdots v_k w_k | x = v_1 \cdots v_k, y = w_1 \cdots w_k, v_i, w_i \in \Sigma^*, 1 \le i \le k\}$ be the *shuffle* of two words $x, y \in \Sigma^*$ and let $\mathcal{L}_1 \sqcup\!\sqcup \mathcal{L}_2 := \bigcup \{w_1 \sqcup\!\sqcup w_2 | w_1 \in \mathcal{L}_1, w_2 \in \mathcal{L}_2\}$ be the *shuffle* of two languages $\mathcal{L}_1, \mathcal{L}_2 \subseteq \Sigma^*$. Furthermore, let $w \in \Sigma^*$ be a word and let $\mathcal{L}(w_1 \sqcup\!\sqcup w_2 \sqcup\!\sqcup \cdots \sqcup\!\sqcup w_n)$ be the *shuffle language* over words $w_1, w_2, \ldots, w_n \in \Sigma^*$. There exists a process tree T only using the sequence and parallel operator with $\mathcal{L}(T) = \mathcal{L}(w_1 \sqcup\!\sqcup w_2 \sqcup\!\sqcup \cdots \sqcup\!\sqcup w_n)$ [cf. 27]. Using the construction in [34], T can be transformed to a safe and sound workflow net that is also a T-system and acyclic. Because MEMBER for a shuffle language (i.e., deciding whether $w \in \mathcal{L}(w_1 \sqcup\!\sqcup w_2 \sqcup\!\sqcup \cdots \sqcup\!\sqcup w_n)$) is already NP-complete [22,32], ALIGN is NP-hard on the class of live and safe (acyclic) T-systems by Lemma 4.

Corollary 3. *On the class of (acyclic) LBFC-systems and on the class of sound (acyclic) free-choice workflow nets,* ALIGN *is NP-complete.*

7 ALIGN on General Safe and Sound Workflow Nets

We finally show that the free-choice assumption is needed to break the PSPACE barrier: In Theorem 8 and Corollary 5, we prove that ALIGN on safe and sound workflow nets is PSPACE-complete.

As a preparatory step, we show that the alignment problem is PSPACE-complete on the class of safe systems (Theorem 7). The usual approach for computing optimal alignments is via a reduction to the reachability problem in Petri nets. Therefore, we express the input trace itself in form of a Petri net:

Definition 22 (Trace System). Let $\sigma \in \Sigma^*$ be a trace over Σ. Its *trace system* $T(\sigma) := (N, M_{init}, M_{final})$ is an accepting system with $N = (P, T, F, \ell)$ where

- $P := \{p_i | 0 \leq i \leq |\sigma|\}$ is the set of places,
- $T := \{t_i | 1 \leq i \leq |\sigma|\}$ is the set of transitions,
- $F := \bigcup_{1 \leq i \leq |\sigma|} \{(p_{i-1}, t_i), (t_i, p_i)\}$ is the flow relation,
- $\ell \colon T \to \Sigma$ is the labeling function such that $\ell(\langle t_i \rangle_{i=1}^{|\sigma|}) = \sigma$,
- $M_{init} = [p_0]$ is the initial marking, and
- $M_{final} = [p_{|\sigma|}]$ is the final marking.

Now, both, the process model and the trace are represented by a Petri net and can be combined using the synchronous product, which was introduced in [3] and is a special case of the product of Petri nets introduced in [33].

Definition 23 (Synchronous Product). Let Σ be an alphabet and let $S_1 = (N_1, M_{1,init}, M_{1,final})$ and $S_2 = (N_2, M_{2,init}, M_{2,final})$ be two accepting systems with $N_1 = (P_1, T_1, F_1, \ell_1)$, $N_2 = (P_2, T_2, F_2, \ell_2)$, $\ell_1 \colon T_1 \to \Sigma \cup \{\tau\}$, and $\ell_2 \colon T_2 \to \Sigma \cup \{\tau\}$ where P_1, T_1, P_2, and T_2 are pairwise disjoint sets. Furthermore, let $\gg \notin \Sigma, T_1, T_2$ be a distinguished no-move symbol. The *synchronous product* of S_1 and S_2, denoted by $S_1 \otimes S_2$, is the accepting system $S_1 \otimes S_2 := (N, M_{init}, M_{final})$ with the Petri net $N := (P, T, F, \ell)$ and labeling function $\ell \colon T \to \Sigma \cup \{\tau\}$ where

- $P := P_1 \cup P_2$,
- $T := \{(t_1, t_2) \in T_1 \times T_2 | \ell_1(t_1) = \ell_2(t_2)\} \cup (T_1 \times \{\gg\}) \cup (\{\gg\} \times T_2)$,
- $F := \quad \{(p, (t_1, t_2)) \in P \times T | (p, t_1) \in F_1 \vee (p, t_2) \in F_2\}$
  $\qquad \cup \{((t_1, t_2), p) \in T \times P | (t_1, p) \in F_1 \vee (t_2, p) \in F_2\}$,
- $(t_1, t_2) \mapsto \ell(t_1, t_2) := \begin{cases} \ell_1(t_1) & t_1 \in T_1, \\ \ell_2(t_2) & t_1 \notin T_1, \end{cases}$
- $M_{init} := M_{1,init} \oplus M_{2,init}$, and $M_{final} := M_{1,final} \oplus M_{2,final}$.

As shown in [3], complete firing sequences in the synchronous product correspond to alignments between the trace and the model.

Proposition 1 ([3, Theorem 4.3.5]). *Given a trace σ and an accepting system S as process model, complete firing sequences of their synchronous product correspond to alignments between σ and S, i.e., $\Gamma_{\sigma,S} = \mathcal{L}(T(\sigma) \otimes S)$.*

Furthermore, the product structure directly transfers to the reachability set:

Proposition 2 ([3, Theorem 4.3.4, 33, Theorem 4.1]). *Given two systems S_1 and S_2, any combination $M_1 \oplus M_2$ of a reachable marking $M_1 \in [S_1\rangle$ and a reachable marking $M_2 \in [S_2\rangle$ is a reachable marking in the synchronous product $S_1 \otimes S_2$, i.e., $\forall M_1 \in [S_1\rangle, M_2 \in [S_2\rangle \colon M_1 \oplus M_2 \in [S_1 \otimes S_2\rangle$ and vice versa.*

Corollary 4. *Given a b_1-bounded system S_1 and a b_2-bounded system S_2, their synchronous product $S_1 \otimes S_2$ is $\max\{b_1, b_2\}$-bounded.*

We now draw the connection to the *reachability problem* and its cost-variant:

Problem 3 (Reachability (REACH)).
   Input: A system (N, M_0) with $N = (P, T, F, \ell)$ and a marking $M \in \mathbb{N}^P$.
   Question: Is $M \in [N, M_0\rangle$?

Problem 4 (Minimum-Cost Reachability (MINCOSTREACH)).
   Input: A system (N, M_0) with a Petri net $N = (P, T, F, \ell)$, a marking $M \in \mathbb{N}^P$, a cost function $c \colon T \to \mathbb{Q}_{\geq 0}$, and a number $k \in \mathbb{Q}_{\geq 0}$.
   Question: Is there a $\sigma \in T^*$ such that $(N, M_0)[\sigma\rangle(N, M)$ and $\sum c(\sigma) \leq k$?

Since we can also assign costs of 0 to any transition, MINCOSTREACH is a generalization of REACH and we have:

Lemma 5. REACH *is polynomial-time reducible to* MINCOSTREACH.

We can now show that MINCOSTREACH on safe systems is in PSPACE:

Theorem 6. *On the class of safe systems,* MINCOSTREACH *can be decided in polynomial space (in short:* MINCOSTREACH $\in$ PSPACE*).*

Proof. To find a *deterministic* PSPACE algorithm for MINCOSTREACH, we use Savitch's Theorem [25]: for every *non-deterministic* PSPACE algorithm, there also exists an equivalent *deterministic* algorithm.
   Let (N, M_0), $N = (P, T, F, \ell)$, c, M, and k be an input of the MINCOSTREACH problem. That is, (N, M_0) is a safe system with a Petri net $N = (P, T, F, \ell)$, $c \colon T \to \mathbb{Q}_{\geq 0}$ is a cost function, $M \in \mathbb{N}^P$ is a marking, and $k \in \mathbb{Q}_{\geq 0}$ is a cost limit. The algorithm stores a marking $\overline{M}$, which is initially set to $\overline{M} = M_0$, and a cost value $\bar{c}$, initially set to $\bar{c} = 0$. Note that since (N, M_0) is safe, a marking of N can be stored in polynomial space. As long as $\overline{M} \neq M$, the algorithm non-deterministically chooses a transition $t \in T$ enabled in marking $\overline{M}$ and computes the marking $\overline{M}'$ after firing t, i.e., $(N, \overline{M})[t\rangle(N, \overline{M}')$. Then, the stored marking is set to $\overline{M} := \overline{M}'$ and the stored cost value is set to $\bar{c} := \bar{c} + c(t)$.
   If M is not reachable, the algorithm does not necessarily terminate. Thus, we add a counter which counts the number of fired transitions. Because (N, M_0) is safe, it has at most $2^{|P|}$ reachable markings. Therefore, we can stop if more than $2^{|P|} - 1$ transitions were fired. If the algorithm reaches the marking M and $\bar{c} \leq k$, it stops and M can be reached within the cost limit. If the counter exceeds $2^{|P|} - 1$ or $\bar{c}$ exceeds k, M cannot be reached within the cost limit.   □

Since REACH on safe systems is PSPACE-complete [10,11], we can conclude:

Lemma 6. *On the class of safe systems,* MINCOSTREACH *is* PSPACE-*complete.*

Proof. On the class of safe systems, REACH is PSPACE-complete [10,11]. By Lemma 5, MINCOSTREACH is PSPACE-hard. In combination with Theorem 6, MINCOSTREACH is PSPACE-complete on the class of safe systems.   □

The next result allows us to transfer the PSPACE-bound to ALIGN.

Lemma 7. ALIGN *is polynomial-time reducible to* MINCOSTREACH.

Proof. Let Σ, $S = (N, M_{init}, M_{final})$, $N = (P, T, F, \ell)$, $\ell \colon T \to \Sigma \cup \{\tau\}$, $\sigma \in \Sigma^*$, $c \colon LM_{\Sigma,S} \to \mathbb{Q}_{\geq 0}$, and k be an input of the ALIGN problem. That is, Σ is an alphabet representing the set of activities, $S = (N, M_{init}, M_{final})$ is an easy-sound system, i.e., $\phi(S) \neq \emptyset$, with the Petri net $N = (P, T, F, \ell)$ and labeling function $\ell \colon T \to \Sigma \cup \{\tau\}$, $\sigma \in \Sigma^*$ is a trace over the alphabet Σ, and $c \colon LM_{\Sigma,S} \to \mathbb{Q}_{\geq 0}$ is a function which assigns costs to each legal move between Σ and S.

According to [3], finding an optimal alignment between σ and S is identical to finding a cost-minimal complete firing sequence in the synchronous product net $\mathcal{T}(\sigma) \otimes S$. Let $\mathcal{T}(\sigma) \otimes S := (N', M'_{init}, M'_{final})$ where $N' := (P', T', F', \ell')$ and in particular $T' \subseteq LM_{\Sigma,S}$. Therefore, a solution to MINCOSTREACH with a system (N', M'_{init}) where $N' = (P', T', F', \ell')$, a marking M'_{final}, a cost function c, and a number k as input is also a solution to ALIGN.   □

Lemma 8. MINCOSTREACH *is polynomial-time reducible to* ALIGN.

Proof. Let (N, M_0), M, c, and k be an input for the MINCOSTREACH problem, i.e., (N, M_0) is a system with Petri net $N = (P, T, F, \ell)$, $M \in \mathbb{N}^P$ the target marking, $c \colon T \to \mathbb{Q}_{\geq 0}$ a cost function, and $k \in \mathbb{Q}_{\geq 0}$ a threshold. It is to decide if M can be reached from M_0 with costs at most k.

To map this to an input of ALIGN, we make use of the empty trace $\sigma = \langle \rangle$. Aligning the empty trace corresponds to finding a firing sequence from the initial marking M_0 to the final marking M with minimal costs. Similarly as in the proof of Lemma 4, we have one technical problem: the system (N, M_0, M) might not be easy-sound. Again, we can solve this by adding a transition $t_{▸◂}$ which allows the system to move from M_0 to M in one step. By making this transition very expensive (at least $k + 1$), we ensure easy-soundness, and, in case M is not reachable from M_0, the optimal alignment has costs at least $k + 1$.   □

Altogether, this yields our first main result of this section:

Theorem 7. *On the class of safe systems,* ALIGN *is* PSPACE-*complete.*

Proof. A trace system is safe by definition, the synchronous product considered in Lemma 7 is also safe according to Corollary 4. Hence, with Lemma 6 we have ALIGN ∈ PSPACE on the class of safe systems. Due to Lemmas 8 and 6, ALIGN is also PSPACE-hard on the class of safe systems and thus PSPACE-complete.   □

We finally turn our attention to safe and *sound* workflow nets, i.e., we add the liveness assumption. The question is whether this property suffices to reduce the complexity of the alignment problem. This turns out not to be the case:

Theorem 8. *There is a polynomial time algorithm which transforms a deterministic Turing machine $\mathcal{M}$ with polynomial space bound $p(n)$ and an input w into a safe and sound workflow net $S = (N, M_{init}, M_{final})$ and a trace σ such that, with respect to the standard cost function, S and σ can be aligned with 0 costs if and only if $\mathcal{M}$ accepts w.*

Proof. We extend the construction in [10, Theorem 4]. First, we make some assumptions on $\mathcal{M}$ which can be guaranteed by preprocessing. When the (deterministic, single tape) Turing machine $\mathcal{M}$ is started with input w, during the computation, the head of $\mathcal{M}$ only moves between positions 0 and $p(n)$, where $n = |w|$, starting at position 0. Moreover, each computation is finite, i.e., the machine $\mathcal{M}$ never repeats a configuration. In particular, $\mathcal{M}$ halts on every input and either accepts or rejects. Finally, there is precisely one accepting and one rejecting configuration of the machine $\mathcal{M}$. To guarantee this, one can implement a subroutine such that, whenever the machine $\mathcal{M}$ enters a final state (accepting or rejecting), then the tape is cleared, the head moves back to position 0, and the machine enters a unique accepting or rejecting state, respectively.

Let $\mathcal{M} = (K, \Sigma, \Gamma, \delta, q_0, q_+, q_-, \bot)$ be a deterministic Turing machine where K is the set of states, Σ the input alphabet, Γ the tape alphabet, $\delta \colon K \backslash \{q_+, q_-\} \times \Gamma \to K \times \Gamma \times \{-1, 1, 0\}$ the transition function (-1 (1) means the head moves one position to the left (right), and 0 means no move), $q_0 \in K$ the initial state, $q_+ \in K$ the unique accepting state, $q_- \in K$ the unique rejecting state, and $\bot \in \Gamma \backslash \Sigma$ the blank symbol. To encode the computation of $\mathcal{M}$ on input w, we define a workflow net $N = (P, T, F, \ell)$ with a set of places P consisting of:

- the initial place p_{init} and the final place p_{final},
- a place p_q^K for each state $q \in K$ (a token in p_q^K indicates that in the current configuration $\mathcal{M}$ is in state q),
- a place p_i^H for each possible head position $i \in \{0, \ldots, p(n)\}$ (a token in p_i^H indicates that in the current configuration the head is at position i),
- a place $p_{i,a}^C$ for each possible tape cell content, i.e., for each combination of a valid position $i \in \{0, \ldots, p(n)\}$ and a tape symbol $a \in \Gamma$ (a token in $p_{i,a}^C$ indicates that in the current configuration tape cell i holds symbol a).

With this preparation, we identify configurations of $\mathcal{M}$ with markings of N. To simulate the computation, we introduce the following set of transitions T:

- One transition $t_{\blacktriangleright}$ that yields the initial configuration, i.e., $\bullet t_{\blacktriangleright} = \{p_{init}\}$ and $t_{\blacktriangleright}\bullet = \{p_{q_0}^K\} \cup \{p_0^H\} \cup \{p_{i,w_i}^C \mid i < |w|\} \cup \{p_{i,\bot}^C \mid |w| \leq i \leq p(n)\}$.
- One transition $t_{\checkmark}$ that finalizes the computation when the (unique) accepting configuration is reached, i.e., $t_{\checkmark}\bullet = \{p_{final}\}$ and $\bullet t_{\checkmark} = \{p_{q_+}^K\} \cup \{p_0^H\} \cup \{p_{i,\bot}^C \mid i \leq p(n)\}$. This transition is labeled by $\checkmark$ and there is no other transition labeled by $\checkmark$. In the same way, we add one transition t_{x} that finalizes the computation when the (unique) rejecting configuration is reached.

– For each possible computational step, we introduce a distinct transition: for each state $q \in K$, symbol $a \in \Gamma$ with $\delta(q, a) = (q', b, d)$, and for each head position $i \in \{0, \ldots, p(n)\}$, we introduce a transition $t[q, a, i]$ that models the configuration change which occurs when $\mathcal{M}$ is in state q, the head is at position i, and reads the symbol a. More precisely, $\bullet t[q, a, i] = \{p_q^K, p_i^H, p_{i,a}^C\}$ and $t[q, a, i] \bullet = \{p_{q'}^K, p_{i+d}^H, p_{i,b}^C\}$.

By construction, we can trigger the simulation from the initial marking $M_{init} = [p_{init}]$ by firing $t_{\blacktriangleright}$. This generates the initial configuration of the computation of $\mathcal{M}$ on w. Since the machine $\mathcal{M}$ is deterministic, from that point onward there is at most one transition of the form $t[q, a, i]$ that can be fired in the current marking/configuration. This transition, in turn, updates the marking/configuration according to the transition function of $\mathcal{M}$. Since we have prepared $\mathcal{M}$ in such a way that the computation is acyclic, we will finally reach the unique accepting or rejecting configuration from which we can fire $t_{\checkmark}$ or t_{x}, respectively, to reach the final marking $M_{final} = [p_{final}]$. Giving the one-to-one correspondence between markings of N and configurations of $\mathcal{M}$, the resulting net is safe (every reachable marking corresponds to a configuration in the sense described above and such markings only hold at most one token per place).

The problem is that N is not sound. In general, it might be that a transition of the form $t[q, a, i]$ can never fire simply because the computation of $\mathcal{M}$ on w does never run into an enabling configuration. Also, we can either fire $t_{\checkmark}$ or t_{x} from the initial marking, but not both. To overcome this, we first add a new place p_{aux} indicating when the net is in *auxiliary* mode. Then, we add two new transitions $t_{\checkmark}^0$ and t_{x}^0 which can fire at the initial marking and activate $t_{\checkmark}$ and t_{x}: we set $\bullet t_{\checkmark}^0 = \{p_{init}\}$ and $t_{\checkmark}^0 \bullet = \bullet t_{\checkmark}$ and analogously for t_{x}^0. In other words, $t_{\checkmark}^0$ and t_{x}^0 produce the two unique markings where precisely $t_{\checkmark}$ or t_{x} is enabled, respectively. So, we can reach the final marking by firing either of them (note that transitions of the form $t[q, a, i]$ are not enabled in these terminal configurations of $\mathcal{M}$). Second, for each transition $t[q, a, i]$ we add two new transitions $t^0[q, a, i]$ and $t^1[q, a, i]$ which activate and deactivate $t[q, a, i]$ from the initial marking, i.e.,

$$\bullet t^0[q, a, i] = \{p_{init}\}, \qquad \text{and} \qquad t^0[q, a, i] \bullet = \{p_{aux}\} \cup \bullet t[q, a, i],$$
$$\bullet t^1[q, a, i] = \{p_{aux}\} \cup t[q, a, i] \bullet, \qquad \text{and} \qquad t^1[q, a, i] \bullet = \{p_{final}\}.$$

In this way, we can move via the sequence $\langle t^0[q, a, i], t[q, a, i], t^1[q, a, i] \rangle$ from the initial to the final marking and fire $t[q, a, i]$ along the way. However, there is one subtlety we have to discuss. In contrast to the case of $t_{\checkmark}$ and t_{x}, the manual activation and firing of $t[q, a, i]$ might enable transitions different from $t^1[q, a, i]$. In fact, if the head does not move in state q while reading a, three tokens would be produced by $t[q, a, i]$ which would allow the net to fire another transition of the form $t[q', b, i]$. However, all such transitions are conservative in the sense that they do not alter the total count of tokens (they all consume and produce three tokens). Thus, the total count of tokens remains three which means that we will never be able to fire one of the final transitions $t_{\checkmark}$ or t_{x}. Since $\mathcal{M}$ does not allow cyclic computations, eventually we get stuck in the simulation component after firing one transition of the form $t[q', b, i]$ which moves the head to the left or right (note that for the new cell, we are missing a token in a place $p^C_{a,i+d}$ for $d \in \{-1, 1\}$). In this setting we can simply fire $t^1[q', b, i]$ which removes the tokens produced by $t[q', b, i]$ and enters the final marking. Finally, all introduced auxiliary transitions get the extra label ▣. Note that we cannot mix a proper simulation of the computation of $\mathcal{M}$ on w triggered by firing $t_▸$ with a transition of the form $t^1[q, a, i]$ since this transition requires a token in p_{aux}. Altogether, by this second extension each transition in the workflow net can be fired from the initial marking and we maintain the property to always reach the final marking.

Finally, let $\sigma := \langle ✔ \rangle$. Then, we claim that σ and the safe and sound workflow net S can be aligned with costs 0 (wrt. the standard cost function) if and only if $\mathcal{M}$ accepts input w. Clearly, if $\mathcal{M}$ accepts w, the simulation triggered by firing $t_▸$ simulates the computation via silent transitions of the form $t[q, a, i]$ until the single transition $t_✔$ with label ✔ can eventually be fired synchronously with the symbol ✔ in σ. This does not incur any costs since we only have one synchronous move and several silent moves in the net. If, on the other hand, the computation of $\mathcal{M}$ on w is rejecting, there is no way to align σ with costs 0. In fact, if any of the auxiliary transitions is used, this will immediately lead to a model move since σ does not contain the symbol ▣. If, on the other hand the proper simulation of $\mathcal{M}$ on w is started via $t_▸$, then this will end up in a completely silent run ending with $t_✗$ and we would require a log move for the symbol ✔ in the trace.   □

Together with PSPACE-membership for safe systems (Theorem 7), we get:

Corollary 5. *On the class of safe and sound workflow nets,* ALIGN *is* PSPACE-*complete.*

8   Conclusion

We proved that the high algorithmic costs for computing alignments are unavoidable: an efficient algorithm for alignments on sound workflow nets (the standard model in process mining) does not exist (assuming P ≠ PSPACE). Furthermore, we derived better algorithmic bounds for important model classes, such as the class of sound free-choice workflow nets which, in turn, includes all process trees.

Our results also show that for a complexity-theoretic understanding of alignments, we cannot simply refer to the reachability problem. For instance, on live, bounded, free-choice systems, the reachability and alignment problem are both NP-complete (see [16] and Sect. 5). However, if we further assume cyclicity (i.e., the initial marking is reachable from any other marking), the complexity of reachability drops to P while the alignment problem remains NP-complete (see [12] and Sect. 5). In particular, this complexity gap also holds for sound free-choice workflow nets as well as simpler model classes like Partially Ordered Workflow Language (POWL) models [20] or process trees [27].

For future research, we want to investigate in how far the bounds of the *Shortest Sequence Theorem* can be further improved on sound free-choice workflow nets. Potentially, this would allow us to generalize our ILP encoding for process trees from [26] to sound free-choice workflow nets and can lead to a much more efficient alignment algorithm on this class in practice. On the more theoretical level, we want to dive deeper into the complexity structure of alignments, in particular, regarding the influence of different parameters of the models.

References

1. van der Aalst, W.M.P.: Verification of workflow nets. In: Azéma, P., Balbo, G. (eds.) ICATPN 1997. LNCS, vol. 1248, pp. 407–426. Springer, Heidelberg (1997). https://doi.org/10.1007/3-540-63139-9_48 isbn: 978-3-540-69187-7
2. van der Aalst, W.M.P.: Workflow verification: finding control-flow errors using Petri-net-based techniques. In: van der Aalst, W.M.P., Desel, J., Oberweis, A. (eds.) Business Process Management. LNCS, vol. 1806, pp. 161–183. Springer, Heidelberg (2000). https://doi.org/10.1007/3-540-45594-9_11 isbn: 978-3-540-45594-3 45594-3 45594-3 45594-3 45594-3 45594-3 45594-3
3. Adriansyah, A.: Aligning observed and modeled behavior. PhD thesis. Technische Universiteit Eindhoven (2014). https://doi.org/10.6100/IR770080. isbn: 978-90-386-3574-3
4. Aho, A.V., Peterson, T.G.: A minimum distance error-correcting parser for context-free languages. SIAM J. Comput. **1**(4), 305–312 (1972). https://doi.org/10.1137/0201022
5. Bloemen, V., van de Pol, J., van der Aalst, W.M.P.: Symbolically aligning observed and modelled behaviour. In: Application of Concurrency to System Design. ACSD 2018, pp. 50–59. IEEE Computer Society (2018). https://doi.org/10.1109/ACSD.2018.00008. isbn: 978-1-5386-7013-2
6. Boltenhagen, M., Chatain, T., Carmona, J.: Generalized alignment-based trace clustering of process behavior. In: Donatelli, S., Haar, S. (eds.) PETRI NETS 2019. LNCS, vol. 11522, pp. 237–257. Springer, Cham (2019). https://doi.org/10.1007/978-3-030-21571-2_14 isbn: 978-3-030-21571-2
7. Boltenhagen, M., Chatain, T., Carmona, J.: Optimized SAT encoding of conformance checking artefacts. Computing **103**(1), 29–50 (2020). https://doi.org/10.1007/s00607-020-00831-8
8. Carmona, J., van Dongen, B.F., Solti, A., Weidlich, M.: Conformance Checking. Relating Processes and Models. Springer, Cham (2018). https://doi.org/10.1007/978-3-319-99414-7. isbn: 978-3-319-99413-0
9. Carmona, J., van Dongen, B.F., Weidlich, M.: Conformance checking: foundations, milestones and challenges. In: van der Aalst, W.M.P., Carmona, J., (eds.) LNBIP, vol. 448, pp. 155–190. Springer, Cham (2022). https://doi.org/10.1007/978-3-031-08848-3_5. isbn: 978-3-031-08847-6
10. Cheng, A., Esparza, J., Palsberg, J.: Complexity results for 1-safe nets. In: Shyamasundar, R.K. (ed.) FSTTCS 1993. LNCS, vol. 761, pp. 326–337. Springer, Heidelberg (1993). https://doi.org/10.1007/3-540-57529-4_66 isbn: 978-3-540-48211-6
11. Cheng, A., Esparza, J., Palsberg, J.: Complexity results for 1-safe nets. Theor. Comput. Sci. **147**(1-2), 117–136 (1995). https://doi.org/10.1016/0304-3975(94)00231-7.
12. Desel, J., Esparza, J.: Reachability in cyclic extended free-choice systems. Theor. Comput. Sci. **114**(1), 93–118 (1993). https://doi.org/10.1016/0304-3975(93)90154-L

13. Desel, J., Esparza, J.: Free Choice Petri Nets. Cambridge Tracts in Theoretical Computer Science, vol. 40. Cambridge University Press, Cambridge (1995). https://doi.org/10.1017/CBO9780511526558. isbn: 978-0-521-01945-3
14. Desel, J., Esparza, J.: Shortest paths in reachability graphs. J. Comput. Syst. Sci. **51**(2), 314–323 (1995). https://doi.org/10.1006/jcss.1995.1070
15. van Dongen, B.F.: Efficiently computing alignments. In: Weske, M., Montali, M., Weber, I., vom Brocke, J. (eds.) BPM 2018. LNCS, vol. 11080, pp. 197–214. Springer, Cham (2018). https://doi.org/10.1007/978-3-319-98648-7_12. isbn: 978-3-319-98647-0 319-98647-0
16. Esparza, J.: Reachability in live and safe free-choice Petri nets is NP-complete. Theor. Comput. Sci. **198**(1–2), 211–224 (1998). https://doi.org/10.1016/S0304-3975(97)00235-1.
17. Esparza, J.: Decidability and complexity of Petri net problems — an introduction. In: Reisig, W., Rozenberg, G. (eds.) ACPN 1996. LNCS, vol. 1491, pp. 374–428. Springer, Heidelberg (1998). https://doi.org/10.1007/3-540-65306-6_20. isbn: 978-3-540-49442-3
18. Esparza, J., Nielsen, M.: Decidability issues for petri nets. BRICS Rep. Ser. **1**(8) (1994). https://doi.org/10.7146/brics.v1i8.21662.
19. Jones, N.D., Landweber, L.H., Lien, Y.E.: Complexity of some problems in Petri nets. Theor. Comput. Sci. **4**(3), 277–299 (1977). https://doi.org/10.1016/0304-3975(77)90014-7
20. Kourani, H., van Zelst, S.J.: POWL: partially ordered workflow language. In: Di Francescomarino, C., Burattin, A., Janiesch, C., Sadiq, S., (eds.) BPM 2023, vol. 14159, pp. 92–108. Springer, Cham (2023). https://doi.org/10.1007/978-3-031-41620-0_6. isbn: 978-3-031-41620-0
21. de Leoni, M., Marrella, A.: Aligning real process executions and prescriptive process models through automated planning. Expert Syst. Appl. **82**, 162–183 (2017). https://doi.org/10.1016/j.eswa.2017.03.047
22. Mansfield, A.: On the computational complexity of a merge recognition problem. Discrete Appl. Math. **5**(1), 119–122 (1983). https://doi.org/10.1016/0166-218X(83)90021-5
23. Pighizzini, G.: How hard is computing the edit distance? Inf. Comput. 165(1), 1–13 (2001). https://doi.org/10.1006/inco.2000.2914
24. Rozinat, A., van der Aalst, W.M.P.: Conformance checking of processes based on monitoring real behavior. Inf. Syst. **33**(1), 64–95 (2008). https://doi.org/10.1016/j.is.2007.07.001
25. Savitch, W.J.: Relationships between nondeterministic and deterministic tape complexities. J. Comput. Syst. Sci. **4**(2), 177–192 (1970). https://doi.org/10.1016/S0022-0000(70)80006-X
26. Schwanen, C.T., Pakusa, W., van der Aalst, W.M.P.: A dynamic programming approach for alignments on process trees. In: Delgado, A., Slaats, T. (eds.) ICPM 2024. LNBIP, vol. 533, pp. 84–97. Springer, Cham (2025). https://doi.org/10.1007/978-3-031-82225-4_7 isbn: 978-3-031-82224-7
27. Schwanen, C.T., Pakusa, W., van der Aalst, W.M.P.: Process tree alignments. In: Borbinha, J., Prince Sales, T., Mira Da Silva, M., Proper, H.A., Schnellmann, M., (eds.) EDOC 2024. LNCS, vol. 15409. Springer, Cham (2025). https://doi.org/10.1007/978-3-031-78338-8_16. isbn: 978-3-031-78337-1
28. Taymouri, F., Carmona, J.: A recursive paradigm for aligning observed behavior of large structured process models. In: La Rosa, M., Loos, P., Pastor, O. (eds.) BPM 2016. LNCS, vol. 9850, pp. 197–214. Springer, Cham (2016). https://doi.org/10.1007/978-3-319-45348-4_12 isbn: 978-3-319-45348-4

29. Taymouri, F., Carmona, J.: Model and event log reductions to boost the computation of alignments. In: Ceravolo, P., Guetl, C., Rinderle-Ma, S. (eds.) SIMPDA 2016. LNBIP, vol. 307, pp. 1–21. Springer, Cham (2018). https://doi.org/10.1007/978-3-319-74161-1_1 isbn: 978-3-319-74160-4

30. Valmari, A.: The state explosion problem. In: Reisig, W., Rozenberg, G. (eds.) ACPN 1996. LNCS, vol. 1491, pp. 429–528. Springer, Heidelberg (1998). https://doi.org/10.1007/3-540-65306-6_21

31. Wagner, R.A.: Order-n correction for regular languages. Commun. ACM **17**(5), 265–268 (1974). https://doi.org/10.1145/360980.360995

32. Warmuth, M.K., Haussler, D.: On the complexity of iterated shuffle. J. Comput. Syst. Sci. **28**(3), 345–358 (1984). https://doi.org/10.1016/0022-0000(84)90018-7

33. Winskel, G.: Petri nets, algebras, morphisms, and compositionality. Inf. Comput. **72**(3), 197–238 (1987). https://doi.org/10.1016/0890-5401(87)90032-0

34. van Zelst, S.J., Leemans, S.J.J.: Translating Workflow nets to process trees: an algorithmic approach. Algorithms **13**(11) (2020). https://doi.org/10.3390/a13110279.

Computing Alignments
for Partially-Ordered Traces Through
Petri Net Unfoldings

Ariba Siddiqui, Wil M. P. van der Aalst⬤, and Daniel Schuster(✉)⬤

RWTH Aachen University, Aachen, Germany
ariba.siddiqui@rwth-aachen.de, {wvdaalst,schuster}@pads.rwth-aachen.de

Abstract. Conformance checking techniques aim to provide diagnostics on the conformity between process models and event data. Conventional methods, such as trace alignments, assume strict total ordering of events, leading to inaccuracies when timestamps are overlapping, coarse, or missing. In contrast, existing methods that support partially ordered events rely upon the interleaving semantics of Petri nets, the reachability graphs, which suffer from the state space explosion problem. This paper proposes an improved approach to conformance checking based upon partially ordered event data by utilizing Petri net unfolding, which leverages partial-order semantics of Petri nets to represent concurrency and uncertainty in event logs more effectively. Unlike existing methods, our approach offers a streamlined one-step solution, improving efficiency in the computation of alignments. Additionally, we introduce a novel visualization technique for partially ordered unfolding-based alignments. We implement unfolding-based alignments with its user-friendly insights in a conformance analysis tool. Our experimental evaluation, conducted on synthetic and real-world event logs, demonstrates that the unfolding-based approach is particularly robust in handling high degrees of parallelism and complexity in process models.

Keywords: Conformance checking · Petri nets unfolding · Partial order semantics · Alignments · Process mining

1 Introduction

Conformance checking, a major task within process mining, deals with comparing event data with process models, typically represented using notations like BPMN [10] or Petri Nets [17]. *Trace alignments* [2] are considered to be state-of-the-art in conformance checking. These techniques aim to search for alignments with minimum deviations, with Adriansyah et al. [2] introducing a method using a shortest-path search using A^* [16] in a Synchronous Product Net (SPN) of the trace net and the model net. Another method [15] relies on a unified representation of process models and event logs based on *event structures*, a well-known model of concurrency. The method returns a set of statements in natural language describing behavior allowed by the model but not observed in the log and vice versa.

© The Author(s), under exclusive license to Springer Nature Switzerland AG 2025
E. Amparore and L. Mikulski (Eds.): PETRI NETS 2025, LNCS 15714, pp. 411–432, 2025.
https://doi.org/10.1007/978-3-031-94634-9_20

One problem with trace alignment algorithms is that they assume totally-ordered event data and produce totally-ordered alignments. In a more recent work by Lu et al. [21], this limitation is overcome. A partially-ordered trace is converted to an occurrence net (i.e., a simple Petri net without choices), and an SPN is computed between the occurrence net and the normative model. From hereon, the shortest path computation is done on the state space of the product net (just like [2]) using the A^* algorithm and a predefined cost function. As a firing sequence with minimum cost is found, it is replayed on the net while unfolding it into a new net. That is, while replaying, whenever a place is seen for the second time (owing to a loop), it is cloned into a new place. The result is an occurrence net, which can be converted into a partially-ordered alignment (*p-alignment*). This results in a two-step computation solely because of the reliance on interleaving semantics of Petri nets, the *reachability graphs*, which additionally suffer from the well-known state space explosion [29], limiting its scalability to larger instances of problems with high concurrency.

To address the issue of state space explosion in the interleaving semantics of Petri nets, researchers in model checking [12], diagnosis [5], and planning [6,18] have extensively explored the Petri net unfolding process. This partial-order semantics, introduced in [23] and further detailed as *branching processes* in [11], offers an alternative to interleaving semantics. A branching process unfolds all partially-ordered runs of a Petri net into a tree-like structure, representing concurrency, causality, and choice points. Although comprehensive, these processes are often infinite. McMillan's seminal work [22] introduced a method to construct a finite prefix of the branching process that retains state reachability information. This prefix, which models concurrency more compactly than the reachability graphs [24], is typically much smaller. However, McMillan's algorithm can produce unnecessarily large prefixes. To address this, Esparza et al. proposed the *ERV unfolding algorithm* [13], which generalizes McMillan's algorithm using *adequate orders*. Since these prefixes encode all information on reachable states of a Petri net, they can effectively be used for shortest-path computation problems. Esparza et al. showed that the reachability problem is NP-complete relative to prefix size [14], with their *OnTheFly ERV* variant proving most efficient amongst all unfolding-based reachability techniques.

Despite advances in alignment computation, partial order-based conformance checking, and Petri net unfolding, these domains have yet to converge. This paper addresses this gap by applying Petri net unfolding to optimal alignment computation based on partially ordered event data. We apply Petri net unfolding to replace reachability graphs to compute *unfolding-based alignments*, using an algorithm we term $ERV[\lhd_c]$. Our method offers a straightforward, one-step solution, avoiding the need to create a partially ordered alignment from execution sequences artificially. A key advantage is faster computation times, particularly when there is high concurrency between events. As a second contribution, we propose a method for visualizing the results of unfolding-based alignments using a technique from [26]. This method visualizes trace variants from partially ordered event data in chevron-based views and is well-suited for graph-based alignments.

Unlike existing methods, our approach makes the relationship between log and model more explicit. We evaluate our algorithm using synthetic event data with varying levels of parallelism, showing that unfolding-based alignments perform better under increasing parallelism. Regarding runtime, the (directed) unfolding-based approach outperforms the classic method by Lu et al. [21]. However, in lower parallelism settings, our method faces higher computational costs. Further experiments with real-world data confirm these findings, demonstrating that unfolding-based algorithms are more robust as model complexity grows.

2   Preliminaries

In process mining, an *event log* captures the occurrence of activities within a business process execution. Logs are analyzed to discover process models, check conformance, or optimize performance. Each event includes *start* and *end* timestamps, a *case identifier*, and a label for the event. Case identifiers group events into a single *case* representing a process execution. Conformance checking compares each case with the normative process model, and deviations are analyzed across the log. A key challenge is representing temporal relationships, especially for concurrently occurring events. To address this, we define *partially-ordered traces (p-traces)* to represent concurrent activities the need for total ordering.

Definition 1 (Partially-ordered Trace (P-Trace)). *For an event log E, in which all events share a common case identifier c, a partially-ordered trace is a labeled directed acyclic graph $\rho_c = ((E, \prec_{\rho_c}), l_{\rho_c})$ where E is a finite set of graph nodes and $\prec_{\rho_c}$ is a partial order over the nodes of the graph such that for any two events a and b, $a \prec_{\rho_c} b$ iff the end timestamp of a is strictly less than the start timestamp of b. $l_{\rho_c} : E \to \mathcal{L}$ is a labeling function for the nodes, where $\mathcal{L}$ is the universe of events labels of E. Each edge in a p-trace is referred to as a dependency between two events e_i and e_j, indicating that e_i has led to the execution of e_j.*

In this paper, we focus on Petri nets as the process modelling formalism. The notations of Petri nets used are largely based upon [7]. In its general, simplest form, Petri nets can be seen as graphs that consist of *places, transitions*, and directed edges between them. Places (depicted as circles) and transitions (depicted as squares) represent the two types of nodes of the graph. An example Petri net $\mathcal{N}_{mgmt}$ is shown in Fig. 1, modeling a software project management process with 11 places and 12 transitions.

Definition 2 (Labeled Petri Net, System Net). *A labeled Petri net (in short, net) is a bipartite graph $\mathcal{N} = (P, T, F, \lambda)$ with a disjoint finite set of nodes consisting of places P and transitions T, a set of (directed) arcs or flow relations $F \subseteq ((P \times T) \cup (T \times P))$, and a labeling function $\lambda : T \to L^{\tau}$ assigning an event label or τ (label of a silent transition) to each transition. A system net $SN = (\mathcal{N}, M_{init}, M_{final})$ is a labeled Petri net with a well-defined initial marking M_{init} and final marking M_{final}.*

Definition 3 (Pre- and Postset). *Let* $\mathcal{N} = (P, T, F, \lambda)$ *be a labeled Petri net and* $x \in P \cup T$ *be a node of* $\mathcal{N}$. *The set* $^\bullet x = \{y \in P \cup T \mid (y, x) \in F\}$ *is called the preset, and* $x^\bullet = \{y \in P \cup T \mid (x, y) \in F\}$ *is called the postset of* x.

We make the following assumptions: the sets of places are always non-empty (i.e., $P \neq \emptyset$), transitions always have incoming and outgoing edges, and the pre- and postsets of transitions are always finite.

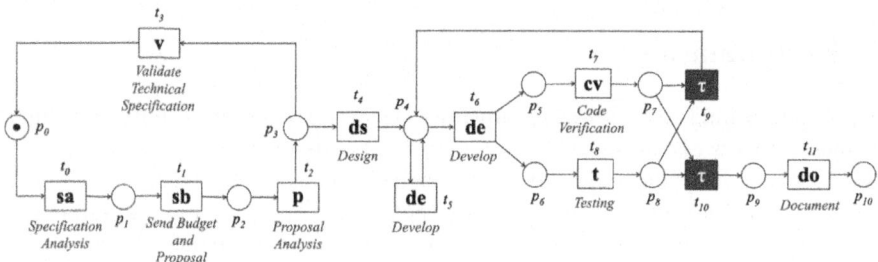

Fig. 1. An example Petri net $\mathcal{N}_{mgmt}$ of a software development management process with the initial marking $M_{init} = [p_0]$ and final marking $M_{final} = [p_{10}]$

To model the dynamic properties of a net, the notion of *marking* exists. It is a multiset over places P of a net, in which the multiplicity of a place denotes the number of *tokens* assigned to it. The net in Fig. 1 has an initial marking $m_0 = [p_0]$. A marking is said to be *reachable* if there exists a sequence of transition *firings* from the initial state of a net, where a transition firing follows the well-known firing semantics of Petri nets [24]. In this paper, we assume all nets to be 1-safe, meaning no place contains more than one token at a time.

The dynamic behavior of Petri nets can be viewed in two ways: the first, called *interleaving semantics*, considers firing sequences as possible execution paths. The second, *true concurrency semantics*, views execution as partially-ordered sequences, representing runs (hereby referred to as *distributed runs*) as partial orders. These partial orders can be modeled with simpler structure nets called *occurrence nets*, obtained by 'unfolding' system nets into a tree-like structure. Occurrence nets represent partially-ordered event sequences from the initial marking, together forming *branching processes*. Stopping the unfolding at different instances yields different processes, but *the* unique *unfolding* is the complete branching process obtained by unfolding as much as possible. The subsequent definitions in this section are largely based upon [13]. Prior to introducing occurrence nets and causal nets, we define the notions of *causality* and *conflict* as follows—1. two nodes x and y are in a *causal relation* with each other if there exists a path between x to y, 2. two nodes are in *conflict* with each other if there exist two different paths to these nodes which start at the same point and diverge immediately afterwards (although later on they can converge again).

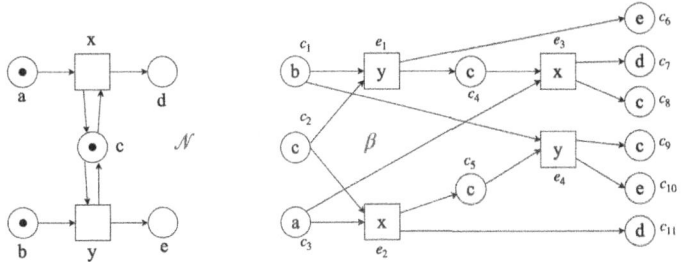

Fig. 2. A system net $\mathcal{N}$ (left) with $M_{init} = [a, b, c]$, $M_{final} = [d, c, e]$, and its branching processes β (right), which is also a complete finite prefix of the unfolding of $\mathcal{N}$. (Color figure online)

Definition 4 (Occurrence Net, Causal Net). *Let $\mathcal{N} = (B, E, F, \lambda)$ be a labeled Petri net. We have an unlabeled net $\mathcal{O} = (B, E, F)$ consisting of the transitions, places and flow relation of $\mathcal{N}$:*

1. *$\mathcal{O}$ is an occurrence net, if the following conditions hold:*
   (a) *$\forall b \in B(|{}^\bullet b| \leq 1)$, no place in $\mathcal{O}$ contains two or more incoming edges;*
   (b) *$\nexists e \in E$ st. it is in conflict with itself;*
   (c) *$F^+ \cap (F^{-1})^+ = \emptyset$, $\mathcal{O}$ is acyclic;*
   (d) *$\mathcal{O}$ is finitely preceded, i.e., for every $x \in B \cup E$, the set of elements $y \in B \cup E$ such that there exists a path from y to x is finite.*
   Places and transitions in an occurrence net are called as conditions B and events E.
2. *$\mathcal{O}$ is a causal net, if (a)-(d) hold, in addition to:*
   (e) *$\forall b \in B(|b^\bullet| \leq 1)$, no place in $\mathcal{O}$ contains two or more outgoing edges*

Those labeled occurrence nets obtained by unfolding system nets are referred to as *branching processes*, formalized as below:

Definition 5 (Branching Process). *A branching process of a system net $SN = ((P, T, F, \lambda), M_{init}, M_{final})$ is a labeled occurrence net $\beta = (\mathcal{O}, h) = (B, E, F, h)$ where the labeling function $h : B \cup E \rightarrow P \cup T$ is a net homomorphism from β to SN such that: $\forall v, v' \in B \cup E.({}^\bullet v = {}^\bullet v' \wedge h(v) = h(v') \implies v = v')$. It is simply a mapping between nets that preserves the nature of nodes and the environment of nodes. It associates the conditions/events of a branching process to the places/transitions of its system net. For a formal definition of a net homomorphism, we refer to [25].*

Figure 2 shows the branching process β of a Petri net $\mathcal{N}$. It consists of two different distributed runs corresponding to firing sequences of x followed by y (highlighted in blue) and y followed by x, respectively. Labels inside each node of β represent the places/transitions it maps to. Notice how no condition (place) in the highlighted run has more than one incoming/outgoing arc, and all events are related in a causal or concurrency relationship. Therefore, it is a causal net. Another thing to notice in the highlighted distributed run of β is that when the run includes conditions c_9 and c_{10}, it has no outgoing arcs that enable any more transitions in the Petri net. Such a distributed run is called *complete*.

The unfolding process of a system net starts with the initial conditions and extends it iteratively with the new possible events and a condition for every output place of the newly added event. This process can typically be infinite. McMillan's [22] (and ERV [13]) algorithm builds a *complete finite prefix* of the unfolding, which terminates when the unfolding represents all of the reachable markings of the Petri net. The foundations of the algorithm are laid by the key concepts of *configurations* and *cuts*. A finite configuration C represents the set of events that have occurred so far in a distributed run. For the highlighted run in β (cf. Fig. 2), $\{e_2, e_4\}$ form a finite configuration while $\{e_1, e_2\}$ does not, since e_1 and e_2 are in conflict with each other. A configuration is related to a marking $Mark(C)$ by identifying which conditions will contain a token after firing all events in configuration C. Thus, $Mark(C)$ corresponds to a final marking in the original system net, reached by firing all transitions labeled by the events in a given finite configuration. When a finite configuration is associated with an event, it is termed as a *local configuration*.

Definition 6 (Local Configuration). *For an occurrence net $\mathcal{O} = (B, E, F)$, a local configuration $[e]$ of an event $e \in E$ is defined by $\{e' \in E \mid (e', e) \in F^*\}$*

It represents a minimal set of all the events denoted by $[e]$ that must have been fired for the event e to be enabled. For example, in β, $[e_4] = \{e_2, e_4\}$. Each local configuration is also associated with the distributed run it represents. It consists of precisely the events in $[e]$, in addition to the conditions and arcs connecting them. We term it as a *local distributed run*.

Definition 7 (Local Distributed Run). *Let $\mathcal{O} = (B, E, F)$ be an occurrence net. A local distributed run of an event $e \in E$ is a causal net $\mathcal{O}_e = (B_e, E_e, F_e)$ where:*

- *$B_e = \{c \in B \mid (c, e) \in F^*\}$ is the set of conditions,*
- *$E_e = [e]$ is the set of events, and*
- *F_e is the flow relation such that $F_e(x, y) = F(x, y), \forall x, y \in B_e \cup C_e$.*

The key to building a complete finite prefix (due to McMillan) is to identify those events where the unfolding process can be stopped without any loss of information. Those events are called *cut-off events* and can be defined in terms of *adequate order* on configurations [8,13,22]. Adequate orders are simply strict partial orders to choose between the extensions of two events when the $Mark$ of their respective local configurations is the same, and one of them precedes the other one with respect to the partial order. The one not chosen to be extended will thus be the *cut-off event*. For a strict partial order to qualify as an adequate order, it must satisfy certain properties (cf. [13] for details). Thus, the unfolding can be ceased as soon as an event e takes a prefix to a marking which has

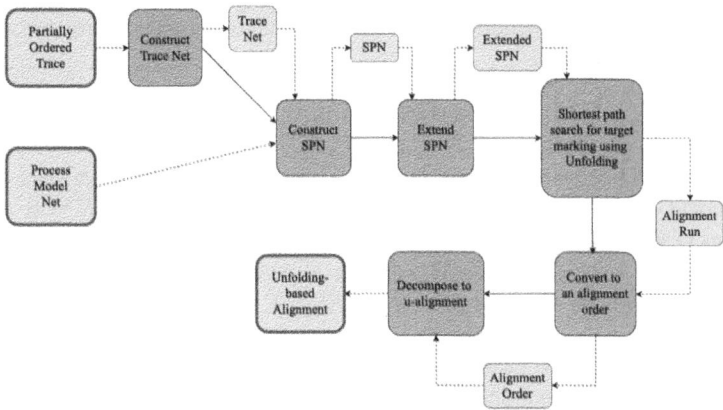

Fig. 3. Overview of the proposed method to compute unfolding-based alignments given a partially-ordered trace and a process model. Input and output of each step are shown through dashed arrows. Various steps are highlighted in blue. (Color figure online)

already been induced by a previously added event e' such that $[e'] \lhd [e]$. The event e is marked as a cutoff event. Assume an adequate order $\lhd$ where $e_1 \lhd e_2$ iff $|[e_1]| < |[e_2]|$ and a current state of marking as $\{d, c, e\}$ in $\mathcal{N}$ of Fig. 2. If there were a transition z from place d to a in $\mathcal{N}$, the event mapped to z in β would have been marked as a cut-off event. This is because it would bring the marking of $\mathcal{N}$ to $\{a, c, e\}$, already represented in β by the local distributed run of e_1, for which $|[e_1]|$ is lesser already.

3   Computation of Unfolding-Based Alignments

We describe our proposed approach as illustrated in Fig. 3. Given a partially ordered trace (p-trace) and a model net, we first model the p-trace as a *trace net* [21], a Petri net capturing the p-trace's behavior. Each event becomes a transition, and places represent dependencies between events. Next, we construct the SPN with the model net. The product is constructed by pairing transitions in one net with transitions in the other net that have the same label [4]. This helps to model all possible alignment moves explicitly. A formal definition of SPN is provided online (Def. 1 in Appendix, [28]). Then we extend it by adding a target transition and place, and revise the final marking to the target place. Using unfolding algorithms $ERV[\lhd_c]$ or $ERV[\lhd_h]$ (cf. Sect. 3.3 and 3.2), we unfold the extended SPN to find a minimum-cost distributed run leading to the target transition, yielding an optimal alignment run. Finally, we use this run to build an alignment order and decompose it into *unfolding-based alignments (u-alignments)*. The goal is to relate p-trace with the closest run through the model net (one with minimum deviations) in a meaningful and interpretable way.

We begin by defining u-alignments, in the context of an alignment function that relates synchronous moves when the alignment is decomposed into its log

and model parts. This refinement allows for a clearer relationship between log
and model visualizations.

Definition 8 (Unfolding-based Alignment (u-alignment)). *Let $\rho_c = ((E,$
$\prec_{\rho_c}), l_{\rho_c})$ denote a p-trace and $\mathcal{N} = ((P, T, F, \lambda), M_{init}, M_{final})$ be a system net
with induced partial order $G_{\mathcal{N}} = ((T', \prec_{\mathcal{N}}), l_{\mathcal{N}})$ over a run of $\mathcal{N}$ (we refer to
it as a model run), where T' is an arbitrary set such that $T' \cap E = \emptyset$ and $l_{\mathcal{N}} :$
$T' \rightarrow rng(\lambda)$ is a labeling function. An unfolding-based alignment (u-alignment)
is a tuple $(\rho_c, G_{\mathcal{N}}, \varphi)$ where $\varphi : E' \rightarrow T'$ is an injective function, we refer to as
the alignment function defined over a set $E' \subseteq E$ and $T' \subseteq T$ such that:*

1. *$\forall e \in E', t \in T'$ st. $(l_{\rho_c}(e) = l_{\mathcal{N}}(t)) \Leftrightarrow \varphi(e) = t$ (nodes with same activity
   labels are mapped together; they are synchronous), and*
2. *A directed graph consisting of nodes and edges from the p-trace ρ_c, model
   run $G_{\mathcal{N}}$ such that there is a single node for each pair of synchronous nodes is
   a strict partial order. $\rho_c \uplus G_{\mathcal{N}}$ denotes a partial order that combines ρ_c and
   $G_{\mathcal{N}}$ in such a way. We refer to it as an alignment order.*

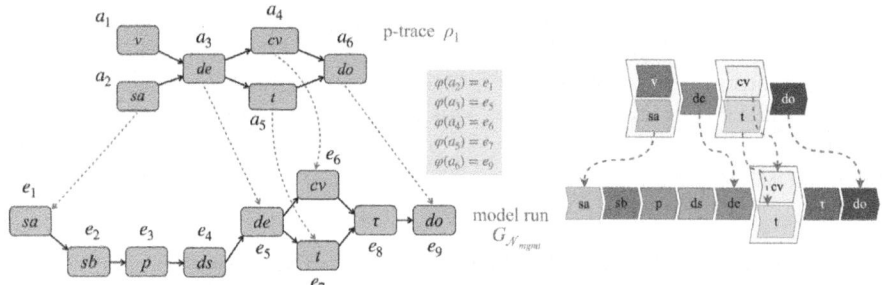

(a) A u-alignment comprising of an arbitrary p-trace ρ_1 and a partial order $G_{\mathcal{N}_{mgmt}}$
induced by an example model run through $\mathcal{N}_{mgmt}$ Fig. 1. Mapping by the alignment
function φ is shown in dashed red arrows. A chevron-based view of the alignment is
visualized on the right

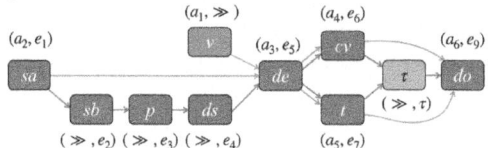

(b) An alignment order $\rho_1 \uplus G_{\mathcal{N}_{mgmt}}$. The partial order contains a single node (a_i, e_j)
for each pair of synchronous nodes, colored in green. All other nodes and dependencies
are colored according to their origin — orange if it originates from ρ_1 (a node $(a_i, \gg)$),
blue if from $G_{\mathcal{N}_{mgmt}}$ (a node $(\gg, e_j)$). A grey node $(\gg, \tau)$ represents an invisible move
originating from model

Fig. 4. An example of a u-alignment along with its corresponding alignment order.
(Color figure online)

Figure 4a shows an illustration of how in a u-alignment, an alignment function maps a p-trace ρ_1 to a partial order induced by an example model run $G_{\mathcal{N}_{mgmt}}$ through $\mathcal{N}_{mgmt}$. The alignment function φ is defined over nodes of p-trace $\{a_2, a_3, a_4, a_5, a_6\}$ such that $\varphi(a_2) = e_1$, $\varphi(a_3) = e_5$, $\varphi(a_4) = e_6$, $\varphi(a_5) = e_7$ and $\varphi(a_6) = e_9$. Note that φ maps only nodes with the same labels; the label for a_2 and e_1 is 'sa', and so on. Figure 4b shows an alignment order for ρ_1 and $G_{\mathcal{N}_{mgmt}}$, a partially ordered structure of *alignment moves*. Nodes in this graph are colored according to where they originate from. An orange node represents a *log move* from the p-trace, while a blue node represents a *model move* from the model run. A green node indicates a *synchronous move*, originating from both the p-trace and the model run. A grey node denotes an invisible move from the model run. Edges represent dependencies, categorized as *model dependencies* (blue) or *log dependencies* (orange), based on their origin. Given such an alignment, the partial orders being directed acyclic graphs, can be visualized in a chevron-based view based upon recursive sequential and parallel partitioning of the graphs, a visualization technique from [26]. The result of the same is illustrated on the right of Fig. 4a. Notice how this view retains the relationship between the log and model parts through the alignment function.

By understanding how to combine a trace with a model run to create an alignment order, we can similarly decompose an alignment order into two partial orders while defining the alignment function simultaneously. This yields a u-alignment. Given an alignment order $G_\varphi = ((M, \prec_\varphi), l_\varphi)$, we denote the two decomposed partial orders as $G_{\varphi \downarrow 1}$ and $G_{\varphi \downarrow 2}$, respectively, i.e., $G_\varphi \rightarrow G_{\varphi \downarrow 1} \uplus G_{\varphi \downarrow 2}$. An alignment function φ is defined for every synchronous move. For our example in Fig. 4, decomposing the alignment order from Fig. 4b results in two partial orders isomorphic to ρ_1 and $G_{\mathcal{N}_{mgmt}}$ in Fig. 4a.

In order to compute u-alignments from an SPN, it is important to realize that each transition of an SPN of a trace net and a model net represents an alignment move [4]. Thus, distributed runs of the product net define a partially-ordered structure of alignment moves, an alignment order. Consequently, the problem of computing alignment(s) for a given p-trace and a model net can be reformulated as the problem of finding complete distributed run(s) in its SPN. Since we seek an *optimal* run corresponding to the fewest log and model moves, we use a *default cost function* [3], which assigns higher costs to log and model moves than to synchronous and invisible moves, reducing the problem to finding minimum-cost complete distributed runs in the product net.

3.1   Finding Optimal Distributed Runs

We already know that a complete finite prefix of a system net encodes all its reachable markings. We exploit this to search for a target marking corresponding to the final marking of an SPN in order to find complete distributed run(s). To enable search for a target marking in an SPN, we apply the standard encoding trick of adding a *target transition* t^* to the product net, which consumes the final marking. To preserve 1-safeness, we also add a target place p^* as a successor to t^*.

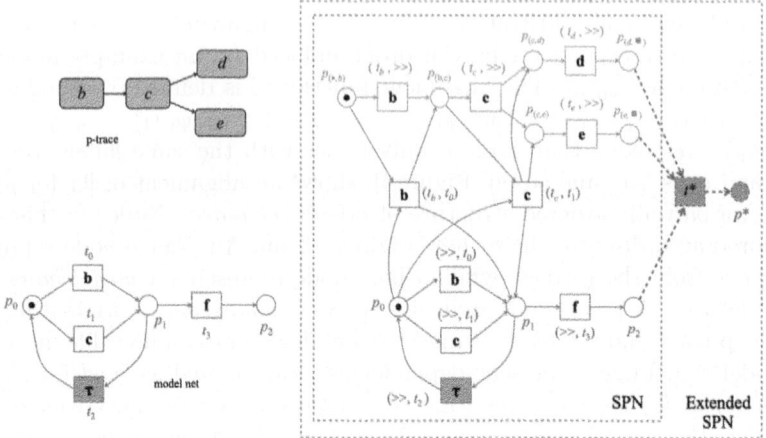

Fig. 5. An extended synchronous product net $SN_\otimes^{ext}$ of an SPN of a trace net and a model net. The target transition is t^*, and a target place p^* is a successor to t^*. For this system net, $M_{init,ext} = [p_{(\blacktriangleright, b)}, p_0]$ and $M_{final,ext} = [p^*]$

Definition 9 (Extended Synchronous Product Net). *Let $SN_\otimes = ((P_\otimes, T_\otimes, F_\otimes, \lambda_\otimes), M_{init,\otimes}, M_{final,\otimes})$ denote a synchronous product net of a trace net and an easy sound[1] model net. An extended sychronous product net of $SN_\otimes$ is another system net $SN_\otimes^{ext} = ((P_\otimes^{ext}, T_\otimes^{ext}, F_\otimes^{ext}, \lambda_\otimes^{ext}), M_{init,\otimes}^{ext}, M_{final,\otimes}^{ext})$ obtained by extending $SN_\otimes$ with a transition t^* and a place p^* such that $^\bullet(t^*) = M_{final,\otimes}$, $^\bullet(p^*) = t^*$ and $(p^*)^\bullet = \emptyset$. For such a system net, $M_{init,\otimes}^{ext} = M_{init,\otimes}$ and $M_{final,\otimes}^{ext} = [p^*]$.*

Figure 5 shows an extended SPN $SN_\otimes^{ext}$ of an SPN $SN_\otimes$ after adding the target transition and place depicted in dotted red. We shall be using this product net of an exemplary trace net (induced by a p-trace) and an easy sound model net shown in Fig. 5) in the remainder of this paper.

Assume a branching process of this extended product net. We define a *target event* e^* in its branching process that maps to the target transition, i.e., for which $h(e^*) = t^*$. The encoding trick ensures that e^* is enabled only when the target marking corresponding to the final marking is reached. Therefore, its local distributed run represents a *complete* distributed run leading to e^*. We refer to such a complete distributed run as an *alignment run*.

Definition 10 (Alignment Run). *Let $SN_\otimes^{ext}$ denote an extended synchronous product net obtained by extending $SN_\otimes$ with a transition t^* and a place p^*. An alignment run $\gamma = (B, E, F, h)$ is a complete distributed run of $SN_\otimes^{ext}$ leading to a target event e^* such that the set $\{b \in B \mid b^\bullet = \emptyset\}$ contains a single condition b^* where $^\bullet b^* = e^*$ and $h(e^*) = t^*$. We denote the universe of all such alignment runs by $\mathcal{U}_{SN_\otimes^{ext}}$.*

---

[1] A net is *easy sound* if its complete finite prefix $\mathcal{P}$ contains at least one complete distributed run.

To ensure optimality, we use a *scoring function over finite configurations*, a monotonic function assigning additive costs. This function allows comparing local configurations of target events, where a configuration with lower cost is preferred, effectively comparing distributed runs based upon costs of their member events. The function is used to define an *optimal alignment run* in the following definition.

Definition 11 (Scoring Function Over Finite Configurations). *Let $\mathcal{C}_{all}$ denote the set of all finite configurations of a branching process $\beta = (B, E, F, h)$ of an extended SPN and cost be a cost function on transitions of the SPN such that it assigns a constant positive cost to log and model transitions, whereas synchronous transitions have a cost of 0, and invisible transitions have a negligible positive cost. The scoring function over finite configurations is defined as $s : \mathcal{C}_{all} \to \mathbb{N}_0$ such that $s(C) = \sum_{e \in C} cost(h(e)), C \in \mathcal{C}_{all}$. A scoring function is monotonic since $\forall C, C' \in \mathcal{C}_{all}, C \subseteq C' \implies s(C) \leq s(C')$.*

Definition 12 (Optimal Alignment Run). *Let $SN_{\otimes}^{ext}$ denote an extended SPN and let β denote its branching process. Let $s : \mathcal{C}_{all} \to \mathbb{N}_0$ be a scoring function over the finite configurations of β. An optimal alignment run $\gamma^{opt} \in \mathcal{U}_{SN_{\otimes}^{ext}}$ is an alignment run of $SN_{\otimes}^{ext}$ leading to a target event $(e^*)^{opt}$ such that for all the other alignment runs $\gamma \in \mathcal{U}_{SN_{\otimes}^{ext}}$ leading to target events $e_1^*, ..., e_n^*$ respectively, it holds that $s([(e^*)^{opt}]) \leq s([e_i^*]), 1 \leq i \leq n$.*

To obtain an optimal alignment run, we unfold $SN_{\otimes}^{ext}$ (using the *ERV* unfolding algorithm described in Sect. 3.2) until a target event is pulled out from the queue of events. At this point, the unfolding process can be terminated and a local distributed run associated with e^* is retrieved. This local distributed run is an alignment run as defined in Definition 10. Its alignment cost is given by $s([e^*])$.

Note that in the unfolding algorithm, an adequate order is used to order events in a queue. Using different adequate orders leads to different results, i.e., different complete distributed runs. Here we choose a *cost-based adequate order* that compares finite configurations based upon their costs.

Definition 13 (Cost-based Partial Order $\vartriangleleft_c$). *Let C, C' be two finite configurations of a branching process β of an extended SPN and s be the scoring function. Let $\gg$ be a total order on the events of β such that for any two events e and e', $e \gg e'$ if e was added earlier to the prefix than e'. Given a set E of events, $\phi(E)$ denotes the sequence of events that is ordered according to $\gg$. It is easy to see that $\gg$ is a well-founded relation. A cost-based partial order $\vartriangleleft_c$ is defined such that $C \vartriangleleft_c C'$ if and only if:*

- $s(C) < s(C')$, *or*
- $s(C) = s(C')$ *and* $|C| < |C'|$, *or*
- $s(C) = s(C')$ *and* $|C| = |C'|$ *and* $\phi(C) \gg \phi(C')^2$.

---

[2] We say $\phi(E_1) \gg \phi(E_2)$ if $\phi(E_1)$ is lexicographically smaller than $\phi(E_2)$ with respect to the total order $\gg$.

That $\triangleleft_c$ is irreflexive, asymmetric and transitive is obvious from the definition. Hence, $\triangleleft_c$ is also a strict partial order. In addition to ordering the priority queue, the chosen adequate order is used to identify cut-off events. For correctness proofs to work, i.e., for the ERV algorithm to correctly identify cut-off events, a cost-based order $\triangleleft_c$ defined as above must be *adequate* [13]. It can be proven that $\triangleleft_c$ satisfies the three properties of adequate orders (cf. Theorem 1 in Appendix, [28]) and hence can be used to correctly unfold extended SPNs with respect to the target marking.

Next, we present the $ERV[\triangleleft_c]$ (along with $ERV[\triangleleft_h]$ as discussed later in Sect. 3.3) *unfolding algorithm*, an on-the-fly variant of ERV algorithm [13] to produce optimal alignment runs. It is implemented as the *UnfoldSyncNet* procedure in Algorithm 1. It differs from the standard ERV unfolding algorithm in that it terminates when a target event is pulled out from the queue (based upon the chosen *stopAtFirst* flag), returning an optimal alignment run as an output. Additionally, it uses $\triangleleft_c$ as its adequate order.

3.2 $ERV[\triangleleft_c]$ Unfolding Algorithm

Given an extended SPN $SN_{\otimes}^{ext} = ((P_{\otimes}^{ext}, T_{\otimes}^{ext}, F_{\otimes}^{ext}, \lambda_{\otimes}^{ext}), M_{init,\otimes}^{ext}, M_{final,\otimes}^{ext})$ and a boolean flag for *stopAtFirst* to indicate if only one optimal alignment run is to be computed or all, we can obtain (all) optimal alignment run(s) (and thus, all optimal alignment orders) from the initial marking $M_{init,\otimes}^{ext}$ to the final marking $M_{final,\otimes}^{ext}$ using Algorithm 1.

Similar to the classic ERV unfolding algorithm, $ERV[\triangleleft_c]$ builds a prefix of events and conditions using a priority queue ordered by an adequate order $(\triangleleft_c)$. Unlike the classic version, it adds steps (lines 8–13) to handle target events enabling the target marking. When such an event is retrieved, its local distributed run is stored in *alignmentRuns* (line 9). If only one alignment run is needed (line 11), the loop ends early, returning the result. Otherwise, the algorithm checks if the event's local configuration contains a cut-off event (line 14) using a lookup table *imarks* for easy O(1) retrieval. If no cut-off exists, the prefix is extended with the event's postset conditions (Lines 15–16). Additionally, if e itself is a cutoff-event, it is added to the set of cutoff events (Lines 17–18). Any new possible extensions are calculated (line 20), added to the prefix, and *imarks* is updated (line 21). The process repeats until the queue is empty.

The calculation of all possible extensions (lines 5 and 20) is the one crucial to implementation. It deals with the problem of finding events that are enabled by the entire set of input conditions while ensuring that they are not in causal or conflict relation with each other. Those identified events are thus added to the prefix and priority queue. Efficient conflict/causal detection while downsizing the number of combinations to check for is key to managing the combinatorial explosion inherent in this part of the algorithm. Roemer in his work [25] already proposed ways to optimize this part using intelligent data structures and procedures. In our work, we implement the same techniques.

Algorithm 1: UnfoldSyncNet - $ERV[\lhd_c]$ (and $ERV[\lhd_h]$, cf. Sect. 3.3)

Input: $SN_\otimes^{ext} = ((P_\otimes^{ext}, T_\otimes^{ext}, F_\otimes^{ext}, \lambda_\otimes^{ext}), M_{init,\otimes}^{ext}, M_{final,\otimes}^{ext}), stopAtFirst \in \mathbb{B}$
Output: $alignmentRuns \subseteq \mathcal{U}_{SN_\otimes^{ext}}, lowestCost \in \mathbb{N}_0$

1   *cut-off, pe, finalEvents, $\beta \leftarrow \emptyset$*   // Initialize the set of cutoff events,
    priority queue ordered by $\lhd_c$ (or $\lhd_h$ for $ERV[\lhd_h]$), the set of target
    events with minimal $s([e])$, and the prefix $\beta = (B, E, F, h)$
2   *imarks $\leftarrow \{M_{init,ext} : \bot\}$*   // Lookup table of induced markings initialized
    with the initial marking induced by a dummy event $\bot$
3   **for** $p \in M_{init,ext}$ **do**   // Add initial conditions to prefix
4 | $B \leftarrow B \cup \{$a new condition mapped to place $p\}$
5   calculate all possible extensions of B
6   **while** $pe \neq \emptyset$ **do**
7 | $e \leftarrow Min\{pe\}$ // Retrieve an event e s.t. $[e]$ is minimal w.r.t. $\lhd_c$
    | (or $\lhd_h$ for the $ERV[\lhd_h]$ variant)
8 | **if** $h(e) = t^*$ **then**   // A target event found
9 | | $alignmentRuns \leftarrow alignmentRuns \cup \{\mathcal{O}_e\}$ // A local distributed run
    | | $\mathcal{O}_e$ of e (cf. Definition 7) is added
10 | | $lowestCost \leftarrow s([e])$
11 | | **if** $stopAtFirst = true$ **then**
12 | | | **break**
13 | | **else continue**
14 | **if** $[e] \cap cut\text{-}off = \emptyset$ **then**
15 | | **for** $p \in h(e)^\bullet$ **do**   // Add e's postset conditions one by one
16 | | | $c \leftarrow$ a new condition mapped to p extend from event e to condition c in the
    | | | prefix β
17 | | **if** e is a cut-off event w.r.t. $\lhd_c$ (or $\lhd_h$ for $ERV[\lhd_h]$) **then**   // check if
    | | $Mark([e])$ is already present in $imarks$
18 | | | $cut\text{-}off \leftarrow cut\text{-}off \cup \{e\}$
19 | | **else**
20 | | | calculate all possible extensions of B
21 | | | add $Mark([e])$ to $imarks$
22 **return** $alignmentRuns, lowestCost$

3.3   Directing the Unfolding

In the context of $ERV[\lhd_c]$ algorithm, we retrieve events from the queue based upon the total cost of mapped transitions in its local configuration, i.e., $s([e])$; equivalent to selecting events based upon the transitions fired so far in its local distributed run. For efficiency, we also consider an *underestimate* of the cost of transitions remaining till the target event is reached. Thus, when selecting events from the queue, we additionally favor those "closer" to t^*, resulting in a *directed unfolding* strategy. To this end, we use a *heuristic function* on configurations to guide the unfolding process towards a target configuration. Its value is an estimate of the distance (where the notion of distance is related to the total cost of transitions fired) between configurations. Thus, in the context of $ERV[\lhd_c]$ algorithm, orderings can be constructed upon the values of a function $f : \mathcal{C}_{all} \rightarrow \mathbb{N}_0$ composed of two parts $f(C) = s([e]) + est(C)$, where $s([e])$ is the total cost of mapped transitions of $[e]$ and $est(C)$, the heuristic value, estimates the distance from $Mark(C)$ to the target marking $[p^*]$. We use a heuristic function that exploits the marking equation [7] to underestimate the remaining cost to $[p^*]$.

Definition 14 (Heuristics-based Partial Order $\lhd_h$). *Let C, C' be two finite configurations of a branching process β of an SPN. Let $f(C) = s([e]) + est(C)$ be a function on finite configurations where s is a scoring function and est is a heuristic function. Let $\lhd_c$ be a cost-based adequate order over finite configurations. A heuristics-based partial order $\lhd_h$ is defined such that:*

$$C \lhd_h C' \text{ iff } \begin{cases} f(C) < f(C'), & \text{if } f(C) \neq f(C') \\ C \lhd_c C', & \text{otherwise.} \end{cases}$$

Given that the first part of f (in our case, s) is a monotonic function and est is an admissible heuristic function, Bonet et al. [6] proved that a partial order defined as above belongs to a family of *semi-adequate* orders, which still enables us to use the $ERV[\lhd_c]$ algorithm with the same definition of cut-off events. Here, we simply replace $\lhd_c$ in $ERV[\lhd_c]$ (Algorithm 1) by $\lhd_h$ to obtain another algorithm variant, referred to as $ERV[\lhd_h]$ *algorithm*. The only changes in $ERV[\lhd_h]$ occur in Lines 7 and 17 where an adequate (semi-adequate, in this case) order is used.

3.4 Building and Visualizing Optimal Alignments from Optimal Alignment Runs

Building a u-alignment from an optimal alignment run, referred to as the *optimal u-alignment*, is a two-step process:

1. Construction of an optimal alignment order by using the events and conditions of the net corresponding to an optimal alignment run.
2. Decomposition of the optimal alignment order to obtain two partial orders, one corresponding to the p-trace and the other to a model run, while setting our alignment function at the same time. This gives us an optimal u-alignment, our final artefact.

In the first step, to build an optimal alignment order G_φ^{opt} from an optimal alignment run γ^{opt}, remove the target event e^*, map each event to a labeled node representing an alignment move, and create dependencies based on conditions, classifying them as log or model dependencies. In the second step, we decompose an alignment order into two partial orders while setting the alignment function. Figure 6 shows the mechanism through an illustration. On top we see an optimal alignment run through the example extended SPN from Fig. 5 using $ERV[\lhd_c]$ or $ERV[\lhd_h]$. Below it, is the transformed optimal alignment order G_φ^{opt}. G_φ^{opt} is decomposed into different partial orders $G_{\varphi\downarrow1}^{opt}$ and $G_{\varphi\downarrow2}^{opt}$, i.e., $G_\varphi^{opt} \rightarrow G_{\varphi\downarrow1}^{opt} \uplus G_{\varphi\downarrow2}^{opt}$. The alignment function φ relates synchronous nodes 'b' and 'c' in the two partial orders. Moreover, notice that $G_{\varphi\downarrow1}^{opt}$ is isomorphic to the p-trace we computed our alignment for (observe the p-trace in Fig. 5). Thus, our final u-alignment is $(G_{\varphi\downarrow1}^{opt}, G_{\varphi\downarrow2}^{opt}, \varphi)$.

The decomposed partial orders $G_{\varphi\downarrow1}^{opt}$ and $G_{\varphi\downarrow2}^{opt}$ are visualized into chevron-based views as exemplified in Fig. 4a. This visualization method effectively highlights alignment between log and model events but falls short in identifying

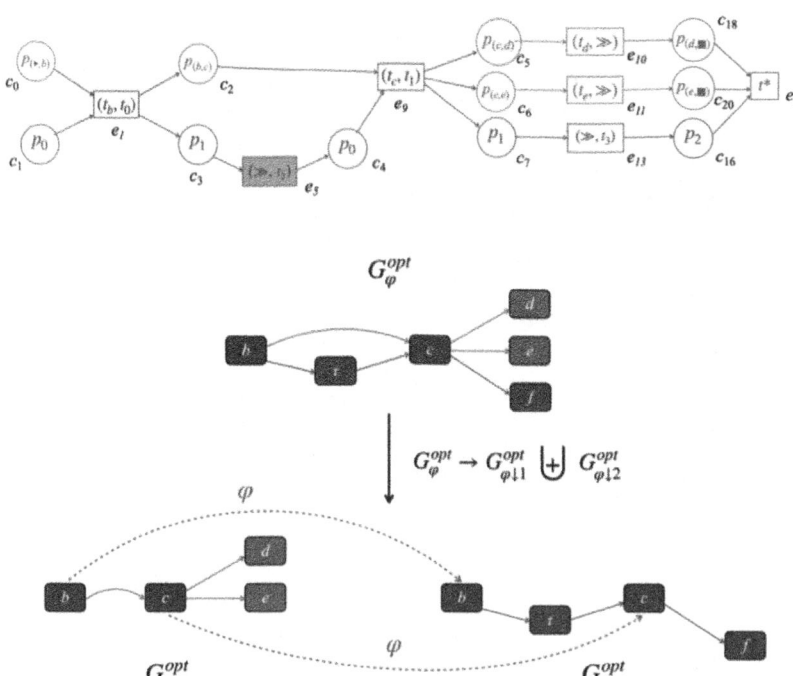

Fig. 6. Optimal alignment order (middle), derived from the alignment run (top) of the extended SPN $SN_{\otimes}^{ext}$, with partial order decomposition showing the u-alignment (bottom). $G_{\varphi\downarrow 1}^{opt}$ is isomorphic to the p-trace and $G_{\varphi\downarrow 2}^{opt}$ is the partial order from the model net run (cf. Fig. 5). Dotted red arrows show the mapping of the alignment function. (Color figure online)

misalignments, particularly regarding dependencies. When comparing a model with a log variant, we aim to identify events present in reality but missing in the model (and vice versa) and to detect violations in event orders. These observations allow us to classify misalignments into two main categories. In the following, we assume a u-alignment $(G_{\varphi\downarrow 1}^{opt}, G_{\varphi\downarrow 2}^{opt}, \varphi)$ where $\prec_{\varphi\downarrow 1}^{opt-}$, $\prec_{\varphi\downarrow 2}^{opt-}$ are the edge relations of $G_{\varphi\downarrow 1}^{opt-}$ and $G_{\varphi\downarrow 2}^{opt-}$, respectively, where G^- denotes the transitive reduction of a DAG G.

1. **Missing events and dependencies**: Events and dependencies in the log but not in the model. Missing dependencies are relations $(m_1, m_2) \in \prec_{\varphi\downarrow 1}^{opt-}$ with $(m_1, m_2) \notin \prec_{\varphi\downarrow 2}^{opt-}$, representing log dependencies without corresponding model dependencies. Missing events are log moves of the form $(m_1, \gg)$.
2. **Undesired events and dependencies**: Events and dependencies in the log but not in the model. Missing dependencies are relations $(m_1, m_2) \in \prec_{\varphi\downarrow 2}^{opt-}$ with $(m_1, m_2) \notin \prec_{\varphi\downarrow 1}^{opt-}$, representing log dependencies without corresponding model dependencies. Missing events are log moves of the form $(\gg, m_2)$.

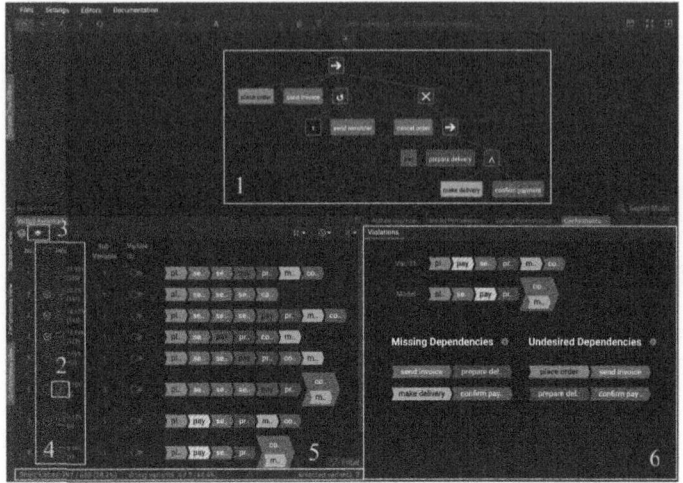

Fig. 7. Overview of the Conformance Analysis interface in Cortado

In $G^{opt}_{\varphi\downarrow1}$ and $G^{opt}_{\varphi\downarrow2}$ from Fig. 6, there are in total three missing dependencies corresponding to all dependencies visible in $G^{opt}_{\varphi\downarrow1}$, three undesired dependencies corresponding to all dependencies in $G^{opt}_{\varphi\downarrow2}$. Moreover, there are two missing events corresponding to 'd' and 'e' (orange nodes) and two undesired events corresponding to 'τ' and 'f' (blue nodes).

4   Tool Support

The $ERV[\lhd_c]$ and $(ERV[\lhd_h])$ algorithms have been implemented[3] in *Cortado* [27], a tool for interactive process discovery. Cortado traditionally checks conformance by computing all possible sequentializations of each partially ordered trace variant and aligning each sequence using the classic sequential alignment method [2]. The final alignments for each variant are aggregates of the sub-alignments. Our implementation replaces this sequentialization-based approach with unfolding-based alignments, adding new features.

We present a visual overview of the modified Conformance Analysis interface in Fig. 7. Once an initial model based on selected variants is available, users can compute unfolding-based alignments for new variants. The model, shown in the process tree editor (1), allows alignment computation by clicking one of the two buttons to (re)calculate conformance statistics (2), (3). After computation, one of three statuses appears (4): 1. variant fits the model, 2. variant does not fit, or 3. variant fits but with order deviations (our contribution). Conformance statistics is shown at the bottom (5). For a more detailed analysis, the Violations tab (6) in 'Conformance View' displays chevron-based concurrency visualizations

---

[3] https://github.com/ariba-work/cortado.

of each selected variant and the lowest cost model run corresponding to that variant. Missing events are highlighted with stripes, and synchronous events can be paired by hovering over them. Missing and undesired dependencies are listed as pairs of chevrons, showing where dependencies should exist but don't in the model (e.g., 'send invoice' followed by 'prepare delivery') or vice versa.

5 Evaluation

We evaluate $ERV[\lhd_c]$ and $ERV[\lhd_h]$ against the classic approach to partial alignments based on A^* algorithm. Since we are particularly interested in the performance of our algorithm in settings of high parallelism, we split our experiments into two: 1). where we generate data artificially in order to control the degree of parallelism and compare different algorithms in the respective settings (Sect. 5.1), and 2). where we evaluate using real-life event logs (Sect. 5.2). In each part, we present our experimental setup and discuss the results. Both $ERV[\lhd_c]$ and $ERV[\lhd_h]$ are implemented in a forked repository[4] of the open-source software tool for process mining, *Cortado* [27]. Furthermore, to compare our results with the classic approach to partial alignments (referred to as *Classic PA* from hereon) [21], we implement their algorithm in a separate Github repository[5].

5.1 Using Artificial Event Logs

To examine the impact of parallelism on computation time, we conducted an experiment to test our hypothesis: as concurrency in models increases, the runtime overhead for exploring interleavings in *Classic PA*'s reachability graph grows larger than the overhead for calculating extensions in $ERV[\lhd_c]$ and $ERV[\lhd_h]$. Consequently, our proposed algorithms are expected to outperform *Classic PA* at higher parallelism levels.

Experimental Setup. The experimental workflow involves generating 8 process trees with varying parallelism levels (0 % to 70%) using *PTAndLogGenerator* [19] such that all trees depict block-structured Petri nets without loops and/or duplicate labels. We then convert the trees into Petri nets, and reduce silent transitions, using different ProM[6] plugins. Event logs with 500 traces are simulated for each net using the plugin *StochasticPetriNets*. Noise in levels of 0 % to 50 % is inserted using an existing noise-insertion technique proposed in [1]. Finally, conformance checking is performed using $ERV[\lhd_c]$, $ERV[\lhd_h]$, and A^*-based *Classic PA* on each of the simulated event log.

---

[4] https://github.com/ariba-work/cortado.

[5] https://github.com/ariba-work/classic-pa.

[6] https://promtools.org/.

Results. In Fig. 8, we compare $ERV[\triangleleft_c]$, $ERV[\triangleleft_h]$, and *Classic PA* for varying noise and parallelism levels, focusing on average computation times. Results are shown for parallelism levels of 30%, 50%, and 70%, where significant differences were observed. The x-axis represents noise percentage in event logs, and the y-axis shows computation times in milliseconds, with linear regression lines illustrating trends. At 70% parallelism, $ERV[\triangleleft_h]$ outperforms other variants, showing robustness. On the other hand, $ERV[\triangleleft_h]$ starts to slow down even at 50%, due to larger prefixes generated, caused by increased events in high-parallelism models and the lack of heuristics to limit prefix expansion. At 30% parallelism, all variants perform almost similarly, with ERV variants excelling under higher noise levels. These findings support the hypothesis that unfolding-based alignments are more robust as parallelism increases.

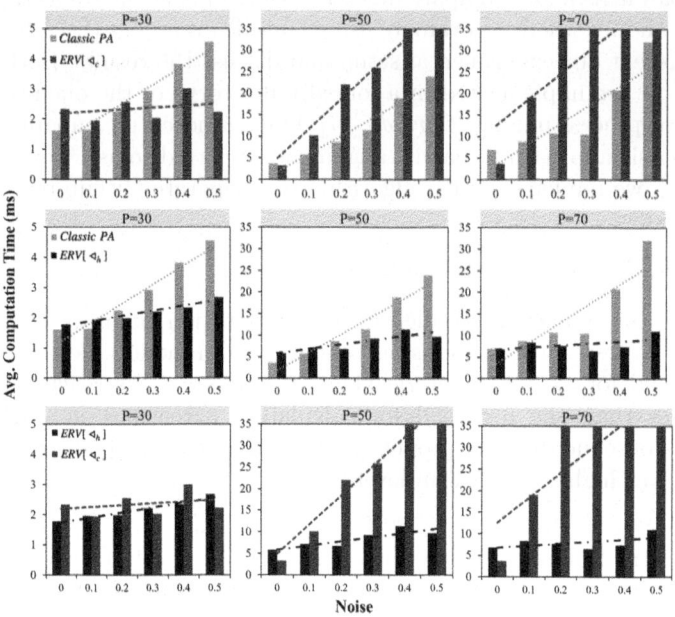

Fig. 8. Average computation time (ms) per percentage of noise, plotted per level of parallelism (30%, 50% and 70%). The level of parallelism is shown on top of the plots. In each row, the comparison is shown for two algorithms at a time.

5.2   Using Real Life Event Logs

Experimental Setup. For this experiment, we selected the BPI Challenge 2012 (BPIC 12) [9] dataset, which contains partial event data suitable for validating our approach. The BPIC 12 dataset contains 262,200 events across 13,087 cases and 4,366 variants. It includes both start- and end-timestamps, with case lengths ranging from 3 to 175 events.

Table 1. Statistics of the models discovered by Infrequent Inductive Miner. Column names are abbreviated (#Pl: number of places, #Tr: number of transitions, #Si: number of silent transitions, ECyM: Extended Cyclomatic Metric)

BPIC 12 Model	#Pl	#Tr	#Si	ECyM
noise thr. 0.05	37	50	33	1899
noise thr. 0.5	26	27	11	289

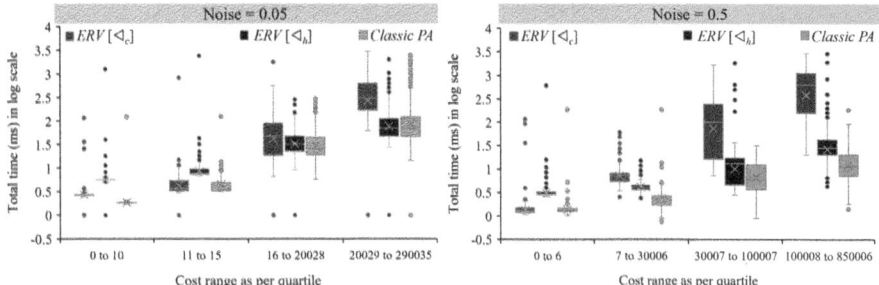

Fig. 9. Runtime comparison regarding deviation costs for different algorithms applied on models with different noise thresholds for BPIC 12 event log. Timeout for individual alignments for single traces is set to 3 s. $ERV[\triangleleft_c]$ generally performs the worst, while $ERV[\triangleleft_h]$ is more robust at the lowest noise threshold.

For conformance checking experiments, process models were discovered using the *Infrequent Inductive Miner* algorithm in ProM with noise thresholds of 5% and 50% where higher thresholds simplify models by filtering out rare traces. Table 1 shows the number of places, transitions, silent transitions and additionally the *Extended Cyclomatic Metric* (ECyM) [20] of the mined models. This metric is calculated using the reachability graph of Petri nets where higher values indicate more reachability paths.

Results. In Fig. 9, we compare the runtime of $ERV[\triangleleft_c]$, $ERV[\triangleleft_h]$, and *Classic PA* across cost ranges and noise thresholds. Box plots show that runtime increases approximately exponentially with deviation costs, with lower noise threshold models taking longer. In both plots, it is apparent that *Classic PA* performs better for lower deviation costs. At 0.5 noise threshold, *Classic PA* still outperforms both *ERV* variants at higher deviation costs. This can be attributed to the higher number of non-synchronous transitions in the SPN of a higher noise threshold model, which necessitates exploring more equal-cost events. This results in broader prefixes and increased computation times. At 0.05 noise, *Classic PA* performs better for low deviation costs (0–10), while $ERV[\triangleleft_h]$ is slower but excels for higher costs. The difference between $ERV[\triangleleft_h]$ and *Classic PA* is not *so* apparent in this plot since runtime in the plots discussed so far does not take into account the traces for which alignment compu-

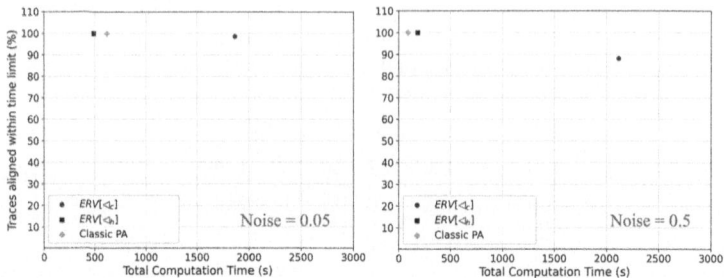

Fig. 10. Percentage of traces aligned within time limit by different algorithms, along with the total computation time. The time limit applies to individual traces, and total time is for aligning the entire BPIC 12 event log with its model at various noise thresholds.

tations timed out (>3 s). However, the total computation for this model differs by nearly 130 s. As shown in Fig. 10, $ERV[\vartriangleleft_h]$ and *Classic PA* align nearly all traces across thresholds, while $ERV[\vartriangleleft_c]$ frequently times out at higher thresholds.

6   Conclusion

This paper presented Petri net unfolding as an efficient solution for computing alignments from partially-ordered event data, overcoming the state space explosion of traditional reachability graph-based methods. Our approach simplifies alignment computation into a single step, avoiding the need for reconstructing alignments from totally-ordered sequences. We also introduce a graph-based visualization for unfolding-based alignments, enhancing the interpretability of log-model relationships. Experimental results show that the computation of unfolding-based alignments outperforms traditional alignment methods in high-parallelism and complex models. While there is some computational overhead due to the calculation of possible extensions in low-parallelism settings, the runtime and robustness of our proposed algorithms in handling concurrency validate its effectiveness. Moreover, with our visualization technique, it is possible to obtain diagnostics over missing and undesired events and dependencies on partially-ordered alignments that have not been possible otherwise. Implemented in *Cortado*, this work provides both theoretical and practical contributions to conformance analysis. Future work will focus on optimizing $ERV[\vartriangleleft_c]$ and $ERV[\vartriangleleft_h]$, exploring heuristics and caching techniques.

References

1. van der Aa, H., Rebmann, A., Leopold, H.: Natural language-based detection of semantic execution anomalies in event logs. Inf. Syst. **102**, 101824 (2021). https://doi.org/10.1016/j.is.2021.101824

2. van der Aalst, W., Adriansyah, A., van Dongen, B.F.: Replaying history on process models for conformance checking and performance analysis. WIREs Data Mining Knowl. Discov. **2**(2), 182–192 (2012). https://doi.org/10.1002/widm.1045
3. Adriansyah, A., van Dongen, B.F., van der Aalst, W.M.: Memory-efficient alignment of observed and modeled behavior. BPM Center Rep. **3**, 1–44 (2013)
4. Adriansyah, A., Munoz-Gama, J., Carmona, J., van Dongen, B.F., van der Aalst, W.: Alignment based precision checking. In: La Rosa, M., Soffer, P. (eds.) BPM 2012. LNBIP, vol. 132, pp. 137–149. Springer, Heidelberg (2013). https://doi.org/10.1007/978-3-642-36285-9_15
5. Benveniste, A., Fabre, E., Haar, S., Jard, C.: Diagnosis of asynchronous discrete-event systems: a net unfolding approach. IEEE Trans. Autom. Control **48**(5), 714–727 (2003). https://doi.org/10.1109/TAC.2003.811249
6. Bonet, B., Haslum, P., Hickmott, S.L., Thiébaux, S.: Directed unfolding of petri nets. Trans. Petri Nets Other Model. Concurr. **1**, 172–198 (2008). https://doi.org/10.1007/978-3-540-89287-8_11
7. Carmona, J., van Dongen, B.F., Solti, A., Weidlich, M.: Conformance Checking - Relating Processes and Models. Springer, Heidelberg (2018). https://doi.org/10.1007/978-3-319-99414-7
8. Chatain, T., Khomenko, V.: On the well-foundedness of adequate orders used for construction of complete unfolding prefixes. Inf. Process. Lett. **104**(4), 129–136 (2007). https://doi.org/10.1016/j.ipl.2007.06.002
9. van Dongen, B.: BPI Challenge 2012 (2012). https://doi.org/10.4121/UUID: 3926DB30-F712-4394-AEBC-75976070E91F. https://data.4tu.nl/articles/_/12689204/1
10. Dumas, M., Rosa, M.L., Mendling, J., Reijers, H.A.: Fundamentals of Business Process Management, 2nd edn. Springer, Heidelberg (2018). https://doi.org/10.1007/978-3-662-56509-4
11. Engelfriet, J.: Branching processes of petri nets. Acta Informatica **28**(6), 575–591 (1991). https://doi.org/10.1007/BF01463946
12. Esparza, J.: Model checking using net unfoldings. Sci. Comput. Program. **23**(2–3), 151–195 (1994). https://doi.org/10.1016/0167-6423(94)00019-0
13. Esparza, J., Römer, S., Vogler, W.: An improvement of McMillan's unfolding algorithm. Formal Methods Syst. Des. **20**(3), 285–310 (2002). https://doi.org/10.1023/A:1014746130920
14. Esparza, J., Schröter, C.: Unfolding based algorithms for the reachability problem. Fund. Informaticae **47**(3-4), 231–245 (2001). http://content.iospress.com/articles/fundamenta-informaticae/fi47-3-4-05
15. García-Bañuelos, L., van Beest, N.R., Dumas, M., Rosa, M.L., Mertens, W.: Complete and interpretable conformance checking of business processes. IEEE Trans. Softw. Eng. **44**(3), 262–290 (2018). https://doi.org/10.1109/TSE.2017.2668418
16. Hart, P.E., Nilsson, N.J., Raphael, B.: A formal basis for the heuristic determination of minimum cost paths. IEEE Trans. Syst. Sci. Cybern. **4**(2), 100–107 (1968). https://doi.org/10.1109/TSSC.1968.300136
17. van Hee, K.M., Sidorova, N., van der Werf, J.: Business process modeling using petri nets. Trans. Petri Nets Other Model. Concurr. **7**, 116–161 (2013). https://doi.org/10.1007/978-3-642-38143-0_4
18. Hickmott, S.L., Rintanen, J., Thiébaux, S., White, L.B.: Planning via petri net unfolding. In: Veloso, M.M. (ed.) IJCAI 2007, Proceedings of the 20th International Joint Conference on Artificial Intelligence, Hyderabad, India, 6–12 January 2007, pp. 1904–1911 (2007). http://ijcai.org/Proceedings/07/Papers/307.pdf

19. Jouck, T., Depaire, B.: Generating artificial data for empirical analysis of control-flow discovery algorithms - a process tree and log generator. Bus. Inf. Syst. Eng. **61**(6), 695–712 (2019). https://doi.org/10.1007/s12599-018-0541-5

20. Lassen, K.B., van der Aalst, W.: Complexity metrics for workflow nets. Inf. Softw. Technol. **51**(3), 610–626 (2009). https://doi.org/10.1016/J.INFSOF.2008.08.005

21. Lu, X., Fahland, D., van der Aalst, W.: Conformance checking based on partially ordered event data. In: Fournier, F., Mendling, J. (eds.) BPM 2014. LNBIP, vol. 202, pp. 75–88. Springer, Cham (2015). https://doi.org/10.1007/978-3-319-15895-2_7

22. McMillan, K.L.: Using unfoldings to avoid the state explosion problem in the verification of asynchronous circuits. In: von Bochmann, G., Probst, D.K. (eds.) CAV 1992. LNCS, vol. 663, pp. 164–177. Springer, Heidelberg (1993). https://doi.org/10.1007/3-540-56496-9_14

23. Nielsen, M., Plotkin, G.D., Winskel, G.: Petri nets, event structures and domains. Part I. Theor. Comput. Sci. **13**, 85–108 (1981). https://doi.org/10.1016/0304-3975(81)90112-2

24. Reisig, W.: Understanding Petri Nets - Modeling Techniques, Analysis Methods, Case Studies. Springer, Heidelberg (2013). https://doi.org/10.1007/978-3-642-33278-4

25. Römer, S.: Theorie und Praxis der Netzentfaltungen als Grundlage für die Verifikation nebenläufiger Systeme. Ph.D. thesis, Technical University Munich, Germany (2000). http://tumb1.biblio.tu-muenchen.de/publ/diss/in/2000/roemer.pdf

26. Schuster, D., Schade, L., van Zelst, S.J., van der Aalst, W.: Visualizing trace variants from partially ordered event data. In: Munoz-Gama, J., Lu, X. (eds.) ICPM 2021. LNBIP, vol. 433, pp. 34–46. Springer, Cham (2022). https://doi.org/10.1007/978-3-030-98581-3_3

27. Schuster, D., van Zelst, S.J., van der Aalst, W.: Cortado—an interactive tool for data-driven process discovery and modeling. In: Buchs, D., Carmona, J. (eds.) PETRI NETS 2021. LNCS, vol. 12734, pp. 465–475. Springer, Cham (2021). https://doi.org/10.1007/978-3-030-76983-3_23

28. Siddiqui, A., van der Aalst, W.M.P., Schuster, D.: Computing alignments for partially-ordered traces through petri net unfoldings (2025). https://arxiv.org/abs/2504.00550

29. Valmari, A.: The State Explosion Problem, pp. 429–528. Springer, Heidelberg (1998). https://doi.org/10.1007/3-540-65306-6_21

Simplifying LTL Model Checking Given Prior Knowledge

Alexandre Duret-Lutz[3] , Denis Poitrenaud[1,2] , and Yann Thierry-Mieg[1(✉)]

[1] Sorbonne Université, CNRS, LIP6, 75005 Paris, France
{denis.poitrenaud,yann.thierry-mieg}@lip6.fr
[2] Université Paris Cité, 75006 Paris, France
[3] EPITA, LRE, Le Kremlin-Bicêtre, France
adl@lrde.epita.fr

Abstract. We consider the problem of the verification of an LTL specification φ on a system S given some prior knowledge K, an LTL formula that S is known to satisfy. The automata-theoretic approach to LTL model checking is implemented as an emptiness check of the product $S \otimes A_{\neg\varphi}$ where $A_{\neg\varphi}$ is an automaton for the negation of the property. We propose new operations that simplify an automaton $A_{\neg\varphi}$ *given* some knowledge automaton A_K, to produce an automaton B that can be used instead of $A_{\neg\varphi}$ for more efficient model checking.

Our evaluation of these operations on a large benchmark derived from the MCC'22 competition shows that even with simple knowledge, half of the problems can be definitely answered without running an LTL model checker, and the remaining problems can be simplified significantly.

1 Introduction—Knowledge is Power

LTL model checking consists in verifying whether all infinite executions of a system S satisfy an LTL formula φ, i.e., $\mathscr{L}(S) \subseteq \mathscr{L}(\varphi)$. In this case we write $S \models \varphi$. In the automata-theoretic approach to model checking [48], this inclusion test is usually implemented as an emptiness check of the product of two automata: $\mathscr{L}(S \otimes A_{\neg\varphi}) = \emptyset$, where $A_{\neg\varphi}$ represents the negation of φ.

The premise of this paper is that we assume to have some additional knowledge K about S. In particular, the knowledge we consider are over-approximations of the system: $\mathscr{L}(S) \subseteq \mathscr{L}(K)$. For instance K might be an LTL formula that has already been proven on S. Of course if K implies φ, i.e. $\mathscr{L}(K) \subseteq \mathscr{L}(\varphi)$, then φ holds as well since $\mathscr{L}(S) \subseteq \mathscr{L}(K)$. And if $\mathscr{L}(K) \subseteq \mathscr{L}(\neg\varphi)$, any run of the system is a counter-example.

But if none of these basic implications hold, we can still benefit from prior knowledge. We show that verifying $S \models \varphi$ *given* K is equivalent to checking $\mathscr{L}(S \otimes B) = \emptyset$ for an automaton B that is *simpler* than $A_{\neg\varphi}$, hopefully allowing a faster exploration of $S \otimes B$.

As an example the automaton $A_{\neg\varphi}$ that is on the left of Fig. 3 can be replaced by the automaton B that is on the right of the same figure. This new automaton is smaller, uses fewer atomic propositions, is now deterministic, and needs

E. Amparore and L. Mikulski (Eds.): PETRI NETS 2025, LNCS 15714, pp. 433–456, 2025.
https://doi.org/10.1007/978-3-031-94634-9_21

fewer acceptance sets because it is now a terminal automaton [9,29]. Using this automaton B should therefore simplify the job of a model checker.

This paper is organized as follows. In Sect. 2 we formalize notion of $S \models \varphi$ *given* K from the point of view of languages, and discuss possible goals when transposing this on automata. In Sect. 3 we pose useful definitions, then Sect. 4 proposes basic and Sect. 5 advanced automata operations that aim to simplify $A_{\neg \varphi}$ based on some given knowledge K. In Sect. 6 we propose costlier automata operations that aim to modify $A_{\neg \varphi}$ to make it stutter-insensitive, within the bounds allowed by some knowledge K. Finally, in Sect. 8 we evaluate the above techniques on a large third-party benchmark provided by the model checking contest [27].

2   Bounding Languages "Given That..."

In this section, we focus on providing justification for our approach at the *language* level. The language $\mathscr{L}(X)$ of a system or property X is a set of infinite words over an alphabet Σ, $\mathscr{L}(X) \subseteq \Sigma^\omega$. We denote $\overline{\mathscr{L}(X)} = \Sigma^\omega \setminus \mathscr{L}(X)$ the complement of the language of X.

A system S satisfies property φ, denoted $S \models \varphi$ if and only if the language $\mathscr{L}(S)$ of the system is a subset of the property language $\mathscr{L}(\varphi)$, *i.e.*, $\mathscr{L}(S) \subseteq \mathscr{L}(\varphi)$. When φ is an LTL formula, the classical automaton-based approach [48] is to test $\mathscr{L}(S) \cap \mathscr{L}(\neg \varphi) = \emptyset$, *i.e.*, perform an emptiness check with the language of the negated property.

In the following, we assume that $\mathscr{L}(S) \neq \emptyset$ since the empty system would satisfy any property and its negation.

Now, consider a property K (a *knowledge*) such that it has already been established that $S \models K$, *i.e.*, we know that $\mathscr{L}(S) \subseteq \mathscr{L}(K)$. This *a priori* knowledge gives us some degrees of freedom when testing whether S satisfies a new property φ. Indeed, we already know that words outside $\mathscr{L}(K)$ are definitely *not* part of $\mathscr{L}(S)$.

The main intuition is given by Fig. 1. Since $\mathscr{L}(S) \subseteq \mathscr{L}(K)$, it is safe to replace the test $\mathscr{L}(S) \cap \mathscr{L}(\neg \varphi) = \emptyset$ by a test $\mathscr{L}(S) \cap \mathscr{L}(B) = \emptyset$ where $\mathscr{L}(B)$ is built from $\mathscr{L}(\neg \varphi)$ by either removing or including words of $\overline{\mathscr{L}(K)}$. Indeed, words in $\overline{\mathscr{L}(K)}$ are not part of the system, so they cannot belong to $\mathscr{L}(S) \cap \mathscr{L}(\neg \varphi)$.

This leads to the following theorem whose proof follows immediately from Fig. 1.

Theorem 1. *Let S be a system, and K a property such that $\mathscr{L}(S) \subseteq \mathscr{L}(K)$. For any property φ, we can define a $\mathscr{L}(B)$ such that $\mathscr{L}(S) \cap \mathscr{L}(\neg \varphi) = \emptyset$ if and only if $\mathscr{L}(S) \cap \mathscr{L}(B) = \emptyset$, by choosing $\mathscr{L}(B)$ between the following bounds:*

$$\mathscr{L}(\neg \varphi) \cap \mathscr{L}(K) \quad \subseteq \quad \mathscr{L}(B) \quad \subseteq \quad \mathscr{L}(\neg \varphi) \cup \overline{\mathscr{L}(K)}$$

The lower bound $\mathscr{L}(\neg \varphi) \cap \mathscr{L}(K)$ is called the *restriction* of $\neg \varphi$: it is constructed from $\mathscr{L}(\neg \varphi)$ by removing words from $\overline{\mathscr{L}(K)}$ (Fig. 1a). The upper bound

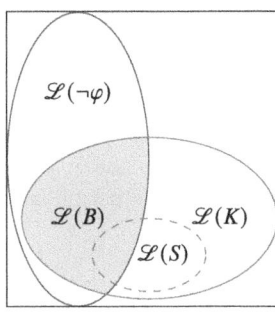 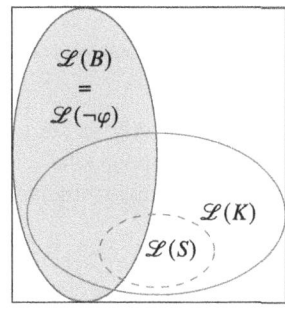 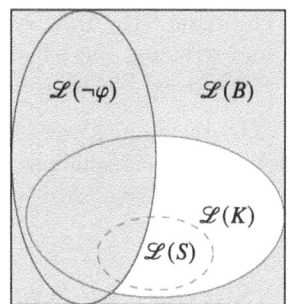

(a) The most restricted $\mathscr{L}(B)$ that can be constructed from K, $\mathscr{L}(B) = \mathscr{L}(\neg\varphi) \cap \mathscr{L}(K)$

(b) Classic approach simply using the language of $\neg\varphi$, $\mathscr{L}(B) = \mathscr{L}(\neg\varphi)$.

(c) The most relaxed $\mathscr{L}(B)$ that can be constructed given K, $\mathscr{L}(B) = \mathscr{L}(\neg\varphi) \cup \overline{\mathscr{L}(K)}$.

Fig. 1. The outside box represents all words in Σ^ω. Each language is depicted as an ellipse, with the language of the system $\mathscr{L}(S)$ inside the knowledge $\mathscr{L}(K)$ but we do not know whether the system language overlaps the negated property language $\mathscr{L}(\neg\varphi)$. The language $\mathscr{L}(B)$ represented in magenta can be chosen anywhere between these extremes to replace $\mathscr{L}(\neg\varphi)$ in the model-checking procedure.

$\mathscr{L}(\neg\varphi) \cup \overline{\mathscr{L}(K)}$ is the *relaxation* of $\neg\varphi$, constructed from $\mathscr{L}(\neg\varphi)$ by adding words from $\overline{\mathscr{L}(K)}$ (Fig. 1c).

The above theorem gives us more freedom in the automata-theoretic approach to LTL model checking. In this context, both the property φ and the knowledge K are expressed as linear-time temporal logic (LTL) formulas which can be converted into automata over infinite words.

Thus, model checking $S \models \varphi$ is implemented as $\mathscr{L}(S \otimes A_{\neg\varphi}) = \emptyset$ where $A_{\neg\varphi}$ is an automaton for $\neg\varphi$, and $\otimes$ is the product of automata [48]. Here, we want to use Theorem 1 to find a *simpler* automaton B such that model checking with $\mathscr{L}(S \otimes B) = \emptyset$ is more efficient. Contrary to intuition, choosing the automaton B with the smallest *language* might be counter-productive, because a small language does not necessarily equate to a small automaton.

To make model checking more efficient, we target the following goals:

smaller or more deterministic Reducing the size of B, or making it more deterministic can often reduce the size of the product $S \otimes B$. (Blahoudek et al. [4] suggest that contrary to previous measurements [40], "smaller" is more important than "more deterministic" for model checking.)

simpler strength class The emptiness check algorithms can be simplified if B belongs to simpler classes of automata, such as *weak*, or *terminal* automata [6, 9, 36].

stutter-insensitive For concurrent systems, many partial-order reductions (POR) techniques [22, 33, 47] and structural reductions [23, 45] can be used when it is known that B is stutter-insensitive.

fewer atomic proposition checks Reducing the number of atomic propositions and the syntactic complexity of the formulas labeling the edges of B

can reduce the time required to build $S \otimes B$ in explicit model checking [5], and reducing the set of observed propositions also helps the aforementioned POR based techniques.

The techniques we will propose mainly attempt to reduce the size of the automata, their number of atomic propositions, and attempt to make them stutter-insensitive. Any determinism improvement or strength reduction is a welcome side effect.

3   Simplifying Automata "Given That..."

We now turn the language bounds of Sect. 2 into automata constructions. We use a variant of Büchi automata called *transition-based generalized Büchi automata* (TGBA). This variant uses accepting transitions instead of accepting states. Additionally, the acceptance condition is generalized: a run has to visit *multiple* accepting sets of transitions infinitely often. This variant is particularly compact to express weak fairness conditions [11], and it also makes our subsequent definitions easier without loss of generality.

3.1   Definitions

The following definitions are freely adapted from the literature.

Let $\mathbb{B} = \{\bot, \top\}$ represent the Boolean set, and let AP represent a set of Boolean atomic propositions. A valuation ℓ is a function from AP to $\mathbb{B}$. The set of valuations is denoted $\mathbb{B}^{AP}$. The set of Boolean formulas over AP is denoted $\mathbb{B}(AP)$. In the following, we consider words that are infinite sequences of valuations, therefore the alphabet Σ of Sect. 2 is $\Sigma = \mathbb{B}^{AP}$. For an atomic proposition $a \in AP$ we use $\bar{a}$ or $\neg a$ interchangeably to represent its negation.

Definition 1 (TGBA). *A Transition-based Generalized Büchi Automaton (TGBA), is a structure $A = \langle AP, Q, \iota, Acc, \delta \rangle$ where*

- *AP is a finite set of Boolean atomic propositions,*
- *Q is a finite set of states,*
- *$\iota \in Q$ is the initial state,*
- *Acc is a finite set of acceptance marks (denoted ⓿, ❶, ❷, etc.)*
- *$\delta \subseteq Q \times \mathbb{B}(AP) \times 2^{Acc} \times Q$ is the transition relation where we use $t = q \xrightarrow{f,a} q'$ to denote an element $t \in \delta$, f is a Boolean formula over AP that we call the label of the transition and a is a set of acceptance marks.*

A run of A on an infinite word $w = \ell_1 \ell_2 \ell_3 \ldots \in (\mathbb{B}^{AP})^\omega$ is an infinite sequence of connected transitions $\rho = q_1 \xrightarrow{f_1, a_1} q_2 \xrightarrow{f_2, a_2} q_3 \xrightarrow{f_3, a_3} q_4 \ldots \in \delta^\omega$ such that $q_1 = \iota$ and for all i, $\ell_i \Rightarrow f_i$. (Recall that ℓ_i is a valuation of all atomic propositions, therefore a conjunction of atomic propositions, in negative or positive form, but f_i is a Boolean formula.) A run is accepting iff for each mark $m \in Acc$ there are infinitely many i such that $m \in a_i$.

The language of A, denoted $\mathscr{L}(A)$, is the set of all infinite words w such that there exists an accepting run of A on w.

Theorem 2 (TGBA for a formula [11,20,21]). *Given an LTL formula φ over AP, one can build a TGBA A_φ with $O(2^{|\varphi|})$ states such that $\mathscr{L}(A_\varphi) = \mathscr{L}(\varphi)$.*

For instance the leftmost automaton of Fig. 3 is a TGBA for $\mathsf{F}(p \wedge r) \vee \mathsf{G}((\mathsf{F}q) \vee (\mathsf{F}\bar{q}))$. An accepting run has to encounter marks ⓪ and ① infinitely often. Therefore, any run reaching state 1 is accepting, and any run reaching state 2 is accepting if both q and $\bar{q}$ hold infinitely often. Note that A_φ is not unique. There is a vast literature on techniques for building and simplifying automata [2,13,16,17,19,43, ...].

An obvious optimization is to discard the useless parts of the automaton by trimming it. The *trim* of an automaton A, denoted *Trim(A)*, is the restriction of A to the transitions and states that appear in at least one accepting run of A. Doing so preserves the language of A. This operation can be done in linear time by studying the strongly connected components of the automaton [16].

The intersection of the languages of two automata A_1 and A_2 is represented by a product $A_1 \otimes A_2$ such that $\mathscr{L}(A_1 \otimes A_2) = \mathscr{L}(A_1) \cap \mathscr{L}(A_2)$.

Definition 2 (*Product of TGBA*). *Given two automata $A_1 = \langle AP_1, Q_1, \iota_1, Acc_1, \delta_1 \rangle$ and $A_2 = \langle AP_2, Q_2, \iota_2, Acc_2, \delta_2 \rangle$, where $Acc_1 \cap Acc_2 = \emptyset$, the product $A_1 \otimes A_2$ is the automaton $\langle AP, Q, \iota, Acc, \delta \rangle$ where:*

- $AP = AP_1 \cup AP_2$
- $Q = Q_1 \times Q_2$
- $\iota = (\iota_1, \iota_2)$
- $Acc = Acc_1 \cup Acc_2$.
- $\delta = \left\{ (q_1, q_2) \xrightarrow{f_1 \wedge f_2, a_1 \cup a_2} (q_1', q_2') \mid q_1 \xrightarrow{f_1, a_1} q_1' \in \delta_1, q_2 \xrightarrow{f_2, a_2} q_2' \in \delta_2 \right\}$

For instance Fig. 2 shows in the bottom right the product $A_{\neg\varphi} \otimes A_K$ of the two surrounding automata. The transitions that would be removed by *Trim* are dashed.

One can also define the sum of two TGBA $A_1 \oplus A_2$ such that $\mathscr{L}(A_1 \oplus A_2) = \mathscr{L}(A_1) \cup \mathscr{L}(A_2)$, and the complement $\overline{A}$ such that $\mathscr{L}(\overline{A}) = (\mathbb{B}^{AP})^\omega \setminus \mathscr{L}(A)$. We omit the precise definition of these operations. While sum and product are cheap operations (at most quadratic in the size of the automata), the complement is worse than exponential [39,49] so it is often desirable to avoid it (e.g., we prefer to compute $A_{\neg\varphi}$ instead of $\overline{A_\varphi}$).

4   Basic Strategies

The simplest way to apply Theorem 1 is to build automata for the most restricted and the most relaxed languages pictured in Fig. 1. Consider the following two definitions:

$$\min{}_{|K}(A_{\neg\varphi}) = A_{\neg\varphi} \otimes A_K \tag{1}$$

$$\max{}_{|K}(A_{\neg\varphi}) = A_{\neg\varphi} \oplus A_{\neg K} \tag{2}$$

When B is chosen as $\min_{|K}(A_{\neg\varphi})$, we are using the most restricted language of Fig. 1a. If B is $\max_{|K}(A_{\neg\varphi})$ we are using the most relaxed language of Fig. 1c.

Note that if $\mathscr{L}(\min_{|K}(A_{\neg\varphi})) = \emptyset$, then it follows from the definition that $\mathscr{L}(K) \subseteq \mathscr{L}(\varphi)$, and since $\mathscr{L}(S) \subseteq \mathscr{L}(K)$ we have $S \models \varphi$. Dually, if $\mathscr{L}(\max_{|K}(A_{\neg\varphi})) = (\mathbb{B}^{AP})^\omega$, then $\mathscr{L}(K) \subseteq (\neg\varphi)$, which means that $S \models \neg\varphi$ (i.e., every run of S is a counterexample of φ) and therefore $S \not\models \varphi$ (because S is nonempty).

While the emptiness check of a TGBA $\mathscr{L}(A) = \emptyset$ can be performed in linear time [11], the universality test $\mathscr{L}(A) = (\mathbb{B}^{AP})^\omega$ requires exponential time [18]. Fortunately the universality test $\mathscr{L}(\max_{|K}(A_{\neg\varphi})) = (\mathbb{B}^{AP})^\omega$ can be avoided by replacing it with $\mathscr{L}(\min_{|K}(A_\varphi)) = \emptyset$ provided a formula for φ is known.

Moreover, the automata products and sums in the above $\min_{|K}$ and $\max_{|K}$ constructions can also be replaced by logical operations on formulas before translating them to TGBA, as in $A_{\neg\varphi\wedge K}$ and $A_{\neg\varphi\vee\neg K}$ respectively.

In the case where the min and max automata are neither empty nor universal, their sizes are unlikely to be smaller than the original $A_{\neg\varphi}$. In a way, using these automata for model checking is similar to asking the model checker to prove K in addition to $\neg\varphi$. As stated in Sect. 2, we would prefer to select a B that is "simpler" than $A_{\neg\varphi}$.

The knowledge K could contain atomic propositions that do not appear in $\neg\varphi$. Let P be the set of atomic propositions that appear in K but not in φ. To avoid introducing needless atomic propositions in P, we suggest to existentially quantify them. This quantification can be done precisely on the automaton A_K by existentially quantifying P from all labels, or it can be over approximated on the LTL formula K by quantifying P from all its Boolean subformulas (considered individually). We note $QE(P,K)$ the latter operation. To show that this is an over-approximation, consider the unsatisfiable formula $K = X(a \wedge b) \wedge X(\bar{a} \wedge b)$ and $P = \{a\}$. We have $(\exists a, a \wedge b) = b$ and $(\exists a, \bar{a} \wedge b) = b$, therefore, $QE(P,K) = X(b) \wedge X(b) = X(b)$ which is satisfiable.

Assuming P contains the atomic propositions of K that are not in φ, let us introduce the following notations:

$$\min_{|K}^{\exists}(\neg\varphi) = A_{(\neg\varphi)\wedge QE(P,K)} \tag{3}$$

$$\max_{|K}^{\exists}(\neg\varphi) = A_{(\neg\varphi)\vee\neg QE(P,K)} \tag{4}$$

5   Using Transition-Based Boolean Bounds on Labels

In this section, we investigate how to leverage Theorem 1 so that given an automaton for a knowledge K, we rewrite the automaton $A_{\neg\varphi}$ into a simpler automaton B.

Simplicity here is measured syntactically on the automaton; we want an automaton that has fewer states, fewer transitions, fewer atomic propositions, fewer acceptance marks, and simpler (smaller) Boolean formulas labeling the transitions of the automaton.

To achieve this, we propose to compute a set of Boolean bounds for each transition of the automaton $A_{\neg\varphi}$. These bounds enable more flexibility in the selection of transition labels by providing the most restrictive and the most relaxed Boolean formulas that can label each transition.

Minato's algorithm [31] is a recursive way to rewrite a Boolean formula as a prime-irredundant cover, which is very compact in general. The algorithm works recursively using formulas in three-valued logic, and Minato [31, Section 4.4] suggests an implementation of this algorithm using Binary Decision Diagrams [8] where a three-valued formula is simply bounded using two Boolean functions: (f_{low}, f_{high}) and the algorithm generates an irredundant sum-of-product f' such that $f_{low} \Rightarrow f' \Rightarrow f_{high}$. In other words, f' is generated as a disjunction of conjunctions of literals, such that no conjunct is uncessary, and no literal can be removed from any conjunct. We use this algorithm to simplify transition labels, as it removes literals that are unnecessary to stay within those bounds.

In Sects. 5.1 and 5.2, we introduce strategies to compute Boolean lower and upper bounds for each label of the automaton. Then, in Sect. 5.3, we show how to simplify transition labels by using Minato's algorithm on the computed bounds. This approach preserves the transition structure of the automaton. It can sometimes remove transitions (if its label becomes $\bot$), it can remove states (when they become unreachable), it can reduce the number of atomic propositions used, and it generally simplifies the expression of the labels. So, contrary to the $\min_{|K}^{\exists}(\neg\varphi)$ and $\max_{|K}^{\exists}(\neg\varphi)$ approaches presented in Sect. 4, this approach always produces a simpler automaton.

5.1 Boolean Upper Bounds

The first step consists in realizing that since $S \models K$, in *every state* of A_K we are over-approximating the state the system S might be in. Some paths in A_K might not be realizable by S, but the system definitely cannot do anything that K does not allow.

We start by building the synchronized product $A_{\neg\varphi} \otimes A_K$ in which every state is a pair (q, k). We can then apply the *Trim* operation to discard any transition that does not belong to an accepting Strongly Connected Component (SCC) or to the prefix of one, and then discard any state of the product unreachable from the initial state.

Now consider for a given state q of $A_{\neg\varphi}$ the set of states Q_q of the knowledge automaton A_K in correspondence with q. The state of the system S in this set of states can be over-approximated as the logical disjunction of the formulas labeling any transition that is outgoing from any state in Q_q.

Definition 3 (Knowledge-based state guarantee). *Given two automata* $A_{\neg\varphi} = \langle AP, Q, \iota, Acc, \delta \rangle$, *and* $A_K = \langle AP, Q_K, \iota_K, \delta_K, Acc_K \rangle$, *let* $Trim(A_{\neg\varphi} \otimes A_K) = \langle AP, Q_P, \iota_P, Acc_P, \delta_P \rangle$ *be the trim product of* $A_{\neg\varphi}$ *and* A_K.

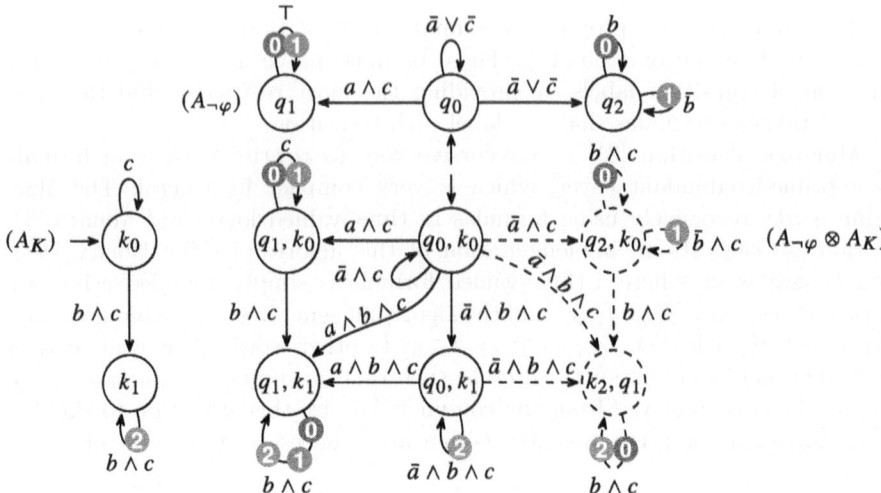

Fig. 2. Example of product of $A_{\neg\varphi} \otimes A_K$ for $\neg\varphi = F(a \wedge c) \vee G((Fb) \wedge (F\bar{b}))$ and $K = FG(b) \wedge G(c)$. The dashed transitions are those removed by *Trim*. We have $SG(q_0) = SG(q_1) = (c) \vee (b \wedge c) \vee (b \wedge c) = c$ because these two states can be synchronized with all the states of A_K, therefore their state guarantee is the disjunction of all labels of A_K. This result indicates that when the system is synchronized with state q_0, it will always satisfy c. We have $TG(q_1 \xrightarrow{T, 0 0} q_1) = (c) \vee (b \wedge c) = c$, which indicates that when a transition of the system is synchronized with this self-loop, it will always satisfy c. Finally, $TG(q_0 \xrightarrow{\bar{a} \vee \bar{c}, \emptyset} q_2) = \bot$ because the only transition synchronizing with this one was trimmed, showing that this transition is not needed.

```
A = spot.translate('F(a & c) | G(Fb & F!b)')
K = 'FG(b) & G(c)'
Ab = spot.update_bounds_given(A, K)
Aminato = spot.bounds_simplify(Ab)
Asimpl = Aminato.postprocess('small')
display_inline(A, Ab, Aminato, Asimpl)
```

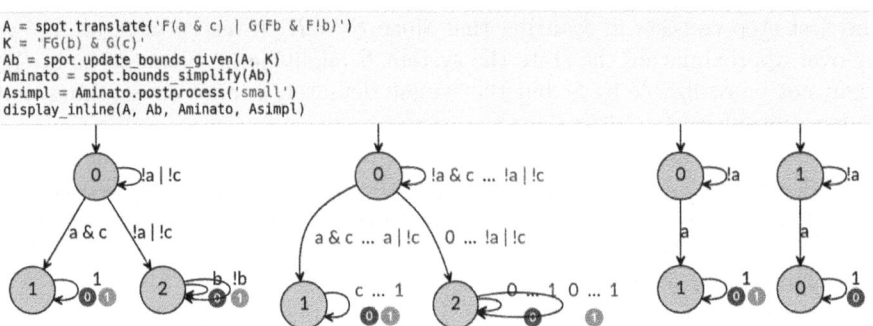

Fig. 3. Use of Spot in a Jupyter notebook to integrate some knowledge $K = FG(b) \wedge G(c)$ into the automaton for $\neg\varphi = F(a \wedge c) \vee G((Fb) \wedge (F\bar{b}))$. Note that this is the same example as Fig. 2 where the construction of the bounds is explained in detail. $A_{\neg\varphi}$ is on the left. The integration of knowledge K is represented as an intermediate "bounded automaton" in which each transition is bounded according to Theorem 3 (second automaton). Applying Minato's algorithm gives the third automaton, which can be further simplified to the rightmost automaton reducing the problem to verification of Fa. Note that Spot's notations differ slightly from those used in the paper, for instance 0, 1, and !a|!c stand for $\bot$, $\top$ and $\bar{a} \vee \bar{c}$ respectively.

For any state $q \in Q$, let $Q_q \subseteq Q_K$ denotes the subset of states of A_K that can synchronize with q in the product:

$$Q_q = \{k \in Q_K \mid (q, k) \in Q_P\}$$

From this set of states, we define the state guarantee *in state q :*

$$SG(q) = \bigvee_{k \in Q_q} \bigvee_{k \xrightarrow{f,a} k' \in \delta_K} f$$

that represents the disjunction of all transition labels of A_K that leave a state of Q_q.

In the trim product $Trim(A_{\neg\varphi} \otimes A_K)$, consider a transition $(q, k) \xrightarrow{f \wedge f_k, a \cup a_k} (q', k')$ that was built as a product of $q \xrightarrow{f,a} q'$ and $k \xrightarrow{f_k, a_k} k'$. Then it is guaranteed that $f_k \Rightarrow SG(q)$ by construction. Hence, when the component $A_{\neg\varphi}$ of the product is known to be in q, this $SG(q)$ is an over approximation of the labels f_k that the transitions in component A_K can satisfy.

Since A_K overapproximates the system S, it is also true that $SG(q)$ will overapproximate the behaviors of the states of S that can synchronize with q. This state guarantee formula thus provides an upper bound or over approximation of the system state when reaching q; therefore, for any transition $q \xrightarrow{f,a} q' \in \delta$, *relaxing* the transition label f to accept $f \vee \neg SG(q)$ would not modify the language of the product with the system S, since the system cannot satisfy $\neg SG(q)$ in this state of the product.

Figure 2 shows examples of computation of SG.

5.2 Boolean Lower Bounds

Let us look at a way to restrict a transition label of $A_{\neg\varphi}$ without limiting the ways in which the system can synchronize with this transition. For this purpose, we introduce $TG(t)$ the *transition guarantee* of a transition $t = q \xrightarrow{f,a} q'$ of $A_{\neg\varphi}$, as the disjunction of all labels of transitions of K that synchronize with t in the trim product.

Definition 4 (Knowledge-based transition guarantee). *Using the same automata as in Definition 3, for any transition $t = q \xrightarrow{f,a} q' \in \delta$, we consider the set of formulas $K_t = \{f_k \mid (q, k) \xrightarrow{f \wedge f_k, a \cup a_k} (q', k') \in \delta_P\}$ that appear on transitions of A_K that synchronize with t in the product. From the disjunction of this set of formulas, we define the* transition guarantee *for transition t:*

$$TG(t) = \bigvee_{f_k \in K_t} f_k$$

Intuitively the label f of t can be restricted to $f' = f \wedge \mathsf{TG}(t)$ since this formula is already enough to match all labels of transitions of K that would synchronize with t in an accepted run. Hence, labeling t with f' is also enough to match all states of the system S that would synchronize with t with its original label f.

Figure 2 shows examples of computation of TG.

5.3   Using the Bounds

Theorem 3. *Using the same automata as in Definition 3, consider a transition* $t = q \xrightarrow{f,a} q' \in \delta$, *and let* $B = \langle AP, Q, \iota, Acc, \delta \setminus \{t\} \cup \{t'\} \rangle$ *be a copy of* $A_{\neg\varphi}$ *where* t *has been replaced by* $t' = q \xrightarrow{f',a} q'$ *where* $f' \in \mathbb{B}(AP)$ *is any formula such that*

$$\underbrace{f \wedge \mathsf{TG}(t)}_{lower\ bound} \quad \Rightarrow \quad f' \quad \Rightarrow \quad \underbrace{f \vee \neg\mathsf{SG}(q)}_{upper\ bound}$$

Then $\mathscr{L}(B \otimes A_K) = \mathscr{L}(A_{\neg\varphi} \otimes A_K)$

The proof, omitted to meet size constraints, can be found in the authors' copy [14].

Our implementation of this construction is an extension of Spot [15] in which the Boolean bounds of Theorem 3 can be represented directly on the automaton. Figure 3 shows our implementation at work. Our knowledge bound integration function `spot.update_bounds_given` can be called repeatedly to integrate multiple knowledge incrementally, as we will discuss in Sect. 7.

To select a simple label compatible with the bounds, we apply Minato's algorithm [31] (introduced at the top of Sect. 5) to compute a simpler label f' such that $f \wedge \mathsf{TG}(t) \Rightarrow f' \Rightarrow f \vee \neg\mathsf{SG}(q)$. Note that when the lower bound is $\bot$, Minato's algorithm will always return $f' = \bot$ (the transition can be removed), else if the upper bound is $\top$, $f' = \top$ will be returned.

If A and K are defined over different sets of atomic propositions, the result of Theorem 3 might include atomic propositions from K that were not in A, which is counterproductive. In this case, we simplify K by existential quantification of the propositions that are not in A to produce an automaton K_{QE}. The language of this automaton contains $\mathscr{L}(K)$, therefore it also contains $\mathscr{L}(S)$ and it can still be used as a knowledge. We denote $BM_{|K}(A)$ the "Bounded by Minato" automaton, i.e., the automaton built from A and K by applying Minato's algorithm on the Boolean bounds computed by Theorem 3 with existential quantification of the atomic propositions not in A.

```
A = spot.translate('XFa'); K = spot.translate('!a', 'Buchi')
sirelax = spot.stutterize_given(A, [K], True).postprocess('small')
sirestrict = spot.stutterize_given(A, [K], False).postprocess('small')
display_inline(A, K, sirestrict, sirelax, per_row=2)
```

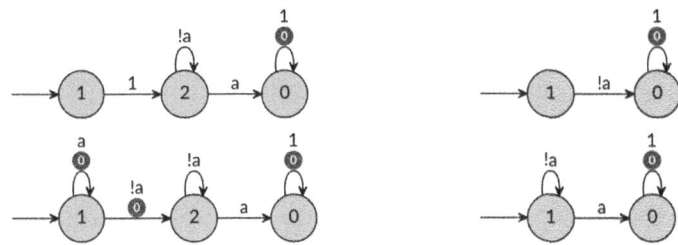

Fig. 4. Use of Spot in a Jupyter notebook to integrate some trivial knowledge $K = \bar{a}$ (top right automaton) into the stutter-sensitive automaton for $\neg\varphi = \mathsf{XF}(a)$ (top-left automaton) and turn it into a stutter-insensitive automata. Simplified automata for $sirestrict_{|K}(A)$ and $sirelax_{|K}(A)$ are given on the bottom left and right respectively.

6 Building Stutter-Insensitive Automata "Given That..."

When model checking a concurrent system S against a formula φ, several advanced and very effective simplification techniques can be used when it is known that $\mathscr{L}(A_{\neg\varphi})$ is stutter-insensitive with reductions up to a factorial factor [22,23,33,45,47].

In this section, we consider the case where the language of $A_{\neg\varphi}$ is *stutter-sensitive*, and, given a knowledge K, we want to replace $A_{\neg\varphi}$ by an automaton B whose language is *stutter-insensitive*. Even if B is "bigger" than $A_{\neg\varphi}$, model checking might become more efficient thanks to the aforementioned simplifications.

Given a word w, let $[w]$ be the set of stutter-equivalent words that can be obtained from w by finitely duplicating letters or removing repetitions.[1] A language $\mathscr{L}(A)$ is stutter-sensitive iff there exists at least one equivalence class $[w]$ that is *only partly* covered by $\mathscr{L}(A)$, i.e., such that $[w] \cap \mathscr{L}(A) \neq \emptyset$ and $[w] \cap \overline{\mathscr{L}(A)} \neq \emptyset$.

As an example, consider the automaton $A_{\mathsf{XF}a}$ *given that* $K = \bar{a}$. Figure 4 shows this example using automata. The knowledge provided here is very simple, but this kind of effect could occur anywhere in the automaton, not just starting in the initial state. Automaton $A_{\mathsf{XF}a}$ is stutter-sensitive: it rejects the word $w = a\bar{a}\bar{a}\bar{a}\cdots$, but it accepts words starting with more than one a. However, notice w is outside $\mathscr{L}(K)$, so by Theorem 1, w could be added to the language of B to make it stutter-insensitive (now accepting $\mathsf{F}a$), giving the bottom-right automaton of Fig. 4. Another strategy would be to remove all words in $[w]$ from B: since all these words all start by a, the whole class $[w]$ is outside $\mathscr{L}(K)$. This second option, corresponding to the formula $\mathsf{G}(a) \vee \mathsf{F}(\bar{a} \wedge \mathsf{F}a)$ gives the bottom-left automaton of Fig. 4. Note that on this example, using the approach based

---

[1] $[w]$ is an equivalence class for the $\sim^{\mathrm{lim}}$ relation of Peled et al. [34, Lemma 3].

on Minato's algorithm described in Sect. 5.3 will only give us bounds $\bar{a}...\top$ on the first transition of $A_{\mathsf{XF}a}$, but since this transition is already labeled by $\top$ it would not be changed.

As seen in this example, we propose two strategies to turn a stutter-sensitive automaton into a stutter-insensitive one. For each partly-covered equivalence class $[w]$, the relaxing strategy consists in adding the rest of $[w]$ to $\mathscr{L}(A)$. Dually, the restricting strategy consists in removing $[w]$ from $\mathscr{L}(A)$. This is legitimated by Theorem 1 provided that the added or removed parts are outside the knowledge.

To realize these strategies on automata, let us equip ourselves with a function $si(A)$ that returns an automaton A' such that $\mathscr{L}(A')$ is the smallest stutter-insensitive language that contains $\mathscr{L}(A)$. Such an operation has already been defined for Büchi automata [24] or TGBA [30]. Intuitively, it consists in two simple syntactic transformations: adding shortcut edges to reduce stutter, and adding states to allow stuttering after traversing any transition. The effect of this operation is that all classes $[w]$ partly covered by $\mathscr{L}(A)$ get fully included into $\mathscr{L}(si(A))$.

Additionally, let us define $ss(A)$ the stutter-sensitive part of A as the automaton that recognizes only the words $w \in \mathscr{L}(A)$ such that $[w] \cap \overline{\mathscr{L}(A)} \neq \emptyset$. While, to our knowledge, this operation does not exist in the literature, it can be defined using si, complement and product as follows:

$$ss(A) = A \otimes si(si(A) \otimes \overline{A})$$

In the above formula, $si(A) \otimes \overline{A}$ accepts exactly the words that should be added to $\mathscr{L}(A)$ to make it stutter-invariant. Therefore, $si(si(A) \otimes \overline{A})$ accepts all the words w such that $[w]$ is partly covered by A.

We now show how to realize our two strategies using these automata operations.

Theorem 4. *Let A be a stutter-sensitive TGBA and K be an LTL formula. We define the SI-relaxation and SI-restriction of A given K as follows.*

$$sirelax_{|K}(A) = \begin{cases} si(A) & \text{if } \mathscr{L}(si(A) \otimes \overline{A} \otimes A_K) = \emptyset \\ A & \text{else} \end{cases}$$

$$sirestrict_{|K}(A) = \begin{cases} A \otimes \overline{ss(A)} & \text{if } \mathscr{L}(ss(A) \otimes A_K) = \emptyset \\ A & \text{else} \end{cases}$$

Then $\mathscr{L}(sirelax_{|K}(A) \otimes A_K) = \mathscr{L}(sirestrict_{|K}(A) \otimes A_K) = \mathscr{L}(A \otimes A_K)$.

Proof. The proof follows from Theorem 1 and the fact that $\mathscr{L}(ss(A)) \subseteq \mathscr{L}(A) \subseteq \mathscr{L}(si(A))$. Indeed, the above relaxation returns $si(A)$ if and only if the words added by si (i.e. $\mathscr{L}(si(A) \otimes \overline{A})$) are outside $\mathscr{L}(K)$ (i.e., $si(A) \otimes \overline{A} \otimes A_K$ has an empty language). Similarly, the restriction removes from $\mathscr{L}(A)$ the words of

$\mathcal{L}(ss(A))$ if and only if they are all outside of $\mathcal{L}(K)$. When the given knowledge does not allow adding or removing those words, the original automaton is returned.[2]   □

Figure 4 shows stutter-insensitive automata obtained with these two constructions.

In practice the complement of A (present in both strategies) can be avoided when an LTL formula for A is known. However, the complement of $ss(A)$, needed only for $sirestrict_{|K}(A)$ can be rather costly, especially considering the current definition of $ss(A)$, which tends to create large automata. Fortunately, the latter complementation need only be performed after it has been checked that the removed words are not part of the knowledge. Our hope is therefore that any stutter-insensitive optimization performed by the model checker will offset the costs incurred by the computation of $sirestrict_{|K}(A)$.

7   Incremental Integration of Knowledge

In this section we show how to integrate knowledge when we know multiple facts about the system. We also discuss some strategies to obtain cheap knowledge tailored to help simplify a given property.

7.1   Working with a Knowledge Base

Previously, in Sect. 3–6, we discussed how to simplify $A_{\neg\varphi}$ given a single knowledge K. We now assume that we have multiple knowledge facts A_{K_1}, A_{K_2}, ..., A_{K_n} about the system S and discuss strategies to integrate them all.

We could simplify $A_{\neg\varphi}$ by applying Theorems 3–4 using $A_K = A_{K_1} \otimes A_{K_2} \otimes \cdots \otimes A_{K_n}$. However, since the product of automata is quadratic in size, this automaton A_K might be very big. Even the translation of the conjunction of all K_i at once $A_K = A_{K_1 \wedge K_2 \wedge \ldots \wedge K_n}$ might be a large automaton.

In the following, we propose techniques to integrate knowledge incrementally, even if this comes with a loss of precision.

For the Boolean Bounds, we suggest applying Theorem 3 using one A_{K_i} at a time, in a loop, and delay the choice of the label (using Minato's algorithm) until the end of the loop. In the syntax of Fig. 3 we do:

```
for k in list_of_facts:
    a = spot.update_bounds_given(a, k)
a_minato = spot.bounds_simplify(a)
```

Here the operation `spot.update_bounds_given(a, k)` is using the lower bounds of each transition of automaton `a` when building the product in the definition of TG and SG. In this loop, each call to `spot.update_bounds_given`

---

[2] The user of these functions may therefore assume that the returned automaton is stutter-insensitive whenever it is different from the input.

may only restrict the lower bounds and relax the upper bounds of the transitions of a.

The incremental construction of $sirestrict_{|K}(A)$ and $sirelax_{|K}(A)$ is handled differently. Since the automata $si(A)$ and $A \otimes ss(A)$ constructed by these techniques are independent of the knowledge that allow to adopt them, we can stop as soon as we find a suitable knowledge. More formally:

$$sirelax_{|K_1, K_2, \ldots, K_n}(A) = \begin{cases} si(A) & \text{if } \exists i, \mathscr{L}(si(A) \otimes \overline{A} \otimes A_{K_i}) = \emptyset \\ A & \text{else} \end{cases}$$

$$sirestrict_{|K_1, K_2, \ldots, K_n}(A) = \begin{cases} A \otimes \overline{ss(A)} & \text{if } \exists i, \mathscr{L}(ss(A) \otimes A_{K_i}) = \emptyset \\ A & \text{else} \end{cases}$$

This generalization, which is what we implement, explains why the single fact k was being passed in an array in Fig. 4. In the implementation the terms $si(A) \otimes \overline{A}$ and $ss(A)$ are of course computed only once, and not for each K_i.

7.2   Seeking Knowledge

The strength of our approach is that it is agnostic to the source or proof method of the knowledge. Of course some facts might simply be other formulas we have already proven, when we are dealing with a set of specification formulas to check against a given system.

We now suggest ways to obtain some cheap knowledge about the system S, tailored to fit the formula φ that we intend to verify. For instance, we can find some simple facts on S using bounded explorations (breadth-first search, bounded model checking [3], . . .), structural analysis of the system, or a decision procedure for reachability. . . and that can help simplify $A_{\neg \varphi}$.

Our implementation currently looks for various kinds of knowledge, using simpler decision procedures than full LTL to prove them.

Initial state First, we can check the label of the initial state of the system, giving us a knowledge of the form $\ell \in 2^{AP}$. While this knowledge is very basic, it is free. Using the initial state of the model to simplify a property has already been proposed for CTL [7].

First steps Similarly, exploring the first steps of the system is cheap. We compute the set of formulas labeling the transitions reachable in the first n steps of $A_{\neg \varphi}$, and check the first n steps of S to check if their values allow us to define a knowledge of the form $Xf, XXf . . .$

We limit our exploration to $n = 2$ in our experiments. We use a breadth-first search with some limits to avoid explosion on models with very large branching factors ($\geq 10^4$). We could also have used any technique based on bounded model checking relying on a SAT or SMT solver.

Invariants Proving some invariants of a system can be delegated to tools that are specialized in reachability analysis and are more effective at this task than LTL model checkers.

We start by looking at the value of each atomic proposition of φ in the initial state of S and try to prove that this value never evolves using a reachability solver. We then try to evaluate compatibility of the atomic propositions, checking given two atomic propositions a and b whether all of $\bar{a}\bar{b}$, $\bar{a}b$, $a\bar{b}$ and ab are possible. For instance, $a = [x > 2]$ and $b = [x > 3]$ have a strong relationship. Knowledge about the exclusions between APs was also used by Blahoudek et al. [5]. We use an SMT solver to check if some of these cases are impossible, not even looking at the system but simply at the atomic proposition definitions. We finally also check if formulas labeling the transitions of $A_{\neg\varphi}$ are invariants. All of these strategies output knowledge of the form G.

Convergent atomic propositions In this approach we try to prove that a given atomic proposition a will eventually converge, providing a knowledge of the form $F(Ga \vee G\bar{a})$.

We use a low complexity structural test based on an analysis of recurring behaviors (SCC in the state graph of the system). Any atomic proposition that only observes variables in the prefix of such SCC must converge; they cannot oscillate indefinitely. We can also, using an SMT solver, try to determine the polarity of atomic propositions at convergence, yielding knowledge of the form FGa (or $FG\bar{a}$).

These strategies are all very basic currently, but show how we can leverage a diversity of decision procedures (with lower complexity than full LTL model-checking) to populate a knowledge base that is tailored for a given formula to assist an LTL model checking step.

8 Experimental Study

Knowledge-based simplifications ("given that") have been implemented in Spot 2.13 [15]. The knowledge collection described in Sect. 7.2 has been implemented in ITS-tools [44], which won the LTL category of the Model Checking Contest in 2023 for the first time, thanks in part to these strategies. The tools to gather the knowledge and integrate it are open source and publicly available. A reproducibility package for the experiments can be found at https://codeocean.com/capsule/1210152/tree/v1.

During the competition, ITS-tools allots a small time slice to incrementally collect and integrate knowledge. After this time, it runs a portfolio of model-checkers, including a symbolic solution [44] and LTSmin [25] configured as an explicit model checker with partial-order reductions.

Measurement of the entire model checking procedure would introduce many biases due to the complex interactions of the portfolio techniques with the main refinement loop of ITS-Tools [45]. Therefore, we focus our evaluation on the knowledge integration step of the procedure, and compare the automata obtained using the strategies introduced in this paper.

8.1   Experimental Setup

The following performance analysis is based on the models and formulas of MCC'22 [26]. The benchmark uses a total of 150 different model families (coming from various domains) configured to build 1617 model instances (some models are scalable).

For each of these (colored) Petri net models, the benchmark contains 32 randomly generated LTL formulas providing a total of 51744 LTL formulas.

For each model instance, we first collected some knowledge using the basic approach of Sect. 7.2 using ITS-Tools [44], setting a generous timeout of 15 minutes to collect it. Obtaining knowledge is cheap (median 0.67 minutes and 75% of cases below 4 minutes) as it leverages low complexity structural and symbolic tests, and in the worst case reachability queries which are much simpler than full LTL. High time usage to collect this basic knowledge correlates with models where LTL model-checking (at least in the empty product case, with no counter-example) is typically prohibitively expensive (huge models with millions of elements), so that the effort is worth it.

After this processing, we obtain some knowledge for 1601 model instances (out of 1617). For each model instance the knowledge is represented as a set of LTL assertions, for a total of 240345 small facts (roughly 150 facts per model instance). From the original set of 51744 LTL formulas we only retain 48975 formulas that intersect the gathered facts.

To add some diversity, we consider the above 48975 formulas and their negations for a total of 97950 formulas. Note that verifying an LTL formula and its negation are two independent chalenges: it can be the case that neither of these formula is verified.

For each formula, we retain only the subset of available facts whose alphabet intersects that of the formula in the experiments.[3]

Our benchmark therefore contains 97950 problems that consist in one specification LTL formula accompanied by a set of knowledge facts (on average 12.7, median 9 facts per formula).

We then proceed to apply each of our strategies to these problems to build an automaton and compute various metrics on its size. The strategies we compare are the following. A "p." used as prefix indicates a *precise* construction that considers all facts $K = \bigwedge_i K_i$ at once. While *precise* variants can pay a significant cost to manipulate the conjunction of known facts, they also benefit from a more precise knowledge so that there is a trade-off between precise and incremental approaches.

raw is formula $\neg\varphi$ translated to a TGBA, without any integration of knowledge

---

[3] This is not necessarily optimal. Consider $\varphi = \mathsf{GF}a$ with alphabet $\{a\}$ and facts $k_1 = \mathsf{G}(b \rightarrow \mathsf{X}a)$ and $k_2 = \mathsf{FG}b$, ignoring k_2 because its alphabet does not intersect φ's is in fact a mistake ; however selecting too many facts can easily overload some approaches particularly those using the conjunction of known facts, and does not help our incremental approaches that consider each fact in isolation since they typically ignore atomic propositions not in φ.

Table 1. Amount of problems (out of 97950 composed of formulas and their negation) that could be shown to be empty or universal using the provided knowledge.

Strategy	Universal	Empty	Total	Strategy	Universal	Empty	Total
p.min	0	25508	25508	p.BM	23258	25508	48766
p.max	24453	0	24453	p.SIrelax	212	0	212
p.min∃	0	25508	25508	p.SIrestrict	0	223	223
p.max∃	25508	0	25508	SIrelax+BM	23286	25091	48377
BM	23080	25095	48175	BM+SIrelax	23164	25095	48259
SIrelax	208	0	208	p.SIrelax+p.BM	23708	25508	49216
SIrestrict	0	219	219	p.BM+p.SIrelax	23344	25508	48852

p.min, p.max are obtained by building $A_{\neg\varphi\wedge K}$ and $A_{\neg\varphi\vee\neg K}$ as discussed in Sect. 4, where K is the conjunction K of all facts.

p.min∃, p.max∃ are the variants with existential quantification shown in equations (3)–(4) from Sect. 4.

p.BM, BM use respectively the strategy $BM_{|K}(A)$ presented in Sect. 5.3, and the incremental strategy described in Sect. 7.1.

p.SIrelax, p.SIrestrict, SIrelax, SIrestrict are the strategies of Sect. 6, and their incremental variants from Sect. 7.1

Finally, we also consider some combination of techniques. For instance "SIrelax+BM" designates the incremental implementation of SIrelax followed by the incremental implementation of BM.

When providing statistics about the automata produced by the above variants, we always assume that those automata have been further simplified using techniques implemented in Spot (notably, removing useless states, useless acceptance marks, and using simulation-based reductions to merge states and prune unnecessary transitions [2]).

The average runtime for solving a problem with any strategy is 35.6 ms. On the 97950 benchmark problems, the only strategies that exceed a very generous timeout of 10 s are "SIrestrict" on 470 problems, and "p.SIrestrict" on 522 problems. Those strategies are occasionally very slow only because of the amount of automata complementations they have to perform. Overall the knowledge integration step is truly negligible before any test involving the actual system. If the knowledge gathering step only uses low complexity procedures or knowledge simply consists of previously proven properties, knowledge integration scales exceptionally well to complex problems.

8.2   Problems Reduced to Empty or Universal

We first study the problems that could be fully solved given the knowledge by reducing the automaton to an empty or universal one. For testing universality, we syntactically check if the resulting automaton has been reduced to a single-state all-accepting automaton.

Table 2. Comparison of the different strategies over 46934 problems that could not be already reduced to false or true by previous methods. 'raw' designates the original automata, for baseline. For each strategy we report various metrics of the produced automata: its number of states and transitions, $\sum |f|$ is the total size of all labels, SI (resp. det) shows the fraction of automata that were stutter-insensitive (resp. deterministic), $|AP|$ is the number of atomic proposition, Time reports the number of milliseconds needed by the strategy, and TO counts the number of timeouts (> 10 seconds). Different statistics are provided for some measurements: 'q95' denotes the 95% quantile (i.e., 95% of all values are below the indicated value), 'geom' denotes the geometric mean. Values within 2% of the best (resp. worse) value of a column, 'raw'excluded, are highlighted in yellow (resp. pink).

Strategy	States				Transitions				$|AP|$	$\sum|f|$	SI	det	Time (ms)		TO
	q95	max	mean	geom	q95	max	mean	geom	mean	mean			mean	geom	
raw	8	73	3.71	3.13	18	286	7.32	5.52	2.14	21.76	49%	50%	20.26	19.95	0
p.min	9	684	5.30	4.79	19	16834	8.71	6.71	3.35	48.36	9%	50%	55.04	49.82	0
p.max	10	76	6.30	5.90	25	295	12.78	11.23	3.35	58.15	9%	43%	55.86	50.27	0
p.min∃	9	684	5.23	4.67	19	16834	8.60	6.55	2.21	35.06	11%	50%	54.18	48.97	0
p.max∃	10	76	6.14	5.73	25	295	12.37	10.77	2.21	42.54	11%	45%	54.91	49.39	0
BM	6	65	3.13	2.68	13	286	5.42	4.24	1.70	13.39	46%	59%	41.46	40.62	0
SIrelax	8	73	3.86	3.16	22	286	8.07	5.82	2.14	26.51	66%	49%	42.19	41.09	0
SIrestrict	10	19463	6.77	3.23	27	373093	52.25	5.96	2.14	335.58	67%	51%	57.65	44.28	122
p.BM	6	65	3.13	2.68	13	286	5.41	4.24	1.69	13.35	46%	59%	46.13	43.50	0
p.SIrelax	9	73	3.93	3.18	24	340	8.47	5.92	2.14	29.15	70%	49%	46.20	43.71	0
p.SIrestrict	11	84249	10.59	3.28	32	1252969	105.47	6.12	2.14	677.23	70%	51%	57.34	45.53	130
SIrelax+BM	6	65	3.07	2.62	13	286	5.38	4.18	1.70	13.49	51%	59%	63.37	61.75	0
BM+SIrelax	7	65	3.19	2.67	15	286	5.93	4.49	1.70	16.21	67%	58%	63.15	61.46	0
p.SIrelax+p.BM	6	65	3.04	2.59	13	286	5.34	4.15	1.69	13.44	52%	59%	72.08	66.89	0
p.BM+p.SIrelax	7	65	3.16	2.63	16	286	5.99	4.46	1.69	16.83	70%	58%	71.85	66.69	0

Table 1 presents those results. While it is certainly due to the random nature of the formulas of the MCC, in total $51016/97950 \approx 52\%$ of the formulas of the MCC benchmark we kept (49% of all formulas of the MCC) could be solved using only the basic approach to glean related knowledge presented in Sect. 7.2, thus avoiding a full LTL model-checking procedure.

We can see that "min∃" and "p.min∃" are the most effective strategies to find empty problems, on par with "p.BM". Dually "p.max∃" is the only most effective at deducing universal problems. The amount of problems reduced to empty by "p.min∃" and to universal by "p.max∃" are identical as hoped, because our benchmark includes both formulas and their negations. Strategy "p.max" is less effective than "p.max∃" because it keeps atomic propositions that are not in φ. Still, while they might produce a larger automaton as discussed in the next section if they can't solve the problem, it is important in a full decision approach involving some knowledge to first test "p.min∃" and "p.max∃" for full solutions.

Generally, strategies based only on restriction (resp. relaxation) can prove only emptiness (resp. universality) and obtaining "empty" seems easier on this benchmark than obtaining "universal" perhaps due to our limited syntactic check for universality.

8.3   Simplifying the Remaining Unsolved Formulas

We now study in Table 2 statistics for all the 46934 problems that could not be proven empty or universal by any strategy.

Since our goal is to reduce the size of the automaton, we first study the number of states and transitions. To better understand the distribution of values, we present the 95% quantile, as well as the arithmetic and geometric means. Cases where the arithmetic mean is much larger than the geometric mean indicate the presence of a few very large outliers.

We observe that basic strategies based on "p.min" or "p.max" are not very good, as feared, doubling the average size. However, all strategies involving "BM" perform well: the average number of state is reduced by 15%, and transitions by 25%. The "SIrelax" strategy produces a moderate size increase that can be further alleviated by combining it with "BM". However, "SIrestrict" can dramatically increase the size of the automaton, and even time out in extreme cases.

The number of atomic propositions (column $|AP|$) and size of the formula labels ($\sum |f|$) is also significantly reduced by all variants using "BM". The average number of atomic propositions is reduced from 2.14 to around 1.7 (a 21% gain), and the average size of formula labels goes from 21.76 to around 13.4 (a 38% gain).

Concerning the stutter insensitivity (column "SI") of the resulting automaton, only 49% of the "raw" problems are stutter-insensitive. "min" and "max" degrade this number significantly. Both of the strategies "SIrestrict" and "SIrelax" developed to optimize this metric are indeed effective (but "SIrestrict" is more expensive and liable to timeout). Combined strategies that finish with a "SIrelax" step lead to the best results, being both small and stutter-insensitive in 70% of cases. The precise variants are a bit better than the incremental constructions.

While our algorithm is not looking to improve determinism (column "det"), this characteristic is nonetheless improved by all variants involving "BM". This is a welcome side-effect since having small *and* deterministic automata can only help model checking [4,40].

On this subset of 46934 cases, the average time to solve a problem is up to two times higher than the average of the 97950 benchmark problems (which was 35.6 ms as mentioned earlier). However it is still very cheap. The fastest strategies to integrate knowledge is "BM" (with an average of 41 ms). "p." precise variants all pay a reasonable time penalty, but the improvement in size is very modest. The only timeouts we observe on these problems that cannot be entirely solved are for "SIrestrict" and its precise variant (in less than 3‰ cases).

In conclusion, combined strategies using "SIrelax" and "BM" produce the smallest automata without real drawbacks apart from moderate increase of the run time. Given the very reasonable run times, it is even feasible to run several of these strategies and then select the most appropriate automaton on a case by case basis.

9   Related Work

Theorem 1 proposes an original framework for exploiting prior knowledge in LTL model checking. This generalizes approaches that only consider invariants [7] or quasi-invariants [28].

Theorem 1 is also related to the problem of language separation: given two languages, a separator is a third language that contains the first one and is disjoint from the second one [35]. In our case, we are looking for an automaton B whose language separates $\mathscr{L}(A) \cap \mathscr{L}(K)$ (which it should include) from $\mathscr{L}(K) \setminus \mathscr{L}(A)$ (which it should not intersect). However, the two languages to separate aren't independent: A is already known to be a separator, and we are trying to find a simpler B by simplifying A.

Blaoudek et al. [5, Section 5] also consider a simplification of labels leveraging Minato's algorithm as we did in Sect. 5. While it is limited to an invariant about mutually exclusive propositions, it did prove to be an effective simplification. Our approach generalizes theirs: if the knowledge encodes mutual exclusion of atomic propositions, we will generate the same bounds, however we can handle arbitrary LTL knowledge, and we take the structure of the automaton into account.

Using Minato's algorithm to find a simple f' such that $f_{low} \Rightarrow f' \Rightarrow f_{high}$ can be related to Coudert and Madre's `restrict` and `constraint` operators [10] that find f' such that $f \wedge c \Rightarrow f' \Rightarrow f \vee \neg c$, where c is a Boolean formula. However, in our case, f_{low} and f_{high} are not limited to this form.

The use of bounded automata in Sect. 5 evokes the notion of incompletely specified Mealy machines used in synthesis, where "don't care" edges are leveraged to produce smaller automata [1,32,37]. The bounded automata we propose can be used for bound-aware simulation-based reductions [42]; this could complement our current approaches.

Dureja and Rozier [12] consider the problem of model checking a single model against a large set of LTL formulas. They compute a matrix of implications between formulas $f_i \Rightarrow f_j$, and they use previously proven formula f_1 to avoid model checking of implied formulas f_2. Such an implication test, between a previously proven formula f_1 (the knowledge) and an unproved formula f_2, is covered in our approach since "f_2 given f_1" will be an empty automaton (see Sect. 8.2). However, we can also obtain a simpler automaton for f_2 even in the absence of full implication. Moreover, we suggest several approaches to leverage *all* accumulated knowledge incrementally.

Our definitions suggest explicit representation of automata, however our approach can be used for symbolic model checking. Instead of using a direct symbolic encoding of Büchi automaton [38], obtained directly from LTL, we can encode the explicit automaton resulting from our knowledge simplifications into a symbolic representation [41]. In fact, ITS-tools uses both a knowledge-based approach and a symbolic encoding.

Although this work is motivated by model checking, our techniques can be used to optimize any inclusion check $\mathscr{L}(A) \subseteq \mathscr{L}(B)$. E.g., in the traditional implementation based on a complementation [46] of B, any knowledge about A, can be used to simplify B before its complementation.

10 Conclusion

We have introduced new operations that help simplify the model-checking of a new formula when we already possess some prior knowledge on the system. Our strategies are automata-based operations, thus capturing any nature of LTL property or prior knowledge. The evaluation of our current implementation on a large benchmark demonstrates the effectiveness of the approach.

Studying the problem of knowledge integration led us to the problem of producing a (small) automaton given bounds on the language it represents. This challenging problem is new to our knowledge and while we have proposed several strategies in this paper, there is a lot of room for more research in this direction. For instance the strategies we presented in Sect. 6 to produce stutter-insensitive automata currently do not take any advantage of the Boolean bounds computed in Sect. 5. Similarly, those Boolean bounds could very likely be used for other kinds of simplifications, such as bound-aware simulation-based reductions [42]. The problem of seeking relevant knowledge by leveraging simpler decision procedures than full LTL is also an avenue for further exploration, paving the way to strategies achieving an incremental verification process.

References

1. Abel, A., Reineke, J.: MeMin: SAT-based exact minimization of incompletely specified Mealy machines. In: Proceedings for the 34th International Conference on Computer-Aided Design (ICCAD'15), pp. 94–101. IEEE Press (2015). https://doi.org/10.1109/ICCAD.2015.7372555

2. Babiak, T., Badie, T., Duret-Lutz, A., Křetínský, M., Strejček, J.: Compositional approach to suspension and other improvements to LTL translation. In: Bartocci, E., Ramakrishnan, C.R. (eds.) SPIN 2013. LNCS, vol. 7976, pp. 81–98. Springer, Heidelberg (2013). https://doi.org/10.1007/978-3-642-39176-7_6

3. Biere, A.: Bounded model checking. In: Biere, A., Heule, M., van Maaren, H., Walsh, T. (eds.) Handbook of Satisfiability - Second Edition, Frontiers in Artificial Intelligence and Applications, vol. 336, pp. 739–764. IOS Press (2021).https://doi.org/10.3233/FAIA201002

4. Blahoudek, F., Duret-Lutz, A., Křetínský, M., Strejček, J.: Is there a best Büchi automaton for explicit model checking? In: Proceedings of the 21th International SPIN Symposium on Model Checking of Software (SPIN'14), pp. 68–76. ACM (2014). https://doi.org/10.1145/2632362.2632377

5. Blahoudek, F., Duret-Lutz, A., Rujbr, V., Strejček, J.: On refinement of Büchi automata for explicit model checking. In: Fischer, B., Geldenhuys, J. (eds.) SPIN 2015. LNCS, vol. 9232, pp. 66–83. Springer, Cham (2015). https://doi.org/10.1007/978-3-319-23404-5_6

6. Bloem, R., Ravi, K., Somenzi, F.: Efficient decision procedures for model checking of linear time logic properties. In: Halbwachs, N., Peled, D. (eds.) CAV 1999. LNCS, vol. 1633, pp. 222–235. Springer, Heidelberg (1999). https://doi.org/10.1007/3-540-48683-6_21

7. Bønneland, F., Dyhr, J., Jensen, P.G., Johannsen, M., Srba, J.: Simplification of CTL formulae for efficient model checking of petri nets. In: Khomenko, V., Roux, O.H. (eds.) PETRI NETS 2018. LNCS, vol. 10877, pp. 143–163. Springer, Cham (2018). https://doi.org/10.1007/978-3-319-91268-4_8

8. Bryant, R.E.: Graph-based algorithms for boolean function manipulation. IEEE Trans. Comput. **35**(8), 677–691 (1986)

9. Černá, I., Pelánek, R.: Relating hierarchy of temporal properties to model checking. In: Rovan, B., Vojtáš, P. (eds.) MFCS 2003. LNCS, vol. 2747, pp. 318–327. Springer, Heidelberg (2003). https://doi.org/10.1007/978-3-540-45138-9_26

10. Coudert, O., Madre, J.C.: A unified framework for the formal verification of sequential circuits. In: Proceedings of the International Conference on Computer-Aided Design (ICCAD'90), pp. 126–129 (1990). https://doi.org/10.1109/ICCAD.1990.129859

11. Couvreur, J.-M.: On-the-fly verification of linear temporal logic. In: Wing, J.M., Woodcock, J., Davies, J. (eds.) FM 1999. LNCS, vol. 1708, pp. 253–271. Springer, Heidelberg (1999). https://doi.org/10.1007/3-540-48119-2_16

12. Dureja, R., Rozier, K.Y.: More scalable LTL model checking via discovering design-space dependencies (D^3). In: Beyer, D., Huisman, M. (eds.) TACAS 2018. LNCS, vol. 10805, pp. 309–327. Springer, Cham (2018). https://doi.org/10.1007/978-3-319-89960-2_17

13. Duret-Lutz, A.: LTL translation improvements in Spot 1.0. Int. J. Crit. Comput.-Based Syst. **5**(1/2), 31–54 (2014). https://doi.org/10.1504/IJCCBS.2014.059594

14. Duret-Lutz, A., Poitrenaud, D., Thierry-Mieg, Y.: Simplifying LTL model checking given prior knowledge (2025). https://hal.science/hal-04999078

15. Duret-Lutz, A., et al.: From Spot 2.0 to Spot 2.10: what's new? In: Proceedings of the 34th International Conference on Computer Aided Verification (CAV'22). Lecture Notes in Computer Science, vol. 13372, pp. 174–187. Springer, Heidelberg (2022). https://doi.org/10.1007/978-3-031-13188-2_9

16. Etessami, K., Holzmann, G.J.: Optimizing Büchi automata. In: Palamidessi, C. (ed.) CONCUR 2000. LNCS, vol. 1877, pp. 153–168. Springer, Heidelberg (2000). https://doi.org/10.1007/3-540-44618-4_13

17. Etessami, K., Wilke, T., Schuller, R.A.: Fair simulation relations, parity games, and state space reduction for büchi automata. In: Orejas, F., Spirakis, P.G., van Leeuwen, J. (eds.) ICALP 2001. LNCS, vol. 2076, pp. 694–707. Springer, Heidelberg (2001). https://doi.org/10.1007/3-540-48224-5_57

18. Fogarty, S., Vardi, M.Y.: Efficient Büchi universality checking. In: Esparza, J., Majumdar, R. (eds.) TACAS 2010. LNCS, vol. 6015, pp. 205–220. Springer, Heidelberg (2010). https://doi.org/10.1007/978-3-642-12002-2_17

19. Fritz, C.: Constructing Büchi automata from linear temporal logic using simulation relations for alternating Büchi automata. In: Ibarra, O.H., Dang, Z. (eds.) CIAA 2003. LNCS, vol. 2759, pp. 35–48. Springer, Heidelberg (2003). https://doi.org/10.1007/3-540-45089-0_5

20. Gastin, P., Oddoux, D.: Fast LTL to Büchi automata translation. In: Berry, G., Comon, H., Finkel, A. (eds.) CAV 2001. LNCS, vol. 2102, pp. 53–65. Springer, Heidelberg (2001). https://doi.org/10.1007/3-540-44585-4_6

21. Giannakopoulou, D., Lerda, F.: From states to transitions: improving translation of LTL formulae to Büchi automata. In: Peled, D.A., Vardi, M.Y. (eds.) FORTE 2002. LNCS, vol. 2529, pp. 308–326. Springer, Heidelberg (2002). https://doi.org/10.1007/3-540-36135-9_20

22. Godefroid, P. (ed.): Partial-Order Methods for the Verification of Concurrent Systems. LNCS, vol. 1032. Springer, Heidelberg (1996). https://doi.org/10.1007/3-540-60761-7

23. Haddad, S., Pradat-Peyre, J.F.: New efficient Petri nets reductions for parallel programs verification. Parallel Process. Lett. **16**(01), 101–116 (2006)

24. Holzmann, G.J., Kupferman, O.: Not checking for closure under stuttering. In: Proceedings of the 2nd workshop on the Spin Verification System (SPIN'96), pp. 17–22. American Mathematical Society (1996)

25. Kant, G., Laarman, A., Meijer, J., van de Pol, J., Blom, S., van Dijk, T.: LTSmin: high-performance language-independent model checking. In: Baier, C., Tinelli, C. (eds.) TACAS 2015. LNCS, vol. 9035, pp. 692–707. Springer, Heidelberg (2015). https://doi.org/10.1007/978-3-662-46681-0_61

26. Kordon, F., et al.: Complete Results for the 2022 Edition of the Model Checking Contest (2022). http://mcc.lip6.fr/2022/results.php

27. Kordon, F., Hillah, L.M., Hulin-Hubard, F., Jezequel, L., Paviot-Adet, E.: Study of the efficiency of model checking techniques using results of the MCC from 2015 To 2019. Int. J. Softw. Tools Technol. Transfer **23**(6), 931–952 (2021). https://doi.org/10.1007/s10009-021-00615-1

28. Larraz, D., Nimkar, K., Oliveras, A., Rodríguez-Carbonell, E., Rubio, A.: Proving non-termination using max-SMT. In: Biere, A., Bloem, R. (eds.) CAV 2014. LNCS, vol. 8559, pp. 779–796. Springer, Cham (2014). https://doi.org/10.1007/978-3-319-08867-9_52

29. Manna, Z., Pnueli, A.: A hierarchy of temporal properties. In: Proceedings of the sixth Annual ACM Symposium on Principles of Distributed Computing (PODC'90), pp. 377–410. ACM, New York (1990)

30. Michaud, T., Duret-Lutz, A.: Practical stutter-invariance checks for ω-regular languages. In: Fischer, B., Geldenhuys, J. (eds.) SPIN 2015. LNCS, vol. 9232, pp. 84–101. Springer, Cham (2015). https://doi.org/10.1007/978-3-319-23404-5_7

31. Minato, S.: Fast generation of irredundant sum-of-products forms from binary decision diagrams. In: Proceedings of the Third Synthesis and Simulation and Meeting International Interchange Workshop (SASIMI'92), Kobe, Japan, pp. 64–73 (1992)

32. Paull, M.C., Unger, S.H.: Minimizing the number of states in incompletely specified sequential switching functions. IRE Trans. Electron. Comput. **EC-8**(3), 356–367 (1959). https://doi.org/10.1109/TEC.1959.5222697

33. Peled, D.: Combining partial order reductions with on-the-fly model-checking. In: Dill, D.L. (ed.) CAV 1994. LNCS, vol. 818, pp. 377–390. Springer, Heidelberg (1994). https://doi.org/10.1007/3-540-58179-0_69

34. Peled, D., Wilke, T., Wolper, P.: An algorithmic approach for checking closure properties of temporal logic specifications and ω-regular languages. Theor. Comput. Sci. **195**(2), 183–203 (1998). https://doi.org/10.1016/S0304-3975(97)00219-3

35. Place, T., Zeitoun, M.: Separating regular languages with first-order logic. Logical Methods Comput. Sci. **12**(1) (2016). https://doi.org/10.2168/lmcs-12(1:5)2016

36. Renault, E., Duret-Lutz, A., Kordon, F., Poitrenaud, D.: Strength-based decomposition of the property Büchi automaton for faster model checking. In: Piterman,

N., Smolka, S.A. (eds.) TACAS 2013. LNCS, vol. 7795, pp. 580–593. Springer, Heidelberg (2013). https://doi.org/10.1007/978-3-642-36742-7_42

37. Renkin, F., Schlehuber-Caissier, P., Duret-Lutz, A., Pommellet, A.: Effective reductions of Mealy machines. In: Proceedings of the 42nd International Conference on Formal Techniques for Distributed Objects, Components, and Systems (FORTE'22). Lecture Notes in Computer Science. Springer, Heidelberg (2022)

38. Rozier, K.Y., Vardi, M.Y.: A multi-encoding approach for LTL symbolic satisfiability checking. In: Butler, M., Schulte, W. (eds.) FM 2011. LNCS, vol. 6664, pp. 417–431. Springer, Heidelberg (2011). https://doi.org/10.1007/978-3-642-21437-0_31

39. Schewe, S., Varghese, T.: Tight bounds for the determinisation and complementation of generalised Büchi automata. In: Chakraborty, S., Mukund, M. (eds.) ATVA 2012. LNCS, pp. 42–56. Springer, Heidelberg (2012). https://doi.org/10.1007/978-3-642-33386-6_5

40. Sebastiani, R., Tonetta, S.: "More Deterministic" vs. "Smaller" Büchi automata for efficient LTL model checking. In: Geist, D., Tronci, E. (eds.) CHARME 2003. LNCS, vol. 2860, pp. 126–140. Springer, Heidelberg (2003). https://doi.org/10.1007/978-3-540-39724-3_12

41. Sebastiani, R., Tonetta, S., Vardi, M.Y.: Symbolic systems, explicit properties: on hybrid approaches for LTL symbolic model checking. In: Etessami, K., Rajamani, S.K. (eds.) CAV 2005. LNCS, vol. 3576, pp. 350–363. Springer, Heidelberg (2005). https://doi.org/10.1007/11513988_35

42. Smolka, D.: Simulation-Based Reduction of Modal Omega-Automata. Bachelor's thesis, Mazaryk University, Faculty of Informatics (2023). https://is.muni.cz/th/qq7ad/?lang=en

43. Somenzi, F., Bloem, R.: Efficient Büchi automata from LTL formulae. In: Emerson, E.A., Sistla, A.P. (eds.) CAV 2000. LNCS, vol. 1855, pp. 248–263. Springer, Heidelberg (2000). https://doi.org/10.1007/10722167_21

44. Thierry-Mieg, Y.: Symbolic Model-Checking Using ITS-Tools. In: Baier, C., Tinelli, C. (eds.) TACAS 2015. LNCS, vol. 9035, pp. 231–237. Springer, Heidelberg (2015). https://doi.org/10.1007/978-3-662-46681-0_20

45. Thierry-Mieg, Y.: Structural reductions revisited. In: Janicki, R., Sidorova, N., Chatain, T. (eds.) PETRI NETS 2020. LNCS, vol. 12152, pp. 303–323. Springer, Cham (2020). https://doi.org/10.1007/978-3-030-51831-8_15

46. Tsai, M.H., Fogarty, S., Vardi, M.Y., Tsay, Y.K.: State of Büchi complementation. Logical Methods Comput. Sci. 10(4) (2014). https://doi.org/10.2168/lmcs-10(4:13)2014

47. Valmari, A.: On-the-fly verification with stubborn sets. In: Courcoubetis, C. (ed.) CAV 1993. LNCS, vol. 697, pp. 397–408. Springer, Heidelberg (1993). https://doi.org/10.1007/3-540-56922-7_33

48. Vardi, M.Y.: Automata-theoretic model checking revisited. In: Cook, B., Podelski, A. (eds.) VMCAI 2007. LNCS, vol. 4349, pp. 137–150. Springer, Heidelberg (2007). https://doi.org/10.1007/978-3-540-69738-1_10

49. Yan, Q.: Lower bounds for complementation of omega-automata via the full automata technique. Logical Methods Comput. Sci. 4(1) (2008)

Failure Resilience of Strongly Synchronized Processes

Rüdiger Valk[(✉)]

Department of Informatics, University of Hamburg, Hamburg, Germany
`ruediger.valk@uni-hamburg.de`

Abstract. Strictly synchronized sequential processes are characterized by a high degree of interdependence. The failure of a single such process can result in the deadlock of a large proportion of the entire system. We model a failure by removing the control token of a sequential process. A system design that limits such failures to the affected process can be very complicated. A method is presented whereby the crash of a sequential process does not affect the other processes.

Keywords: sequential processes · failure tolerance · structure of Petri nets · Petri net foldings · cycloids

1 Introduction

Consider a distributed system of a finite number of circular and sequential processes. The processes are synchronized by uni-directional one-bit channels in such a way that they behave like a circular traffic queue when folded together. To give an example, Fig. 1a) shows three such sequential circular processes a_0, a_1, a_2, each of length 7. In the initial state the control is in positions 0, 1 and 2, respectively. The synchronization, realized by the connecting channels, should be a queue when folded together, as shown in Fig. 1b). This means, that the controls of processes a_0 and a_1 cannot make a step until process a_2 makes a step itself, while the control of a_2 can make four steps until a_0 makes a step. Following [6] this behaviour is realized by the net of Fig. 1c). In this article we denote a process a_j by transitions $[t_0, a_j], \cdots, [t_{p-1}, a_j]$, where p is the process length. The output place of $[t_i, a_j]$ is denoted $[s_i, a_j]$, while the output place of the communication channel is $[s_i', a_j]$ (see $[s_3', a_0]$ for example in Fig. 1c).

Since all transitions in this model depend on other processes, the stop of one process brings the entire system to a standstill. Section 5 will show how the introduction of a net folding on the channel places changes the behaviour of the system in such a way that the non-failed processes continue unchanged. If, for example, the transition $[t_{stop}, a_j]$ in Fig. 2 stops the process a_j by removing the control token, the other processes continue to run normally. Furthermore, if the failed process is removed, a net of the same structure is obtained and a repeated failure elimination is adequately represented. As introduced in Sect. 4, this behaviour

E. Amparore and L. Mikulski (Eds.): PETRI NETS 2025, LNCS 15714, pp. 457–477, 2025.
https://doi.org/10.1007/978-3-031-94634-9_22

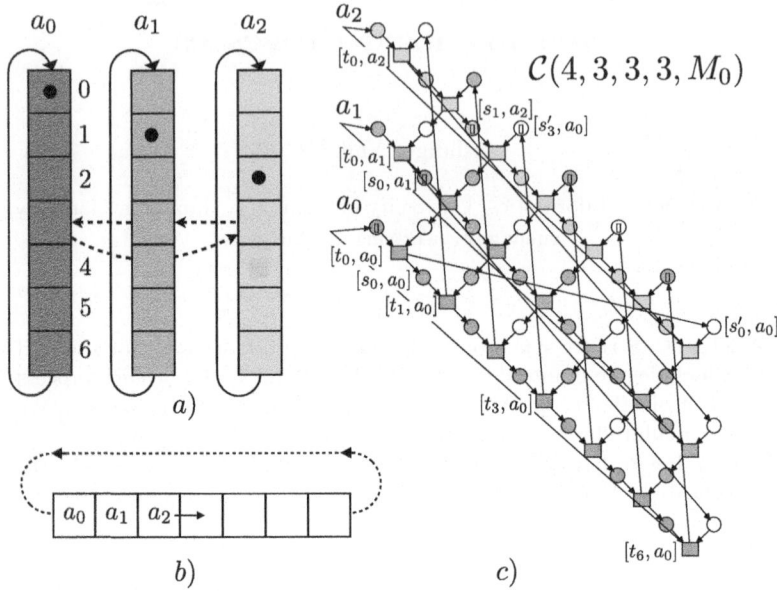

Fig. 1. Three sequential processes synchronized by single-bit channels.

is based on a previously introduced folding of the communication places, which do not change the behaviour of the system. In the graphical representation in Fig. 2 the seven folded places are identified by the name $[\![S_i]\!], 0 \leq i \leq 6$. Obviously, this modelling changes from synchronization by message passing to shared data synchronization. The results presented here are based on the now relatively well-developed theory of cycloids, which is briefly introduced in Sect. 2. These apply to the subclass of so-called regular cycloids, which is better adapted to the representation of co-operating processes (Sect. 3). The results shown for the net in Fig. 2 apply to all so-called canonical regular cycloids. However, an important contribution of this article is that they apply in modified form to the more general class of regular cycloids. It was also possible to prove for them which cycloid structure the process components have that remain alive after failure of some processes. These extended results are not yet included in the preprint [9]. For an initial review, Sect. 2 on cycloids can be skipped, as the rest of the article only deals with the coordinates for regular cycloids, as discussed in Sect. 3.

We recall some standard notations for set theoretical relations. If $R \subseteq A \times B$ is a relation and $U \subseteq A$ then $R[U] := \{b \mid \exists u \in U : (u, b) \in R\}$ is the *image* of U and $R[a]$ stands for $R[\{a\}]$. R^{-1} is the *inverse relation* and R^+ is the *transitive closure* of R if $A = B$. Also, if $R \subseteq A \times A$ is an equivalence relation then $[\![a]\!]_R$ is the *equivalence class* of the quotient A/R containing a. Furthermore $\mathbb{N}$, $\mathbb{N}_+$, $\mathbb{Z}$ and $\mathbb{R}$ denote the sets of integers, positive integer, integer and real numbers, respectively. For integers: $a|b$ if a is a factor of b. The *modulo* function is used in the form $a \bmod b = a - b \cdot \lfloor \frac{a}{b} \rfloor$, which also holds for negative integers $a \in \mathbb{Z}$. As a short notation we write $x \oplus_b y$ for $(x + y) \bmod b$ and $x \ominus_b y$ for $(x - y) \bmod b$. From the above equation follow $-a \bmod b = b - a$ for $0 < a \le b$ and

$$a \ominus_b b = \begin{cases} 0 & \text{if } a = b \\ a & \text{if } a < b \end{cases}.$$

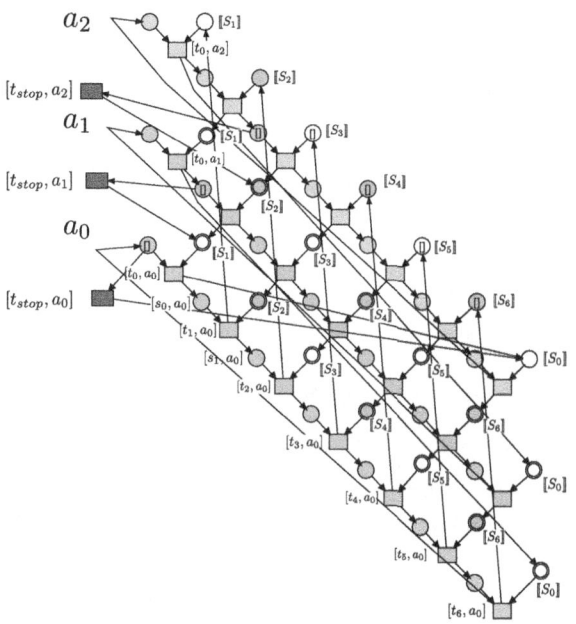

Fig. 2. Stop resilient cycloid $\mathcal{C}_{bf}^{stop}(4, 3, 3, 3, \llbracket M_0 \rrbracket)$.

2 Petri Space and Cycloids

We define (Petri) nets as they will be used in this article.

Definition 1 ([5]). *As usual, a net* $\mathcal{N} = (S, T, F)$ *is defined by non-empty, disjoint sets* S *of places and* T *of transitions, connected by a flow relation* $F \subseteq (S \times T) \cup (T \times S)$ *and* $X := S \cup T$. *A transition* $t \in T$ *is* active *or* enabled *in a marking* $M \subseteq S$ *if* $^\bullet t \subseteq M$. *In this case we obtain* $M \xrightarrow{t} M'$ *if* $M' = (M \backslash {}^\bullet t) \cup t^\bullet$, *where* $^\bullet x := F^{-1}[x]$, $x^\bullet := F[x]$ *denote the sets of input and output elements of an element* $x \in X$, *respectively.* $\xrightarrow{*}$ *is the reflexive and transitive closure of* $\to$.

After many relevant lectures, Carl Adam Petri published the formalism of cycloids in his article *Nets, Time and Space* [3]. Based on formal descriptions in [2] and [1] more elaborate formalizations are given in [5]. Petri started with an event-oriented version of the Minkowski space which is called Petri space now. Cycloids were defined by him as a folding of the unbounded Petri space into a substructure that is finite with respect to space and time. More detailed motivation can be found in [5–8,10].

Figure 3b) shows the coordinate grid of the Petri space with the axes ξ for the time dimension and η for the space dimension. Part a) of this figure describes the system of naming the transition elements of the Petri space. The places obtain their names by their input transitions.

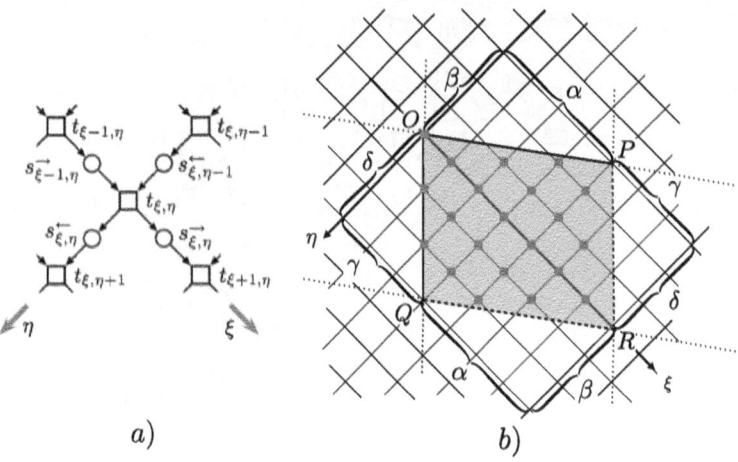

a)   b)

Fig. 3. a) Petri space coordinates and b) Fundamental parallelogram of $\mathcal{C}(4,3,3,3)$.

Definition 2 ([5]). *A Petri space is defined by the net* $\mathcal{PS}_1 := (S_1, T_1, F_1)$ *where* $S_1 = S_1^{\rightarrow} \cup S_1^{\leftarrow}$, $S_1^{\rightarrow} = \{s_{\xi,\eta}^{\rightarrow} \mid \xi, \eta \in \mathbb{Z}\}$, $S_1^{\leftarrow} = \{s_{\xi,\eta}^{\leftarrow} \mid \xi, \eta \in \mathbb{Z}\}$, $S_1^{\rightarrow} \cap S_1^{\leftarrow} = \emptyset$, $T_1 = \{t_{\xi,\eta} \mid \xi, \eta \in \mathbb{Z}\}$, $F_1 = \{(t_{\xi,\eta}, s_{\xi,\eta}^{\rightarrow}) \mid \xi, \eta \in \mathbb{Z}\} \cup \{(s_{\xi,\eta}^{\rightarrow}, t_{\xi+1,\eta}) \mid \xi, \eta \in \mathbb{Z}\} \cup \{(t_{\xi,\eta}, s_{\xi,\eta}^{\leftarrow}) \mid \xi, \eta \in \mathbb{Z}\} \cup \{(s_{\xi,\eta}^{\leftarrow}, t_{\xi,\eta+1}) \mid \xi, \eta \in \mathbb{Z}\}$. $S_1^{\rightarrow}$ *is the set of forward places and* $S_1^{\leftarrow}$ *the set of backward places.* $^{\rightarrow}t_{\xi,\eta} := s_{\xi-1,\eta}^{\rightarrow}$ *is the forward input place of* $t_{\xi,\eta}$ *and in the same way* $^{\leftarrow}t_{\xi,\eta} := s_{\xi,\eta-1}^{\leftarrow}$, $t_{\xi,\eta}^{\rightarrow\bullet} := s_{\xi,\eta}^{\rightarrow}$ *and* $t_{\xi,\eta}^{\leftarrow\bullet} := s_{\xi,\eta}^{\leftarrow}$ *(Fig. 3a).*

In two steps, by a twofold folding with respect to time and space, Petri defined the cyclic structure of a cycloid. One of these steps is a folding f with respect to space with $f(\xi,\eta) = f(\xi+\alpha, \eta-\beta)$, fusing all points (ξ,η) of the Petri space with $(\xi+\alpha, \eta-\beta)$ where $\xi, \eta \in \mathbb{Z}, \alpha, \beta \in \mathbb{N}_+$ ([3], page 37). As in the next definition, the second step then consists of a folding of the time dimension using γ and δ. Transitions folded to the same element are equivalent with respect to the relation $\equiv$.

Definition 3 ([5]). *A cycloid is a net* $\mathcal{C}(\alpha, \beta, \gamma, \delta) = (S, T, F)$, *defined by para-meters* $\alpha, \beta, \gamma, \delta \in \mathbb{N}_+$, *by a quotient of the Petri space* $\mathcal{PS}_1 := (S_1, T_1, F_1)$ *with respect to the equivalence relation* $\equiv\,\subseteq X_1 \times X_1$ *with* $X_1 = S_1 \cup T_1$, $\equiv[S_1^{\rightarrow}] \subseteq S_1^{\rightarrow}, \equiv[S_1^{\leftarrow}] \subseteq S_1^{\leftarrow}, \equiv[T_1] \subseteq T_1$, *defined by* $x_{\xi,\eta} \equiv x_{\xi+m\alpha+n\gamma, \eta-m\beta+n\delta}$ *for all* $\xi, \eta, m, n \in \mathbb{Z}$ *and the conditions* $X = X_1/_{\equiv}$ *and* $[\![x]\!]_{\equiv} F [\![y]\!]_{\equiv} \Leftrightarrow$ $\exists\, x' \in [\![x]\!]_{\equiv} \exists\, y' \in [\![y]\!]_{\equiv} : x' F_1 y'$ *for all* $x, y \in X_1$. *The matrix* $\mathbf{A} = \begin{pmatrix} \alpha & \gamma \\ -\beta & \delta \end{pmatrix}$ *is called the matrix of the cycloid. Petri denoted the number* A *of transitions as the area of the cycloid and proved in* [3] *its value to* $A = \alpha\delta + \beta\gamma$ *which equals the determinant* $A = det(\mathbf{A})$. *The embedding of a cycloid in the Petri space is called* fundamental parallelogram *(see Fig. 3b). The term* $n := \alpha + \beta$ *is called* standard initial number, *or shortly* number *of* $\mathcal{C}(\alpha, \beta, \gamma, \delta)$.

Figure 3b) gives the fundamental parallelogram of the cycloid $\mathcal{C}(\alpha, \beta, \gamma, \delta) = \mathcal{C}(4, 3, 3, 3)$ within the Petri space. All coordinate elements within the parallelo-gram and on solid edges represent transitions. Of the corners, only the origin O belongs to it, but not P, Q and R. The latter belong to neighbouring parallelo-grams that coincide due to the folding. Figure 4b) gives the same cycloid in net representation.

Theorem 4 ([5,8]). *The following cycloids are isomorphic[1] to* $\mathcal{C}(\alpha, \beta, \gamma, \delta)$:

a) $\mathcal{C}(\beta, \alpha, \delta, \gamma)$,   (*The* dual cycloid *of* $\mathcal{C}(\alpha, \beta, \gamma, \delta)$.)
b) $\mathcal{C}(\alpha, \beta, \gamma - q \cdot \alpha, \delta + q \cdot \beta)$ *if* $q \in \mathbb{N}_+$ *and* $\gamma > q \cdot \alpha$,
c) $\mathcal{C}(\alpha, \beta, \gamma + q \cdot \alpha, \delta - q \cdot \beta)$ *if* $q \in \mathbb{N}_+$ *and* $\delta > q \cdot \beta$.

For proving the equivalence of two points in the Petri space the following procedure[2] is useful.

Theorem 5 ([6,9]). *With respect to a cycloid* $\mathcal{C}(\alpha, \beta, \gamma, \delta)$ *two points* $\boldsymbol{x}_1, \boldsymbol{x}_2 \in X_1$ *of the Petri space are equivalent* $\boldsymbol{x}_1 \equiv \boldsymbol{x}_2$ *if and only if for the difference* $\boldsymbol{v} := \boldsymbol{x}_2 - \boldsymbol{x}_1$ *the parameter vector* $\pi(\boldsymbol{v}) = \frac{1}{A} \cdot \mathbf{B} \cdot \boldsymbol{v}$ *has integer values, where* A *is the area and* $\mathbf{B} = \begin{pmatrix} \delta & -\gamma \\ \beta & \alpha \end{pmatrix}$. *In analogy to Definition 3 we obtain* $\boldsymbol{x}_1 \equiv \boldsymbol{x}_2 \Leftrightarrow$ $\exists\, m, n \in \mathbb{Z} : \boldsymbol{x}_2 - \boldsymbol{x}_1 = \mathbf{A} \begin{pmatrix} m \\ n \end{pmatrix}$.

The transitions connected via forward places allow cycloids to be structured into cyclic sequential processes that are synchronized by channels via the back-ward places.

---

[1] By a net isomorphism [4].

[2] The algorithm is implemented under https://cycloids.de.

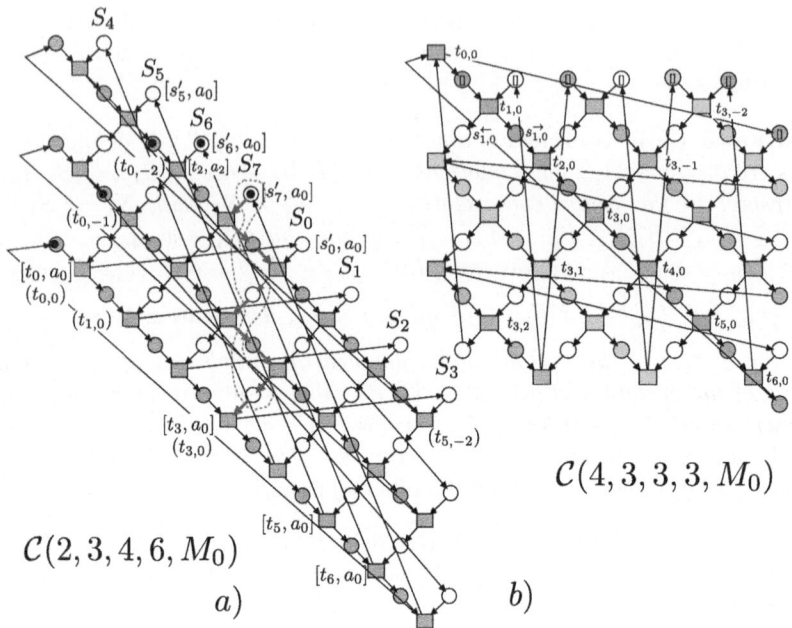

Fig. 4. Cycloids in regular a) and Fundamental Parallelogram b) representation.

Definition 6. *If, starting from a transition $t_{\xi,\eta}$ of a cycloid $\mathcal{C}(\alpha,\beta,\gamma,\delta)$, the forward output place $t_{\xi,\eta}^{\bullet\rightarrow} := s_{\xi,\eta}^{\rightarrow}$ (Definition 2) is selected repeatedly, a cycle called* forward cycle *or* forward process *is obtained.*

Since it is customary to call nets with initial marking net systems, we introduce the term cycloid system accordingly.

Definition 7 ([5]). *For a cycloid $\mathcal{C}(\alpha,\beta,\gamma,\delta)$ we define a cycloid system $\mathcal{C}(\alpha,\beta,\gamma,\delta,M_0)$ or $\mathcal{C}(\mathcal{N},M_0)$ by adding the so-called* standard initial marking:
$$M_0 = \{s_{\xi,\eta}^{\rightarrow} \in S_1^{\rightarrow} \mid \beta\xi + \alpha\eta \leq 0 \ \wedge \ \beta(\xi+1) + \alpha\eta > 0\}/_\equiv \ \cup$$
$$\{s_{\xi,\eta}^{\leftarrow} \in S_1^{\leftarrow} \mid \beta\xi + \alpha\eta \leq 0 \ \wedge \ \beta\xi + \alpha(\eta+1) > 0\}/_\equiv.$$
Note, that by [5] *a standard initial marking contains β tokens in forward and α tokens in backward places.*

The motivation of this definition is given in [5]. See Fig. 4b) for an example of a cycloid with standard initial marking. The following definition of a regular initial marking in Petri space coordinates is relevant for Sect. 3. For an example in Fig. 4a) some Petri space coordinates are added in round brackets: $[t_2, a_2]$ in regular coordinates is $t_{0,-2}$ in Petri space coordinates. $s_{-1,i}^{\rightarrow}$ is the forward output place of $t_{-1,i}$. Regular initial markings will be defined for Petri space coordinates in Definition 8 and translated to regular coordinates in Corollary 13. They are motivated by the observation that the process a_j starts in position $j-1$ (see Figs. 1 and 4a).

Definition 8. *For a cycloid $C(\alpha, \beta, \gamma, \delta)$ a regular initial marking is defined by a number of β forward places $\{s^{\rightarrow}_{-1,i}|\ 0 \geq i > -\beta\}$ and a number of α backward places $\{s^{\leftarrow}_{i,-\beta}|\ 0 \leq i < \alpha\}$ all in Petri space coordinates. Note, that a regular initial marking contains β tokens in forward and α tokens in backward places.*

3   Regular Cycloids

Many applications, like circular traffic queues, are composed by a number of sequential and interacting processes of equal length, each of which has a single control token. In the formalism of cycloids this corresponds to a number of β disjoint forward cycles of equal length p. Cycloids with such a property are called *regular*. The widespread type of cycloids as shown in Fig. 1c) form important subclasses and are called *canonical regular*.

Definition 9. *A cycloid system $C = C(\alpha, \beta, \gamma, \delta, M_0)$ is called* regular *if β divides δ. Furthermore C is called* canonical regular *if $\beta = \gamma = \delta$.*

Theorem 10. *A regular cycloid system $C = C(\alpha, \beta, \gamma, \delta, M_0)$ with area A is composed of β disjoint processes (Definition 6), each of length $p = \frac{A}{\beta}$ and containing exactly one token.*

Proof. The proof is done by considering a path in the Petri space, which is folded into the cycloid by the equivalence $\equiv$. We follow a path from a transition $t_{\xi,\eta}$ with coordinates (ξ, η) by p steps only via forward output places by proceeding in the ξ−dimension until a point $(\xi + p, \eta)$ is obtained, which is, by closing the cycle, equivalent to (ξ, η). By Theorem 5 a necessary and sufficient condition

for $\begin{pmatrix} \xi + p \\ \eta \end{pmatrix} \equiv \begin{pmatrix} \xi \\ \eta \end{pmatrix}$ is $\pi(\begin{pmatrix} \xi + p \\ \eta \end{pmatrix} - \begin{pmatrix} \xi \\ \eta \end{pmatrix}) = \pi(\begin{pmatrix} p \\ 0 \end{pmatrix}) = \frac{1}{A} \begin{pmatrix} \delta & -\gamma \\ \beta & \alpha \end{pmatrix} \begin{pmatrix} p \\ 0 \end{pmatrix} =$

$\frac{1}{A} \begin{pmatrix} \delta \cdot p \\ \beta \cdot p \end{pmatrix} \in \mathbb{Z}^2$. This is equivalent to $\frac{\delta}{A} \cdot p \in \mathbb{Z} \wedge \frac{\beta}{A} \cdot p \in \mathbb{Z}$. For the first part

of the conjunction we obtain $\frac{\delta}{A} \cdot p = \frac{\delta}{A} \cdot \frac{A}{\beta} = \frac{\delta}{\beta} \in \mathbb{Z}$ since $\beta | \delta$. In a similar

way for the second part: $\frac{\beta}{A} \cdot p = \frac{\beta}{A} \cdot \frac{A}{\beta} = 1 \in \mathbb{Z}$. For a smaller value $p' < p$ in the second part, a value smaller 1 is obtained. Therefore all points before on the process from (ξ, η) to $(\xi, \eta + p)$ are distinct. Hence, we have exactly a number of β processes of equal length $p = \frac{A}{\beta}$ sharing exactly one of the β tokens in forward places of the initial marking.   □

To exploit the structure of a regular cycloid we define specific coordinates, called *regular coordinates*. The process a_0 starts with transition $t_{0,0}$ which is denoted $[t_0, a_0]$, having the input place $[s_{p-1}, a_0]$. The next transitions are $[t_1, a_0]$ up to $[t_{p-1}, a_0]$ and then returning to $[t_0, a_0]$. The other processes a_1 to $a_{\beta-1}$ are denoted in the same way (see Fig. 1c). As the process a_j starts in position $j - 1$ of the queue, its initial token is in $[s_{j \ominus_p 1}, a_j]$.

Definition 11. *Given a regular cycloid* $\mathcal{C}(\alpha, \beta, \gamma, \delta)$, *regular coordinates are defined as follows: transitions of an* a_j-*process* $0 \le j < \beta$, *each with length* p, *are denoted by* $\{[t_0, a_j], \cdots, [t_{p-1}, a_j]\}$. *For each transition* $[t_i, a_j]$ *we define the output places by* $[t_i, a_j]^{\overset{\bullet}{\rightarrow}} := [s_i, a_j]$ *and* $[t_i, a_j]^{\overset{\bullet}{\downarrow}} := [s'_i, a_j]$ *and the output transition by* $[s_i, a_j]^\bullet := [t_{i \oplus_p 1}, a_j]$ *for* $0 \le i < p, 0 \le j < \beta$. *Regular coordinates are related to standard coordinates of the Petri space by defining the following initial condition* $\mathrm{stand}([t_0, a_j]) := t_{-j,-j}$ *for* $0 \le j < \beta$ *(taking the equivalent transition of* $t_{-j,-j}$ *in the fundamental parallelogram).*

For instance, in Fig. 4a) we obtain for the last formula in Definition 11: $[t_0, a_2] := t_{-2,-2}$. While the output place $[s'_i, a_j]$ in regular coordinates takes its name from the input transition in Definition 11, it remains to determine its output transition according to the corresponding regular coordinates.

Lemma 12. *In a regular cycloid the injective mapping* stand *from regular to Petri space coordinates is given by* $\mathrm{stand}([t_i, a_j]) = t_{i-j,-j}$ *for* $0 \le i < p$ *and* $0 \le j < \beta$ *(modulo equivalent transitions). The output transition of* $[s'_i, a_j]$ *is*

a) $[s'_i, a_0]^\bullet = [t_{(i+n-1) \bmod p}, a_{\beta-1}]$ *for* $j = 0$ *and*
b) $[s'_i, a_j]^\bullet = [t_{i \ominus_p 1}, a_{j \ominus_\beta 1}]$ *for* $0 < j < \beta$.
c) *If* $p = n$ *the two cases coincide.*

Proof. For a given j by Definition 11 we have $\mathrm{stand}[t_0, a_j] := t_{-j,-j}$. Adding a value $i \in \{0, \cdots, p-1\}$ to the index of t_0 we obtain the index of t_i, hence $\mathrm{stand}([t_i, a_j]) := t_{-j+i,-j}$.

a) By the preceding result $\mathrm{stand}([t_i, a_0]) := t_{i,0}$. To prove a) we observe that in the fundamental parallelogram from $[t_i, a_0] = {}^\bullet[s'_i, a_0]$ we should come to $[s'_i, a_0]^\bullet = [t_{(i+n-1) \bmod p}, a_{\beta-1}]$ by taking one step in the η-direction. Therefore it is sufficient to prove $\mathrm{stand}([t_i, a_0]) + (0,1) \equiv \mathrm{stand}([t_{(i+\beta+\alpha-1)}, a_{\beta-1}])$

or $\begin{pmatrix} i \\ 0 \end{pmatrix} + \begin{pmatrix} 0 \\ 1 \end{pmatrix} \equiv \begin{pmatrix} i + \beta + \alpha - 1 - (\beta-1) \\ -(\beta-1) \end{pmatrix}$ or $\begin{pmatrix} i \\ 1 \end{pmatrix} \equiv \begin{pmatrix} i + \alpha \\ 1 - \beta \end{pmatrix}$. By

Theorem 5 we derive $\pi(\begin{pmatrix} i \\ 1 \end{pmatrix} - \begin{pmatrix} i + \alpha \\ 1 - \beta \end{pmatrix}) = \pi(\begin{pmatrix} -\alpha \\ \beta \end{pmatrix}) = \frac{1}{A} \begin{pmatrix} \delta & -\gamma \\ \beta & \alpha \end{pmatrix} \begin{pmatrix} -\alpha \\ \beta \end{pmatrix} =$

$\frac{1}{A} \begin{pmatrix} -A \\ 0 \end{pmatrix} \in \mathbb{Z} \times \mathbb{Z}$.

b) The same method results in the following equivalence to be proved:

$\mathrm{stand}([t_i, a_j]) + (0,1) \equiv \mathrm{stand}([t_{(i-1)}, a_{j-1}])$ or $\begin{pmatrix} i - j \\ -j \end{pmatrix} + \begin{pmatrix} 0 \\ 1 \end{pmatrix} \equiv$

$\begin{pmatrix} i - 1 - (j-1) \\ 1 - j \end{pmatrix}$ which is obvious (without using Theorem 5).

c) If $p = \alpha + \beta = n$ then $(i + \beta + \alpha - 1) \bmod p = (i - 1) \bmod p$. $\qquad\square$

Corollary 13. *a) The regular initial marking of a regular cycloid system* $\mathcal{C}(\alpha, \beta, \gamma, \delta, M_0)$ *with process length* p *in regular coordinates is* $M_0 = \{[s_{p-1}, a_0]\} \cup \{[s_i, a_{i+1}] | 0 \le i < \beta - 1\} \cup \{[s'_i, a_0] | p - \alpha \le i < p\}$.
b) From the regular initial marking M_0 *for each* $k \in \{0, \cdots, p-1\}$ *the marking* $M_k = \{[(s_{p-1+k}) \bmod p, a_0]\} \cup \{[s_{i \oplus_p k}, a_{i+1}] | 0 \le i < \beta - 1\} \cup \{[s'_{i \oplus_p k}, a_0] | p - \alpha \le i < p\}$ *is reachable by* $k \cdot \beta$ *transition occurrences.* M_k *is called a* k-*regular or simply* regular *marking of the cycloid.*

Proof. a) Since $p - 1 = (-1) \bmod p$ we can write $\{[s_i, a_{i+1}]| -1 \le i < \beta - 1\}$ for the forward places of M_0. As the mapping *stand* is defined on transitions, we go to the input transitions and apply *stand* to obtain $\{[t_i, a_{i+1}]| -1 \le i < \beta - 1\}$ and $\{stand([t_i, a_{i+1}])| -1 \le i < \beta - 1\} = \{t_{-1,-(i+1)}| -1 \le i < \beta - 1\} = \{t_{-1,i}| 0 \ge i > -\beta\}\}$, which is the same as the set of input transitions of the forward places in Definition 8. To deal with the third part of the union we prove that $U := \{[s_i', a_0]| p - \alpha \le i < p\}$ maps to $V := \{s_{i,-\beta}^{\leftarrow}| 0 \le i < \alpha\}$. To simplify the notation we start from the index $i = p - \alpha$ in U. In order to be able to use the map *stand*, we consider the output transition of $[s_{p-\alpha}', a_0]$, which is by Lemma 12 a): $[s_{p-\alpha}', a_0]^\bullet = [t_{(p-\alpha+\alpha+\beta-1) \bmod p}, a_{\beta-1}] = [t_{(p+\beta-1) \bmod p}, a_{\beta-1}] = [t_{\beta-1}, a_{\beta-1}]$ and $stand([t_{\beta-1}, a_{\beta-1}]) = t_{0,1-\beta}$. Now going back to the backward input place in the Petri space $s_{0,-\beta}^{\leftarrow}$ is obtained, which is the value of V for $i = 0$. By gradually increasing the index from $i = 0$ to $i = \alpha - 1$, the remaining values of V are obtained.

b) In the marking M_k transitions $[t_i, a_{\beta-1}], [t_{i\ominus_p 1}, a_{\beta-2}], \cdots, [t_{i\ominus_p \beta}, a_0]$ for some $0 \le i < p$ can occur and the indices of the forward markings are increased by 1. The token in the first place $[s_{(p-\alpha+k) \bmod p}', a_0]$ in the backward places of M_k is removed by $[t_i, a_{\beta-1}]$. The last transition $[t_{i\ominus_p \beta}, a_0]$ adds a token to $[s_{(p-1+k+1) \bmod p}, a_0]$ and therefore also to $[s_{(p-1+k+1) \bmod p}', a_0]$. Therefore the set $\{[s_{i\oplus_p k}', a_0]| p - \alpha \le i < p\}$ of M_k is transformed to the corresponding set $\{[s_{(i+k+1) \bmod p}', a_0]| p - \alpha \le i < p\}$ of M_{k+1}. $\qquad\square$

In the regular cycloid system $\mathcal{C}(2, 3, 4, 6, M_0)$ in Fig. 4a) we obtain for the last part in the proof of Corollary 13 a) $[s_{p-\alpha}', a_0]^\bullet = [s_{8-2}', a_0]^\bullet = [t_2, a_2]$ and $stand([t_2, a_2]) = t_{0,-2}$.

Lemma 14. *In a regular cycloid for* $0 \le j < p$:

a) $^\bullet[t_i, a_{\beta-1}] = [s_{i\ominus_p(n-1)}', a_0]$

b) $^\bullet[t_i, a_j] = [s_{i\oplus_p 1}', a_{j+1}]$ *for* $0 \le j < \beta - 1$.

Proof. a) By Lemma 12 a) $[s_k', a_0]^\bullet = [t_{(k+n-1) \bmod p}, a_{\beta-1}]$, hence $^\bullet[t_i, a_{\beta-1}] = [s_k', a_0]$ with $i = (k + n - 1) \bmod p$ and $k = (i - n + 1) \bmod p = i \ominus_p (n - 1)$.

b) By Lemma 12 b) $[s_k', a_j]^\bullet = [t_{k\ominus_p 1}, a_{j\ominus_\beta 1}]$ for $0 < j < \beta$. It follows $^\bullet[t_i, a_j] = [s_k', a_{j+1}]$ with $i = k \ominus_p 1$, hence $k = i \oplus_p 1$. $\qquad\square$

4   Backward Foldings and Cycloid Hierarchies

Considered as cooperating processes, cycloids perform a strong synchronization regimen. Therefore it is surprising that by a small extension we can model these processes to be stoppable or failing without stopping the other processes. As a by-product it is proved that by eliminating the process with the highest index $\beta - 1$ from a canonical regular cycloid $\mathcal{C}(\alpha, \beta, \beta, \beta)$ we obtain $\mathcal{C}(\alpha+1, \beta-1, \beta-1, \beta-1)$. If the illustration of the cycloid $\mathcal{C}(4, 3, 3, 3)$ in Fig. 1 is considered by the circular queue in part b) of this figure, this means that after the elimination of a_2 instead of $\beta = 3$ queue elements and $\alpha = 4$ empty slots, there are now only $\beta - 1 = 2$ queue elements and $\alpha + 1 = 5$ empty slots. The results are not limited to such canonical regular cycloids, but apply to all regular cycloids. As this construction can be repeated, a hierarchy of regular cycloids results.

Since the extension is defined by a folding of the backward places only, the forward places and transitions of the processes are not modified and the cycloid algebra (Theorem 5 and [9]) can be applied. To obtain a live system when one process is stopped we require that at least $\beta > 1$ processes are present. As a first step, we prove in this section that the failure of the last process $a_{\beta-1}$ is again a cycloid and calculate its parameters.

Definition 15. *For a given regular cycloid system* $C = C(\alpha, \beta, \gamma, \delta, M_0)$ *with* $\beta > 1$, *process length* p *and a fixed set* $D \subseteq \{0, \cdots, \beta - 1\}$ *with* $|D| > 1$, *called the set of* back indices, *we define the* backward folding $C_{bf(D)}(\alpha, \beta, \gamma, \delta, [\![M_0]\!]_D)$ *by a relation* $\equiv_{bf(D)}$ *on the backward places* $[t_i, a_j]^{\bullet\rightarrow} = [s_i', a_j]$ *by*

$$[s_i', a_j] \quad \equiv_{bf(D)} \quad [s_r', a_s] \quad \Leftrightarrow \quad i = r \ \wedge \ \{j, s\} \subseteq D \quad \text{for} \quad 0 \le i, r < p \quad (1)$$

The folding is extended to markings by $[\![M]\!]_D := \{[\![s]\!]_{bf(D)} | s \in M\}^3$, *where* $[\![s]\!]_{bf(D)}$ *denotes the class containing* s. *If* $D = \{0, \cdots, \beta - 1\}$ *the folding and the equivalence relation are called* total *and denoted by* $C_{bf}(\alpha, \beta, \gamma, \delta, [\![M_0]\!])$ *and* $\equiv_{bf}$, *respectively. Many properties of a cycloid, like number* n, *area* A, *process length* p, *regular or canonical regular are transferable to its backward folding.*

The folding is defined on backward places modelling the channels of the cooperating processes. Therefore by the folding we switch from a message oriented synchronization to a shared variable synchronization mechanism. If $j \in D$ then the process $a_{j \ominus_c 1}$ is sharing its backward input places.

Lemma 16. *a) For* $0 \le i < p$ *the class* S_i^{bf} *(or shorter* S_i*) of* $[s_i', a_0]$ *with respect to* $\equiv_{bf}$ *(and total folding) is* $S_i^{bf} := \{[s_i', a_0]\} \cup \{[s_{i \oplus_p n}', a_j] | 0 < j < \beta\}$ *with* $n = \alpha + \beta$. *When used as the name of a (fused) place we write* $[\![S_i]\!]_{bf}$ *or shorter* $[\![S_i]\!]$ *for this class. The input and output transitions of such a place are* $^\bullet[\![S_i]\!] = \{[t_i, a_0]\} \cup \{[t_{i \oplus_p n}, a_j] | 0 < j < \beta\}$ *and* $[\![S_i]\!]^\bullet = \{[t_{(i+n-1) \bmod p}, a_{\beta-1}]\} \cup \{[t_{(i+n-1) \bmod p}, a_{j \ominus_\beta -1}] | 0 < j < \beta\}$.
b) The classes of the relation $\equiv_{bf(D)}$ *on the backward places are* $S_i^D :=$
$$\begin{cases} \{[s_i', a_0]\} \cup \{[s_{i \oplus_p n}', a_j] | j \in D \setminus \{0\}\} & \text{if } 0 \in D \\ \{[s_{i \oplus_p n}', a_j] | j \in D\} & \text{if } 0 \notin D \end{cases} \quad \text{for} \quad 0 \le i < p.$$

Proof. a) $[s_i', a_0] \in S_i^{bf}$ by definition. The output transition of $[s_i', a_0]$ is by Lemma 12 $[s_i', a_0]^\bullet = [t_{(i+n-1) \bmod p}, a_{\beta-1}]$. By Definition 15 the remaining elements of S_i^{bf} are $^{\bullet\rightarrow}[t_{(i+n-1) \bmod p}, a_j]$ for $0 < j < \beta$. By Lemma 14 b) we obtain for second set of S_i^{bf} in this lemma : $\{^{\bullet\rightarrow}[t_{(i+n-1) \bmod p}, a_j] | 0 \le j < \beta - 1\} = \{[s_{(i+n-1+1) \bmod p}', a_{j+1}] | 0 \le j < \beta - 1\} = \{[s_{(i \oplus_p n)}', a_j] | 0 < j < \beta\}$.
b) This follows directly from a) as $\equiv_{bf(D)}$ is equal or finer than $\equiv_{bf}$. $\qquad\square$

---

[3] We will show that in each reachable marking of the cycloids under investigation each class contains at most one token.

For the cycloid $\mathcal{C}(2, 3, 4, 6, M_0)$ from Fig. 4a) and total D we obtain $A = 24, p = 8, n = \alpha + \beta = 5$ and $S_0 = \{[s'_0, a_0], [s'_5, a_1], [s'_5, a_2]\}^4$, $S_1 := \{[s'_1, a_0], [s'_6, a_1], [s'_6, a_2]\}, \cdots, S_7 := \{[s'_7, a_0], [s'_4, a_1], [s'_4, a_2]\}$ (dotted line for S_7). Note that for canonical regular cycloids (where $p = n$) $S_i^{bf} = \{[s'_i, a_j] | 0 \leq j < \beta\}$. For the definition of place invariants the following Lemma will be used.

Lemma 17. *The places of a class S_i^{bf} of a backward folding are included in a path, called bf-path or i-bf-path, as follows: $[s'_i, a_0], [t_{(i+n-1) \bmod p}, a_{\beta-1}], \cdots,$ $[t_{(i+n-1) \bmod p}, a_j], [s_{(i+n-1) \bmod p}, a_j]^*,$ $[t_{(i+n) \bmod p}, a_j], [s'_{(i+n) \bmod p}, a_j],$ $[t_{(i+n-1) \bmod p}, a_{j-1}], \cdots, [t_{(i+n-1) \bmod p}, a_0]$. A i-bf-path and a a_j-process share the place $[s_{(i+n-1) \bmod p}, a_j]$ $(0 \leq i < p, 1 \leq j < \beta)$. For $p - n + 1 \leq i < p - \alpha$ which is i $= p - n + 1 + k$ $(0 \leq k < \beta - 1)$ the bf-path starting in $[s'_i, a_0]$ contains exactly one token in the regular initial marking.*

Proof. By Lemma 12 the output transition of $[s'_i, a_0]$ is $[s'_i, a_0]^\bullet = [t_{(i+n-1) \bmod p}, a_{\beta-1}]$. By induction, from $[t_{(i+n-1) \bmod p}, a_j]$ $(0 < j < \beta)$ we come in 4 steps to $[t_{(i+n-1) \bmod p}, a_{j-1}]$ which gives for $j = 1$ the end of the bf-path $[t_{(i+n-1) \bmod p}, a_0]$. The place shared with the a_j-process is marked in the lemma by an asterisk. The place $[s_k, a_{k+1}]$ $(0 \leq k < \beta - 1)$ contains the single token the bf-path starting in $[s'_i, a_0]$. $\qquad\square$

For an example of a bf-path, see the highlighted path from $[s'_7, a_0]$ to $[t_3, a_0]$ in Fig. 4a) with $n = 5, p = 8$. It is sharing the place $[s_3, a_1]$ with the a_1-process. It is important to prove that under a mild restriction the backward folding of a cycloid is safe and live.

Theorem 18. *The backward folding $\mathcal{C}_{bf(D)}(\alpha, \beta, \gamma, \delta, [M_0])$ of a regular cycloid system $\mathcal{C}(\alpha, \beta, \gamma, \delta, M_0)$ with regular M_0 and $n - 1 \leq p$ is a safe net, i.e. in each reachable marking each place contains at most one token.*

Proof. Since the cycloid system $\mathcal{C}(\alpha, \beta, \gamma, \delta, M_0)$ is safe (Theorem 5.4 of [5]), it is sufficient to prove that each equivalence class S_i^{bf} (Lemma 16) is contained in a S-invariant containing exactly one token. Thus the same is holding for the classes S_i^D of $\equiv_{bf(D)}$. As all places of a cycloid have exactly one input and output transition the S-invariants can be defined by cycles. We distinguish two cases, namely those where $[s'_i, a_0]$ is marked in the regular initial marking $p - \alpha \leq i < p$ and the complementary case $0 \leq i < p - \alpha$.

Case 1: $[s'_i, a_0]$ is marked. Starting from $[s'_i, a_0]$ the cycle to be constructed initially is the bf-path until the a_0-process is reached and then follows this process until the input-transition of $[s'_i, a_0]$ is reached. Formally, since $[s'_i, a_0]$ is marked, we have $p - \alpha \leq i < p$ by Corollary 13 and i can be represented as $i = p - \alpha + k$ with $0 \leq k < \alpha$. Then the input transition of $[s'_i, a_0]$ is $[t_i, a_0] = [t_{p-\alpha+k}, a_0]$. The cycle is starting in $[s'_{p-\alpha+k}, a_0]$ and follows the bf-path which ends by Lemma 17 in $[t_{(i+n-1) \bmod p}, a_0] = [t_{(p-\alpha+k+\alpha+\beta-1) \bmod p}, a_0] = [t_{(p+k+\beta-1) \bmod p}, a_0] = [t_{k+\beta-1}, a_0]$. From this transition the cycle follows the a_0-process until the input

---

[4] See the dashed line below S_0 in Fig. 4.

transition of $[s'_i, a_0]$ namely $[t_i, a_0] = [t_{p-\alpha+k}, a_0]$ is reached. This is possible without passing the end of the a_0-process to $[s_{p-1}, a_0]$ if the inequality $k+\beta-1 \leq p - \alpha + k$ is holding. The inequality follows from the condition $n - 1 \leq p$ of the theorem by $n-1 \leq p \Leftrightarrow \alpha+\beta-1 \leq \alpha+p-\alpha \Leftrightarrow \beta-1 \leq p-\alpha \Leftrightarrow \beta-1+k \leq p-\alpha+k$. The cycle contains a token in $[s'_i, a_0]$. The remaining forward places are unmarked since by Lemma 17 the i-bf-path is sharing the place $[s_{(i+n-1) \bmod p}, a_j]$ with the a_j-process, which contains a single token in the place $[s_{j-1}, a_j]$. The places are different since $[s_{(i+n-1) \bmod p}, a_j] = [s_{(p-\alpha+k+\alpha+\beta-1) \bmod p}, a_j] = [s_{k+\beta-1}, a_j]$ and we prove $j - 1 < k + \beta - 1$. This applies because $j < \beta \Rightarrow j - 1 < \beta - 1 + k$ since $k \geq 0$. Also the places of the a_0-process are unmarked since the place $[s_{p-1}, a_0]$ is not contained,

Case 2: $[s'_i, a_0]$ is unmarked, hence $0 \leq i < p - \alpha$. Here we consider two subcases: $0 \leq i < p - n + 1$ and $p - n + 1 \leq i < p - \alpha$.

Case 2.1: $0 \leq i < p - n + 1$. Again by Lemma 17 the bf-path starting in $[s'_i, a_0]$ ends in $[t_{(i+n-1) \bmod p}, a_0]$ with $0 \leq i \leq p - n$. Here, contrary to case 1, we are passing the marked place $[s_{p-1}, a_0]$ to reach the input transition $[t_i, a_0]$ of $[s'_i, a_0]$. The remaining places of the cycle are unmarked by the same arguments as before.

Case 2.2: $p - n + 1 \leq i < p - \alpha$ or $i = p - n + 1 + k$ with $0 \leq k < \beta - 2$. By the last sentence in Lemma 17 the bf-path starting in $[s'_i, a_0]$ contains exactly one token in the regular initial marking. This bf-path is extended to a cycle as in Case 1. □

By the cycloid systems of Fig. 5 the proof of Theorem 18 is illustrated and it is shown that the assumption $n - 1 \leq p$ of the theorem is sharp. For the cycloid system in a) the equivalence class S_3^{bf} is given by places contained in the dotted line. They are connected by the 3-bf-path in bold arrows, which is contained in a single marked cycle. This does not hold for the 1-bf-path in part b) of the figure. A cycle contains two tokens in $[s_3, a_0]$ and $[s_2, a_3]$ in the given regular initial marking. The backward folding $\mathcal{C}_{bf}(2, 4, 2, 4, [M_0])$ is not safe since the places marked by a star represent a reachable marking where $[S_1^{bf}]$ contains two tokens.

Theorem 19. Let be $\mathcal{C}_{bf(D)} = \mathcal{C}_{bf(D)}(\alpha, \beta, \gamma, \delta, [M_0]_D)$ the backward folding of a regular cycloid system $\mathcal{C} = \mathcal{C}(\alpha, \beta, \gamma, \delta, M_0) = (S, T, F, M_0)$ with $n - 1 \leq p$.

a) For each transition $t \in T$ and all markings M, M' of $\mathcal{C}$

$$M \xrightarrow{t} M' \quad \Leftrightarrow \quad [M]_D \xrightarrow[bf(D)]{t} [M']_D$$

where $\xrightarrow{t}$ and $\xrightarrow[bf(D)]{t}$ are the transition relation of the net $\mathcal{C}$ and $\mathcal{C}_{bf(D)}$, respectively.

b) $\mathcal{C}_{bf(D)}$ is live.

Proof. a) We first prove the equivalence of the activations $M \xrightarrow{t} \quad \Leftrightarrow$
$[M]_D \xrightarrow[bf(D)]{t}$.

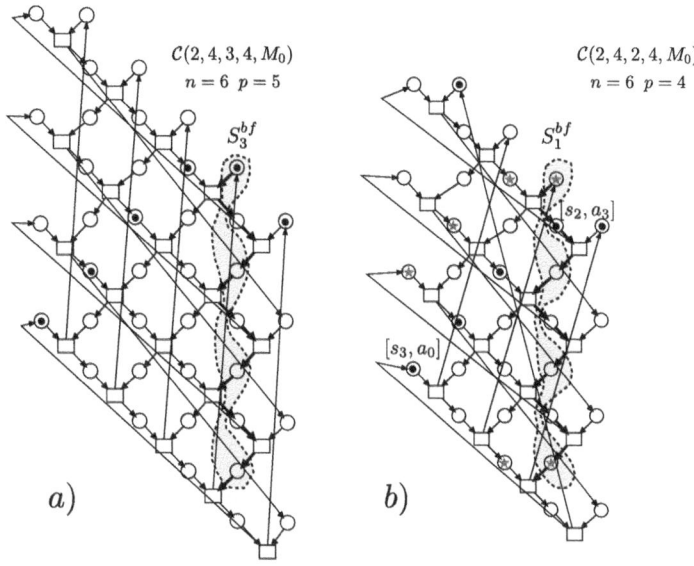

$\mathcal{C}(2,4,3,4,M_0)$
$n=6 \quad p=5$

$\mathcal{C}(2,4,2,4,M_0)$
$n=6 \quad p=4$

S_3^{bf}

S_1^{bf}

$[s_2,a_3]$

$[s_3,a_0]$

$a)$

$b)$

Fig. 5. Cycloids with $n-1=p$ and $n-1>p$.

Let be $t=[t_i,a_j]$ a transition of $\mathcal{C}$. If $M \xrightarrow{t}$ then the input places $[s_{i\ominus_p 1},a_j]$ and $[s'_h,a_{j\oplus_\beta 1}]$ are marked. It follows that the input places $[s_{i\ominus_p 1},a_j]$ and $[\![s'_h,a_{j\oplus_p 1}]\!]_D$ of $[t_i,a_j]$ in $\mathcal{C}_{bf(D)}$ are marked and $[\![M]\!]_D \xrightarrow[bf(D)]{t}$.

The inverse implication is proved by contradiction: $\neg M \xrightarrow{t} \Rightarrow \neg [\![M]\!]_D \xrightarrow[bf(D)]{t}$.

If $\neg M \xrightarrow{t}$ there are two cases:

Case 1: $[s_{i\ominus_p 1},a_j] \notin M$. As $[s_{i\ominus_p 1},a_j]$ is unchanged in $\mathcal{C}_{bf(D)}$ also $\neg [\![M]\!]_D \xrightarrow[bf(D)]{t}$.

Case 2: $[s_{i\ominus_p 1},a_j] \in M$ but $[s'_h,a_{j\oplus_\beta 1}] \notin M$, where $[s'_h,a_{j\oplus_\beta 1}]$ is the backward input place of $t=[t_i,a_j]$ (see Fig. 5). To prove $\neg [\![M]\!]_D \xrightarrow[bf(D)]{t}$ we deduce that the class $[\![s'_h,a_{j\oplus_\beta 1}]\!]_{bf(D)}$ of $[s'_h,a_{j\oplus_\beta 1}]$ is not in $[\![M_0]\!]_D$. This is the case if all elements of the class $S_h^{bf} = \{[s'_h,a_0]\} \cup \{[s'_{h\oplus_p n},a_j]|0<j<\beta\}$ with $0 \le i < p$ are unmarked. For the element $[s'_h,a_{j\oplus_\beta 1}]$ of this set, this property is part of the assumption of Case 2. To prove it for the remaining elements of this set consider the bf-path of $[\![s'_h]\!]_D$, as defined in Lemma 17 containing all these elements (highlighted by bold edges in Fig. 6). As shown in Lemma 17 this bf-path is extended to a S-invariant which contains exactly one token. We define a re-route of the corresponding cycle by replacing the sub-path $[t_{h\ominus_p 1},a_{j\oplus_\beta 1}],[s_{h\ominus_p 1},a_{j\oplus_\beta 1}],[t_h,a_{j\oplus_\beta 1}],[s'_h,a_{j\oplus_\beta 1}],[t_i,a_j]$ by $[t_{h\ominus_p 1},a_{j\oplus_\beta 1}],[s'_{h\ominus_p 1},a_{j\oplus_\beta 1}],[t_{i\ominus_p 1},a_j],[s_{i\ominus_p 1},a_j],[t_i,a_j]$ (see Fig. 6) which is also a S-invariant containing the place $[s_{i\ominus_p 1},a_j]$ which is marked in Case 2. As

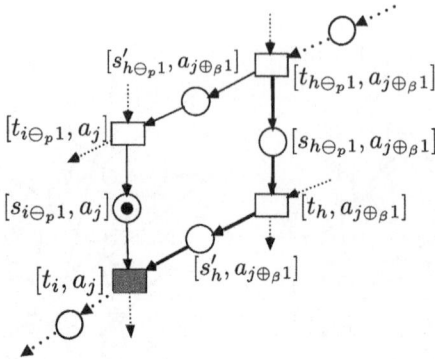

Fig. 6. Case 2 in the proof of Theorem 19.

this is the only token in the re-routed S-invariant all places different to $[s_{i\ominus_p 1}, a_j]$ are unmarked, including all the elements of S_h^{bf}.

b) By part a) of the proof $\mathcal{C}_{bf(D)}$ is behavioural equivalent to $\mathcal{C}$ which is live by Theorem 5.4 of [5]. $\qquad\square$

Lemma 20. *a) For a given regular cycloid $C = \mathcal{C}(\alpha, \beta, \gamma, \delta, M_0)$ with $\beta > 1$ (and process length p) the cycloid $C' = \mathcal{C}(\alpha + 1, \beta - 1, p - (\alpha + 1), \beta - 1, M')$ has the same process-length as C.*
b) Each regular cycloid $C'' = \mathcal{C}(\alpha + 1, \beta - 1, \gamma'', \delta'', M'')^5$ with the same process-length p is isomorphic to C'.

Proof. a) If A' is the area of C' the process-length of C' is $p' = \frac{A'}{\beta-1} = \frac{1}{\beta-1} \cdot$
$((\alpha + 1) \cdot (\beta - 1) + (\beta - 1) \cdot (p - (\alpha + 1))) = p.$
b) To prove that C' and C'' are isomorphic, by Theorem 4 b) it is sufficient to show that for some $q \in \mathbb{N}_+$ we have

$$\gamma'' = \gamma' - q \cdot (\alpha + 1) \quad \text{and} \quad \delta'' = (\beta - 1) + q \cdot (\beta - 1) \quad \text{for} \quad \gamma' = p - (\alpha + 1) \quad (2)$$

Since C'' is supposed to be regular we have $\delta'' = r \cdot (\beta - 1)$ for some $r \in \mathbb{N}_+$. C'' is also required to have the same process-length $p'' = \frac{1}{\beta-1} \cdot ((\alpha+1) \cdot \delta'' + (\beta - 1) \cdot \gamma'') = \frac{1}{\beta-1} \cdot ((\alpha+1) \cdot r \cdot (\beta-1) + (\beta-1) \cdot \gamma'') = (\alpha+1) \cdot r + \gamma''$ as $p = \gamma' + (\alpha + 1)$.
$p = p''$ gives $\gamma' + (\alpha + 1) = (\alpha + 1) \cdot r + \gamma''$ and $\gamma'' = \gamma' + (\alpha + 1) - (\alpha + 1) \cdot r = \gamma' - (r - 1) \cdot (\alpha + 1)$, hence $q = r - 1$ as required in Eq. (2). The observation $\delta'' = r \cdot (\beta - 1) = (q + 1) \cdot (\beta - 1) = (\beta - 1) + q \cdot (\beta - 1)$ satisfies the second part of Eq. (2). $\qquad\square$

A bf-folding with $D = \{0, \beta - 1\}$ is merging the backward input places of the processes of the processes $a_{\beta-1}$ and $a_{\beta-2}$. Then the $a_{\beta-1}$-process can be

---

[5] M_0, M', M'' are the corresponding regular initial markings.

stopped or eliminated such that the remaining cycloid is still live and safe with the same process length. This is done with the cycloid C_0 in Fig. 7. Via the intermediate step of the bf-folding C_1 a cycloid C_2 is obtained, where the a_2-process is eliminated.

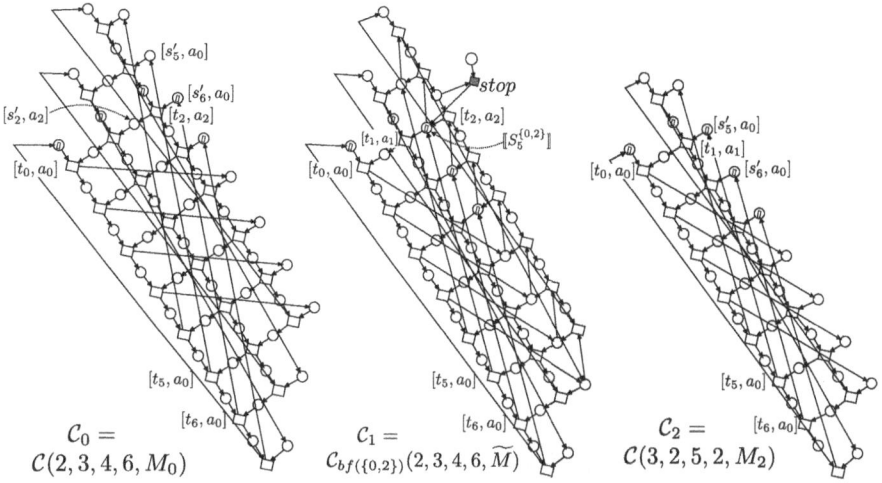

Fig. 7. Two cycloids and bf-folding $C_{bf(2,1)}(2,3,4,6,\widetilde{M})$ as intermediate step.

Definition 21. *For a bf-folding $C_1 = C_{bf(D)}(\alpha, \beta, \gamma, \delta, [\![M_0]\!]_D)$ with $j \in \{0, \cdots, \beta - 1\}$ we define $[C_1] \downarrow (j)$ by deleting all transitions $\{[t_i, a_j] | 0 \leq i < p\}$ of the j-process, together with its forward output places (and their tokens). The initial marking is $[\![M_0]\!]_D$ after adding a token in the backward output place of $[t_j, a_j]$ if $\beta > 0$ or adding $p - n + 1$ tokens in the places $[\![S_0]\!], \cdots, [\![S_{p-n}]\!]$ (Definition 16) if $\beta = 0$.*

Note that the cases coincide for canonical regular cycloids where $p = n$.

Theorem 22. *Let be $C_0 = C(\alpha, \beta, \gamma, \delta, M_0)$ a regular cycloid with $\beta > 1$ and $n \leq p$.*

a) $C_2 = C(\alpha + 1, \beta - 1, p - (\alpha + 1), \beta - 1, M_2)^6$ is behaviour equivalent to $C_1 = C_{bf(\{0, \beta-1\})}(\alpha, \beta, \gamma, \delta, \widetilde{M})$ where $\widetilde{M} = [\![M_0]\!] - \{[s_{\beta \ominus_p 2}, a_{\beta-1}]\} \cup \{[\![S_{p-(\alpha+1)}^{\{0, \beta-1\}}]\!]\}$
b) C_2 is isomorphic to $C_3 = [C_1] \downarrow (\beta - 1)$.

Proof. a) By the elimination of the token in $[s_{\beta \ominus_p 2}, a_{\beta-1}] \in M$ all transitions of the $a_{\beta-1}$-process are dead. Therefore it is sufficient to prove part b) of the theorem.

---

[6] M_0, M_2 are the corresponding regular initial markings.

b) As a first step, note that in C_3 the process $a_{\beta-1}$ is deleted. Therefore in formulas relating to C_3 the value of $\beta - 1$ is the be replaced by $\beta'' = \beta - 2$. By the folding $bf(D)$ with $D = \{0, \beta - 1\}$ the difference between C_3 and C_2 consists in the replacement of the two element class $[\![S_i^D]\!] = \{[s_i', a_0], [s_{i\oplus_p n}', a_{\beta-1}]\}$ with $0 \leq i < p$ (by Lemma 16 b) in C_3 by the place $[s_i', a_0]$ in C_2. The corresponding isomorphism from C_3 to C_2 is $\phi([\![S_i^D]\!]) = [s_i', a_0]$ and the identical map for all different places and transitions of C_3. The structure of C_3 is inherited from C_0. Since the elements of $[\![S_i^D]\!]$ obtain their names from their input transitions, these input transitions are (I) $[t_i, a_0]$ and (II) $[s_{i\oplus_p n}', a_{\beta-1}]\}$, respectively. The output transitions of the elements of $[\![S_i^D]\!]$ are computed by Lemma 12 with respect to C_1 by (III) $[t_{(i+n-1) \bmod p}, a_{\beta-1}]$ and (IV) $[t_{(i+n-1) \bmod p}, a_{\beta-2}]$, respectively. By the definition of C_3 the process $a_{\beta-1}$ is eliminated and thereby also the elements (II) and (III). This means that for $[\![S_i^D]\!]$ in C_3 there remain the input and output transitions (I) $[t_i, a_0]$ and (IV) $[t_{(i+n-1) \bmod p}, a_{\beta-2}]$. In remains to prove that these are the same under the mapping $\phi([\![S_i^D]\!]) = [s_i', a_0]$. In fact, one receives in C_2 the values ${}^\bullet[s_i', a_0] = [t_i, a_0]$ and $[s_i', a_0]^\bullet = [t_{(i+n'-1) \bmod p'}, a_{\beta'-1}]$, which matches with (IV) since for C_2 we have $n' = n, p' = p$ and $\beta' = \beta - 1$. It remains to prove that the initial marking M_3 of C_3 maps to M_2. The marking M_3 is defined from M_0 of C_0 via $\widetilde{M}$ of C_1. From Corollary 13 we obtain $M_0 = M_0^1 \cup M_0^2 \cup M_0^3$ with $M_0^1 = \{[s_{p-1}, a_0]\}$, $M_0^2 = \{[s_i, a_{i+1}] | 0 \leq i < \beta - 1\}$ and $M_0^3 = \{[s_i', a_0] | p - \alpha \leq i < p\}$. By the transition from C_0 to C_1 M_0^1 and M_0^2 are not changed, but the places $[s_i', a_0]$ in M_0^3 are changed to $[\![S_i^D]\!]$. Furthermore by adding a token to $\{[\![S_{p-(\alpha+1)}^{\{0,\beta-1\}}]\!]\}$ the marking M_0^3 is changed to $\{[\![S_i^D]\!] | p - \alpha' \leq i < p\}$ with $\alpha' = \alpha + 1$. Finally with the transition from C_1 to C_3 with the $(\beta - 1)$-process the place $[s_{\beta-1}, a_{\beta-1}]$ is eliminated and M_0^2 becomes $\{[s_i, a_{i+1}] | 0 \leq i < \beta' - 1\}$ with $\beta' = \beta - 1$. Collecting these changes of the M_0^i we obtain $\{[s_{p-1}, a_0]\} \cup \{[s_i, a_{i+1}] | 0 \leq i < \beta' - 1\} \cup \{[\![S_i^D]\!] | p - \alpha' \leq i < p\}$ with $\beta' = \beta - 1$ and $\alpha' = \alpha + 1$ which is the initial marking of C_2. $\square$

Figure 7 shows the cycloid $C_0 = \mathcal{C}(\alpha, \beta, \gamma, \delta, M_0) = \mathcal{C}(2, 3, 4, 6, M_0)$ and its folding $\mathcal{C}_{bf(\{0,2\})}(2, 3, 4, 6, \widetilde{M})$ as intermediate step to the cycloid $C_2 = \mathcal{C}(\alpha + 1, \beta - 1, p - (\alpha + 1), \beta - 1, M_2) = \mathcal{C}(3, 2, 5, 2, M_2)$. The transition $stop$ in C_1 simulates the change from the marking M_0 to $\widetilde{M}$. The map ϕ in the proof for $i = 5$ is $\phi([\![S_5^{\{0,2\}}]\!]) = [s_5', a_0]$. Theorem 22 can be applied again to the cycloid system $C_2 = \mathcal{C}(3, 2, 5, 2, M_2)$ to obtain a cycloid with only one process: $C_3 = \mathcal{C}(4, 1, 4, 1, M_2)$. In this way, a small hierarchy of regular cycloids is created.

5   Failure-Resilient Regular Cycloids

To model resilience to failure of *any* process, we consider total backward foldings here. The failure or stop of any process a_k is represented by the removing of its control token in the initial regular marking in the place $[s_i, a_{i+1}]$ with $0 \leq i < \beta - 1$. By this deletion of a token the cycle as described in the proof of Theorem

18 becomes unmarked and the net is not longer safe and live. Therefore a token is added to the backward output transition of $[t_i, a_{i+1}]$. The addition of one single token does not work for the process a_0, as, again by the proof of Theorem 18 a number of $p - n + 1$ cycles become unmarked. Therefore such a number of tokens will be added when the process a_0 is stopping. The proof for processes a_1 to $a_{\beta-1}$ is similar to that in Theorem 22. Since we need to make some renaming here, we formulate again a corresponding isomorphism. We ask what are the parameters of a (backward folded) cycloid $\mathcal{C}'_{bf} = \mathcal{C}_1(\alpha', \beta', \gamma', \delta', M'_0)$ that is isomorphic to $\mathcal{C} = \mathcal{C}(\alpha, \beta, \gamma, \delta, M_0)$ when the process a_i is cancelled. $\mathcal{C}'_{bf}$ has one less process, hence $\beta' = \beta - 1$ is expected.

Theorem 23. *Let $\mathcal{C}_1 = \mathcal{C}_{bf}(\alpha, \beta, \gamma, \delta, [\![M_0]\!])$[7] be a total backward folding (Definition 15) of a cycloid $\mathcal{C}_0 = \mathcal{C}_0(\alpha, \beta, \gamma, \delta, M_0)$ with process length p and $n \leq p$.*

a) *For $0 < k < \beta$ the net $\mathcal{C}_2 := [\mathcal{C}_1] \downarrow (k)$ (Definition 21) is isomorphic to $\mathcal{C}_3 = \mathcal{C}_{bf}(\alpha + 1, \beta - 1, p - \alpha - 1, \beta - 1, [\![M_2]\!])$.*
b) *For $j = 0$ the net $\mathcal{C}_4 = [\mathcal{C}_1] \downarrow (0)$ is isomorphic to $\mathcal{C}_5 = \mathcal{C}_{bf}(p - \beta + 1, \beta - 1, \beta - 1, \beta - 1, [\![M_5]\!])$.*
c) *If $\mathcal{C}_0$ is a canonical regular cycloid (i.e. $n = p$) then the cases a) and b) coincide and $\mathcal{C}_3 = \mathcal{C}_5 = \mathcal{C}(\alpha + 1, \beta - 1, \beta - 1, \beta - 1)$ is again canonical regular.*

Proof. a) If the stopping process is a_k with $0 \leq k < \beta$ we define the isomorphism from $\mathcal{C}_2$ to $\mathcal{C}_3$ for $0 \leq i < p$ and $0 \leq j < \beta$ by $\phi([t_i, a_j]) = \begin{cases} [t_i, a_{j-1}] & \text{if } j > k \\ [t_i, a_j] & \text{otherwise} \end{cases}$, and $\phi([\![S_i]\!]) = [\![S_i]\!]$ for $k > 0$ or $k = 0 \wedge n = p$. The values of $n = \alpha + \beta$ of $\mathcal{C}_0$ remain unchanged in $\mathcal{C}_1$ and $\mathcal{C}_3$. Also by Lemma 20 the process length p is unchanged. Therefore it remains to prove that the backward output places of $\mathcal{C}_2$ and $\mathcal{C}_3$ are in the same relation to the transitions. This follows immediately from the formulas for ${}^\bullet[\![S_i]\!]$ and $[\![S_i]\!]^\bullet$ in Lemma 16 as they depend only on p and n (with the exception of β). The regular initial marking of $\mathcal{C}_3$ is given in Corollary 13. After the occurrence of $[t_{\beta-2}, a_{\beta-2}], \cdots, [t_j, a_j]$ in $\mathcal{C}_3$ the backward output place $[s'_j, a_j]$ is marked. This is exactly the token added to the backward output place of $\phi([t_j, a_j])$ in the definition of $\mathcal{C}_2$ (Definition 21).

b) If $k = 0$ and $n < p$ the isomorphism from $\mathcal{C}_4$ to $\mathcal{C}_5$ is defined for $[t_i, a_j]$ and $[s_i, a_j]$ as in case a), but $\phi([\![S_i]\!])$ may be different since the value of n may be different for $\mathcal{C}_1$ and $\mathcal{C}_5$. Recall that the names of the places of $[\![S_i]\!]$ are defined by the transitions $[t_i, a_0]$. We determine the relation of the indices by comparison with their output transitions $[t_0, a_{\beta-1}], \cdots, [t_{p-1}, a_{\beta-1}]$ which are in a total ordering. We define the position of $[\![S_i]\!]$ by $pos([\![S_i]\!]) = j$ if $[\![S_i]\!]^\bullet = [t_j, a_{\beta-1}]$. Then $pos([\![S_0]\!]) = n - 1$ in $\mathcal{C}_1$ by Lemma 12 and $pos([\![S_0]\!]) = n' - 1$ in $\mathcal{C}_5$ with $n' = p - \beta + 1 + \beta - 1 = p$ by the same lemma (this is for instance in Fig. 8 $pos([\![S_0]\!]) = 4$ in $\mathcal{C}_1$ and $pos([\![S_0]\!]) = 7$ in $\mathcal{C}_5$). Therefore by the isomorphism from $\mathcal{C}_4$ to $\mathcal{C}_5$ the position of $[\![S_0]\!]$ switches from $n - 1$ to $p - 1$ and $[\![S_i]\!]$ becomes $\phi([\![S_i]\!]) = [\![S_{(i+n-p) \bmod p}]\!]$. Because of the precondition

---

[7] $[\![M_0]\!]$ and in the following $[\![M_2]\!]$ and $[\![M_5]\!]$ are as defined in Definition 15.

$n \leq p$ of the theorem and the formula at the end of Sect. 1 we have two cases: $n \ominus_p p = 0$ if $n = p$ and $n \ominus_p p = n$ if $n < p$. Since the first case has already been dealt with in part a) of the proof, it is now sufficient to deal with the second case and we define $\phi([S_i]) = [S_{(i+n-p) \bmod p}] = [S_{i \oplus_p n}]$. Again it remains to prove that the backward output place of $[t_i, a_1]$ in $\mathcal{C}_4$ maps to the corresponding backward output place of $\phi([t_i, a_1])$ in $\mathcal{C}_5$. The place $[t_i, a_1]^{\bullet} = [s_i', a_1]$ is in the class $[S_{i \ominus_p n}]$ since $[s_{i \oplus_p n}', a_1] \in [S_i]$ by Lemma 16 b). Therefore $\phi([S_{i \ominus_p n}]) = [S_{(i-n+n) \bmod p}] = [S_i]$ in $\mathcal{C}_4$. The latter is also obtained in $\mathcal{C}_5$ since $n' = p$ and $\phi([t_i, a_1])^{\bullet} = [t_i, a_0]^{\bullet} = [S_i]$.

From Definition 13 we obtain for M_0 in $\mathcal{C}_0$ the formula $M_0 = M_0^1 \cup M_0^2 \cup M_0^3$ as defined in the proof of Theorem 22. $[M_0]$ of $\mathcal{C}_1$ is defined as the backward folding of M_0. Therefore $[M_0] = [M_0^1] \cup [M_0^2] \cup [M_0^3]$ with $[M_0^1] = M_0^1 = \{[s_{p-1}, a_0]\}$, $[M_0^2] = M_0^2 = \{[s_i, a_{i+1}] | 0 \leq i < \beta - 1\}$ and $[M_0^3] = \{[S_i] | p - \alpha \leq i < p\}$. $\mathcal{C}_4 = [\mathcal{C}_1] \downarrow (0)$ is obtained from $\mathcal{C}_1$ by deleting all transitions and places of the a_0-process and adding the following set R of $p-n+1$ tokens to the places $R := \{[S_i] | 0 \leq i \leq p - n\}$ (Definition 21). Therefore $[M_0^1]$ is deleted. By the isomorphism ϕ we obtain $\phi([M_0^2]) = \{[s_i, a_i] | 0 \leq i < \beta - 1\}$ and $\phi([M_0^3]) = \{[S_i] | (p - \alpha + n) \bmod p \leq i < (p + n) \bmod p\} = \{[S_i] | (p - \alpha + \alpha + \beta) \bmod p \leq i < n\} = \{[S_i] | \beta \leq i < n\}$ and $\phi(R) = \{[S_i] | n \leq i \leq p - n + n\} = \{[S_i] | n \leq i \leq p\}$. Bringing these results together we finally obtain $\phi([M_0]) = M_0^2 \cup \{[S_i] | \beta \leq i \leq p\}$, which is

$$\phi([M_0]) = \{[s_i, a_i] | 0 \leq i < \beta - 1\} \cup \{[S_i] | \beta \leq i \leq p\} \tag{3}$$

in $\mathcal{C}_4$. This initial marking of $\mathcal{C}_4$ is proved to be equal with the initial marking of $\mathcal{C}_5$ after the occurrence of β' transitions $[t_{\beta'-1}, a_{\beta'-1}], [t_{\beta'-2}, a_{\beta'-2}], \cdots, [t_0, a_0]$ where we denote $\mathcal{C}_5$ by $\mathcal{C}_5 = \mathcal{C}(\alpha', \beta', \gamma', \delta')$ with $\alpha' = p - \beta + 1$, $\beta' = \gamma' = \delta' = \beta - 1$. To compute this follower marking we apply Theorem 13 b) to the cycloid $\mathcal{C}_6 = \mathcal{C}(\alpha', \beta', \gamma', \delta')$. This marking is denoted M_1 in Theorem 13 b), which becomes for $\mathcal{C}_5$ the marking $[M_1] = \{[s_{(p-1+1) \bmod p}, a_0]\} \cup \{[s_{i \oplus_p 1}, a_{i+1}] | 0 \leq i < \beta' - 1\} \cup \{[s_{i \oplus_p 1}', a_0] | p - \alpha' \leq i < p\} = \{[(s_0, a_0]\} \cup \{[s_{i+1}, a_{i+1}] | 0 \leq i < \beta - 2\} \cup \{[S_{i+1}] | p - p + \beta - 1 \leq i < p\} = \{\{[s_i, a_i] | 0 \leq i < \beta - 1\} \cup \{[S_{i+1}] | \beta - 1 \leq i < p\} = \{\{[s_i, a_i] | 0 \leq i < \beta - 1\} \cup \{[S_i] | \beta \leq i \leq p\}$, which is the same as in equation (3).

c) If $n = p = \alpha + \beta$ in $\mathcal{C}_0$ we obtain $\gamma' = p - \alpha - 1 = \alpha + \beta - \alpha - 1 = \beta - 1$ in a) and $\alpha' = p - \beta + 1 = \alpha + \beta - \beta + 1 = \alpha + 1$ in b). $\qquad \square$

To give an example for the proof of part b) of Theorem 23 consider the bf-foldings of Fig. 8. On the left hand side the net $\mathcal{C}_1 = \mathcal{C}_{bf}(2, 3, 4, 6, [M_0])$ is given if the transitions $[t_{stop}, a_j]$ are considered non-existent at this point. Then $\mathcal{C}_5 = \mathcal{C}_{bf}(\alpha', \beta', \gamma', \delta', [M_5]) = \mathcal{C}_{bf}(6, 2, 2, 2, [M_5])$ is given on the right of this figure, since $p - \beta + 1 = 8 - 3 + 1 = 6$. $\mathcal{C}_4 = [\mathcal{C}_1] \downarrow (0)$ is obtained from $\mathcal{C}_1$ by deleting all transitions $[t_i, a_0]$ and places $[s_i, a_0]$. For the isomorphism ϕ we obtain $\phi([S_i]) = [S_{i \oplus_p n}] = [S_{i \oplus_8 5}]$ since $p = 8$ and $n = 5$ for $\mathcal{C}_1$, e.g. $\phi([S_0]) = [S_5]$. With respect to the the backward output transitions in $\mathcal{C}_4$ we observe for

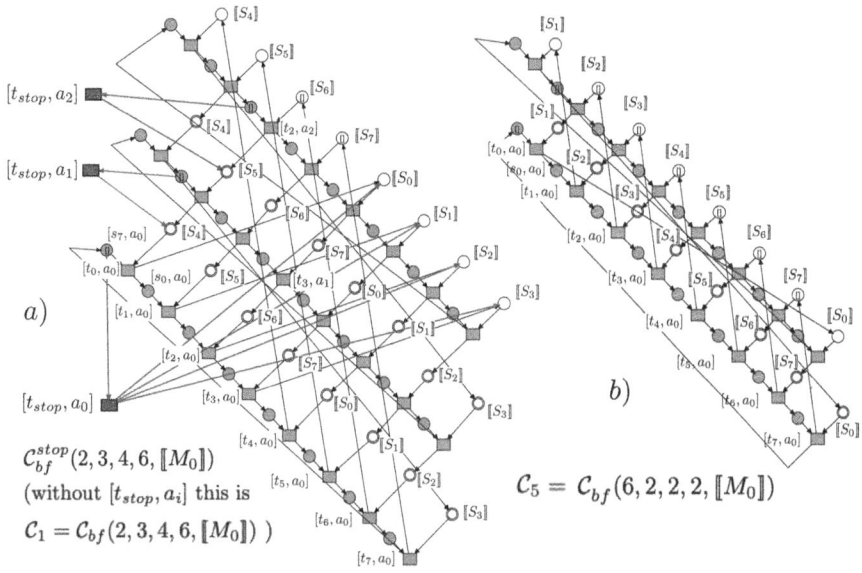

a)

$\mathcal{C}_{bf}^{stop}(2,3,4,6,[\![M_0]\!])$
(without $[t_{stop},a_i]$ this is
$\mathcal{C}_1 = \mathcal{C}_{bf}(2,3,4,6,[\![M_0]\!])$)

b)

$\mathcal{C}_5 = \mathcal{C}_{bf}(6,2,2,2,[\![M_0]\!])$

Fig. 8. The stop resilient cycloid $\mathcal{C}_{bf}^{stop}(2,3,4,6,[\![M_0]\!])$ and $\mathcal{C}_{bf}(6,2,2,2,[\![M_0]\!])$.

example $[t_3,a_1]^{\leftarrow\bullet} = [\![S_{3\ominus_8 5}]\!] = [\![S_{(-2)\ mod\ 8}]\!] = [\![S_6]\!]$ and $\phi([\![S_6]\!]) = [\![S_{6\oplus_8 5}]\!] = [\![S_3]\!]$. This result is proved now to be equal in the image of ϕ by $\phi([t_3,a_1])^{\leftarrow\bullet} = [t_3,a_0]^{\leftarrow\bullet}[\![S_{3\ominus_8 n'}]\!] = [\![S_{3\ominus_8 8}]\!] = [\![S_3]\!]$. The image of the initial marking of $\mathcal{C}_4$ (augmented by R) is $\phi(\{[s_0,a_1],[s_1,a_2],[\![S_6]\!],[\![S_7]\!]\})\cup\phi(\{[\![S_0]\!],[\![S_1]\!][\![S_2]\!],[\![S_3]\!]\}) = \{[s_0,a_0],[s_1,a_1],[\![S_3]\!],[\![S_4]\!]\} \cup \{[\![S_5]\!],[\![S_6]\!][\![S_7]\!],[\![S_0]\!]\}$. This is the initial marking of $\mathcal{C}_5 = \mathcal{C}_{bf}(6,2,2,2,[\![M_5]\!])$ after the occurrence of $[t_1,a_1]$ and $[t_0,a_0]$ (as given in Eq. (3).

Definition 24. *Given a regular and total backward folding $\mathcal{C}_{bf}(\alpha,\beta,\gamma,\delta,[\![M_0]\!])$ (Definition 15), a stop resilient cycloid $\mathcal{C}_{bf}^{stop}(\alpha,\beta,\gamma,\delta,[\![M_0]\!])$ is defined by adding transitions $[t_{stop},a_j],(0 \le j < \beta)$ with $^\bullet[t_{stop},a_j] := \{[s_{j\ominus_p 1},a_j]\}$ and one output place $[t_{stop},a_j]^\bullet := \{[t_j,a_j]^{\leftarrow\bullet}\}$ for $j > 0$. If $j = 0$ we define $^\bullet[t_{stop},a_0] := \{[s_{p-1},a_0]\}$ and $p-n+1$ output places $[t_{stop},a_0]^\bullet := \{[t_k,a_k]^{\leftarrow\bullet} | 0 \le k \le p-n\}$. Then the process of a_j is called a stoppable process. After the occurrence of $[t_{stop},a_j]$ the process a_j is denoted as stopped.*

A transition $[t_{stop},a_j]$ deletes the control token in $[s_{j\ominus_p 1},a_j]$ of the a_j-process, but in same time it generates permit signals by one or more tokens in the backward output places of $[t_j,a_j]$. In the particular case of a canonical regular backward folded cycloid this is just as the alternative transition $[t_j,a_j]$ would have done in its next step. In terms of a queue of cars and gaps in a lane, this means that the stopping car is still able to leave the lane giving place to the next car, which are able to proceed. We now formulate the final result of this section.

Corollary 25. *Let $C_0 = C_{bf}^{stop}(\alpha, \beta, \gamma, \delta, [\![M_0]\!])$ be a stop resilient cycloid with $\beta > 1$ and $n \le p$.*

a) *Stopping any number of $s \in \mathbb{N}_+$ $(0 < s < \beta)$ processes of C_0 results in a safe net, where all transitions $[t_i, a_j]$ from not stopped processes are live.*

b) *Stopping C_0 by a transition $[t_{stop}, a_k]$ and after deleting this transition and all transitions $[t_i, a_k]$ and forward places $[s_i, a_k]$ results in the stop resilient cycloid $C_1 = C_{bf}^{stop}(\alpha + 1, \beta - 1, p - \alpha - 1, \beta - 1, [\![M_0]\!])$ if $0 < k < \beta$ and $C_2 = C_{bf}^{stop}(p - \beta + 1, \beta - 1, \beta - 1, \beta - 1, [\![M_0]\!])$ if $k = 0$.*

The corollary follows from Theorem 23 since the transitions $[t_{stop}, a_j]$ implement the marking change as given in Definition 21. In Fig. 8 a) the regular stop resilient cycloid $C_{bf}^{stop}(2, 3, 4, 6, [\![M_0]\!])a$) is given, while Fig. 2 shows the construction for the special case of the canonical regular stop resilient cycloid $C_{bf}^{stop}(4, 3, 3, 3, [\![M_0]\!])$.

Since the result of a stopping is a regular cycloid again (when removing the dead parts), this can happen in any order until only one process remains: $\beta = 1$. The two transformations from the corollary commute: stopping for $i = 0$ and stopping for $i > 0$, in any order, result in the same cycloid. If only one type of these transformations is used, the result may be different. To give an example removing only process a_0 results in the following chain (the initial markings are omitted): $C(2, 3, 4, 6), C(6, 2, 2, 2), C(7, 1, 1, 1)$. Compare with the case where always $a_{\beta-1}$ is stopped (see Fig. 7): $C(2, 3, 4, 6), C(3, 2, 5, 2), C(4, 1.4, 1)$. The cycloid $C(7, 1, 1, 1)$ is isomorphic to $C(4, 4, 1, 1)$ but, by the methods introduced in [10], not to $C(4, 1, 4, 1)$. In addition, we conjecture that the construction can be extended so that a process can fail at other or all points. Also to the transitions $[t_{stop}, a_j]$ an inverse transition $[t_{resume}, a_j]$ to restart a failed process can be added.

6   Conclusion

Despite the tight synchronization of sequential processes in the form of regular cycloids, it has been possible to extend the formalism in such a way that individual processes can fail without hindering the other processes. In particular, it was shown that after the failure of a process, the residual system can be characterized by a known class of cycloids. This was achieved by a folding of places, which did not change the behaviour of the process system. Initially, this method was used to remove the process with the highest index, allowing a regular cycloid to be constructed without folded places. By iterating this construction, a hierarchy of regular cycloids is obtained that includes all subsystems down to a single process. In a further step, this was extended to the elimination, stopping or failure of any processes. Surprisingly, there was a special case for the failure of the process with the lowest index, which, however, does not occur for the important class of canonical regular cycloids.

References

1. Fenske, U.: Petris Zykloide und Überlegungen zur Verallgemeinerung. Diploma Thesis, Department of Informatics, University of Hamburg (2008)
2. Kummer, O., Stehr, M.-O.: Petri's axioms of concurrency a selection of recent results. In: Azéma, P., Balbo, G. (eds.) ICATPN 1997. LNCS, vol. 1248, pp. 195–214. Springer, Heidelberg (1997). https://doi.org/10.1007/3-540-63139-9_37
3. Petri, C.A.: Nets, time and space. Theor. Comput. Sci. (153), 3–48 (1996)
4. Smith, E., Reisig, W.: The semantics of a net is a net - an exercise in general net theory. In: Voss, K., Genrich, J., Rozenberg, G. (eds.) Concurrency and Nets, pp. 461–479. Springer, Berlin (1987). https://doi.org/10.1007/978-3-642-72822-8_29
5. Valk, R.: Formal properties of Petri's cycloid systems. Fund. Inform. **169**, 85–121 (2019)
6. Valk, R.: Circular traffic queues and Petri's cycloids. In: Janicki, R., Sidorova, N., Chatain, T. (eds.) PETRI NETS 2020. LNCS, vol. 12152, pp. 176–195. Springer, Cham (2020). https://doi.org/10.1007/978-3-030-51831-8_9
7. Valk, R.: Deciphering the co-car anomaly of circular traffic queues using Petri nets. In: Buchs, D., Carmona, J. (eds.) PETRI NETS 2021. LNCS, vol. 12734, pp. 443–462. Springer, Cham (2021). https://doi.org/10.1007/978-3-030-76983-3_22
8. Valk, R.: Analysing cycloids using linear algebra (2024). https://arxiv.org/abs/2402.07303
9. Valk, R.: Modelling cooperating failure-resilient processes (2024). https://arxiv.org/abs/2409.18318
10. Valk, R., Moldt, D.: On reduction and parameter recovering of Petri's cycloids (2024). https://arxiv.org/abs/2405.21025v2

Symbolic Model Checking in the Modular State Space Using Binary Decision Diagrams

Lukas Zech(✉)

University of Rostock, Schwaansche Str. 2, 18055 Rostock, Germany
lukas.zech@uni-rostock.de

Abstract. A *modular Petri net* is composed of individual Petri nets, the modules. Modules are composed by fusing their *interface* transitions. The behavior of *internal* transitions is unrelated to other modules and is recorded in *local reachability graphs* for each module. The behavior of interface transitions is recorded in a single *synchronization graph*. The local reachability graphs and the synchronization graph form the *modular state space* [18].

In this paper, we study the reachability problem in the composed Petri net using the modular state space instead of the state space of the composed Petri net. We analyze how the reachability of markings that satisfy a *state predicate* can be verified. Local reachability graphs are encoded as Binary Decision Diagrams [9,23,29]. We present algorithms to solve the mentioned reachability problem in this setting.

Finally, we compare the implementation with traditional state space exploration in the non-modular state space.

Keywords: Petri nets · Modular · Model Checking · Symbolic · Binary Decision Diagrams · Verification and model checking using nets · Regular Paper

1 Introduction

Compared to regular Petri nets, a modular Petri net is composed of multiple components, the modules. Many models are inherently compositional, making verification using modular Petri nets interesting. This is further illustrated by modeling tools like CPN Tools [35] supporting compositional or hierarchical models. Compositional verification in general is a well researched topic [7,20,22, 33]. Additionally, proposals exist to automatically separate nets without a user-defined modular structure into modules [8,17]. The composed Petri net can be obtained by fusing the modules along their *interface* transitions. In this paper we assume the notation used in [18,39] to describe the modular state space. The latter was introduced in [10]. The modular state space consists of the local reachability graphs of the modules and a single global synchronization graph. The former model the internal behavior in a module, while the latter models the firing of interface transitions. A modular Petri net may consist of multiple

© The Author(s), under exclusive license to Springer Nature Switzerland AG 2025
E. Amparore and L. Mikulski (Eds.): PETRI NETS 2025, LNCS 15714, pp. 478–500, 2025.
https://doi.org/10.1007/978-3-031-94634-9_23

instances of the same module, which are differentiated by their initial markings. In this case, we assume those instances that share a module to behave similarly. Therefore, data structures used to store reachable markings can be used more efficiently. In addition to computing multiple smaller graphs instead of a single, potentially large, reachability graph, this helps to reduce the state explosion problem [34].

The local reachability graphs of the modules describe the projection of the composed reachability graph to a single module. By exploring the local reachability graphs and the synchronization graph together, the behavior in a local reachability graph is just the behavior of the module in the composed Petri net. Therefore, local reachability graphs are always finite as long as the reachability graph of the composed Petri net is finite. This is the main difference to aforementioned compositional verification, which assumes local state spaces to be finite and explores them independently, before composing the global state space. In particular, exploration in the modular state space does not build the composed state space at all.

In a local reachability graph, *segments* abstract away the behavior between two consecutive firings of interface transitions. A vertex in the synchronization graph is then a tuple of segments for each module. Edges in the synchronization graph describe interface transitions. Therefore, the synchronization graph describes how the modules move from one segment to another synchronously.

In this paper, we deal with symbolic verification in the modular state space. Compared to traditional verification, (part of) the state space is encoded to allow for faster verification using specific techniques [11,13,26] or to allow for further compression of the state space [9,26,30] to combat the state explosion problem. We encode the local reachability graphs of the modules using binary decision diagrams (BDDs) [6,29] with the aim to further compress the local state spaces [9]. We restrict ourselves to the verification of *state predicates*, which are conjunctions of inequalities over places. Many properties can be described through state predicates. Deadlocks can be searched by asserting that every pre-place of a transition should have fewer tokens than the arc weight. The fireability of a transition can be asserted by searching for markings where every pre-place of the transition has tokens greater than or equal to the arc weight. Lastly, k-boundedness can be verified by asserting that every place has at most k tokens.

The paper is structured as follows. Section 2 introduces the basic notation for modular Petri nets, the modular state space and BDDs. In Sect. 3, we present two methods for the mentioned symbolic verification. First, we adapt results from [1] to convert state predicates to BDDs. Second, we present an algorithm to directly evaluate a state predicate on a BDD. Finally, in Sect. 4, we discuss an implementation of our findings and compare its performance with traditional state space exploration of the composed Petri net.

2 Preliminaries

The definitions for modular Petri nets and the modular state space follow existing notation introduced in [18,39].

2.1 Modular Petri Nets

Modular Petri nets consist of multiple individual Petri nets called *modules*. They alone do not model behavior. This is added through initial markings to form *instances*.

Definition 1 (Module). *A place/transition Petri net $N = [P, T, F, W]$, consisting of places P, transitions T, arcs F and weight function W, where T is partitioned into subsets $T_{internal}$ of internal transitions and $T_{interface}$ of interface transitions is called a* module.

Definition 2 (Instance). *An* instance *is a Petri net system $[N, m_0]$ where N is a module and m_0 is the initial marking of N. In general, a marking is a mapping $m\colon P \to \mathbb{N}$.*

Instances can be composed along their interface transitions. This is modelled through fusion vectors.

Definition 3 (Fusion Vector). *Let $\{[N_1, m_{01}], \ldots, [N_\ell, m_{0\ell}]\}$ be a set of instances. A fusion vector $f \in (T_{1|interface} \cup \{\bot\}) \times \ldots \times (T_{\ell|interface} \cup \{\bot\})$ is a vector of interface transitions of the instances or the $\bot$-symbol. If $f[j] = t$ for $t \in T_{j|interface}$, then instance $[N_j, m_{0j}]$ participates in this fusion with t. If $f[j] = \bot$, then instance $[N_j, m_{0j}]$ does not participate in this fusion.*

By explicitly marking non-participating instances with $\bot$, we define transition fusions over all instances. This removes ambiguities, since multiple instances could be of the same module. Further, multiple interface transitions of the same instance cannot participate in a single fusion simultaneously. The above definitions describe the base of the *modular structure*, a system composed of instances and fusion vectors.

Definition 4 (Modular Structure). *A modular structure is a tuple $\mathcal{M} = [\mathcal{I}, \mathcal{F}]$, where*

- *$\mathcal{I} = \{[N_1, m_{01}], \ldots, [N_\ell, m_{0\ell}]\}$ is a set of instances with disjoint modules and*
- *$\mathcal{F} \subseteq (T_{1|interface} \cup \{\bot\}) \times \ldots \times (T_{\ell|interface} \cup \{\bot\})$ is a set of fusion vectors.*

The components of a modular structure are fused together into a single Petri net, the *modular Petri net*.

Definition 5 (Modular Petri Net). *From a modular structure $\mathcal{M} = [\mathcal{I}, \mathcal{F}]$, we can derive a Petri net system $N = [P, T, F, W, m_0]$, where*

- $P = \bigcup_{j \in \{1,\ldots,\ell\}} P_j,$
- $T = \bigcup_{j \in \{1,\ldots,\ell\}} T_{j|internal} \cup \{t_f \mid f \in \mathcal{F}\},$ where t_f is a freshly introduced fusion transition *for fusion vector* f,
- $F = \bigcup_{j \in \{1,\ldots,\ell\}} (F_j \cap (P_j \times T_{j|internal} \cup T_{j|internal} \times P_j))$

$$\cup \{(p, t_f) \mid f \in \mathcal{F}, \exists j \in \{1,\ldots,\ell\} : f[j] = t, (p,t) \in F_j\}$$

$$\cup \{(t_f, p) \mid f \in \mathcal{F}, \exists j \in \{1,\ldots,\ell\} : f[j] = t, (t,p) \in F_j\},$$

- $W(t,p) = W_j(t,p)$ for $(t,p) \in F_j, t \in T_{j|internal},$
- $W(t_f, p) = W_j(t^*, p)$ for $(t^*, p) \in F_j, t^* = f[j], f \in \mathcal{F}, j \in \{1,\ldots,\ell\},$
- $W(p,t) = W_j(p,t)$ for $(p,t) \in F_j, t \in T_{j|internal},$
- $W(p, t_f) = W_j(p, t^*)$ for $(p, t^*) \in F_j, t^* = f[j], f \in \mathcal{F}, j \in \{1,\ldots,\ell\},$
- $m_0 = \bigcup_{j \in \{1,\ldots,\ell\}} m_{0j}.$

For every fusion vector f, a new transition t_f is introduced (called *fusion transition*). These fusion transitions combine the behavior of the interface transitions of the individual instances. Interface transitions, that do not participate in any fusion, are not represented in the modular Petri net. Since an instance only participates in a fusion vector with at most a single transition, the weight function is well-defined. Similarly, since the domains of the initial markings m_{0j} are pairwise disjoint, m_0 is also well-defined.

Example 1 (Modular Petri net). Figure 1 describes a modular structure $\mathcal{M} = (\{[N_1, p_1], [N_2, p_3]\}, \{(t_{13}, t_{21})\})$ consisting of two instances for two modules N_1 and N_2. The single fusion vector (t_{13}, t_{21}) fuses them together into a single Petri net. The initial markings are written in multiset notation.

In the following, unless stated otherwise, $\mathcal{M}$ shall denote a modular structure $\mathcal{M} = [\mathcal{I}, \mathcal{F}]$, while N denotes the corresponding modular Petri net $N = [P, T, F, W, m_0]$.

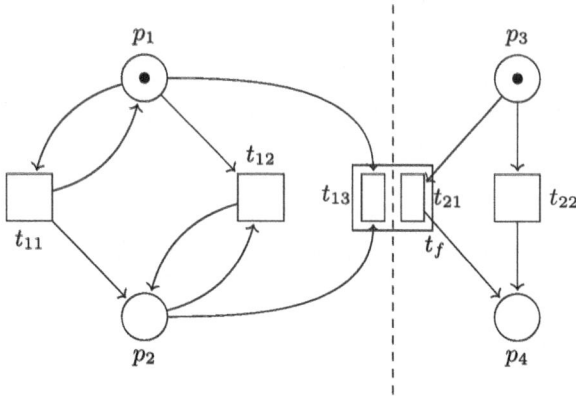

Fig. 1. Modular Petri net

2.2   Modular State Space

Obviously, the state space of a modular structure can be described by the composed state space of the corresponding modular Petri net. It is however more natural to describe the modular state space without constructing the composed state space. For this, [18,39] introduce the notions of *local reachability* and *synchronization*.

Since the instances that make up a modular structure are themselves Petri net systems, existing notation is used to describe their local behavior. The basis of this is the activation and firing of transitions.

Definition 6 (Transition Activation and Firing). *Let m be a marking and $t \in T$ be a transition of Petri net N. Transition t is* activated *or* fireable *in m (denoted by $m \xrightarrow{t}$) iff, $\forall p \in P : W(p,t) \leq m(p)$. t can then* fire *in m and lead to marking m' (denoted $m \xrightarrow{t} m'$) if $m \xrightarrow{t}$ and $m' = m + \Delta_t$ with $\Delta_t(p) = W(t,p) - W(p,t)$. We assume $W(x,y) = 0$ for $(x,y) \notin F$ for convenience. If m does not activate t we denote this by $m \not\xrightarrow{t}$.*

The firing of a single transition can be extended to transition sequences $\omega \in T^*$, saying $m \xrightarrow{\epsilon} m$ for the empty sequence and $m \xrightarrow{\omega t} m''$, if $m \xrightarrow{\omega} m' \xrightarrow{t} m''$. Similarly, since we use $\perp$ in fusion vectors like transitions, we say $m \xrightarrow{\perp} m$. By extension, we define that marking m' is *reachable* from marking m (denoted $m \rightarrow m'$), if a sequence $\omega \in T^*$ exists such that $m \xrightarrow{\omega} m'$.

With this, we can define the state space of an instance. This follows similar notions regarding reachability graphs, but we make one restriction to produce the *local* reachability graph. First, we can calculate the *reachability set* $RS_N(M)$ for a given Petri net system N and a set of markings M. We define $RS_N(M) = \{m' \mid m \rightarrow m', m \in M\}$.

Definition 7 (Projection). *Let $m \in RS(\{m_0\})$ be a reachable marking of a modular Petri net N. For a given instance $[N_j, m_{0j}]$ for $j \in \{1, \ldots, \ell\}$ of $\mathcal{M}$, let $\pi_j(m) = m \cap (P_j \times \mathbb{N})$ be the projection of m to $[N_j, m_{0j}]$.*

Definition 8 ((Local) Reachability Graph). *The* reachability graph *of a Petri net system N is the directed labeled graph $R = [V^R, E^R]$, where $V^R = RS_N(\{m_0\})$ and $(m,t,m') \in E^R$ iff $m \xrightarrow{t} m'$ for some $t \in T$.*

Given a modular structure $\mathcal{M} = [\mathcal{I}, \mathcal{F}]$ and with N as its corresponding modular Petri net, we can define the local reachability graph *$L_j = [V_j^L, E_j^L]$ for an instance $[N_j, m_{0j}]$ with $j \in \{1, \ldots, \ell\}$ based on R:*

- *$V_j^L = \{\pi_j(m) \mid m \in V^R\}$*
- *$E_j^L = E_{j|internal}^L \cup E_{j|interface}^L$ with*
  - *$(\pi_j(m), t, \pi_j(m')) \in E_{j|internal}^L$ iff $(m,t,m') \in E^R, t \in T_{j|internal}$*
  - *$(\pi_j(m), t, \pi_j(m')) \in E_{j|interface}^L$ iff $(m, t_f, m') \in E^R, f[j] = t, f \in \mathcal{F}$*

By projecting the behavior of the modular Petri net to the instances, we avoid the case where the isolated behavior of an instance is infinite, while the composed system has finite behavior. This is because the projection disallows arbitrary firing of interface transitions. However, we still restrict ourselves to local reachability graphs that are finite after this restriction, since infinite local reachability are still infeasible to handle.

Since fusion transitions only appear in the composed modular Petri net, the local reachability graphs depend on the reachability graph of the modular Petri net. In order to reason about the latter and to build the local reachability graphs without it, we notice that fusion transitions only appear in the composed reachability graph when every interface transition that participates in the fusion is activated. By capturing these *synchronization* points in the *synchronization graph*, we can avoid constructing the composed reachability graph and explore the local reachability graphs immediately.

From the perspective of a single instance, it does not matter which exact markings from other instances activate interface transitions that participate in a fusion together. It only matters if they can be activated without having to fire another interface transition along the way. This way, we can abstract single markings to sets of markings when dealing with synchronization. We call these sets *segments*.

Definition 9 (Segment). *Let $L_j = [V_j^L, E_j^L]$ be the local reachability graph of instance $[N_j, m_{0j}]$ for $j \in \{1, \ldots, \ell\}$. A segment is a set of markings $O \subseteq V_j^L$, which is forwardly closed in the following way:*
If $m \in O$ and $(m, t, m') \in E_{j|internal}^L$, then $m' \in O$.

For a set of markings $M \subseteq V_j^L$, let $\mathring{M}$ be the smallest segment that contains M, i.e. the *closure* of M regarding internal transitions. We call set M a *generator* of $\mathring{M}$.

Markings in a segment are not necessarily connected or strongly connected, as generators of a segment are not required to be connected. Segments can lie arbitrarily to each other, i.e. they can intersect, be disjoint or contain each other.

Segments describe internal behavior of an instance between occurrences of interface transitions. For a fixed interface transition and a fixed segment, we define the successor segment.

Definition 10 (Successor Segment). *Let O be a segment of the local reachability graph L_j of instance $[N_j, m_{0j}]$ and $t \in T_{j|interface}$ be an interface transition with $j \in \{1, \ldots, \ell\}$. The successor segment is defined as*

$$O^{+t} = \mathring{M}, \text{ where } M = \{m' \mid m \in O, (m, t, m') \in E_{j|interface}^L\}$$

As we permit only one interface transition per instance in a fusion vector, the successor segment is unambiguous. From time to time, we will abuse the notation for a fusion transition $t_f \in T$ where $f \in \mathcal{F}$ is a fusion vector as $O^{+t_f} = O^{+f[j]}$ for $f[j] \neq \perp$ and $O^{+t_f} = O$ for $f[j] = \perp$ for $j \in \{1, \ldots, \ell\}$.

After abstracting the markings of an instance to segments, the next step is to abstract the activation of interface transitions, from a single marking in the reachability graph of the modular Petri net to a tuple of segments for every instance.

Definition 11 (Global Activation). *A fusion transition t_f for fusion vector $f \in \mathcal{F}$ is globally activated in a tuple of segments $<O_1, \ldots, O_\ell>$ if for all instances $[N_j, m_{0j}]$ with $j \in \{1, \ldots, \ell\}$ and $f[j] \neq \perp$, there exists a $m_j \in O_j$ with $m_j \xrightarrow{f[j]}$.*

With this abstraction, we are ready to define the synchronization graph. A vertex in the synchronization graph is an ℓ-tuple of segments of the local reachability graphs, one per instance.

Definition 12 (Synchronization Graph). *The synchronization graph $S = [V^S, E^S]$ of modular structure $\mathcal{M} = [\mathcal{I}, \mathcal{F}]$ is inductively defined as follows:*
Base: $< \{\overset{\circ}{m}_{01}\}, \ldots, \{\overset{\circ}{m}_{0\ell}\} > \in V^S$
Step: if $v = < O_1, \ldots, O_\ell > \in V^S$ and fusion transition t_f for fusion vector $f \in \mathcal{F}$ is globally activated in v, then $v' = < O_1^{+t_f}, \ldots, O_\ell^{+t_f} > \in V^S$ and $(v, t_f, v') \in E^S$.

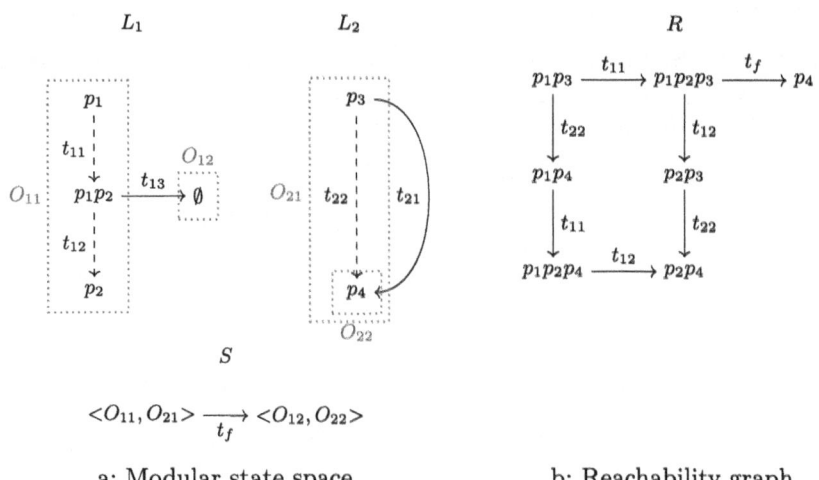

a: Modular state space   b: Reachability graph

Fig. 2. State spaces for Example 1

Example 2 (Modular state space). The graphs in Fig. 2a depict the above definitions applied to the modular structure from Example 1. As before, markings are written in multiset notation. Internal transitions in the instances are depicted through dashed lines. Segments are denoted through dotted borders. For comparison, Fig. 2b depicts the reachability graph of the corresponding modular Petri net. Notice that p_4 is part of both O_{21} and O_{22}, since the marking can be reached through both internal and interface transitions.

Since we assumed local reachability graphs of the modules to be finite, the synchronization graph will in turn also be finite. Based on these definitions, the modular state space is well-defined for a given modular structure. Since the modular Petri net resulting from that modular structure is also well-defined (per Definition 5), a modular structure producing finite local reachability and synchronization graphs would also produce a finite reachability graph of the modular Petri net.

In the following we shall write R for the reachability graph of a modular Petri net N and S for the synchronization graph of the underlying modular structure. When relating markings m of R to vertices $v = <O_1, \ldots, O_\ell>$ of S in the following sections, we shall write $m \in v$ to mean $m \in O_1 \times \ldots \times O_\ell$.

2.3   Binary Decision Diagrams

Binary Decision Diagrams (BDDs) [6,23] are a method to describe boolean functions $b : \{0,1\}^n \to \{0,1\}$ over boolean variables $x_i, i \in \{0, \ldots, n\}$. They are described through a compressed decision tree over the variables.

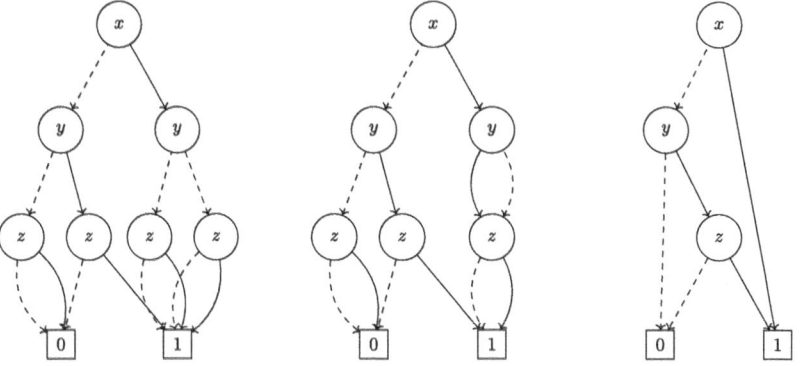

a: Decision Tree   b: Merge equivalent sub- c: Reduced ordered BDD
                        trees

Fig. 3. BDD representing $x \vee (y \wedge z)$

Beginning from the root node, a path in the BDD leading to a 0-node (*False* in the following) describes a variable mapping where b evaluates to 0, while a path leading to a 1-node (*True* in the following) describes a mapping where b evaluates to 1. Compared to simple binary decision trees, BDDs can be compressed, leading to *reduced* BDDs by merging equivalent subtrees and removing nodes where both children are the same (thereby connecting that single child with the parent of the removed node).

For a given variable ordering (meaning that variables x_i are passed in the same order along every path in the BDD), the resulting reduced *ordered* BDD

(ROBDD) has a canonical form for a given boolean function b [5]. In the following, we shall mean ROBBDs when talking about BDDs.

Example 3 (BDD reduction). Figure 3 depicts the reduction process to obtain a BDD representing the boolean function $x \vee (y \wedge z)$ with variables x, y and z. 0-Decisions are marked through dashed lines, while 1-decisions are marked through solid lines. Beginning from a binary decision tree, equivalent subtrees are merged and nodes with single children eliminated. In theory, both steps would be repeated until a fixed point is reached.

To encode the reachability graph of a Petri net as a BDD, we will assume 1-safe (modular) Petri nets. In this special case, the "state" of a place in a marking can be described through a single boolean variable. In the following, we will refer to these variables using the names of places p_i. The characteristic function of the reachability set of the Petri net can therefore be described as a boolean function over those variables. This way, the state space of the Petri net can be encoded in a single BDD. In [29], methods for this are described as well as methods to transform a BDD describing a set of markings M to a BDD describing another set of markings $M' = \{m' \mid \exists m \in M : m \xrightarrow{t} m'\} \cup M$ for some transition t. This allows the construction of the state space using a BDD describing just $\{m_0\}$ and symbolic operations that mimic transition firings.

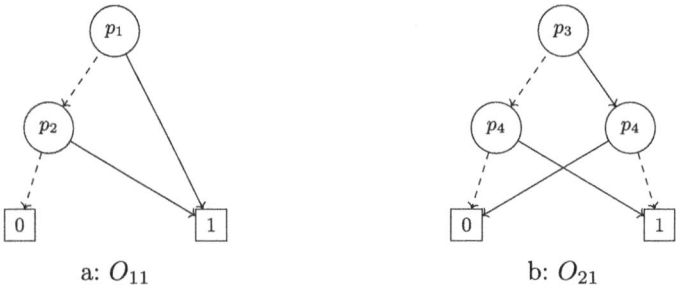

a: O_{11} b: O_{21}

Fig. 4. Segments from Example 2 as BDDs

Example 4 (BDD reachability sets). The two BDDs in Fig. 4 apply the symbolic state space exploration from [29] to the two segments $O_{11} = \{p_1, p_2, p_1 p_2\}$ and $O_{21} = \{p_3, p_4\}$ from Example 2 that make up the initial vertex in the synchronization graph.

We apply the symbolic representation to segments in the local reachability graphs. Since segments make up vertices in the synchronization graph, they are more relevant than the local reachability graphs as a whole. Further, since segments already abstract away the internal behavior and only describe sets of markings, encoding these sets as BDDs carries no loss in information.

We do not encode the synchronization graph using BDDs since an encoding is not as obvious as for 1-safe Petri nets and would require a separate discussion in itself.

Encoding the modular state space using BDDs introduces several BDDs instead of a single one for the whole state space. Similarly to explicit exploration, the goal is that the set of BDDs is still smaller than the one describing the composed state space. This can make operations on BDDs faster, since their complexity scales with the size of the BDDs [6].

Since we apply the findings of [29] to the modular state space in the following, we assume 1-safe reachability sets (segments, local reachability graphs, generators, etc.).

3 Symbolic Reachability Analysis in the Modular State Space

Reachability analysis deals with the *reachability problem*, i.e. the question if a given marking m is reachable in a given Petri net. This idea can be broadened to the reachability of markings that satisfy *state predicates*, which are made up of *atomic* state predicates.

Definition 13 ((Atomic) State Predicate). *In a Petri net N, an atomic state predicate $\varphi_k(m) := \sum_i \lambda_i \cdot m(p_i) \leq \lambda$ describes an inequality over places $P = \{p_1, \ldots, p_{|P|}\}$, with $\lambda, \lambda_i \in \mathbb{Z}$ and $i \in \{1, \ldots, |P|\}$. A marking m satisfies an atomic state predicate if the inequality is true in m. For atomic state predicate φ, the support $supp(\varphi)$ describes the places p_i where $\lambda_i \neq 0$. A state predicate is a conjunction of atomic state predicates $\varphi = \bigwedge_k \varphi_k$. In the following, let the atomic state predicate $\varphi_k := \sum_i \lambda_i \cdot p_i \leq \lambda$ mean $\varphi_k(m)$ and let the formal sum $\sum_i \lambda_i \cdot p_i$ be shortened to s_k.*

$$\varphi := \begin{array}{rl} & p_1 + 2p_2 \leq 1 \\ \wedge & 3p_1 + 4p_2 - 3p_3 \leq 6 \\ \wedge & -2p_1 + 4p_3 - 2p_4 \leq 1 \end{array} \tag{1}$$

Example 5 (State predicate). In the following, Eq. 1 will be used as a running example of a state predicate. It consists of three atomic state predicates $\varphi_1 := p_1 + 2p_2 \leq 1$, $\varphi_2 := 3p_1 + 4p_2 - 3p_3 \leq 6$ and $\varphi_3 := -2p_1 + 4p_3 - 2p_4 \leq 1$. Looking at Example 2, one can see that φ can be satisfied in R, since both p_1p_4 and p_4 satisfy φ. Since $p_1p_4 \in < O_{11}, O_{21} >$ and $p_4 \in < O_{12}, O_{22} >$, both vertices of the synchronization graph should also satisfy φ. We discuss the methods to obtain that result in the following.

In [18], a method is presented to check whether a state predicate is satisfied in a modular Petri net by analyzing the underlying modular structure and without constructing the composed state space. The satisfiability of state predicates depends on the satisfiability of the atomic state predicates. For this, the problem is divided into two parts: (1) atomic state predicates φ_k where every $p \in supp(\varphi_k)$ belongs to the same instance and (2) atomic state predicates φ_k where places $p \in supp(\varphi_k)$ belong to different instances. In order to check (2), vertices v in the synchronization graph need to be checked. In the following, we describe how both parts are solved when the segments that make up v are encoded using BDDs.

3.1   Atomic State Predicates φ_k with All Places of $supp(\varphi_k)$ from the Same Instance

When every place $p \in supp(\varphi_k)$ belongs to the same instance, the satisfiability of the atomic state predicate can be checked using just the state space of said instance. In theory, if this is the case for all atomic state predicates, and it is the same instances for all of them, the modular approach would not be needed. Since this is not the case in general and the local behavior of an instance describes just the behavior possible in the composed system, this edge case will not be covered. But, checking such a single atomic state predicate does not depend on the modular approach. Therefore, existing techniques for non-modular Petri nets can be reused. In the case of BDD-encoded state spaces, evaluating an atomic state predicate φ_k on a single marking is hard. Instead, φ_k can be encoded as a BDD that represents the set of markings that satisfy it. The conjunction of the resulting BDD and a BDD representing a set of reachable markings represents the intersection of the two sets, i.e. the set of reachable markings that satisfy φ_k. If that results in a non-*False* BDD, one can conclude that φ_k can be satisfied.

In [1], a method is presented to convert a *pseudo-boolean constraint* into a BDD. *Pseudo-boolean constraints* are constraints of the form $\sum_i a_i \cdot x_i \# K$, with $a_i, K \in \mathbb{Z}$, $\# \in \{<, >, \leq, \geq, =\}$ and boolean variables x_i. The algorithm presented in [1] assumes $a_i, K \in \mathbb{N}$ and that $\#$ is always $\leq$. In the following, we present adaptions to the algorithm that easily allow integer coefficients and constants, allowing the application to our atomic state predicates. We take the liberty to introduce the required concepts already adapted to our notion of atomic state predicates.

First, [1] introduces the notion of *intervals* for an atomic state predicate.

Definition 14 (Interval of an atomic state predicate and BDD) [1]**.** *For a given atomic state predicate $\varphi_k = \lambda_1 \cdot p_1 + \ldots + \lambda_{|P|} \cdot p_{|P|} \leq \lambda$ the interval of φ_k describes the set of integers $[\beta, \gamma]$ such that $\lambda_1 \cdot p_1 + \ldots + \lambda_{|P|} \cdot p_{|P|} \leq h$ for a given $h \in [\beta, \gamma]$ is true iff φ_k is true. Similarly, for a given BDD $\mathcal{B}$ rooted at variable p_i, its interval includes all integers h such that $\lambda_i \cdot p_i + \ldots + \lambda_{|P|} \cdot p_{|P|} \leq h$ is represented by $\mathcal{B}$.*

In a BDD $\mathcal{B}$ representing φ_k, every node can be annotated with its interval by calculating them bottom up.

Proposition 1 (Computation of intervals [1]). *For a given node $\mathcal{V}$ with selector variable p_i in BDD $\mathcal{B}$ representing an atomic state predicate φ_k, the* *interval* $[\beta, \gamma]$ *of* $\mathcal{V}$ *can be calculated as follows based on the intervals* $[\beta_f, \gamma_f]$ *and* $[\beta_t, \gamma_t]$ *of the false child* $\mathcal{V}_f$ *with selector variable* p_f *and true child* $\mathcal{V}_t$ *with selector variable* p_t:

$$\beta = \max(\beta_f, \beta_t + \lambda_i),$$
$$\gamma = \min(\gamma_f, \gamma_t + \lambda_i)$$

The above definition assumes the simple case $p_f = p_t = p_{i+1}$, i.e. that no variables in the order are skipped when traversing the BDD downwards. This may happen in a reduced BDD however, but only because the skipped variable would be represented by a node $\mathcal{V}'$ where both children are the same. In this case, $[\beta_f, \gamma_f] = [\beta_t, \gamma_t]$, therefore the resulting interval can easily be calculated without having a node there.

Proof. Proposition 1 is adapted from Proposition 7 of [1], which deals with the same problem. Their proof however never requires every λ_i with $i \in \{1, \ldots, |P|\}$ and λ to be positive integers, which is why their arguments still hold.

Since the $\textit{False}$-BDD represents the contradiction for $0 \leq \lambda$, its interval is $(-\infty, -1]$. Similarly, the $\textit{True}$-BDD represents the tautology for $0 \leq \lambda$, therefore its interval is $[0, \infty)$. With this, the interval of every node in the BDD $\mathcal{B}$ representing φ_k can be calculated, which is used in Algorithm 1 and 2 to construct $\mathcal{B}$.

Proposition 2 (Correctness of Algorithm 1 [1]). *Algorithm 1 is correct in the following sense:*

1. λ' *belongs to the interval returned by* $\text{BDDCONSTRECUR}(i, \lambda_i \cdot p_i + \ldots + \lambda_{|P|} \cdot p_{|P|} \leq \lambda', \mathcal{K})$.
2. *The tuple* $([\beta, \gamma], \mathcal{B})$ *returned by* BDDCONSTRECUR *consists of a BDD* $\mathcal{B}$ *and its interval* $[\beta, \gamma]$.
3. *If* BDDCONSTRECUR *returns, the returned BDD* $\mathcal{B}$ *is reduced.*

Proof. Similarly to Proposition 1, Algorithm 1 and 2 are adapted from Algorithms 1 and 2 presented in [1] to support negative integer coefficients and constants. The three statements are proven by induction on i.
Base: The base case is $i = |P| + 1$. $\mathcal{K}_{|P|+1}$ contains the intervals $(-\infty, -1]$ and $[0, \infty)$, therefore the SEARCH at line 2 of Algorithm 2 succeeds and the first statement holds. For the second statement, there are definite minimum and maximum values of $\lambda_i \cdot p_i + \ldots + \lambda_{|P|} \cdot p_{|P|}$ if places are assumed to be 1-safe. Markings that result in values less than the minimum are therefore impossible, and all possible values are less than or equal to the maximum. Therefore, the initial tuples in $\mathcal{K}$ are correct. Further, all of these tuples are reduced.
Step: For $i < n + 1$, we can assume that $\lambda' \in [\beta_f, \gamma_f]$ and $\lambda' - \lambda_i \in [\beta_t, \gamma_t]$. It is easy to see that $\lambda' \in [\beta_f, \gamma_f] \cap [\beta_t + \lambda_i, \gamma_t + \lambda_i]$ (which is implied by line 13 of Algorithm 2). Therefore, the first statement holds. Assuming previously inserted intervals and BDDs to be correct, Proposition 1 (applied in lines 11 and 13) implies the current interval and BDD to be correct. This way, the second

---

Algorithm 1. Construction of BDD [1]

1: **procedure** BDDCONSTRUCTION($\lambda 1.p_1 + \ldots + \lambda_{|P|} \cdot p_{|P|} \leq \lambda$)
2:     **for all** $i \in \{1, \ldots, |P| + 1\}$ **do**
3:         min $\leftarrow$ sum$\{\lambda_j \mid \lambda_j < 0, j \in \{i, \ldots, |P|\}\} - 1$
4:         max $\leftarrow$ sum$\{\lambda_j \mid \lambda_j > 0, j \in \{i, \ldots, |P|\}\}$
5:         $K_i \leftarrow \{((-\infty, \min], False), ([\max, \infty), True)\}$ ▷ Initialize Cache
6:     **end for**
7:     $\mathcal{K} \leftarrow (K_1, \ldots, K_{|P|+1})$
8:     $([\beta, \gamma], \mathcal{B}) \leftarrow$ BDDCONSTRECUR$(1, \lambda_1 \cdot p_1 + \ldots + \lambda_{|P|} \cdot p_{|P|} \leq \lambda, \mathcal{K})$
9:     **return** $\mathcal{B}$
10: **end procedure**

Algorithm 2. Recursive BDD construction [1]

1: **procedure** BDDCONSTRECUR$(i, \lambda_i \cdot p_i + \ldots + \lambda_{|P|} \cdot p_{|P|} \leq \lambda, \mathcal{K})$
2:     $([\beta, \gamma], \mathcal{B}) \leftarrow$ SEARCH(λ', K_i)
3:     **if** $[\beta, \gamma] \neq$ null **then**
4:         **return** $([\beta, \gamma], \mathcal{B})$ ▷ Cache hit
5:     **end if**
6:     $([\beta_f, \gamma_f], \mathcal{B}_f) \leftarrow$ BDDCONSTRECUR$(i + 1, \lambda_{i+1} \cdot p_{i+1} + \ldots + \lambda_{|P|} \cdot p_{|P|} \leq \lambda', \mathcal{K})$
7:     $([\beta_t, \gamma_t], \mathcal{B}_t) \leftarrow$ BDDCONSTRECUR$(i+1, \lambda_{i+1} \cdot p_{i+1} + \ldots + \lambda_{|P|} \cdot p_{|P|} \leq \lambda' - \lambda_i, \mathcal{K})$
8:     **if** $[\beta_t, \gamma_t] = [\beta_f, \gamma_f]$ **then**
9:         $\mathcal{B} \leftarrow \mathcal{B}_t$
10:     **else**
11:         $\mathcal{B} \leftarrow$ ITE$(p_i, \mathcal{B}_t, \mathcal{B}_f)$ ▷ Root p_i, true child $\mathcal{B}_t$, false child $\mathcal{B}_f$
12:     **end if**
13:     $[\beta, \gamma] \leftarrow [\max(\beta_f, \beta_t + \lambda_i), \min(\gamma_f, \gamma_t + \lambda_i)]$
14:     INSERT$(([\beta, \gamma], \mathcal{B}), K_i)$
15:     **return** $([\beta, \gamma], \mathcal{B})$
16: **end procedure**

---

statement holds. Assuming previous BDDs to be reduced, $\mathcal{B}_t$ and $\mathcal{B}_f$ only imply different BDDs if their intervals are different. Therefore, lines 8–12 guarantee that the current BDD is also reduced.

Example 6 (BDD conversion). Looking at Example 5, the atomic state predicate φ_1 is the only one where every place in $supp(\varphi_1)$ belongs to the same instance. Figure 5 depicts the application of Algorithm 1 to φ_1, resulting in a single BDD. For every call of BDDCONSTRECUR, the recursive sub-calls are depicted. Every call is also numbered in the order they happen in. Finally, if no recursive calls split from a call, the cache delivered the result.

3.2   Atomic State Predicates φ_k with Places in $supp(\varphi_k)$ from Different Instances

If $supp(\varphi_k)$ includes places from more than one instance, [18] proposes to split the formal sum s_k according to the instance affiliation of places, resulting in partial

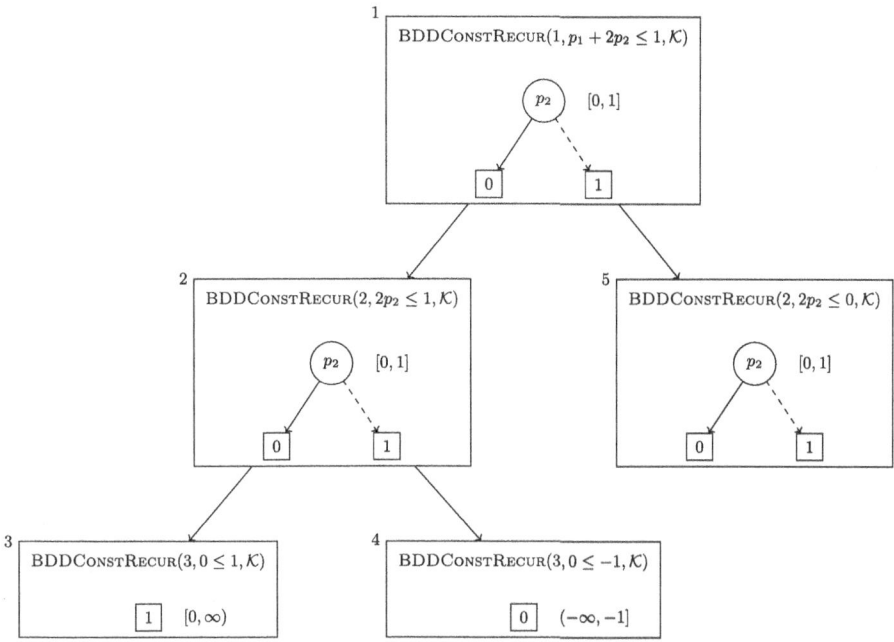

Fig. 5. Conversion of $p_1 + 2p_2 \leq 1$ with recursive calls of BDDCONSTRECUR

sums $s_{k,j} = \sum_i \lambda_i \cdot p_i$ for $i \in P_j$. With all atomic state predicates $\varphi_1, \ldots, \varphi_n$ that make up φ, a marking $m_j \in V_j^L$ implies a vector of partial sums $\mathbf{c}_{m_j} = (s_{1,j}, \ldots, s_{n,j})$. A segment $O_j \subseteq V_j^L$ therefore implies a set $C_j = \{\mathbf{c}_{m_j} \mid m_j \in O_j\}$ of such vectors. For a given vertex $< O_1, \ldots, O_\ell >$ in the synchronization graph the resulting vector of sets can be used to check the satisfiability of φ by searching for a combination $\mathbf{c}_{m_1} \in C_1, \ldots, \mathbf{c}_{m_\ell} \in C_\ell$ such that $\mathbf{c}_{m_1} + \ldots + \mathbf{c}_{m_\ell} \leq (\lambda_1, \ldots, \lambda_n)$. If there is such a combination and for segment O_j there is a marking m_j' such that $\mathbf{c}_{m_j'} \leq \mathbf{c}_{m_j}$, the combination with $\mathbf{c}_{m_j'}$ also satisfies φ. If $\mathbf{c}_{m_j'}$ is found in the search, any $\mathbf{c}_{m_j}$ fitting this example does not need to be stored. Therefore, the sets C_j can be reduced to $C_j^{min} = \{\mathbf{c}_{m_j} \mid \nexists \mathbf{c}_{m_j}' \in C_j : \mathbf{c}_{m_j}' \leq \mathbf{c}_{m_j}\}$. In the following, the condition "$\nexists \mathbf{c}_{m_j}' \in C_j : \mathbf{c}_{m_j}' \leq \mathbf{c}_{m_j}$" may be referred as the *minimization criterion*.

In the explicit exploration of the modular state space, the sets C_j^{min} can be aggregated using Algorithm 3 while exploring the segments that make up a vertex v in the synchronization graph, while the satisfiability can be checked after completing v. As already mentioned in Sect. 3.1, evaluating an atomic state predicate on a single marking is hard when segments are encoded using BDDs. Algorithm 4 describes a method to calculate C_j^{min} for a given segment O_j in $v = < O_1, \ldots, O_\ell >$ with $j \in \{1, \ldots, \ell\}$ when O_j is given as BDD $\mathcal{B}$. The algorithm works by traversing $\mathcal{B}$ depth-first and building C_j^{min} bottom up, annotating nodes with sets of vectors of partial sums while backtracking. To describe the idea, let j be fixed to some instance $[N, m_0]$. The idea is, that any path in $\mathcal{B}$

Algorithm 3. Insert vector of partial sums

```
 1: procedure INSERT(c, C)
 2:     for all c' ∈ C do
 3:         if c < c' then
 4:             C ← C \ {c'}                    ▷ New element smaller? Remove old
 5:         else if c ≥ c' then
 6:             return C                        ▷ New element bigger? Don't insert
 7:         end if
 8:     end for
 9:     return C ∪ {c}
10: end procedure
```
---

leading to the $True$ node implies a (set of) reachable marking(s) and an assignment of variables $\{p_1 = v_1, \ldots, p_{|P|} = v_{|P|}\}$. For a given atomic state predicate φ_k, the partial sum s_k can be evaluated on this assignment, resulting in a partial sum for the corresponding set of markings. Looking at the path in reverse, going from the $True$ node to the root of $\mathcal{B}$, one can notice that when reaching node $\mathcal{V}$ with selector variable p_i with $i \in \{1, \ldots, |P|\}$, s_k describes the partial sum if we "ignore" all places before p_i in the order. Therefore, the BDD rooted at $\mathcal{V}$ can be associated with all possible partial sums when ignoring places p_o with $1 \leq o < i$. This set can be calculated based on the children $\mathcal{V}_f$ and $\mathcal{V}_t$ of $\mathcal{V}$ by adding the effect λ_i to every element in the set of $\mathcal{V}_t$ and merging both sets together, while respecting the minimization criterion. Keep in mind that the set associated with $\mathcal{V}_f$ does not need to be changed before merging, since no longer ignoring p_i would add $\lambda_i \cdot 0 = 0$ to every element.

This leaves only the terminal cases, which are made up of the $False$ and $True$ nodes, which can be annotated with $\emptyset$ and $\{\mathbf{0}\}$ respectively. The empty set is reasonable for the $False$ node, since any path in $\mathcal{B}$ that leads to $False$ implies an empty set of reachable markings. The zero vector as a single element for the $True$ node is also reasonable, since any path leading to $True$ implies (1) some reachable marking(s) and (2) every path starting there when backtracking ignores all places at this point, so s_k is still 0 for every k.

This merging of children while backtracking is described in Algorithm 4 and 5. Similar to Algorithm 2, a cache can be utilized to not calculate the set of partial sums for each node visited twice. The operation COFACTOR in lines 6 and 7 of Algorithm 5 takes a BDD $\mathcal{B}$ and returns a BDD $\mathcal{B}'$ representing the function of $\mathcal{B}$ where p_i is assumed $False$ (line 6) or assumed $True$ (line 7). In this case, since p_i is the selector variable of the root of $\mathcal{B}$, COFACTOR returns the $False$- and $True$-children of the root respectively. This allows for recursive descend into the BDD.

Proposition 3 (Correctness of Algorithm 4). *Algorithm 4 is correct in the following sense: The set C returned by* CALCULATESUMS$(\mathcal{B}, \varphi)$ *describes the set C^{min} for the segment O described by $\mathcal{B}$.*

Proof. The correctness can be shown inductively on the length of the path while backtracking.

Algorithm 4. Calculate set of partial sum vectors

1: **procedure** CALCULATESUMS($\mathcal{B}, \varphi$)
2:     **for all** $i \in \{1, \ldots, |P| - 1\}$ **do**
3:         $K_i \leftarrow \{(\emptyset, False)\}$ ▷ Initialize cache
4:     **end for**
5:     $K_{|P|} \leftarrow \{(\emptyset, False), (\{\mathbf{0}\}, True)\}$
6:     $\mathcal{K} \leftarrow (L_1, \ldots, L_{|P|})$
7:     **return** CALCSUMSRECUR($1, \mathcal{B}, \varphi, \mathcal{K}$)
8: **end procedure**

Algorithm 5. Recursive partial sum vector calculation

1: **procedure** CALCSUMSRECUR($i, \mathcal{B}, \varphi, \mathcal{K}$)
2:     $C \leftarrow$ SEARCH($\mathcal{B}, K_i$)
3:     **if** $C \neq$ null **then**
4:         **return** C ▷ Cache hit
5:     **end if**
6:     $C_f \leftarrow$ CALCSUMSRECUR($i) + 1,$ COFACTOR($\mathcal{B}, \neg p_i), \varphi, \mathcal{K}$ ▷ Check false child
7:     $C_t \leftarrow$ CALCSUMSRECUR($i) + 1,$ COFACTOR($\mathcal{B}, p_i), \varphi, \mathcal{K}$ ▷ Check true child
8:     **for all** $c \in C_t$ **do**
9:         $c' \leftarrow c +$ EFFECTS(p_i, φ)   ▷ Update c with every λ_i
10:        $C_f \leftarrow$ INSERT(c', C_f)
11:     **end for**
12:     INSERT($(\mathcal{B}, C_f), K_i$)
13:     **return** C_f
14: **end procedure**

Base: The *False*-node is correctly annotated with $\emptyset$, because any path leading to this node implies a marking that is not reachable. The *True*-node is correctly annotated with $\{\mathbf{0}\}$, because when backtracking starts in this node, every place is "ignored" and s_k is 0 for every k. This set should also not be empty, since reaching *True* implies some reachable markings.

Step: Assuming C_f and C_t from lines 6 and 7 in Algorithm 5 to be correct, lines 8-11 apply Algorithm 3 to insert every element of C_t into C_f. Algorithm 3 respects the minimization criterion, since a new vector is only added when it does not compare less-or-equal with any existing vector. Further, any vector that compares less-than a new vector is removed, leaving any remaining vector incomparable.

Example 7 (C_j^{min} *calculation*). Figure 6 applies Algorithm 4 to O_{11} from Example 2 (Fig. 4a) and φ from Example 5. Like with Example 6, recursive calls of CALCSUMSRECUR are depicted and numbered in order. Notice that only φ_2 and φ_3 have places from differing instances. Therefore, any c is 2-dimensional. The operation "merge" describes the merge respecting the minimization criterion. Any operation $\{\ldots\} + (\ldots)$ shall mean "for every element in $\{\ldots\}$, add $(\ldots)$". Further, notice that calls 6 and 7 are the same (just happen twice), which is why they are grouped together.

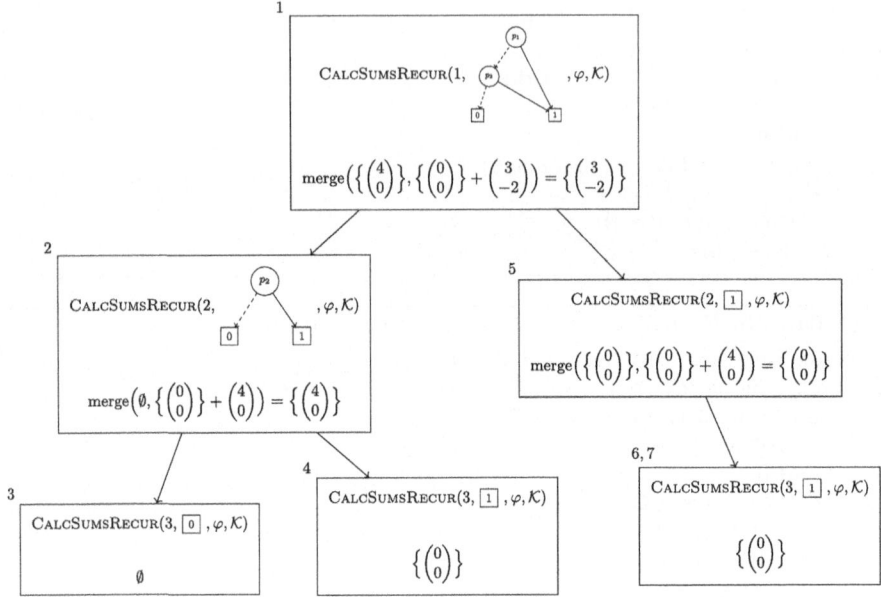

Fig. 6. C_1^{min} calculation for O_{11} with recursive calls of CALCSUMSRECUR

4   Experimental Results

The algorithms introduced in Sect. 3 were implemented and tested in LoLA [37]. We will refer to atomic state predicates with places belonging to a single module as "category 1" and others as "category 2" in the following. For the implementation of BDDs, CUDD [31] was used. Other libraries were considered [14, 25, 27], but the main reasons for using CUDD were the following:

- The concept of *managers* allows multiple variable spaces in the same program. This allows the complete separation of BDDs relating to different modules. The manager based approach allows easier, lock-free multithreading, exploring local reachability graphs of different modules in parallel.
- BDDs are kept unique via an internal hash table. BDDs can then be identified using just the memory address where the root of the BDD is stored. Since a memory address can be represented as a bit vector and its length is generally independent of the modular structure and property we want to check, lookup of a BDD becomes $\mathcal{O}(1)$ in a prefix tree data structure in terms of our inputs.

In addition to the above, CUDD supports a wide variety of variable reordering algorithms. The variable order of a BDD is very important for its size [4, 16]. The modular exploration and Algorithm 1 work with any variable order in practice and the order could also dynamically change (the order $p_1, \ldots, p_{|P|}$ was chosen for simplicity). Algorithm 4 however does not work with a changing variable order. The calculations that happen while backtracking in Algorithm 4 "ignore" variables in the order. If the order changes, some variables that were ignored at some point should not have been ignored. This invalidates the cache. This is fine for a given vertex in the synchronization graph (since the calculation is not changing any BDDs), but reusing results between vertices (i.e. when revisiting a segment) is not possible. Therefore, static variable order heuristics as compared in [3] or introduced in [21] may be applied to find good initial orderings. Dynamic reordering algorithms can be applied when cache between vertices of the synchronization graph is disabled. The comparison of different algorithms were outside the scope of this paper but remain an area for future research.

We used models provided by the Model Checking Contest [2] to evaluate time and memory performance. Information on compilation and runtime parameters, handling of the inputs as well as hardware used and the resulting output files can be found in the `eval`-folder at the LoLA git repository at [37].

In the 2019 edition of the contest, 4896 inputs exist for the reachability of state predicates with our atomic state predicates in safe Petri nets. We choose the 2019 edition because prior analysis processed the result table from [2] to allow for correctness checking. Out of those 4896 inputs, 2862 were suitable for our method, i.e. the state predicate was a conjunction of atomic state predicates. All models were modularized using hypergraph partitioning presented in [17] to achieve variations having two to five instances with no replications of modules. The verification of each input therefore happened with the following configurations:

1. Traditional depth-first search in the non-modular state space. Reduction techniques using stubborn sets [32], net reduction rules [24] and the state equation [36] were applied
2. Symbolic depth-first search in the non-modular state space by applying [29] and the presented algorithms. No further reduction techniques were applied.
3. Symbolic depth-first search in the modular state space by splitting the original net into $2\ldots5$ instances ($2\ldots5$ modules respectively).

We use the first configuration as a baseline and present results relative to this. The different configurations may be abbreviated as "1", "2" and "3.{2...5}". In cases where multithreading was applicable, multithreading was applied as good as possible. Meaning, when concurrently computing the local state space of five modules, the computation was done over five cores, not less. Further, to avoid outliers, each input was repeated ten times and the results were averaged.

Table 1 shows information about the runtime performance. The second column shows, for every configuration, how many inputs of the baseline were solved. The third column relates the total time the latter took versus the total time of the baseline for the same inputs. The last column relates the same time, just now versus the previous configuration, i.e. how long 3.2 took versus 2. Over all configurations, two conclusions can be drawn regarding runtime performance.

First, the baseline performance is significantly faster than any symbolic verification. This result may be heavily biased towards the traditional depth-first search, since reduction techniques were not applied to the symbolic verification. This is because their implementations were not available.

Table 1. Number of inputs solved

Config	# inputs (of base)	Time (vs. base)	Time (vs. prev)
1	4653	1	-
2	82	~1100	-
3.2	90	~978	~2.1
3.3	105	~331	~0.34
3.4	339	~106	~0.42
3.5	414	~122	~0.86

Table 2. BDD memory usage relative to markings needed

Config	Markings		
	< 1000	< 10000	> 10000
2	~325 MB	~109 MB	~589 MB
3.2	~117 MB	~320 MB	~145 MB
3.3	~109 MB	~274 MB	~270 MB
3.4	~101 MB	~219 MB	~173 MB
3.5	~136 MB	~217 MB	~175 MB

Second, modularization improved the symbolic verification (i.e. more instances led to faster verification). The configuration splitting a Petri net into 5 instances proved to be the fastest configuration for symbolic verification. The speedup gained by modularization reduces with higher modularization, however. With higher modularization, more atomic state predicates fall into category 2. We suspect this to have a higher performance impact than category 1. This is further supported by the fact, that configuration 2 (no modularization) performed faster than modularization into two modules. For configuration 2, every atomic state predicate falls in category 1. This cannot be clearly verified, however. The methods presented in [17] only perform modularization based on net structure. A clear conclusion can only be drawn when modularization based on state predicates is available.

Table 2 relates the average memory usage of BDDs as reported by CUDD for each configuration with the orders of magnitudes of markings needed in the baseline configuration. Here, two conclusions can be drawn again. First, the baseline outperforms the symbolic verification, but memory consumption in the symbolic verification does not directly relate to the number of explored markings. Second, modularization in the symbolic verification scales appears to not impact memory consumption much. Because of the default cache size of 262144 entries used internally by CUDD, a lot of overhead is introduced even for simple inputs requiring few markings. While memory consumption does grow for "harder" inputs (i.e. where traditional state space exploration explores many markings), it actually stabilizes or reduces for even harder inputs. This could derive from more efficient usage of the (per default) large cache. In theory, since every module introduces a new manager, higher modularization should result in higher memory consumption from more caches. The data does not seem to indicate that, which would imply that more resizing is required with less modularization. The relatively high memory consumption of the non-modularized configuration (configuration 2) seems to support that with a lot of unnecessary resizing. Therefore, higher modularization also seems to be beneficial for memory consumption.

Measuring the memory consumption of the cache used in Algorithm 4 was not feasible, since the prefix tree data structure implemented in LoLA cannot measure this. This only impacts atomic state predicates falling into category 2. Therefore, the suspected higher performance impact of the method stands.

5   Conclusions

We analyzed the verification of state predicates in the modular state space. For this, we implemented the method presented in [18]. To achieve further compression, we encoded the local reachability graphs of instances using binary decision diagrams. We adapted work from [1] to convert atomic state predicates into BDDs and developed a way to calculate the formal sums of atomic state predicates that span over multiple instances based on BDD-encoded segments. This way, atomic state predicates can be verified in the synchronization graph.

Performance-wise, traditional depth-first search still outperforms both methods. Modularization itself gives a speedup, however. Therefore, testing against large and complex models with complex state predicates appears a natural field for research. This is further supported by the fact that memory consumption in the symbolic verification does not appear to scale directly with the number of markings required for traditional verification, implying higher efficiency for large models.

The presented methods allow for high flexibility. Converting an atomic state predicate into a BDD happens when the formers formal sum can be calculated using just one instance. The representation of other instances does not matter. Similarly, for atomic state predicates whose formal sums span over multiple instances, the calculation of the partial sums per instance is independent. This allows the mix of symbolic representation and explicit representation. Other symbolic representations like SAT-based encoding [28] are also possible. This implies a natural field for future research, i.e. to find heuristics to determine an optimal representation for a local reachability graph.

In contrast to this, sacrificing flexibility could improve performance. The algorithm presented in Sect. 3.2 uses caching to improve performance. Such caching does not work with dynamic variable reordering. The latter is beneficial for BDD performance in general, so a comparison is again a natural field for future research. Ideally, a different method to calculate the sets without dependence on the variable order would be the best of both worlds. The existence of this method is a field for future research.

Currently, the only state predicates we looked at were conjunctions of atomic state predicates. While other forms, namely disjunctions or a mix of both, can be converted to conjunctions, the resulting state predicate could explode in the number of atomic state predicates. Accepting only conjunctions simplifies the theory of atomic state predicates with participating places in multiple instances, therefore the theory for that case needs to be adjusted.

Finally, comparisons with different implementations and combinations with different techniques appear possible. Comparisons with not just the traditional state space exploration, but with explicit modular state space exploration is interesting, an explicit exploration is not implemented yet however. Traditional state space exploration in Petri nets benefits strongly from partial order reductions [19,32,38], net reductions [12,24] and optimizations using the state equation [15,36]. Currently, there is nothing stopping one from adapting or outright implementing those methods in the modular state space to achieve similar benefits.

References

1. Abío, I., Nieuwenhuis, R., Oliveras, A., Rodríguez-Carbonell, E., Mayer-Eichberger, V.: A new look at BDDs for pseudo-boolean constraints. J. Artif. Intell. Res. **45**, 443–480 (2012). https://doi.org/10.1613/jair.3653

2. Amparore, E., et al.: Presentation of the 9th edition of the model checking contest. In: Beyer, D., Huisman, M., Kordon, F., Steffen, B. (eds.) TACAS 2019. LNCS, vol. 11429, pp. 50–68. Springer, Cham (2019). https://doi.org/10.1007/978-3-030-17502-3_4

3. Amparore, E.G., Donatelli, S., Beccuti, M., Garbi, G., Miner, A.: Decision diagrams for petri nets: a comparison of variable ordering algorithms. In: Koutny, M., Kristensen, L.M., Penczek, W. (eds.) Transactions on Petri Nets and Other Models of Concurrency XIII. LNCS, vol. 11090, pp. 73–92. Springer, Heidelberg (2018). https://doi.org/10.1007/978-3-662-58381-4_4

4. Breitbart, Y., Hunt, H., Rosenkrantz, D.: On the size of binary decision diagrams representing Boolean functions. Theoret. Comput. Sci. **145**(1), 45–69 (1995). https://doi.org/10.1016/0304-3975(94)00181-H

5. Bryant, R.E.: Symbolic Boolean manipulation with ordered binary-decision diagrams. ACM Comput. Surv. **24**(3), 293–318 (1992). https://doi.org/10.1145/136035.136043

6. Bryant, R.E.: Graph-based algorithms for boolean function manipulation. IEEE Trans. Comput. **C-35**, 677–691 (1986). https://doi.org/10.1109/TC.1986.1676819

7. Buchholz, P., Kemper, P.: Efficient computation and representation of large reachability sets for composed automata. Discrete Event Dyn. Syst. **12**(3), 265–286 (2002). https://doi.org/10.1023/A:1015669415634

8. Buchholz, P., Kemper, P.: Hierarchical reachability graph generation for petri nets. Formal Methods Syst. Des. **21**(3), 281–315 (2002). https://doi.org/10.1023/A:1020321222420

9. Burch, J.R., Clarke, E.M., McMillan, K.L., Dill, D.L., Hwang, L.J.: Symbolic model checking: 10^{20} states and beyond. Inf. Comput. **98**(2), 142–170 (1992). https://doi.org/10.1016/0890-5401(92)90017-A

10. Christensen, S., Petrucci, L.: Modular analysis of petri nets. Comput. J. **43**(3), 224–242 (2000). https://doi.org/10.1093/comjnl/43.3.224

11. Ciardo, G., Lüttgen, G., Siminiceanu, R.: Saturation: an efficient iteration strategy for symbolic state-space generation. In: Goos, G., Hartmanis, J., Van Leeuwen, J., Margaria, T., Yi, W. (eds.) Tools and Algorithms for the Construction and Analysis of Systems, vol. 2031, pp. 328–342. Springer, Heidelberg (2001)

12. Clarke, E.M., Enders, R., Filkorn, T., Jha, S.: Exploiting symmetry in temporal logic model checking. Formal Methods Syst. Des. **9**(1), 77–104 (1996). https://doi.org/10.1007/BF00625969

13. Classen, A., Heymans, P., Schobbens, P.Y., Legay, A.: Symbolic model checking of software product lines. In: Proceedings of the 33rd International Conference on Software Engineering, ICSE 2011, pp. 321–330. Association for Computing Machinery, New York (2011). https://doi.org/10.1145/1985793.1985838

14. van Dijk, T., van de Pol, J.: Sylvan: multi-core decision diagrams. In: Baier, C., Tinelli, C. (eds.) TACAS 2015. LNCS, vol. 9035, pp. 677–691. Springer, Heidelberg (2015). https://doi.org/10.1007/978-3-662-46681-0_60

15. Esparza, J., Melzer, S.: Verification of safety properties using integer programming: beyond the state equation. Formal Methods Syst. Des. **16**(2), 159–189 (2000). https://doi.org/10.1023/A:1008743212620

16. Friedman, S.J., Supowit, K.J.: Finding the optimal variable ordering for binary decision diagrams. In: Proceedings of the 24th ACM/IEEE Design Automation Conference, DAC 1987, pp. 348–356. Association for Computing Machinery, New York (1987). https://doi.org/10.1145/37888.37941

17. Gaede, J., Overath, J., Wallner, S.: Automatic modularization of place/transition nets. In: Köhler-Bussmeier, M., Moldt, D., Rölke, H. (eds.) Proceedings of the International Workshop on Petri Nets and Software Engineering 2024 co-located with the 45th International Conference on Application and Theory of Petri Nets and Concurrency (PETRI NETS 2024), 24–25 June 2024, Geneva, Switzerland. CEUR Workshop Proceedings, vol. 3730, pp. 53–73. CEUR-WS.org (2024). https://ceur-ws.org/Vol-3730/paper03.pdf

18. Gaede, J., Wallner, S., Wolf, K.: Modular state spaces - a new perspective. In: Kristensen, L.M., van der Werf, J.M. (eds.) Application and Theory of Petri Nets and Concurrency, pp. 312–332. Springer, Cham (2024). https://doi.org/10.1007/978-3-031-61433-0_15

19. Godefroid, P.: Using partial orders to improve automatic verification methods. In: Clarke, E.M., Kurshan, R.P. (eds.) CAV 1990. LNCS, vol. 531, pp. 176–185. Springer, Heidelberg (1991). https://doi.org/10.1007/BFb0023731

20. Graf, S., Steffen, B.: Compositional minimization of finite state systems. In: Clarke, E.M., Kurshan, R.P. (eds.) CAV 1990. LNCS, vol. 531, pp. 186–196. Springer, Heidelberg (1991). https://doi.org/10.1007/BFb0023732

21. He, L., Liu, G.: Petri net based CTL model checking: using a new method to construct OBDD variable order. In: 2021 International Symposium on Theoretical Aspects of Software Engineering (TASE), pp. 159–166 (2021). https://doi.org/10.1109/TASE52547.2021.00033

22. Le Cornec, Y.S.: Compositional analysis of modular Petri nets using hierarchical state space abstraction. In: Joint 5th International Workshop on Logics, Agents, and Mobility, LAM 2012, the 1st International Workshop on Petri Net-Based Security, WooPS 2012 and the 2nd International Workshop on Petri Nets Compositions, CompoNet 2012. CEUR Workshop Proceedings, Hamburg, Germany, vol. 853, pp. 119–133 (2012). https://hal.archives-ouvertes.fr/hal-00785569

23. Lee, C.Y.: Representation of switching circuits by binary-decision programs. Bell Syst. Tech. J. **38**, 985–999 (1959). https://doi.org/10.1002/j.1538-7305.1959.tb01585.x

24. Lee-Kwang, H., Favrel, J., Baptiste, P.: Generalized petri net reduction method. IEEE Trans. Syst. Man Cybern. **17**(2), 297–303 (1987). https://doi.org/10.1109/TSMC.1987.4309041

25. Lind-Nielsen, J.: BuDDy: A binary decision diagram package. Report, Department of Information Technology Denmark (1999). https://buddy.sourceforge.net/manual/main.html

26. Miner, A.S., Ciardo, G.: Efficient reachability set generation and storage using decision diagrams. In: Donatelli, S., Kleijn, J. (eds.) ICATPN 1999. LNCS, vol. 1639, pp. 6–25. Springer, Heidelberg (1999). https://doi.org/10.1007/3-540-48745-X_2

27. Mrena, M., Kvassay, M., Zaitseva, E.: TeDDy: templated decision diagram library. SoftwareX **26**, 101715 (2024). https://doi.org/10.1016/j.softx.2024.101715
28. Ogata, S., Tsuchiya, T., Kikuno, T.: SAT-based verification of safe petri nets. In: Wang, F. (ed.) ATVA 2004. LNCS, vol. 3299, pp. 79–92. Springer, Heidelberg (2004). https://doi.org/10.1007/978-3-540-30476-0_11
29. Pastor, E., Cortadella, J., Roig, O.: Symbolic analysis of bounded Petri nets. IEEE Trans. Comput. **50**, 432–448 (2001). https://doi.org/10.1109/12.926158
30. Shi, G.: A survey on binary decision diagram approaches to symbolic analysis of analog integrated circuits. Analog Integr. Circ. Sig. Process **74**(2), 331–343 (2013). https://doi.org/10.1007/s10470-011-9773-8
31. Somenzi, F.: CUDD: CU decision diagram package. Public Software, University of Colorado (1997). https://cir.nii.ac.jp/crid/1570009749922215168
32. Valmari, A.: Stubborn sets for reduced state space generation. In: Rozenberg, G. (ed.) ICATPN 1989. LNCS, vol. 483, pp. 491–515. Springer, Heidelberg (1991). https://doi.org/10.1007/3-540-53863-1_36
33. Valmari, A.: Compositional analysis with place-bordered subnets. In: Valette, R. (ed.) ICATPN 1994. LNCS, vol. 815, pp. 531–547. Springer, Heidelberg (1994). https://doi.org/10.1007/3-540-58152-9_29
34. Valmari, A.: The state explosion problem. In: Reisig, W., Rozenberg, G. (eds.) ACPN 1996. LNCS, vol. 1491, pp. 429–528. Springer, Heidelberg (1998). https://doi.org/10.1007/3-540-65306-6_21
35. Westergaard, M.: CPN tools 4: multi-formalism and extensibility. In: Colom, J.-M., Desel, J. (eds.) PETRI NETS 2013. LNCS, vol. 7927, pp. 400–409. Springer, Heidelberg (2013). https://doi.org/10.1007/978-3-642-38697-8_22
36. Wimmel, H., Wolf, K.: Applying CEGAR to the petri net state equation. In: Abdulla, P.A., Leino, K. (eds.) TACAS 2011. LNCS, vol. 6605, pp. 224–238. Springer, Heidelberg (2011). https://doi.org/10.1007/978-3-642-19835-9_19
37. Wolf, K.: Petri net model checking with LoLA 2. In: Khomenko, V., Roux, O.H. (eds.) Application and Theory of Petri Nets and Concurrency, pp. 351–362. Springer, Cham (2018). https://doi.org/10.1007/978-3-319-91268-4_18. https://git.informatik.uni-rostock.de/theo/lola-2
38. Wolper, P., Godefroid, P.: Partial-order methods for temporal verification. In: Best, E. (ed.) CONCUR 1993. LNCS, vol. 715, pp. 233–246. Springer, Heidelberg (1993). https://doi.org/10.1007/3-540-57208-2_17
39. Zech, L., Wolf, K.: Verifying temporal logic properties in the modular state space. In: Kristensen, L.M., van der Werf, J.M. (eds.) Application and Theory of Petri Nets and Concurrency, pp. 333–354. Springer, Cham (2024). https://doi.org/10.1007/978-3-031-61433-0_16

Author Index

E. Amparore and Ł. Mikulski (Eds.): PETRI NETS 2025, LNCS 15714, pp. 501–502, 2025.
https://doi.org/10.1007/978-3-031-94634-9

The manufacturer's authorised representative in the EU is Springer
Nature Customer Service Centre GmbH, Europaplatz 3, 69115 Heidelberg,
Germany. If you have any concerns regarding our products, please
contact ProductSafety@springernature.com

Printed and bound by CPI Group (UK) Ltd, Croydon, CR0 4YY

24/04/2026

02096367-0020